Super Hybrid Rice Breeding and Cultivation

CNS | K
Hunan Science & Technology Press

Editor-in-Chief Yuan Longping

Executive Editor-in-Chief Liao Fuming

Deputy Editors-in-Chief He Qiang, Deng Qiyun, Ma Guohui, Liu Aimin, Xu Qiusheng, Zhao Bingran, Yang Yuanzhu, Xin Yeyun

Translators Luo Zhongwu, Xu Qiliang, Liu Xiangyang, Yue Ling

Proofreader Xie Fangming

Authors (in the order of the Chinese Pinyin of surnames)

Bai Bin	Chang Shuoqi	Deng Qiyun	Fu Chenjian	Guo Xiayu
He Qiang	He Jiwai	Hu Zhongxiao	Huang Min	Huang Zhinong
Li Jianwu	Li Xiaohua	Li Xinqi	Li Yali	Liao Fuming
Liu Aimin	Liu Shanshan	Long Jirui	Lv Qiming	Ma Guohui
Mao Bigang	Qin Peng	Wang Kai	Wang Weiping	Wei Zhongwei
Wen Jihui	Wu Jun	Wu Xiaojin	Xiao Cenglin	Xie Zhimei
Xin Yeyun	Xu Qiusheng	Yang Yishan	Yang Yuanzhu	Yao Dongping
Yuan Longping	Zhang Haiqing	Zhang Qing	Zhang Yuzhu	Zhao Bingran
Zhu Xinguang	Zhuang Wen	Zou Yingbin		

图书在版编目（CIP）数据

超级杂交水稻育种栽培学 = Super Hybrid Rice Breeding and Cultivation : 英文 / 袁隆平主编 ;骆忠武等译. — 长沙 : 湖南科学技术出版社, 2023.7
ISBN 978-7-5710-2103-0

Ⅰ. ①超… Ⅱ. ①袁… ②骆… Ⅲ. ①水稻—杂交育种—英文 Ⅳ. ①S511.035.1

中国国家版本馆CIP数据核字(2023)第047153号

Super Hybrid Rice Breeding and Cultivation

主　　编：袁隆平

执行主编：廖伏明

译　　者：骆忠武 等

责任编辑：欧阳建文

特约编辑：侯晓卿

出版发行：湖南科学技术出版社

社　　址：长沙市芙蓉中路416号泊富国际广场40楼

网　　址：http://www.hnstp.com

印　　刷：长沙超峰印刷有限公司

厂　　址：宁乡市金州新区泉洲北路100号

邮　　编：410600

版　　次：2023年7月第1版

印　　次：2023年7月第1次印刷

开　　本：889mm×1194mm　1/16

印　　张：46.5

字　　数：1378千字

书　　号：ISBN 978-7-5710-2103-0.

定　　价：300美元

Catologing In Publication (CIP)

Super Hybrid Rice Breeding and Cultivation: English Version / Yuan Longping: Editor-in-Chief ; Luo Zhongwu, etc. : Translators. — Changsha: Hunan Science and Technology Pressing House, July, 2023.
ISBN 978-7-5710-2103-0

I. ①Super… II. ①Yuan… ②Luo… III. ①Rice—Hybrid Breeding—English IV. ①S511.035.1

China National Archives of Publications and Culture — CIP Number (2023) No. 047153

Super Hybrid Rice Breeding and Cultivation

Editor-in-Chief: Yuan Longping

Executive Editor-in-Chief: Liao Fuming

Translators: Luo Zhongwu, etc.

Editor: Ouyang Jianwen

Assistant Editor: Hou Xiaoqing

Publication&Distribution: Hunan Science and Technology Pressing House

Address: 40F, BOFO International Plaza, No.416, First Section, Furong Middle Road, Changsha

Website: http://www.hnstp.com

Printing House: Changsha Chaofeng Printing Co., LTD

Address: No. 100 Quanzhou North Road, Jinzhou New District, Ningxiang City, Hunan Province

Postcode:410600

Edition: The first edition, July, 2023.

Printing: The first printing, July, 2023.

Format: 889mm×1194mm 1/16

Sheets: 46.5

Words: 1378,000

ISBN 978-7-5710-2103-0

Price: $300

Preface

As former US Secretary of State Henry Kissinger once said, "control oil and you control nations; control food and you control people." Similarly, a Chinese saying goes, "with sufficient food in hand, we feel assured." This shows the importance of food to the security of a country and its people.

Rice is the most important food crop in the world, and it is the staple food of more than half of the world's population which has exceeded 7.4 billion up to now and is still increasing, predicted to reach 9.3 billion by 2050. In contrast, the area of cultivated land is shrinking. Therefore, increasing rice yield per unit area is of great significance to global food security.

China took the lead in making use of rice heterosis, which greatly improved rice yield. In 1976, China began to plant hybrid rice on a large scale, and the yield per unit area was about 20% higher than that of conventional rice varieties. In 1995, two-line hybrid rice was successfully put into production in China and the average yield reached 7,240 kg/ha, 10.2% higher than that of three-line hybrid rice (6,569 kg/ha) (National Agro-Tech Extension and Service Center, 2003).

The breeding of super high-yielding rice has been the most important research project of many countries and research institutes since the 1980s. In 1996, the Ministry of Agriculture (MOA) of China launched China's super rice breeding program. Through the cultivation of super hybrids with high-yielding plant types and the utilization of intersubspecific heterosis, China succeeded in the breeding of super high-yielding rice, and achieved its goals in three phases. Large-scale planting showed that the yield of super hybrid rice per hectare was more than 750 kg higher than that of common high-yielding hybrid rice.

In order to better promote super hybrid rice, we have compiled this book which consists of 20 chapters in five parts: The Basics, Breeding, Cultivation, Seeds and Achievements. The first part discusses the heterosis and male sterility of rice, and the theory and strategies of traditional and molecular breeding. The second part examines the breeding of male sterile lines and restorer lines, the selection of super hybrid rice combinations, and the outlook of breeding third-generation hybrid rice. The third part discusses the ecological adaptability, growth and development, cultivation physiology and cultivation techniques, as well as the control of main diseases and pests of super hybrid rice. The fourth part introduces the production techniques for male sterile lines, hybrids, and seed quality control of super hybrid rice. The last part introduces the production extension of super hybrid rice, describes the backbone parents, combinations and award-winning achievements in super hybrid rice development. This book presents a systematic discussion of the basic principles and techniques regarding the breeding, seed production and cultivation of super hybrid rice, which we hope will be helpful to those engaged in the research and promotion of agricultural science and technology.

There are always secrets of nature for mankind to explore and there is no end to scientific development. Therefore, we will never stop striving for progress in regard to super hybrid rice. We firmly believe that by making use of advanced biological and information technology, and combining traditional and molecular breeding, the yield, quality and resistance of super hybrid rice will be further improved and we will have a brighter future in the promotion of super hybrid rice.

Despite efforts by the authors, there might still be some errors in the book. Readers are welcomed to point out any such errors so that we can make corrections in the second edition.

Editors

Contents

Part I

The Basics

Chapter 1

Heterosis of Hybrid Rice

He Qiang / Lv Qiming

Section 1 Introduction

Ⅰ. Discovery of Heterosis

Heterosis is a common phenomenon in the biological world. As early as 2,000 years ago, *Qimin Yaoshu* (*Important Arts for the People's Welfare*), a book on agricultural science written by Jia Sixie of the Northern Wei Dynasty (386 – 534) in ancient China, recorded the fact that a mule is the product of a horse crossed with a donkey, which marked the start of human observation and utilization of heterosis. Research on crop heterosis first started in Europe in the mid-18th century when Kolreuter, a German scientist, observed heterosis in the interspecific hybridization of Carnation, *mirabilis jalapa* and Tobacco. In the 1920s and 1930s, the United States adopted a suggestion made by D. F. Jones, a maize geneticist and breeder, to carry out maize double cross breeding, which made the utilization of heterosis among maize inbred lines practical and feasible. The planting area of hybrid maize reached 0.1% (about 3,800 ha) of the total corn planting area in the United States and marked the start of heterosis utilization in cross-pollinated crops. In 1937, American scholar J. C. Stephens proposed to utilize heterosis in the breeding of male sterile lines in sorghum. In 1954, it was reported that the sorghum male sterile line 3197A was developed by crossing West African sorghum with South African sorghum, and a restorer line was selected from sorghum Leytbaying 60. Hybrid sorghum was prepared with the three-line method, and its application in production set a good example for the utilization of heterosis in often cross-pollinated crops.

Before the 1960s, it was generally believed that inbreeding in cross-pollinated and often cross-pollinated crops such as maize and sorghum would result in inbreeding depression, while hybridization would bring heterosis. However, because of long-term natural and artificial selection, the lethal genes of self-pollinated crops such as rice and wheat have been eliminated and favorable genes have been accumulated and preserved, so there is neither inbreeding depression nor heterosis in self-pollinated crops. Given the view that self-pollinated crops have no heterosis, there had rarely been studies on the heterosis of self-pollinated crops in the past. Moreover, to make use of rice heterosis, there must be all three lines available but this is all the more difficult because rice is a self-pollinated crop with small floral organs. In 1964, Yuan Longping initiated research on the utilization of rice heterosis. In 1973, he successfully completed the three-line set and developed Nanyou 2 and other hybrid rice combinations with strong heterosis, which were soon put into large-scale production. This success in China firmly proves that besides cross-pollinated and often cross-pollinated crops, self-pollinated crops also have strong heterosis.

Research on the utilization of rice heterosis first began in the 1920s. In 1926, American scholar J. W. Jones first proposed rice heterosis and found that some rice hybrids had obvious advantages in tillering ability and yield compared with their parents, which attracted the attention of rice breeders all over the world. After that, B. S. Kadam (1937) of India, F. B. Brown (1953) of Malaysia, A. Alim (1957) of Pakistan and Hiroko Okada (1958) of Japan all studied rice heterosis. The utilization of rice heterosis began with the breeding of male sterile lines. In 1958, Katsuo and Mizushima from Tohoku University, Japan, crossed Japanese rice Fujisaka 5 with Chinese wild rice Hongmang, which combined the nucleic gene of cultivated rice with the cytoplasm of wild rice, and obtained homozygous and stable male sterile materials by continuous backcross, and finally succeeded in breeding the Fujisaka 5 male sterile line. In 1966, Shinjyo C. and O'mura T. from the University of the Ryukyus, Japan, crossed and backcrossed *indica* rice Chinsurah Boro Ⅱ with Taichung 65, a *japonica* rice variety from Taiwan, China, and developed the BT-type Taichung 65 male sterile line with the cytoplasm of Chinsurah Boro Ⅱ. Most *japonica* rice varieties have the ability to maintain sterility, but it is difficult to find the corresponding restorer lines (Shinjyo C., 1966; 1969; 1972a, b).

In 1968, Watanabe of Japan Institute of Agricultural Technology crossed Burmese *indica* rice Lead Rice with Fujisaka 5 and developed the Fujisaka 5 male sterile line with the cytoplasm of Lead Rice. Although a small number of homo-cytoplasmic restorer lines were found and a three-line set was completed, it could not be planted on a large scale because of weak heterosis and difficulty in seed production. In 1970, Li Bihu, an assistant of Yuan Longping, and Feng Keshan, a technician from Nanhong Farm, discovered a plant of typical abortive male sterility with thin, yellow anthers and indehiscent pollen in a common wild rice community in Yaxian county, southern China's Hainan province, which was named Yebai or wild abortion (WA). In 1972, rice breeders in Jiangxi and Hunan provinces bred male sterile lines of the WA type such as Zhenshan 97 A and Erjiunan 1A, as well as their maintainer lines. In 1973, rice breeders in Guangxi and Hunan successively selected such strong and excellent restorer lines as IR24 through test crossing, and completed the three-line set.

Thus, the utilization of rice heterosis in production became a reality. In 1976, China began to plant hybrid rice on a large scale, and by 2013, more than 4,000 hybrid rice varieties were developed with a total planting area of about 500 million hectares. Up to now, Chinese hybrid rice has been demonstrated and planted in more than 30 countries in Southeast Asia, South Asia, Africa and America. The annual planting area outside China has reached more than six million hectares. The successful utilization of rice heterosis is very important for China's food self-sufficiency and the alleviation of food shortage in the world.

Ⅱ. Measurements and Manifestations of Heterosis

1. Definition of heterosis

Heterosis refers to the superiority of the hybrid generation produced by the crossing of two genetically different parents over their parents in such traits as growth vigor, viability, reproduction rate, stress resistance, adaptability, yield and quality. In 1908, American scholar G. H. Shull put forward the concept of heterosis and applied it to maize cross breeding. It is called heterosis utilization when the first-generation hybrid was put into agricultural production for maximal economic benefit.

Heterosis in rice refers to the phenomenon that the first-generation (F_1) rice hybrid produced by crossing two parents with different genetic compositions outperforms its parents in vitality, growth vigor, adaptability, stress resistance and/or yield. From the perspective of agricultural production, rice heterosis is ultimately reflected in yield (Xie Huaan, 2005). Since the successful completion of the three-line set in 1973, Chinese scientists have been carrying out studies on the utilization of rice heterosis, including the relevant genetic mechanism and techniques, parent and hybrid breeding, resource mining, seed production, hybrid rice physiology and ecology, cultivation, and seed testing and processing. These studies have become an important part of rice breeding and cultivation in China, and a comprehensive and systematic discipline has been formed.

2. Measurements of heterosis

Heterosis, though common in the biological world, is a quite complex biological phenomenon. There are many forms of heterosis, but on the whole, they can be divided into positive heterosis and negative heterosis. When the trait value of the F_1 hybrid is higher than that of its parents, it is called positive heterosis, while when the trait value of the F_1 hybrid is lower than that of its parents, it is called negative heterosis. The yield of most hybrid rice varieties, as well as the plant height, spike length and effective tiller generally show positive heterosis for both hybrids between varieties and hybrids across subspecies. In terms of interspecific hybridization, the flag leaf area, root activity, spike weight and grain weight generally show positive heterosis, while chlorophyll content, amylose content and gel consistency tend to show negative heterosis. In the varietal crosses, the probability of negative heterosis is much higher than that of positive heterosis. Since what man wants is not exactly the same as what is beneficial to living organisms, some of the heterosis are favorable for the living organism, but unfavorable for human beings. Depending on the different perspectives of research, evaluation and utilization of heterosis, the commonly used measurements of heterosis are mid-parent heterosis, over-parent heterosis, standard heterosis, relative heterosis and the heterosis index.

(1) Mid-parent heterosis

Mid-parent heterosis refers to the ratio of the measured value of a trait in the F_1 hybrid to the average value of the same trait in its parents.

$$V=\frac{F_1-MP}{MP}\times100\%$$

where F_1 is the trait value of the single cross hybrid and MP is the average value of the parents ($MP=$

$\frac{P_1+P_2}{2}$). The greater the difference between F_1 and MP is, the stronger the heterosis will be.

(2) Over-parent heterosis

Over-parent heterosis, including high-parent heterosis and low-parent heterosis, refers to the ratio of one trait value of the F_1 hybrid deviating from the value of the same trait of its high-value or low-value parent.

$$V=\frac{F_1-HP(\text{or } LP)}{HP(\text{or } LP)}\times 100\%$$

where F_1 is the trait value of the single cross hybrid and HP or LP is the trait value of the high-value or low-value parent. The greater the difference between the two is, the stronger the high-parent or low parent heterosis will be.

(3) Standard heterosis

Standard heterosis refers to the deviation ratio of a certain trait value of the F_1 hybrid from the value of the same trait in the control variety.

$$V=\frac{F_1-CK}{CK}\times 100\%$$

where F_1 is the trait value of the single cross hybrid and CK is the value of the control (check) variety. The greater the difference is, the stronger the heterosis will be.

(4) Relative heterosis

Relative heterosis refers to the deviation ratio of the trait value of the single cross hybrid from the average value of the same trait of the parents to half of the difference of the same trait of the parents.

$$hp=\frac{F_1-MP}{\frac{1}{2}(P_1-P_2)}$$

where F_1 is the trait value of the single cross hybrid, P_1 and P_2 are the trait values of the two parents, and MP is the average trait value of the two parents.

If $hp=0$, then there is no vigor (no heterosis); if $hp=\pm 1$, there is positive or negative heterosis; if $hp>1$, there is positive over-parent heterosis; if $hp<-1$, there is negative over-parent heterosis; if $-1<hp<0$, there is partial negative heterosis; and if $0<hp<1$, there is partial positive heterosis.

(5) Heterosis index

The heterosis index is the ratio of one trait value of the single cross hybrid to the value of the same trait of its parents.

$$a_1=\frac{F_1}{P_1} \qquad a_2=\frac{F_1}{P_2}$$

The higher the heterosis index is, the stronger the heterosis will be. The possibility of developing a heterotic hybrid will grow as the heterosis index increases. There are great differences in such quality traits of hybrid rice as chalkiness size, chalkiness rate, gel consistency and head rice rate, both in high-parent and low-parent heterosis indices. Among the agronomic traits, the difference in the number of grains per panicle is the greatest, followed by seed setting rate, effective panicle rate and number of effective panicles per plant.

All the measurements mentioned above are valuable for analyzing the heterosis in biological traits. However, standard heterosis is of greater practical value.

3. Manifestations of heterosis

Rice heterosis is manifested in many aspects, and it is obvious in many traits. Heterosis can be found in both external morphology and internal structure, in physiological and biochemical indicators and processes, and in enzyme activity. From the perspective of economic traits, the heterosis of hybrid rice is mainly shown in nutrition, reproduction, resistance and quality.

(1) Heterosis in nutrition

The F_1 hybrid has great growth vigor and strong heterosis in nutrition. Its heterosis in nutrition over conventional rice is shown in the following aspects.

1) Tall plants

Hybrid rice generally has obvious and mostly positive heterosis in plant height. Researchers at Jiangxi Academy of Agricultural Sciences (JAAS) carried out studies on 29 hybrid rice combinations, and found positive heterosis in plant height in 27 of them. At present, the parents with the semi-dwarf trait are generally selected for the breeding of hybrids so as to control the plant height of the hybrid.

2) Faster seeds germination with early and strong tillering

Researchers at Hunan Agricultural University (HAU) carried out studies on the seed germination rate of hybrid rice Nanyou 2 and its parents, and found that the hybrid germinated much faster than its male sterile parent. The Shanghai Institute of Plant Physiology and Ecology (SIPP) observed that single-cropping rice hybrids Nanyou 2 and Nanyou 6 began to tiller 12 days after sowing, 6 – 8 days earlier than their male parents. Researchers at Guangxi Academy of Agricultural Sciences (GAAS) investigated the highest seedling numbers of the hybrid Nanyou 2, the maintainer Erjiunan 1, the restorer IR24 and the control variety Guangxuan 3 under the same condition, and the results showed that the number of seedlings of Nanyou 2 reached 4.24 million per hectare, an increase of 0.29 – 1.25 million per hectare over its parents and the control variety.

3) Well-developed root system with deep and wide distribution and active absorption and synthesis

According to researchers at Hunan Academy of Agricultural Sciences (HAAS) and SIPP, Nanyou 2 has obvious heterosis in root number and root weight compared with its parents and conventional rice. Researchers at Wuhan University studied the characteristics of the root growth and respiratory metabolism of four rice hybrids and their parents, and found that the hybrids were superior to their parents in terms of the weight and volume of the root system. The root protein content reached its peak from heading to grain filling. The length and diameter of the roots, lateral roots and surface roots of the hybrids had the characteristics of both of their parents with a high increment. Under various seedling ages and nitrogen levels, Shanyou 63 had obvious heterosis over its parents Minghui 63 and Zhenshan 97A in terms of total dry weight per plant, dry weight of root system, relative growth rate of root system and nitrogen absorption rate (Xie Huaan, 2005).

4) More green leaves per plant with thicker, larger canopy and enhanced photosynthetic function.

According to Wuhan University (1977), under the same cultivation conditions, the leaf area per plant of the hybrid Nanyou 1 was 58.8% and 80.6% larger than those of its male parent IR24 at the heading and maturity stages respectively. Hybrid rice is significantly superior to conventional rice in terms of the accumulation of organic matter in the early growth stage and the transportation of organic matter in the late growth stage (Xie Huaan, 2005).

(2) Heterosis in reproduction

The huge heterosis in vegetative growth lays a good foundation for reproductive growth. Compared with conventional rice, the F_1 hybrid has obvious heterosis in yield, which is resulted from large panicles, more and larger grains, as well as high grain yield. Hybrid rice has a significant mid-parent heterosis and over-parent heterosis in total number of grains per panicle, grain weight, number of panicles and yield per plant (S. S. Virmani et al., 1981).

1) Larger panicles with more grains

Hybrid rice has significantly larger panicles with more grains and balances well between the size and number of panicles. When the effective panicles per hectare reach 2.7 million, the total number of grains per panicle can reach 150 or even up to 200. JAAS surveyed 29 rice hybrids and found 89.75% of them to have positive heterosis in terms of total grains per panicle. An experiment showed that among the yield components of hybrid rice, the heterosis in total grains per panicle was the highest, and the average heterosis of all tested hybrids was higher than that of their parents (Zhang Peijiang, 2000). Sichuan Academy of Agricultural Sciences studied the grain structure of different rice varieties in China before 1980, and the results showed that the yield of the conventional dwarf variety in the 1960s was 31.3%–98.5% higher than that of the conventional tall variety in the 1950s with a 67.5%–77.7% increase in the number of panicles, but minor difference in the total grains per panicle and grain weight. The yield of hybrid rice in the 1970s increased by 11.2%–32.1% compared with that of conventional dwarf rice, mainly because of the 18.0%–30.9% increase in the total number of grains per panicle. The heterosis of hybrid rice in yield was realized through the heterosis of large panicles on the basis of a stable number of panicles. Intersubspecific hybrid rice has more prominent heterosis in that they have larger panicles and more grains. Zhu Yunchang (1990) observed 44 intersubspecific combinations, and found that 33 of them (or 75.0%) had more than 180 grains per panicle, 25 (or 56.82%) had more than 200 grains, and nine of them had more than 250 grains.

2) Big grains and high 1,000-grain weight

The grain weight of hybrid rice is generally higher than that of its parents. According to research findings, the 1,000-grain weight of hybrid rice varieties of the 1970s was 9.2%–12.0% higher than that of conventional dwarf rice varieties. Zeng Shixiong (1979) studied the heterosis of grain weight of 34 rice hybrids and found that 23 combinations had heterosis over their high-value parents, and 31 combinations exceeded the average value of their parents. JAAS studied the grain weight of 400 hybrids and their parents and found that 67.75% of the hybrids showed positive heterosis.

3) High grain yield

The yield increase of hybrid rice is realized through the increase in panicles per unit area, total grains per panicle and grain weight. According to statistics, the average yield of hybrid rice in China between 1986 and 1992 increased by 25.6%-45.4% over that of conventional varieties (Xie Huaan, 2005). A large number of studies on the yield of hybrid rice showed that the standard yield heterosis of hybrid rice ranged from 1.9% to 157.4%, and the yield heterosis of hybrid rice over their parents ranged from 1.9% to 386.6%. At present, super hybrid rice of *indica*, *japonica* and their intersubspecific hybrids planted on a large scale in China generally yields over 5.0% more than hybrid rice control varieties, and the increase is more obvious in comparison with conventional rice varieties.

4) Long growth duration

Generally, growth duration is an inherited quantitative trait, which is greatly influenced by the ecotype of the parents. In intervarietal crosses, when both parents are early maturing varieties, the heading date of the hybrid is generally earlier than that of their parents. For a hybrid between early and medium maturing parents, the heading date is around the mid-value of the parents'. For a hybrid of an early or medium maturing parent and a late maturing parent, the heading date is close to that of the late maturing parent. In intersubspecific hybridization, it is common for the growth duration of the F_1 hybrid to be longer than that of the late maturing parent. Luo Yuehua et al. (1991) studied the maturity of more than 30 intersubspecific hybrids of *indica* and *japonica* varieties derived from W6154S and a number of typical *japonica* rice varieties, and found that only one combination had a short growth duration, while the other combinations all matured later than their late maturing parents, and 92.86% of the hybrids had a longer growth duration than the control variety Shanyou 63.

(3) Heterosis in resistance

F_1 hybrids often have stronger resistance and adaptability to adverse environmental conditions than their parents due to their growth heterosis. Research shows that the F_1 hybrids of crops like rice, maize and rape oilseeds present obvious heterosis in resistance to lodging, diseases, drought and low temperature.

The resistance of hybrid rice combinations to pests and diseases depends on their parents' genetics of resistance. If the resistance is controlled by a single gene and subject to quality trait inheritance, the resistance performance of the hybrid depends on the dominant and recessive resistance genes of both parents. If the resistance is controlled by multiple genes and subject to quantitative trait inheritance, the hybrid will have the mid-value of its parents, or one similar to that of the more resistant parent, but there are always some exceptions. Most of the genes for resistance to rice blast and bacterial blight are dominant, while only a few are recessive. Researchers from Hunan Agricultural University in Chenzhou analyzed rice blast resistance in 224 rice hybrids and their parents and found that 102 F_1 hybrids had dominant resistance genes, 31 had recessive resistance genes, 15 had partially dominant genes, and 18 appeared to have new types of resistance. Some researchers studied the correlation between the blast resistance of hybrid rice combinations and that of their restorer lines and found a highly positive correlation between them. The blast resistance of male sterile lines also has a certain influence on the blast resistance of the hybrid, which indicates that the blast resistance of the hybrid is affected by both the restorer line and the male sterile line. However, if the male sterile line is susceptible to rice blast, the blast resistance of the hybrid is mainly affected by the blast resistance of the restorer line (Huang Fu et al., 2007).

Hybrid rice combinations have strong adaptability. Shanyou 63, a hybrid rice variety with wide adaptability, high heterosis and the largest planting area in China, has an outstanding heterosis in not only yield, but also ecological adaptability. It was approved for planting in more than 10 provinces (autonomous regions and municipalities) with a total planting area of over 60 million hectares in China. In a study of yield using 140 hybrids planted in different seasons and at different nitrogen application levels, it was found that all the hybrids had heterosis in yield (Yang Jubao et al., 1990). In hybrid rice yield trials of India, Malaysia, the Philippines, Vietnam and some other countries from 1980 to 1986, the average maximal yield of different hybrids was between 4.7 t/ha and 6.2 t/ha, and the average heterosis over the control varieties was from 108% to 117%.

(4) Heterosis in quality

The quality of a hybrid rice combination is mainly determined by the quality traits inherited from the parents and there is a good correlation between the hybrids and their parents. Zhang Xueli et al. (2017) studied 12 quality traits and found that only the trait of amylose content is affected more by the female parent, while all of the other 11 traits are less affected by the female, and all are subject to both the additive and non-additive effects of genes. The parents have obvious interaction effects on protein content, brown rice and milled rice, transparency and alkali spreading value. Li Shigui et al. (1996) believed that it is easier to realize heterosis in grain weight and grain width, but more difficult to have heterosis in length/width ratio, chalkiness rate and chalkiness area. The grain weight, grain width, chalkiness rate and chalkiness area of a rice hybrid are between the levels of the parents and close to that of the high-value parent; while the grain length and length/width ratio are between the parents' levels and close to that of the low-value parent. The MOA had approved 108 super hybrid rice combinations as of 2017, and among them, more than 45% are high-quality varieties above Grade 3 and 20% are above Grade 2.

Ⅲ. Prediction of Heterosis

The utilization of heterosis is an effective way to improve crop yield, quality and resistance. After more than a century's development, great achievements have been made in utilizing the heterosis of major crops such as rice, maize, rape oilseeds, cotton, soybean and vegetables. However, so far, we are still in the stage of in-depth exploration of the mechanism of heterosis, and have not succeeded in fully revealing it at the molecular level. This leads to a lack of an effective prediction model for the utilization of heterosis, which, in turn, results in a long breeding cycle, heavy workload, and blind and inefficient hybrid breeding. Therefore, the quick and accurate prediction of heterosis has become a hot topic and key issue in the research of crop heterosis.

So far, there have been many reports on heterosis prediction, mainly including population genetic prediction, physiological and biochemical prediction, molecular genetic prediction and so on.

1. Population genetic prediction

(1) Heterotic group prediction

The term "heterotic group", which first appeared in maize breeding, refers to a group of inbred lines with rich genetic basis, close co-ancestral relationship, similar main traits and strong general combining ability. A hybrid from individuals within one heterotic group generally does not show significant heterosis, but crosses among individuals from different heterotic groups tend to have high heterosis. The crossing between two heterotic groups, which is likely to produce strong heterosis, is called a heterosis utilization mode. The theory of heterotic grouping in maize, known as the third great leap in the hybrid maize breeding theory, contributes to the rapid development of modern hybrid maize breeding (Lu Zuomei et al., 2010). In the 1940s, a high heterosis mode for maize hybrids was found between Reid's Yellow Dent of the South and Lancaster of the North through pedigree analysis. This heterosis utilization mode became widely known and has been in use since its discovery. With the development of breeding technology, the heterotic groups of maize inbred lines in the United States are mainly classified by ecological region, pedigree, combining ability and molecular marker. No matter which classification is adopted, the genetic difference between parents is the fundamental principle of heterosis.

Research of heterotic grouping in rice lags far behind due to the less attention paid to it. Existing research in this aspect has been quite fuzzy and shallow with unclear conclusions. The rice species cultivated in China is *Oryza sativa*, which can be divided into two major subspecies, *indica* and *japonica*. *Indica*, with a longer history and wider distribution, has become the major heterotic group for rice heterosis utilization in China due to its great adaptability and stress resistance, as well as other excellent traits such as the semi-dwarf plant, high photosynthetic efficiency, heat resistance, and growth vigor (Wang Xiangkun et al., 1997; 1998). A five-level classification system composed of species, subspecies, ecogroups, ecotypes and varieties was proposed for *Oryza sativa* based on origin and evolution, morphological traits, daylength and temperature, isozyme and hybridization compatibility. This classification is of certain theoretical and practical significance for studying the relationship between genetic differences and heterosis, as well as the heterotic grouping of *indica* rice (Wang Xiangkun et al, 2000). Zhang et al. (1995) believed that there are two heterotic groups of rice varieties in South China and Southeast Asia. Xie et al. (2015) found that the male sterile lines and restorer lines of the three-line and two-line hybrid rice combinations planted on a large scale in China are from two different *indica* rice subgroups. Chen Liyun et al. (1992) studied heterosis in hybrids among the different ecotypes of early, middle and late rice, concluding that the hybrids of mid-season rice have a short growth duration and short stem, but more grains per plant; therefore, it has a high value for production and utilization. Sun Chuanqing et al. (1999) found that hybrids between the partial *japonica* variety N422S and American rice, African *japonica*, Yunnan-Guizhou *japonica* and improved varieties of North China *japonica* are dominant ecotypes, while the hybrids between the partial *indica* variety Peiai 64S and American rice, improved varieties of North China *japonica* and African *japonica* are dominant ecotypes.

(2) Combining ability prediction

Griffing (1956) proposed a linear model to predict heterosis in crops based on combining ability, which was the first time that combining ability was used to predict crop heterosis. According to the Griffing model, the heterosis of F_1 hybrids can be predicted by the general combining ability (GCA) of the two parents when the special combining ability (SCA) is not significant.

Most studies concluded that both the GCA and the SCA play important roles in a hybrid's yield and other important traits. The high yield heterosis of a hybrid is only possible when both the GCA of the parents and the SCA of the hybrid are high. Zhou Kaida et al. (1982) studied the incomplete diallel crosses of six male sterile lines and five restorer lines and found that the yield and other traits of the hybrids are affected by the GCA of both parents and the SCA of the hybrid. The GCA of the parents is more important than the SCA of the hybrid and heterosis can be roughly predicted through the total effect of parental GCA and the hybrid's SCA. Gordon (1980) proposed a method for parent selection and heterosis prediction based on the estimation of the GCA with computer simulation of incomplete diallel crosses. It was believed that if a trait has large additive genetic variance, the hybrid's heterosis in the trait can be predicted through the estimation of the GCA. Ni Xianlin et al. (2009) used incomplete diallel crosses of five male sterile lines and four restorer lines to study the correlation between SCA and heterosis in yield of hybrid rice and found a significant positive correlation between the SCA and the standard heterosis or mid-parent heterosis. Therefore, to some extent, or within a certain range, heterosis can be predicted through the estimation of the SCA.

(3) Genetic distance prediction

The genetic difference between the parents is the genetic basis of heterosis. In 1970, G. M. Bhat first applied genetic distance to parent selection for wheat hybridization. Generally speaking, in a certain range, the greater the genetic distance between the parents, the higher the probability of high heterosis in the hybrid.

Genetic distance has been used to guide parent selection and the prediction of hybrid combinations since the 1970s (G. M. Bhat, 1973; Liu Laifu, 1979; Huang Qingyang, 1991; He Zhonghu, 1992; Wang Yiqun et al., 1998). Hou Heting et al. (1995) found that there is a very significant parabolic regression between genetic distance and the SCA of sorghum hybrids, and it can be used to predict the SCA of a hybrid, which can improve the predictability in breeding hybrids with high heterosis. Wang Yiqun et al. (2001) studied the relationship between genetic distance and heterosis on the basis of the quantitative traits of 70 sweet corn inbred lines and the results showed that genetic distance can be used to predict the heterosis in the yield of hybrids because of a significant quadratic relationship between the standard heterosis and the genetic distance between the parents. Xu Jingfei et al. (1981) and Li Chengquan et al. (1984) attempted to determine the genetic distance of quantitative traits related to yield through multivariate analysis, and found a significant linear regression relationship between the genetic distance and yield heterosis in *indica* and in *japonica* rice hybrids. However, some studies found no direct correlation between genetic distance and heterosis (N. M. Cowen, 1987; A. K. Sarawgi et al., 1987; M. L. Sarahe et al., 1990).

2. Physiological and biochemical prediction

(1) Yeast prediction method

In 1962, Matzkov and Manzyuk proposed the yeast method for crop heterosis prediction. Yeast was cultured in the leaf extracts of one parent, both parents and the hybrid respectively and the results showed that 76% of the both-parent extracts had a promoting effect on yeast growth similar to that of the extracts of the hybrid, better than the effect of the single-parent extract. Li Jigeng et al. (1964) adopted the yeast method to predict maize heterosis, and the accuracy rate was 82.9%. Guan Chunyun et al. (1980) used the yeast method to predict the heterosis of *Brassica napus* hybrids and the accuracy rate was 66.7%.

(2) Cell-level prediction

Mitochondrial complementation, chloroplast complementation and cell homogenate complementation were used in the prediction of crop heterosis at the cell level involving both the parents and the hybrids.

In 1966, McDaniel and Sarkissian found for the first time that heterosis in the oxidative activity of mitochondrial mixture between maize inbred lines was higher than that of a single parent, and called this phenomenon "mitochondrial heterosis". In 1972, McDaniel adopted the method of mitochondrial complementation to predict barley heterosis and found a significant correlation between barley yield heterosis and mitochondrial heterosis.

Hybrid crops generally have a large photosynthetic leaf area, which can significantly improve photosynthetic intensity. There are some reports that the chloroplast mixture from the parents of heterotic hybrids has high photosynthetic activity due to *in vitro* chloroplast complementation. On the contrary, non-heterotic hybrids show no such complementation and no increase in the photosynthetic activity of the chloroplast mixture. In a research on three-line hybrid rice, Li Liangbi et al. (1978) found that the Hill reaction photosynthetic activity of the chloroplast mixture of the male sterile and restorer lines of a heterotic hybrid is higher than that of any one of the parents. In 1972, McDaniel found a correlation between yield heterosis and chloroplast heterosis in barley.

The concept of cell homogenate complementation was first proposed by Chinese scholars and used to predict rice heterosis (Yang Fuyu et al., 1978). The accuracy can be as high as 85%. Zhu Peng et al. (1987) compared the effects of chloroplast complementation and cell homogenate complementation in predicting the heterosis of hybrid rice at the seedling stage, and concluded that cell homogenate complementation produces more consistent and reliable results.

(3) Isozyme prediction method

D. Schwartz initiated a study on the prediction of maize heterosis by isozyme in 1960 and found a "hybrid enzyme" band in the isozyme spectrum of heterotic hybrids, which was speculated to be related to heterosis. Li Jigeng et al. (1979, 1980) studied the relationship between isozyme and maize heterosis and found that the zymograms of heterotic hybrids and their parents are different or complementary, while the zymograms of non-heterotic hybrids or low-heterotic hybrids are generally the same as those of their parents. Many studies found new bands of esterase or peroxidase isozymes in organs or tissues of rice hybrids, such as seedlings, young leaves, pollen and stamens. A heterotic hybrid has the enzyme bands of both parents or has relatively high enzyme activity, while non-heterotic hybrids or low-heterotic hybrids have the same enzyme bands as one of their parents.

Zhu Peng et al. (1991) studied the relationship between rice heterosis and the activities of malate dehydrogenase (MDH) and glutamic dehydrogenase (GDH), and believed that MDH can be used for early prediction of rice heterosis. Sun Guorong et al. (1994) studied the activity of glutamine synthetase during the growth of hybrid rice and its parents, and found that the activity of glutamine synthetase during the reproductive stage reflected the level of heterosis in yield to a certain extent. Zhu Yingguo et al. (2000) believed that there is a significant correlation between rice heterosis and the isozyme diversity index.

3. Molecular genetic prediction

(1) DNA molecular marker prediction

DNA molecular markers are used to measure the molecular genetic distance between the parents, which, in turn, is used to predict crop heterosis. At present, restriction fragment length polymorphism (RFLP), random amplified polymorphism DNA (RAPD), amplified fragment length polymorphism (AFLP), simple-sequence repeat (SSR), sequence-tagged microsatellites (STMs), single nucleotide polymorphism (SNP), intron fragment length polymorphism (IFLP) and single strand conformation polymorphism (SSCP) are widely used in the prediction of crop heterosis.

RFLP has long been used to predict crop genetic diversity and heterosis. A high correlation between the genetic distance based on RFLP markers and the heterosis of F_1 maize hybrids was found in a study of maize inbred lines and crosses, and it was thus believed that genetic distance based on RFLP markers can be used to predict heterosis (M. Lee et al., 1989; O. S. Smity et al., 1990); Zhang Peijiang et al. (2001) found a significant correlation between the RFLP genetic distance and mid-parent heterosis of rice in terms of total grains per panicle, and a very significant correlation with standard heterosis.

With the development of molecular biotechnology, new DNA molecular markers have been used in the research of crop heterosis. Peng Zebin et al. (1998) found significant correlations between the genetic distance of RAPD molecular markers and yield, mid-parent heterosis, and the SCA of the hybrids derived from 15 maize inbreds of six types. Zhang Peijiang et al. (2000) showed that the mid-parent heterosis in total grains per panicle is significantly correlated with the genetic distance of the parents, and a highly significant correlation between the mid-parent heterosis and the standard heterosis in total grains per panicle in rice based on RAPD molecular markers. Fu Hang et al. (2016) used SSR molecular markers closely linked to functional genes associated with various traits of rice hybrids to predict hybrid heterosis in Southwest China's Sichuan province and showed that the heterozygosity rate of SSR molecular markers associated with grain weight and grains per panicle have a highly significant correlation with yield per plant, and concluded that SSR molecular markers can be used to effectively predict the over-parent heterosis in yield per plant. Zhang et al. (1994, 1996) proposed measuring parental genotype heterozygosity in research on the screening of molecular markers for heterosis-related traits in rice using general heterogeneity (genetic differences between two parents estimated by all molecular markers) and special heterogeneity (genetic differences between parents estimated by molecular markers with significant effects on a single trait determined by one-way analysis of variance). It was found that usually there is a low correlation between the general heterogeneity of the parents and the performance of F_1 hybrids, while there is a significantly positive correlation between the special heterogeneity of the parents and the performance of F_1 hybrids.

However, some scholars believe that the genetic distances of RFLP, RAPD, SSR and other molecular markers have lower value in the effective prediction of hybrid heterosis because of the low correlation with heterosis (E. B. Godshalk et al., 1990; J. W. Dudley et al., 1991; J. Boppmmaier et al., 1993; J. Xiao et al., 1996; S. P. Joshi et al., 2001). The conclusion from known studies show that the correlation, ranging from low to high, between the genetic distance of DNA molecular markers and heterosis is still not enough for accurate prediction of hybrid heterosis.

(2) Prediction of QTL genetic information

Many traits of crops are quantitative traits controlled by polygenes with minor effects. With the rapid development of molecular genetics, the heterozygosity of quantitative trait locus (QTL) and relevant interactions are used for predicting crop heterosis. Xiao et al. (1995) used the recombinant inbred line F_7's obtained from a cross between *indica* and *japonica* rice to backcross with its parents respectively to detect and analyze the molecular marker polymorphism of 12 quantitative traits related to yield. It was found that 60% of the 37 QTLs showed dominant effect and 27% of them showed the partially dominant effect, and concluded that the dominant complementary degree of QTLs that affect the yield of parents can be used to predict heterosis. R. Bernardo (1992) used mathematical modeling and field data analysis to reach the conclusion that the accuracy of predicting heterosis by molecular markers mainly depends on the coverage of QTLs related to heterosis and the proportion of markers linked to heterosis-related QTLs. He believed that at least 30%-50% of QTLs were linked to molecular markers. Wu Xiaolin et al. (2000) classified the markers into specific markers significantly correlated with QTLs and non-specific markers uncorrelated with QTLs based on the correlation between molecular marker loci and heterosis and showed that increased coverage of QTLs can greatly improve the accuracy of heterosis prediction, while heterosis prediction by discrete markers unlinked to QTLs is of low effectiveness. Gang et al. (2009) designed a genome-wide oligonucleotide microarray based on those known and predicted *indica* rice genes to study hybrid rice Liangyoupeijiu and its parents and examined leaves at the seedling and tillering stages, flag leaves at the booting, heading, flowering and filling stages, and panicles at the filling stage. There are 3,926 genes with differential expression during these seven different stages and those involved in energy metabolism and transportation are only differentially expressed between F_1 hybrids and their parents, and most of them are located in QTLs related to yield, which provides candidate genes for heterosis prediction.

(3) Best linear unbiased prediction

Best linear unbiased prediction (BLUP), also known as genomic hybrid breeding, is a method proposed and verified by Xu et al. (2014, 2016) for the prediction of the yield of hybrid rice. Transcriptome and metabolomic data are used as potential resources for yield prediction. The performance of all possible hybrid combinations (21,945) from 210 inbreds was successfully predicted based on the phenotypic data of 278 randomly selected training set hybrids. The yield of the first 100 high-yielding combinations was 16% higher than the average yield of all possible hybrid combinations. This method is more effective for traits with high heritability. The yield of the top 10 high-yielding hybrids selected from the top 100 high-yielding combinations by metabolite prediction was about 30% higher. Compared with genome prediction, the predictability of hybrid yield is nearly twice as high when metabolomics data was used.

This prediction method provides technical support for quick and effective identification of the best hybrid combination among numerous hybrid combinations.

Section 2 Genetic Mechanism of Heterosis

Ⅰ. Genetic Basis of Heterosis

Heterosis is a complex phenomenon in biological genetics. The genetic causes of heterosis were studied and discussed as early as the end of the 19th century. With wide application of heterosis in crop production, a lot of studies have been carried out on the genetic mechanism of heterosis from different aspects and perspectives in the last century, resulted in various hypotheses. At present, with the rapid development of molecular biotechnology, evidence has been found for these various hypotheses from molecular genetics, but the genetic mechanism of heterosis is still under exploration.

1. Major hypotheses of the genetic mechanism of heterosis and their verification

(1) Dominance hypothesis

The dominance hypothesis was first proposed by Charles Davenport in 1908 and was later developed by A. B. Bruce (1910). The basis of the hypothesis is that dominant traits are favorable in most cases, while recessive traits are harmful as a result of long natural selection and adaptation. Heterosis exists because F_1 hybrids inherit the favorable genes from both parents and partially dominant genes cover up the relatively recessive unfavorable genes. In other words, the dominant effect is the complementary effect caused by the aggregation of dominant genes from both parents in the hybrid. In 1917, Donald Jones further developed the hypothesis by introducing the concepts of linkage gene and additive effect.

One of the most powerful genetic evidence came from the experiment carried out by Frederick Keeble and Caroline Pellew in 1910. Two pea varieties 1.5 – 1.8 m in height were crossed; one variety had more nodes and shorter internodes, the other had fewer nodes and longer internodes. The dominant genes of both parents for nodes and internodes were aggregated in the F_1 hybrid with an obvious heterosis shown in the plant height of 2.1 – 2.4 m. Since the 21st century, the development of molecular biotechnology has provided molecular evidence for the dominance hypothesis. Chromosome segment substitution lines covering the whole rice genome derived from a cross of Zhenshan 97 and Minghui 63 was used for the study of plant height, a trait with high heritability and the results showed that all genes related to plant height show dominance, and most of the loci scattered between the parents are genes with a synergistic dominant effect (Shen et al., 2014). A high correlation between yield heterosis and the number of dominant alleles was found on the basis of genome-wide re-sequencing of 1,495 hybrid rice combinations. Most parents have only a few superior alleles, while the high-yielding hybrids have more superior alleles, indicating the aggregation of many rare superior alleles with positive dominance to form rice yield heterosis (Huang et al., 2015). Hybrid lineages, genome-wide re-sequencing and phenotypic data were used to study 10,074 F_2 individuals of 17 rice hybrids from the three groups of three-line, two-line and intersubspecific hybrids. The results showed that a small number of loci from the female parent's genome were the causes of the over-male parent heterosis in yield; that most of the genomic loci related to heterosis were positively dominant; and that partial dominance in some genomic loci was the cause of yield heterosis (Huang et al., 2016). The major genes related to yield traits, such as *Ghd7*, *Ghd7.1*, *Ghd8* and *Hd1*, cloned in the past decade have confirmed the important contribution of the dominance effect to rice heterosis (Xue et al., 2008; Yan et al., 2011; Yan et al., 2013; Garacia, 2008).

Though it is supported by a wide range of evidence, the dominance hypothesis clearly failed to explain the necessary association of dominant or recessive genes related to favorable or unfavorable traits. In fact, some recessive genes also play an important role in organisms, while some dominant genes are not the prerequisite of the growth and development of organisms.

(2) Overdominance hypothesis

The overdominance hypothesis, also known as the allelic heterozygosity theory, was first proposed separately by George Shull and Edward East in 1908. Then, in 1936, Edward East supplemented the overdominance hypothesis with the function of multiple allele accumulation. The theory believes that a hybrid's heterosis is caused by the interaction between heterozygous alleles of the parental genotypes and the degree of heterosis is closely associated with the degrees of heterozygosity, rather than dominant or recessive genes or the coverage by dominant genes or the accumulation of dominant genes in F_1 hybrids.

This theory has found evidence in some traits controlled by single genes. For example, Berger (1976) found that the alcohol dehydrogenase gene of maize clearly functions better in the heterozygous state. With molecular and genetic evidence, Krieger et al. (2010) proved that SFT is an overdominant locus contributing to the heterosis in the yield of tomato. In addition, the theory is supported by more molecular evidence in traits controlled by multiple genes. For example, Stuber et al. (1992) used recombinant inbred lines derived from maize inbred lines to construct backcross introgression lines with their parents. Drawing on genome-wide molecular marker data, they found that all QTLs are overdominant, and yield is highly correlated with marker heterozygosity. Li et al. (2001) used genome-wide molecular markers to analyze recombinant inbred lines (RILs) of intersubspecific hybrid combinations of Lemont and Teqing, crossing them with parents and two test cross lines respectively to obtain two backcross populations and two test cross populations for the study of the genetic mechanism of heterosis. Analysis showed that 90% of the heterosis-related QTLs of most traits are overdominant. In the components of rice yield, most of the major QTLs for tillers per plant and grains per panicle are overdominant (Luo et al., 2001).

Molecular biotechnology has provided more evidence in support of the overdominance hypothesis. However, it fully ignores the dominance effect in heterosis and the existence of dominant and recessive alleles and interactions between them.

(3) Epistasis hypothesis

The epistasis hypothesis was proposed by A. K. Sheridan in 1981. It believes that heterosis is the result of not only alleles, but also non-allelic interactions at different loci.

This hypothesis was supported by evidence at the molecular level. Li et al. (1997) analyzed the F_4 population derived from the intersubspecific rice hybrid of Lemont and Teqing, and found that the disharmonious interaction between the parents' alleles led to hybrid depression. A large number of interactions affecting yield were detected in the F_4 population, and more than 70% of the interactions occurred between non-major loci. The interaction of traits with low heritability such as grains and panicle weight was more important. An experiment using 34 reciprocal crosses from 17 rice inbred lines and one test cross showed that the yield heterosis is not prominent in general, but the product of those factors resulted in huge heterosis in yield. By the hierarchical additive effect model of heterosis, traits are divided into single-locus, component and complex traits. Single-locus traits, controlled by additive effects, can be involved in different component traits. The product of the additive effects that regulate the process factor is the source of heterosis in complex traits (Dan et al., 2015).

The above three hypotheses are classical genetic hypotheses for heterosis. However, we still have a very limited understanding of the genetic mechanism of rice heterosis, and there is no unified theory to explain it. The existing theoretical hypotheses and evidence explain the complex molecular genetic mechanism of heterosis from different aspects, but they fail to provide a full explanation and are not mutually exclusive. There is molecular evidence that dominance, overdominance and epistasis jointly regulate heterosis in a well-coordinated manner. Different studies on the hybrids of Shanyou 63 and its related populations found that dominance, overdominance and epistasis all played a role in heterosis. Through the analysis of yield traits in the $F_{2:3}$ population of Shanyou 63, it was found that most yield QTLs and a few yield component QTLs are overdominant but also important in interactions for yield heterosis. An analysis of 33 heterosis loci detected in the permanent F_2 population of Shanyou 63 revealed that various single-locus effects, such as partial dominance, complete dominance and overdominance, contribute significantly to heterosis, and the three types of epistasis is also an important part of the genetic basis of heterosis. Analysis of high-density gene mapping of the permanent F_2 population of Shanyou 63 showed that the accumulation of dominance and overdominance can well explain heterosis in grains per panicle, grain weight and yield, and that the dominant interaction contributes greatly to heterosis in the number of effective panicles (Yu et al., 1997; Hua et al., 2003; Zhou et al., 2012). Some researchers studied intersubspecific hybrid combinations of *indica* and *japonica* and also found the importance of dominance, overdominance and epistasis for intersubspecific rice hybridization (Xiao et al., 1995; Li et al., 2008; Wang et al., 2012).

2. Other hypotheses on the genetic mechanism of heterosis

(1) Gene network system hypothesis

The gene network system hypothesis was raised by Bao Wenkui (1990) on the basis of triticale distant hybridization. It believes that all organisms with different genotypes have a set of genetic information to ensure individual growth and development, including all coding genes, functional genes, regulatory sequences that control gene expression, and components that coordinate the interaction between different genes. Genome encodes the invisible information on DNA, forming a network for orderly genetic expression. Mutation in some genes affects other members of the network, further expands its influence through the network, and finally develops into visible variations.

According to this hypothesis, heterosis occurs in the new network system (F_1) formed with the combination of two different genetic groups, and allele members are in the best working state and the whole genetic system thus delivers its best performance.

(2) Genetic equilibrium hypothesis

The genetic equilibrium hypothesis, first put forward by K. Mather (1942), considers that the development of any trait is the result of genetic equilibrium. Turbon (1964, 1971) improved and supplemented the hypothesis by proposing that heterosis is a complex genetic phenomenon in a polygenic system, which is based on the interaction between genetic factors, the interaction between the cytoplasm and the nucleus, the relationship between individuals and phylogeny, and the influence of the environment on the development of traits. It believes that the loss of genetic equilibrium results in poor development of inbred lines of cross-pollinated plants and that hybrids of pure parents carefully selected have heterosis because heterozygous systems with genetic equilibrium are formed in them.

It is generally believed that the genetic equilibrium hypothesis only provides a conceptual explanation for the source of heterosis, but fails to explain the role and contribution of specific genes in the formation of heterosis.

(3) Active gene hypothesis

The active gene hypothesis suggests that heterosis is the result of the additive and interaction effects of active genes. The main arguments are that there are active and inactive genes in the gene pool because of genomic imprinting, and the activity of genes is temporary and not hereditary. The genes that produce heterosis are active minor effect genes which cannot be divided into dominant and recessive ones, and are only different in a cumulative effect. When the alleles that produce heterosis are homozygous, only one of them can be active due to genomic imprinting, and it plays a role in the formation of a phenotype; while the other is inactive, or has no effect on the formation of a phenotype. However, when these genes are heterozygous, they do not produce genomic imprinting, and can all be active and have their effect. The heterozygotes of hybrids have a greater number of genes not imprinted than homozygotes, and the additive and interaction effects are accumulatively great to produce heterosis (Zhong Jincheng, 1994).

(4) Polygene hypothesis

The polygene hypothesis, suggested by Swedish scholar Nilsson Ehle (1909), is used to explain the inheritance of quantitative traits. It points out that one quantitative trait is controlled by many genes that have small and almost equal effects on the trait. The effect of minor genes that control a quantitative trait is generally additive and not obviously dominant or recessive. Quantitative traits are usually the result of the accumulation of multiple minor effect genes.

Ⅱ. Effect of Cytoplasm on Heterosis

Heterosis is not only controlled by karyogene, but also influenced by cytoplasmic genes. In the male sterile system of rice caused by nucleo-cytoplasmic interaction, the sterile cytoplasm not only results in male sterility, but also has effects on agronomic traits, which not only directly reflects the different genetic basis of different cytoplasmic sources and their sterile lines, but also relates to the quality of economic traits and the prospect of heterosis utilization of hybrid rice.

F_1 hybrids with the same karyogene but different cytoplasmic backgrounds have significantly different heterosis, indicating the effect of nucleo-cytoplasmic interaction. A study of 12 male sterile lines from the eight different cytoplasmic sources of WA, Liubai, Shenqi, Gangxing, Hongye, Baotai, Dian 1, and Dian 3, as well as the corresponding maintainer lines crossed with the same restorer line to examine the effects of the male sterile lines on plant height, panicle neck length, days of heading, maximal tiller number, effective panicle number, panicle rate, total grains per panicle, filled grains per panicle, seed setting rate, grain weight, yield per panicle and plot yield. The results showed a minor positive effect of sterile cytoplasm on the traits of days of heading, maximal tiller number, total grains per panicle and grain weight, and a significant negative effect on all other traits. Comparing the hybrids of a male sterile line and a restorer line with the hybrids of a maintainer line and a restorer line, male sterile cytoplasm has the tendency of short plant and short panicle neck, delayed heading, low seed setting rate and low yield per plant despite the cytoplasmic sources. In addition, reciprocal hybrids of five homo-plasmic restorer lines for three typical cytoplasmic male sterile maintainer lines (WA, Hongye and Baotai) were compared with the hybrids derived from corresponding maintainers and restorers, and the results also show the negative effects of cytoplasm on heterosis, further proving a universal negative effects of the sterile cytoplasm on the main economic traits of F_1 hybrids. However, such negative effects on heterosis are only relative, because it will not change the direction and performance of heterosis in F_1 hybrids for hybrids derived from parents with high genetic diversity, high combining ability and good restoration.

Ⅲ. Heterosis Research Methodology

In the early days of research on the genetic mechanism of heterosis, classical genetic methodology was used to study the allelic interaction and non-allelic interaction within the nucleus and the nucleo-cytoplasmic interaction based on phenotype data that became available as science and technology develop generally. Such efforts were able to explain the three classical genetic hypotheses of heterosis to a certain, and sometimes a large, extent, but not to give a full explanation. With the development of molecular genetics and the application of new technologies and methods, explorations deepened and widened regarding the hypotheses of the heterosis mechanism, and the prediction models for heterosis found powerful molecular evidence.

1. QTL location and analysis

The location and systematic analysis of the loci that control quantitative traits make it possible to analyze the genetic mechanism of heterosis at the molecular level. Stuber et al. (1992) used isozyme markers and RFLP molecular markers to locate QTLs for grain yield and other traits in single-cross maize hybrids and successfully identified QTLs related to heterosis. Devicente et al. (1993) studied the QTLs related to heterosis in tomato hybrids, and identified no trait with heterosis in F_1 hybrids, but found significant overdominant effect in QTLs related to traits with heterosis.

2. Epigenetics

Epigenetics, as a branch of genetics, is the study of heritable phenotype changes that do not involve alterations in the DNA sequence. It covers many phenomena, such as DNA methylation, genomic imprinting, maternal effect, gene silencing, nucleolar dominance, dormant transposon activation and RNA editing. The most important function of DNA methylation is to inhibit the activity of transposons so as to maintain the stability of the genome. Some DNA methylation located in the gene promoter and genomic region may be related to the transcriptional activity of the gene. Therefore, DNA methylation has become an important research topic in epigenetics and epigenomics. Recent studies have shown that DNA methylation and high expression of small RNAs are correlated with heterosis, and verified the correlation between DNA methylation and heterozygote formation, thus conveying the idea that DNA methylation plays an important role in the formation of heterozygote (Hofmann, 2012). Some studies revealed that the DNA methylation level of rice hybrid chromosomes from their parents is similar to that of the corresponding parents. Further analysis showed that allele-specific DNA methylation occurs in hybrids, and it is speculated that allele-specific DNA methylation, together with allele DNA sequence differences, may play a role in the allele-specific expression of hybrids (Chodavarapu et al., 2012).

3. Transcriptomics

Hybrids tend to have different allelic expression from their parents and the transcriptome activity of the hybrids is therefore also different from that of their parents. By analyzing the differences in transcriptome activity between hybrids and their parents, identifying the regulatory factors of these differences, and establishing correlation with phenotypic differences, it is possible to explain the mechanism of heterosis formation at the molecular level (He Guangming et al., 2016). Research shows that some genes that express differently between hybrids and their parents might be associated with known QTLs for the yield trait in rice, and it was speculated that these genes might be candidates for yield heterosis in rice. Peng et al. (2014) used genome-wide microarray to measure and compare the transcripts of the hybrid rice Liangyoupeijiu, Liangyou 2163 and Liangyou 2186 and their parents at the flowering and grain filling stages. The results showed that there are a large number of differentially expressed genes between the hybrids and their parents. Functional analysis showed that there are many of these genes in carbohydrate and energy metabolism, especially in carbon fixation, and 80% of the differentially expressed genes are located in rice QTLs in the Gramene database, while 90% of the differentially expressed genes are located in yield-related QTLs.

4. Molecular markers

At present, researchers pay close attention to using molecular markers to study the mechanism of heterosis. As mentioned above, molecular markers are mainly used to predict heterosis. Because molecular markers exist in all parts of a crop genome, the genetic distance between different varieties can be well understood by analyzing the polymorphism of molecular markers in a crop genome, and hybrids with strong heterosis can be obtained. Previous studies have shown that the genetic distance between parents determines the level of genotype heterozygosity of hybrids, and ultimately affects the level of heterosis in the hybrids (Kang Xiaohui et al., 2015). Stuber et al. (1992) used 76 RFLP molecular markers that cover 90%–95% of the maize genome to analyze the relationship between the polymorphic composition of loci and the heterosis of maize varieties. They believed that heterosis in quantitative traits that determine the yield of maize is positively correlated with the heterozygosity of loci (the correlation coefficient is 0.68); the phenotype of traits controlled by single loci has a weaker correlation with heterozygosity, yet the correlation coefficient increases with the increase of the number of loci involved in determining the trait.

5. Proteomics

As an important component of cells and tissues, protein is the main undertaker of life activities, and has many important functions. For example, structural proteins participate in the construction of tissue structures, and functional proteins participate in matter transport, catalyze biochemical reactions, and send signals. Therefore, proteomics provides an important means for further study on the genetic mechanism of heterosis. The variation of differential proteins between hybrids and their parents was analyzed for their roles in the formation of biological heterosis (Zhang Yuanyuan et al., 2016). Many proteins expressed differentially are involved in stress response and metabolic processes in two-dimensional electrophoresis and mass spectrometry of rice hybrids and their parents (Wang et al., 2008). Marcon et al. (2010) used maize reciprocal hybrids and their parents to construct differential expression profiles of proteome by two-dimensional electrophoresis, and revealed that there are three types of non-additive proteins in the hybrids, namely, the dominant, the overdominant and the partially dominant, accounting for 21% of all proteins. Mass spectrometry also showed that some proteins related to the glucose metabolism pathway play an important role in the formation of heterosis in immature embryos. Fu et al. (2011) studied the heterotic proteome of five hybrid maize seeds at germination by two-dimensional electrophoresis and mass spectrometry, and found many heterosis-related proteins, most of which show non-additive expression patterns, and most of these expression patterns show high affinity and super high affinity expression. Therefore, crop heterosis results from the interaction between many proteins, and polygenic heterosis results from the difference in protein metabolism between hybrids and parents.

Section 3 Approaches and Methods of Rice Heterosis Utilization

The utilization of rice heterosis can be divided into three categories based on the heterosis level, i. e. intervarietal, intersubspecific and distant heterosis.

Ⅰ. Categories of Rice Heterosis Utilization

1. Intervarietal heterosis

At present, most three-line hybrids, either *indica* or *japonica*, can be regarded as representatives of intervarietal heterosis utilization. In the early 1970s, three-line hybrid rice, with the set successfully completed, was widely planted in China, and it served as a typical example of *indica* intervarietal heterosis with a yield increase of more than 20% compared with conventional *indica* varieties at that time. In the middle and late 1970s, BT-type and Dian-type *japonica* hybrid rice, with the set successfully completed, was planted in China and represented *japonica* intervarietal heterosis with a yield increase of more than 10% compared with conventional *japonica* varieties.

2. Intersubspecific heterosis

Rice hybrids between the two subspecies of rice, *indica* and *japonica*, generally has a high heterosis and yields 15% more than intervarietal hybrids due to the diverse genetic distance. Rice breeders have long been trying to utilize intersubspecific heterosis to improve rice yield. Chinese scientists so far have made great efforts and progress in the discovery of genes for early maturity, dwarf plant and wide compatibility, as well as fertility loci and relevant molecular breeding for developing intersubspecific hybrid rice. The utilization of intersubspecific heterosis is the most promising and effective way to achieve high rice yield in a short time.

Scientists from the Republic of Korea have made remarkable achievements in conventional breeding through intersubspecific hybridization between *indica* and *japonica* since the end of the 1960s. High-yielding dwarf rice lineages such as Tongil, Suwon, Milyang and Iri have been developed and the yield of these varieties is 20%–40% higher than that of conventional *japonica*. Japan launched a super high-yielding rice breeding project in 1981 using intersubspecific hybridization between *indica* and *japonica*. The main concept was to use a large number of *indica* rice varieties from China and the Republic of Korea to cross with *japonica* rice varieties from Japan so as to increase the number of spikelets and improve stress resistance and yield stability. The goal was a 50% yield increase, and intersubspecific hybridization was the only way possible to attain it. By the end of the 20th century, Japan had bred a number of super high-yielding varieties of brown rice with a yield of 10 t/ha, such as Chugoku 91, Hokuriku 125, Chugoku 96, Hokuriku 129, Hokuriku 130, Oryza 331, etc. There is no doubt that intersubspecific F_1 hybrids between *indica* and *japonica* rice have higher heterosis than intervarietal hybrids. In theory, the yield of intersubspecific hybrid rice could be 30%–50% higher than that of high-yielding intervarietal hybrid rice.

China started the cross breeding between *indica* and *japonica* subspecies in the 1950s with the belief that positive heterosis thus obtained could be stabilized. In the 1980s, China initiated the breeding of intersubspecific hybrids. High-yielding hybrids such as Chengte 232/26 Zhaizao (Hunan Hybrid Rice Research Center, 1987), 3037/02428, and W6154S/Vary Lava (Gu Fulin et al., 1988) yielded up to 10.50–11.25 t/ha, which was more than 20% higher than Shanyou 64.

At present, intersubspecific hybrids such as the Yongyou series and the Chunyou series are planted on a large scale in China. Those hybrids were mainly derived from *japonica* sterile lines and *indica-japonica* intermediate restorer lines with wide compatibility. Although those hybrids are not typical *indica-japonica* hybrids, they show high heterosis and traits of intersubspecific hybrids, with a yield increase of more than 15% compared with conventional hybrid rice. For example, in a regional trial, the yields of Chunyou 927 and Yongyou 540 increased by 18.1% and 19.0% respectively over that of the control variety. The utilization of intersubspecific heterosis became possible as continuous exploration was made regarding genes associated with various superior traits such as growth duration, plant height, compatibility and fertility in rice germplasm, and more progress was made in the molecular breeding technology. For example, the seed setting rate of *indica-japonica* hybrids derived from 509S is more than 85% because of allele replacement in which multiple fertility genes of *indica* are integrated into the *japonica* sterile line with a high compatibility.

3. Distant heterosis

Distant hybridization can, to a certain extent, break the reproductive isolation and promote genetic exchange between different species. As a breeding method, it is mainly used to introduce useful genes from a different species to improve the existing varieties. There are many examples of direct utilization of heterosis produced through distant hybridization to cultivate new varieties. For example, in some countries, emmer wheat, aegilops and thinopyrum have been used as parents to breed wheat hybrids resistant to rust. The inter-generic hybridization between maize and gramineae produced maize varieties with high protein and fat content. China has reported the use of more than 10 families and genera of sorghum, maize, wheat, reed, zizania latifolia, Chinese pennisetum, coix lachryma, echinochloa crusgalli and even bamboo as male parents in distant hybridization of rice. Some of the hybrids have excellent agronomic traits and show certain heterosis. In-depth research on hybridization of rice, maize and sorghum has been widely carried out. Compared with rice, sorghum and maize have many excellent agronomic traits as they are C4 plants, with high photosynthetic efficiency, smooth transportation of photosynthate, good seed setting, good plant and leaf morphology, thick and hard stems, well-developed root systems, great fertilizer and lodging resistance, large panicles with more grains, strong adaptability, greater drought and waterlogging resistance, and high and stable yield. Fu Jun et al. (1994) developed the distant hybrid of Chaofengzao 1 with favorable genes from 89Zao281 rice and Qingkeyang sorghum. Compared with 89Zao281, Chaofengzao 1 shows heterosis in having large panicles, more grains, good seed setting, a high photosynthetic rate and a high growth rate. Chen Shanbao et al. (1989) developed the heterotic Zhongyuan rice hybrids through distant sexual hybridization with Yinfang rice varieties as the female parent and Henrijia and other sorghum varieties as the male parent. Liu Chuanguang et al. (2003) carried out distant hybridization using the dual-purpose genic male sterile line D1S as the female parent and super sweet corn Jinyinsu as the male parent, and bred the dual-purpose genic male sterile line Yu - 1S with large panicles and big grains. The hybrids derived from this line showed super heterosis and had large panicles and big grains, as well as high seed setting rates with a yield significantly higher than that of the control variety Peizashuangqi.

Biotechnology has provided new methods to introduce foreign genes, such as using total foreign DNA to create new rice germplasm resources, including pollen tube introduction, ear stem injection, embryo soaking, etc.; and introducing distant favorable genes with the transgenic technology to obtain rice breeding materials with stable inheritance and expression of foreign genes.

Ⅱ. Methods of Rice Heterosis Utilization

1. Three-line method

This is an effective and classic method for the utilization of heterosis in rice. At present, most of the intervarietal hybrid rice varieties planted on a large scale is three-line hybrids. It is also used for intersubspecific heterosis utilization, introducing the gene for wide compatibility into the male sterile line, maintainer line or restorer line. Three-line hybrids have been successfully planted in China since the 1970s, with a total planting area of over 400 million hectares.

2. Two-line method

Photo-and thermo-sensitive genic male sterile (PTGMS) lines can not only reproduce PTGMS seeds by selfing, but also be used for hybrid seed production. Two-line heterosis utilization based on PTGMS reduces the steps required in seed production. More importantly, all normal rice varieties can be used as the restorer line, and the probability of breeding heterotic hybrids is higher than with the three-line method. In addition, the two-line hybrid rice technology can avoid the negative effects of the sterile cytoplasm and the simplification of the genetic basis.

The two-line method can be used not only for intervarietal hybrids but also for intersubspecific hybridization. After more than 20 years of efforts, a lot of scientific and technological innovations and new technology integration have been applied to dealing with key issues such as the fertility conversion of PTGMS lines, the breeding of practical PTGMS lines and their hybrids, and safe and efficient breeding and seed production. Practical two-line male sterile lines such as Peiai 64S, Guangzhan 63S, C815S, Y58S and Longke 638S, and two-line rice hybrids such as Liangyoupeijiu, Yangliangyou 6, Y Liangyou 1, and Longliangyou-Huazhan have been developed and planted in 16 provinces in China. As of 2012, the total planting area of two-line hybrid rice had reached more than 33 million hectares.

By using the technology of two-line super hybrid rice, China has achieved the goals of the first, second, third and fourth phases of its super hybrid rice breeding program, which are respectively 10.5t/ha, 12.0 t/ha, 13.5 t/ha and 15.0 t/ha in yield respectively in 2000, 2004, 2012 and 2014. Two-line hybrid rice, initiated in China with independent intellectual property rights, provides new theory and method for crop genetic improvement and ensures China's leading position in hybrid rice research and application in the world. Based on the theory and experience of the two-line hybrid rice technology, China has succeeded in the breeding of two-line rape oilseed, sorghum and wheat, which opens up new possibilities for crops that have difficulties in heterosis utilization with three-line hybrids.

3. One-line method

The one-line method is to cultivate non-segregated F_1 hybrids of fixed heterosis without the need to produce seeds annually. Apomixis, a kind of asexual reproduction without fertilization, is considered to be the most promising method.

In apomixis, heterosis is fixed without genotype change or segregation of traits in the offspring and large-scale planting can be achieved through the reproduction of its own hybrid seeds. Gramineae is one of the families with the most apomictic genera and species, and it is theoretically speculated that apomictic genes may exist in *Oryza*. Meanwhile, heterologous apomixis genes can be introduced into rice through distant hybridization or genetic engineering. Back in the 1930s, Navashin, Karpachenko and Stebbins proposed the idea of using apomixis to fix heterosis. Some experiments were carried out in this aspect on sorghum, maize and other crops, and Bashaw successfully bred an apomictic bafel forage grass variety. Explorations are still being made in the breeding of rice using this method. Apomixis is a valuable, promising but difficult research topic for hybrid rice breeding.

4. Chemical emasculation

Chemical emasculation is a method of heterosis utilization where the pollen of one parent (as the female parent) is deprived of fertility through chemical treatment and the other parent (as the male parent) is selected to pollinate and produce hybrid seeds. It can be regarded as a different two-line method. Such efforts started abroad in the early 1950s and China has been doing chemical emasculation since 1970. A hybrid created via chemical emasculation that has been applied in rice production is Ganhua 2.

Free from the influence of genetic factors, chemical emasculation offers more options for parent selection and allows for the utilization of a wider range of heterosis than the three-line method. However, at present, rice hybrids bred through this method have not been planted on a large scale due to poor yield and low purity of seeds as a result of poor emasculation effectiveness and asynchronous development of rice tillers. Ineffective chemical hybridizing agents usually cause different degrees of female organ damage and poor flowering, and also it is difficult to control the time of emasculation.

Ⅲ. Problems in Rice Heterosis Utilization

1. Level of heterosis utilization

The genetic relationship between varieties is close with a relatively small genetic difference. The genetic diversity within the varieties of rice, maize, cotton and other crops is usually poor and reducing with relatively simple breeding technology. Most of the hybrids have high yield but low quality, or high quality but low yield, or high yield but no resistance. Therefore, they are not suitable for large-scale commercial planting. The range of heterosis utilization is also limited, and there is no hybrid which has a high heterosis with many other superior traits, resulting in the small increase in the yield of intervarietal hybrid rice and not a significant progress in the yield per unit area.

Intersubspecific heterosis utilization aims mainly at making use of the large genetic difference between subspecies so that the different superior traits of the two subspecies can complement each other. However, typical *indica-japonica* hybridization has an obviously negative effect on heterosis such as long growth duration, low seed setting rate, high plant height and poor grain filling. Therefore, they are not suitable for direct production with existing breeding technology. Rice breeders need to find ways to make good use of the genetic difference between *indica* and *japonica* rice, or to use suitable breeding techniques for better utilization of the positive intersubspecific heterosis.

Direct utilization of hybrids of distant hybridization may result in unimaginable and unpredictable heterosis, and it is also very difficult to achieve because of reproductive isolation. Hybrids from distant hybridization are usually sterile, which may be caused by genetic or chromosomic sterility and the seed setting rate is very low. Moreover, these hybrids have irregular segregation with various plant types, long segregating generations and slow stabilization of the progenies. Despite all these difficulties, scientists have been looking for ways to solve problems, and some remarkable results have been achieved.

2. Methods of heterosis utilization

With the three-line method, the utilization scope of rice germplasm is very limited due to the cytoplasmic effect and the limitation of the relationship between the restorer line and the maintainer line. As far as three-line *indica* hybrid rice is concerned, only 0.1% and 5.0% of the existing *indica* varieties can be directly bred into sterile lines and restorer lines respectively. Moreover, the breeding procedure and seed production process are complicated, resulting in a long time, low efficiency and multiple steps for breeding new hybrids and promoting them.

As for the two-line method, fertility can be restored for more than 95% of the rice germplasm with more hybridization choices. However, the fertility of PTGMS lines is subject to the influence of day-length and temperature, which can be unstable, resulting in risks in parent reproduction and hybrid seed production.

In the long run, the classic three-line and two-line methods will eventually be replaced by more advanced technology. For example, Yuan Longping's research team developed the technology of breeding third-generation hybrid rice on the basis of genetic engineering of common genic male sterility in 2017. It is an advanced new technology for heterosis utilization in rice. This technology does not only have the advantages of stable fertility of three-line male sterile lines and free hybridization of two-line male sterile lines, but also overcome the hybridization limitation of the sterile lines in the three-line method, and eliminates the possibility of fertility recovery of PTGMS lines in seed production under adverse climate conditions.

References

[1] YUAN LONGPING. Hybrid rice science[M]. Beijing: China Agricultural Press, 2002.

[2] YUAN LONGPING, CHEN HONGXIN. Breeding and cultivation of hybrid rice[M]. Changsha: Hunan Science and Technology Press, 1996.

[3] XIE HUAAN. Theory and practice of Shanyou 63 breeding[M]. Beijing: China Agricultural Press, 2005.

[4] JONES D F. Dominance of linked factors as a means of accounting for heterosis[J]. Genetics, 1917, 2: 466－479.

[5] JONES J W. Hybrid vigor in rice[J]. J Am Soc Agron, 1926, 18: 423－428.

[6] YUAN LONGPING. Male sterility of rice[J]. Chinese Science Bulletin, 1966, 17:185－188.

[7] KADAM B S, PATIL G G, PATANKAR V K. Heterosis in rice[J]. Indian J Agric Sci, 1937, 7: 118－126.

[8] BROWN F B. Hybrid vigor in rice[J]. Malay Agric, 1953, 36: 226－236.

[9] WEERARATNE H. Hybridization technique in rice[J]. Trop Agric, 1954, 110: 93－97.

[10] SAMPATH S, MOHANTY H K. Cytology of semi-sterile rice hybrids[J]. Curr Sci, 1954, 23: 182－183.

[11] KATSUO K, MIZUSHIMA U. Studies on the cytoplasmic difference among rice varieties, *Oryza sativa* L. 1. On the fertility of hybrids obtained reciprocally between cultivated and wild varieties[J]. Japan J Breed, 1958, 8 (1): 1－5.

[12] SHINJYO C, O'MURA T. Cytoplasmic male sterility in cultivated rice, *Oryza sativa* L. I. Fertility of F_1, F_2, and offspring obtained from their mutual reciprocal backcrosses; and segregation of completely male sterile plants[J]. Japan J Breed, 1966, 16 (suppl. 1): 179－180.

[13] SHINJYO C. Cytoplasmic-genic male sterility in cultivated rice, *Oryza sativa* L. [J]. J Genet, 1969, 44 (3): 149－156.

[14] SHINJYO C. Distribution of male sterility inducing cytoplasm and fertility restoring genes in rice. I. Commercial lowland rice cultivated in Japan[J]. Japan J Genet, 1972, 47: 237－243.

[15] SHINJYO C. Distribution of male sterility inducing cytoplasm and fertility restoring genes in rice. Varieties introduced from sixteen countries[J]. Japan J Breed, 1972, 22: 329－333.

[16] SHULL G H. The composition of a field of maize[J]. Am Breed Assoc Rep, 1908, 4: 296－301.

[17] VIRMANI S S, Chaudhary R C, Khush G S. Heterosis breeding in rice (*Oryza sativa* L.)[J]. Theor Appl

Genet, 1981, 63: 373 – 380.

[18] DENG HUAFENG, HE QIANG. Study on plant type model of super hybrid rice with wide adaptability in the Yangtze river basin[M]. Beijing: China Agricultural Press, 2013.

[19] DENG HUAFENG. Encyclopedia of hybrid rice[M]. Beijing: China Science and Technology Press, 2014.

[20] ZHANG PEIJIANG, CAI HONGWEI, LI HUANCHAO, et al. Genetic distance by using RAPD markers and its relationship with heterosis in rice[J]. Journal of Anhui Agricultural Sciences, 2000, 28 (6): 697 – 700.

[21] ZHU YUNCHANG, LIAO FUMING. Research progress on heterosis of two-line hybrid rice[J]. Hybrid Rice, 1990 (3): 32 – 34.

[22] ZENG SHIXIONG, LU ZHUANGWEN, YANG XIUQING. Studies on heterosis in F_1 intervatietal hybrids and its relationship with parents[J]. Acta Agronomica Sinica, 1979, 3 (5): 23 – 34.

[23] YANG JUBAO, LU HAORAN. Review on the development of rice heterosis utilization home and abroad[J]. Fujian Science and Technology of Rice and Wheat, 1990 (2): 1 – 5, 31.

[24] HUANG FU, XIE RONG, LIU CHENGYUAN, et al. Effect of blast resistance of parents on blast resistance of hybrid rice combinations[J]. Hybrid Rice, 2007 (2): 64 – 68.

[25] ZHANG XUELI, ZHANG ZHENG, HU ZHONGLI, et al. Study on combining ability and heritability of quality traits in hybrid rice [J/OL]. Molecular Plant Breeding, 2017 (10): 4133 – 4142 [2017 – 09 – 19]. http://kns. cnki. net/kcms/detail/46. 1068. s. 20170919. 0836. 002. html

[26] LI SHIGUI, LI HANYUN, ZHOU KAIDA, et al. Genetic correlation analysis of appearance quality traits in hybrid rice[J]. Southwest China Journal of Agricultural Sciences, 1996, 9 (Special Edition): 1 – 7.

[27] LU ZUOMEI, XU BAOQIN. On the guiding significance of heterosis group theory for hybrid rice breeding [J]. China Journal of Rice Science, 2010 (1): 1 – 4.

[28] WANG XIANGKUN, LI RENHUA, SUN CHUANQING, et al. Identification and classification of subspecies and intersubspecific hybrids of *Oryza sativa* [J]. Chinese Science Bulletin, 1997, 42 (24): 2596 – 2602.

[29] WANG XIANGKUN, SUN CHUANQING, CAI HONGWEI, et al. Origin and evolution of rice cultivation in China[J]. Chinese Science Bulletin, 1998, 43 (22): 2354 – 2363.

[30] WANG XIANGKUN, SUN CHUANQING, LI ZICHAO. Origin and evolution of biodiversity and classification of rice cultivated in Asia[J]. Plant Genetic Resources Science, 2000,1 (2): 48 – 53.

[31] ZHANG Q F, GAO Y J, SAGHAI M A, et al. Molecular divergence and hybrid performance in rice[J]. Mol Breed, 1995, 1: 133 – 142.

[32] XIE WB, WANG G W, YUAN M, et al. Breeding signatures of rice improvement revealed by a genomic variation map from a large germplasm collection[J]. Proc Natl Acad Sci USA, 2015, 112: 5411 – 5419.

[33] CHEN LIYUN, DAI KUIGEN, LI GUOTAI, et al. Comparative study on different types of *indica-japonica* F_1 hybrid[J]. Hybrid Rice, 1992 (4): 35 – 38.

[34] SUN CHUANQING, CHEN LIANG, LI ZICHAO, et al. Tentative study on the dominant ecotype of two-line hybrid rice[J]. Hybrid Rice, 1999, 14 (2): 34 – 38.

[35] GRIFFING B. A generalized treatment of the use of diallel crosses in quantitative inheritance[J]. Heredity, 1956, 10: 31 – 50.

[36] ZHOU KAIDA, LI HANYUN, LI RENRUI, et al. Preliminary study on combining ability and heritability of main traits in hybrid rice[J]. Acta Agronomica Sinica, 1982,8 (3): 145 – 152.

[37] GORDON G H. A method of parental selection and cross prediction using incomplete partial diallels[J]. Theor Appl Genet, 1980, 56: 225 – 232.

[38] NI XIANLIN, ZHANG TAO, JIANG KAIFENG, et al. Correlation of SCA with heterosis and genetic distance between parents in hybrid rice[J]. Hereditas, 2009, 31 (8): 849 - 854.

[39] BHAT G M. Multivariate analysis approach to selection of parents for hybridization aiming at yield improvement in self-pollinated crops[J]. Aust J Agric Res, 1970, 21: 1 - 7.

[40] BHAT G M. Comparison of various method of selecting parents for hybridization in common wheat[J]. Aust J Agric Res, 1973, 24: 257 - 264.

[41] LIU LAIFU. Genetic distance of crop quantitative traits and its determination[J]. Acta genetics Sinica, 1979, 6 (3): 349 - 355.

[42] HUANG QINGYANG, GAO ZHIREN, RONG TINGZHAO. Relationship between genetic distance, yield heterosis and hybrid yield of maize inbred lines[J]. Journal of Genetics and Genomics, 1991, 18 (3): 271 - 276.

[43] HE ZHONGHU. Application of distance analysis method in wheat parent selection[J]. Acta Agronomica Sinica, 1992, 18 (5): 359 - 365.

[44] WANG YIQUN, ZHAO RENGUI, WANG YULAN, et al. Relationship among distance, heterosis and special combining ability of sweet corn[J]. Journal of Jilin Agricultural Sciences, 1998, 92 (3): 17 - 19.

[45] HOU HETING, DU ZHIHONG, ZHAO GENDI. Relationship between genetic distance, heterosis and special combining ability of sorghum parents[J]. Hereditas, 1995, 17 (1): 30 - 33.

[46] WANG YIQUN, ZHAO RENGUI, WANG YULAN, et al. Study on distance analysis and heterosis of sweet corn[J]. Journal of Jilin Agricultural Sciences, 2001, 26 (3): 16 - 20.

[47] XU JINGFEI, WANG LUYING. Heterosis and genetic distance of rice[J]. Journal of Anhui Agricultural Sciences, 1981 (rice quantitative genetics): 65 - 71.

[48] LI CHENGQUAN, ANG SHENGFU. Studies on heterosis and genetic distance of japonica rice[C]//International Symposium on Hybrid Rice, 1986.

[49] COWEN N M, FERY K J. Relationships between three measures of genetic distance and breeding behavior in oats[J]. Genome, 1987, 29: 97 - 106.

[50] SARAWGI A K, SHRIVASTANA. Heterosis in rice under irrigated and rain-fed situations[J]. Oryza, 1987, 25: 20 - 25.

[51] SARATHE M L, PERRAJU P. Genetic divergence and hybrid performance in rice[J]. Oryza, 1990, 27: 227 - 231.

[52] GUAN CHUNYUN, WANG GUOHUAI, ZHAO JUNTIAN. Preliminary study on heterosis and early prediction of heterosis in brassica napus[J]. Journal of Genetics and Genomics, 1980, 7 (1): 55 - 63

[53] LI LIANGBI, ZHANG ZHENGDONG, TAN KEHUI, et al. Studies on chloroplast complementarity in plants Ⅰ. Chloroplast complementarity in hybrid parents[J]. Journal of Genetics and Genomics, 1978, 5 (3): 196 - 203

[54] YANG FUYU, XING JINGRU, SHI BAOSHENG, et al. Study on heterosis test by homogenate complementation method (Ⅰ)[J]. Chinese Science Bulletin, 1978, 23 (12): 752 - 755.

[55] ZHU PENG, LIU WENFANG, XIAO YIHUA. Hill reaction activity of chloroplast in hybrid rice[J]. Wuhan Botanical Research, 1987, 5 (3): 257 - 266.

[56] SCHWARTZ D. Genetic studies of mutant isozymes in maize[J]. Proc Natl Acad Sci VSA, 1960, 88: 1202 - 1206.

[57] LI JIGENG, YANG TAIXING, ZENG MENGQIAN. Studies on isozyme and heterosis of maize Ⅰ. Compari-

son between hybrids and parents at vegetative growth stage[J]. Heredita, 1979 (3): 8 - 11.

[58] LI JIGENG, YANG TAIXING, ZENG MENGQIAN. Studies on isozymes and heterosis of maize Ⅱ. Types of complementary enzymes and their distribution in different organs[J]. Hereditas, 1980, 2 (4): 4 - 6.

[59] ZHU PENG, SUN GUORONG, XIAO YIHUA, et al. MDH and GDH activities and heterosis prediction of rice[J]. Journal of Wuhan University (Natural Science Edition), 1991 (4): 89 - 94.

[60] ZHU YINGGUO. Biology of male sterility in rice[M]. Wuhan: Wuhan University Press, 2000.

[61] LEE M, GODSHALK E B, LAMKEY K R, et al. Association of restriction fragment length polymorphisms among maize inbreds with agronomic performance of their crosses[J]. Crop Sci, 1989, 29: 1067 - 1071.

[62] SMITH O S, SMITH J S C, BOWEN S L, et al. Similarities among a group of elite maize inbreds as measured by pedigree, F_1 grain yield, grain yield heterosis, and RFLPs[J]. Theor Appl Genet, 1990, 80: 833 - 840.

[63] ZHANG PEIJIANG, CAI HONGWEI, YUAN PINGRONG, et al. Genetic distance by RFLP marker and its relationship with heterosis in rice[J]. Hybrid Rice, 2001, 16 (5): 50 - 54.

[64] PENG ZEBIN, LIU XINZHI. Studies on the relationship between F_1 yield, heterosis, parents' SCA and RAPD genetic distance in maize[C]//Wang Lianzheng, Dai Jingrui. Proceedings of National Symposium on Crop Breeding. Beijing: China Agricultural Science and Technology Press, 1998: 221 - 226.

[65] FU HANG, XIANG XUCHAO, XU SHUNJU, et al. A method for predicting heterosis of hybrid rice in sichuan using molecular marker heterozygosity[J]. Journal of China Agricultural University, 2016, 21 (9): 40 - 48.

[66] ZHANG Q F, GAO Y J, YANG S H, et al. A diallel analysis of heterosis in elite hybrid rice based on RFLPs and microsatellites[J]. Theor Appl Genet, 1994, 89: 185 - 192.

[67] ZHANG Q F, ZHOU Z Q, YANG G P, et al. Molecular marker heterozygosity and hybrid performance in *indica* and *japonica* rice[J]. Theor Appl Genet, 1996, 92: 637 - 643.

[68] GODSHALK E B, LEE M, LAMKEY K R. Relationship of restriction fragment length polymorphisms to single cross hybrid performance of maize[J]. Theor Appl Genet, 1990, 80: 273 - 280.

[69] DUDLEY J W, SAGHAI M A, RUFENER G K. Molecular markers and grouping of parents in maize breeding programs[J]. Crop Sci, 1991, 31: 718 - 723.

[70] BOPPENMAIER J, MELCHINGER A E, SEITZ GETAL. Genetic diversity for RFLPs in European maize inbreds performance of crosses within between heterotic group for grain trait[J]. Plant Breeding, 1993, 111: 217 - 226.

[71] XIAO J, LI J, YUAN L, et al. Genetic diversity and its relationship to hybrid performance and heterosis in rice as revealed by PCR-based markers[J]. Theor Appl Genet, 1996, 92: 637 - 643.

[72] JOSHI S P, BHAVE S G, GHOWDARL K V, et al. Use of DNA markers in prediction of hybrid performance and heterosis for a three-line hybrid system in rice[J]. Biochemical Genetics, 2001, 39 (5 - 6): 179 - 200.

[73] JINHUA XIAO, JIMING LI, LONGPING YUAN, et al. Dominance is the major genetic basis of heterosis in rice as revealed by QTL analysis using molecular markers[J]. Genetics, 1995, 140: 745 - 754.

[74] BERNARDO R. Relationship between single-cross performance and molecular marker heterozygosity[J]. Theor Appl Genet, 1992, 83: 628 - 643.

[75] WU XIAOLIN, XIAO BINGNAN, LIU XIAOCHUN, et al. Searching and locating QTLs affecting heterosis [J]. Animal Biotechnology Bulletin, 2000, 7 (1): 116 - 122.

[76] GANG W, YONG T, LIU G Z, et al. A transcriptomic analysis of super hybrid rice Liangyoupeijiu and its parents[J]. Proceedings of the National Academy of Sciences of the United States of America, 2009, 106 (19): 7695 - 7701.

[77] XU S, ZHU D, ZHANG Q. Predicting hybrid performance in rice using genomic best linear prediction[J]. Proceedings of the National Academy of Sciences of the United States of America, 2014, 111 (34): 12456 - 12461.

[78] XU S, XU Y, GONG L, et al. Metabolomic prediction of yield in hybrid rice[J]. Plant Journal, 2016, 88 (2): 219 - 227.

[79] DAVENPORT C B. Degeneration albinism and inbreeding[J]. Science, 1908, 28: 454 - 455.

[80] BRUCE A B. The Mendelian theory of heredity and the augmentation of vigor[J]. Science, 1910, 32: 627 - 628.

[81] JONES D F. Dominance of linked factors as a means of accounting for heterosis[J]. Genetics, 1917, 2: 466 - 479.

[82] KEEBLE J, PELLEW C. The mode of inheritance of stature and of time of flowering in peas (Pisum sativum)[J]. J Genet, 1910, 1: 47 - 56.

[83] SHEN G, ZHAN W, CHEN H, et al. Dominance and epistasis are the main contributors to heterosis for plant height in rice[J]. Plant Science, 2014, s215 - 216 (2): 11 - 18.

[84] HUANG X, YANG S, GONG J, et al. Genomic analysis of hybrid rice varieties reveals numerous superior alleles that contribute to heterosis[J]. Nature Communications, 2015, 6: 6258.

[85] HUANG X, YANG S, GONG J, et al. Genomic architecture of heterosis for yield traits in rice[J]. Nature, 2016, 537 (7622): 629 - 633.

[86] XUE W Y, XING Y Z, WENG X Y, et al. Natural variation in *Ghd7* is an important regulator of heading date and yield potential in rice[J]. Nat Genet, 2008, 40: 761 - 767.

[87] YAN W H, WANG P, CHEN H X, et al. A major QTL, *Ghd8*, plays pleiotropic roles in regulating grain productivity, plant height, and heading date in rice[J]. Mol Plant, 2011, 4: 319 - 330.

[88] YAN W H, LIU H Y, ZHOU X C, et al. Natural variation in *Ghd7. 1* plays an important role in grain yield and adaptation in rice[J]. Cell Res, 2013, 23: 969 - 971.

[89] GARCIA A A F, WANG S C, MELCHINGER A E, et al. Quantitative trait loci mapping and the genetic basis of heterosis in maize and rice[J]. Genetics, 2008, 180: 1707 - 1724.

[90] SHULL G H. The composition of a field of maize[J]. Ann Breed Assoc Rep, 1908, 4: 296 - 301.

[91] EAST E M. Inbreeding in corn [R]. In: Reports of the Connecticut Agricultural Experiment Station for Years 1907—1908, 1908: 419 - 428.

[92] BERGER E. Heterosis and the maintenance of enzyme polymorphism[J]. Am Nat, 1976, 11: 823 - 839.

[93] KRIEGER U, LIPPMAN Z B, ZAMIR D. The flowering gene single flower truss drives heterosis for yield in tomato[J]. Nat Gene, 2010, 42: 459 - 463.

[94] STUBER C W, LINCOLN S E, WOLFF D W, et al. Identification of genetic factors contributing to heterosis in a hybrid from two elite maize inbred lines using molecular markers[J]. Genetics, 1992, 132: 823 - 839.

[95] LI Z K, LUO L J, MEI H W, et al. Overdominant epistatic loci are the primary genetic basis of inbreeding depression and heterosis in rice. Ⅰ. Biomass and grain yield[J]. Genetics, 2001, 158: 1737 - 1753.

[96] LUO L J, LI Z K, MEI H W, et al. Overdominant epistatic loci are the primary genetic basis of inbreeding depression and heterosis in rice. Ⅱ. Grain yield components[J]. Genetics, 2001, 158: 1755 - 1771.

[97] LI Z, PINSON S R M, PARK W D, et al. Genetics of hybrid sterility and hybrid breakdown in an intersubspecific rice (*Oryza sativa* L.) population[J]. Genetics, 1997, 145: 1139 – 1148.

[98] YU S B, LI J X, XU C G, et al. Importance of epistasis as the genetic basis of heterosis in an elite rice hybrid[J]. Proc Natl Acad Sci USA, 1997, 94: 9226 – 9231.

[99] HUA J, XING Y, WU W, et al. Single-locus heterotic effects and dominance by dominance interactions can adequately explain the genetic basis of heterosis in an elite rice hybrid[J]. Proc Natl Acad Sci USA, 2003, 100: 2574 – 2579.

[100] ZHOU G, CHEN Y, YAO W, et al. Genetic composition of yield heterosis in an elite rice hybrid[J]. Proc Natl Acad Sci USA, 2012, 109 (39): 15847 – 15852.

[101] XIAO J, LI J, YUAN L, et al. Dominance is the major genetic basis of heterosis in rice as revealed by QTL analysis using molecular markers[J]. Genetics, 1995, 140: 745 – 754.

[102] LI L Z, LU K Y, CHEN Z M, et al. Dominance, overdominance and epistasis condition the heterosis in two heterotic rice hybrids[J]. Genetics, 2008, 180: 1725 – 1742.

[103] WANG Z, YU C, LIU X, et al. Identification of *indica* rice chromosome segments for the improvement of *japonica* inbreds and hybrids[J]. Theoretical & Applied Genetics, 2012, 124 (7): 1351 – 1364.

[104] BAO WENKUI. Opportunity and risk- reflections on 40 years of breeding research[J]. Plants, 1990 (4): 4 – 5.

[105] MATHER K. The balance of polygenic combinations[J]. Journal of Genetics, 1942, 43 (3): 309 – 336.

[106] ZHONG JINCHENG. Hypothesis of active gene effect[J]. Journal of Southwest University for Nationalities (Natural Science Edition), 1994, 20 (2): 203 – 205.

[107] NILSSON-EHLE H. Kreuzungsuntersuchungen an Hafer und Weizen[M]. Lunds Universitets Arsskrift, East E M, 1909.

[108] DEBICENTE M C, TANKSLEY S D. QTL analysis of transgressive segregation in an interspecific tomato cross[J]. Genetics, 1993, 134 (2): 585 – 596.

[109] HOFMANN N R. A global view of hybrid vigor: DNA methylation, small RNAs and gene expression[J]. The Plant Cell, 2012, 24 (3): 841.

[110] CHODAVARAPU R K, FENG S, DING B, et al. Transcriptome and methylene interactions in hybrids[J]. Proc Natl Acad Sci USA, 2012, 109: 12040 – 12045.

[111] HE GUANGMING, HE HANG, DENG XINGWANG. Transcriptome basis of rice heterosis[J]. Chinese Science Bulletin, 2016, 65 (35): 3850 – 3857.

[112] PENG Y, WEI G, ZHANG L, et al. Comparative transcriptional profiling of three super-hybrid rice combinations[J]. Int J Mol Sci, 2014, 15: 3799 – 3815.

[113] KANG XIAOHUI, PENG YUJIAO, FU JUMEI, et al. SSR analysis of stripe rust resistance genes in sichuan wheat[J]. Journal of Natural Science of Hunan Normal University, 2015, 38 (3): 11 – 15.

[114] WANG W, MENG B, GE X, et al. Proteomic profiling of rice embryos from a hybrid rice cultivar and its parental lines[J]. Proteomics, 2008, 8 (22): 4808 – 4821.

[115] MARCON C, SCHUCTZENMEISTER A, SCHUTZ W, et al. Non-additive protein accumulation patterns in maize (*Zea mays* L.) hybrids during embryo development[J]. Journal of Proteome Research, 2010, 9 (12): 6511 – 6522.

[116] FU Z, JIN X, DING D, et al. Proteomic analysis of heterosis during maize seed germination[J]. Proteomics, 2011, 11 (8): 1462 – 1472.

[117] FU JUN, XU QINGGUO. Study on distant hybridization between rice and sorghum[J]. Journal of Hunan Agricultural University, 1994, 20 (1): 6-12.

[118] LIU CHUANGUANG, JIANG YIJUN, LIN QINGSHAN, et al. Improvement of *indica* photoperiod and temperature sensitive genic male sterile lines by rice-maize distant hybridization[J]. Guangdong Agricultural Sciences, 2003 (4): 7-9.

Chapter 2
Rice Male Sterility

Liao Fuming

Rice is a typical hermaphroditic, self-pollinated crop, which means that the offspring is reproduced by pollination and fertilization in the same flower. Male sterility refers to the degeneration of male organs, which makes the plant unable to form pollen or form only unviable pollen, and therefore, unable to bear self-pollinated seeds; while the female organs are normal and once pollinated with normal fertile pollen, can be fertilized and set seed. Lines featuring male sterility are called male sterile lines. At present, hundreds of rice cytoplasmic male sterile (CMS) lines and a large number of photo-and thermo-sensitive genic male sterile (PTGMS) lines from a wide range of sources and of various different types have been developed in China.

Section 1 Classification of Rice Male Sterility

There are two types of male sterility in rice, genetic sterility and non-genetic sterility. Genetic sterility refers to sterility that is controlled by genetic factors and exhibits inheritable characteristics, such as the sterility of the male sterile lines widely used in three-line and two-line hybrid production. Non-genetic sterility means that the sterility is caused by abnormal external conditions without the involvement of any sterility gene and is non-inheritable, such as the sterility caused by abnormally high or low temperatures, induced by male gametocides, or realized in other ways. From the perspective of genetics and breeding, genetic male sterility has high practical value and is the focus of research and utilization.

Rice genetic male sterility can generally be classified into cytoplasmic sterility, genic sterility and cytoplasmic-genic sterility.

1) Cytoplasmic sterility means that sterility is solely controlled by cytoplasmic genes and is not associated with the nucleus. In this case, a restorer line is difficult to find, and the sterility has no practical value in production.

2) Genic male sterility (GMS) refers to the sterility that is controlled only by nuclear genes whose action is not influenced by the cytoplasm, which is more common in nature. The earliest rice male sterile material discovered in China, namely the natural mutant pollen-free male sterile material discovered by Yuan Longping from Shengli *indica* in 1964 ("non-pollen *indica*" for short), is an example of genic male sterility. This type of sterility is generally controlled by a pair of recessive nuclear

genes, and all varieties with normal fertility are its restorer lines. However, it has no maintainer line, which means its sterility cannot be fully maintained, so it cannot be used directly in production. Although breeders proposed some ideas and made attempts to use this kind of sterility, all were unsuccessful. For example, the Wuhu Institute of Agricultural Sciences in Anhui tried to produce hybrid seeds using a highly sterile line with a restorer line which had marked traits, and then screen the hybrid and the sterile plants based on the marked traits (the "two-line method") in the next season in the rice seedling field, thus making use of the heterosis. In 1974, they produced a hybrid using a dual-purpose sterile line with a sterile plant rate of 98% and a sterility degree of over 90%, and crossed it with a restorer line with a stable purple trait, thus completing a two-line set. However, because the sterility of this sterile line is greatly affected by the environment, which resulted in an unstable ratio of hybrids and self-pollinated inbred plants, the hybrid could not be widely used in production.

3) Cytoplasmic-genic male sterility (CMS) is the sterility controlled by both cytoplasmic genes and nuclear genes, in which case, sterility occurs only when sterile genes are present in both the cytoplasm and the nucleus. This type of sterility has not only the maintainer lines (those with fertile cytoplasmic genes and sterile nuclear genes) to maintain its sterility, but also the restorer lines (fertile or sterile cytoplasmic genes and fertile nuclear genes) to restore fertility in F_1 hybrids, thus completing a three-line set. Therefore, it can be used directly in production. The three-line hybrid rice was widely and successfully put into rice production in the 1970s in China.

However, long-term rice breeding practice has revealed that among the above three types of rice male sterility, the sterility controlled solely by the cytoplasm has not actually been discovered. For example, the *japonica* wild abortive sterile line was once considered to be a cytoplasmic sterile line because no restorer line was found for a long time; but later, the Academy of Agricultural Reclamation Sciences of the Xinjiang Construction Corps found a restorer lines in the progenies of early *japonica* 3373 × IR24 hybrids introduced from Chinese Academy of Agricultural Sciences. Another example is that restorers had not been found for a long time for the male sterile line of Chinese wild rice × Fujisaka 5 produced in Japan, but was later discovered in the progenies of Fujisaka 5 × *indica* hybrid in Hubei. Therefore, in practical application, cytoplasmic sterility in actually refers to cytoplasmic-genic male sterility.

In 1973, Shi Mingsong from Mianyang county of Hubei province discovered a PTGMS strain, Nongken 58S, in a population of the *japonica* late-season variety Nongken 58. Subsequently, more PTGMS materials were discovered, such as those *indica* types of Annong S-1, discovered by Deng Huafeng (1988) from Anjiang Agricultural School of Hunan Hengnong S-1, discovered by Zhou Tingbo (1988) from Hengyang Agricultural Science Institute of Hunan; and 5460S, discovered by Yang Rencui (1989) from Fujian Academy of Agricultural Sciences. After extensive and in-depth research, this new category of genic male sterility was found to be controlled by recessive nuclear genes and not associated with the cytoplasm. Although it still belongs to the category of GMS, it is different from the previously found cases of general GMS, because the expression of its fertility is mainly regulated by daylength and temperature: At a certain stage of development, long daylength and high temperature lead to sterility, while short daylength and low temperature lead to fertility, showing obvious fertility conversion. This is a typical case of ecological genetics where sterility is controlled by both nuclear sterility genes, and daylength and temperature, so it is called PTGMS. A PTGMS line can be used to produce hybrid seeds for field production during the sterile period, and for self-reproduction to maintain its sterility during the fertile period. Thus, one line serves dual purposes. Practice has showed broad application prospects for using this type of sterility to breed two-line hybrid rice. In addition, the United States, Japan and other countries have also bred some sterile lines featuring fertility conversion.

In summary, from a practical perspective, male sterility in rice can be classified into two categories, namely CMS and GMS. PTGMS is a special type of GMS.

Ⅰ. Classification of Cytoplasmic Male Sterility

After successfully completing the three-line set of hybrid rice in 1973, Chinese scientists have made more detailed studies on the classification of CMS for different research purposes, which is summarized in the following five classification methods.

1. Classification by restorer—maintainer relationship

CMS lines can be classified into three types, wild abortion (WA), Honglian (HL) and BT, based on the differences between the maintainer and restorer lines of a sterile line.

(1) Wild abortive type

This type of male sterile lines was developed through nuclear substitution using WA rice from Yaxian county as the female parent and dwarf early *indica* Erjiuai 4, Zhenshan 97, Erjiunan 1, 71 – 72, V41, etc. as the male parents. Most dwarf early *indica* varieties from the Yangtze River Basin are its maintainer lines; while its restorer lines come from Peta (a Southeast Asian variety) and Indonesian Paddy rice, as well as Peta-derived low-latitude *indica* rice varieties such as Taiyin 1, IR24, IR661, IR26, etc. and Indonesian Paddy rice-related South China late *indica* varieties such as Shuangqiuai 2, Qiushuiai, etc. In terms of the restoration rate and degree, *indica* is greater than *japonica*, late-season *indica* is greater than early- season *indica*, late-maturing varieties are greater than mid-maturing varieties, mid-maturing varieties are greater than early-maturing varieties, and *indica* at low latitudes is greater than *indica* at high latitudes. CMS lines with a restorer-maintainer relationship similar to that of the WA CMS include those of the Gang type, D type, dwarf abortive type, and Yezai Guangxuan 3A.

(2) Honglian type

The Honglian (HL) type of male sterile lines was developed through nuclear substitution using red awn wild rice as the female parent and the high stalk early-season *indica* variety Liantangzao as the male parent. Further development resulted in HL Huaai 15A. The relationship between HL-type restorers and maintainers is obviously different from that of the WA type. For instance, the dwarf *indica* rice varieties Erjiuai 4, Zhenshan 97, Jinnante 43, Bolizhan'ai, Xianfeng 1, Zhulian'ai, Erjiuqing, Wenxuanzao, Longzi 1, etc. from the Yangtze River Basin are the maintainers of WA CMS lines, but are restorers to HL CMS lines. Similarly, Taiyin 1, which is a restorer to WA CMS lines, can maintain the sterility of HL CMS lines. IR24, IR26 and some other varieties can restore the fertility of WA CMS lines, but are semi-restorative for HL CMS lines. Generally, the restoration spectrum of the HL CMS is wider, but the restorability is lower than that of the WA CMS. Tianjidufuyu 1A belongs to this type too.

(3) BT type

The BT type of sterile lines was developed by Japanese scholar Choyu Shinjyo through an intersubspecific cross using spring *indica* from India as the female parent and Taichung 65, a *japonica* variety from Taiwan, China, as the male parent. Most *japonica* rice varieties are their maintainers, but it is difficult to find a restorer. High-altitude *indica* rice and Southeast Asian *indica* rice varieties can be used as its restore lines; however, the resulted hybrids have low seed setting rates due to the intersubspecific incompatibility between *indica* and *japonica*, so it is not suitable for large-scale production. The breeding of the restorers is more complicated. CMS lines bred in China, such as Dian -1, Dian -3 and Lead, as well as BT-derived CMS lines such as Liming A, Nonggui 6A, Akihigari A, and other *japonica* CMS lines, belong to this type.

2. Classification by anther and pollen morphology

According to the difference in anther and pollen morphology, CMS can be divided into five types, anther-free, pollen-free, typical (mononuclear abortion), spherical (binuclear abortion), and stained (trinuclear abortion) abortion.

(1) Anther-free type

Song Deming et al. (1998, 1999) found the anther-free sterile materials M01A, M02A and M03A in the F_3 population of the distant cross of Dongxiang Wild Rice × M872 (*indica*), the F_3 population of an *indica-japonica* cross, and the F_4 population of 02428 (*japonica*) × Milyang 46, respectively. The anthers of these CMS lines were completely degraded. The morphology of the anthers and pollen of the offspring, derived from the above CMS lines as female parents, varies with the male parent, from being completely anther-free to having severely degraded anthers (no pollen or very few typically abortive pollen) or incomplete anthers (a few typically abortive pollen), or indehiscent anthers (with stained or normal pollen).

(2) Pollen-free type

The pollen-free type is aborted in various periods prior to the mononuclear stage. The development of the sporogenous cells is blocked and they fail to form pollen mother cells, or abnormal meiosis of the pollen mother cells results in failure to form tetrads, or tetrad development is blocked and pollen grains cannot be formed. The characteristic is that amitosis is extremely common and the abortion pathway varies greatly, which eventually leads to no pollen in the anther sac, leaving only residual pollen walls. Examples of this type of sterility include the sterile plants of Nanguangzhan (the C system), Jingyin 63, Nanluai, and the Jiangxi O type.

(3) Typical abortion (mononuclear abortion)

Pollen is mainly aborted at the mononuclear stage, and pollen that develops to the binuclear stage is not filled with contents and cannot be stained by iodine-potassium iodide. The morphology of the empty-shell pollen is very irregular. Examples include the CMS lines of the WA type, the Gang type and the dwarf abortive type.

(4) Spherical abortion (binuclear abortion)

In the pollen development of this type of sterile lines, most of the pollen can pass through the mononuclear stage, but the reproductive nucleus and the vegetative nucleus successively disintegrate, which leads to abortion after entering the binuclear stage. Some pollen aborted in the late binuclear stage can be stained and most aborted pollen are spherical. Examples include the CMS of the HL and Dian − 1 types.

(5) Stained abortion (trinuclear abortion)

These sterile lines have the latest pollen abortion period, with most pollen aborting only after the early trinuclear stage. Most of the pollen has normal external morphology, accumulates starch, and can be stained by iodine-potassium iodide, but the reproductive and nutritional nuclei are not developed properly, resulting in sterility. Example of the sterile lines belong to this category are BT type and Lead type CMS.

3. Classification by nuclear substitution type

CMS is generally derived from nuclear substitution through distant hybridization, and it thus can be divided into three categories based on the forms of nuclear substitution − interspecific, intersubspecific and intervarietal.

(1) Interspecific nuclear substitution

Interspecific nuclear substitution includes nuclear substitution between *Oryza sativa* F. Spontanea and *Oryza sativa* L., and between *Oryza glaberrima* and *Oryza sativa* L. Examples of the former include Erjiuai 4A bred through nuclear substitution between pollen-aborted wild rice male sterile strains (as the female parent) and the dwarf early-season *indica* Erjiuai 4 (as the male parent); Guangxuanai 3A bred through nuclear substitution between Hainan common wild rice (as the female parent) and the dwarf *indica* rice Guangxuan 3 (as the male parent); HL A bred through nuclear substitution between red awn wild rice (as the female parent) and the tall-stalk early-season *indica* Liantangzao (as the male parent); and O-type sterile materials bred through nuclear substitution between cultivated rice (as the female parent) and South China wild rice (as the male parent). Examples of the latter include the Huaai 15 sterile material bred through nuclear substitution between *Oryza glaberrima* "DAN BOTO" (as the female parent) and *Oryza sativa* dwarf early-season *indica* Huaai 15 (as the male parent).

(2) Intersubspecific nuclear substitution

Intersubspecific nuclear substitution refers to the nuclear substitution between *indica* and *japonica* rice. For example, Hongmaoying A of the Dian − 1 type was bred through nuclear substitution between a sterile plant resulted from natural crossing between Yunnan high-altitude *indica* rice and *japonica* rice Taipei 8 (as the female parent) and *japonica* Hongmaoying (as the male parent); Hongmaoying A of the Dian − 5 type was bred through nuclear substitution between South China late *indica* Baotaiai (as the female parent) and *japonica* Hongmaoying (as the male parent); and Hongmaoying A of the Dian − 7 type was bred through nuclear substitution between Indian spring *indica* 190 (as the female parent) and *japonica* Hongmaoying (as the male parent). In addition, there is also the *indica* Taichung 1A of the Dian − 8 type, which was bred through nuclear substitution between a highly sterile F_1 plant of *japonica* rice [(Keqing 3 × Shanlan 2) F_2 × Taichung 31] (as the female parent) and *indica* rice Taichung 1 (as the male parent).

(3) Intervarietal nuclear substitution

Intervarietal nuclear substitution refers to nuclear substitution between *indica* or between *japonica* varieties that are geographically distant or of different ecotypes. Examples include the Gang-type Chaoyang 1A bred through nuclear substitution between West African late *indica* Gambiaka (as the female parent) and Chinese dwarf *indica* (as the male parent); and Keqing 3A of the Dian－4 type bred through nuclear substitution between Yunnan high-altitude *japonica* Zhaotongbeizigu (as the female parent) and *japonica* Keqing 3 (as the male parent).

4. Classification by cytoplasmic source

Based on the sources of the cytoplasm, CMS can generally be divided into the following four categories.

(1) Cytoplasm from *Oryza rufipogon Griff.*

This category includes sterile lines bred through nuclear substitution with common wild rice (*Oryza rufipogon* Griff.), including WA, as the female parent and cultivated rice (*Oryza sativa*) as the male parent. Examples include WA, HL, and dwarf abortive CMS.

(2) Cytoplasm from *Oryza glaberrima*

Backcrossed progenies with 100% sterility were either obtained by crossing *Oryza glaberrima* as the female parent and a common cultivated *japonica* variety as the male parent in the United States, or by crossing *Oryza glaberrima* as the female parent with a common cultivated rice variety in India. Guangshenhuaai 15A was bred in Hubei of China by crossing and backcrossing *Oryza glaberrima* (female) with the early-season *indica* Guangshenhuaai 15.

(3) Cytoplasm from *indica*

This is the type of sterile lines produced by nuclear substitution with *indica* as the female parent and *japonica* as the male parent, as well as between *indica* varieties that are geographically distant or of different ecotypes. The former includes BT CMS lines bred through nuclear substitution between *indica* rice Boro Ⅱ as the female parent and *japonica* rice Taichung 65 as the male parent. The latter includes the Gang-type Chaoyang 1A and the D-type D-Shan A, as well as the CMS lines of the Dian－1, Dian－5, and Lead types and Indonesian Paddy rice.

(4) Cytoplasm from *japonica*

This includes the Dian－8 type Taichung 1A bred with a *japonica* female parent and an *indica* male parent, and the Dian－4 type Keqing 3A bred with *japonica* varieties of different ecotypes.

5. Classification by CMS genetic characteristics

Based on genetic characteristics, CMS can be classified as sporophyte sterility and gametophyte sterility.

(1) Sporophyte sterility

The pollen fertility of sporophyte male sterility is controlled by the genotype of the sporophyte (plant that produces the pollen), independent of the genotype of the pollen (gametophyte) itself, and pollen abortion occurs in the sporophyte stage. When the genotype of the sporophyte is S(rr), all pollen is aborted; when the genotype is N(RR) or S(RR), all pollen is fertile; when the genotype is S(Rr), male gametes of two different genotypes, S(R) and S(r), can be produced, but their fertility is determined by the dominant fertile gene in the sporophyte, so both pollen can be fertile. When such sterile lines are crossed with restorer lines, the pollen of an F_1 plant is normal without fertility segregation, but F_2 plants will have fertility segregation with a certain proportion of sterile plants (Fig. 2 – 1). The pollen of sporophyte sterility is mainly aborted at the mononuclear stage with irregular forms such as boat shape, spindle shape, or triangle; while the anthers are milky white, water-stained and indehiscent. The sterility is relatively stable and less affected by external environmental factors. The CMS plants have short panicle necks and the panicles are partially inside the leaf sheath. Sterile lines of the WA, Gang, D, and dwarf abortive types belong to this category.

(2) Gametophyte sterility

The pollen fertility of gametophyte male sterility is directly controlled by the genotype of the gametophyte (pollen), independent of the genotype of the sporophyte, and its genetic characteristics are shown in Fig. 2 – 2. Pollen with the gametophyte genotype S(r) is sterile, while that with S(R) is fertile. When such sterile lines are crossed with restorer lines, the F_1 pollen has both S(R) and S(r) genotypes, equally represented. Since fertility depends on the genotype of the gametophyte, S(r) pollen is aborted while S(R) pollen is fertile. Although only half of the pollen is fertile, they can pollinate and set seeds normally so that the entire F_2 population is fully fertile and set seeds normally without sterile plants. The pollen of gametophyte sterile lines is mainly aborted after the binuclear stage. The abortive pollen is spherical and some can be stained by iodine-potassium iodide. The anthers are milky yellow, small and indehiscent. The sterility is generally less stable and susceptible to high temperature and low humidity, causing some anthers to dehisce and disperse pollen and some self-pollinated seeds to be set. The panicles of the gametophyte sterile lines can be fully out of the leaf sheath. Examples of the gametophyte sterile lines include those of the BT, HL, Dian – 1 and Lead types.

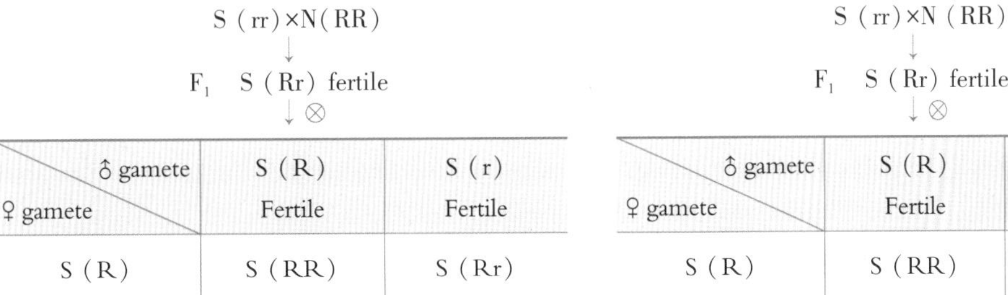

♂ gamete / ♀ gamete	S (R) Fertile	S (r) Fertile
S (R) Fertile	S (RR) Fertile	S (Rr) Fertile
S (r) Fertile	S (Rr) Fertile	S (rr) Sterile

Fig. 2 – 1 Genetic model of sporophyte sterile line

♂ gamete / ♀ gamete	S (R) Fertile	S (r) Abortif
S (R) Fertile	S (RR) Fertile	
S(r) Fertile	S(Rr) Fertile	

Fig. 2 – 2 Genetic model of gametophyte sterile line

Ⅱ. Classification of Genic Male Sterility

In recent years, various new types of rice genic male sterility (GMS), especially photo-and thermo-sensitive genic male sterility (PTGMS), have been discovered, which greatly enriched the types of rice nuclear male sterility. Its classification is summarized as follows.

First of all, rice genic male sterility can be classified as dominant and recessive GMS according to the dominant-recessive genetic characteristics of the genes that control nuclear sterility. Recessive nuclear sterility means that the sterility is controlled by recessive genes, while dominant nuclear sterility means that the sterility is controlled by dominant genes. Most of the genic sterility discovered so far is recessive genic sterility, such as those ordinary GMS and PTGMS mutated naturally or artificially. Dominant GMS has also been reported, such as that of the Pingxiang dominant GMS line discovered in 1978 by Yan Longan et al. (1989) from the Pingxiang Academy of Agricultural Sciences, Jiangxi, and the dominant thermo-sensitive GMS line 8987 discovered in 1989 by Deng Xiaojian et al. (1994) from Sichuan Agricultural University. Currently, recessive rice GMS is more commonly used for rice heterosis utilization, while dominant GMS is mainly used in recurrent selection and population improvement. However, the genetic interaction-based dominant GMS controlled by two pairs of independently inherited dominant genes have also been used for completing three-line or two-line sets, thus for the purpose of using heterosis. For example, Pingxiang dominant GMS rice is temperature-sensitive, and some rice varieties with a pair of dominant epistatic gene can inhibit the expression of the sterility gene (Ms-p) and restore the hybrid fertility. Therefore, it is possible to complete a two-line set using a homozygote sterile line as the female parent, which is obtained through continuous selfing for multiple generations under a specific high temperature, and a restorer line with dominant epistatic gene as the male parent. In other crops such as rapeseed, three-line sets have been completed based on gene-interaction dominant GMS.

Secondly, rice GMS can be classified into environmentally sensitive GMS and ordinary GMS according to whether the sterility is sensitive to environmental factors. The former type of sterility is affected by environmental factors, and the sterility or fertility is convertible regularly under certain environmental conditions. The latter type of sterility is generally not affected by external environmental factors, and as long as the plant has the nuclear sterility gene, it will always be sterile no matter how the environment conditions change.

Environmentally sensitive GMS, currently mainly refers to PTGMS, where sterility/fertility is regulated by external daylength and temperature. There are currently different opinions on the classification of PTGMS. Some divide it into the two types of PGMS and TGMS; others divide it into the three types of PGMS, TGMS and PT-interactive GMS (PTGMS). Still, there is another classification as PTGMS, TGMS, etc. In addition, Sheng Xiaobang et al. (1993) categorized PTGMS into four genetic types, low-temperature and strong-photo sensitivity, high-temperature and weak-photo sensitivity, high-temperature and strong-photo sensitivity, and low-temperature and weak-photo sensitivity. Zhang Ziguo et al. (1993, 1994) classified PTGMS into four types based on the range of critical daylength and temperature required for the sterility/fertility conversion, the high-low type (i. e., high upper limit of sterility critical temperature and low lower limit of fertility critical temperature, with a wide critical temperature range), the low-low type (i. e., low upper limit and low lower limit of critical temperatures with a narrow critical temperature range), as well as the high-high and low-high types. Chen Liyun (2001) proposed to classify PTGMS into four groups, long daylength and high temperature (PTGMS), high temperature (thermo-sensitive MS), short daylength and low temperature (inverse PTGMS), and low temperature (inverse thermo-sensitive MS), and proceeded to set specific daylength and temperature requirements for each type when they are used in production.

There is a general consensus that neither pure photo-sensitive nor pure thermo-sensitive GMS has yet been found based on the discoveries and studies of existing photo- and thermo-sensitive GMS materials. For example, the sterility of Nongken 58S, which was originally considered to be solely photo-sensitive, is also found to be subject to the influence of temperature after further study; while the fertility of Annong S-1, Hengnong S-1, 5460S, etc., which had generally been regarded as typical thermo-sensitive, is also affected by the daylength. Sun Zongxiu (1991) found that at the same temperature (25.8 ℃), the selfed seed setting rates of Annong S-1, Hengnong S-1 and 5460S were significantly higher with a short daylength of 12 hours than under a long daylength of 15 hours. Therefore, existing PTGMS is actually affected by both daylength and temperature, and the roles of daylength and temperature vary with the materials. In view of this, PTGMS is generally classified into two main types, photo-sensitive and thermo-sensitive GMS, based on the major or minor influence of daylength and temperature on fertility. For the former type, fertility is mainly regulated by daylength, while temperature plays a minor role. Examples are Nongken 58S, N5088S and 7001S. For the latter, fertility is mainly controlled by temperature, while daylength has a minor effect, such as those *indica* sterile lines of Annong S-1, Hengnong S-1, 5460S, Peiai 64S, as well as the *japonica* lines of Nonglin PL12, Diannong S-1 and Diannong S-2. So far, PGMS has been more common in *japonica* varieties and TGMS is found mainly in *indica* varieties.

Regarding how to determine the major and minor effects of daylength and temperature on the fertility of a PTGMS line, Chinese researchers have made many useful attempts which are summarized into the four methods below.

1) Under certain temperature conditions, use red and far-red light (R-FR) intermittently during long dark periods (short daylength) to test the difference of fertility and the conversion effect of R-FR, thus determining whether fertility is regulated by photo- or thermo-sensitivity.

2) Under the conditions of artificially controlled daylength and temperature, study the fertility difference of a sterile line statistically, and determine whether the sterile line is photo- or thermo-sensitive, and their interaction at the sensitive stages.

3) Under natural conditions, split seeding to observe the effects of different daylength and temperature on fertility during the sensitive period to determine the PTGMS's response to daylength and temperature.

4) Allelic test among different sterile lines to determine whether a sterile line is photo- or thermo-sensitive.

With the above methods, the classification of a particular PTGMS line as decided on by different researchers could be consistent, inconsistent or even exactly opposite. For example, the classification of the sterile lines Nongken 58S, N5088S, Annong S-1 and 5460S are generally consistent. The first two are photo-sensitive, while the latter two are thermo-sensitive. However, the classification of the *indica* sterile lines derived from Nongken 58S, such as W6154S, was quite inconsistent as it was categorized as thermo-sensitive in some cases, but photo-sensitive in some others. The reason for such discrepancy may be related to the inconsistency of the daylength and temperature conditions used by different researchers. It is feasible to reach a general agreement for a particular PTGMS line if a certain set of daylength and temperature conditions, their ranges, experiment method and classification criteria are adopted as standards.

PTGMS can be divided into four types, long daylength, short daylength, high temperature and low temperature, based on the different directions of daylength and temperature effects on sterility. Long daylength sterility refers to the case where a long daylength induces sterility, while a short daylength induces fertility for a PTGMS line, while the short daylength type is just the opposite, i. e. a short daylength induces sterility and a long daylength induces fertility. The high temperature type means a high temperature leads to sterility and a low temperature leads to fertility, while the low temperature type is just the opposite, i. e. a low temperature induces sterility and a high temperature induces fertility. These four types are all found in nature, but the long daylength and high temperature types are more common, while the short daylength and low temperature types are rare. Long daylength sterile lines include Nongken 58S and its derivatives, and short daylength sterile lines include Yi DS1, 5201S, etc. The high temperature type includes W6154S, Peiai 64S, Annong S-1, Hengnong S-1 and 5460S; while the low temperature type includes go543S, IVA, Diannong S-1, Diannong S-2, etc.

In a breeder's point of view, the above PTGMS can be classified further based on the critical daylength and temperature required for fertility conversion. For example, the high temperature type can be further divided into high critical temperature (high thermo-sensitivity) and low critical temperature (low thermo-sensitivity) groups. The former includes Hengnong S-1 and 5460S, and the latter includes Peiai 64S, Guangzhan 63S, Y58S, etc., which are a group of newly bred sterile lines. For another example, the long daylength type can be further divided into long critical daylength and short critical daylength groups. The former includes Nongken 58S, and the latter includes HS-1.

In recent years, with the rapid development of molecular biology and other basic sciences, molecular techniques such as RFLP and RAPD markers have been used in the classification of rice GMS in more scientific and accurate manners, better serving production practice.

The following chart is a summary of the classification of rice GMS.

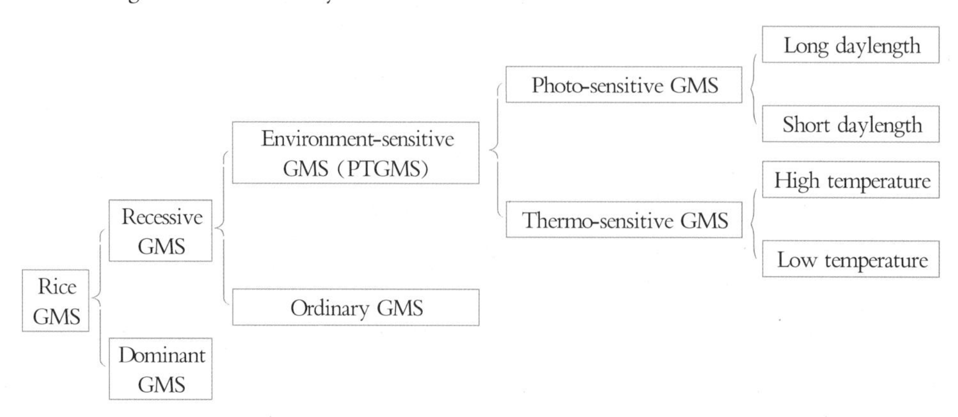

Section 2 Cell Morphology of Rice Male Sterility

Ⅰ. Normal Rice Pollen Development

1. Anther development and structure

After the differentiation of the pistil and the stamen during reproductive development, the stamen further differentiates into anthers and filaments. The anther structure is simple at the early stage of formation, with the outermost layer being the epidermis, and the inside composed of basic tissue cells with the same morphology and structure. Later, a row of archesporial cells form immediately under the epidermis at the four corners of the anther. They then divide into two layers of cells – the outer layer is the parietal cells, while the inner layer is the sporogenous cells. The parietal cells then further divide into three layers, the outer layer of which, next to the epidermis, is the endothecium, whose cell walls are unevenly thickened and will lose its protoplasm with a function related to the dehiscence of pollen sacs when the anther matures. The layer inside of the endothecium is the middle lamella, which gradually fades in the process of anther development and no longer exists in a mature anther. The innermost layer is the tapetum which is composed of large cells rich in nutrients. It encloses the periphery of the sporogenous tissue and plays an important role in pollen development. The tapetum cells will gradually vanish after playing their role of providing nutrients for pollen development when the pollen develops to a certain stage (Fig. 2 – 3).

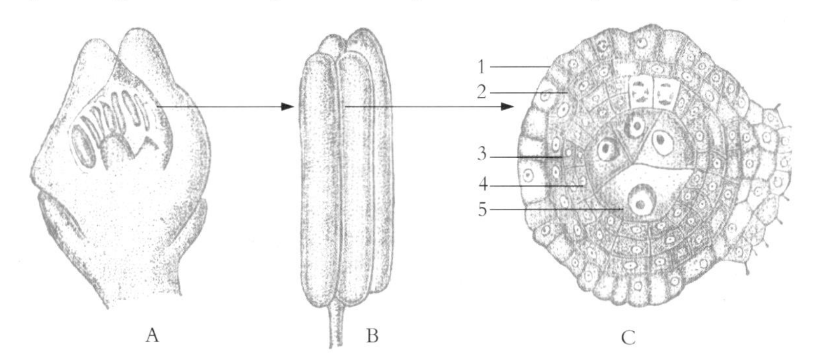

Fig. 2 – 3 Rice anther development (source – *Hybrid Rice Research and Practice*, 1982)

When pollen development is completed and the anthers are fully mature, the only parts of the pollen sac that actually remain are a layer of epidermis and endothecium cells. Once the endothecium shrinks, the pollen sac dehisces and pollen spreads out.

2. Formation and meiosis of pollen mother cells

Pollen mother cells develop from sporogenous cells. As parietal cells divide and change, sporogenous cells undergo multiple mitoses with increased cell numbers and grow into pollen mother cells (microsporocytes). Meanwhile, colloidal callose is gradually formed in the center of the anther chamber, which extends to the intercellular space of the pollen mother cells, encloses the mother cells, and then forms a transparent callose wall on its outer periphery. Before this, pollen mother cells are connected to one another, and the cells are polyhedral. After the callose wall is formed, the pollen mother cells separate from each other, and the cells change from a polyhedral to a round or oval shape.

The nucleus of a pollen mother cell is large and obvious with a large nucleolus. The chromatin is filamentous and faintly visible. The pollen mother cells begin to undergo meiosis at a certain stage of development. The meiosis of rice pollen mother cells includes two successive nuclear divisions, which are called the first meiotic division and the second meiotic division (or division Ⅰ, division Ⅱ) respectively. Both divisions go through the stages of prophase, metaphase, anaphase and telophase, and finally form four daughter cells with haploid chromosomes (n).

3. Development process of pollen grains

The transparent callose wall remains in place until the tetrad is formed, so the four cells of the tetrad will not separate from one another. Yet, shortly after that, the callose wall begins to disintegrate, and the tetrad spores begin to separate, turning into microspores. The microspores gradually change from a fan shape to a round shape, then undergo three developmental stages, namely the mononuclear, binuclear and trinuclear stages, and finally form mature pollen grains (Fig. 2-4).

(1) Mononuclear pollen

The round microspores have thin cell walls with a nucleus in the center and no vacuole (Fig. 2-4, 1). Soon, the outer periphery shrinks, and then the shrinkage intensifies to the maximal extent. The cell appears radially polygonal and this is the first contraction phase (Fig. 2-4, 2). Shortly after the cells shrink intensively, a transparent pollen intine begins to form on the periphery of the cells, and then an exine appears on the intine. Meanwhile, a germinal aperture appears on the exine and the cells return to a round shape. Then, the entire pollen soon shrinks to a spindle shape or boat shape, which is the second contraction phase (Fig. 2-4, 3 and 4). These two contraction phases constitute the initial stage of pollen wall formation, during which uneven growth of the pollen wall causes shrinkage. Then, the pollen grains return to the round shape and get enlarged, the center of the cells occupied by a large vacuole, the cytoplasm becoming a very thin layer clinging to the pollen wall, and the nucleus squeezed to one side of the pollen grain. This is known as the late mononuclear stage (Fig. 2-4, 5 and 6).

(2) Binuclear pollen

When mononuclear pollen develops to a certain point, the nucleus moves along the pollen wall to the opposite side of the germinal aperture, and the first mitotic division of the pollen grain takes place there. Generally, the long axis of the spindle in the mitotic phase is perpendicular to the perisporium (Fig. 2-4, 7) and two daughter nuclei are formed, one close to the pollen wall and the other on its inside. They are identical in form and size at the beginning, but soon get separated, and the morphological differentiation occurs during the separation process. Due to the uneven distribution of cytoplasm, the cell close to the pollen wall, called the reproductive cell, is smaller and its nucleus is called the reproductive nucleus. The inner cell is larger, called the vegetative cell, and its nucleus is called the vegetative nucleus. The two nuclei are separated by a thin membrane. At this time, pollen grains are binuclear, and microspores are in the initial phase of male gametophyte (Fig. 2-4, 8, 9, and 10). The reproductive nucleus is in the shape of biconvex lens, close to the pollen wall and stays in place; while the vegetative nucleus moves quickly away from the reproductive nucleus along the pollen wall toward the germinal aperture. When the vegetative nucleus sets in place close to the germinal aperture, both the nucleus and the nucleolus

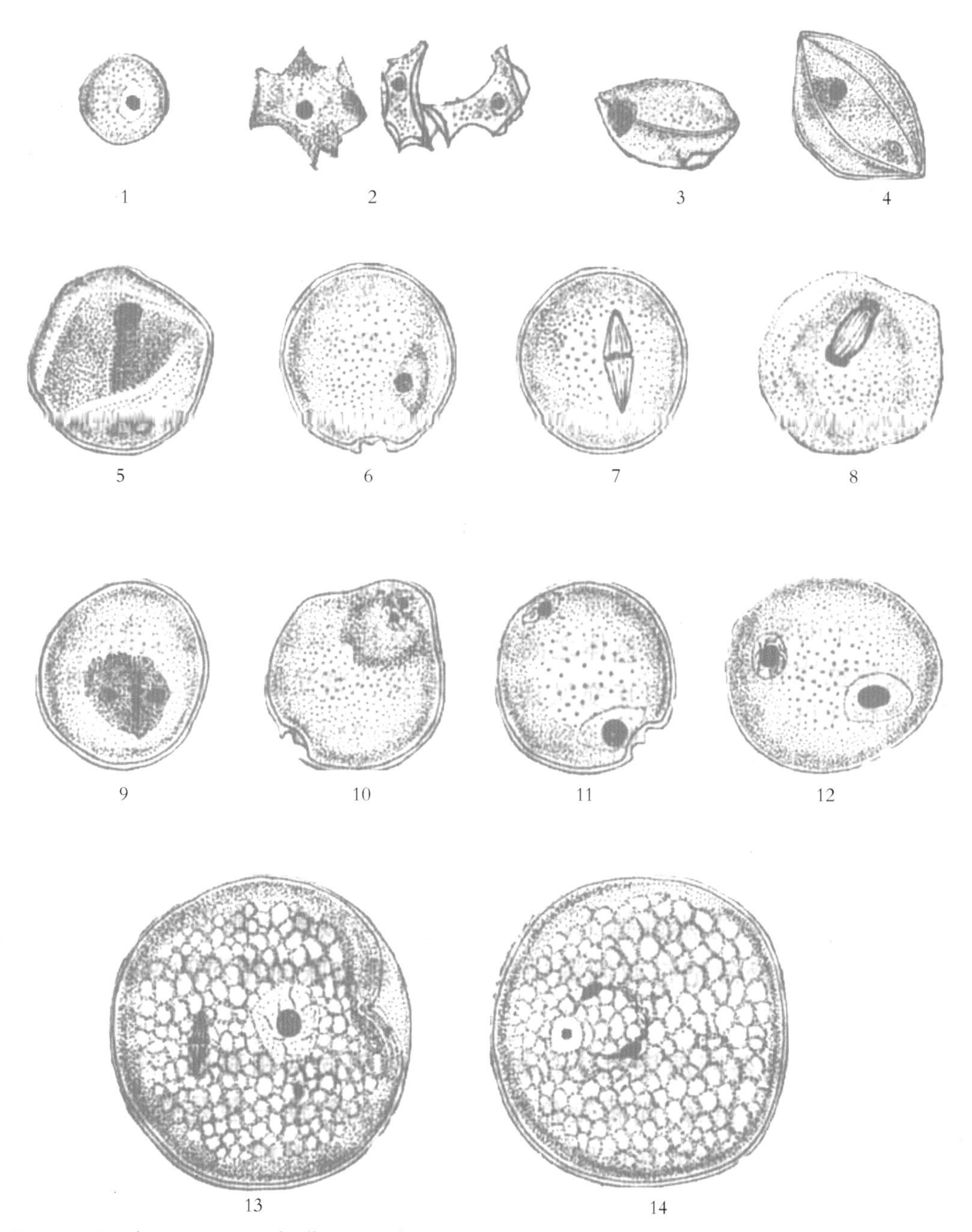

Fig. 2 – 4 Development process of pollen grain of rice Erjiunan 1 (Department of Biology, Hunan Normal University, 1973)

1. Microspore 2. Germinal aperture formed during the first contraction phase 3 – 4. Spindle-shaped pollen grain during the second contraction phase 5. Mononuclear pollen grain 6. Mononuclear pollen grain, nucleus moves to the opposite side of the germinal aperture 7. Metaphase of the first mitosis of pollen grain 8. Anaphase of the first mitosis of pollen grain 9. Telophase of the first mitosis of pollen grain 10. Binuclear pollen 11. Vegetative nucleus moves to the vicinity of the germinal aperture during the binuclear pollen stage, and both are close to the wall 12. Two nuclei get close to each other during the binuclear pollen stage 13. Metaphase of the second mitosis of pollen grain 14. Mature pollen with two sesame-shaped sperm cells

enlarge significantly. At this time, the vegetative and reproductive nuclei stand far apart facing each other, both close to the wall (Fig. 2-4, 11). After maintaining this state for a period of time, the cell membrane of the two nuclei dissolves and the two nuclei get immersed in the same cytoplasm. The reproductive nucleus moves toward the vegetative nucleus, and the vegetative nucleus begins to move to the opposite side of the germinal aperture as the reproductive nucleus approaches. The reproductive nucleus follows the vegetative nucleus to the opposite side of the germinal aperture. Both nuclei move in the manner of amoeba movement during the approach so they are both radial, with the protrusions on the side of the advancing direction tending to get longer than other protrusions. After the two are close, the reproductive nucleus undergoes mitosis (the second mitosis of pollen), and produce two daughter nuclei, i. e. the sperm cells or male gametes (Fig. 2-4, 12 and 13).

(3) Trinuclear pollen

When at the telophase of mitosis and the two sperm cells begin to form, the nuclei of the sperm cells are approximately spherical, with an obvious nucleolus in the center and inconspicuous cytoplasm on the periphery. Later, the sperm cells' nuclei gradually change to a rod shape, with slightly pointed ends, and the cytoplasm extends to both ends, one connected to the vegetative nucleus, and the other to the extended cytoplasm of the other sperm cell. As the pollen matures further, this ribbon-like connection disappears and the sperm nucleus becomes a sesame punctate (Fig. 2-4, 14). A nucleolus is visible in the sperm nucleus under high-power microscope, and many chromatin granules are scattered in the nuclear cavity. When sperm cells undergo the above morphological changes, the nucleolus of the vegetative nucleus shrinks further, and finally becomes very small.

At the anaphase of binuclear pollen development, starch grains begin to form in the pollen grains. When entire pollen grains are filled with starch grains, the pollen is fully matured.

A summary of anther and pollen development is given below.

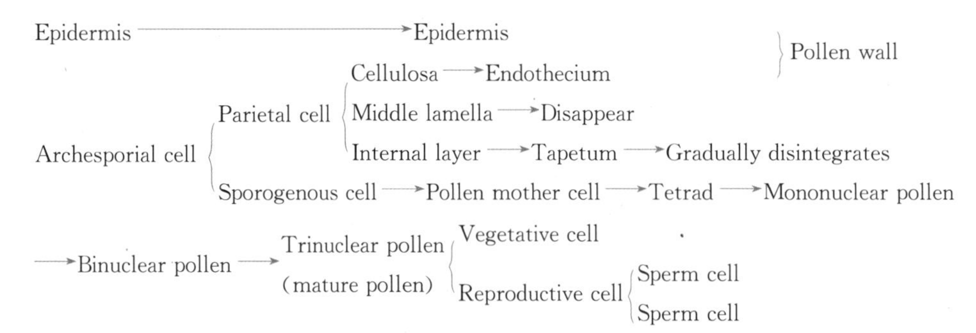

(4) Relationship between pollen development and panicle development

There is a certain correlation between the processes of rice pollen development and panicle development. It is the transition period from the meiosis stage to the mononuclear stage when the flag leaf collar comes out from the penultimate leaf (the first leaf after the flag leaf) collar. The majority of the spikelets are in the peak stage of meiosis when the flag leaf collar is about 3 cm below the penultimate leaf collar. When the panicle top is close to the penultimate leaf collar, it is the transition period from the mononuclear stage to the binuclear stage, i. e. the spikelets in the upper part of the main axis are at the binuclear stage while the rest are at the mononuclear stage. When the panicle top is close to the flag leaf collar, it is the transition period from the binuclear stage to the trinuclear stage, i. e. the spikelets on the main axis are at the trinuclear stage while the rest are at the binuclear stage. When panicles gradually exsert from the flag leaf sheath, spikelets in the middle and lower parts gradually enter the trinuclear stage, while the pollen in the spikelets in the upper part of the panicle is mature.

This timeline may vary with rice varieties and across main and tiller panicles. For instance, the distance between the collars of the flag leaf and the penultimate leaf is greater than 3 cm at the meiosis stage for a tall-stalk variety; on the contrary, it may be less than 3 cm for some dwarf varieties. Therefore, pollen should be collected at the meiosis, mononuclear, binuclear and trinuclear stages for preparation, preferably at peak times of meiosis, i. e. 06:00—07:00 and 16:30—17:30. Materials collected during 16:30—17:30 also show more active pollen mitoses.

Ⅱ. Traits of Pollen Abortion for Male Sterile Lines

The pollen abortion pathway of male sterile rice is intricate, and the most important difference is in the stage at which the pollen is aborted. The development of rice pollen can generally be divided into four stages, the proliferation of sporogenous cell to the meiosis stage, the mononuclear pollen stage, the binuclear pollen stage, and the trinuclear pollen stage. The abortion of rice pollen can also be divided into four types accordingly, 1) the pollen-free type (abortion before mononuclear pollen formation); 2) mononuclear abortion; 3) binuclear abortion; 4) trinuclear abortion.

In 1977, the Institute of Genetics of Chinese Academy of Sciences selected 17 male sterile lines with 13 different cytoplasm sources to observe and compare pollen development. The results are shown in Table 2-1.

Table 2-1 Abortion stages of different male sterile lines

(Institute of Genetics, Chinese Academy of Sciences, 1977)

Sterile line	Type	Mononuclear stage	Binuclear stage	Trinuclear stage	Starch accumulation
Taichung 65A	BT	—	—	++++	++
Baijin A	BT	—	—	++++	++
Erjiuai 4A	WA	+++		—	—
Guangxuan 3A	WA		++	++	++
Erjiuai 4A	Gang	++	++	—	—
Chaoyang 1A	Gang	++++	—	—	—
New Zealand A	Nan	++++	—	—	—
Guoqing 20A	Nan	++++	—	—	—
Taichung 65A	Lead			++++	++

Continued

Sterile line	Type	Mononuclear stage	Binuclear stage	Trinuclear stage	Starch accumulation
Sanqizao A	Yangye		+	+++	++
Liming A	Dian - 1		+++	+	++
Liming A	Dian - 2	+	++	+	+
Nantaigeng A	Jing	+	+++	—	+
Liantangzao A	Hongye	++	++	—	—
Guangxuan 3A	Haiye	+	++	+	+
Erjiuqing A	Tengye	+++	+	—	—
Nongken 8A	Shen	++++	—	—	—

Note. "+" indicates the quantity, while "—" means none.

Table 2 - 1 shows that different sterile lines have different stages of pollen abortion. Some have a more concentrated abortion stage, such as the WA-type Erjiuai 4A; while some have scattered periods such as the Haiye-type Guangxuan 3A. Yet they are always aborted mainly in certain stages. Pollen aborted at an early stage does not contain starch grains, while in the case of a late abortion (after the late binuclear stage), the pollen will contain different quantities of starch grains.

The main cytological characteristics of the four types of rice pollen abortion are detailed below.

1. Pollen-free abortion

Chen Meisheng et al. (1972) from the Department of Biology, Hunan Normal University, studied the three pollen-free *indica* sterile materials Nanguangzhan C35171, Nanluai D31134, and 68 - 899, as well as one pollen-free *japonica* sterile line Jingyin 63, and concluded that pollen-free sterility can be classified into three types.

(1) Abnormal development of sporogenous cells. In this case, the sporogenous cells fail to develop into normal pollen mother cells, but continue to proliferate through amitosis, which is accomplished through nucleolus budding. The nucleolus grows quickly after budding, and it separates to form two new nuclei when it grows to the size of a mother nucleolus. Then a septum is formed between the two nuclei to form two cells. When the spikelets are elongated to 2 - 5 mm, these cells divide in a knife-like manner to gradually form many very irregular sheet-shaped cells of different sizes, and then gradually grow longer, finally becoming filamentous and disintegrated. By the time the spikelets stretch to 6 mm, nothing but a pack of liquid remains in the anther sac.

(2) Abnormal development of pollen mother cells. In this case, sporogenous cells can develop into pollen mother cells and seem to be able to undergo meiosis, but the size and shape of the pollen mother cells are extremely inconsistent, with round and long forms being more common. These cells show no typical prophase change during the first meiotic division, and the shape of the chromosomes is very irregular. This abnormal phenomenon makes it difficult to distinguish between the metaphase and the anaphase

for some cells. When they enter the telophase to form a dyad, their development is not like normal division. The two half-moon-shaped cells of the dyad are connected at both ends, not forming a tetrad, but continuing to undergo mitosis after two divisions. The cells get smaller and smaller, and finally disappear. Chen Zhongzheng et al. (2002) studied the pollen formation and development process of WS-3-1, a new male sterile germplasm produced with space mutagenesis and found that WS-3-1 is a pollen-free sterile germplasm. The anther middle layer begins to vacuolize at the early phase of microspore mother cell development, and degrades prematurely, causing the tapetum to degenerate prematurely and making the tapetum unable to function normally, and the microspore mother cell to adhere and disintegrate at the dyad stage, failing to form pollen. Huang Yuxiang et al. (2000) observed Annong S-1 and found that under high temperature conditions with an average daily temperature higher than 30 ℃, the plants produced no pollen and the microspore mother cells were aborted at the metaphase and anaphase. The sticky edge of the mother cells disintegrated immediately, losing the cell structure and becoming an irregular protoplasmic mass. Finally, the protoplasmic masses gradually disappeared and no pollen existed in the anthers until the flowering stage.

(3) Abnormal development after tetrad. In this case, some of the above-mentioned pollen mother cells of various sizes and shapes can form tetrads through meiosis. When these tetrads develop into tetrad spores, some will undergo amitosis through nucleolar budding and form many cells of different sizes which will then gradually disappear; while others proceed to the first and second contraction periods. After entering the second contraction period, the cells stay shrunk with the protoplasm in them gradually disappearing, and the shrinking continues till there are only residual pollen walls of different sizes and shapes.

In any of the above three cases, no normal microspore or pollen can be formed, thus the name "pollen-free abortion".

2. Mononuclear pollen abortion

Researchers at the Department of Biology, Hunan Normal University, (1973, 1977), through observation of sterile materials from the low generations ($B1F_1$-$B3F_1$) and high generations ($B15F_1$-$B17F_1$) of the original WA strain and the WA-type Erjiunan 1, found the following conspicuous abnormalities in the original WA plants. During the telophase of the first meiotic division, in some cells, one pair of the chromosomes fails to form bivalents, but form a trivalent with the other two bivalents instead; in some other cells, a quadrivalent is formed with two bivalents combined; yet for more of the cells, normal nuclear plates fail to take shape at the metaphase of the second meiotic division, forming only a loose and irregular structure; when entering anaphase I, a pair of homologous chromosomes will go ahead or lag behind; and during the second meiotic division, the two mitosis phases of the dyad are unparalleled, sometimes perpendicular to for a T-shaped tetrad, and sometimes lined up horizontally to form a straight tetrad. Materials from low-generation WA-type sterile lines tend to be aborted at an earlier stage, that is, through abnormal meiosis. Yet for $B3F_1$, abnormal meiosis becomes far less common. Meiosis happens normally in most young panicles and it is more common to see a pair of homologous chromosomes go ahead or lag behind between metaphase I and anaphase I. Abnormalities can still be found in some spikelets of a small number of young panicles. For example, homologous chromosomes can fail to pair up normally, metaphase I chromosomes can fail in forming a neat nuclear plate, or the second meiotic division may result in T-shaped or straight tetrads instead of parallel ones. However, most pollen is aborted at the mononuclear stage with only a small share aborted after entering the binuclear stage. For the higher generations, pollen abortion occurs more stably at the mononuclear stage. Some are aborted at the end of the second contraction and after pollen grains become spherical; while others are aborted when shrinking occurs.

The genetics research group of the Department of Biology at Sun Yat-sen University (1976) revealed through observation of the WA-type Erjiuai 4A, Zhenshan 97A and Guangxuan 3A that most pollen mother cells went through normal meiotic processes with only a small number of Erjiuai 4A plants showing abnormal meiosis. Multipolar spindle fibers appeared at telophase Ⅱ, and chromosomes were divided unevenly into four groups; at the dyad and tetrad phases, complete cell walls could not be formed between daughter cells to complete the division normally. Instead, the two daughter cells were separated where there was a cell wall but connected where there was not, forming O-shaped, X-shaped or T-shaped gaps between them. At the tetrad phase, some cells undergo amitosis or unequal mitosis. Amitosis is where the nucleolus buds or fragments into multiple daughter nucleoli, each of which then form a new nucleus that will later be separated by a cell wall from other new nuclei. Unequal mitosis mostly occurs at the prophase of division. Filamentous chromosomes divide into several masses of different sizes in the cytoplasm, between which cell walls will then form to create microspores, triad spores and polyspores of different sizes. Binuclear and multinuclear spores were observed. Two or more nucleoli can be seen in one nucleus and the cell is large. Only some of such abnormal microspores can develop into mononuclear pollen, with various sizes though. In mature anthers, there is mostly late mononuclear pollen with no cellular content and some have unclear nuclear membranes. They are all aborted pollen. The meiosis and microspore development of Zhenshan 97A are normal, but at the mononuclear anaphase, pollen stick together to form several masses in the anther sac, failing to separate and get released from the anther sac. When the anther is separated with a needle, most of the pollen is found to adhere to tapetum cells, or to stick together with several or a dozen other pollen cells. Pollen spore walls, including intine and extine, are found collapsed or degenerated at the adhesion points, while where this does not happen, the intine and extine are found to be very thick. For a small number of pollen that do get released from the anther sac, the cell walls are incomplete or thinned and the germinal aperture is obscure. Most of this mononuclear pollen, adhesive or not, has no cellular content, with degenerated small nuclei or nucleoli seen in a few of them, with collapsed nuclear membranes. The cytoplasm of some agglutinates into large or small masses that become quite dark after being stained. For Guangxuan 3A, abortion happens as follows. In some microspores the cytoplasm is vacuolated, yet more of them stay at the mononuclear pollen stage and a few at the binuclear stage. The cytoplasm is thin in both cases, and most of the pollen is transparent with nuclei seen in only a small number of them, in which case the nucleoli are small or degenerated completely with the germinal aperture still visible.

Wu Hongyu et al. (1990) revealed through observation of the pollen abortion of the PTGMS line Nongken 58S under long daylength conditions that some pollen mother cells showed abnormalities at the zygotene and pachytene stages of meiosis, such as pollen mother cell adhesion, abnormal cellular structures, cytoplasmic disintegration, and nuclear disappearance; and there were also tripod-shaped and straight tetrads. Pollen abortion mainly occurs at the metaphase and anaphase of the mononuclear stage, when the cell walls shrink severely, cellular content disintegrates, and the nuclei decompose into chromatin masses in some cases.

Liang Chengye et al. (1992) observed from the cytological perspective the seven PTGMS lines of Nongken 58S, W6154S, W6417 Xuan S, 31111S, Annong S – 1, KS – 14 and Peiai 64S at different stages of anther development, and found that pollen abortion for all these lines occurred at the mononuclear stage, but the main specific time point of the abortion differed slightly. W6154S, W6417 Xuan S, and Annong S – 1 are aborted rather late, at the late mononuclear stage (when the single nucleus moves to the periphery), while for Nongken 58S, 31111S, KS – 14 and Peiai 64S, the pollen are aborted earlier at the mononuclear stage (median of the mononuclear stage). The abortion shows the following features. The pollen content disintegrates, pollen grains shrink till only the walls and a small quantity of the cytoplasm and pollen nuclei remain, and the nuclei also disappear in some cases.

Li Rongqian et al. (1993) and Sun Jun et al. (1995) found that pollen abortion of Nongken 58S occurred at the late mononuclear stage under long daylength conditions. The ribosomes are in aggregation, organelles like the endoplasmic reticulum and mitochondrion gradually disintegrate, starch accumulation is low, phagophores increase, and the cytoplasm is thin.

Feng Jiuhuan et al. (2000) studied pollen formation and development of the PTGMS line Peiai 64S and found that its microspore mother cells developed normally before meiosis, but some abnormal changes occurred afterwards. At the prophase of meiosis, the cytoplasm of about half of the microspore mother cells becomes abnormal. There are fewer free ribosomes, the mitochondria are underdeveloped, and many of the endoplasmic reticulum is vesicular, gradually vacuolating and disintegrating. After the early microspores are formed, the extine of almost all microspores develops abnormally with unclear boundaries and a lack of the intermediate zona pellucida. The intine fails to form and abortion occurs early at the binuclear pollen stage. Yang Liping (2003) compared the cytological changes of the pollen of Jiyu *japonica* and D18S at various stages of their formation and development, and found that for the latter, abortion mainly occurred from the late microsporous stage to the early binuclear pollen stage. Huang Xingguo et al. (2011) made cytological observations on the pollen development of seven constructed isonuclear alloplasmic male sterile lines and found that the typical abortion rate of all lines was above 92%, with pollen abortion starting at the mononuclear stage. Guo Hui et al. (2012) made cytological observations on the pollen of CMS lines from five different sources, i. e., the D type, K type, Gang type, WA type and Indonesian paddy rice type, and found that pollen abortion of the WA and D types occurred at the transition from the mononuclear to the binuclear stage, pollen abortion of the Gang type, K type and Indonesian paddy rice type was slightly earlier, at the telophase of the mononuclear stage.

In conclusion, pollen of the mononuclear abortion type is mostly aborted at the mononuclear pollen stage, and the aborted pollen show various irregular shapes. Thus, mononuclear abortion is also commonly referred to as typical abortion.

3. Binuclear pollen abortion

HL-type sterile lines are representatives of this type. According to cytological observation on the pollen of five generations ($B2F_1$ –$B7F_1$) of red-awn wild rice × Liantangzao made by the Genetic Research Laboratory of Wuhan University (1973), most aborted pollen is spherical, while part of it is irregular in shape; up to 98.2% of the pollen did not turn blue when tested with iodine. Jiang Jiliang et al. (1981) observed the pollen abortion process of HL sterile line $B26F_1$ and found that pollen abortion mainly occurred at the binuclear stage (80.3%), with 12.8% occurring at the mononuclear stage. Pathways of binuclear pollen abortion include nucleolus deformation and then nucleus dissolution. The two nuclei get connected and nuclear materials disperse in the cytoplasm and form irregular masses, finally disappearing; some nucleoli germinate and form many more small nucleoli, the nuclear membrane dissolves, and nuclear materials disperse in the cytoplasm and gradually shrink till they disappear; some generative nuclei disintegrate, the nuclear membrane dissolves, and abortion occurs after the vegetative nucleus deforms, thus forming a spherical empty pollen.

Xu Shuhua (1980) made cytological observations on the pollen of $B8F_1$ and $B9F_1$ of Huaai 15A trans-bred from HL type sterile lines, and found that most pollen was aborted at the binuclear stage. The pathway of abortion is that the generative nucleus disintegrates and many chromatins appear around the nucleus to form chromatin masses. Then the vegetative nucleus disintegrates and produces chromatin masses too, which are gradually absorbed and disappear. Meanwhile, the cytoplasm disintegrates, leaving only spherical empty pollen with a germinal aperture. In some cases, this may not happen until when the generative nucleus is about to divide in the binuclear stage.

Pollen aborted at the binuclear stage is spherical, so this type of abortion is also known as spherical abortion.

4. Trinuclear pollen abortion

Trinuclear pollen abortion mainly occurs early in the trinuclear stage. As pollen grains have accumulated much starch by this time, easy to be stained by iodine-potassium iodide solution, it is also known as stained abortion. Actually, stained abortion can happen from the late binuclear stage to the trinuclear stage. The BT type is a typical representative. According to observation on Taichung 65A bred by the genetics research group of the Department of Biology at Sun Yat-sen University (1976), from the meiosis of pollen mother cells to the trinuclear pollen stage, the appearance of most pollen is normal, showing no obvious abnormality compared with the counterparts in the maintainer line along the way. For only a very small number of the cells, the nucleolus of the generative nucleus becomes smaller during the binuclear and trinuclear stages, some vegetative nuclei degenerate at the trinuclear stage with the nucleoli becoming smaller and the nuclear membrane disappearing; and some pollen are two-thirds smaller than normal ones. Jiang Jiliang et al. (1981) observed Nongjin 2A and Fuyu 1A trans-bred from BT-type sterile lines and found that when the pollen developed to the binuclear stage, 88% and 93% of the normal pollen entered the trinuclear stage where they were aborted. The abortion pathway is as follows. In Nongjin 2A, many chromatin granules were thrown out when the generative nuclei are in the anaphase of mitosis. The thrown out chromatin granules were very small, and would disappear later. The ratio of these cells to normal cells was 39 : 59. This abnormality is quite rare in the maintainer line. In Fuyu 1A, it was found that at the binuclear anaphase, the vegetative nuclei of many pollen grains were as large as the generative nuclei, and the nucleoli of the generative nuclei germinated to produce many small nucleoli, which were scattered in the cytoplasm and finally disintegrated. Some vegetative nuclei that entered the trinuclear stage had two isometric nucleoli and some nucleoli germinated. The size of the sperms formed through the division of the generative nuclei differed. These phenomena are extremely rare in the maintainer line.

The stage at which abortion actually occurs and the pathway of pollen abortion of rice male sterile lines are much more complicated than what is described above. They can differ for sterile lines of different types, for isoplasmic allonuclear and isonuclear alloplasmic lines, and even for different backcrossed generations of the same line, different plants, different spikelets and different environmental conditions.

Ⅲ. Characteristics of the Tissue Structure of Rice Male Sterile Lines

A normal rice anther has four anther chambers, two on each side of the anther connective vascular bundle in the center, bilaterally symmetric. There is a line of dehiscence between the two anther chambers on each side and a dehiscence cavity below the line. The filament is single-veined, and there are more than one annular vessels in the vascular bundle and two or more annular vessels or tracheids in anther connective. The anther wall is composed of four layers of cells which are the epidermis, the endothecium, the middle lamella and the tapetum (Fig. 2 - 5).

After the pollen mother cells enter the meiosis phase, the cells of the middle lamella begins to degenerate, becoming indistinguishable in the trinuclear pollen stage. After the mononuclear microspores are formed, tapetum cells start to disintegrate too and finally disappear at the trinuclear pollen stage, leaving only the epidermis and the endothecium (Fig. 2 - 6).

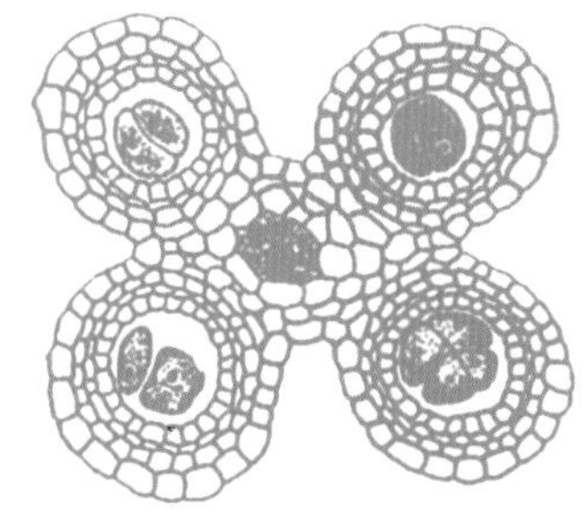

Fig. 2 - 5 Cross section of a rice anther and the formation of pollen mother cells
(Jiangsu Agricultural College, 1977)

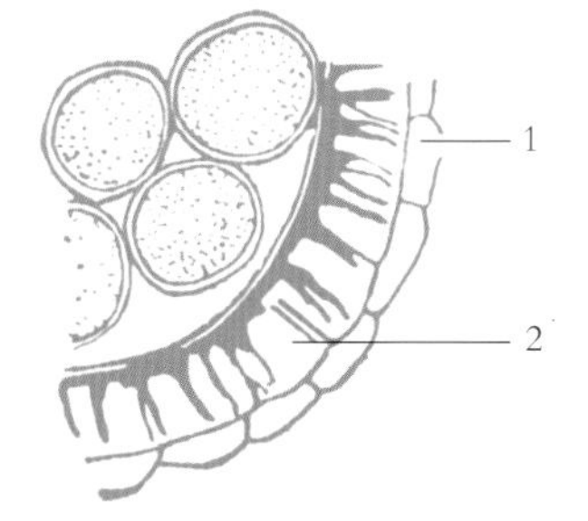

Fig. 2 - 6 Structure of a mature rice anther wall
1. Endothecium 2. Epidermis

Cellular walls of the fibrous layer form a "spring" through uneven ring proliferation. When the flower blooms, the anther wall cells lose water and the extine shrinks, making the elater stretch out and the anther wall dehisce to eject the pollen, thus completing the flowering and pollen release process (Fig. 2 - 7).

Compared with the above, a rice male sterile line usually show abnormalities in the development of its anthers, which are inherently related to pollen abortion and the difficulty of anther dehiscence.

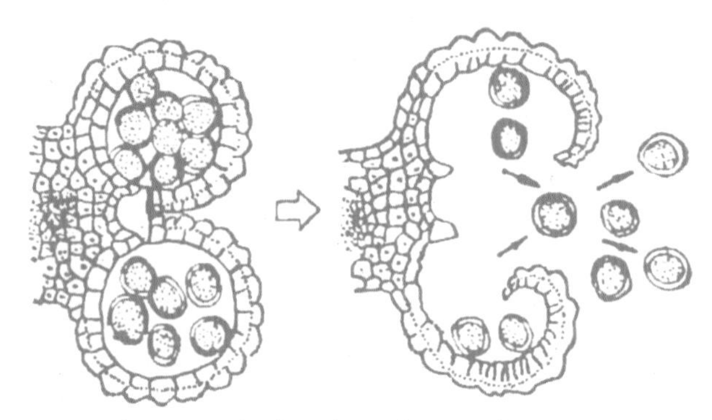

Fig. 2 - 7 Dehiscence and pollen release of a rice anther (Xing Chuan, 1975)

1. Relationship between tapetum & middle lamella development and pollen abortion

It is generally acknowledged that the tapetum is the nutritive tissue for pollen development and that its functions include- 1) to decompose callose enzyme and thus control the synthesis and decomposition of the callose walls of microspore mother cells and microspores; 2) to provide sporopollenin that constitutes the pollen extine; 3) to provide protective pigments (carotenoid) and lipids that constitute the extine of mature pollen grains; 4) to provide extine protein, i. e. the recognition protein controlled by the sporophyte; and 5) to transport nutrients to ensure microsporous development. The products of tapetum degradation can serve as raw materials for the synthesis of pollen DNA, RNA, proteins and starch. Therefore, abnormal development of the tapetum (early or delayed disintegration) is considered the inducement of various types of pollen abortion in rice male sterility.

Xu Shuhua (1980) found in the HL-type Huaai 15A that tapetum cells proliferate abnormally to form peripheral tapetum periplasmodium, pushing pollen mother cells to the center of the anther chamber and causing all pollen mother cells in the same anther chamber to disintegrate. In the WA-type Huaai 15A, at the mononuclear pollen stage, in some anthers, the epidermis and the fibrous layer abruptly start to expand radially, destroying the tapetum and pushing the cells to the center of the anther chamber. Thus the tapetum quickly disintegrates and disappears, and the pollen is aborted. Pan Kunqing (1979) found in the WA-type Erjiuai 4A and Erjiunan 1A sterile lines that tapetum cells underwent rapid cytoclasis and disappeared quickly during the mononuclear pollen stage, resulting in pollen abortion. Guo Hui et al. (2012) observed five CMS lines with different cytoplasmic sources, i. e. the D-type D62A, K-type K17A, Gang-type Gang 46A, WA-type Zhenshan 97A and Indonesian paddy rice Ⅱ-32A5, and found that the five male sterile lines had little difference in their abortion processes, with only slight difference as to the specific stage when the abortion occurs. Specifically, abortion occurs mostly during the transition from the mononuclear to the binuclear stage for the WA type and the D type, when the pollen is ready for mitosis and pollen content is about to form. At this phase, the volume of pollen cells increases rapidly, requiring lots of nutrients which come mainly from the disintegration of the tapetum. However, the anther tissues undergo abnormal changes at this point and the tapetum undergoes rapid cytoclasis and decomposition, resulting in the loss of the nutrients required for microsporous development. This may be the reason why the pollen of WA-type and D-type sterile lines is aborted at the late mononuclear stage or early binuclear stage. For the other three types of CMS lines, i. e. the Gang type, K type and Indonesian paddy rice type, pollen abortion occurs slightly earlier, generally at the mononuclear telophase. Tapetum decomposition peaks at the mononuclear anaphase. Analysis reveals that the complete decomposition of the tapetum in a short period of time prevents the microspores from getting the required nutrient supply, resulting in the abortion of microspores. Hu Lifang et al. (2015) studied tissue sections of the rice male sterile mutant *tda* obtained from Songxiang early *japonica*, a *japonica* rice variety created through radiation with ^{60}Co-γ rays and found that it started to show abnormalities during microsporous development as the tapetum degraded earlier than usual and the microspores deformed and failed to form normal pollen grains.

The Department of Biology at Guangxi Normal University (1975) found in Guangxuan 3A that when the pollen developed to the trinuclear stage, its tapetum cells failed to disintegrate and the nucleolus still existed, keeping pollen development stagnated at a certain stage and getting the pollen aborted. Wang Tai et al. (1992) found in Nongken 58S that the inner tangential walls of tapetum cells in sterile anthers decomposed during meiosis and the cells began to separate. At the early mononuclear stage, tapetum cells separated but did not disintegrate, while at the late mononuclear stage, the walls of the tapetum cells would disintegrate and the cytoplasm would merge, resulting in two different outcomes. In some of the cells, the cytoplasmic mass extends toward the center of the anther chamber and fills the space between microspores at the late mononuclear stage while in other cells, the cytoplasm forms a complete protoplasm layer around the anther chamber, with uneven inner surfaces and partial disintegration. Li Rongqian et al. (1993) observed the ultrastructure of Nongken 58S anthers under long daylength conditions and found the tapetum cells of sterile anthers to have a complete structure all along the way, containing nuclei and abundant cytoplasm with well-developed endoplasmic reticulum, a small number of vacuoles, large spheroplasts (lipids), plastids and other organelles. Multiple circular vesicles were seen inside the tangential walls of tapetum cells, formed with secretion. The tapetum cells generally were found to be very active metabolically without any sign of disintegration, while the nearby pollen was already aborted. Observation on Nongken 58S made by Sun Jun et al. (1995) also showed that the disintegration of tapetum cells in sterile anthers was delayed under long daylength conditions. Yang Liping (2003) systematically observed the pollen formation and development process of the normal Jiyu *japonica* line and the PTGMS line D18S. By comparing and analyzing the cytological changes of the pollen of the two lines at different stages, it was found that a series of abnormalities appeared in the sterile D18S at the microspore metaphase. Specifically, the cytoplasm shrank, the content degraded to vesicles, and the cytoplasm and the nuclei degenerated to about 2/3 of the normal pollen volume. The tapetum of the fertile Jiyu *japonica* degraded from the meiosis period on and basically disintegrated at the late microspore stage, while that of the sterile line D18S did not show clear signs of disintegration during the meiosis period, with cavities formed only in some cells. By the microspore metaphase, the basic feature is amalgamation into hill-like structures, and thick belts remained even in the microspore anaphase. Thus, compared with Jiyu *japonica*, D18S showed delayed disintegration of its various structures at basically all stages. Peng Miaomiao et al. (2012) observed tissue sections of sterile mutant TP79 derived from naturally mutated rice variety Taipei 309, and found that TP79 showed abnormalities at the prophase of microspore formation, i. e. the tapetum failed to degrade normally, the microspores developed abnormally, and the tapetum was still in a condensed state at the final stage of pollen maturity, forming shriveled and inactive pollen. However, according to Xu Hanqing et al. (1981), Lu Yonggen et al. (1988) and Feng Jiuhuan et al. (2000), tapetum development in the anthers of male sterile lines was not abnormal, with no obvious difference compared with normal fertile anthers. So it is still difficult to determine whether abnormal tapetum development is the real cause of pollen abortion.

In addition, the anther wall structure generally has starch or other storage material in the middle lamella, which gradually tends to disintegrate and be absorbed during the development of the microspores. Abnormal cellular structure in the middle lamella may impair normal pollen development. For example, Pan Kunqing (1979) found in paraffin sections of Erjiuai 4A, Erjiunan 1A and other WA-type sterile lines at the mononuclear pollen stage that cells of the middle lamella began to thicken and vacuolize along the radial direction of the anther, pushing the tapetum cells to the center. At this point, many vacuoles appeared in the cytoplasm of tapetum cells, substantially thinning the cytoplasm which can be only very lightly stained. Afterwards, cells of the middle lamella continued to vacuolize and enlarge, and all of them would have become vacuoles and fully enlarged by the binuclear pollen stage (the pollen would have already been aborted and could not proceed to the binuclear stage). As the anthers grew, the middle lamella cells enlarged, but did not continue to thicken, tending to shrink instead. On the cross section, the original approximate square has become a narrow and elongated shape, and the linearly curved nuclei are still visible. At this point, a secondary tapetum wall can be seen clearly outside the middle lamella. Sun Jun et al. (1995) observed anther wall development of the PTGMS line Nongken 58S under long daylength conditions and found delayed disintegration for the middle lamella of sterile anthers.

Chen Zhongzheng et al. (2002) studied the pollen formation and development process of the new non-pollen type male sterile germplasm WS-3-1 produced through space mutation, its parent line Texianzhan 13 and the normal line IR36. The findings revealed that the male sterility in WS-3-1 is caused by abnormalities of the middle lamella (premature vacuolization and disintegration). Such abnormalities caused premature degradation and loss of the normal functions of the tapetum, as a result of which microspore mother cells failed to get sufficient nutrients to initiate and proceed with meiosis and are starved to consume its own cytoplasm. As such, lots of vacuoles would appear in the cells and microtubules would be disarranged in microspore mother cells during meiosis. With constant nutrient consumption in the cytoplasm of the microspore mother cells, vacuoles enlarge and cavities form, hampering normal meiosis. Thus, irregular spindle apparatuses will be formed at metaphase I, which will be aborted and disintegrate shortly after dyads are formed, ultimately producing no pollen. Additionally, due to the premature degradation of the tapetum, callose enzyme cannot be secreted normally and microspore mother cells stick together for a longer time. This may accelerate abortion.

2. Relationship between filament & anther connective vascular bundle development and pollen abortion

The filament and anther connective vascular bundle of rice stamen are channels for the absorption of water and the transport of nutrients to the anther chambers, providing nutrients required for pollen development, which is highly important. Poor differentiation and development thereof may impair substance transport and hamper normal pollen development.

Pan Kunqing (1979) compared filament tissues of common WA-type sterile lines, non-pollen sterile lines, and Zhenzhu'ai A, Erjiunan 1A, Erjiuai 4 A and Lushuang 101A of the WA type, as well as their maintainer lines, and found that vessels were completely degenerated in WA and wild non-pollen lines. For WA lines, in particular, the degree of vessel degradation in the filament is correlated to the backcrossing generation, i. e. the higher the generation, the higher the degree of degeneration. Generally, degeneration starts from $B1F_1$, and usually from the middle section of the filament. More than half of the vessels degenerate in $B2F_1$ and most of them are degenerated by $B3F_1$. The degree of filament degeneration is positively correlated with the proportion of fertile and aborted pollen in the stamen anther chamber: When 100% of the pollen in an anther chamber is sterile, no fully developed vessel can be seen in the filament. When 50% are sterile, the vessels are interrupted so there can be two vessels, one vessel or no vessel at different points of the filament. Some may have only a small section of a vessel at the base of the anther connective. When 20% are sterile, vessels in the filament develop normally or degenerate only slightly.

Xu Shuhua (1980, 1984) found in the HL-type Huaai 15A that adjacent filaments in some spikelets tend to merge at the base. Sometimes two merge into one, and sometimes three merge into one. In addition, the WA type and HL type male sterile lines differ in the development of filament vascular bundles. Both show traits of the original female parent. In the WA type, the filament vascular bundles are degenerated severely, while those in the HL type are better developed. More specifically, Huaai 15A and Huaai 15B, both of the HL type, mainly show differences in their anther connective vascular bundle. In integral samples of the maintainer line, it can be seen that the anther connective vascular bundles differentiated well, featuring uniform wall thickness and regular arrangement; cells stand in compact lines to form vessels; in paraffin sections, the vascular bundles also show uniform thickness and sound differentiation. In comparison, in integral samples of the sterile line, vessels in the anther connective vascular bundle generally show poor development before pollen abortion, as the number, width and ring spacing of vessels in the upper part of the anther connective differ greatly, and the tubular cells leading into to anther wall are also less differentiated; in paraffin sections, vascular bundles are found underdeveloped and poorly differentiated, thinner than those of the maintainer line and sometimes with obscure structures and uneven thickness. Underdevelopment of anther connective vascular bundles was extremely common and serious in the WA-type Huaai 15A as sections show serious or extremely serious vascular bundle degeneration, and even missing or interrupted vascular bundles in some cases. Similar conclusions have also been reached in comparative studies of the anther connective vascular bundles of the WA-type Erjiunan 1A, Xiang'aizao 4A, Bolizhan'ai A and their maintainer lines made by the Department of Biology at Hunan Normal University (1975), and those of the WA-type Erjiuai 4A and Erjiunan 1A and their maintainer lines made by the Department of Biology at Sun Yat-sen University (1976).

Wang Tai et al. (1992) found through observation that the anther connective tissue development of fertile anthers of the PTGMS line Nongken 58S was similar to that of common rice varieties, while for sterile anthers, at the microspore mother cell phase, parenchymal cells in the vascular bundle developed poorly with thin walls, and invisible vessels and sieve tubes. At meiosis, vascular parenchymal cells increased, but failed to develop normally. At the early mononuclear stage, differentiated vessels and sieve tubes were visible, but cell walls of the sieve tubes were thin, and the parenchymal cells had little cytoplasm. At the late mononuclear stage, there can be three types of abnormal vascular bundles. The first type shows crimpled sheath cells, poorly developed parenchymal cells, a complete xylem with two or three thin vessels and a complete phloem with three to four thin sieve tubes. The second type has crimpled sheath cells, underdeveloped parenchymal cells, a xylem with poorly differentiated vessels, and degenerated sieve tubes. The third type has severely crimpled sheath cells, incomplete xylem and phloem, and severely degenerated parenchymal cells. Huang Xingguo et al. (2011) made cytological observation on seven isonuclear alloplasmic male sterile lines and found abnormal vascular tissue development in all of the lines when the pollen was aborted. According to their observation, there were no or only one to two vessels or sieve tubes in the anther connective of the sterile lines, and the vascular parenchymal cells enlarged abnormally. In comparison, the corresponding maintainer lines had perfect vascular tissues, so it was speculated that underdevelopment of the anther connective vascular tissue was the major cause of pollen abortion in the sterile lines.

3. Relationship between development of anther dehiscence structure and sterile line

Whether and how well an anther dehisces has direct bearings on pollination and fertilization, and thus the seed setting rate. In addition to abnormalities in the development of the tapetum, filament and anther connective vascular bundle, sterile lines also show different degrees of abnormalities in the development of their anther dehiscence structure.

Zhou Shanzi (1978) and Pan Kunqing et al. (1981) described rice anther dehiscence and its mechanism by comparing the tissue structures of a three-line set. Into the binuclear pollen stage, a dehiscence cavity is formed on each side of the anther connective, under the epidermal cells at the bottom of the depression between the two anther chambers. Between the openings of the cavities is a row of four to six small epidermal cells. On the opposite side are parenchymal cells of the anther connective tissue. There can be one or two fibrous layer parenchymal cells on the left and right sides, which remain parenchymal all through the process, without fibrosis, and shrink further before the anther dehisces, clearly different from the fibrous layer cells that have gone through fibrillation. The dehiscence cavity has only one layer of cells on three sides except for the anther connective, and this one layer can even be shrunken parenchymal cells. Therefore, this is the weakest point of the anther and the place that opens when the anther dehisces (Fig. 2 - 8).

The dehiscence cavity forms first on one side of the middle section of the anther, and then extends to the other side, going from small to large, spanning the two ends of the anther, and connecting the two sides. At the same time, the fibrous layer of the anther wall begins to fibrillate. Thickened rings form along the anticlinal direction on the cellular walls and get connected laterally to create a "spring" perpendicular to the vertical axis of the anther, connected to both sides of the dehiscence cavity. The thickened rings are the strongest at the upper and lower ends of the anther, and gradually get weaker toward the middle where there are no such rings. The "springs" are best developed by the sides of the split, and be-

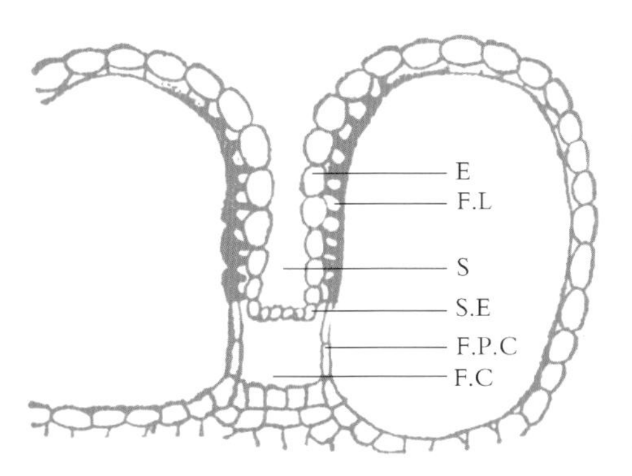

E: Epidermal cells; F. L: Fibrous layer cell; S: Split; S. E: Split small epidermal cells; F. P. C: Fibrous parenchymal cells; F. C: Fracture (dehiscence) cavity

Fig. 2-8 Cross section of a rice anther at dehiscence (Pan Kunqing et al., 1981)

come weaker toward the back. Such a structure determines that a rice anther dehisces first at the two ends and the opening will then extend toward the middle until the dehiscence is completed. During glume opening, as the anther wall loses water and crimples, fibrous cells will have vertical tension which is higher for stronger "springs", so the anther dehisces from the ends toward the middle. The anther chambers dehisce before the part between anther chambers. The exact place where the anther chambers dehisce is between the fibrous layer cells on both sides of the dehiscence cavity and the fibrous parenchymal cells while dehiscence between the two anther chambers happens between the small epidermal cells at the split.

Zhou Shanzi (1978) observed the WA-type Bolizhan'ai A and Nantai 13A, the BT-type Liming A and their corresponding maintainer lines and found that the "springs" formed by the fibrous layer cells of Bolizhan'ai A plants were not strong, having no dehiscence cavity or only one. The elasticity of the "springs" is low, so the dehiscence cavity fails to open and the anther cannot dehisce. Nantai 13A has strong "springs" but no dehiscence cavity, so the two anther chambers are firmly bonded and the "springs" are not strong enough to break the anther chamber and dehisce the anther. Liming A also has strong "springs", but some anthers fail to form dehiscence cavities, while some only have a cavity on one side, and others have it on both sides. Therefore, the anther may fail to dehisce, dehisce on one side, or on both. Thus, anther dehiscence can be quite different for different types of sterile lines. This is because abnormally developed anther walls impair the formation and development of strong "springs" or dehiscence cavities. Pan Kunqing and He Liqing (1981) observed the anther dehiscence of the WA-type Erjiuai 4A, Zhenshan 97A and their corresponding maintainer lines, and found that the reason why the anthers of the sterile lines failed to dehisce or had difficulty dehiscing was not the fibrillation level of their fibrous layer cells. Instead, the reason lies with the split, especially the dehiscence cavity that fails to differentiate properly.

Section 3 Physiological and Biochemical Characteristics of Rice Male Sterility

There is a large amount of literature on the physiological and biochemical differences between sterile and fertile male rice lines. The results showed that rice male sterile lines, be it CMS or PTGMS lines, have their own characteristics in substance transport and energy metabolism, the composition of amino acids and proteins, enzyme activity, hormone level and so on. Study on the physiological and biochemical functions and physical characteristics of rice male sterile lines can help us understand the causes of male sterility as well as the expression and regulation of male sterility genes.

I. Substance Transport and Energy Metabolism

Shanghai Institute of Plant Physiology (1977) measured the starch content of rice pollen of some three-line sets and found that for the WA type, the pollen of the sterile line contained no starch, that of the maintainer line contained some starch, and that of the restorer line had the most starch; for the HL type and the BT type, the pollen of the sterile lines had little starch, with only a few small starch grains. Wang Tai et al. (1991) compared carbohydrates in the leaves of the PTGMS line Nongken 58S and its fertile counterpart Nongken 58 under long daylength and short daylength conditions, and found that changes in the content of starch, sucrose and reducing sugar were consistent between the sterile and fertile lines before the pistil and stamen were formed; but after that, the sucrose content in the leaves of the fertile plants decrease and the starch and reducing sugar increase gradually, while for sterile plants, the content of all three increased significantly. Based on this, it is speculated that the transport of carbohydrates from the leaves to the stamen is hampered and the microspores are thus aborted due to a lack of nutrient supply. Wang Zhiqiang et al. (1993) studied the transport and distribution of assimilates during spikelet development of Nongken 58S and found that assimilate input was significantly lower under long daylength conditions than under short daylength conditions.

Researchers from the Department of Biology at Sun Yat-sen University (1976) measured the absorption and distribution of ^{32}P, ^{14}C and ^{35}S in WA-type and BT-type sterile lines and their maintainer lines. The results show that the absorption intensity of the male sterile pollen, vascular bundles of the spike branches and spikelets was lower than that of the maintainer lines, while the absorption of the ovary is the same as that of the maintainer lines, indicating that the ovary has normal substance metabolism, but the anther metabolism is impaired.

Chen Cuilian et al. (1990) studied the phosphorus metabolism of Nongken 58S using the ^{32}P tracer technique, and found that the radioactive intensity of ^{32}P-hexose phosphate in flag leaves under short daylength conditions was less than 1/3 that under long daylength conditions; meanwhile the radioactive intensity of ^{32}P-hexose phosphate in the anthers (pollen) under short daylength conditions was more than three times that under long daylength conditions. This indicates that the accumulation of large amounts of hexose phosphate under long daylength conditions slows down transport along the central metabolic pathway, and this in turn affects normal oxidative phosphorylation and the conversion of carbohydrates to nitrogenous compounds and phospholipids. As a result, the content of DNA, RNA, phospholipids, high-energy phosphate and other compounds are significantly lower than under short daylength conditions. In plants under long daylength conditions, the content of phospholipids, RNA and DNA in the anthers is also significantly lower than in those under short daylength conditions, indicating that low levels of substance and energy conversion is a major factor causing pollen abortion.

He Zhichang et al. (1992) studied the distribution of ^{32}P in Nongken 58S. Under short daylength conditions, most ^{32}P absorbed by fertile plants, or specifically 78.86% of the total, was transported to the panicles, while the penultimate leaf received only 8.07%; in comparison, under long daylength conditions, sterile plants absorbed 9.23% more ^{32}P than fertile plants, but only 4.15% of the total absorption was transported to their panicles and 75.89% accumulated in the penultimate leaf. Therefore, it is believed that inadequate supply of nutrients for pollen development is one of the reasons for pollen abortion in Nongken 58S under long daylength conditions.

Xia Kai et al. (1989) measured the changes in the ATP content of the leaves of Nongken 58S and found that when the materials were moved from the scotophase to the photophase, the ATP content was lower under long daylength conditions than under short daylength conditions, while the opposite was true when the materials moved from the photophase to the scotophase. In comparison, the control variety Nongken 58 shows no significant difference under long and short daylength conditions. It is speculated that Nongken 58S has different metabolic pathways under different daylength conditions. Deng Jixin et al. (1990) studied the ATP content of Eyi 105S during pollen development. Sterile plants of pollen abortion had far lower ATP content than fertile plants from the early mononuclear stage on, with the specific content only 1/7 – 1/4 that of fertile plants under short daylength conditions. This indicates that ATP content is closely related to fertility. Chen Xianfeng et al. (1994) studied the ATP content of Nongken 58S sterile anthers and found it was significantly lower than that of fertile anthers at the early mononuclear, late mononuclear, binuclear and trinuclear stages. As the pollen of Nongken 58S are aborted at the mononuclear stage, it is believed that the occurrence of male sterility is related to the lack of energy (ATP) supply for normal morphogenesis of the anther caused by abnormal energy metabolism.

Zhou Hantao et al. (1998) found through microcalorimetric analysis of the mitochondria of Maxie CMS line and its maintainer line that the energy release from the mitochondria was significantly lower in the sterile line than in the maintainer line. This may be related to the insufficient substance involved in energy metabolism in the mitochondria of the sterile line. Wei Lei et al. (2002) made microcalorimetric analysis on the anthers of four male sterile lines, i. e. Zidao A, Zhenshan 97A (WA type), Yuetai A (HL type) and Maxie A (Maxie type), and one maintainer line Zidao B, and proved that anther energy metabolism was closely related to CMS. Zhou Peijiang et al. (2000) studied the mitochondrial in vitro energy release thermogram and differential scanning calorimetry (DSC) curve of rice CMS lines Zhenshan 97A (WA type), Guangcong 41A (HL type) and Maxie A (Maxie type) as well as their corresponding maintainer lines, and found the mitochondria of the sterile lines to release more heat and have higher energy levels and more complex mechanism than those of their maintainer lines, so their energy release rates were lower.

In conclusion, impaired substance transport and abnormal energy metabolism are closely related to male sterility.

Ⅱ. Changes in Free Amino Acid Content

Amino acids are raw materials for protein synthesis and also products of the decomposition of proteins. The Shanghai Institute of Plant Physiology (1977) measured free amino acid content in the anthers of rice three-line sets and found that the anthers of sterile plants contained more amino acids than their maintainers and restorers. It suggests that protein decomposition is greater than protein synthesis as a result of pollen abortion. Among the various types of amino acids, the contents of proline and asparagine have the greatest difference between fertile and sterile lines. The proline content in the anthers of a sterile line is very low, at only 5.6% of the total amino acids content, while the asparagine content is quite high, accounting for 59.2% of the total. Conversely, the proline content in fertile pollen is very high, while the asparagine content is very low. Similar measurements done by the Chemistry Teaching and Research Section of Hunan Agricultural University (1974), Crop Ecological Genetics Laboratory of Guangdong Agriculture and Forestry University (1975), and Department of Biology at Guangxi Normal University (1977) on different three-line sets all obtained similar results. Shen Yuwei et al. (1996) studied the content of free amino acids in the anthers of R-type fertile revertant T24 obtained from the γ-ray mutated sterile *indica* rice line Ⅱ-32A. The results also showed that the free proline content of the sterile line was much lower than that of the fertile line, while asparagine was accumulated in large quantities in the sterile line. It was also found that the free arginine content of the sterile line was 6-10 times higher than that of the fertile line, indicating that free arginine may also be associated with male sterility.

Xiao Yihua et al. (1987) measured the free amino acid content in anthers at different pollen development stages of Nongken 58S, Nongken 58, V20A and V20B under artificially controlled long and short daylength conditions. Among the 17 amino acids whose content was measured, proline was the most closely related to pollen abortion, followed by alanine. During pollen development, the proline content in sterile pollen declines continuously, down to only 0.1% of the dry weight of the anther at the trinuclear stage; while the proline content in fertile pollen rises gradually, reaching up to 1.0% of the dry weight of the anther at the trinuclear stage. Alanine content follows a trend opposite to that of proline content, but the slope is not that steep.

According to Wang Xi et al. (1995) and Liu Qinglong et al. (1998), the proline content of anthers aborted through chemical gametocide is also significantly lower than those of their fertile counterparts, indicating that lower proline content is common across different types of rice sterile lines.

Proline, as a storage form of amino acid, can be converted into other amino acids. Working with carbohydrates in the pollen, it can provide nutrition to promote pollen development, germination and pollen tube elongation. Decrease in proline content can cause nutritional disorder and thus male abortion.

Ⅲ. Changes in Proteins

As important components of pollen, proteins play a very important role in the development and biological activity of pollen. Shanghai Institute of Plant Physiology (1977) compared the protein content of the anthers of Erjiunan 1A, Erjiunan 1B, and the restorer line IR661 and found the sterile line to have the lowest anther protein content, followed by the maintainer line, and the restorer line to have the highest. In terms of protein per 100 mg of fresh anthers, the maintainer line has 2. 65 times as much protein as the sterile line, and the restorer line has 2. 94 times as much. For every 100 anthers, the maintainer line has 4. 14 times as much protein as the sterile line and the restorer line has 10. 89 times as much. Dai Yaoren et al. (1978) analyzed the free histone content of Erjiunan 1A and Erjiunan 1B through disc electrophoresis, and found the content of various free histones in the sterile line to be significantly lower than in the maintainer line at various pollen development stages. At the critical stage (mononuclear stage) of pollen abortion, in particular, one of the free histones tended to disappear in the pollen of the sterile line, while it was clearly present in that of the maintainer line. The same histone completely disappeared from the binuclear stage to the trinuclear stage in the sterile line and this must be closely related to pollen abortion, probably through its involvement in the suppression of the expression of certain genes in the nuclei, thus controlling specific transcription processes and affecting pollen development. Zhu Yingguo et al. (1979) analyzed free histones in the anthers of various sterile lines and maintainer lines, such as Zhenshan 97, and confirmed that the content is lower in sterile lines than in maintainer lines.

Ying Yanru et al. (1989) compared the component I protein (RuBP carboxylase/oxygenase) of CMS lines and their maintainer lines of rice (Zhenshan 97), wheat (Fan 7), rapeseed (Xiang'aizao) and tobacco (G28) with immunochemical methods and amino acid composition analysis. Little difference was found. Therefore, it is deduced that the chloroplast gene product, i. e. the large subunit of component I protein has little to do with CMS.

Xu Renlin et al. (1992) compared the proteins and polypeptides of the chloroplast, mitochondrion and cytoplasm of the WA-type CMS line Zhenshan 97A and its maintainer line Zhenshan 97B through one-way SDS-PAGE and protein chromium silver staining, and found significant differences between the two, with disparities more clearly seen in the reproductive organ (panicles) than in the vegetative organ (leaves). For mature panicles of a sterile plant, soluble proteins in the chloroplast have 25 bands, while for the maintainer line there are only 16 bands. Differences were observed on 19 polypeptides. As for the mitochondrion, soluble proteins in the sterile line have 28 bands while the maintainer line lack two polypeptides, 30. 1 kD and 21. 8 kD, compared with the sterile line. The soluble protein component of protein acetone precipitates of the cytoplasm of the sterile line have 24 bands, while its counterpart of the maintainer line have 29 bands, with differences seen on seven polypeptides. The SDS-solubilized protein component of the soluble protein acetone precipitates of the sterile line cytoplasm have 18 bands, while its counterpart of the maintainer line have only 11 bands, with differences seen on seven polypeptides. Based on these, it is believed that the expression of WA-type CMS phenotypes may require multiple genes to be enabled or disabled, involving not only the chloroplast and the mitochondrion, but also the nuclear genome.

Zhang Mingyong et al. (1999) measured the soluble protein content in the seedlings, flag leaves, young panicles and anthers of Zhenshan 97A and Zhenshan 97B plants, and found that results were basically the same between the two lines for the first three, while the sterile line had far lower soluble protein content in the last than the maintainer line.

Wen Li et al. (2007) separated total proteins from the pollen of the HL-type CMS line YTA and its maintainer line YTB at the binuclear stage. Comparing the two, YTA lacks or has lower expression of some proteins involved in substance and energy metabolism, namely the α chain of mitochondrion H^+-transporting ATPase (H^+-ATPase), salt-inducible annexin, mitochondrion NAD^+-dependent malic enzyme and phosphoribosyl pyrophosphate synthase. The reduced or lack of expression of these proteins may be related to the abnormal pollen development resulted from insufficient energy supply from the mitochondrion. The increased expression of mitochondrial voltage-dependent anion-selective channel (VDAC), which is an important protein in YTA, may be related to programmed cell death during pollen abortion.

Wen Li et al. (2012) assessed and analyzed the total pollen proteins in the HL-type CMS line Yuetai A, its maintainer line Yuetai B and its F_1 hybrid (Hongliangyou 6) at the mononuclear stage, and found in the sterile line a lack or decreased expression of some proteins involved in substance and energy metabolism, cell cycle, transcription, and cellular transporters compared with the fertile line, namely the K^+/H^+ cotransport protein, zinc finger protein and WD-repeat protein. The decreased expression or lack of these proteins may be related to the abnormal pollen development caused by insufficient mitochondrial energy supply.

Cao Yicheng et al. (1987) compared proteins in the PTGMS line Nongken 58S and the corresponding fertile line Nongken 58 at major stages of young spike development under different daylength conditions through polyacrylamide gel IEF-SDS two-dimensional electrophoresis, and found that the proteins differed most at the stage when the pistil and stamen were formed. They then proceeded to postulate that these differences in proteins were probably related to the fertility conversion of PTGMS lines. Deng Jixin et al. (1990) measured the protein synthesis dynamics of the PTGMS line 105S at various stages of pollen development under different daylength conditions, and found that protein synthesis activity was very low for 105S with long daylength at various pollen development stages, without obvious peaks.

Wang Tai et al. (1990) analyzed changes in leaf protein of Nongken 58S and the conventional late *japonica* variety Nongken 58 under different daylength conditions through two-dimensional gel electrophoresis, and found that changes at points with a molecular weight of 23 – 35 kD may be associated with the fertility conversion. Cao Mengliang et al. (1992) compared the young spike protein of Nongken 58S and the control variety Nongken 58 at the secondary branch primordium differentiation stage, pistil and stamen formation stage, pollen mother cell formation stage and meiotic stage. Out of the 17 specific protein components observed, 11 showed the difference of existence or non-existence between fertile and sterile lines, three differed in expression quantity, and the other three showed position translation on the two-dimensional electrophoresis map. Huang Qingliu et al. (1994) studied changes in anther protein of the PTGMS line 7001S, and found that the soluble protein content of fertile anthers was higher than that of sterile anthers; also in terms of the components, the protein in fertile anthers had two more bands, 43 kD and 40 kD, than that of sterile anthers according to SDS-PAGE analysis.

Liu Lijun et al. (1995) used Nongken 58S, N5088S, 8902S and Nongken 58 as materials under different daylength conditions to perform two-dimensional electrophoresis on leaf soluble proteins, and found that there were various numbers of PTGMS-related protein plaques and tangles in each group of materials from the secondary branch primordium differentiation metaphase to spikelet primordium differentiation anaphase; among them, 63 kD and PI6.1 – 6.4 products only appeared in the short daylength groups of PTGMS lines, not in the long daylength groups. Huang Qingliu et al. (1996) studied the PTGMS lines Erjiuai S and Annong S – 1 and found that both were sterile under a high temperature (30 ℃), and the soluble protein content in the sterile anthers increased significantly, which was contrary to what was observed in the *japonica* PTGMS line 7001S. The protein map also showed that some bands were missing in the sterile anthers under a high temperature, which meant that the protein components differed. Li Ping et al. (1997) observed that soluble protein content in the sterile spikelets and anthers of the PTGMS line Peiai 64S was significantly lower.

Shu Xiaoshun et al. (1999), with conventional Zike as the control variety, measured the soluble protein content in the leaves and young spike anthers of Zike, high temperature sensitive GMS material 1356S, and Hengnong S – 2 at the pollen mother cell formation stage and pollen mother cell meiosis stage under different fertility conditions. The results showed that the soluble protein content in the young spike anthers of sterile 1356S and Hengnong S – 2 plants was very significantly lower than in the fertile plants during these two stages critical to fertility, their figures were respectively 44.0%, 45.1%, 35.8% and 42.6% of the figures of the fertile plants. Such a significantly low supply of proteins in young spike anthers of the sterile plants is bound to affect a series of life activities of the pollen, and ultimately impair normal pollen development.

Chen Zhen (2010) studied the differences in fertility-related proteins in young panicles between long daylength and short daylength PTGMS lines, and obtained a large number of spectrum showing differences in protein expression under sterile and fertile conditions. His comparative analysis also revealed some co-variation of EIF3, glycometabolism, and energy metabolism-related proteins with the difference in fertility, which meant that these proteins may be related to pollen sterility.

The above studies show that there exists certain relationship among the content and components of proteins and fertility. However, more needs to be done to ascertain any connection between the various specific protein components and fertility.

Ⅳ. Changes in Enzyme Activity

The complex biochemical reactions in plants are completed with the catalysis of enzymes, so changes in enzyme activity in pollen can reflect pollen development to a certain extent. Studies have revealed that the activities of various enzymes in male sterile rice lines change during pollen development.

Hunan Normal University (1973) compared the activities of related enzymes in fertile and sterile lines of 68 – 899 and the C system (Nanguangzhan) during the pollen development process (Table 2 – 2), and found peroxidase to be more active in abortive and non-pollen sterile plants than in normal fertile plants. In sterile plants, enzyme activity increases from the tetrad period to the binuclear stage of pollen development, then gradually decreases and disappears completely by maturity. In comparison, in fertile plants, enzyme activity first grows, then drops, but never show full inactivity. Meanwhile, polyphenol oxidase, acid phosphatase, alkaline phosphatase, ATPase, succinic dehydrogenase and cytochrome oxidase showed a different pattern with their activities higher in normal plants than in abortive and non-pollen sterile plants. Generally, in normal plants, enzyme activity gradually increases with pollen development, while in sterile plants it decreases and finally disappears. Meanwhile, the activity of catalase was also measured, and the results showed that the activity was higher in normal plants than in sterile plants, specifically, by 48.91%–57.66% at the mature phase. Research work done by Jiangxi Communist Labor University (1977), Hunan Agricultural College (1977), and Dai Yaoren et al. (1978) on rice three-line sets all obtained similar results.

Table 2 – 2 Comparison of enzyme activity in the pollen of 68 – 899 plants with different fertility conditions (Hunan Normal University, 1973)

Enzyme	Fertility condition	Pollen development process				
		Tetrad stage	Mononuclear stage	Binuclear stage	Binuclear anaphase	Mature stage
Peroxidase	Normal plant	+	+++	++	+	+
	Abortive type	+	+++	+++	++	(+) 0
	Non-pollen type	+	+++	+++	++	0
Cytochrome oxidase	Normal plant	+	++	++	++	+++
	Abortive type	+	++	+	+	(+) 0
	Non-pollen type	+	+	+	+	0
Polyphenol oxidase	Normal plant	+	++	++	+++	+++
	Abortive type	+	++	++	+	(+) 0
	Non-pollen type	+	+	+	+	0
Acid phosphatase	Normal plant	+	++	++	+++	+++
	Abortive type	+	++	++	+	(+) 0
	Non-pollen type	+	+	+	+	0
Alkaline phosphatase	Normal plant	+	++	++	+++	+++
	Abortive type	+	++	++	+	(+) 0
	Non-pollen type	+	++	+	+	0
ATPase	Normal plant	+	++	++	+++	+++
	Abortive type	+	++	++	+	(+) 0
	Non-pollen type	+	++	++	+	0

Continued

Enzyme	Fertility condition	Pollen development process				
		Tetrad stage	Mononuclear stage	Binuclear stage	Binuclear anaphase	Mature stage
Succinic dehydrogenase	Normal plant	+	++	++	++	+++
	Abortive type	+	++	+	+	(+) 0
	Non-pollen type	+	+	+	+	0

Note. Enzyme activity is compared according to the relative levels of staining or fading. "+" indicates staining or fading, "0" indicates no staining or fading, and "(+) 0" means no or trace staining or fading.

Peroxidase is an important oxidoreductase. Its main physiological function is to eliminate toxic substances produced in the body, and it plays an important role in the electron transport in the respiratory chain and multiple other pathways. During pollen development, the activity of peroxidase drops sharply from a relatively high level and can even disappear, which is detrimental to the respiratory function, substance transformation and self-detoxification. Cytochrome oxidase and polyphenol oxidase are two main terminal oxidases whose activity is low, reflecting the weakening of the metabolic function of pollen respiration. Moreover, reduced ATPase activity is even more detrimental to the energy metabolism of pollen cells, affecting the absorption, transport and transformation of substances and the synthesis of biomacromolecules. Catalase activity is an indicator of metabolic intensity. Increased catalase activity indicates more active physiological functions and higher metabolic intensity, while decreased catalase activity indicates less active physiological functions and lower metabolic intensity. Sterile plants generally have lower catalase activity than fertile plants, so their metabolic intensity is lower.

Chen Xianfeng et al. studied changes in the activity of peroxidase (POD), catalase (CAT) and superoxide dismutase (SOD) in the anthers of the sterile and maintainer lines of rice 7017 and Erjiuai varieties and found the anther enzyme activity to be basically the same for the sterile line and the maintainer line at the early mononuclear stage and then at the late mononuclear stage, binuclear stage and trinuclear stage, the activity of the sterile anthers became significantly lower than that of the fertile anthers. In the sterile anthers, the Cu-ZnSOD isozyme band was missing, and the efficiency of O^{2+} generation was 4. 1 – 5. 5 times that of fertile anthers, with H_2O_2 and MDA accumulated. H_2O_2 accumulation and increased membrane lipid peroxidation in the sterile anthers may be associated with pollen abortion.

There are numerous studies on the changes in enzyme activity of PTGMS rice lines. Chen Ping et al. (1987) found that the change in peroxidase activity of the sterile anthers of the PTGMS line Nongken 58S was similar to that of the CMS line V20A. Early in anther development, enzyme activity in sterile anthers is much higher than in fertile anthers. Throughout the anther development process, peroxidase activity goes from high to low in sterile anthers, but from low to high in fertile anthers. Mei Qiming et al. (1990) used Nongken 58S to measure the activities of various enzymes and isozymes in leaves, young panicles and anthers under long daylength (LD), short daylength (SD), red light interrupted dark period (R) and far-red light (FR) treatments, and found that the activity of RuBPCase, GOD, NR, PAL,

ADH, DAO and PAO in Nongken 58S decreased under LD and R treatments with isozymes missing, while the activity of COD, SOD, ADC and SAMDC enhanced with isozymes increased. These abnormal changes in enzymes corresponded to changes in pollen fertility-the pollen was sterile under LD and R treatments, and fertile under SD and FR treatments (Table 2 − 3), indicating that changes in enzyme activity and isozymes are associated with fertility conversion of PTGMS rice lines.

Table 2 − 3 Effects on enzyme activity and isozymes during Nongken 58S young panicles development under LD and R treatments (Mei Qiming et al., 1990)

Enzyme	Secondary branch and spikelet primordium differentiation stage	Pistil and stamen primordium formation stage	Pollen mother cell formation stage	Pollen mother cell meiosis stage	Late mononuclear pollen stage	Trinuclear pollen stage
RuBPC	−	−	−	−	−	−
GOD	±	−	−	−	−	−
NR	−	−	−	−	−	−
PAL	±	±	±	−	−	−
ADH	±	−	−	−	−	−
EST	±	±	±	±	±	−
COD	±	+	+	+	+	−
SOD	±	+	+	+	+	+
POD	±	+	±	±	±	+
ADC	+	+	±	+	+	+
SAMDC	±	±	+	+	+	+
DAO	±	−	±	−	−	−
PAO	±	±	±	−	−	−

Notes. (1) RuBPC: 1,5 − ribulose bisphosphate carboxylase; GOD: glycolate oxidase; NR: nitrate reductase; PAL: phenylalanine ammonialyase; ADH: alcohol dehydrogenase; EST: esterase; COD: cytochrome oxidase; SOD: superoxide dismutase; POD: peroxidase; ADC: arginine decarboxylase; SAMDC: S-adenosylmethionine decarboxylase; DAO: diamine oxidase; PAO: polyamine oxidase. (2) The increase and decrease of enzyme activity and isozymes are measured with the figures of Nongken 58S and Nongken 58 under SD or FR treatment at the corresponding stages as the benchmark figures. "+", "−" and "±" respectively represent increase, decrease and insignificant difference in enzyme activity and isozymes.

Zhou Hantao et al. (2000) used the *indica* CMS line Maxie A and its maintainer line Maxie B to examine the peroxidase isozyme, cytochrome oxidase isozyme electrophoresis, and enzyme activity of anthers at the pollen mother cell formation stage, meiosis tetrad stage, mononuclear stage, binuclear stage and trinuclear stage. They found that for the sterile line, the peroxidase isozyme had a wide variety of banding patterns on the zymogram of electrophoresis and a high activity, while the cytochrome oxidase isozyme had only limited banding patterns and a low activity. Specifically, these phenomena first appeared in the anthers at the mononuclear stage, and became more obvious with anther development, which is consistent with cytological observations on the abortion of Maxie-type sterile lines.

Chang Xun et al. (2006) used the leaves and young panicles at different development stages of the HL-type CMS line Yuetai A and its maintainer line Yuetai B to analyze the changes in tissue transglutaminase (tTG) activity and found that tTG activity increased with pollen development from the tetrad stage to the binuclear stage in Yuetai A and peaked at the binuclear stage; while tTG activity did not change significantly with pollen development in the maintainer line Yutai B. It was thus inferred that tTG is related to programmed cell death during pollen abortion.

Chen Xianfeng et al. (1992) found that the gross activity of cytochrome oxidase, ATPase, peroxidase, catalase, and superoxide dismutase in sterile anthers from the mononuclear to the trinuclear stage in Nongken 58S and W6154S was generally lower than in fertile anthers; and that the sterile anthers lacked 1 – 5 bands of cytochrome oxidase, one band of superoxide dismutase and 1 – 2 bands of Cu-ZnSOD isozyme later in the development process and showed higher superoxide anion production efficiency and H_2O_2 and MDA accumulation. It indicates that the membrane lipid peroxidation intensifies in the sterile anthers as the pollen abortion process proceeds. Liang Chengye et al. (1995) found that when the anthers of Nongken 58S developed from the mononuclear to the trinuclear stage, the content of ascorbic acid (ASA) and reduced glutathione (GSH) in fertile anthers were high, while the content of the two in the sterile anthers were only 35%–58% and 22%–32% the figures of the fertile anthers, with lipid hydroxide accumulation. With anther development, the activities of ascorbate peroxidase (ASA-POD), glutathione reductase and glucose – 6 – phosphate dehydrogenase in fertile anthers gradually increase, peaking at the trinuclear stage; while with pollen abortion, the activities of these enzymes in sterile anthers decrease gradually from the early mononuclear to the trinuclear stage, down to 26%, 22% and 19% the figures of fertile anthers respectively by the trinuclear stage. The activities of malic enzyme and malate dehydrogenase in sterile anthers are also lower than those in fertile anthers. Therefore, low cell reduction potential was deemed as one of the characteristics of sterile anthers, and such a low reduction potential may cause reactive oxygen metabolic disorder and anther abortion. Lin Zhifang et al. (1993), Zhang Mingyong et al. (1997) and Li Meiru et al. (1999) studied CMS lines V20A and Zhenshan 97A and PTGMS lines Nongken 58S, W6154S, GD1S and N19S and obtained similar results.

Li Ping et al. (1997) studied the *indica* PTGMS line Peiai 64S and found that its sterility was closely related to the significant reduction in NAD+-MDH activity in the spikelets and anthers at the time when the pollen was fully mature, while from pollen mother cell formation to the meiosis stage, it is closely related to changes in spikelet AP activity and isozyme composition. Based on this, it is believed that the fertility expression of Peiai 64S may be related to fat metabolism early in the pollen development process and respiratory metabolism late in the pollen development process.

Du Shiyun et al. (2012) measured the activities of superoxide dismutase (SOD), peroxidase (POD), catalase (CAT) and other antioxidant enzymes and the content of malondialdehyde (MDA) in the anthers and flag leaves of sterile rice and normal fertile *japonica* rice at the later stage of panicle development under different daylength and temperature conditions. The results showed that rice anthers were more sensitive to daylength and temperature stress than leaves, and there were significant differences in reactive oxygen metabolism between sterile and fertile anthers. Different types of sterile rice abortive physiology were different. The environment changes of the daylength and temperature had more obvious stressful effects on daylength and temperature sensitive nuclear sterile rice 2310S. The above three antioxidant enzymes could not act synergistically in the late stage of sterile pollen development, with high SOD activity, low POD activity, and high degree and earlier time of membrane lipid peroxidation. The POD activity was also consistently lower in the other two types of sterile lines, indicating that it may be more relevant to the formation of sterile pollen in rice.

In conclusion, for both CMS lines and PTGMS lines, the relationship between enzyme activity change and fertility expression is quite complex. The results can vary when researchers use different materials and methods. Gene controls zymoprotein synthesis, enzyme regulates metabolic reactions, and the concentrated expression of multiple metabolic reactions is the properties and physiological functions. Therefore, change in fertility is also one of the results of a series of changes in enzyme activity.

V. Changes in Plant Hormones

Plant hormones are trace amounts of physiologically active substances in plants and the products of normal plant metabolism. They regulate and control the growth and development of plants.

Huang Houzhe et al. (1984) used rice three-line sets and semi-sterile *indica-japonica* hybrid plants plus their parents to study the relationship between indoleacetic acid (IAA) content and male sterility, and found that combined-state IAA (C-IAA) and fertility degree decreased in parallel and that the activity of IAA oxidase and peroxidase in sterile anthers increased by several or even dozens of times as the sterility increased. Based on this, it is believed that male sterility is caused by the depletion of the IAA pool of sterile anthers due to damages related to oxidases, which inevitably leads to abnormal anther metabolism and microspore development, thus resulting in pollen abortion.

Xu Mengliang et al. (1990) studied changes in IAA content under long daylength and short daylength conditions during the young spike development of the PTGMS line Nongken 58S and the control line Nongken 58 through enzyme-linked immunosorbent assay (ELISA). The results show that IAA content was closely related to fertility expression. Under long daylength conditions, free IAA (F-IAA) in the leaves of Nongken 58S accumulated in great quantities during pollen mother cell formation, meiosis and pollen content enrichment, while F-IAA in young panicles and anthers was severely depleted. This did not happen to Nongken 58S under short daylength treatment or the control variety Nongken 58. C-IAA measurements show none of the above accumulation or depletion under either long or short daylength conditions in Nongken 58S, and changes in the C-IAA content of Nongken 58S under long daylength conditions was unrelated to F-IAA accumulation or depletion. Thus, it was speculated that the F-IAA content in leaves was regulated by daylength and its transport is impaired by long daylength, resulting in F-IAA accumulation in leaves and depletion in young panicles.

Yang Daichang et al. (1990) analyzed changes in the content of four endogenous hormones in the leaves of Nongken 58S under different daylength treatments. Under long daylength treatment, IAA was severely depleted, the content of gibberellins (GAs) was obviously higher than under short daylength treatment, abscisic acid (ABA) content rose sharply at the meiosis stage, and the content of the four hormones was extremely low under long daylength treatment when the pollen was fully mature. The time sequence of changes of the four endogenous hormones is as follows, IAA before zeatin (ZT), ZT before GAs, and GAs before ABA. Based on this, it is believed that IAA depletion is leading factor for the changes in the content of the four hormones, while changes in the others are metabolic adjustments caused by IAA depletion. Thus, the low hormone levels after pollen abortion is not the cause but a result.

Zhang Nenggang et al. (1992) studied the relationship between three endogenous hormones and the fertility conversion of Nongken 58S and Shuang 8 – 14S, and found that during the sensitive period for fertility conversion (from secondary branch primordium differentiation to pollen mother cell formation), IAA content in the penultimate leaf and young spike under long daylength treatment was lower than under short daylength treatment, while IAA oxidase activity was higher under long daylength treatment than under short daylength treatment; the GA_{1+4}/ABA value in the penultimate leaf was lower than under short daylength treatment, negatively correlated with the activity of IAA oxidase in the leaf. Thus, it is believed that sterility induced by long daylength is related to endogenous IAA depletion, while IAA depletion may be caused by the activation of IAA oxidase and the decrease of GA_{1+4}/ABA in the functional leaf.

Tong Zhe et al. (1992) applied a variety of hormones to the leaves or roots of rice plants during the sensitive period for fertility conversion, and found that a certain dose of gibberellin GA3 and GA4 can restore fertility partially in the PTGMS line Nongken 58S under long daylength treatment, while auxin, cytokinin and abscisic acid did not restore the fertility. Under long daylength conditions, the content of active gibberellin decreased sharply in the sterile leaves and gibberellin biosynthesis inhibitor also reduced the seed setting rate of Nongken 58S under short daylength treatment. So it is believed that the photoperiod is the second messenger to regulate the development of male organs in young panicles via its influence on the increase and decrease of gibberellin in leaves.

Huang Shaobai et al. (1994) compared the contents of endogenous GA_{1+4} and IAA in young panicles and penultimate leaves of WA-type and BT-type CMS lines and their maintainer lines (Zhenshan 97A, Zhenshan 97B, Hua 76 – 49A, and Hua 76 – 49B), and found that the sterile lines had lower contents of these hormones than the maintainer lines. It is believed that GA_{1+4} and IAA depletion is a physiological cause of CMS in rice.

Tang Risheng et al. (1996) measured the content of endogenous ABA, IAA and GAs in young panicles and other organs of TO3 (a gametocide)-treated rice with ELISA to analyze the relationship between changes in the contents of the three hormones and TO3-induced rice male sterility. The study showed that TO3 can significantly increase the content of endogenous ABA in young panicles, anthers and other organs of rice; and IAA and GAs were obviously depleted; the ratio of IAA+GAs to ABA was obviously lowered. This is believed to be one of the leading causes of the impaired fertility expression and the final abortion of rice male organs.

Luo Bingshan et al. (1990, 1993) used the ethylene biosynthesis inhibitor $CoCl_2$ to treat the PTGMS line Shuang 8 -2S and found out that the plants turned fertile under long daylength sterile conditions after the treatment and its fertility expression was greatly promoted at the critical daylength for fertility conversion. The ethylene release of the young panicles of Nongken 58S and its derived sterile lines under long daylength conditions was 2.5 -5.0 times higher than that of the control variety Nongken 58, while it was close to the low ethylene release level of the control Nongken 58S at the critical daylength for fertility conversion. It indicates that PTGMS may involve an ethylene metabolism system regulated by daylength, and its fertility conversion was regulated by ethylene metabolism. Li Dehong et al. (1996) found through further study that the young panicles of Nongken 58S, when within the range of suitable temperature for fertility conversion, showed a significantly higher ethylene release rate under long daylength treatment than under short daylength treatment; yet the long daylength release rate was greatly reduced when the temperature was low and it remained high even under short daylength conditions when the temperature was high. The ethylene release rate of young panicles was negatively correlated with pollen fertility. When treating the plants with the ethylene metabolism inhibitor aminoethoxyvinylglycine (AVG) under sterile conditions, obvious pollen fertility expression can be induced, but when using 1 -aminocyclopropane -1 -carboxylic acid (ACC) to promote ethylene production, the fertility level of Nongken 58S under short daylength conditions can be sharply reduced. So it is believed that the ethylene release rate of the young panicles of Nongken 58S was jointly regulated by daylength and temperature, and highly consistent with how daylength and temperature impacts fertility conversion; and that ethylene is involved in fertility conversion regulation and may play a key role in pollen abortion.

Tian Chang'en et al. (1999) found that treating a maintainer line (Zhenshan 97B) with the ethylene biosynthesis precursor ACC can reduce pollen fertility; decrease the content of proteins, DNA and RNA in young panicles; and lower the activity of protease, RNase and DNase; it can also increase the O^{2-} generation rate and MDA content, lower the activity of CAT and SOD, and increase the activity of POD. When the sterile line (Zhenshan 97A) is treated with the ethylene synthesis inhibitor AVG, it can partially improve pollen fertility; increase the content of proteins, DNA and RNA in young panicles; and lower the activity of protease, RNase and DNase; and it can also lower the O^{2-} generation rate and MDA content, and increase the activity of CAT, SOD and POD. Ethylene may affect the expression of pollen fertility by regulating macromolecular synthesis and reactive oxygen metabolism.

Zhang Zhanfang et al. (2014) studied the differences and changes in the content of endogenous jasmonic acid (JA) and methyl jasmonate (MeJA) during young spike development of Xieqingzao A and Xieqingzao B, and found that a low content of endogenous JAs (JA+MeJA) in the pollen mother cells of Xieqingzao A during the meiosis stage may be the cause of CMS.

In conclusion, the relationship between plant hormones and male sterility in rice is complex. Though regular patterns have been found, there are still inconsistencies. Meanwhile, it should be noted that there are many types of plant hormones and no one alone can control fertility. Growth and development are almost always regulated by a series of different endogenous hormones in coordination.

Ⅵ. Changes in Polyamines

Polyamines, ubiquitous in higher plants, have certain regulatory effects on many development processes, such as stem and bud growth, germination of dormant buds, leaf senescence, flower induction, pollen germination and embryogenesis. Polyamines can exist in free and bound forms in leaves. There are three kinds, putrescine (Put), spermidine (Spd) and spermine (Spm). It has been reported that maize male sterility is related to a significant reduction in polyamine content. In the past years, Chinese scholars have done some studies on the relationship between polyamine changes and the pollen development of PTGMS rice, and found that polyamines in young panicles are closely related to pollen fertility conversion.

Feng Jianya et al. (1991, 1993) measured changes in polyamines during young spike development of Nongken 58S and Nongken 58, and found the polyamine content per spike of Nongken 58S to increase slowly with spike development under long daylength conditions, while it doubles with spike development under short daylength conditions, with spermidine increasing particularly sharply from the secondary branch differentiation phase to the pistil and stamen formation phase. Polyamines in rice panicles are closely related to anther development. Polyamine changes in Nongken 58S under short daylength conditions result in normal pollen development, while reduced polyamine content under long daylength conditions results in pollen abortion. Based on further studies on the characteristics of polyamine changes in the panicles of the PTGMS line C407S, it was found that the change in polyamine content per gram of fresh weight of C407S showed an unimodal curve as young panicles develop. The total content of polyamines was close between long daylength and short daylength conditions, yet more specifically, the content of spermidine and spermine was higher with short daylength than with long daylength. After the differentiation of young panicles, the induction of pollen abortion was not determined by the total amount of polyamines in young panicles under different daylength conditions, but may be related to the content of spermidine and spermine, especially the peak content of spermidine. Under long daylength conditions, the contents of spermidine and spermine in young panicles reduced and the pollen developed abnormally and became completely sterile.

Mo Leixing et al. (1992) measured endogenous polyamines in the leaves and panicles of Nongken 58S treated with red light and far-red light and applied polyamine biosynthesis inhibitor to study the role of polyamines in fertility conversion. The results showed that there was no obvious correlation between the polyamine changes in Nongken 58S leaves and fertility conversion; that the polyamine content of young panicles was indirectly regulated by phytochrome and closely related to fertility conversion. During different development stages, phytochrome has different regulating effects on polyamines, and the types of polyamines playing the major role in fertility may differ too. The regulating effect of polyamines on the fertility conversion of Nongken 58S is related to the polyamine content and the share of different types of polyamines. Specifically, the content of spermine and spermidine, and the spm/spd ratio are important factors. The polyamine biosynthesis inhibitor methylglyoxal-double amidine hydrazone (MGBG) can partially reverse red light-induced sterility, and promote Nongken 58S to produce a few fertile pollen and self-seeding under long daylength and complete sterile conditions. Li Rongwei et al. (1997) also found that polyamine content in young panicles of PTGMS rice under sterile conditions was significantly lower than that of fertile conditions.

Additionally, Liang Chengye et al. (1993) and Tian Chang'en et al. (1998) found that the polyamine content in the anthers and young panicles of CMS lines was also obviously lower than in those of corresponding maintainer lines, and proved that polyamine depletion was one of the causes of male sterility by adding polyamine biosynthesis inhibitor or polyamine. Applying additional polyamine to a sterile line can partially restore pollen fertility, applying polyamine synthesis inhibitor to a maintainer line can reduce pollen fertility, and polyamine can partially offset the reducing effect of polyamine synthesis inhibitor on pollen fertility. Slocum et al. (1984) and Smith (1985) comprehensively reviewed the action mechanism of polyamine in plants and believed that polyamine can promote macromolecular synthesis in plant tissues and inhibit macromolecular degradation. Li Xinli et al. (1997) believed that the regulating effect of polyamine on rice pistil development may be realized by promoting nucleic acid synthesis and protein translation. Tian Chang'en et al. (1999) studied rice CMS lines Zhenshan 97A and Zhenshan 97B, and found that applying additional polyamine can slightly increase the content of DNA, RNA and protein in the young panicles of sterile lines, and decrease the activities of DNA enzyme, RNA enzyme and protease, which indicates that polyamine can increase the content of protein and nucleic acid by reducing the activities of the said enzymes. Applying additional inhibitor D-Arg+MGBG can slightly lower the content of DNA, RNA and protein in the maintainer line, and also reduce the activity of DNase, RNase and protease (it may be related to inhibited polyamine synthesis by the inhibitor and the reduced synthesis of proteins including the above enzymes). Polyamine replenishment can eliminate the effect of the inhibitor on the protein and nucleic acid content, and further reduce the activity of the above enzymes, which is consistent with the results of applying polyamine to sterile lines. It indicates that polyamine may react by promoting the synthesis of proteins and nucleic acids. Polyamine depletion in sterile lines may lower protein and RNA synthesis in young panicles, or accelerate decomposition, thus resulting in insufficient protein and nucleic acid content, affecting the morphogenesis of the anther and pollen, and finally causing male sterility.

Ⅶ. Ca^{2+} Messenger and Rice Male Sterility

Ca^{2+}, as the second messenger, plays extensive roles in plant cells. Intracellular reactions caused by various external and internal signal factors (such as touch, light, chilling stress, salt-alkali stress, thermosensitive reaction, and hormones) have been proved to be related to changes in Ca^{2+} concentration. Recent studies have shown that Ca^{2+} signal transduction is closely related to rice male sterility.

Chen Zhangliang et al. (1991) pointed out that a Ca^{2+} flow may exist in some plant cells and organs after exposure to light, and compounds that can change intracellular Ca^{2+} concentration have clear effects on the photo-stimulation response of plants. Yu Qing et al. (1992) measured changes in the total Ca^{2+} content in leaves, Ca^{2+} content in leaf cells and CaM content of Nongken 58S from phase Ⅲ to phase Ⅶ of young panicle development, and found that the pistil and stamen formation phase (phase Ⅳ) and the pollen mother cell formation phase (phase Ⅴ) were the two photosensitive phases of Nongken 58S, a long daylength caused the Ca^{2+} content in leaves to change, and especially increased the intracellular Ca^{2+} content, indicating that Ca^{2+} was closely related to photo signals. Under short daylength treatment, the CaM content in leaves began to rise from phase V, and maintained at phases VI and Ⅶ a level higher than in leaves under long daylength treatment at the same phases. Such a higher CaM content may be necessary for normal development of Nongken 58S at these phases, so obviously insufficient CaM content in leaves under long daylength treatment may be inherently related to pollen abortion. Li Hesheng et al. (1993) took Nongken 58S as material to feed $^{45}Ca^{2+}$ to plants under long daylength and short daylength treatments through leaf coating and root irrigation, and found that $^{45}Ca^{2+}$ can be transferred from leaves or roots into the functional organs, with the transferred amount increasing with marker strength; after absorbing $^{45}Ca^{2+}$, the fertility rate and selfing seed setting rate obviously increased under long daylength treatment, while the opposite occurred under short daylength conditions, indicating that absorbing Ca^{2+} from the outside may affect pollen fertility, thus proving Ca^{2+} related to rice male sterility. Wu Wenhua et al. (1993) found that under red light and long daylength treatment, the soluble Ca^{2+} content in leaves increased and the Ca^{2+}-ATPase activity in chloroplast decreased from phase Ⅳ to phase V of young panicle development of Nongken 58S, showing a trend completely opposite to that under short daylength and far-red light treatment. It indicates that the soluble Ca^{2+} content in leaves and changes in chloroplast Ca^{2+}-ATPase activity are related to fertility conversion in Nongken 58S. Tian et al. (1993) carried out calcium localization study on fertile and sterile anthers of Hubei Photoperiod-sersitive Genic Male Rice (HPGMR) using the potassium pyroantimonate precipitation method and found that the abnormal distribution of accumulated calcium accumulation was related to retard pollen development and abortion.

Xia Kuaifei et al. (2005) studied tapetum cell development and changes in the distribution of Ca^{2+} in cells of the WA-type sterile line Zhenshan 97A and its maintainer line Zhenshan 97B and found that tapetum cells of the maintainer line began to rapidly disintegrate at the late mononuclear pollen stage, while tapetum cells of the sterile line began to undergo nuclear membrane and cell membrane disintegration at the pollen mother cell stage and the process continued into the binuclear pollen phase. In Zhenshan 97A tapetum cells, small Ca^{2+} precipitation granules started to be seen in the cytoplasm in the pollen mother cell stage; during meiosis, a lot of large Ca^{2+} precipitation granules existed on the inward tangential wall surface of the tapetum cells; and during the mononuclear pollen stage, a layer of Ca^{2+} precipitation gathered around the tapetum cells. While there was no Ca^{2+} precipitation in the tapetum cells of the maintainer line at the pollen mother cell and meiosis stages, Ca^{2+} precipitation in tapetum cells at the mononuclear pollen stage is mainly distributed in the disintegrated cytoplasm. It is speculated that abnormal development of the tapetum cells and abnormal Ca^{2+} distribution may be related to pollen abortion.

Hu Chaofeng et al. (2006), in order to accurately understand the distribution and its changes of Ca^{2+} in anthers, especially anther mitochondria of the HL-type Yuetai sterile and maintainer lines, established a system for efficiently observing the dynamic changes of Ca^{2+} in living cells, and monitored the Ca^{2+} distribution and dynamic changes in the microspores at different development stages based on the FRET principle. Preliminary results showed that the Ca^{2+} concentration at the mononuclear stage was low in the sterile line, but it massively accumulated at the binuclear stage, resulting in a much higher concentration. While on the contrary, the Ca^{2+} concentration in the maintainer line at the mononuclear stage was higher than that in the sterile line, and decreased at the binuclear stage. Zhang Zaijun et al. (2007) took the pollen of the HL-type male sterile line Yuetai A and Yuetai B at different development stages to analyze the dynamic changes in Ca^{2+}-ATPase activity in the plasma membrane, mitochondrial membrane and tonoplast of the pollen. It showed that the Ca^{2+}-ATPase activity changes in the plasma membrane, mitochondrial membrane and tonoplast of the pollen of Yuetai A and Yuetai B differed significantly. From the tetrad stage to the early mononuclear stage, the activity of Ca^{2+}-ATPase in the plasma membrane of the pollen of the sterile line Yuetai A is slightly higher than that of Yuetai B; but it is just the opposite at the late mononuclear and binuclear stages, i. e. the activity of Ca^{2+}-ATPase of the plasma membrane in the pollen of Yuetai A is significantly lower than that of Yuetai B. Only at the late mononuclear stage, the activity of tonoplast Ca^{2+}-ATPase of Yuetai A pollen is significantly higher than that of Yuetai B pollen and the activity of tonoplast Ca^{2+}-ATPase of Yuetai A and Yuetai B at the other three stages shows only insignificant differences. The activity of mitochondrial membrane Ca^{2+}-ATPase in Yuetai A decreases gradually from the tetrad stage to the binuclear stage. At the tetrad stage, the activity of mitochondrial membrane Ca^{2+}-ATPase in Yuetai A pollen is significantly higher than that in Yuetai B pollen, and the two show no significant differences at the early mononuclear stage, while at the late mononuclear and binuclear stages, the activity of Ca^{2+}-ATPase in Yuetai A is significantly lower than that in Yuetai B. According to comprehensive analysis, during the dynamic process of pollen development of Yuetai A, in order to maintain calcium homeostasis in the pollen cytoplasm, first the Ca^{2+}-ATPase on the plasma membrane transfers calcium out of the pollen cytoplasm, mitochondrial membrane Ca^{2+}-ATPase transfers the excess calcium ions in the cytoplasm to the mitochondrion; then with pollen development, the activity of plasma membrane and mitochondrial membrane Ca^{2+}-ATPase continuously decreases at the late mononuclear stage, failing to further maintain calcium homeostasis, thus, large vacuoles are formed in the center of pollen cells and tonoplast Ca^{2+}-ATPase rises sharply to maintain pollen calcium homeostasis; yet still, the vacuoles fail to withstand the excessive calcium accumulation and rupture, and excessive calcium ions accumulate in the cytoplasm, destroying calcium homeostasis and finally resulting in pollen abortion.

Xia Kuaifei et al. (2009) studied changes in Ca^{2+} distribution for male sterility induced by high temperature and normal fertile anthers of the PTGMS line Peiai 64S with the potassium pyroantimonate precipitation method, and found that when Peiai 64S was sterile under a high temperature, there were more vacuoles, more Ca^{2+} precipitation and fewer mitochondria in its pollen mother cells compared with fertile anthers, and more Ca^{2+} precipitation existed in the middle lamella, epidermis and tapetum of sterile anthers. At the tetrad and mononuclear pollen stages, there were more Ca^{2+} precipitation on the secondary thickened wall of the xylem cells of sterile anthers, and the Ca^{2+} precipitation in the conjunctive tissues also increased greatly, the extines of the sterile pollen were thick and abnormally developed. The Ca^{2+} content in sterile anthers was higher than that in fertile anthers at all anther development stages, indicating that the increase of Ca^{2+} precipitation in anthers under high temperature conditions may be related to pollen abortion in Peiai 64S.

Ouyang Jie et al. (2011) studied anther development and Ca^{2+} distribution changes in cells of the non-pollen CMS line G37A and its maintainer line G37B with the potassium pyroantimonate precipitation method and found that calcium distribution in anthers of the two lines differed greatly. Little Ca^{2+} precipitation existed in the fertile anthers of G37B at the pollen mother cell and dyad stages; while at the mononuclear pollen stage, Ca^{2+} precipitation increased sharply and was mainly found in tapetum cells, the outer layer of the pollen extine and the surface of Ubisch bodies; then Ca^{2+} precipitation on the anther wall reduced while many Ca^{2+} precipitation still existed on the outer layer of the pollen extine. On the contrary, sterile anthers of G37A had large amounts of Ca^{2+} precipitation in the microsporous mother cells and the anther wall during the pollen mother cell and dyad stages, particularly abundant in the middle layer and tapetum. After the dyad stage, Ca^{2+} precipitation in sterile anthers reduced, especially near the inward tangential wall plasma membrane in the tapetum where precipitation almost disappeared. In comparison, in fertile anthers at the same stage, a mass of Ca^{2+} precipitation existed in the tapetum. So it is speculated that there is a certain connection between more calcium ions at the early stage of sterile anther development and pollen abortion.

Ⅷ. Role of Plant Phytochrome in Fertility Conversion of Photosensitive Sterile Rice

As the first messenger to receive and transmit light information in plants, phytochrome is closely related to various morphogenesis regulation mechanisms in plants. Its physiological functions involve seed germination, photoperiod-induced flowering, fertility conversion, auxin transport, metabolism of ethylene and carotenoid and so on. It has been preliminarily confirmed that the fertility of photosensitive sterile rice is regulated by phytochrome.

Li Hesheng et al. (1987, 1990) and Tong Zhe et al. (1990, 1992) found that a long dark period with intermittent red light may result in pollen abortion in Nongken 58S and reduce the natural seed setting rate. The typical abortion rate of the pollen could reach 86% and the natural seed setting rate was 3%-7%. If far-red light followed, the results could be reversed to a typical abortion rate of 11%-12% and a natural seed setting rate of 49%-50%; and this can again be reversed by subsequent red light, pushing the typical abortion rate back to more than 80% and the natural seed setting rate down to less than 10%. In conclusion, as a photoreceptor, phytochrome participates in the fertility conversion process of Nongken 58S.

Section 4 Genetic Mechanism of Rice Male Sterility

Ⅰ. Hypotheses of the Genetic Mechanism of Male Sterility

1. Three-type theory

The Three-type theory was proposed by American scientist Ernest Robot Sears (1947) based on previous research work. He sorted plant male sterility into genic male sterility, cytoplasmic male sterility and genic-cytoplasmic male sterility, thus the name "three type".

(1) Genic male sterility

Genetic male sterility (GMS) is controlled by a pair of sterility genes in the nucleus and is unrelated to the cytoplasm. For this type of male sterility, there are restorer lines but no maintainer line, so three-line sets cannot be completed. In GMS lines, there are homozygous sterile genes (rr) in the nucleus, while in the corresponding normal lines, there are homozygous fertile genes (RR) in the nucleus. When a GMS plant is crossed with a normal plant, fertility is restored in the F_1 population, and about one in every four F_2 plants is sterile. When a sterile F_2 plant is crossed with a fertile F_1 plant, about half of the hybrids are fertile, while the other half is sterile (Fig. 2 - 9).

Fig. 2 - 9 Inheritance of genic male sterility

Fig. 2 - 10 Inheritance of cytoplasmic male sterility

(2) Cytoplasmic sterility

Cytoplasmic sterility is completely controlled by cytoplasmic genes and unrelated to the nucleus. When a cytoplasmic sterile plant is crossed with a fertile plant, the F_1 plants will remain sterile and if back-crossing continues, the progenies will remain sterile. This means that a CMS line can easily find a maintainer line, but not a restorer line (Fig. 2 - 10).

The basis of Sears's proposal of cytoplasmic male sterility is Rhoades's study (1933). In that study, 10 pairs of chromosomes in a male sterile line of maize was replaced one by one with 10 pairs of marked chromosomes of a male fertile line, but fertility was not restored. Thus, it was believed that this type of maize male sterility is unrelated to any of the chromosomes, and is controlled by the cytoplasm.

(3) Genic-cytoplasmic male sterility

Genic-cytoplasmic male sterility (CMS) is jointly controlled by cytoplasmic and nuclear genetic materials. When a CMS plant is pollinated with a certain type of pollen, the F_1 population will continue to show male sterility; while when it is pollinated with some other type of pollen, the fertility of F_1 plant will be restored.

Suppose *S* is the gene for cytoplasmic male sterility, *N* is the gene for cytoplasmic fertility, *R* is the dominant nuclear gene for fertility, and *r* is the recessive nuclear gene for sterility, and then when a CMS plant is crossed with fertile plants of different types, there can be five different results in the F_1 population (Fig. 2 - 11).

Cytoplasmic genes can only be passed on to offspring through the egg cells of the female parent. Only those plants whose cytoplasmic and nuclear genes are both for sterility [S(rr)] will show male sterility. When the cytoplasmic gene is for sterility, and the nuclear gene is homozygously fertile [S(RR)] or

A.	S(rr)	×	N(rr)	→	S(rr)		
	Sterile		Fertile		Sterile		
B.	S (rr)	×	S (RR)	→	S (Rr)		
	Sterile		Fertile		Fertile		
C.	S (rr)	×	N (RR)	→	S (Rr)		
	Sterile		Fertile		Fertile		
D.	S (rr)	×	S (Rr)	→	S (Rr)	and	S (rr)
	Sterile		Fertile		Fertile		Sterile
E.	S (rr)	×	N(Rr)	→	S(Rr)	and	S (rr)
	Sterile		Fertile		Fertile		Sterile

Fig. 2 - 11 Inheritance of genic-cytoplasmic male sterility

heterozygously fertile [S(Rr)], the plant will be fertile. If the cytoplasmic gene is for fertility, the plant will always be fertile regardless of the nuclear gene.

For male sterility [S(rr)] × male fertility [N(rr)], F_1 hybrids will show male sterility [S(rr)], and the male parent that maintains male sterility [N(rr)] is called the maintainer line.

For male sterility [S(rr)] × male fertility [N(RR) or S(RR)], F_1 hybrids will show heterozygous fertility [S(Rr)], and both of the male parents are the restorer lines.

Obviously, CMS lines have both the maintainer lines and the restorer lines, forming a complete three-line set.

2. Two-type theory

This is Edwardson's (1956) modification of Sears' Three-type theory. He merged cytoplasmic and genic-cytoplasmic sterility in the Three-type framework into one category, thus forming a binary system for male sterility comprising genic male sterility and genic-cytoplasmic male sterility as cytoplasmic male sterility was later proved to be genic-cytoplasmic sterility while pure cytoplasmic male sterility does not exist in nature.

3. Correspondence theory of multiple genic-cytoplasmic genes

The genic-cytoplasmic male sterility discussed above is for cases where there is only one pair of genic-cytoplasmic fertility genes. That is to say, there must be only one fertility gene (RR, Rr, rr) in the nucleus and one fertility gene (S, N) in the cytoplasm. However, according to the study of Kihara, Maan et al., on genic-cytoplasmic correspondence of common wheat, male sterility in a plant was not so simple as to involve the correspondence of only one pair of genic-cytoplasmic fertility genes, but more complex as to involve multiple pairs of genic-cytoplasmic fertility genes. Thus, the theory of correspondence among multiple genic-cytoplasmic fertility genes was proposed (Fig. 2 - 12).

For example, when common wheat is used as the male parent to cross with five different varieties of wild *aegilops tauschiis* and five more primitive varieties of wheat, 10 different genic-cytoplasmic sterile lines were obtained with different cytoplasmic sources and the same nuclear source. It indicates that there are at least 10 different genes for genic sterility in the nucleus of the common wheat plants that served as the male parent and in each hybrid, one of these nuclear genes interacted with the cytoplasmic sterility gene to pro-

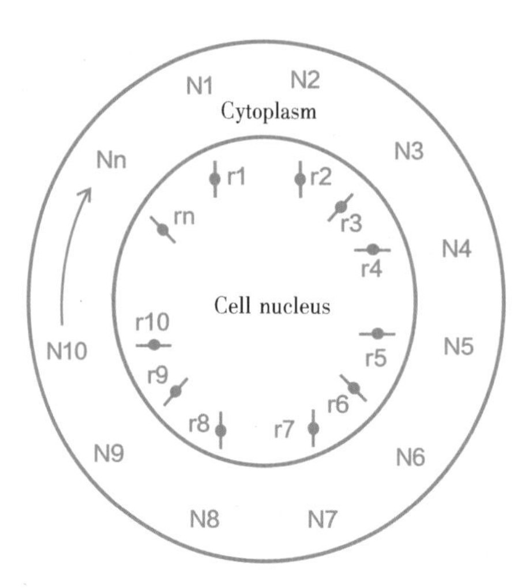

Fig. 2 – 12 Correspondence of genic-cytoplasmic fertility factors

duce male sterility. However, when reciprocal crossing was carried out instead, i. e. using common wheat as the female parent to cross and back cross with the five more primitive wheat varieties and the five varieties of *aegilops tauschiis*, all the progenies turned out fertile. It indicates that there are no cytoplasmic sterility genes corresponding to the above 10 nuclei of common wheat, but rather, there exist 10 different fertile genes to cover up and inhibit the effect of the nuclear sterile genes and make the common wheat fertile.

In general, genic-cytoplasmic fertility genes interact in pairs, without mutual interference between non-corresponding fertility genes. For example, there are interactions between the pair $N_1 - r_1$ or the pair $N_2 - r_2$, but not between $N_1 - r_2$ or $N_2 - r_1$. The theory reflects more complicated situations in the male sterility mechanism.

4. Pathway theory

After summarizing the practice of male sterility research in China, Wang Peitian et al. of the Institute of Genetics, Chinese Academy of Sciences proposed a schematic diagram of the evolution of cytoplasmic and nuclear genes controlling pollen formation, which is known as the "pathway" theory (Fig. 2 – 13).

The upper part of Fig. 2 – 13 is the evolutionary tree of rice varieties, and the lower part is the evolutionary process of fertility genes. The evolutionary process of pollen fertility genes is depicted with changes in the nature and quantity of the cytoplasmic gene *N* for fertility, cytoplasmic gene *S* for sterility, nuclear gene (+) for fertility and nuclear gene (−) for sterility.

When rice evolves from a lower stage to an advanced stage, there are more normal genes (*N*) and fewer sterile genes (*S*) for pollen fertility in the cytoplasm, and fewer normal genes (+) and more sterile genes (−) in the nucleus. Namely, some processes which used to be completed mainly by nuclear genes are gradually shifted to the corresponding cytoplasmic genes.

Suppose that both "*N*" and "+" represent a "pass" while "*S*" and "−" represent an interruption, then normal pollen can be formed as long as there is one pathway out of the three possible options controlling pollen formation (one in the cytoplasm and two in the nucleus); otherwise it will not be formed.

(South) → (North)

Lower stage of evolution (low latitude) (Low altitude) → (high latitude) (high altitude) Higher stage of evolution

Several processes of pollen formation	Primitive wild rice ↓ Other wild rice	Common wild rice ↓ New types of wild rice	Late-season *indica* ↓ Wild *indica*	Med-season *indica* ↓ Early-season *indica*	Late-season *japonica* ↓ Wild *japonica*	Med-season *japonica*	Early-season *japonica*	
	S1 ++	S1 ++	S1 or	S1 ++	S1 ++	S1 ++	S_1	++Pollen mother cell
	S2 ++	S2 ++	S2 ++	S2 ++	S2 ++	S2 ++	S_2	++Sporoblast
	S3 ++	S3 ++	S3 ++	S3 ++	S3 ++	S3 ++	S_3	++Dyad
	N4 ++	N4 --	N4 --	N4 --	N4 --	N4 --	N_4	Tetrad
or	S4 ++	N5 ++	N5 --	N5 --	N5 --	N5 --	N_5	--Early mononuclear stage
	S5 ++	or S5 ++	N6 ++	N6 --	N6 --	N6 --	N_6	--Middle mononuclear stage
	S6 ++	S6 ++	or S6 ++	N7 ++	N7 --	N7 --	N_7	-- Late mononuclear stage
	S7 ++	S7 ++	S7 ++	or S7 ++	N8 ++	N8 --	N_8	--Early binuclear stage
	S8 ++	S8 ++	S8 ++	S8 ++	or S8 ++	N9 ++	N_9	--Late binuclear stage
	S9 ++	S9 ++	S9 ++	S9 ++	S9 ++	or S9 ++	N_{10}	++Trinuclear stage
	S10 ++	S10 ++	S10 ++	S10 ++	S10 ++	S10 ++	or S_{10}	++

Fig. 2 - 13 Evolution of cytoplasmic and nuclear genes that control pollen formation

1. Normal factor in cytoplasm (*N*) = pass; sterility factor in cytoplasm (*S*) = interruption.

2. Normal factor in nucleus (+) = pass; sterility factor in nucleus (−) = interruption.

3. When one of the three lines is a "pass" (one in cytoplasm and two in nucleus) a complete pathway can be formed to produce normal pollen, while an interruption produces sterile pollen.

4. "or" indicates that the cytoplasmic factor that controls a certain process can be *N* or *S*.

During pollen formation, if an interruption occurs at an earlier stage, sterility usually shows earlier and is more serious; while sterility resulted from more interruptions along the process tends to be more difficult to convert back to fertility.

Thus, the pathway theory can be summarized as follows.

1) It is easier to obtain sterile lines but more difficult to get restorer lines through hybridization between varieties with distant genetic relationships (Example *a*); while it is difficult to obtain sterile lines by crossing more closely related varieties but easier to get restorer lines due to fewer interruptions (Example *b*).

Example *a*:	++++ ++++	×	---+ ---+	Multiple backcrossing -------→	---+ --+
	$N_5S_6S_7S_8$		$N_5N_6N_7N_8$		$N_5S_6S_7S_8$
	Common wild rice		Early *indica* or late *japonica*		Typical abortion (the 6th and 7th processes are interrupted)
Example *b*:	+++ +++	×	-++ -++	-------→	-++ -++
	$N_5S_7S_8$		$S_6S_7S_8$	Nuclear substitution	$S_6S_7S_8$
	Late *indica*		Medium *indica*		Normal and fertile

2) It is easier to obtain sterile lines through hybridization between a southern variety which is at a lower evolutionary stage as the female parent and a northern variety as the male parent (Example *c*); while on the contrary, it is difficult to obtain sterile lines through hybridization between a northern variety as the female parent and a southern variety as the male parent (Example *d*).

Example *c*:	++ ++	×	-- --	-------→	-- --
	S_6S_7		N_6N_7	Nuclear substitution	S_6S_7
	Late *indica*		Late *japonica*		Sterile
Example *d*:	-- --	×	++ ++	-------→	++ ++
	N_6N_7		S_6S_7	Nuclear substitution	N_6N_7
	Late *japonica*		Late *indica*		Fertile

3) When sterile lines are produced by nuclear substitution, varieties distributed in the south have more restorative lines of the maternal (cytoplasmic donor) type than other southern varieties that produce sterile lines; varieties distributed in the north have more maintainers of the paternal (nuclear donor) type than other northern varieties that produce sterile lines.

4) Sterile lines can be obtained from reciprocal crossing between collateral distant varieties.

5) The type of a sterile line is related to the stage at which the interruption occurs. For example, if a certain *indica* rice variety suffers interruption of the nuclear pathway at the first possible stage and fails to complete the first process of pollen formation, showing the features of non-pollen sterility (S_1/--) with severe anther degeneration, it will be easy to find a restorer line (S_1/++) in normal varieties, but difficult to find a maintainer line (N_1/--).

S_1/++ -------→ S_1/-- × S_1/++ -------→ S_1/++

Nuclear mutation　　　　Restore fertility

If interruption occurs during the sixth process along the cytoplasmic pathway, preventing the sixth process from being completed, the result will be typical abortion. For this type of sterile lines, it is easy to find maintainer lines in middle and early *indica* varieties but difficult to find restorer lines, which may be find in late *indica* varieties.

5. Relative theory

Proposed by Pei Xinshu et al. from Hunan Agricultural College, this theory believes that male sterility in plants is a quantitative character controlled by multiple genes, rather than a qualitative character. The reason is that F_1 hybrids show fertility segregation, while F_2 ones show a continual variation of fertility, making it impossible to strictly divide between sterility and fertility. Sterility is unstable and subject to the influence of environmental factors, especially temperature. Generally, the theory divides the causes of male sterility into the following two categories. 1) In distant hybridization, genetic materials from the two parents cannot be coordinated due to the large genetic distance between the parents, thus resulting in sterility; 2) Male sterility can be caused by changes in the chromosome in the nucleus or the cytoplasm. So, sterility or fertility depends on the genetic difference between the parents. Male sterility is caused by the combination of genetic materials with distant parents, while fertility restoration is the result of the closeness of the genetic materials of the parents. In order to obtain male sterility, it is necessary to select parents that are genetically distant; while when aiming to restore fertility, parents genetically close should be used for crossing.

6. Ca^{2+}-CaM system regulation hypothesis

Yang Daichang et al. (1987) found with the X-ray energy spectrum method that the Ca^{2+} content differed greatly between long daylength and short daylength conditions during the fertility conversion of photosensitive genic sterile rice, and thus concluded that there was a "phytochrome-Ca^{2+} regulatory system" in the fertility conversion of photosensitive genic sterile rice. Studies have shown that Ca^{2+} acts as a second messenger molecule in cells. After Ca^{2+} combines with calmodulin (CaM), it can regulate gene expression and enzyme activity in multiple ways, and also regulate the membrane potential difference between the inside and outside of cells. Thus, a regulatory model for fertility conversion in photosensitive genic sterile rice is outlined. During the fertility conversion of photosensitive genic sterile rice, a series of cascade reactions are generated centering on the "phytochrome-Ca^{2+}" regulatory system to enable or disable relevant genes through modification of the nucleoprotein or regulatory protein, thus expressing or inhibiting a group of genes that control fertility according to the spatial-temporal sequence of development.

In this model, red-light state phytochrome (P_r) is converted to far-red light state phytochrome (P_{fr}) under long daylength conditions, so the P_r/P_{fr} balance is tilted toward P_{fr}, thus changing the membrane permeability, activating the Ca^{2+} pathway and increasing Ca^{2+} concentration in the cytoplasm; then protein kinase is activated by CaM, phosphorylating the regulatory protein of the repressor gene, thus activating the repressor gene to produce repressor protein, and then disable the regulator gene. As activation of fertility genes requires action of regulatory protein, after disabling the regulator genes, fertility genes fail to express themselves and sterility occurs. Under short daylength conditions, P_r cannot be converted to P_{fr}, and the balance is tilted toward P_r, decreasing membrane permeability; Ca^{2+} goes outward along the pathway, which decreases Ca^{2+} concentration in cells, inactivating protein kinase, activating phosphorylase to dephosphorylate the regulatory protein of the repressor gene, thus disabling the repressor gene. As the regulator gene is not repressed by the repressor protein, it is expressed and produces a regulatory protein to activate the fertility genes, so a group of fertility genes are expressed in the spatial-temporal sequence through cascade reactions, showing normal fertility. However, this group of fertility genes is affected by both physiological factors and environmental conditions.

7. Light-temperature promoter hypothesis

This hypothesis was proposed by Zhou Tingbo (1992, 1998) and its main points are as follows. 1) Pollen development is achieved through a series of physiological and biochemical processes that occur in sequence over time and are controlled and coordinated by nuclear sequence genes, cytoplasmic sequence genes and external light and temperature conditions. 2) Nuclear sequence genes are a group of genes performing some physiological functions for a certain process of pollen development. It includes light-temperature sensing genes, integrator genes, multiple producer genes and related promoter genes. 3) The correspondence between the nuclear sequence genes and cytoplasmic sequence genes for every process of pollen development is species-specific and is established over a long process of natural evolution. 4) Cutting off any of the connections-that between nuclear sequence genes, between nuclear and cytoplasmic sequence genes, or between nuclear and/or cytoplasmic sequence genes and the external light and temperature conditions-will cause male sterility.

The light-temperature promoter hypothesis believes that physiological sterility and genetic sterility are unified. As long as conditions for male sterility remain unchanged, the plants will remain sterile year after year and this trait is inheritable. Similarly, photo-thermo-sensitive sterility is obviously inheritable, but it is indeed physiological in nature. The hypothesis also believes that male sterility is a quantitative character in general, because it is controlled by a series of genes, but it does not exclude the possibility that in a specific hybrid combination, it may be the result of differences in a few nuclear sequence gene loci, which makes the fertility change a qualitative character.

Ⅱ. Molecular Mechanism of Rice Male Sterility

In recent years, with rapid development of molecular biology, numerous useful explorations have been made regarding the mechanism of rice male sterility at the molecular level and many meaningful results have been achieved, deepening our understanding of male sterility and contributing to the utilization of rice heterosis. Studies on the molecular mechanism of rice male sterility mainly involve two aspects, cytoplasmic male sterility and nuclear male sterility.

1. Cytoplasmic male sterility

The cytoplasm contains two relatively independent genetic systems, the mitochondria and the chloroplasts. Cytoplasmic male sterility is supposed to be closely related to both of them. Kadowaki et al. (1986), Liu Yansheng et al. (1988) and Zhao Shimin et al. (1994) studied cytoplasmic male sterility of rice and found that chloroplasts were not directly related to male sterility, while mitochondria may play a more important role.

A comparative study on the mitochondria of cytoplasmic male sterile lines and the corresponding maintainer lines revealed that the two differed obviously in mitochondrial genome, plasmid-like DNA in mitochondria and mitochondrial gene translation products, suggesting that cytoplasmic male sterility is related to mitochondria.

Molecular hybridization of mitochondrial DNA restriction enzyme digestion sections revealed that cytoplasmic male sterile lines and maintainer lines of rice differed in the position or copy number of mitochondrial *Cox Ⅰ*, *Cox Ⅱ*, *Cox Ⅲ*, *atp6*, *atp9*, *atpA*, *Cob* and other genes. Liu Yansheng et al. (1988) compared the electrophoresis banding patterns of the mitochondrial DNA restriction enzyme digestion sections of the sterile line Zhenshan 97A and its maintainer line, as well as the positions of cytochrome C oxidase subunit I (*Cox Ⅰ*) and subunit Ⅱ (*Cox Ⅱ*) in the sections, and found significant differences. Li Dadong et al. (1990) found that a BT-type sterile line had two copies of the *atpA* gene, while the maintainer line had only one. Kaleikau et al. (1992) found that a WA-type sterile line had only one *Cob* copy, while the maintainer line had two, including one pseudogene *Cob2*, which was produced by recombination or insertion between *Cob1* and a 192bp fragment. Yang Jinshui et al. (1992, 1995) found a duplicate *atp6* copy in BT-type lines, the sterile line had two *atp6* copies, while the maintainer line had only one. In the WA-type sterile line Digu A and its maintainer line, different numbers of *atp9* copies were found. The sterile line had only one *atp9* copy, while the maintainer line had two. Kadowaki et al. (1989) used a synthetic mitochondrial gene sequence as the probe to compare the mitochondrial DNA of a BT-type sterile line and the corresponding maintainer line through RFLP and molecular hybridization. The findings were that the sterile line had twice as many *atp6* and *Cob* copies in the mitochondrial DNA as the maintainer line, that is, in addition to one normal *atp6* gene copy, the sterile line also had one additional chimeric *atp6* gene copy. Further studies showed that the cytoplasm of the sterile line contained a chimeric *atp6* gene (*urf-rmc*) and a normal *atp6* gene, while the cytoplasm of the fertile line had only a normal *atp6* gene. After a restorer gene was introduced, the length of the *urf-rmc* gene transcript was changed, while the length of the normal *atp6* gene transcript remained unchanged. It implies that the chimeric gene is related to cytoplasmic male sterility (Kadowaki et al., 1990). Iwahashi et al. (1993) also obtained similar results. In conclusion, correct RNA processing and editing may play an important role in the expression and fertility restoration of cytoplasmic rice male sterility.

Xu Renlin et al. (1995) obtained a specific amplified fragment R_{2630}WA from the mitochondrial DNA of a WA-type male sterile line with the arbitrary single-primer polymerase chain reaction technology. The fragment was used as a probe to carry out Southern hybridization analysis, and mitochondrial DNA polymorphism was then detected across the male sterile cytoplasm and the normal fertile cytoplasm. The sterile line Zhenshan 97A and its F_1 hybrids had the same hybrid map, while the maintainer line Zhenshan 97B and the restorer line Minghui 63 had the same hybrid map. The full length of the fragment was determined to be 629bp and the sequence contained an inverted repeat sequence 5'-ACCATATGGT-3' with a length of 10bp, located in fragments 262 – 272. Additionally, its segments 379 – 439 can encode a short peptide containing 20 amino acid residues. So it is believed that R_{2-630}WA is closely related to male sterility in WA-type rice, and the inverted repeat sequence 5'-ACCATATGGT-3' may play an important role in the formation of cytoplasmic male sterility. Liu Jun et al. (1998) studied the mitochondrial genetic composition of a Maxie-type sterile line and its maintainer line through RFLP analysis and molecular hybridization experiments. Tu Jun (1999) analyzed the mitochondrial DNA of HL-type sterile line Congguang 41A and its maintainer line through RFLP analysis and compared their mitochondrial DNA restriction maps, finding that the mitochondrial genome differed greatly between the sterile line and the maintainer line.

Reports on the relationship between plasmid-like DNA and cytoplasmic male sterility of rice include the following. Yamaguchi et al. (1983) found B_1 and B_2 plasmid-like DNA in BT-type rice sterile lines, respectively 1.5kb and 1.2kb; while no B_1 or B_2 existed in the corresponding maintainer lines. Then, Kadowaki et al. (1986) confirmed that B_1 and B_2 existed in Taichung 65A, but not in the corresponding maintainer line. Nawa et al. (1987) also reported that changes in B_1 and B_2 in the mitochondria were related to rice cytoplasmic male sterility. Mignouna et al. (1987) found that in addition to the main mitochondrial DNA contained in the mtDNA of the WA-type sterile line Zhenshan 97A, there were also four covalently closed circular (ccc) plasmids, while in the maintainer line there were only three of them, leaving a 2.1kb plasmid-like DNA unique to the sterile line. Shikanai et al. (1988) also found that the mtDNA of CMS-A58 in a BT-type sterile line contained four ccc plasmid-like DNAs, while A58 with normal cytoplasm did not contain these DNAs. Mei Qiming et al. (1990) compared the mtDNA of the HL-type Qingsiai A and the WA-type Zhenshan 97A and their maintainer lines through agarose electrophoresis and observation through an electron microscope, and found that the sterile lines had micro molecular mtDNA, while the corresponding maintainer lines had none. Tu Jun et al. (1997) also found the HL-type maintainer line Congguang 41 to have three more mitochondrial DNAs, 6.3kb, 3.8kb, and 3.1kb respectively, than the corresponding sterile line. The above findings show that differences exist in the plasmid-like DNAs of the BT-type, WA-type and HL-type sterile lines and maintainer lines. However, Saleh et al. (1989) extracted and analyzed mtDNAs from the leaves of V41A and V41B and the results showed that both the sterile line and the maintainer line contained four types of low-weight mtDNA molecules, and that there was no difference in their plasmid-like DNAs. Thus, it was believed that there was no simple connection between plasmid-like DNA and cytoplasmic male sterility. In addition, Nawa et al. found that the iso-cytoplasmic restorer line of the BT-type Taichung 65A also had plasmid-like B_1 and B_2, clearly supporting Saleh's conclusion. Liu Zuochang et al. (1988) studied WA-type sterile lines and found two plasmid-like DNAs, 3.2 kb and 1.5 kb respectively, not only in the sterile line, but also in the maintainer line and the F_1 hybrid, thus proposing that these plasmid-like DNAs may not be related to male sterility.

Studies on the *in vitro* translation products of mitochondrial genes also showed differences between the sterile lines and the maintainer lines. Liu Zuochang et al. (1989) studied the mitochondrial protein and *in vitro* translation products of WA-type sterile lines and maintainer lines, and found an extra 20 kD polypeptide in the *in vitro* translation products of the mitochondrial genes of the sterile lines in comparison with the maintainer lines and restorer lines. Zhao Shimin (1994) also found a unique 70.8 kD polypeptide in a D-type sterile line, and believed it was related to sterility and a product of sterility genes. Liu Zuochang et al. found that the *in vitro* translation product of the mitochondrial genome of BT-type sterile lines lacked one 22 kD polypeptide in comparison with the maintainer line, but its restorer line and F_1 hybrids had a nucleus-encoded 22 kD polypeptide, which compensated for the lack of the 22 kD polypeptide in the cytoplasm of the sterile line and restored fertility. This reflects that the fertility-related mutations in the mitochondrial genome of the cytoplasm of BT-type sterile lines may be related to a certain physiological process in the formation of microspores and the lack of this polypeptide may have affected the formation and normal development of microspores. Based on this, the researchers called the mitochondrial gene that encoded the 22 kD polypeptide the fertility gene. Moving forward from here, Zhao Shimin et al. proposed two hypotheses regarding fertility restoration in sterile rice lines featuring cytoplasmic male sterility. One is that a defect causes sterility and compensation restores fertility. The sterility of BT-type rice fits this. The other is that addition causes sterility and inhibition results in fertility restoration. The WA type and D type fit here. Defect-caused male sterility may be the result of the lack of a specific protein (22 kD) in the cytoplasm, which ultimately interrupts pollen development. This is controlled by the mitochondrial genome, but the whole process of fertility expression is not determined by the mitochondria. One 22 kD polypeptide may be necessary for a certain step in the pollen development process, and sterility occurs when it is absent. In the case of addition-induced sterility, the sterile cytoplasm has one more polypeptide than the normal cytoplasm. This polypeptide may inhibit a certain step of pollen development, thus resulting in male sterility. It is slightly expressed in the mitochondrial genome translation in the F_1 hybrids, so it is speculated that the nuclear genome of the restorer line inhibits the synthesis of the polypeptides that hinder fertility expression.

Ren Juansheng et al. (2009) did sequencing and analysis on the differential fragments of the mitochondrial genomes of the Wanhui 88-type cytoplasmic male sterile line Neixiang 2A and its corresponding maintainer line Neixiang 2B, and speculated that male sterility was caused by a point of mutation in the core promoter region of the mitochondrial *atp6* gene of the male sterile line, preventing proper transcription of the *atp6* gene, causing insufficient energy supply for the mitochondria, and finally resulting in pollen abortion.

The mitochondrial genome of BT-type sterile lines has two copies of the *atp6* gene, and the 3' end of one of the two copies also has a predicted ORF (*orf79*). This *orf79* and *atp6* co-transcribe to form the *B-atp6/orf79* mRNA. Liu Yaoguang et al. (2005, 2006) found that when *orf79*, which expresses a toxic protein, was introduced into normal rice varieties, gametophyte-type male sterility can be produced. Two closely linked homologous restorer genes *Rf1a* and *Rf1b* were isolated from the restorer locus *Rf1* through localization, cloning and other methods. The two genes encoded pentatricopeptide repeat (PPR) protein and located in the mitochondria. *Rf1a* can mediate specific dissection of the *B-atp6/orf79* mRNA, while *Rf1b* can mediate the full degradation of this RNA when *Rf1a* is absent. So they can silence the sterility gene *orf 79* through different channels, thus restoring normal pollen development. When *Rf1a* and *Rf1b* coexist, *Rf1a* preferentially dissects the target RNA and expresses epistasis, while *Rf1b* fails to mediate the degradation of the RNA fragments produced by the dissection controlled by *Rf1a*. By comparing the structure of the whole genome of the mitochondrial DNA and the transcription profile of the WA-type sterile line and maintainer line, it was found that the sterile line had two specific mRNAs and one was degraded through post-transcriptional processing of the restorer gene *Rf4*. Thus, it may be a transcript of the sterile gene, but the mitochondrial DNA section (including the promoter region) encoding this mRNA exist in all rice varieties, including wild rice lines (20 species) and has the same sequence, with transcription occurring only in the sterile cytoplasmic materials. Based on this, it was speculated that there was a specific transcription factor in the mitochondrial genome of the sterile line that controlled the expression of the sterile gene, and since the mitochondrial genome of normal cytoplasm did not have this transcription factor, the sterile gene was present but not expressed. Another specific mRNA of the sterile line was produced by the transcription of a DNA fragment specific to the mitochondria of the sterile line, and this mRNA was not affected by *Rf4*. Thus, it was speculated that this mRNA was probably a transcript of this transcription factor gene. The two mRNAs were not affected by *Rf3*, suggesting that *Rf3* may restore fertility through post-translational modification of the proteins of the sterile gene. It indicates that cytoplasmic male sterility produced by the expression of the sterile gene in WA-type rice was regulated by the transcription level of the mitochondrial-encoded gene, and its fertility restoration was controlled by post-transcriptional and post-translational levels of the nuclear restorer genes. On this basis, WA-type cytoplasmic male sterility gene WA352 and restorer gene *Rf4* were successfully cloned, which revealed that the WA352 protein interacted with the mitochondrial localization protein *COX11* expressed by nuclear gene and induced abnormal anther tapetum degradation and pollen sterility, *Rf4* restored fertility by lowering transcript level of WA352, thus clarifying molecular mechanism of controlling the male sterility occurrence and fertility restoration through genic-cytoplasmic interaction of rice CMS/Rf system (Chen Letian et al., 2016).

2. Genic male sterility

The nuclear genome plays an important role in the production of rice male sterility, but related research literature is still limited due to the huge size of the nuclear genome. Studies on the relationship between the nuclear genome and male sterility mainly focus on the localization, cloning and functional analysis of the sterile genes in common genic sterile mutants and photo-thermo-sensitive genic sterile lines, and some progress has been made.

It is well known that male sterile rice plants can be obtained through both natural and artificial mutations. Mostly, such genic male sterility features simple sterility genetic behavior mainly controlled by a single recessive gene. In recent years, with the completion of rice genome sequencing and rapid development of the molecular marker technology, studies on the localization, cloning and functional analysis of the sterile genes of numerous sterile mutants or the discovered photo-thermo-sensitive genic sterile lines have found dozens of sterile genetic loci (Tables 2 - 4 and 2 - 5) and some genes have been cloned (Table 2 - 6). Current research findings show that the genetic loci of genic male sterility are widely distributed on chromosomes, covering all 12 chromosomes of rice.

In fact, during the male reproductive development of rice plants, any gene mutation involved in the whole processes of stamen development, sporogonium differentiation, pollen mother cell meiosis, microspore mitosis, pollen development or flowering may cause abnormal anther or pollen development, and

Table 2 – 4 Localization and cloning of sterile genes of rice nuclear male sterile mutants (lines)

Sterile mutant (line)/gene	Sterility type	Pathway	Number of gene pairs	Dominant or recessive	Chromosome	Cloning/ localization	First author	Publishing time
Nongken 58S, Shuang 8 – 2S, N98S	Photo-thermo-sensitive sterility		2	Recessive	3, 11	Localization	Hu Xueying	1991
32001S	Photosensitivesterility		2	Recessive	3, 7	Localization	Zhang	1994
tms1	Thermo-sensitive sterility		1	Recessive	8	Localization	Wang	1995
Nekken2 / *tms2*	Thermo-sensitive sterility		1	Recessive	7	Localization	Yamagushi	1997
IR32364 / *tms3*	Thermo-sensitive sterility		1	Recessive	6	Localization	Subudhi	1997
Nongken 58S	Photosensitive sterility		1	Recessive	12	Localization	Li Ziyin	1999
IR32364 / *tms3* (*t*)	Thermo-sensitive sterility		1	Recessive	6	Localization	Lang	1999
Nongken 58S / *pms3*	Photosensitive sterility		1	Recessive	12	Localization	Mei	1999
Annong S – 1/ *tms5*	Thermo-sensitive sterility		1	Recessive	2	Localization	Wang	2003
Gene *APRT*	Thermo-sensitive sterility				4	Cloning	Li Jun	2003
OsMS-L	Non-pollen sterility	Radiation mutation	1	Recessive	2	Localization	Liu Haisheng	2005
OsMS121	Male sterility	Radiation mutation	1	Recessive	2	Localization	Jiang Hua	2006
Osms2	Male sterility	Radiation mutation	1	Recessive	3	Localization	Chen Liang	2006
Osms3	Male sterility	Radiation mutation	1	Recessive	9	Localization	Chen Liang	2006
Peiai 64S	Thermo-sensitive sterility		2	Recessive	7, 12	Localization	Zhou Yuanfei	2007
osms7	Male sterility		1	Recessive	11	Localization	Zhang Hong	2007
ms gene	Non-pollen type		1	Recessive	1	Localization	Cai Zhijun	2008
XS1	Pollen abortion	Natural mutation	1	Recessive	4	Localization	Zuo Ling	2008
ohs1 (*t*)	Female and male gamete sterility	Transgenosis	1	Recessive	1	Localization	Liu Xiaoling	2009
ms-np	Non-pollen sterility	Natural mutation	1	Recessive	6	Localization	Chu Mingguang	2009
D52S[*rpms3*(*t*)]	Short photosensitive sterility		1	Recessive	10	Localization	Ma Dong	2010
sms1	Severe sterility of anther	Natural mutation	1	Recessive	8	Localization	Yan Wenyi	2010

continued

Sterile mutant (line)/gene	Sterility type	Pathway	Number of gene pairs	Dominant or recessive	Chromosome	Cloning/ localization	First author	Publishing time
Guangzhan 63S/*ptgms 2-1*	Photo-thermo-sensitive sterility		1	Recessive	2	Localization	Xu	2011
tms7	Thermo-sensitive sterility	Wild planting	1	Recessive	9	Localization	Zou Danni	2011
802A[*ms92(t)*]	Typical abortive type pollen		1	Recessive	3	Localization	Sun Xiaoqiu	2011
tms7	Inverse thermo-sensitive sterility	Wild planting	2	Recessive	9, 10	Localization	Hu Hailian	2011
rtms2	Inverse thermo-sensitive sterility	Wild planting	1	Recessive	10	Localization	Xu Jiemeng	2012
rtms3	Inverse thermo-sensitive sterility	Wild planting	1	Recessive	9	Localization	Xu Jiemeng	2012
Zhu 1S/*tms9*	Thermo-sensitive sterility		1	Recessive	2	Localization	Sheng	2013
tms9-1	Photo-thermo-sensitive sterility		1	Recessive	9	Localization	Qi Yongbin	2014
osms55	Male sterility	Chemical mutation	1	Recessive	2	Cloning	Chen Zhufeng	2014
IR64 mutant	Pollen abortion		1	Recessive	11	Localization	Hong Jun	2014
012S-3	Non-pollen type	Natural mutation	1	Recessive	7	Localization	Ouyang Jie	2015
D63	Non-pollen type	Chemical mutation	1	Recessive	12	Localization	Zhu Baiyang	2015
Osgsl5	Male low-fertility	T-DNA insertion			6	Cloning	Shi Xiao	2015
oss125	Male fertility	Chemical mutation	1	Recessive	2	Localization	Zhang Wenhui	2015
gamyb5	Non-pollen type	Radiation mutation	1	Recessive	1	Map-based cloning	Yang Zhengfu	2016
cyp703a3-3	Non-pollen type	Radiation mutation	1	Recessive	8	Map-based cloning	Yang Zhengfu	2016
9522 mutant	Male fertility	Radiation mutation	1	Recessive	4	Localization	Yang Zhen	2016
mil3	Non-pollen type	Radiation mutation	1	Recessive		Map-based cloning	Feng Mengshi	2016
D63	Non-pollen type	Chemical mutation	1	Recessive	2	Localization	Jiao Renjun	2016
OsDMS-2	Less pollen	Natural mutation	1 *	Dominant	2, 8	Localization	Min Hengqi	2016
whf41	Non-pollen type	Radiation mutation	1	Recessive	3	Localization	Xuan Dandan	2017

Note. * Genetic analysis is for a pair of dominant genes, but gene mapping found two loci, on chromosome 2 and 8 respectively.

Table 2-5 Information related to photo-thermo-sensitive male sterile genes located in rice (Fan Yourong et al., 2016)

Locus	Chromosome	Sterility parent	Candidate interval	Function
pms1	7	32001S	85 kb	—
pms1 (t)	7	Peiai 64S	101.1 kb	—
pms2	3	32001S	17.6 cM	—
pms3	12	Nongken 58S	LDMAR	long non-coding RNA
pms4	4	Mian 9S	6.5 cM	—
p/tms12-1	12	Peiai 64S	osa-smR5864m	small RNA
CSA	1	csa mutant	LOC_Os01g16810	Participate in sugar distribution
rpms1	8	Yi D1S	998 kb	—
rpms2	9	Yi D1S	68 kb	—
ptgms2-1	2	Guangzhan 63S	50.4 kb	—
tms1	8	5460S	6.7 cM	—
tms2	7	Norin PL12	1.7 cM	—
tms3 (t)	6	IR32364TGMS	2.4 cM	—
tms4 (t)	2	TGMS—VN1	3.3 cM	—
tms5	2	Annong S-1, Zhu 1S	LOC_Os02g12290	RNase Z
tmsX	2	*Indica* S	183 kb	—
tms6	5	Sokcho—MS	2.0 cM	—
tms9	2	Zhu 1S	107.2 kb	—
tms9-1	9	Hengnong S-1	162 kb	—
TGMS	9	SA2	11.5 cM	—
Ugp1	9	Ugp1 co-inhibition	LOC_Os09g38030	UDP glucose pyrophosphorylase
tms6 (t)	10	G20S	1455 kb	—
rtms1	10	J207S	7.6 cM	—

Table 2-6 Cloned rice recessive genic male sterility genes (Ma Xiqing et al., 2012)

Genic male sterility gene	Corresponding protein encoded by fertility gene	Corresponding fertility gene function
msp1	LRR receptor-like kinase	Early development of microspores
pair1	Coiled-coil domain protein	Homologous chromosomes synapsis

Continued

Genic male sterility gene	Corresponding protein encoded by fertility gene	Corresponding fertility gene function
pair2	HORMA domain protein	Homologous chromosomes synapsis
zep1	Coiled-coil domain protein	Meiosis phase synaptic complex formation
mel1	ARGONAUTE (AGO) family protein	Cell division before germ cell meiosis
pss1	Kinesin family protein	Dynamic changes in male gamete meiosis
tdr	bHLH	Degradation of tapetum
udt1	bHLH	Degradation of tapetum
gamyb4	MYB transcription factor	Aleurone layer and pollen sac development
ptc1	PHD-finger transcription factor	Tapetum and pollen grain development
api5	Anti-apoptotic protein 5	Delayed tapetum degradation
wda1	Carbon lyase	Lipid synthesis and pollen grain extine formation
cyp704B2	Cytochrome P450 gene family	Development of pollen sac and pollen exine
dpw	Fatty acid reductase	Development of pollen sac and pollen exine
mads3	Homeosis	Pollen sac late development and pollen development
osc6	Lipid transfer family protein	Liposome and pollen exine development
rip1	WD40	Pollen maturation and germination
csa	MYB	Allocation of pollen and pollen sac sugar
id1	MYB	Pollen sac dehiscence

eventually result in male sterility (Ma, 2005; Glover et al., 1988). In recent years, with the completion of rice genome sequencing and the building of rice mutant libraries and gene expression profiling, progress has been made in research on the molecular mechanism of rice pollen development. According to Zhang Wenhui et al. (2015), some genes that control the number of floral organs in rice, such as *FON1 - 4* and *OsLRK1*, have been discovered, as well as genes that control the division and differentiation of pollen sac cells, *MSP1* and *OsTDL1A*; genes that control male meiosis, *PAIR1*, *PAIR2*, *PAIR3*, *MEL1*, *MIL1*, *DTM1*, and *OsSGO1*; key genes that promote pollen grain development, *CYP703A3*, *CYP704B2*, *WDA1*, *OsNOP*, *DPW*, *Ugp2*, and *MTR1*, etc. These genes are involved in many development processes, including the meiosis of microspore mother cells, tapetum development and degradation, and pollen cell wall formation. According to the differences in function and regulation period of sterility genes, they are sorted into three types. 1) sterility genes that work during microspore mother cell development; 2) sterility genes that work during tapetum development; 3) sterility genes that work during pollen sac and pollen exine development.

MSP1 (multiple sporocyte), which encodes the leucine-rich receptor protein kinase and regulates early microspore cell development, is the first rice fertility gene cloned. The *msp1* mutant produces excessive female and male spore mother cells, resulting in disorder in the development of the pollen sac cell wall and tapetum, keeping the development of microspore mother cells at the stage of the first meiotic division, while not affecting the development of megaspore mother cells. The result is complete pollen abortion with normal female organ development (Nonomura, 2003). Zhang Wenhui et al. (2015) found *OsRPA1a* to be the gene controlling sterile mutant *oss125* phenotype and found that the A663 locus in the *OsRPA1a*-coded region mutated to C, resulting in abnormal pollen development. *OsRPA1a* was involved in male and female gamete development, and was necessary for rice meiosis and somatic cell DNA repair. Shi Xiao (2015) found a low-fertility mutant from the library of mutants inserted with T-DNA and *Tos17*, and concluded that the reason for the formation of this mutant was that the T-DNA fragment was inserted on the fifth intron of the glucan synthase-like 5 (*GSL5*) gene on chromosome 6. The gene was expressed in all parts of the plant throughout the whole growth period, with the highest expression period and position being the microspore mother cells at the dyad and tetrad stages in the meiosis of rice male gametophyte. *GSL5* in rice is homologous to gene *At GSL2* in Arabidopsis and belongs to the *OsGSL* gene family. It is the first gene in rice reported to encode callose synthase, and the main function of the encoded *OsGSL5* is to control callose synthesis. During the meiosis of microspore mother cells, after *GSL5* expression was silenced, the callose synthesis controlled by *GSL5* reduced dramatically. At this stage, the role of callose was to form cell plates, pollen mother cells, and the callose wall of dyad and tetrad. As further development of male gametes was severely affected by callose deletion, the number of active and mature pollen formed was down to only about 3% of the level of wild plants, and the spike seed setting rate was only about 10% of that of the wild type. Most cloned fertility-related rice genes associated with tapetum development are those encoding transcription factors. For example, undeveloped tapetum 1 (*UDT1*) regulates early gene expression for the tapetum and the meiosis of pollen mother cells, which is necessary for secondary wall cells to differentiate into mature tapetum. The tapetum of *udt1* mutant becomes vacuolated during meiosis, the middle lamella fails to degrade timely, and the auxocyte fails to develop into pollen, thus resulting in complete pollen abortion (Jung et al., 2005). Both the tapetum degeneration retardation (*TDR*) and *UDT1* genes encode bHLH-like transcription factors. Studies found that *TDR* can be directly bound to the *Os-CP1* and *OsC6* promoter regions in the PCD gene and positively regulates the PCD process of tapetum and the formation of pollen cell walls. Degradation of the tapetum and middle lamella in *tdr* mutant was delayed while the microspores rapidly degraded after release, thus resulting in complete male sterility (Li et al., 2006). Chen Liang (2006) found that during pollen wall development of mutant *Osms3*, the tapetum suffered premature vacuolization and abnormal degradation, thus failing to form normal pollen grains. Zhang Hong (2007) studied rice male sterile mutant *osms7*, which was recessive genic male sterility, and believed that the gene may be the negative regulatory factor that controls programmed cell death of the tapetum. Feng Mengshi (2016) located precisely the sterile gene of the non-pollen mutant *mil3* obtained from ^{60}Co-γ ray radiation-induced sterile line Zhongxian 3037 to be between STS markers S10 and S11. The physical distance between the two markers is about 40kb, with nine open reading frames (ORFs). One of the ORFs has a single-base insertion at nucleotide 496, and single-base mutations at nucleotides 497 and 499, changing the amino acid sequence that comes after them. The gene was thus deemed a candidate gene and also an oxidoreductase-related gene. Abnormal anther development in *mil3* mutant resulted in microspore degradation, with no pollen produced. qPCR analysis on the five genes of *MSP1*, *UDT1*, *TDR*, *DTC1 and OsCP1* that affected tapetum development showed that *MIL3* may be located downstream of *MSP1*, *UDT* and *TDR*, and upstream of *DTC1* and *OsCP1*, and that it played an important role in tapetum development. Yang Zhengfu (2016) discovered a non-pollen male sterile mutant *gameb5* in the mutant library of ^{60}Co-γ ray radiation-induced sterility from Zhonghui 8015. Anther semi-thin section showed that the meiosis of microspore mother cells of mutant *gamyb5* was abnormal, making it unable to form normal tetrads and microspores, and the tapetum elongated abnormally with programmed death delayed. The mutant gene was finely located to be between markers ZF - 29 and ZF - 31 on the long arm of chromosome 1, the physical distance between which was about 16.9kb. Sequencing analysis on two complete ORFs in this region found that the second exon of the MYB transcription factor gene *Os01G0812000* whose coding was induced by gibberellin had a deletion of eight bases, resulting in premature termination of the translation. *qRT-PCR* detected that the expression quantity of regulatory factors *UDT1*, *TDR*, *CYP703A3* and *CYP704B2* that affect anther development in the mutant was significantly lower than that in wild type plants, further proving that GAMYB played a key role in anther meiosis and programmed tapetum death.

The outermost layer of a rice pollen sac wall is composed of a layer of waxy cuticle, which can protect the pollen sac during development by resisting various adversity stresses, and preventing bacterial infections and water loss. Genic sterility genes that affect wax formation have been reported. For example, wax-deficient anther 1 (*Wda1*) is involved in the long-chain fatty acid synthesis pathway that regulates lipid synthesis and pollen wall development. It is mainly expressed in the epidermal cells of the pollen sac. The cuticle waxy crystal of the pollen sac is absent in the *wda1* mutant, and microspore development was severely delayed, thus resulting in male sterility (Jung et al., 2006). *Cyp704B2* belongs to the cytochrome P450 gene family and plays an important role in the hydroxylation pathway of fatty acid. It is mainly expressed in tapetum cells. The tapetum of the *cyp704B2* mutant suffered developmental defect, and the development of the pollen sac and pollen exine was impaired, thus resulting in pollen abortion (Li et al., 2010). Xuan Dandan et al. (2017) isolated and identified a non-pollen male sterile mutant *whf41* from the library of radiation-induced mutants of the *indica* line Zhonghui 8015. Phenotypic analysis showed that the anthers of the *whf41* mutant were thin, transparent and opalescent, and contained no pollen grain; semi-thin sections showed that the microspores of the mutant failed to form normal pollen extine, tapetum cells expanded abnormally but failed to proceed to programmed death, and finally, the expanded tapetum and pollen cell fragments merged and filled the anther chamber. Electron microscope observation further revealed that the intine and extine of the mutant's anthers were smooth but lacking lipids, so the pollen cells gradually disrupted and degraded. This gene was located between markers XD - 5 and XD - 11 on the short arm of chromosome 3, with a physical distance of 45.6kb between them, containing nine ORFs. Sequencing analysis showed that there were one single-base replacement and a deletion of three bases at the fourth exon of the cytochrome P450 gene *Locos03G07250*, leading to the substitution of one amino acid (aspartic acid substituted by methionine) and the deletion of one amino acid (valine) in the translated sequence, thus resulting in functional change for the phenotype. Results of qRT-PCR testing showed that the expression levels of *CYP704B2* and a series of genes related to anther lipid synthesis and transport in the *whf41* mutant were significantly down-regulated. It was inferred that *OsWHF41* was a new allele of *CYP704B2*. Other related results further proved the important role of *CYP704B2* in rice anther lipid synthesis and pollen wall formation. Yang Zhengfu (2016) located the sterile gene of mutant *cyp703a3* - *3* from the library of radiation-induced mutants of the *indica* line Zhonghui 8015 between the two markers S15 - 29 and S15 - 30 on chromosome 8, with a physical distance of about 47.78kb. Sequencing revealed that there were three bases deleted in the first exon of *CYP703A3*, which caused frameshift mutation, thus resulting in a mutant phenotype. Through transgenic complementary verification, the mutant phenotype was restored by transferring wild-type *CYP703A3* gene into the mutant, indicating that *CYP703A3* was the target gene that played an important role in anther and pollen wall development.

Chen Liang (2006) located the *Osms2* mutant sterile gene to be between InDel markers CL6 - 4 and CL7 - 4 on chromosome 3, 0.1 cM from the former and 0.04 cM from the latter, making a physical distance between the two of about 100kb. Thirteen genes existed within the range; one was similar to the *MS2* gene and homologous to the Arabidopsis male sterility gene *MS2*. Sequencing showed that the eighth exon of this gene had a base deletion, which led to frameshift mutation. *OsMS2* was preliminary determined to be a candidate gene. Observation revealed that the microsporous extine developed abnormally, which caused degradation of microspores, thus preventing the formation of mature pollen grains. Jiang Hua et al. (2006) obtained a rice male sterile mutant *OsMS121* from the seeds of the radiation-induced *japonica* 9522 mutant, and located the gene between molecular markers R2M16 - 2 and R2M18 - 1 on chromosome 2, 200kb apart, through map-based cloning. Analysis indicated that abnormality of the germinal aperture of pollen during development may be the cause of pollen abortion.

In addition, Zhang et al. (2010) isolated and identified a carbon-starved anther (*CSA*) mutant, which had increased sugar content in the stem and leaves, but reduced sugar and starch content in floral organs. In particular, the carbohydrate level in the anther was low at the anaphase, showing male sterility. After map-based cloning, the *CSA* gene was preferentially expressed in tapetum cells and vascular tissues responsible for sugar transport, and encoded the R2R3 MYB transcription factor. The *CSA* gene was closely related to the MST8 promoter that encoded the monosaccharide transporters. In the anther of the *CSA* mutant, the expression quantity of MST8 reduced greatly. Analysis showed that *CSA* played a key role in regulating the transcription of genes responsible for sugar distribution in rice male sterility development.

In recent years, some progress has also been made in studies of PTGMS genes. Li Jun et al. (2003) reported for the first time that the adenine ribose phosphate transferase gene *APRT* that causes plant male sterility in Arabidopsis was cloned from *Oryza sativa subsp. indica*, and mapped a BAC clone on chromosome 4. The gene was 4,220bp long, the encoded APRT protein of 212 amino acid residues, and AP-

RT catalytic structural domain existed in the protein. Further study showed that expression changes in APRT gene in the PTGMS line Annong S－1, induced by temperature, may be related to PTGMS phenotype. Zhou Yuanfei (2007) found that PTGMS line, Peiai 64S, was controlled by two pairs of overlapping recessive genes, and mapped the two, *pms1* and *pms3*, on chromosomes 7 and 12 respectively. The *pms1* was co-segregated with the SSR marker RM6776, and the genetic distance between *pms1* and the linkage marker RM21242 and YF11 was both 0.2 cM, making a physical distance of 101.1kb. The interval had 14 predicted genes, among which *LOCOs07g12130* locus encoded a MYB-like protein containing a DNA binding domain. The gene product was related to a sensitive response to thermal stimulus, and it was speculated that *LOCOs07g12130* was the most likely candidate gene for *pms1* fertile allele. Qi Yongbin (2014) located the gene that controlled thermo-sensitive male sterility of Hengnong S－1 to be between the two dCAPS markers on chromosome 9, 162kb in between, and named it *tms9－1*. Results of BAC function annotation of the candidate gene and sequencing of the candidate gene confirmed that *OsMS1* gene had a base mutation from C to T at its third exon, so *OsMS1* was speculated as the candidate gene of *tms9－1*. Point mutation in Hengnong S－1 happened to occur in the central region of the predicted transcription factor S－Ⅱ, regulating the transcription elongation controlled by RNA polymerase Ⅱ. Ding et al. (2012) discovered a 1,236bp long non-coding long-chain RNA (lncRNA), and named it LDMAR, which regulated the photosensitive sterility in Nongken 58S. Normal pollen development required sufficient LDMAR transcript under long daylength conditions. Study showed that sterile mutant and the wild type differed only in single nucleotide polymorphism (SNP), the difference changed LDMAR's secondary structure, which increased the degree of methylation of the LDMAR promoter region, thus resulting in transcriptional down-regulation of LDMAR. Insufficient LDMAR transcription led to premature programmed cell death during anther development, which eventually resulted in sterility. Ding et al. (2012) and Zhou (2012) successively reported cloning and function analysis of photosensitive genic male sterile gene *pms3*, and revealed that the sterile gene *pms3* located in Nongken 58S by Ding et al. and the sterile gene *p/tms12－1* located in Peiai 64S by Zhou et al. were actually the same gene. However, as the two sterile lines made different responses to light and temperature, two function analyses on this gene differed slightly. Yet it was still confirmed that Nongken 58's conversion to the photosensitive sterile Nongken 58S and Peiai 64's conversion to the thermo-sensitive Peiai 64S were caused by the same base at *pms3* locus which turned from G to C. This base mutation was mapped at locus 11 of a small RNA (*osa-smR5864*) of 21nt. Ding et al. found that the mutation increased CG methylation in DNA methylation of the promoter region of the gene, and inhibited expression quantity of the gene in young panicles under long daylength conditions, resulting in sterility, and confirmed that the regulation was typical RNA-mediated DNA methylation. Zhou et al. believed that both small RNA *osa-smR5864m* and its wild-type counterpart *osa-smR5864w* were preferentially expressed in young panicles, but expression difference between the two was not regulated by daylength and temperature, it was speculated that regulation of the fertile gene was not reflected by expression quantity of small RNA, but the regulation on downstream target genes bound by small RNA.

Annong S－1, as a major source of PTGMS gene resources currently used in production, is widely used in the breeding of PTGMS lines. South China Agricultural University and the Institute of Genetics and Developmental Biology of Chinese Academy of Sciences jointly made positional cloning for the thermo-sensitive sterile gene *tms5* from Annong S-1 and Zhu 1S, and revealed the molecular mechanism of the gene controlling thermo-sensitive sterility (Zhou et al., 2014). *TMS5* encoded a conserved short-version RNase Z homologous protein named RNase Z^{S1}. The 71st base in the *TMS5* coding region in Annong S-1 and Zhu 1S mutated from C to A, resulting in premature termination of RNase Z^{S1} protein translation. The RNase Z^{S1} protein had precursor tRNA 3'-end incision enzyme activity. Zhou Hai et al. (2014) speculated that the protein can regulate thermo-sensitive male sterility by processing mRNA transcribed from ubiquitin-ribosome fusion protein L40 (UbL_{40}). In wild type rice, induced by high temperature, mRNA transcribed by UbL_{40} gene can be normally degraded by RNase Z^{S1}, and the fertility was normal; in *tms5* thermo-sensitive sterile line, due to deletion of RNase Z^{S1} function, UbL_{40} mRNA induced and expressed by high temperature cannot be normally degraded and excessively accumulated, thus resulting in pollen abortion (Fig. 2-14).

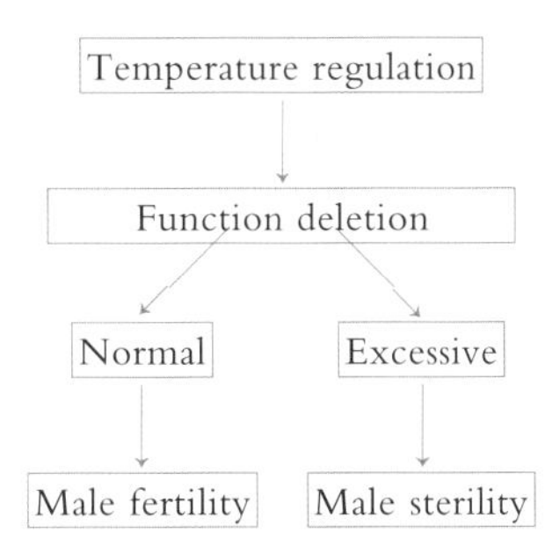

Fig. 2-14 Mechanism of RNase Z^{S1} regulating thermo-sensitive male sterility

In conclusion, with rapid development of modern biotechnology, cloning and functions of more rice genic male sterile genes have been clarified. In the future, the genetics and molecular regulatory mechanisms of rice male sterile genes will be more clearly understood, providing better opportunities for effective use of sterile genes for rice heterosis utilization. Meanwhile, it will also provide more genetic resources for rice molecular design and breeding, thus pushing rice heterosis utilization to a new stage of development.

References

[1] CAI YAOHUI, ZHANG JUNCAI, LIU QIUYING. Study on relationship between Pingxiang genic male sterile rice apiculus character inheritance and temperature sensitivity[J]. Jiangxi Agricultural Science and Technology, 1990 (1): 14-15.

[2] CAO MENGLIANG, ZHENG YONGLIAN, ZHANG QIFA. Comparative analysis on two-dimensional electrophoresis of protein in photosensitive genic male sterile rice Nongken 58S and Nongken 58[J]. Journal of Huazhong Agricultural University, 1992, 11(4): 305-311.

[3] CAO YICHENG, FU BINYING, WANG MINGQUAN, et al. Preliminary analysis on two-dimensional electrophoresis of protein in photosensitive genic male sterile rice[J]. Journal of Wuhan University (HPGMR Special

Issue), 1987: 73 - 80.

[4] CHEN CUILIAN, SUN XIANGNING, ZHANG ZIGUO, et al. Preliminary study on phosphorus metabolism of Hubei photosensitive genic male sterile rice[J]. Journal of Huazhong Agricultural University, 1990, 9(4): 472 - 474.

[5] CHEN PING, XIAO YIHUA. Comparative study on anther peroxidase activity during pollen abortion of photosensitive genic male sterile rice[J]. Journal of Wuhan University (HPGMR Special Issue), 1987: 39 - 42.

[6] CHEN XIANFENG, LIANG CHENGYE. Anther energy and reactive oxygen metabolism of Hubei photosensitive genic male sterile rice[J]. Chinese Bulletin of Botany, 1992, 34(6): 416 - 425.

[7] CHEN XIONGHUI, WAN BANGHUI, LIANG KEQIN. Study on sensitivity of fertility to photoperiod and thermo period response of photo-thermo-sensitive genic male sterile rice[J]. Journal of South China Agricultural University, 1997, 18(4): 8 - 11.

[8] CHENG SHIHUA, SUN ZONGXIU, SI HUAMIN, et al. Study on classification of fertility conversion photoperiod and thermo period response types of rice dual-purpose genic male sterile lines[J]. Chinese Agricultural Sciences, 1996, 29(4): 11 - 16.

[9] DENG HUAFENG. Discovery and preliminary study on Annong S - 1 photosensitive sterile rice[J]. Adult Higher Agricultural Education, 1988 (3): 34 - 36.

[10] DENG JIXIN, LIU WENFANG, XIAO YIHUA. Study on anther ATP content and synthesis of nucleic acid and protein during HPGMR pollen development[J]. Journal of Wuhan University (Natural Science Edition), 1990 (3): 85 - 88.

[11] DENG XIAOJIAN, ZHOU KAIDA. Study on fertility conversion and inheritance of low thermo-sensitive dominant genic male sterile rice "8987"[J]. Journal of Sichuan Agricultural University, 1994, 12(3): 376 - 382.

[12] FENG JIANYA, CAO DAMING. Characteristics of fertility and changes of polyamines in spike of photosensitive genic male sterile rice C407S[J]. Journal of Nanjing Agricultural University, 1993, 16(2): 107 - 110.

[13] FENG JIANYA, YU BINGGAO, CAO DAMING. Changes in polyamines during young spike development of photosensitive genic male sterile rice[J]. Journal of Nanjing Agricultural University, 1991, 14(1): 12 - 16.

[14] FENG JIUHUAN, LU YONGGEN, LIU XIANGDONG. Cytological mechanism of pollen abortion of rice photo-thermo-sensitive genic male sterile line Peiai 64S (in English)[J]. Chinese Journal of Rice Science, 2000, 14(1): 7 - 14.

[15] HE HAOHUA, ZHANG ZIGUO, YUAN SHENGCHAO. Preliminary study on effects of temperature on development and fertility conversion of photo-induced photosensitive genic male sterile rice[J]. Journal of Wuhan University (HPGMR Special Issue), 1987: 87 - 93.

[16] HU XUEYING, WAN BANGHUI. Genetic relationship and chain determination of rice photo-thermo-sensitive genic male sterile genes and isozyme genes[J]. Journal of South China Agricultural University, 1991, 12 (1): 1 - 9.

[17] HUANG HOUZHE, LOU SHILIN, WANG HOUCONG, et al. Auxin deletion and occurrence of male sterility [J]. Journal of Xiamen University (Natural Science Edition), 1984, 23(1): 82 - 97.

[18] HUANG QINGLIU, TANG XIHUA, MAO JIANLEI. Photoperiod and thermo period response characteristics of *japonica* type photosensitive genic male sterile rice 7001s as well as pollen fertility conversion and changes in anther protein during the process[J]. Acta Agronomica Sinica, 1994, 20(2): 156 - 160.

[19] HUANG QINGLIU, TANG XIHUA, MAO JIANLEI. Effects of temperature on pollen fertility and anther pro-

tein of thermo-sensitive genic male sterile rice[J]. Plant Physiology Journal, 1996, 22(1): 69 - 73.
[20] HUANG SHAOBAI, ZHOU XIE. Relationship between rice cytoplasmic male sterility and endogenous GA_(1+4) and IAA[J]. Acta Agriculturae Boreali-sinica, 1994, 9(3): 16 - 20.
[21] JIANG YIMING, RONG YING, TAO GUANGXI, et al. Breeding of *japonica* rice new cytoplasm source thermo-sensitive genic male sterile line Diannong S - 2[J]. Southwest China Journal of Agricultural Sciences, 1997, 10(3): 21 - 24.
[22] JIANG YIMING, RONG YING, TAO GUANGXI, et al. Breeding and performance of new cytoplasm source *japonica* rice thermo-sensitive genic male sterile line Diannong S - 1[J]. Hybrid Rice, 1997, 12(5): 30 - 31.
[23] LI SHILING, GAO YIZHI, LI HUIRU, et al. Preliminary study on observation of outcrossing of short photo-sensitive sterile rice Yi DS1[J]. Hybrid Rice, 1996 (1): 32.
[24] LI DADONG, WANG BIN. Cloning of rice mitochondrial *aptA* gene and its relationship with cytoplasmic male sterility[J]. Hereditas (Beijing), 1990, 12(4): 1 - 4.
[25] LI HESHENG, LU SHIFENG. Preliminary study on correlation between fertility transfer and phytochrome of Hubei photosensitive genic male sterile rice[J]. Journal of Huazhong Agricultural University, 1987, 6(4): 397 - 398.
[26] LI MEIRU, LIU HONGXIAN, WANG YIROU, et al. Changes in oxygen metabolism during fertility conversion of *indica* type dual-purpose genic male sterile rice[J]. Chinese Journal of Rice Science, 1999, 13(1): 36 - 40.
[27] LI PING, LIU HONGXIAN, WANG YIROU, et al. Fertility expression of *indica* type dual-purpose genic male sterile line rice Peiai 64S-changes in NAD^+-MDH and AP isozyme during young spike development[J]. Chinese Journal of Rice Science, 1997, 11(2): 83 - 88.
[28] LI PING, ZHOU KAIDA, CHEN YING, et al. Mapping of restorer genes of rice wild abortive type genic-cytoplasmic male sterility with molecular markers[J]. Journal of Genetics and Genomics, 1996, 23(5): 357 - 362.
[29] LI RONGWEI, LI HESHENG. Changes in polyamine content in PTGMS rice fertility conversion (Briefing) [J]. Plant Physiology Communications, 1997, 33(2): 101 - 104.
[30] LI ZEBING. Preliminary study on classification of rice male sterility in China[J]. Acta Agronomica Sinica, 1980, 6(1): 17 - 26.
[31] LI ZIYIN, LIN XINGHUA, XIE YUEFENG, et al. Mapping of photosensitive genic male sterile gene of Nongken 58S with molecular markers[J]. Chinese Bulletin of Botany, 1999, 41(7): 731 - 735.
[32] LI RONGQIAN, WANG JIANBO, WANG XIANGMING. Effect of photoperiod on ultra-structure of microsporogenesis and pollen development of photosensitive genic male sterile rice[J]. Chinese Journal of Rice Science, 1993, 7(2): 65 - 70.
[33] LIANG CHENGYE, MEI JIANFENG, HE BINGSEN, et al. Cytological observation on main phases of microspore abortion of photo-thermo-sensitive genic male sterile rice[C]//Current Status of Two Line Hybrid Rice Research. Beijing: China Agriculture Press, 1992:141 - 149.
[34] LIANG CHENGYE, CHEN XIANFENG, SUN GUCHOU, et al. Some biochemical metabolism characteristics in anther of Hubei photosensitive genic male sterile rice Nongken 58S[J]. Acta Agronomica Sinica, 1995, 21(1): 64 - 70.
[35] LIN ZHIFANG, LIANG CHENGYE, SUN GUCHOU, et al. Microspore abortion and organic free radical lev-

el in anther of male sterile rice[J]. Chinese Bulletin of Botany, 1993, 35(3): 215 -4221.

[36] LIU JUN, ZHU YINGGUO, YANG JINSHUI. Study on mitochondrial DNA of Maxie type cytoplasmic male sterile rice[J]. Acta Agronomica Sinica, 1998, 24(3): 315 -319.

[37] LIU LIJUN, XUE GUANGXING. Preliminary study on protein products related to rice photosensitive genic male sterile gene[J]. Acta Agronomica Sinica, 1995, 21(2): 251 -253.

[38] LIU QINGLONG, PENG LISHA, LU XIANGYANG, et al. Study on purification of two-line hybrid rice with chemical hybridization agent Ⅱ. Effect of Baochunling treatment on physiology and biochemistry of photo-thermo-sensitive genic male sterile rice[J]. Journal of Hunan Agricultural University, 1998, 24 (5): 345 - 350.

[39] LIU YANSHENG, WANG XUNMING, WANG YUNZHU, et al. Analysis on mitochondrial *CO Ⅰ* and *CO Ⅱ* gene structure differences between rice cytoplasmic male sterile line and maintainer line[J]. Journal of Genetics and Genomics, 1988, 15(5): 348 -354.

[40] LIU ZUOCHANG, ZHAO SHIMIN, ZHAN QINGCAI, et al. Rice mitochondrial genome translation products and cytoplasmic male sterility[J]. Journal of Genetics and Genomics, 1989, 6(1): 14 -19.

[41] LU XINGGUI, YUAN QIANHUA, XU HONGSHU. Practice and experience of pilot development of two-line hybrid rice in China[J]. Hybrid Rice, 1998, 13(5): 1 -3.

[42] LU XINGGUI. Review on breeding of rice photo-thermo-sensitive male sterile lines in China[J]. Hybrid Rice, 1994(3 -4): 27 -30.

[43] LUO XIAOHE, QIU ZHIZHONG, LI RENHUA. Dual-purpose sterile line Peiai 64S that causes low critical temperature of sterility[J]. Hybrid Rice, 1992 (1): 27 -29.

[44] LUO BINGSHAN, LI WENBIN, QU YINGLAN, et al. Preliminary study on fertility conversion mechanism of Hubei photosensitive genic male sterile rice[J]. Journal of Huazhong Agricultural University, 1990, 9 (1): 7 -12.

[45] LUO BINGSHAN, LI DEHONG, QU YINGLAN, et al. Relationship between ethylene and fertility conversion of photosensitive genic male sterile rice[J]. Chinese Journal of Rice Science, 1993, 7(1): 1 -6.

[46] MEI QIMING, ZHU YINGGUO, ZHANG HONGJUN. Study on reaction characteristics of enzyme in Hubei photosensitive genic male sterile rice[J]. Journal of Huazhong Agricultural University, 1990, 9(4): 469 - 471.

[47] MEI QIMING, ZHU YINGGUO. Comparative study on mitochondrial DNA (mtDNA) of HL type and wild abortive type rice cytoplasmic male sterile lines[J]. Journal of Wuhan Botanical Research, 1990, 8(1): 25 - 32.

[48] MO LEIXING, LI HESHENG. Role of polyamine in fertility conversion of Hubei photosensitive genic male sterile rice[J]. Journal of Huazhong Agricultural University, 1992, 11(2): 106 -114.

[49] SHEN YUWEI, GAO MINGWEI. Analysis on isozyme and amino acid in R-type fertility revertant of rice cytoplasmic male sterile line[J]. Acta Agronomica Sinica, 1996, 22(2): 241 -246.

[50] SHENG XIAOBANG, DING SHENG. Discussion on several issues of breeding and utilization of photosensitive genic male sterile rice[J]. Hybrid Rice, 1993(3): 1 -3.

[51] SHI MINGSONG. Discovery and preliminary study on recessive male sterile rice sensitive to light length[J]. Scientia Agricultura Sinica, 1985(2): 44 -48.

[52] SHI MINGSONG. Preliminary report on breeding and application of late *japonica* natural dual-purpose line [J]. Hubei Agricultural Sciences, 1981(7): 1 -3.

[53] SHU XIAOSHUN, CHEN LIANGBI. Changes in total RNA content in young panicles and leaves during fertility sensitive period of high temperature sensitive sterile rice (Briefing)[J]. Plant Physiology Communications, 1999, 35(2): 108 - 109.

[54] SONG DEMING, WANG ZHI, LIU YONGSHENG, et al. Discovery of rice non-pollen sterile materials and preliminary observation on fertility of its trans-breeding offspring[J]. Chinese Bulletin of Botany, 1998, 40 (2): 184 - 185.

[55] SONG DEMING, WANG ZHI, LIU YONGSHENG, et al. Study on rice antherless sterile materials[J]. Journal of Sichuan Agricultural University, 1999, 17(3): 268 - 271.

[56] SUN JUN, ZHU YINGGUO. Study on ultra-structure of pollen and anther wall during development of Hubei photosensitive genic male sterile rice[J]. Acta Agronomica Sinica, 1995, 21(3): 364 - 367.

[57] SUN ZONGXIU, CHENG SHIHUA, SI HUAMIN, et al. Fertility response of early indica photosensitive sterile line under artificially controlled light and temperature conditions[J]. Acta Agriculturae Zhejiangensis, 1991, 3(3): 101 - 105.

[58] SUN ZONGXIU, CHENG SHIHUA. Hybrid rice breeding from three-line, two-line to one-line[M]. Beijing: China Agricultural Science and Technology Press, 1994.

[59] TANG RISHENG, MEI CHUANSHENG, ZHANG JINYU, et al. Relationship between TO3-induced rice male sterility and endogenous hormones[J]. Jiangsu Journal of Agricultural Sciences, 1996, 12(2): 6 - 10.

[60] TIAN CHANG'EN, DUAN JUN, LIANG CHENGYE. Effects of ethylene on protein, nucleic acid and reactive oxygen metabolism of rice CMS line and its maintainer line[J]. Scientia Agricultura Sinica, 1999, 32 (5): 36 - 42.

[61] TIAN CHANG'EN, LIANG CHENGYE, HUANG YUWEN, et al. Primary exploration of relationship between polyamine and ethylene during young spike development of rice cytoplasmic male sterile line (English) [J]. Plant Physiology Journal, 1999, 25(1): 1 - 6.

[62] TIAN CHANG'EN, LIANG CHENGYE. Effects of polyamine on protein, nucleic acid and reactive oxygen metabolism in young panicles of rice CMS line and its maintainer line[J]. Plant Physiology Journal, 1999, 25(3): 222 - 228.

[63] TONG ZHE, SHAO HUIDE, ZHAO YUJIN, et al. The second messenger that regulates fertility in photosensitive genic male sterile rice[C]//Current status of two line hybrid rice research. Beijing: China Agriculture Press, 1992: 170 - 175.

[64] TU JUN, ZHU YINGGUO. Research progress on rice mitochondrial genome and cytoplasmic male sterility [J]. Hereditas (Beijing), 1997, 19(5): 45 - 48.

[65] TU JUN. Mitochondrial DNA enzyme digestion analysis on HL type rice sterile line and maintainer line[J]. Journal of Zhongkai Agrotechnical College, 1999, 12(3): 11 - 14.

[66] WAN BANGHUI, LI DINGMIN, QI LIN. Classification of rice cytoplasmic-genic male sterile cytoplasm [C]//Proceedings of the international symposium on hybrid rice. Beijing: Academic Journal Press, 1988: 345 - 351.

[67] WAN BANGHUI. Classification and utilization of rice cytoplasmic-genic male sterility[C]//Study on utilization of rice heterosis. Beijing: China Agriculture Press, 1980.

[68] WANG HUA, TANG XIAOHUA, DAI FENG. Review of study on heterosis utilization of genic male sterile materials in cabbage type rape[J]. Guizhou Agricultural Sciences, 1999, 27(4): 63 - 66.

[69] WANG JINGZHAO, WANG BIN, XU QIONGFANG, et al. Analysis on rice photosensitive genic sterile gene

through RAPD method[J]. Journal of Genetics and Genomics, 1995, 22(1): 53-58.

[70] WANG TAI, TONG ZHE. Microstructure changes of sterile anthers in photosensitive genic male sterile rice Nongken 58S[J]. Acta Agronomica Sinica, 1992, 18(2):132-136.

[71] WANG TAI, XIAO YIHUA, LIU WENFANG. Changes in carbohydrates in leaves during fertility induction and conversion of photosensitive genic male sterile rice[J]. Acta Agronomica Sinica, 1991, 17(5):369-375.

[72] WANG TAI, XIAO YIHUA, LIU WENFANG. Study on changes of photo-induced HPGMR leaf protein[J]. Journal of Huazhong Agricultural University, 1990, 9(4): 369-374.

[73] WANG XI, YU MEIYU, TAO LONGXING. Effect of male gamete attractant CRMS on rice anther protein and free amino acid (English)[J]. Chinese Journal of Rice Science, 1995, 9(2): 123-126.

[74] WU HONGYU, WANG XIANGMING. Effect of photoperiod length on microsporogenesis of Nongken 58S [J]. Journal of Huazhong Agricultural University, 1990, 9(4): 464-465.

[75] XIA KAI, XIAO YIHUA, LIU WENFANG. Analysis on ATP content and RuBPCase activity in leaves during photosensitive phase of Hubei photosensitive genic male sterile rice[J]. Hybrid Rice, 1989 (4): 41-42.

[76] HE ZHICHANG, XIAO YIHUA, FENG SHENGYAN. Distribution of 32P in HPGMR (58s) plants[J]. Journal of Wuhan University (Natural Science Edition), 1992 (1): 127-128.

[77] XIAO YIHUA, CHEN PING, LIU WENFANG. Comparative analysis on free amino acids during anther abortion of photosensitive genic male sterile rice[J]. Journal of Wuhan University (HPGMR Special Issue), 1987: 7-16.

[78] XIE GUOSHENG, YANG SHUHUA, LI ZEBING, et al. Discussion on classification of photo-sensitivity and thermo-sensitivity of dual-purpose genic male sterile rice[J]. Journal of Huazhong Agricultural University, 1997, 16(5):311-317.

[79] XU HANQING, LIAO QINLIN. Cell morphology observation on zinc methylarsenate's androcidal action against rice[J]. Acta Agronomica Sinica, 1981, 7(3): 195-200.

[80] XU MENGLIANG, LIU WENFANG, XIAO YIHUA. Changes in IAA during young spike development of Hubei photosensitive genic male sterile rice[J]. Journal of Huazhong Agricultural University, 1990, 9(4): 381-386.

[81] XU RENLIN, JIANG XIAOHONG, SHI SUYUN, et al. Comparative study on protein polypeptide of rice wild abortive type cytoplasmic male sterile line and maintainer line[J]. Journal of Genetics and Genomics, 1992, 19(5): 446-452.

[82] XU RENLIN, XIE DONG, SHI SUYUN, et al. Cloning and sequence analysis on specific fragments of rice mitochondrial DNA male sterility[J]. Chinese Bulletin of Botany, 1995, 37(7): 501-506.

[83] YAN LONGAN, CAI YAOHUI, ZHANG JUNCAI, et al. Study and application prospects of dominant male genic sterile rice[J]. Acta Agriculturae Jiangxi, 1997, 9(4): 61-65.

[84] YAN LONGAN, ZHANG JUNCAI, ZHU CHENG, et al. Preliminary report on identification of rice dominant male sterile genes[J]. Acta Agronomica Sinica, 1989, 15(2): 174-181.

[85] YANG HUAQIU, ZHU JIE. Study on breeding of indica rice short daylength low temperature sterility genic male sterile line go543S[J]. Hybrid Rice, 1996(1): 9-13.

[86] YANG DAICHANG, ZHU YINGGUO, TANG LUOJIA. Content and fertility conversion of four endogenous hormones in HPGMR leaf[J]. Journal of Huazhong Agricultural University, 1990, 9(4): 394-399.

[87] YANG JINSHUI, VIRGINIA W. Mitochondrial DNA restriction enzyme digestion map analysis on rice wild a-

bortive sterile line and maintainer line[J]. Acta Agronomica Sinica, 1995, 21(2): 181 - 186.

[88] YANG JINSHUI, GE KOULIN, VIRGINIA W. Mitochondrial DNA enzyme digestion electrophoresis banding pattern of rice BT type sterile line and maintainer line[J]. Acta Agriculturae Shanghai, 1992, 8(1): 1 - 8.

[89] YANG RENCUI, LI WEIMING, WANG NAIYUAN, et al. Discovery and preliminary study on indica rice photosensitive genic male sterile germplasm 5460ps[J]. Chinese Journal of Rice Science, 1989, 3(1): 47 - 48.

[90] YING YANRU, NI DAZHOU, CAI YIXIN. Comparative analysis on component I protein in rice, wheat, rape and tobacco cytoplasmic male sterility system[J]. Journal of Genetics and Genomics, 1989, 16(5): 362 - 366.

[91] YUAN SHENGCHAO, ZHANG ZIGUO, LU KAIYANG, et al. Basic characteristics and adaptability to different ecological types of photosensitive genic male sterile rice[J]. Journal of Huazhong Agricultural University, 1990, 9(4): 335 - 342.

[92] YUAN LONGPING. Male sterility of rice[J]. Chinese Science Bulletin, 1966, 17(4): 185 - 188.

[93] YUAN LONGPING. Overview of hybrid rice breeding in China[C]//Study on utilization of rice heterosis. Beijing: China Agriculture Press, 1980: 8 - 20.

[94] YUAN LONGPING, CHEN HONGXIN. Hybrid rice breeding and cultivation[M]. Changsha: Hunan Science & Technology Press, 1988.

[95] YUAN LONGPING. Progress in study on two-line hybrid rice [G]//Current status of two line hybrid rice research. Beijing: China Agriculture Press, 1992: 6 - 12.

[96] ZENG HANLAI, ZHANG ZIGUO, LU XINGGUI, et al. Discussion on classification of photo-sensitivity and thermo-sensitivity of W6154S type rice[J]. Journal of Huazhong Agricultural University, 1995, 14(2):105 - 109.

[97] ZHANG MINGYONG, LIANG CHENGYE, DUAN JUN, et al. Membrane lipid peroxidation levels in different organs of CMS rice[J]. Acta Agronomica Sinica, 1997, 23(5): 603 - 606.

[98] ZHANG NENGGANG, ZHOU XIE. Relationship between three endogenous acidic plant hormones and fertility conversion of Nongken 58S[J]. Journal of Nanjing Agricultural University, 1992, 15(3): 7 - 12.

[99] ZHANG XIAOGUO, LIU YULE, KANG LIANGYI, et al. Construction of rice male sterility and its fertility restoration expression vector[J]. Acta Agronomica Sinica, 1998, 24(5): 629 - 634.

[100] ZHANG ZHONGTING, LI SONGTAO, WANG BIN. Application of RAPD in study on rice thermo-sensitive genic male sterility[J]. Journal of Genetics and Genomics, 1994, 21(5): 373 - 376.

[101] ZHANG ZIGUO, ZENG HANLAI, YANG JING, et al. Study on photoperiod and temperature stability of fertility conversion of rice photo-thermo-sensitive genic male sterile lines[J]. Hybrid Rice, 1994(1): 4 - 8.

[102] ZHAO SHIMIN, LIU ZUOCHANG, ZHAN QINGCAI, et al. Analysis and study on translation products of cytoplasmic genome of WA, BT and D type rice male sterile lines[J]. Journal of Genetics and Genomics, 1994, 21(5): 393 - 397.

[103] Chinese Academy of Agricultural Sciences, Hunan Academy of Agricultural Sciences. Development of hybrid rice in China[M]. Beijing: China Agriculture Press, 1991.

[104] ZHOU TINGBO, CHEN YOUPING, LI DUANYANG, et al. Comparative observation on photoperiod and temperature induction of rice forward and reverse photo-thermo-sensitive sterile lines[J]. Hunan Agricultural Sciences, 1992 (5): 6 - 8.

[105] ZHOU TINGBO, XIAO HENGCHUN, LI DUANYANG, et al. Breeding of indica type photosensitive sterile line 87N123[J]. Hunan Agricultural Sciences, 1988 (6): 17-18.

[106] ZHOU TINGBO. Hypothesis of photoperiod and temperature promoter of rice male sterility inheritance[J]. Hereditas (Beijing), 1998, 20 (Supplement): 143.

[107] ZHU YINGGUO. Study on male sterile lines of different cytoplasmic types of rice[J]. Acta Agronomica Sinica, 1979, 5(4): 29-38.

[108] ZHOU KAIDA, LI HANYUN, LI RENDUAN. Breeding and utilization of D-type hybrid rice[J]. Hybrid Rice, 1987(1): 11-16.

[109] HUANG SHENGDONG, LI YUSHENG, YANG JUAN. Mapping of fertility genes of *japonica* type shortday-length sterile new germplasm 5021S[C]//Proceedings of the first China hybrid rice conference. Changsha: Hybrid Rice Editorial Office, 2010: 268-272.

[110] ZENG HANLAI, ZHANG ZIGUO, YUAN SHENGCHAO, et al. Study on fertility conversion conditions of low thermo-sensitive rice sterile line IVA[J]. Journal of Huazhong Agricultural University, 1992, 11(2): 101-105.

[111] LI XUNZHEN, CHEN LIANGBI, ZHOU TINGBO. Preliminary identification of fertility of new low temperature sterile rice (N-10S, N-13S)[J]. Journal of Natural Science of Hunan Normal University, 1991, 14(4): 376-378.

[112] YANG ZHENYU, ZHANG GUOLIANG, ZHANG CONGHE, et al. Breeding of medium indica high quality photo-thermo-sensitive genic male sterile line Guangzhan 63S[J]. Hybrid Rice, 2002(4): 8-10.

[113] DENG QIYUN. Breeding of eurytopic rice photo-thermo-sensitive sterile line Y58S[J]. Hybrid Rice, 2005, 20(2): 18-21.

[114] GUO HUI, LI SHUXING, XIANG GUANLUN, et al. Comparative study on cell biology of pollen abortion of different cytoplasmic male sterile lines of rice[J]. Seed, 2012, 31(5): 30-33.

[115] HU LIFANG, SU LIANSHUI, ZHU CHANGLAN, et al. Genetic and cytological analysis on radiation-induced rice male sterile mutant *tda* [J]. Journal of Nuclear Agricultural Sciences, 2015, 29(12): 2253-2258.

[116] YANG LIPING. Study on cytological characteristics of rice photo-thermo-sensitive genic male sterile lines-comparative study on Jiyu *japonica* and sterile line D18S[D]. Yanji: Yanbian University, 2003.

[117] PENG MIAOMIAO, DU LEI, CHEN FAJU, et al. Genetic analysis and cytological study on rice male sterile mutant TP79[J]. Chinese Journal of Tropical Crops, 2012, 33(1): 59-62.

[118] CHEN ZHONGZHENG, LIU XIANGDONG, CHEN ZHIQIANG, et al. Cytological study on spacial mutated male sterile new germplasm of rice[J]. Chinese Journal of Rice Science, 2002, 16(3): 199-205.

[119] HUANG XINGGUO, WANG GUANGYONG, YU JINHONG, et al. Cytoplasmic genetic effects and cytological study on rice homonuclear heterogeneous male sterile lines[J]. Chinese Journal of Rice Science, 2011, 25(4): 370-380.

[120] ZHOU HANTAO, ZHU YINGGUO. *In vitro* thermal analysis on mitochondria of Maxie cytoplasmic male sterile line and its maintainer line of rice[J]. Journal of Xiamen University, 1998, 37(5): 757-762.

[121] WEI LEI, DING YI, LIU YI, et al. Microcalorimetric analysis on anther of rice male sterile lines[J]. Journal of Wuhan Botanical Research, 2002 (4): 308-310.

[122] ZHOU PEIJIANG, LING XINGYUAN, ZHOU HANTAO, et al. Thermodynamic and kinetic characteristics of mitochondrial energy release of cytoplasmic male sterile rice[J]. Acta Agronomica Sinica, 2000(6):

818 – 824.

[123] ZHOU PEIJIANG, ZHOU HANTAO, LIU YI, et al. Characteristics of mitochondrial energy release of Maxie cytoplasmic male sterile rice[J]. Journal of Wuhan University (Natural Science Edition), 2000 (2): 222 – 226.

[124] WEN LI, LIU GAI, WANG KUN, et al. Preliminary comparative analysis on total protein in HL-type rice cytoplasmic male sterile pollen[J]. Journal of Wuhan Botanical Research, 2007 (2): 112 – 117.

[125] WEN LI, LIU GAI, WANG KUN, et al. Analysis on differential expression of pollen protein at mononuclear phase in HL type cytoplasmic male sterile rice[J]. Chinese Journal of Rice Science, 2012, 26(5): 529 – 536.

[126] CHEN ZHEN. Differential analysis on young spike fertility-related proteins of long and short photosensitive male sterile rice[D]. Wuhan: Huazhong Agricultural University, 2010.

[127] ZHOU HANTAO, ZHENG WENZHU, MEI QIMING, et al. Analysis on isozyme during microspore development of rice cytoplasmic male sterile lines[J]. Journal of Xiamen University (Natural Science Edition), 2000(5): 676 – 681.

[128] CHANG XUN, ZHANG ZAIJUN, LI YANGSHENG, et al. Comparison of tissue transglutaminase activity during young spike development of rice HL-type cytoplasmic male sterility[J]. Chinese Journal of Rice Science, 2006 (2): 183 – 188.

[129] DU SHIYUN, WANG DEZHENG, WU SHUANG, et al. Changes in antioxidase activity in anthers and leaves of three types of male sterile rice[J]. Plant Physiology Journal, 2012, 48(12): 1179 – 1186.

[130] LI DEHONG, LUO BINGSHAN, QU YINGLAN. Ethylene production and fertility conversion of young panicles of photosensitive genic male sterile rice[J]. Plant Physiology Journal, 1996, 22(3): 320 – 326.

[131] ZHANG ZHANFANG, LI RUI, ZHONG TIANTING, et al. Differences in response to exogenous jasmonic acid and endogenous jasmonic acid synthesis between rice cytoplasmic male sterile line and maintainer line [J]. Journal of Nanjing Agricultural University, 2014, 37(6):7 – 12.

[132] CHEN ZHANGLIANG, QU LIJIA. Regulation of gene expression in higher plants[J]. Chinese Bulletin of Botany, 1991, 33(5): 390 – 405.

[133] YU QING, XIAO YIHUA, LIU WENFANG. Preliminary study on role of Ca^{2+}-CaM system in HPGMR fertility conversion[J]. Journal of Wuhan University (Natural Science Edition), 1992 (1): 123 – 126.

[134] LI HESHENG, WU SUHUI, MA PINGFU. Relationship between fertility and ^{45}Ca in photosensitive genic male sterile rice (Briefing)[J]. Plant Physiology Communications, 1998, 34(3): 188 – 190.

[135] WU WENHUA, ZHANG FANGDONG, LI HESHENG. Effect of light length and quality on activity of Ca^{2+}-ATP in Nongken 58S chloroplast[J]. Journal of Huazhong Agricultural University, 1993, 12(4): 303 – 306.

[136] XIA KUAIFEI, WANG YAQIN, YE XIULIN, et al. Changes in Ca^{2+} distribution during tapetum development of rice cytoplasmic male sterile line Zhenshan 97A and its maintainer line Zhenshan 97B[J]. Acta Botanica Yunnanica, 2005 (4): 413 – 418.

[137] HU CHAOFENG. Detection of Ca^{2+} dynamic changes during microspore development of HL-type cytoplasmic male sterile rice [A]//Abstracts of 2006 academic annual conference and academic symposium. Genetics Society of Hubei Province, Genetics Society of Jiangxi Province, 2006: 1.

[138] ZHANG ZAIJUN. Analysis on Ca^{2+}-ATP enzyme activity during microspore development of HL-type cytoplasmic male sterile rice[C]//Compilation of papers on "Symposium on HL-Type Hybrid Rice" held at

2007 China Association for Science and Technology Annual Conference. China Association for Science and Technology, Hubei Provincial People's Government, 2007: 12.

[139] XIA KUAIFEI, LIANG CHENGYE, YE XIULIN, et al. Changes in calcium during anther development of thermo-sensitive male sterile rice Peiai 64S (English) [J]. Journal of Tropical and Subtropical Botany, 2009, 17(03): 211-217.

[140] OUYANG JIE, ZHANG MINGYONG, XIA KUAIFEI. Changes in Ca^{2+} distribution during anther development of rice non-pollen cytoplasmic male sterile line and its maintainer line (in English) [J]. Plant Science Journal, 2011, 29(01): 109-117.

[141] REN JUANSHENG, LI SHIGUI, XIAO PEICUN, et al. A cytoplasmic male sterility type caused by mutation of rice mitochondrial gene *atp6* promoter[J]. Southwest China Journal of Agricultural Sciences, 2009, 22(03): 544-549.

[142] LIU YAOGUANG. Molecular basis of rice cytoplasmic male sterility and its restoration[C]//Proceedings of 2005 international symposium on plant molecular breeding. China Association of Agricultural Science Societies, Guangxi Academy of Agricultural Sciences, Sichuan Academy of Agricultural Sciences, Hainan Institute of Tropical Agricultural Resources Development and Utilization, 2005: 1.

[143] CHEN LETIAN, LIU YAOGUANG. Discovery, utilization and molecular mechanism of rice wild abortive cytoplasmic male sterility[J]. Chinese Science Bulletin, 2016, 61(35): 3804-3812.

[144] ZHANG WENHUI, YAN WEI, CHEN ZHUFENG, et al. Genetic analysis and gene mapping of rice male sterile mutant *oss125* [J]. Scientia Agricultura Sinica, 2015, 48(4): 621-629.

[145] LI JUN, LIANG CHUNYANG, YANG JILIANG, et al. Cloning of rice gene *APRT* and its relationship with thermo-sensitive genic male sterility (English)[J]. Acta Botanica Sinica, 2003(11): 1319-1328.

[146] LIU HAISHENG, CHU HUANGWEI, LI HUI, et al. Genetic and localization analysis on rice male sterile mutant *OsMS-L* [J]. Chinese Science Bulletin, 2005(1): 38-41.

[147] JIANG HUA, YANG ZHONGNAN, GAO JUFANG. Genetic and mapping analysis on rice male sterile mutant *OsMS121* [J]. Journal of Shanghai Normal University (Natural Science Edition), 2006 (6):71-75.

[148] CHEN LIANG. Genetic and mapping analysis on rice male sterile mutants *Osms2* and *Osms3* [D]. Xiamen: Xiamen University, 2006.

[149] ZHOU YUANFEI. Study on inheritance and gene mapping of rice photo-thermo-sensitive genic male sterility [D]. Hangzhou: Zhejiang University, 2007.

[150] ZHANG HONG. Morphological observation and gene mapping of rice male sterile mutant *Osms7* [D]. Shanghai: Shanghai Jiaotong University, 2007.

[151] CAI ZHIJUN, YAO HAIGEN, YAO JIAN, et al. Genetic relationship analysis and mapping of rice non-pollen type male sterile genes[J]. Molecular Plant Breeding, 2008(5): 837-842.

[152] ZUO LING. Morphological characteristics and genetic mapping of a rice male sterile mutant *XS1* [D]. Ya'an: Sichuan Agricultural University, 2008.

[153] CHU MINGGUANG, LI SHUANGCHENG, WANG SHIQUAN, et al. Genetic analysis and gene mapping of a rice male sterile mutant[J]. Acta Agronomica Sinica, 2009, 35(6): 1151-1155.

[154] MA DONG. Mapping of rice short-photosensitive male sterile genes with SSR molecular markers[D]. Wuhan: Huazhong Agricultural University, 2010.

[155] YAN WENYI. Physiological characters and gene mapping of rice senescence and male sterile mutants[D]. Shanghai: Shanghai Normal University, 2010.

[156] ZOU DANNI. Fine mapping of thermo-sensitive genic male sterile gene *tms7* from wild planted distant hybridization of rice[D]. Haikou: Hainan University, 2011.

[157] SUN XIAOQIU, FU LEI, WANG BING, et al. Genetic analysis and gene mapping of rice male sterile mutant *802A* [J]. Scientia Agricultura Sinica, 2011, 44(13): 2633-2640.

[158] TAI DEWEI, YI CHENGXIN, HUANG XIANBO, et al. Study on fertility inheritance of new rice male sterile material SC316[J]. Journal of Nuclear Agricultural Sciences, 2011, 25(3): 416-420.

[159] HU HAILIAN. Genetic mapping of sterile gene of reverse thermo-sensitive genic male sterile line Tb7S from wild planted hybridization of rice[D]. Haikou: Hainan University, 2011.

[160] XIAO RENPENG, ZHOU CHANGHAI, ZHOU RUIYANG. Abortion characteristics and genetic analysis of rice 60Co-γ radiation induced male sterile mutant[J]. Crops, 2012 (4): 75-78, 163.

[161] XU JIEMENG. Identification and molecular mapping of reverse thermo-sensitive genic male sterile lines from wild planted distant hybridization of rice[D]. Haikou: Hainan University, 2012.

[162] QI YONGBIN. Mapping of rice thermo-sensitive male sterile gene tms9-1 and effect of monosaccharide transporter gene on fertility and grain filling[D]. Hangzhou: Zhejiang University, 2014.

[163] CHEN ZHUFENG, YAN WEI, WANG NA, et al. Cloning of rice male sterile genes with improved MutMap method[J]. Hereditas (Beijing), 2014, 36(1): 85-93.

[164] HONG JUN. Cytological study and sterile gene mapping of a rice male sterile mutant[D]. Nanjing: Nanjing Agricultural University, 2014.

[165] OUYANG JIE, WANG CHUTAO, ZHU ZICHAO, et al. Genetic analysis and gene mapping of rice male sterile mutant 012S-3[J]. Molecular Plant Breeding, 2015, 13(6): 1201-1206.

[166] ZHU BAIYANG. Genetic analysis and gene mapping of rice male sterile mutant D60 and yellow-green leaf mutant 5043ys[D]. Ya'an: Sichuan Agricultural University, 2015.

[167] SHI XIAO. Study on gene cloning and sterility mechanism of rice male sterile mutant gsl5[D]. Beijing: Chinese Academy of Agricultural Sciences, 2015.

[168] ZHANG WENHUI, YAN WEI, CHEN ZHUFENG, et al. Genetic analysis and gene mapping of rice male sterile mutant *oss125* [J]. Scientia Agricultura Sinica, 2015, 48(4): 621-629.

[169] YANG ZHENGFU. Map-based cloning of two rice male sterile genes[D]. Beijing: Chinese Academy of Agricultural Sciences, 2016.

[170] YANG ZHEN. Genetic and mapping analysis on rice male sterile mutants[D]. Yinchuan: Ningxia University, 2016.

[171] FENG MENGSHI. Map-based cloning and functional study on rice male sterile gene *MIL3* [D]. Yangzhou: Yangzhou University, 2016.

[172] JIAO RENJUN, ZHU BAIYANG, ZHONG PING, et al. Genetic analysis and fine mapping of fertility genes of rice male sterile mutant *D63* [J]. Journal of Plant Genetic Resources, 2016, 17(3): 529-535.

[173] MIN HENGQI. Preliminary mapping of rice dominant male sterile gene *OsDMS-2* [D]. Chongqing: Southwest University, 2016.

[174] XUAN DANDAN, SUN LIANPING, ZHANG PEIPEI, et al. Identification and gene mapping of rice non-pollen type genic male sterile mutant *whf41* [J]. Chinese Journal of Rice Science, 2017, 31(3): 247-256.

[175] FAN YOURONG, CAO XIAOFENG, ZHANG QIFA. Research progress in photo-thermo-sensitive male sterile rice[J]. Chinese Science Bulletin, 2016, 61(35): 3822-3832.

[176] MA XIQING, FANG CAICHEN, DENG LIANWU, et al. Research progress and discussion on breeding application of rice recessive genic male sterile gene[J]. Chinese Journal of Rice Science, 2012, 26(5): 511 - 520.

[177] IWASHI M, KYOZUKA J, SHIMAMOTO K. Processing followed by complete editing of an altered mitochondrial atp6 RNA restores fertility of cytoplasmic male sterile rice[J]. Embo J, 1993, 12 (4): 1437 - 1446.

[178] KADOWAKI K, ISHIGE T, SUZUKI S, et al. Differences in the characteristics of mitochondrial DNA between normal and male sterile cytoplasms of *japonica* rice[J]. Jpn J Breed, 1986, 36: 333 - 339.

[179] KADOWAKI K, et al. Differential organization of mitochondrial genes in rice with normal and male-sterile cytoplasms[J]. Jpn J Breed, 1989, 30: 179 - 186.

[180] KADOWAKI K, SUZUKI T, KAZAMA S. A chimeric gene containing the 5'portion of atp 6 is associated with cytoplasmic male sterility of rice[J]. Mol Gen Genet, 1990, 224 (1): 10 - 15.

[181] KALEIKAU E K, ANDRÉ C P, WALBOT V. et al. Structure and expression of the rice mitochondrial apocytochromb gene (*cob1*) and pseudogene (*cob2*)[J]. Curr Genet, 1992, 22: 463 - 470.

[182] KATO H, MURUYAMA K, ARAKI H. Temperature response and inheritance of a thermo-sensitive genic male sterility in rice[J]. Jpn J Breed, 1990, 40 (Suppl. 1): 352 - 369.

[183] LANG N T, SUBUDHI P K, VIRMANI S S, et al. Development of PCR-based markers for thermo-sensitive genetic male sterility gene *tms3* (*t*) in rice (*Oryza sativa L.*)[J]. Hereditas, 1999, 131 (2): 121 - 127.

[184] LIU ZUOCHANG. Hybrid rice[M]. Manila: International Rice Research Institute, 1988: 84.

[185] MIGNOUNA H, et al. Mitochondrial DNA modifications associated with cytoplasmic male sterility in rice [J]. TAG, 1987, 74: 666.

[186] NAWA S, SANO Y, YAMADA MA, et al. Cloning of the plasmids in cytoplasmic male sterile rice and changes of organization of mitochondrial and nuclear DNA in cytoplasmic reversion[J]. Jpn J Genet, 1987, 62: 301.

[187] OARD J H, HU J, RUTGER J N. Genetic analysis of male sterility in rice mutants with environmentally influenced levels of fertility[J]. Euphytica, 1991, 55 (2): 179 - 186.

[188] RUTGER J N, SCHAEFFER G W. An environmental sensitive genetic male sterile mutant in rice [C]. Proceedings twenty-third rice technical working group in USA, 1990: 25.

[189] SALEH N M, et al. Small mitochondrial DNA molecules of wild abortive cytoplasm in rice are not necessarily associated with CMS[J]. TAG, 1989, 77: 617.

[190] SHIKANAI T, YAMADO Y. Properties of the circular plasmid-like DNA B1 from mitochondria of cytoplasmic male-sterile rice[J]. Gurr Genet, 1988, 13 (5): 441 - 443.

[191] YAMAGUCHI M, et al. Electrophoretic analysis of mitochondrial DNA from normal and male sterile cytoplasms in rice[J]. Jpn. J Genet, 1983, 58: 607 - 611.

[192] YOUNG J, VIRMANI S S, KHUSH G S. Cyto-gènic relationship among cytoplasmic-genetic male sterile, maintainer and restorer lines of rice[J]. Philip J Crop Sci, 1983, 8: 119 - 124.

[193] ZHANG Q F, SHEN B Z, DAI X K, et al. Using bulked extremes and recessive class to map genes for photoperiod-sensitive genic male sterility in rice[J]. Proc Natl Acad Sci USA, 1994, 91 (18): 8675 - 8683.

[194] ZHANG Z G, ZENG H L, YANG J. Identifying and evaluating photoperiod sensitive genic male sterile (PGMS) lines in China[J]. IRRN, 1993, 18 (4): 7 - 9.

[195] TIAN H Q, KUANG A, MUSGRAVE ME, et al. Calcium distribution in fertile and sterile anthers of a pho-

toperiod-sensitive genic male-sterile rice[J]. Planta, 1998, 204 (2): 183-192.

[196] WANG Z, ZOU Y, LI X, et al. Cytoplasmic male sterility of rice with Boro Ⅱ cytoplasm is caused by a cytotoxic peptide and is restored by two related PPR motif genes via distinct modes of mRNA silencing[J]. Plant Cell, 2006, 18: 676-687.

[197] MA H. Molecular genetic analysis of microsporogenesis and microgametogenesis in flowering plants[J]. Annual Review of Plant Biology, 2005, 56: 393-434.

[198] GLOVER J, GRELON M, CRAIG S. Cloning and characterization of *MS5* from Arabidopsis, a gene critical in male meiosis[J]. The Plant Journal, 1988, 15: 345-356.

[199] NONOMURA K I, MIYOSHI K, EIGUCHI M, et al. The *MSP1* gene is necessary to restrict the number of cells entering into male and female sporogenesis and to initiate anther wall formation in rice[J]. Plant Cell, 2003, 15 (8): 1728-1739.

[200] JUNG K H, HAN M J, LEE D Y, et al. Wax-deficient anther is involved in cuticle and wax production in rice anther walls and is required for pollen development[J]. Plant Cell, 2006, 18 (11): 3015-3032.

[201] LI H, PINOT F, SAUVEPLANE V, et al. Cytochrome P450 family member CYP704B2 catalyzes the ω-hydroxylation of fatty acids and is required for anther cutin biosynthesis and pollen exine formation in rice[J]. Plant Cell, 2010, 22 (1): 173-190.

[202] ZHANG H, LIANG W Q, YANG X J, et al. Carbon starved anther encodes a MYB domain protein that regulates sugar partitioning required for rice pollen development[J]. Plant Cell, 2010, 22: 672-689.

[203] ZHOU H, LIU Q J, LI J, et al. Photoperiod and thermo-sensitive genic male sterility in rice are caused by a point mutation in a novel non-coding RNA that produces a small RNA[J]. Cell Research, 2012, 22: 649-660.

[204] ZHOU H, ZHOU M, YANG Y Z, et al. RNase ZS1 processes UbL40 mRNAs and controls thermo-sensitive genic male sterility in rice[J]. Nature Communications, 2014, 5: 4884.

[205] DING JH, LU Q, OUYANG Y D, et al. Along non-coding RNA regulates photoperiod-sensitive male sterility, an essential component of hybrid rice[J]. PNAS, 2014, 109 (7): 2654-2659.

[206] ZHOU H, HE M, LI J, et al. Development of commercial thermo-sensitive genic male sterile rice accelerates hybrid rice breeding using the CRISPR/Cas9-mediated *TMS5* editing system[J]. Scientific Reports, 2016, 6: 373.

Chapter 3 Theory and Strategy of Super High-yielding Breeding of Hybrid Rice

Yuan Longping / He Qiang

Section 1 High Yield Potential of Rice

Ⅰ. Theoretical Yield Potential of Rice

The theoretical yield potential of rice refers to the potential of a rice population per unit area of land to convert solar energy into chemical energy stored in carbohydrates to produce rice yield under very ideal ecological conditions (including optimal ecological environment and cultivation conditions without interference from drought, salinity, barren adversity, or disease and pest stresses) during its growth period. Different scholars have used different methods to estimate the yield potential of rice. The estimation by scholars such as Xue Derong (1977), Li Mingqi (1980), Liu Zhenye (1984), Zhang Zhengxian (1992) and Murata Yoshino (1975) is based on the total solar radiation energy during the rice growing period, the percentage of effective physiological radiation used for photosynthesis, the share of losses due to reflection, light leakage, transmission, light saturation and respiration consumption of the rice population, the rate of conversion of photosynthetic energy, the energy contained in photosynthetic products, as well as economic coefficients.

Rice accumulates 90%–95% of its dry matter during its lifetime from photosynthetic products produced by its own photosynthesis. Since the 1960s, many scholars at home and abroad have estimated the light energy utilization rate of rice and the highest theoretical yield potential of rice. The light energy utilization rate is the chemical energy stored through photosynthesis as a percentage of the total light energy input. The essence of increasing the yield per unit area of rice is to improve the light utilization efficiency of the rice population, which involves the solar radiation energy received and the biomass produced by the rice population per unit area during the growth period. Most scholars estimate that the light energy utilization rate can theoretically reach 5%, but at present, the actual rate of high-yielding rice is generally 1%–3%, while that of low-yielding rice is only about 0.5% (Chen Wenfu et al., 2007).

1. Estimation of theoretical yield potential of crops by foreign scholars

The theoretical biological yield of crops at different latitudes was estimated by Hechiborovitz, a plant physiologist of the former Soviet Union, with a light energy utilization rate of 5% (Yuan, 2002) (Table 3-1).

Table 3-1 Theoretical biological yield at different geographical latitudes at 5% light energy utilization

Latitude	Total Radiation Energy (100 million kJ/ha)	Theoretical Biological Yield (t/ha, absolute dry weight)
60°–70°	83.68–41.84	25–12
50°–60°	146.44–83.68	45–25
40°–50°	209.20–146.44	70–40
30°–40°	251.04–188.28	75–55

Continued

Latitude	Total Radiation Energy (100 million kJ/ha)	Theoretical Biological Yield (t/ha, absolute dry weight)
20°-30°	376.56 - 251.04	110 - 75
0°-20°	418.40 - 376.56	125 - 110

Japanese scholars estimated the yield potential of Japanese rice based on the average daily solar radiation from August to September in Japan and the solar radiation from 10 days before to 30 days after heading, which plays a key role in the formation of rice grain yield, and the results showed that the maximum yield potential of brown rice in Japan could reach 24.02 t/ha (Murata Yoshino, 1975).

2. Estimation of theoretical yield potential of rice by Chinese scholars

Chinese researchers have estimated the maximum theoretical yield potential of rice from different perspectives based on relevant parameters such as light energy resources and light energy utilization of rice in various rice cropping areas across the country.

(1) Estimation of maximum theoretical yield potential on the basis of solar radiation energy in whole growth period of rice

Chinese scholars estimated the highest theoretical yield potential of rice light energy utilization during the whole growth period of rice, in different ecotypes of rice cropping areas in Northeast, North, Central, South and Southwest China, as shown in Table 3-2.

Table 3-2 Estimation of the highest theoretical yield potential of rice in the whole growth period in China

		Light energy utilization rate (%)	Highest theoretical yield potential (kg/ha)
Northeast	Single-cropping *japonica*	3.41	28,215
North	Single-cropping *japonica*	5.00	36,630
Central	Early-season rice	2.50	15,375
	Late-season rice	2.50	17,250
	One Mid-season rice	3.80 - 5.50	22,178 - 36,960
South	Early-season rice	4.89	16,335
	Late-season rice	4.89	23,685
	Mid-season rice	4.89	17,115
Southwest	One Mid-season rice	4.90	23,340

Note. Data in the table are quoted from Zhang Xianzheng (1992), Yuan Longping (2002), Qi Changhan (1985), Xue Derong (1977) and Liu Zhenye (1984).

The theoretical yield potential estimated by light energy utilization during the whole growth period of rice was the highest for single-cropping *japonica* in the North China rice region, followed by the single-cropping *japonica* in the Northeast China rice region.

(2) Estimation of maximum theoretical yield potential of rice by solar radiation energy during rice grain formation

The maximum theoretical yield potential of different ecotypes of rice was estimated using the light energy utilization during the rice grain formation period in the southern and northern regions of China, as shown in Table 3 - 3.

Table 3 - 3 Estimation of maximum theoretical yield potential of rice during rice grain formation in the northern and southern rice regions of China

		Light energy utilization rate (%)	Maximum theoretical production potential (kg/ha)	Yield formation period
Northern	Single-cropping *japonica*	5.2	22,500 - 26,250	From 10 days before to 30 days after heading (Lu Qiyao, 1980)
Southern	Double-cropping early-season rice		20,625 - 24,375	
	Double-cropping late-season rice		15,000 - 19,875	
	Single-cropping mid-season rice		20,625 - 26,250	
	Single-cropping late-season rice		15,000 - 18,750	

The significant difference in the theoretical yield potential of rice in different ecological regions and planting seasons are due to the different solar radiation received during the yield formation period of rice. In the northern rice region, the yield formation period of single-cropping *japonica* rice is at the turn of summer and autumn when the climate is dry and sunny and solar radiation is high, and its theoretical yield potential is the highest among all rice regions. The yield formation period of double-cropping early-season rice and single-cropping mid-season rice in southern rice regions is in summer when solar radiation is the highest, while the yield formation period of double-cropping late-season rice and single-cropping late-season rice is in autumn when solar radiation decreases, resulting in the difference in the theoretical yield potential of rice in southern rice regions across different planting seasons. The above estimates of maximum theoretical yield potential of rice in different rice regions does not take into account the effect of temperature on rice photosynthesis, so it is called photosynthetic yield potential. For more realistic estimation, the effects of maximum and minimum temperatures on seed setting during rice grain formation should also be considered.

(3) Estimation of the maximum theoretical yield potential on the basis of climate ecological model of the different growth stages of rice

The maximum theoretical yield potential of different ecological models in each rice area estimated based on the light and temperature conditions and the contribution rate of photosynthesis in the growth period to rice yield in different rice areas across the country are shown in Table 3 - 4.

Table 3 - 4 Estimation of the maximum theoretical yield potential of ecological models in China's rice-growing regions

		Light energy utilization rate (%)	Maximum theoretical production potential (kg/ha)	Factors considered
Northeast	Single-cropping *japonica*	2.9 - 3.5	18,750 - 22,500	Light and temperature conditions, and the contribution rate of photosynthesis at different growth stages (Gao Liangzhi et al., 1984)
North	Single-cropping *japonica*	2.9 - 3.9	18,750 - 24,000	
Central	Double-cropping early-season rice	4.1 - 4.5	16,125 - 17,625	
	Double-cropping late-season rice	4.1 - 4.5	15,375 - 16,875	
	Single-cropping rice	2.7 - 3.2	19,500 - 21,000	
South	Double-cropping early-season rice	3.7 - 4.1	16,875 - 18,375	
	Double-cropping late-season rice	3.7 - 4.1	16,875 - 18,375	
	Single-cropping rice	3.5 - 3.7	19,500 - 21,000	
Southwest	Single-cropping rice	2.9 - 3.7	16,125 - 21,000	

The theoretical yield potential of single-cropping rice is the highest in the North, due to strong solar radiation, low average temperature, high light energy utilization, and long effective grain-filling period. The middle and lower reaches of the Yangtze River and southern China are next because of the long growing season and strong radiation, and the relatively low light energy utilization efficiency of single-cropping rice. The rice growing area of Southwestern China's Yunnan Plateau has a higher theoretical yield potential due to the high radiation intensity, big temperature difference between day and night, low average temperature, and high light energy utilization; while the low theoretical yield potential of the Sichuan-Guizhou rice growing area is due to the low solar radiation caused by more cloudy and foggy days, coupled with high temperature and humidity during the rice growth period, and small temperature difference between day and night which is not conducive to the high yield potential of rice.

Ⅱ. Actual Yield Potential of Rice

The theoretical yield potential of rice is estimated under the premise that all external conditions affecting rice growth are idealized, while the realistic yield potential of rice refers to the actual level of rice output, i. e., the actual rice yield, that has been achieved or attained in specific ecological regions and farming conditions within a certain period of time, including the realistic yield level (average rice yield) that has been achieved in a specific ecological region and the highest record yield in a given ecological zone (Chen Wenfu et al., 2007).

Japanese scholars have estimated that the maximum yield of dry-season rice can reach 15.9 t/ha in tropical areas, and up to 18.0 t/ha in temperate areas (Yoshida Shoichi, 1980). According to relevant data, the world record of rice yield of 21 t/ha was set in Madagascar, Africa, in 1999 (Yuan Longping,

2002), and a super-high yield of 17. 8 t/ha was achieved in India (Suetsugu, 1975).

In a certain ecological region, the continuous exploitation of the extreme yield potential of rice through breeding and cultivation is a technical reserve, a possibility and a path to explore the continuous improvement of rice yield in large areas. So far in China, different rice cropping areas, different rice cropping systems in the same rice cropping area, and different rice types, all have their respective high yield records (Table 3 - 5). The date in the below table shows that the rice yield per unit of small area or contiguous area of 6. 67-ha has reached or even exceeded the realistic productivity of rice estimated by relevant scholars, and the rice demonstration yields of some rice-growing areas is close to the highest theoretical production potential of rice.

Table 3 - 5 Records of the highest yield per unit area in different periods in China

Rice growing area		Year	Demo area (m^2)	Yield (kg/ha)	Variety	Rice type
Northeast	Jilin	2010	11,333	12,741	Dongdao 1	Single-cropping *japonica*
	Heilongjiang	2007	High-yielding plots	12. 600		Single-cropping *japonica*
North	Shandong	2016	66,667	15,207	Chaoyouqian	Single-cropping mid-season *indica*
	Hebei	2016	66,667	16,232	Chaoyouqian	Single-cropping mid-season *indica*
		2017	68,000	17,235	Chaoyouqian	Single-cropping mid-season *indica*
East	Fujian	2004	66,667	13,925	Ⅱ Youhang 1	Single-cropping mid-season *indica*
	Jiangsu	2009	High-yielding plots	14,058	Yongyou 8	*Indica-japonica* hybrid
	Zhejiang	2004	High-yielding plots	12,282	Zhongzheyou 1	Single-cropping mid-season *indica*
		2007	High-yielding plots	11,021	Zhongzao 22	Double-cropping early-season *indica*
		2015	70,000	15,233	Chunyou 927	*Indica-japonica* hybrid
		2016	78,000	15,362	Yongyou 12	*Indica-japonica* hybrid
Central	Hunan	2004	66,667	12,149	88S/0293	Single-cropping mid-season *indica*
		2009	70,667	12,540	D Liangyou 15	Single-cropping late-season *indica*
		2010	66,667	13,080	Guangliangyou 1128	Single-cropping mid-season *indica*
		2011	72,000	13,899	Y Liangyou 2	Single-cropping mid-season *indica*
		2014	68,400	15,401	Y Liangyou 900	Single-cropping mid-season *indica*
	Hubei	2007	High-yielding plots	12,351	Luoyou 8	Single-cropping mid-season *indica*
	Henan	2007	66,667	12,891	Y Liangyou 1	Single-cropping mid-season *indica*

Continued

Rice growing area		Year	Demo area (m^2)	Yield (kg/ha)	Variety	Rice type
South	Guangdong	1990	High-yielding plots	12,863		Double-cropping late-season *indica*
		2016	72,000	12,482	Chaoyouqian	Double-cropping early-season *indica*
		2016	68,000	10,586	Chaoyouqian	Double-cropping late-season *indica*
	Guangxi	1991	994	12,375	Teyou 63	Double-cropping early-season *indica*
		1991	669	11,919	Teyou 63	Double-cropping late-season *indica*
		2016	66,667	21,723	Chaoyouqian	Single-cropping ratooning rice
Southwest	Yunnan	1983	967	16,131	Guichao 2	Single-cropping *indica*
		1987	667	17,011	Sixizhan	Single-cropping *indica*
		1999	5,493	17,081	Peiai 64S/E32	Single-cropping mid-season *indica*
		2001	747	17,948	Ⅱ Youming 86	Single-cropping mid-season *indica*
		2004	713	18,299	Ⅱ You 6	Single-cropping mid-season *indica*
		2005	767	18,450	Ⅱ You 28	Single-cropping mid-season *indica*
		2006	767	19,196	Ⅱ You 4886	Single-cropping mid-season *indica*
		2006	753	19,305	Xieyou 107	Single-cropping mid-season *indica*
		2015	68,000	16,013	Chaoyouqian	Single-cropping mid-season *indica*
		2016	67,334	16,320	Chaoyouqian	Single-cropping mid-season *indica*
	Sichuan	2015	71,667	15,708	Deyou 4727	Single-cropping mid-season *indica*
	Guizhou	2007	High-yielding hilly plots	15,662	Qianyou 88	Single-cropping mid-season *indica*
	Chongqing	2006	1,000	12,258		Single-cropping mid-season *indica*

Note. The data in the table are from the Internet.

The highest rice yield record in China was 19.3 t/ha obtained in 2006 in a small-area of high-yielding demonstration of Xieyou 107 planted as a single-cropping mid-season crop in Taoyuan township, Yongsheng county, Yunnan province. Taoyuan township has a typical low-latitude plateau south subtropical climate, located in a dry and hot-river valley, with many hours of light, high solar radiation, big temperature difference between day and night, no extreme high temperature during the whole rice growing season, low humidity, fertile and permeable soil, and low probability of pests and diseases, which makes the temperature and light conditions ideal for rice growth with a high photosynthetic rate of strong solar radiation and moderate temperature during the day and low respiration consumption at night. Thus, this ecological region has excellent ecological environment and unique natural conditions to fully exploit the highest theoretical yield potential of rice, and such conditions are conducive to the simultaneous increase of the number of panicles per unit area, the total number of grains per panicle and the seed setting rate, so that the three yield components, i. e. panicle number, grain number and grain weight, can be better coordinated. From the early 1980s to the beginning of the 21st century, the highest rice yield records in China and even in the world have been broken one after another. In 2015 and 2016, the average yield per unit area in the super high-yielding demonstration plot with single cropping mid-season Chaoyouqian over 6.67 hectares in Datun town, Gejiu city, Yunnan province, was 16.01 t/ha and 16.32 t/ha, respectively, which were the world records of rice yield in a large area of tropical rice area. Datun town is a low-latitude subtropical mountainous monsoon climate, with four seasons like spring, no extreme temperature and relatively sufficient rainfall during the rice growing period, with a large area of arable land, suitable for the establishment of large-scale super high-yielding demonstration to explore the extreme yield potential of rice.

The theoretical yield potential of the high-latitude one-season rice area in North China is high, and if suitable rice ecological zones, suitable varieties and suitable cultivation techniques are selected, the real productivity of rice in this area can be close to the theoretical yield potential. In 2016, the average yield per unit area of the super high-yielding demonstration of Chaoyouqian cultivated as single-cropping mid-season rice over 6.67 hectares in Yongnian district, Hebei province, reached 16.23 t/ha, which set the world record of rice yield in a large area at high latitudes. In 2017, the region continued to carry out the super high-yielding demonstration with the same variety, and the average yield exceeded 17.0 t/ha over 6.67 hectares, which again set a new record for rice yield in high latitudes.

At present, there is still a considerable gap between the average actual yields and the real productivity of large rice areas in China, and in some cases, the difference is nearly double (Table 3-6). Through the analysis of the relationship between the highest yield records of small areas and the real productivity in each rice crop area, this gap will become smaller and smaller by cultivating and selecting rice varieties that are suitable for each rice crop area in terms of plant type and adaptability to environmental changes, improving cultivation techniques of the corresponding varieties, and the production conditions such as soil and irrigation. Data show that in 2018, the realistic production level of rice yield in China was 6,562 kg/ha, which was 4,670 kg/ha higher than that in 1949, an increase of nearly 250%, and the estimated rice productivity capacity in that period is 7,735 kg/ha, a difference of only 1,173 kg/ha (Fang Fuping, 2009). The yield per unit area of rice can be increased from the current level of large-scale production to a level close to realistic productivity by cultivating better varieties with wide-adaptability, improving the soil of low- and medium-yield fields, improving cultivation techniques, and adjusting cropping system.

Table 3-6 Comparison of realistic yield and realistic productivity of rice in various rice regions of the country

		Realistic yield (kg/ha)	Realistic productivity (kg/ha)
Northeast	Single-cropping *japonica*	7,227.0	10,125-12,000
	Single-cropping *japonica*	7,053.0	10,125-12,750

Continued

		Realistic yield (kg/ha)	Realistic productivity (kg/ha)
North	Double-cropping early-season rice	5,571.0	8,625 - 9,375
	Double-cropping late-season rice	6,027.0	8,250 - 9,000
Central	Single-cropping rice	7,357.5	10,500 - 11,250
	Double-cropping early-season rice	5,386.5	9,000 - 9,750
	Double-cropping late-season rice	5,235.0	9,000 - 9,750
South	Single-cropping rice	5,811.0	10,500 - 11,250
Southwest	Single-cropping rice	6,667.5	10,500 - 11,250

Note. Realistic production levels in the table are the average yields of each rice area from 2005 to 2009; realistic productivity data are cited from research literature (Gao, Liangzhi et al., 1984).

From the breeding point of view, breeding plant types and root types that are adapted to the climatic conditions of each rice crop area and the physiological functions that are compatible with them, selecting and breeding panicle types and grain types that are compatible with the plant types and physiological functions, and selecting and breeding varieties that are adapted to extreme climate change, drought and flood tolerance, and infertile soil and pest resistance are the fundamental ways to improve the realistic productivity of rice.

Section 2 Concept and Goals of Super Rice

Ⅰ. Background of Researches on Super Rice

Since the mid-20th century, the world's major rice producing countries have made high yield the primary goal of rice breeding and tried to use various means to further increase rice yield. South Korea proposed *indica-japonica* cross breeding for increase in rice yield in the 1960s in pursuit of rice self-sufficiency. Japan launched a super yield breeding program in 1981 to increase rice yield by 50% within 15 years. The International Rice Research Institute (IRRI) launched the New Plant Type program to increase rice yield by more than 20%.

1. Super high yield rice breeding in South Korea

Until the 1960s, the goal of rice breeding in South Korea had been to select high-yielding *japonica* varieties that were resistant to blast, stripe virus disease and lodging, but the results had not been satisfactory. Later, with the assistance of IRRI, the Suweon series of semi-dwarf high-yielding varieties with multiple tillers, large panicles and erect leaves were developed over a period of five years. These varieties were named Tongil-type varieties and the yield was higher by 20%-30% compared with original Korean *japonica* varieties (Park et al., 1990), making the country self-sufficient in rice since the 1970s. The Tongil type varieties are triple crosses produced by crossing the intermediate materials of single-cross progenies of *indica* and *japonica* subspecies with IR8 from IRRI. However, this series of rice varieties have many problems such as long growth duration, poor adaptability to the climate conditions in high-latitude cold rice areas, poorer cold tolerance than *japonica* varieties, and poor rice quality not meeting consumer's demand. After the 1990s, the goal of South Korea in rice breeding shifted from pure ultra-high yield to both high yield and high quality, marking the cessation of the production of *indica-japonica* intersubspecific varieties represented by the Tongil type varieties. However, in its super high-yielding breeding program, the goal set for rice varieties intended for food processing was still high yield, and a number of super high-yielding varieties such as Suweon 431 and Milyang 160 were developed with the yield reaching 9 t/ha.

2. Super high-yielding breeding program in Japan

The super high-yielding rice breeding program of Japan was first proposed by Japanese scholars in 1981 as the Reverse 753 Program, which mainly focused on the breeding of rice varieties with high yield potential, and required the breeding of super high-yielding varieties with 10% , 30% , and 50% yield increase over the control varieties in three stages, respectively three, five, and seven years. This means that over the 15 years from 1981 to 1995, Japan would sought to increase the brown rice yield from 5. 0 – 6. 5 t/ha to 7. 5 – 9. 75 t/ha (equivalent to 9. 38 – 12. 19 t/ha of paddy rice) (Sato, 1984; Nakada, 1986). The strategy of the super high-yielding rice breeding program set different objectives for different stages. Based on the results of a previous stage, the objective of the next stage was to enrich the genetic base of Japanese *japonica* rice and expand the range of its genetic variation mainly by using foreign rice resources with elite traits and super high yield and conducting *indica-japonica* intersubspecific hybridization or geographically distant intervarietal hybridization. Within eight years after the implementation of the program, a number of varieties with super high-yielding potential, such as Star, Akihito, Oyu 326, Tatsumaki, Dali, and Sho were developed and the yield of these rice varieties in small trial areas has reached about 12 t/ha, basically achieving the goal of a 30% yield increase set for the second stage of the program. However, due to poor seed setting rate, and poor quality and adaptability, the varieties were not promoted into large-scale production (Xu Zhengjin et al. , 1990; Chen Wenfu et al. , 2007).

3. IRRI's New Plant Type (NPT) breeding program for super high-yielding rice

Since IRRI developed the new high-yielding semi-dwarf rice variety IR8 (referred to as "miracle rice" by the media) , with strong tillering, strong stems and a high harvest index, which marked the beginning of first green revolution lasting from 1966 till the end of the 1990s, a number of new rice varieties, such as IR24, IR36 and IR72, were released and widely planted in tropical rice ecological areas. These varieties were significantly improved in terms of resistance and daily yield, but the yield per unit area hovered at the level of 8 – 9 t/ha without any significant breakthrough. The breeders at IRRI believed that if a new and substantial breakthrough in rice yield is to be achieved, NPT rice must be developed. Therefore, in 1989, they proposed to develop "super rice", which was later named "New Plant Type Super High-Yield Breeding Program". The aim was to breed NPT rice different from the previous multi-tiller and semi-dwarf plant type and to make a significant breakthrough in yield. More specifically, the project intended to develop NTP rice suitable for cultivation in tropical rice ecological areas in 8 – 10 years with a yield potential 30% to 50% higher than that of dwarf varieties at that time. According to that the timeline set for the program, NPT rice varieties should be developed by 2000 with a yield of 12 t/ha. The design of NPT has been described in detail in terms of tillering, panicle shape, growth duration, plant height, stem thickness, harvest index, canopy leaf characteristics, grain size and grain density (Khush, 1990; Peng et al., 1995). After nearly five years of research, IRRI informed the world in 1994 of the success in research on the use of NPT and specific germplasm resources to select NPT of super high-yielding rice, which had significant yield potential. At that time, foreign news media reported that the new "super rice" would help feed nearly 500 million people, thus attracting great attention from the world's major rice producing countries. The term "super rice" has since replaced "NPT rice" or "super high-yielding rice breeding" and spread around the world, becoming a hot topic in the research work of rice breeders.

The breeding strategy of IRRI's NPT breeding program is to select tropical *japonica* rice (*javanica*) with weak tillering ability, hard stems, large panicles with more grains and cross them with *japonica* rice that does not have hybrid compatibility barriers, so as to develop new varieties with the target traits in the NPT design with high-yielding potential. The yield of this NPT rice with tropical *japonica* rice background reached 12.5 t/ha in small area trials, but no significant breakthrough in yield has been achieved in large-scale planting so far, not even with the second and third generations of NPT varieties improved with *indica-japonica* cross. The design and breeding strategy of the first generation of NPT rice have obvious defects. First, the strong heterosis of *indica-japonica* intersubspecific hybrids was intentionally avoided in the pursuit of compatibility, which is not conducive to the expansion of the genetic base of the varieties and causes homogeneity in the genetic background. Second, the design of fewer tillers results in serious waste of solar radiation energy during the growth period which is not conducive to the improvement of photosynthetic potential. Third, NPT rice breeding pursued solely a high harvest index and neglected the high biological yield which is the prerequisite of super high yield. Although the original NPT super rice breeding design was later revised to allow the introduction of favorable *indica* genes to improve grain filling, tillering and plant height, and the utilization of wild rice resources to make up for the defects, it still failed to achieve a substantial breakthrough. Regardless of the reasons, the following issues need to be considered for NPT rice at IRRI. Can varieties with less tillering and larger panicles perform well in the tropical rice ecological regions? Can varieties with heavy panicles perform normally in the tropical rice ecological regions in terms of seed setting? Can varieties with heavy panicles solve the lodging problem in the tropical rice ecological areas? (Yuan Longping, 2011)

Ⅱ. Super Rice Research in China

There is no clear and authoritative definition of super rice. As the name implies, super rice is a kind of rice that significantly exceeds the level of control varieties in terms of yield, quality, resistance and other major indicators. Some scholars have defined super hybrid rice as new hybrid rice varieties with good quality and strong resistance bred through a combination of plant type improvement and utilization of hybrid vigor, morphology and physiological functions, morphologically featured by moderate tillering, straight flag leaves, a moderate plant height, strong stems for better lodging resistance, and large panicles with many grains; delivering a yield increase of 15%–30% over common hybrid rice or conventional rice varieties (Deng Huafeng et al., 2009).

1. Proposal and goal of super hybrid rice in China

At the end of the 1980s, a Chinese scholar presented a paper entitled "New Trends in Super High-Yielding Rice Breeding-Combining Ideal Plant Type and the Utilization of Advanced Heterosis" at the International Rice Research Conference organized by IRRI, the Chinese Academy of Agricultural Sciences (CAAS) and China Rice Research Institute (CRRI). This showed that China started to explore and research the breeding of super high-yielding rice. During the Seventh and Eighth Five-Year Plan periods, China was facing the severe reality of continuous population growth and sharp decline in the total area of arable land. In response, to further improve the rice yield per unit area, the research on super rice breeding was included in the national key science and technology research projects. In 1996, the MOA held China Super Rice Research Seminar at Shenyang Agricultural University (SAU), and decided to start China Super Rice Research Program. In 1998, Academician Yuan Longping submitted a proposal to the State Council to launch a super hybrid rice breeding program, and the State Council decided to allocate the Premier's Fund to support it. In the same year, the MOA and the Ministry of Science and Technology (MOST) jointly launched the Super Hybrid Rice Breeding Program.

The New Century Agricultural Dawn Plan established by the MOA in 1996 initially defined the technical route, research contents, and demonstration and promotion plan of China Super Rice Research project, and set the first and second goals as achieving a yield of 10 t/ha by 2000 and 12 t/ha by 2005 in 6.67-ha continuous demonstration. The yield targets of super rice in different ecological regions are shown in Table 3-7. The table shows the absolute yield targets, while the relative target is an 8% or higher yield increase over the control variety in yield trials at all levels. In addition to this, the rice quality requirement was MOA's standard for Grade Ⅱ rice or above plus resistance to one or two major local pests and diseases. The third phase of the super rice breeding program aimed at a yield of 13.5 t/ha by 2015 on the basis of the second phase.

Table 3-7 Yield targets for different types and stages of super rice development

Year	Conventional rice (t/ha)				Hybrid rice (t/ha)			Increase (%)
	Early-season *indica*	Early-, mid- and late-season *indica*	Single-cropping South	Single-cropping North	Early-season *indica*	Single-cropping rice	Late-season rice	
Current	6.75	7.50	7.50	8.25	7.50	8.25	7.50	0
2000	9.00	9.75	9.75	10.50	9.75	10.50	9.75	15
2005	10.50	11.25	11.25	12.00	11.25	12.00	11.25	30

Note. The data in the table are the average yields on two 6.67-ha demonstration land plots within one ecological area in two consecutive years.

Source – Department of Science, Technology and Quality Standards, MOA, 1996.

In 1997, Yuan Longping made a comprehensive analysis of the yield targets of super high-yielding rice breeding proposed by various rice breeding research institutes at home and abroad, and concluded that the target yield of super rice should vary with the time, ecological region and planting season. He also proposed that in addition to the absolute yield targets, targets should also be set for daily yield per unit area, which is more reasonable because the growth duration is closely related to the yield. On the basis of the yield and breeding level of hybrid rice in China at that time, the target of super hybrid rice breeding in the Ninth Five-Year Plan period was set at 100 kg/ha/day.

In August 2006, the MOA General Office issued China Super Rice Research and Promotion Plan (2006—2010), which clearly stated that during the Eleventh Five-Year Plan period, China's super rice research and promotion should focus on the strategic objective of national food security, adhere to the principle of mainly promoting Phase I results, deepening Phase Ⅱ research and exploring ways towards Phase Ⅲ goals, and seek to accelerate the selection and breeding of new super rice varieties, aggregate favorable genes, innovate in breeding methods, strengthen the integration of cultivation techniques, and expand demonstration. It pursued high yield, high quality and wide adaptability with the support of good seeds, good cultivation methods, and the integration of research, demonstration and promotion. Specifically, the goals were to deliver 20 leading varieties of super rice by 2010, promote them to 30% of the total rice planting area in China (about eight million hectares), and achieve an average yield increase of 900 kg/ha, so as to significantly push up the rice yield nationwide and maintain China's leading position in the world in rice breeding. At the same time, the development targets for different ecological rice planting areas were proposed (Table 3 - 8).

Table 3 - 8 Yield, quality and resistance targets of Chinese super rice varieties

Ecological region		Early-season rice in the Yangtze River Basin	Early-season *japonica* in the Northeast; mid-maturing late-season rice in the Yangtze River Basin	Early-and late-season rice in the South and late-maturing rice in the Yangtze River Basin	Single-cropping rice in the Yangtze River Basin and mid-maturing *japonica* in the Northeast	Late-maturing single-cropping rice in the Yangtze River Basin and late-maturing *japonica* in the Northeast
Growth period (day)		102 - 112	110 - 120	121 - 130	135 - 155	156 - 170
Yield (t/ha)	Fertilizer Tolerant type	9.00	10.20	10.80	11.70	12.75
	Eurytopicity type	Yield increase of more than 8% above provincial yield trials, with a growth duration similar to the control variety				
Quality		Northern *japonica* rice reaches the MAO standard for Grade Ⅱ rice or higher, Southern late-season *indica* reaches the MAO standard for Grade Ⅲ rice or higher, and Southern early-season *indica* and single-cropping rice reaches the MAO standard for Grade Ⅳ rice or higher.				
Resistance		Resistant to one or two major local diseases				

During the 12th Five-Year Plan period, the MOA Super Rice Program proposed to develop 30 new super rice varieties by 2015, and to promote them to 10 million hectares of land nationwide, with an average yield increase of 750 kg and cost saving of RMB 1,500 per hectare (referred to as the "3151" Project). Its development should be in accordance with the idea of expanding the application of Phase I results, deepening the research and promotion of the Phase Ⅱ project, and striving to achieve the Phase Ⅲ goals. More specifically, for Phase-I super rice varieties with a yield of 10,500 kg/ha, wide adaptability and high quality should be pursued in addition to high yield, so demonstration and promotion should shift from high-yielding land plots to medium- and low-yielding ones to explore the yield potential of super rice on medium- and low-yielding land plots. For Phase Ⅱ super rice varieties with a yield of 1,200 kg/ha, more efforts should be made in variety selection and cultivation techniques so as to make them the leading varieties in China's rice production as soon as possible. Research on the Phase Ⅲ super rice varieties should be accelerated to achieve the goal of 13,500 kg/ha by the end of the 12th Five-Year Plan period. As the Phase Ⅲ goals were achieved ahead of time in 2011, the MOA launched the forth phase of the super rice program with a target yield of 15,000 kg/ha in the spring of 2014, aiming to achieve the goal by 2020.

2. Achievements of super rice research in China

Super rice, as one of the China's cutting-edge agricultural technology, including both three-line and two-line hybrids, are new hybrid rice varieties that combine super high yield, high quality and multiple resistances and has shown great vitality since their inauguration. In 2000, the pioneer combination Liangyoupeijiu, which achieved the Phase I goal, yielded more than 10. 5 t/ha on 16 6. 67-ha demonstration plots in Hunan. In 2004, the pioneer two-line super hybrid rice, 88S/0293, which achieved the Phase Ⅱ goal one year ahead of schedule, yielded more than 12. 2 t/ha in 6. 67-ha demonstration plots in three counties in Hunan for two consecutive years, and Y Liangyou 1 yielded 12. 9 t/ha in 6. 67-ha demonstration plots in Henan. In 2011, Yuan Longping's research team succeeded in breeding the super hybrid rice combination Y Liangyou 2, which delivered a yield of 13. 9 t/ha in continuous 6. 67-ha land plots in Longhui county, Hunan province, achieving the Phase Ⅲ goal of 13. 5 t/ha. In 2014, Yuan Longping's research team developed super hybrid rice Y Liangyou 900, which yielded 15. 4 t/ha in Xupu county, Hunan province, exceeding the 15. 0 t/ha target of Phase Ⅳ. In 2015 and 2016, the pioneer hybrid of Phase Ⅳ, Chaoyouqian, set the new world records of 16. 01 t/ha and 16. 32 t/ha, breaking the internationally recognized limit of rice yield in the tropics (15. 9 t/ha). In 2017, super hybrid rice Chaoyouqian was planted on a trial basis in Yongnian district, Handan city, Hebei province, and yielded 17. 24 t/ha on 6. 67-ha continuous land plots, setting a world record for large-area rice yield. Among the three-line super hybrid rice combinations, Zheyou 1 and Deyou 4727 yielded 12. 3 t/ha and 15. 7 t/ha on 6. 67-ha continuous land plots in Zhejiang and Sichuan provinces respectively.

According to statistics, the accumulated area of promotion for super rice in China exceeded 80 million hectares from 2003 to 2006, and its proportion in the total rice planting area increased year by year, from 9. 8% in 2003 to 30. 2% in 2014 (Table 3-9). From 2014 to the present, the annual planting area of super rice nationwide has been stable above 8. 7 million hectares. In addition to high yield, super hybrid rice such as Zhongzheyou 1, Ⅲ-You 98, Wuyou 308, Tianyou 122, and Tianyouhuazhan also shows good quality, with their main quality indicators meeting the national rice quality standard for Grade I. Shenliangyou 5814, Chuanxiangyou 2 and Guangliangyouxiang 66 are up to the national quality standard for Grade Ⅱ, and Liangyou 287 exceeds the national rice quality standard for Grade I hybrid early rice in the Yangtze River Basin.

Table 3-9 Demonstration and promotion of super rice in China

Year	Planting area (10,000ha)	Proportion to China's total rice planting area (%)
2003	266.67	9.8
2004	320.00	11.7
2005	380.00	13.3
2006	433.33	15.1
2007	533.33	18.6
2008	556.13	19.2
2009	606.67	21.2
2010	673.33	23.5
2011	733.33	24.5
2012	800.00	26.6
2013	873.33	29.1
2014	906.67	30.2
2015	873.33	30.0
2016	873.33	30.0

Note. Data of planting area come from the Internet or relevant news reports.

According to statistics, from the first official recognition of super rice varieties by the MOA in 2005, China's super rice research program had produced 166 recognized super rice varieties by 2017 (Table 3-10), including 108 super hybrid rice varieties, accounting for 65.1%, covering the rice growing areas of the Yangtze River Basin, South China, Southwest China, and Northeast China. The success of Super Rice research and promotion has enabled China's grain production to increase for 13 consecutive years after 7 years of yield reduction (1997—2003) in the past 20 years, and the annual comprehensive grain production capacity has reached over 600 million tons, which has made an important contribution to ensuring China's food security, improving China's grain yield level and promoting the strategic adjustment of agricultural structure.

Table 3 - 10 Super rice varieties recognized by the MOA from 2005 to 2017

Year	Super hybrid rice variety	Super conventional rice variety
2005	Xieyou 9308, Guodao 1, Guodao 3, Zhongzheyou 1, Fengyuanyou 200, Jinyou 299, Ⅱ - Youming 86, Ⅱ - Youhang 1, Teyouhang 1, D-You 527, Xieyou 527, Ⅱ - You 162, Ⅱ - You 7, Ⅱ - You 602, Tianyou 998, Ⅱ - You 084, Ⅱ - You 7954, Liangyoupeijiu, Zhunliangyou 527, Liaoyou 5218, Liaoyou 1052, Ⅲ - You 98	Shengtai 1, Shennong 265, Shennong 606, Shennong 016, Jijing 88, Jijing 83
2006	Tianyou 122, Yifeng 8, Jinyou 527, D - You 202, Q - You 6, Qiannanyou 2058, Y Liangyou 1, Liangyou 287, Zhuliangyou 819, Peizataifeng, Xinliangyou 6, Yongyou 6	Zhongzao 22, Guinongzhan, Wujing 15, Tiejing 7, Jijing 102, Songjing 9, Longjing 5, Longjing 14, Kenjing 11
2007	Xinliangyou 6380, Nei2You 6 (Guodao 6), Ganxin 688, Fengliangyou 4, Ⅱ - Youhang 2	Ningjing 1, Huaidao 9, Qianchonglang 2, Liaoxing 1, Chujing 27, Longjing 18, Yuxiangyouzhan
2009	Yangliangyou 6, Luliangyou 819, Fengliangyou 1, Luoyou 8, Rongyou 3, Jinyou 458, Chunguang 1	Longjing 21, Huaidao 11, Zhongjiazao 32
2010	Guiliangyou 2, Peiliangyou 3076, Wuyou 308, Wufengyou T025, Xinfengyou 22, Tianyou 3301	Xindao 18, Yangjing 4038, Ningjing 3, Nanjing 44, Zhongjiazao 17, Hezhanmei
2011	Yongyou 12, Lingliangyou 268, Zhunliangyou 1141, Huiliangyou 6, 03 - You 66, Teyou 582	Shennong 9816, Wuyunjing 24, Nanjing 45
2012	Zhunliangyou 608, Shenliangyou 5814, Guangliangyouxiang 66, Jinyou 785, Dexiang 4103, Qyou 8, Tianyou-Huazhan, Yiyou 673, Shenyou 9516	Chujing 28, Lianjing 7, Zhongzao 35, Jinnong
2013	Y Liangyou 087, Tianyou 3618, Tianyou-Huazhan, Zhong-9-You 8012, H-You 518, Yongyou 15	Longjing 31, Songjing 15, Zhendao 11, Yangjing 4227, Ningjing 4, Zhongzao 39
2014	Y Liangyou 2, Y Liangyou 5867, Liangyou 038, C Liangyou-Huazhan, Guangliangyou 272, Liangyou 6, Liangyou 616, Wufengyou 615, Shengtaiyou 722, Neiwuyou 8015, Rongyou 225, F You 498	Longjing 39, Lianda 1, Changbai 25, Nanjing 5055. Nanjing 49, Wuyujing 27
2015	H Liangyou 991, N Liangyou 2, Yixiangyou 2115, Shenyou 1029, Yongyou 538, Chunyou 84, Zheyou 18	Yangyujing 2, Nanjing 9018, Zhendao 18, Huahang 31
2016	Huiliangyou 996, Shenliangyou 870, Deyou 4727, Fengtianyou 553, Wuyou 662, Jiyou 225, Wufengyou 286, Wuyouhang 1573	Jijing 511, Nanjing 52

Continued

Year	Super hybrid rice variety	Super conventional rice variety
2017	Y Liangyou 900, Longliangyou-Huazhan, Shenliangyou 8386, Y Liangyou 1173, Yixiangyou 4245, Jifengyou 1002, Wuyou 116, Yongyou 2640	Nanjing 0212, Chujing 37

Note. The Confirmation Method of Super Rice Varieties was released in 2008; the 45 varieties (combinations) underlined in the table were later removed from the super rice list due to insufficient planting area.

Section 3 Theory and Technology of Super Hybrid Rice Breeding

Ⅰ. Plant Type Model of Super Hybrid Rice

Rice is a thermophilic crop, and its distribution depends on certain natural ecological conditions. China has the conditions for a wide distribution of rice, but the ecological conditions, including light, temperature and topography, are very complex across the country, and different rice growing areas have their own specific requirements for morphological traits such as height, tillering, plant compactness, leaf traits such as length, width, arrangement and canopy leaf structure, and panicle traits. Super high-yielding rice is no exception. The plant type and the formation of various traits of super high-yielding rice in different rice growing areas are closely related to the local evolutionary environment. For example, under high temperature and low light conditions, the plants tend to form large and thin leaves with a loose leaf structure, which is conducive to capturing solar energy for better solar energy utilization, laying the foundation for higher rice yield potential. The high-yielding plant types also differ at different growth stages. For example, there can be early tillering and rapid development, relatively loose plant shape, and unfolded leaf posture in the early stage of growth, and the plant type gradually turns compact, with upright leaf posture and a high number of green leaves in the late stage of growth, which is also conducive to the capture of more solar energy and for better utilization at different growth stages for higher yield potential.

1. Intersubspecific heavy-panicle model of hybrid rice in the upper Yangtze River rice area

The Sichuan basin in the upper reaches of the Yangtze River is limited in increasing rice yields by increasing density and leaf area under the special ecological conditions of cloudy and wet conditions, high cloudiness, low sunlight and high temperature. Based on this, Zhou Kaida et al. (1995, 1997) believe that the main focus of yield improvement should be increasing photosynthetic efficiency and increasing the weight of each panicle. Therefore, the heavy-panicle model was proposed for super high-yielding rice breeding. The theoretical basis of this model is to use intersubspecific heavy-panicle hybrids to breed higher sink weights and select heavy-panicle plant types for better solar energy utilization.

The main breeding objectives for intersubspecific heavy-panicle hybrids are as follows, 1) a yield potential of 15 t/ha, up by more than 10% over that of current hybrids, with panicle length at 29 – 30 cm, 200 or more grains per panicle, a seed setting rate above 80% and a harvest index of 0.55; 2) in terms of plant morphology, the plant height should be 125 cm, with strong lodging resistance and a panicle setting rate above 70% (15 panicles per plant); slightly scattered plant type in the early growth stage and erect leaves after node elongation with moderately compact structure; flag leaves 40 – 45 cm in length, slightly rolled in; strong roots without early senescence; 3) wide adaptability and disease resistance, and stable seed setting rates.

By the heavy-panicle model, hybrid combinations such as Ⅱ – You 6087, Ⅱ – You 162, Gangyou 160, D You 613 and Deyou 4727 were developed. Ⅱ – You 6087 has a long growth period and can only be planted in areas with good heat conditions in the Sichuan Basin, with a maximum yield of 13.5 t/ha and an average yield of 9.4 t/ha in small-area trials. It is one of the main varieties promoted in Chongqing. Ⅱ – You 162 is a super rice combination with good quality and wide adaptability, designated as a breakthrough rice variety by Sichuan province. Deyou 4727's yield exceeded 15.0 t/ha in a 6.67-ha demonstration plot in Hanyuan, Sichuan. in 2015, setting a record for rice yield per unit area in a large area in the upper reaches of the Yangtze River.

2. Panicles-under-flag-leaf model of mid-season hybrid rice in the middle and lower reaches of the Yangtze River

This is a model of super high-yielding rice plant morphology proposed by Yuan Longping (1997) for single-cropping mid-season rice varieties in the middle and lower reaches of the Yangtze River, which is characterized by improved solar energy utilization and increased biological yield in the late growth stage. The strategy is to breed super hybrid rice with high canopy, low panicle layer, low center of gravity, large and uniform sink, and high resistance to lodging by combining morphological improvement and a higher level of hybrid vigor, mainly using two-line and three-line intersubspecific hybrids.

In 1998, the plant type for super high-yielding rice breeding was specified in line with the principle of "seeking nearness from distance and height from shortness". and specific characteristics of super high-yielding rice in terms of plant height, leaf morphology, yield factors and population structure were proposed. 1) The plant height is about 100cm (culm length 70 cm and panicle length 25 cm); 2) the functional leaves should be long, straight, narrow, concave and thick; 3) the plant should be moderately compact with moderate tillering ability and strong lodging resistance with the panicle layer drooping after grain filling to the point where the panicle tip is about 60 cm above the ground; 4) the panicle size should be medium or big with a weight of 5 g per panicle; 5) the harvest index should be above 0.55. The designed yield potential is about 30% higher than that of intervarietal hybrid rice.

The first batch of China's typical super hybrid rice varieties developed based on this model mainly includes Peiliangyou E32, Liangyoupeijiu and others. The former delivered a maximum yield of 17 t/ha in small-area trials in Yunnan, and the latter was planted on a cumulative area of more than 70 million hectares nationwide for five consecutive years from 2000 to 2004, with an average yield of 9.2 t/ha. The second batch of super hybrid rice varieties developed according to this model includes 88S/0293, Y Liangyou 1, Zhunlianyou 527 and others. 88S/0923 yielded more than 12 t/ha in four 6.67-ha demonstration plots in Hunan for two consecutive years from 2003 to 2004, achieving the Phase Ⅱ goal of China's super rice program in the Yangtze River Basin one year ahead of schedule. The third batch of varieties were also developed with this model and examples include Guangzhan 63S/R1128 and Y Liangyou 2. The former yielded 13.08 t/ha on a 6.67-ha demonstration plot in Hunan in 2010, and the latter achieved a yield of 13.5 t/ha on a 6.67-ha demonstration land plot in Hunan in 2011, achieving the Phase Ⅲ goal.

3. late-stage-function model of super hybrid rice

By comparing the plant type of Xieyou 9308 with that of 65396 (Peiai 64S/E32) and Shanyou 63, researchers at CRRI proposed an ideal plant type for super hybrid rice featuring late-stage function, with a yield of above 12 t/ha (Cheng Shihua et al., 2005). Its specific traits are as follows. 1) Balanced grain and panicle numbers with 12 – 15 effective panicles per plant, 190 – 220 grains per panicle and a moderate seeding density; 2) 115 – 125 cm in plant height, strong stems with high lodging resistance; 3) long, rolled, erect and slightly inward functional leaves with leaf angles of 10°, 20° and 30°, leaf length of 45, 50 – 60 and 55 – 60 cm, leaf width of 2.5, 2.1 and 2.1 cm for the flag leaf, second and third leaf from top respectively, and 250 cm^2 of the total area of the first three leaves from top; 4) high functions for the plant and roots, high photosynthetic capacity of the first three leaves from the top, green stems and yellow ripe without early senescence in the late stage.

The model was proposed to address the phenomena of early root and leaf senescence, low seed setting rates, poor grain filling, and poor overall performance that often occur in the progenies of *indica-japonica* crosses to breed super hybrid rice, with the goal of improving photosynthetic capacity in the late growth stage of rice. This model has produced a series of super hybrid rice combinations such as Xieyou 9308, Guodao 1, Guodao 3 and Guodao 6.

4. New plant type development model for super hybrid rice

After successfully achieving the Phase I and Phase Ⅱ goals of super rice breeding in China following the technical route of panicles under flag leaf in the middle and lower reaches of the Yangtze River, Yuan Longping (2012) proposed a new idea regarding how to further exploit the yield potential of rice by gradually increasing the plant height on the basis of maintaining the existing harvest index with a goal of continuous increasing rice yield. When rice yield is low, increasing both biological yield and harvest index can significantly increase rice yield; when yield reaches a relatively high level, increasing plant height and thus biological yield may be an important ways to further exploit the yield potential of rice.

Development usually follows a spiral path and high-yielding rice breeding is no exception. It has gone from tall stem to short stem, and then semi-dwarf stem, semi-tall stem, new tall stem, and super-high stem (Fig. 3 – 1).

A comparative analysis of the changes in plant height and yield of varieties planted on a large scale before and after dwarfing in China, modern varieties, and representative varieties of hybrid rice that have achieved the target yields of the first three phases of China's super rice breeding program (Table 3 – 11) shows a trend basically in line with the above-mentioned pattern in plant height changes, which fully indicates that the way to increase biological yield by increasing plant height and thus rice yield is practical and feasible.

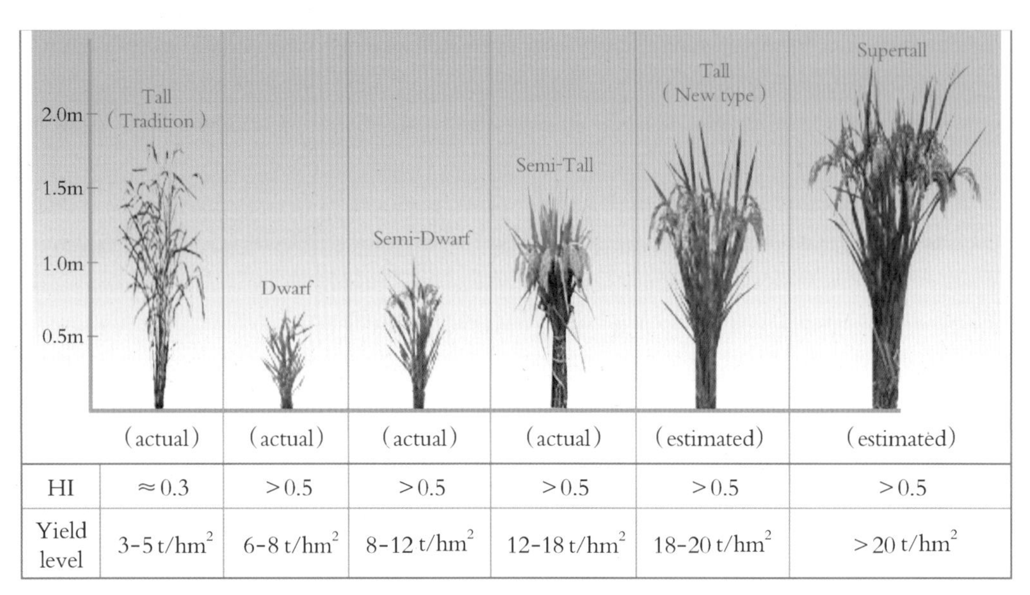

Fig. 3 - 1 New plant type development of super hybrid rice

Table 3 - 11 Changes in plant height and yield of representative varieties of high-yielding rice in the middle and lower reaches of the Yangtze River at different periods

Representative variety	Period	Plant height (cm)	Harvest index	Yield (t/ha)
Shenglixian	Before the 1960s	140 - 160	0. 3	3. 75
Aijiaonante	1960s and 1970s	80	0. 5	6. 0
Shanyou 63	From 1985 to 2005	105	0. 5	8. 0
Liangyoupeijiu	From 2000 to 2010	115	0. 5	10. 5 (6. 67-ha demonstration field)
Y Liangyou 1	Since 2008	120	0. 5	12. 0 (6. 67-ha demonstration field)
Y Liangyou 2	Since 2012	130	0. 5	13. 5 (6. 67-ha demonstration field)

Therefore, in order to further enhance the yield potential of rice, Yuan Longping (2012) proposed a new plant type model for super hybrid rice based on the bold idea of "high biological yield, high harvest index, and high resistance to lodging" with the goal of "increasing sink, expanding source and smoothening flow". The proposed new theory considered the combination of an ideal plant type with intersubspecific hybrid vigor to produce new super hybrid rice with semi-tall stem, good plant type, moderate compactness, strong tillering ability and less difference between the main and tillering panicles. The new plant type has the following specific technical indicators. 1) 120 cm or more in plant height; 2) a growth duration of 145 - 155 days; 3) 2. 7 million effective panicles per hectare, more than 300 grains per panicle, and a seed setting rate of more than 90% ; 4) above 27 g per 1,000 grains; 5) 8 - 10 mm in diameter at the fifth stem node from the top, with high resistance to lodging; 6) strong physiological functions at the late stage and a well-developed root system; 7) above 0.50 in harvest index.

Under the guidance of this model, the average yield of Y Liangyou 900 reached 15.4 t/ha in a 6.67-ha field in Xupu, Hunan, in 2014, achieving the Phase Ⅳ goal of 15 t/ha. In 2015, the average yield of Chaoyouqian reached 16.01 t/ha in 6.67-ha fields in Gejiu, Yunnan, breaking the internationally recognized yield limit of 15.9 t/ha in tropical rice growing areas and setting a new record for large-area rice yield in tropical regions. In 2017, the average yield of Chaoyouqian exceeded 17.0 t/ha in 6.67-ha fields in Yongnian, Hebei, setting a new record for rice yield in a large area at high latitude.

The above-mentioned plant types of super hybrid rice have their own ecological traits. The trend in the plant type design is generally moderately increased plant height, reduced number of tillers, increased panicle weight, improved biological yield and economic coefficient, etc. Therefore, the plant type of super hybrid rice can be summarized as follows. Moderately taller plant and moderately compact plant with intermediate-to-strong tillering ability, well-developed underground root systems, thick above-ground culm, thick culm wall, long and straight canopy of functional leaves, 5 g of grains per panicle, high lodging resistance, good color change at the late stage, large biomass, high grains-to-straw ratio, and a harvest index of 0.55 or higher.

Ⅱ. Technology Route of Super Hybrid Rice Breeding

Breeding practice shows that, up to now, there are only two breeding methods to effectively improve crop yield, morphological improvement and heterosis utilization. The potential of morphological improvement alone is limited; if heterosis is not combined with morphological improvement, the effect will be poor. Other breeding methods and technologies, including genetic engineering and other biotechnological techniques, must ultimately be implemented to achieve excellent morphology and hybrid heterosis; otherwise, they will not contribute to higher yields. However, the development of breeding to a higher level must rely on progress in biotechnology.

1. Morphological improvement

(1) Ideal plant type

The ideal plant type, also called the ideotype, is a concept first proposed by C. M. Donald of Australia in 1968. It refers to an idealized plant type that is conducive to crop photosynthesis, growth and development and grain yield formation. It can maximize a population's solar energy utilization, give full play to the photosynthetic potential, increase the biological yield and improve the economic coefficient. Research on ideotype is to pursue the best combination of many traits in order to maximize the solar energy utilization rate and dry matter production capacity of the population.

As early as the early 1920s, Engledow et al. (1923) proposed that the best synthetics for high yield could be obtained from various high-yielding traits in one through appropriate hybridization and optimal combinations of yield factors. In the 1950s, Japanese scholar Akita Takazaburo (1985) studied the relationship between fertilizer tolerance and plant type in rice, soybean and sweet potato, and put forward the ideal plant type theory. Leaves should be upright, thick and dark, with less early senescence, short and strong stems and leaf sheaths, and moderate tillering.

In the 1960s, C. M. Donald of Australia first introduced the term "ideotype" (ideal type), proposing to find the ideal type in crops with minimum competition among individuals, arguing that the competitiveness within genotype is weak, appropriately dense planting is possible, and that each plant can effectively use the limited favorable conditions above and below the ground to receive photosynthetic products without limiting the ability to enter the economic part. The ideal plant type of wheat was also designed. The theory of minimum competition is still useful and informative for current high-yielding rice breeding and the creation of new plant types. In 1973, Matsushima, a Japanese scholar, emphasized that the ideal plant type for high-yielding rice is "multiple panicles, short stalks and short stems" with the upper two or three leaves short, thick and erect, and the color of the leaves fading slowly after heading and more green leaves. In 1973, Chinese scientist Yang Shouren summarized that dwarf rice varieties have three major characteristics which are tolerance to fertilizer and resistance to lodging, suitability for dense planting, and high grain-to-straw ratio; and he suggested the importance of plant type in solar energy utilization and the yield potential of dwarf rice varieties with large panicles. In 1977, he proposed the model of ideal rice plant type and raised the three requirements for breeding accordingly. The plant should have 1) tolerance to fertilizer and resistance to lodging; 2) large growth; 3) high grain-to-straw ratio. In 1994, he also put forward the "three good theory". 1) A plant height of 90 cm±10 cm; 2) an appropriate panicle size, specifically in the case of downward adjustment of the number of panicles per unit area, the panicle size should be appropriate (blindly increasing the panicle size could result in poor tillering); 3) appropriate tillering ability – too strong tillering will lead to too small panicles, while too weak tillering will result in too few panicles. The essence of the "three good theory" is to balance between biological yield, number of panicles and number of grains for the highest yield possible.

(2) Development of ideal rice plant type in China

The history of rice breeding shows that every major breakthrough in rice production is inseparable from the development of plant type breeding and change. From the first global green revolution in food production in the mid-20th century featured by dwarfing, to hybrid rice breeding and now ideotype breeding, every step sees increase in rice yield as a result of plant type improvement. Many researchers at home and abroad have devoted themselves to research and practice in this area, delivering ground-breaking innovative works, and great achievements which constitute important contributions to the development of rice breeding theory and to global food security. Rice plant type breeding is a dynamic process and the development and formation of plant types are closely linked to the goals of delivering high and super-high yield, and each step in plant type improvement has led to substantial increase in rice yield per unit area.

In terms of breeding practice, the development of ideal rice plant type breeding in China can be summarized into the following three stages.

1) Dwarf breeding stage. Since rice hybridization breeding started in the beginning of the 20th century, there have been effective genetic improvement of rice varieties in various countries, but in general, rice breeding is still limited to trait improvement for tall varieties. Since modern times, the fertilizer industry has been prosperous and the amount of fertilizer applied to rice fields has increased dramatically, so tall plants are often prone to collapse and the yield cannot be increased significantly. Until the 1960s, the

problem of poor fertilizer tolerance and lodging resistance still remained unsolved in *indica* rice breeding in many countries, and the yield per unit area of rice remained low. Therefore, dwarf breeding with the goal of reducing plant height and preventing lodging became the focus of the first stage of rice plant type breeding.

The breeding of dwarf plant types, as the beginning of plant type breeding, is to turn high-stalk varieties into dwarf ones.

Dwarf breeding started as plant type breeding to turn tall varieties into dwarf varieties. Recent studies have found that humans selected an important gene associated with high yield when they started domesticating rice about 10,000 years ago when humans knew nothing about genetics, and scientists have confirmed that this gene is the semi-dwarf gene *SD1*, which was initially present only in *japonica* rice, not in *indica* rice or wild rice. The semi-dwarf gene *SD*1 was the key element in dwarf breeding in rice in the mid-20th century, and in the late 1950s, the Guangdong Academy of Agricultural Sciences (GAAS) of China was the first in the world to start dwarf breeding to solve the problem of rice lodging. The breeding of Aijiaonante and Gaungchang'ai in 1956 started the era of dwarf breeding in the history of China's rice breeding efforts, making China the first country in the world to select and promote dwarf rice varieties. In 1966, IRRI developed the first dwarf rice variety IR8, which was the first semi-dwarf, high-yielding, fertilizer-tolerant, lodging-resistant rice variety with large panicles and many grains. It was developed by introducing the dwarf gene from Dee-geo-woo-gen (DGWG) from Taiwan, China, into the high-yielding Indonesian rice variety Peta. This marked the beginning of the Green Revolution in Southeast Asia, and got dwarf breeding further established. This was an important milestone in the history of rice breeding, and together with Mexican dwarf wheat (semi-dwarf or dwarf wheat varieties developed by crossing Japanese Norin 10 that has dwarf genes with rust-resistant Mexican wheat), triggered the first global green revolution. The breakthrough in rice dwarf breeding is inseparable from the discovery, screening and utilization of dwarf genetic resources. The dwarf varieties have not only shorter plants, high tillering ability, high leaf area coefficients, and well-developed root systems which enhances lodging resistance; but also significantly higher grain-to-straw ratio and economic coefficient than high-stalk varieties. The planting of these varieties basically solves the problems of lodging and yield reduction caused by high planting density, the high demand for fertilizer and strong wind. The yield was increased by 30%-40% compared with ordinary high-stalk varieties. Although efforts to breed plant types conducive to solar energy utilization did not bear fruit at that time, rice yield was still raised to a new level due to three major traits of dwarf varieties-fertilizer tolerance and lodging resistance, suitability for dense planting, and high harvest index.

2) Breeding for ideal plant types. Breeding for ideal plant types kicked off after dwarf breeding, and many breeders and scholars explicitly proposed to take solar energy utilization into consideration in the breeding of new plant types, constantly envisioning and creating breeding models for ideal plant types in different rice ecological zones. This stage can be further divided into two parts. The first features the emergence of ideal plant type breeding. Dwarf breeding solved the problem of fertilizer tolerance and lodging resistance of rice varieties and led to a substantial increase in the yield of *indica* rice. By the 1970s, on the basis of dwarf varieties, further breeding efforts focused on selecting plant and leaf morphology that

could make full use of solar energy, i. e. , morphological breeding for high photosynthetic efficiency, to create a more reasonable comprehensive configuration of roots, stems, leaves and panicles. Its main feature is to change drooping leaves into erect ones, turning variety-oriented breeding to a plant type-oriented approach aimed at more effective utilization of solar energy, thus further raising rice yield. In *japonica* rice, the representative varieties include Liaojing 5 and Shennong 1033 from Liaoning province, as well as Nanjing 35 from Jiangsu province. In *indica* rice, the representative variety is Guichao 2 developed in Guangdong province. Compared with the dwarf varieties developed in the 1960s, the above-mentioned varieties have significantly improved plant types and have achieved high yields due to their thick, narrow, erect and dark green leaves, large leaf area of the population, high photosynthetic production, and more heavy panicles. The second part features improvement of ideal plant types. By the early 1980s, new progress had been made in improving the high-yielding rice plant type, mainly through dwarf and upright stalks, appropriately increased plant height, increased biomass, and balance between sink, source and flow. Such improvements further pushed up rice yield. Rice breeders and scholars at home and abroad made active efforts in this regard and achieved considerable success. The representative varieties include Guichao 1 (representative of cluster breeding) bred by GAAS, which has the traits of a dwarf variety plus a good plant type and further improved photosynthetic productivity of the population. In addition, the variety also has physiological and ecological advantages, thus solving the conflict between panicle number and per panicle weight, which is conducive to the formation of large and heavy panicles, so that the high yield potential can be given full play. In terms of the breeding of *indica* hybrid rice, the representative varieties include Shanyou 63 and Ganhua 2, who has greater plant height than dwarf varieties, but is resistant to lodging because of its abundant accumulation of dry matter at the base of the stems. Its biological yield is higher than that of dwarf varieties, and the number of grains per panicle is obviously increased. Its economic coefficient is close to that of dwarf varieties, so its yield potential is greater. In terms of *japonica* rice breeding, the representative variety BG910 has excellent plant structure and physiological traits, which enables it to obtain the maximum yield potential of improved rice varieties and high social and economic benefits when planted on a large scale.

3) This is a stage of super high-yielding rice breeding that combines ideal plant types with the use of hybrid vigor. This stage is developed on the basis of ideal plant type breeding and has become a popular topic of international research. Rice breeders have noticed the use of hybrid vigor in ideotype breeding and the important trend of ideotype in hybrid vigor utilization, and many varieties developed have shown high yield potential. So far, super high-yielding rice breeding has entered the phase of steady progress towards technical maturity thanks to continuous exploration. Since the 1980s, some important rice producing countries and research institutes in the world have been actively exploring breeding models for super high-yielding rice. Japan was the first to propose a research program called "Ni 753" for super high-yielding rice breeding in 1981. IRRI also started breeding research on NPT, or super rice, in 1989. So far, a number of NPT rice varieties have been developed (Zhuanglu, IR65598 - 1122, etc.), which have shown the potential of super rice yield.

In order to ensure China's food security in the 21st century and maintain the country's leading position in hybrid rice breeding in the world, China started the research on super rice in the mid-1990s. Based on the breeding of dwarf varieties and hybrid heterosis utilization, China took up a technical research route with Chinese characteristics, which combined the ideotype with hybrid heterosis and physiological functions. It was incorporated into the key research projects of the Ninth Five-Year Plan period and the National High-Tech R&D Program (863 Program) during the Tenth Five-Year Plan period, and received the support from MOA, with China Super Rice Research Program officially launched in 1996, the Premier's Fund of the State Council and MOST (after Yuan Longping proposed the super hybrid rice breeding program in 1998). At the same time, a number of research institutes have been organized to carry out joint research. After more than a decade's research and practice, great progress was made and a number of core lines (such as Shennong 265, Shennong 89368, Teqing and Shengtai) with large panicles and high rate of panicle-bearing tillers and large biomass were developed. A number of new super rice hybrids such as Peiai 64S/E32 and 88S/0293 bred by HHRRC and Liangyoupeijiu bred jointly by the HHRRC and the Jiangsu Academy of Agricultural Sciences (JAAS) were also developed. Super hybrid rice combinations were successfully demonstrated and widely applied, delivering significant yield increase. The objectives of the first, second, third and fourth phases of China's super hybrid rice breeding program were achieved in 2001, 2004, 2011 and 2014, respectively. By now, China's super high-yielding rice breeding entered the stage of steady development. By 2017, the yield per unit area of super hybrid rice has exceeded 17 t/ha.

In summary, the improvement of plant and leaf morphology played a great role in further improving the yield potential of rice. Dwarf breeding is the first stage of ideotype breeding, the core of which is to enhance fertilizer tolerance and lodging resistance by reducing plant height, thus improving rice yield. Ideotype breeding is a further step based on dwarf breeding, the key of which is to focus on changing stem and leaf traits, improving leaf posture and leaf quality, requiring the leaves, especially the upper leaves, to be curled, short, thick and erect, with good light exposure, so as to develop high light efficiency and further increase the potential for high rice yield. The modern super high-yielding, new plant type breeding started on the basis of ideotype breeding, not simply to increase leaf area, but to improve population photosynthesis and dry mass production through the comprehensive improvement of population structure and light exposure potential. Its development direction is the combination of ideotype and hybrid heterosis utilization. The key point is to create a new plant type with superior overall traits by optimizing the trait combination through intersubspecific hybridization between *indica* and *japonica*, recombining plant types, and utilizing heterosis (i. e., combining morphology and functions). It is the only way to obtain super high yield in rice, and the mainstream of plant type breeding development today.

(3) Methods to improve super hybrid rice plant type

The ideotype is the morphological basis for super high yield. The purpose of plant type improvement is to adjust the geometric configuration and spatial arrangement of individuals, improve the population structure and light exposure, balance between leaf area, unit photosynthetic efficiency and canopy duration to the maximum extent, and achieve a dynamic equilibrium at a higher photosynthetic efficiency and material production level so as to finally deliver super high yield. Morphological improvement means coordinating and improving the morphological and physiological functions of individuals, so that individuals can give full play to their dry mass production under more adapted population conditions (Chen Wenfu et al., 2007).

Based on domestic and foreign research of super high-yielding rice breeding, the ideal super high-yielding plant type generally has the following characteristics. 1) Moderate plant height and high mechanical strength to ensure sufficient biomass and high lodging resistance; 2) uplifted leaves, curled and erect, high specific leaf weight (SLW), and high chlorophyll content, long functional period and no premature senescence; 3) moderate tillering ability, loose in the early stage of growth and compact in the late stage, and high ratio of panicle-bearing tillers; 4) well-developed vascular bundles of panicles and stems, and high numbers of primary branches; 5) well-developed root systems without premature senescence. There are three main ways to improve the morphology of the new plant type of super high-yielding hybrid rice.

Firstly, innovation in and utilization of specific rice germplasm resources based on exploration and use of good breeding materials and the breeding of new ideotype for super high yield. Crop variety improvement mainly depends on the development and utilization of germplasm resources, and rice ideotype improvement is no exception. The three major breakthroughs in the history of rice breeding all began with the discovery of specific germplasm resources, For example, the discovery of Aijiaonante in Guangdong pioneered dwarf breeding in China, the discovery of the WA plant in Hainan paved the way for successful hybrid rice research; the discovery of PTGMS (Nongken 58S and Annong S-1) and the wide compatibility gene (S5N) has expanded the scope of hybrid rice heterosis utilization and provided material basis for the breeding of super high-yielding rice plant types, especially, the materials of PTGMS lines and parental lines from *indica-japonica* cross with wide compatibility have opened up new ways for the breeding of super high-yielding hybrid rice ideotype. The widely compatible PTGMS line Peiai 64S and widely-adaptive PTGMS line Y58S, as well as their representative hybrids Liangyouteqing and Y Liangyou 1 bred by the HHRRC, the dwarf, strong, long-leaved, large panicle variety Shennong 89366 (became the core material for IRRI's super rice ideotype), Shennong 127, and erect panicle, semi-dwarf, compact, and high individual competitive variety Shennong 159 bred by Shenyang Agricultural University have become representative models of rice plant type breeding in different rice regions, and also the core resources for rice improvement with remarkable results. Thus, the exploring and utilization of, and innovation in specific germplasm resources have provided technical support for super hybrid rice ideotype improvement.

Secondly, the use of hybrid vigor to breed for super high-yielding ideotype. Heterosis utilization is the best combination of favorable traits obtained through the complementation of two parents, while the ideotype for super high-yielding rice refers to the best combination of various favorable traits related to yield under specific ecological conditions. Therefore, research on the use of hybrid heterosis based on the results of ideotype breeding is one of the important ways to breed super high-yielding rice ideotypes. In particular, the utilization of hybrid heterosis between distant species and *indica-japonica* intersubspecific crosses to improve the plant type will further increase the yield of super hybrid rice, due to the great variation caused by high segregation in interspecific or intersubspecific hybrid progenies, which provides the possibility for creating and selecting new plant types.

The methods include 1) combining the high heterosis of *indica-japonica* intersubspecific cross and ideotype through two-line hybrid rice breeding, represented by 65396 (Peiai 64S/E32) and 65002 (Liangyoupeijiu) bred jointly by HHRRC and the Jiangsu Academy of Agricultural Sciences, and 88S/0293 bred by HHRRC. 2) To realize the high heterosis of having heavy or large panicles by *indica-japonica* intersubspecific cross with the three-line method to breed three-line hybrids with super high-yielding ideotype, represented by Ⅱ-You 162 bred by the Rice Institute of Sichuan Agricultural University and Xieyou 9308 bred by CRRI. 3) Create new plant type variations and strong heterosis through *indica-japonica* intersubspecific crosses or through geographically distant hybridization, and then combining the ideotype and heterosis through optimized combination of traits to breed new super high-yielding plant types, represented by Shennong 265 and Shennong 606 with erect and large panicles in the north, and the *indica-japonica* hybrid rice Yongyou, Chunyou and Zheyou varieties bred in the middle and lower reaches of the Yangtze River. These are examples of the high yield resulted from a combination of morphological characteristics and physiological functions. 4) Create core germplasm through continuous improvement to achieve kinship penetration across ecotypes, over geographical distance, and between species and subspecies to strengthen the input of high-quality genes. On this basis, select new resources with target traits, special genes or intermediate materials, or geographically and ecologically distant new materials that combine high-quality ideotype with advantaged lines created by distant hybridization. Examples include the giant-panicle rice line R1128 bred by HHRRC. R900, a strong restorer line from the progenies of *indica-japonica* cross bred by Hunan Yuanchuang Super Rice Technology Co., Ltd., which were selected for the breeding of new super hybrid rice varieties suitable for large-scale planting. The representative varieties include Guangliangyou 1128, Liangyou 1128, Y Liangyou 900 and Xiangliangyou 900.

Thirdly, combine biotechnology and conventional breeding to create new high light-efficient ideotype of super hybrid rice. The rapid development of biotechnology in recent years has promised a bright future for super rice breeding and opened a new way for the creation of new plant types with super-high yield and high photosynthetic efficiency. The potential rice cultivars are limited, so we should focus on the discovery of high yield genes or other genes related to important agronomic traits from wild rice and other rice relatives, and introducing high photosynthetic efficiency genes from C4 crops such as maize through transgenic technology, or combining molecular markers with conventional breeding methods, so as to select new plant types and varieties with super high yield. It has been reported that Japan has successfully introduced CO_2 fixase (PEPC) into rice from highly photosynthetic C4 plants, offering the possibility of improving photosynthesis in C3 plants and breeding new ideotypes.

2. Improvement of the heterosis level

(1) Intersubspecific heterosis of rice

According to Yuan Longping, the trend of rice heterosis is distant interspecific heterosis > intersubspecific heterosis > intervarietal heterosis. Many studies have shown that there is a general heterosis trend, *indica / japonica* > *indica / javanica* > *japonica / javanica* > *indica / indica* > *japonica / japonica*, and this trend roughly indicates that the more distant the two parents are, the stronger the hybrid heterosis is.

Yang Zhenyu (1991) selected different *indica* and *japonica* parents to study the relationship between the genetic distance of the parents and heterosis. The results showed that a highly significant positive correlation between the difference of the Cheng's index of the parents and the dry matter weight of the whole plant of intersubspecific F_1 hybrids ($r=0.87$). When the Cheng's index difference of the parents is greater than 14, the biological heterosis of the intersubspecific F_1 hybrids is very strong, while when the Cheng's index difference of the parents is between 7 and 13, the biological heterosis is moderately strong. Researchers at Anhui Academy of Agricultural Sciences (1977) used genetic correlation matrix and principal component analysis to study the genetic distance and heterosis of *indica-japonica* intersubspecific crossings, and also concluded that the heterotic hybrids were generally derived crosses of distant parents, the closer the relationship between the parents, the smaller the difference in traits. It was also concluded that when the genetic distance is small, the hybrid vigor will be weak or none. Yan Qinquan (2001) studied the relationship between the *indica-japonica* degree and heterosis by crossing four PTGMS lines of different *indica-japonica* degrees with 11 male parents and found that parental *indica-japonica* degree was significantly correlated with hybrid heterosis, which also confirmed that the greater the genetic distance between the parents, the higher the hybrid heterosis.

(2) Expression of intersubspecific hybrid heterosis in rice subspecies

The advantage of rice intersubspecific heterosis in economic traits is mainly in the number of spikelets per panicle and the total number of spikelets per unit area. Among the rice yield components, the prominent heterosis of intersubspecific F_1 hybrids is mainly shown in the average number of spikelets per panicle and the total number of spikelets per unit area.

Yuan Longping et al. (1986) studied the yield potential of intersubspecific F_1 hybrids, and found that the number of spikelets per panicle and the total number of spikelets per plant of the F_1 hybrids from the typical *japonica* line Chengte 232 and the typical *indica* line 26 Zhaizao were 162.8% and 122.4% more than those of the hybrid Weiyou 35 at the same growth duration, respectively. The number of spikelets per hectare was more than 660 million, and the yield was almost equal to that of the control due to the hybrid's high spikelet number per unit area, although the seed setting rate was only 54% (Table 3-12). If the seed setting rate of intersubspecific F_1 hybrids can be increased to 80%, there should be a potential yield increase of 30% over intervarietal hybrids.

Table 3-12 Yield potential of F_1 hybrids between *indica* and *japonica* lines

(Yuan Longping et al., 1986)

	Plant height (cm)	Number of spikelets per panicle	Number of spikelets per plant	Seed setting rate (%)	Actual yield (t/ha)
Chengte 232 (*japonica*)/ 26 Zhaizao (*indica*)	120	269.4	1,779.4	54.0	8.33
Weiyou 35 (CK)	89	102.6	800.3	92.9	8.71
Heterosis (%)	34.8	162.8	122.4	-41.9	-4.3

In 1989, Yuan Longping et al. also reported the heterosis performance of the intersubspecific hybrid Erjiuqing S/DT713. The results showed that the highest heterosis among all yield traits remained in the number of spikelets per panicle and the number of spikelets per unit area. The average heterotic rates of spikelets per panicle and spikelets per unit area were 82. 94% and 59. 86%, respectively (Table 3 - 13).

Table 3 - 13 Comparison of economic traits between *indica-japonica* hybrid Erjiuqing S/DT713 and the control line (Yuan Longping et al., 1989)

	Plant height (cm)	Heading days	Panicles per plant	Panicles per hectare (10,000)	Spikelets per panicle	Spikelets per hectare (10,000)	Seed setting rate (%)	1,000 - grain weight (g)	Grain weight per plant (g)	Theoretical yield (t/ha)
Erjiuqing S/DT713	110. 0	88. 0	15. 06	282. 45	205. 89	58155. 75	75. 40	27. 80	65. 00	12,190. 2
Weiyou 6	103. 0	88. 0	17. 24	323. 25	112. 54	36378. 60	83. 15	27. 38	44. 17	8,282. 1
Heterosis over control (%)	106. 8	100. 0	87. 35	87. 35	182. 94	159. 86	90. 68	101. 53	147. 18	147. 2

(3) Problems in the utilization of intersubspecific heterosis in rice

Although the F_1 hybrid of *indica-japonica* subspecies of rice has strong biological advantages, generally showing tall plants, high number of grains per spike and high number of spikelets per unit area, the hybrid has the potential to increase the yield by 30%-35% (Chen Liyun, 2001). However, it has not been used widely in production due to the excessive genetic differences between the two parents and the existence of genetic barriers, which were shown in the disadvantages of low seed setting rates, super high plants, extra-long growth duration and poor grain filling. In addition to the main problems mentioned above, there are also some other problems that can even restrict application. The vulnerability of the seed setting of the intersubspecific hybrid to changes in environmental conditions (especially temperature) leads to an unstable seed setting rate; low hybrid seed production yield due to the low seed setting rate caused by the difference in flowering habits for both *indica* / *japonica* and *japonica* / *indica* combinations. The intersubspecific hybrid also lacks moderate threshing performance.

(4) Main ways to improve heterosis level

1) Utilization of intersubspecific heterosis

It is an indisputable fact that the intersubspecific F_1 hybrids between *indica* and *japonica* lines have strong heterosis, which is shown in their vigorous growth, more spikelets per panicle and more total spikelets per unit area. However, its strong heterosis cannot be utilized effectively due to the large genetic differences between the two subspecies and the low seed setting rate. Many scholars at home and abroad have done a lot of research on how to exploit the pathway of intersubspecific heterosis. So far, there are mainly four ways. Firstly, use *indica-japonica* intersubspecific crosses to select conventional varieties and strong restorer lines to achieve partial utilization of intersubspecific hybrid heterosis. Secondly, use wide compatibility genes and PTGMS lines to achieve direct utilization of *indica-japonica* intersubspecific hybrid heterosis. An example is the successful breeding of Peiai 64S, a sterile line for two-line hybrids. Thirdly, use intermediate breeding materials with mixed affinity of *indica-japonica* progenies to utilize the intersubspecific hybrid heterosis. For example, a restorer line of *indica-japonica* intermediate type can be combined with a *japonica* sterile line with wide compatibility, for example, C57, a restorer line created by " *indica-japonica* bridge" breeding. Fourthly, exploit *indica-japonica* intersubspecific hybrid heterosis through molecular breeding techniques such as gene fragment replacement and polymerization.

So far, significant progress has been made in the utilization of *indica-japonica* intersubspecific heterosis. 1) Typical examples of the utilization of *indica-japonica* intersubspecific heterosis is the three-line hybrid series of Yongyou, Chunyou and Zheyou, which were bred by combining *japonica* male sterile lines and *indica-japonica* intermediate widely compatible restorer lines. The main features of these hybrids are tall plants, big panicles, long growth duration, and high seed setting rates, showing high heterosis and great yield potential. In 2012, the average yield of Yongyou 12 planted on 6. 67-ha demonstration plots in Zhejiang reached 14. 45 t/ha. In 2015, the average yield of Chunyou 927 planted on 6. 67-ha demonstration plots in Zhejiang reached 15. 23 t/ha. 2) A typical representative of the utilization of *indica-japonica* intersubspecific heterosis through two-line hybrid rice is the hybrid series derived from Peiai 64S and R900. These hybrids were bred by combining *indica* male sterile lines which carry wide compatibility genes and have certain genetic relationship with *japonica* rice, with *indica* or *indica-japonica* intermediate restorer lines. In 2000, the average yield of Liangyoupeijiu planted on 6. 67-ha demonstration plots in Hunan exceeded 10. 5 t/ha, realizing the goal of the first phase of China's super rice breeding project. In 2014, the average yield of Y Liangyou 900 planted on 6. 67-ha demonstration plots in Hunan achieved 15. 4 t/ha, realizing the goal of the fourth phase of China's super rice breeding project.

2) Utilization of distant hybridization heterosis

Despite the incompatibility, sterile hybrids and low seed setting rates, fertility, irregular segregation and difficulty in stabilization, and very difficult and complicate in direct utilization of rice distant hybridization, the breeding of distant cross has been traditionally emphasized both at home and abroad. IRRI's G. S. Kusch believes that one of the ways to improve the yield potential of crops is distant hybridization. Yuan Longping proposed three stages for the strategic development of hybrid rice, the third of which is the use of distant hybrid heterosis.

To date, there is no shortage of examples of direct application of distant hybridization in breeding. In rice distant hybridization, plants from more than a dozen families such as sorghum, maize, Z *izania latifolia*, *Pennisetum alopecuroides*, *Coix lacryma-jobi*, barnyard grass and even bamboo (*Bambusoideae*) have been used as the male parents for sexual hybridization. Some progenies produced through these distant hybridization had excellent agronomic traits and showed strong heterosis according to reports of trial planting. For example, wild rice resources have been the most crucial and most frequently used in the process of three-line hybrid heterosis utilization in China. Yuan Longping et al. (1964) discovered WA rice with pollen abortion, which was a breakthrough in hybrid rice breeding in China. Zhongshan 1, a variety through distant hybridization between common wild rice and cultivated rice, produced a great number of varieties and breeding lines that have been planted on a large scale over the years and contributed greatly to rice production which is rare in the world history of rice breeding.

3. Application of molecular biotechnology

Research of molecular genetics and functional genomics at the molecular and DNA level provides a greater platform for rice hybrid heterosis utilization. As Yuan Longping pointed out, the combination of conventional breeding and molecular biotechnology is the future direction of crop breeding, and this is also a way to select and breed super hybrid rice with great potential. With the help of molecular biotechnology, we can make use of favorable genes or create new germplasm from cultivated rice, wild rice within the gramineae species or from other species.

(1) Utilization of favorable genes in wild rice

In general, the favorable genes in wild rice are exploited through a combination of distant hybridization and molecular marker-assisted selection, and the method is mainly applied to breeding for important traits in rice such as bacterial blight resistance, blast resistance, yield, quality traits, etc. IRRI has been using molecular biotechnology to transfer genes of disease and pest resistance from wild rice since 1969, and has successfully transferred excellent non-AA genes from wild rice, such as the genes of blast and bacterial blight resistance from *O. minuta*, and the genes for brown planthopper and bacterial blight resistance from *O. australiensis*, into cultivated varieties, resulting in a number rice varieties with target traits (A. Amante Bordeos et al., 1992; D. S. Multani et al., 1994).

Yuan Longping (1997) reported that two important QTLs (*yld1. 1* and *yld2. 1*) were identified in Malaysian wild rice *O. rufipongon L.*, located on chromosomes 1 and 2, respectively, and each locus had an effect of yield increase of about 20%. These two QTLs were successfully transferred into the hybrid parent Ce64 - 7 through distant hybridization combined with molecular marker-assisted selection, and the large-panicle restorer line Q611 was selected to carry these two QTLs with a yield increasing effect. The super hybrid rice derived from Q611, such as Jingyou 611, Fengyuan 611, etc., were demonstrated in 6.67-ha trials in Lingling, Hunan, in 2005. The theoretical yield of Fengyuanyou 611 reached 11,502 kg/ha and the actual yield was 10,575 kg/ha, which was more than 3000 kg/ha higher than that of local late-season hybrid rice. Some researchers transferred yield increasing QTLs *yld1. 1* and *yld2. 1* to the rice core parent 9311 by molecular marker-assisted selection and obtained a number of breeding lines carrying one or two of these QTLs. A total of 15 hybrids were created using Peiai 64S and these QTL lines to study the yield increasing effect. The results showed that the theoretical yields of these 15 hybrids were higher than that of the control variety, with an average yield increase of 16.41%.

(2) Utilization of favorable genes in other species within gramineae

Compared with rice, the distant species such as sorghum and maize have many excellent agronomic traits. They are C4 plants with high photosynthetic efficiency, smooth flow of photosynthetic products, high seed setting rates, good plant and leaf morphology, thick stems, hard stalks, well-developed root systems, large panicles with many grains, high adaptability, high tolerance to fertilizer, lodging, drought and water logging, and high and stable yield. The non-AA type wild rice *Oryza officinalis* Wall has the traits of perennial overwintering roots, resistance to senescence, strong ratooning ability, and cold and drought tolerance. *Echinochloa beauv* in Gramineae is a C4 plant with high photosynthetic efficiency, fast grain filling, early maturity and high probability of large panicles.

The direct use of favorable genes for superior traits in distant crosses of rice with these species has little chance of success due to the interspecific reproductive isolation barrier. To exploit these superior traits, it is necessary to introduce distant favorable genes into rice recipients with the help of molecular biotechnology to obtain stable genetically improved lines. At present, two main approaches have been applied for this purpose. 1) Create rice germplasm resources by introducing total DNA from distant species and using their favorable genes. This includes pollen tube introduction, panicle and stem injection, embryo soaking, etc. Li Daoyuan et al. (1990) introduced the total DNA of non-AA, *Oryza officinalis* Wall into *11 indica* and one *japonica* rice varieties through pollen tubes and bred the line of Gui-D1, which showed some special traits of the wild rice, such as hard, straight and thick green leaves, tolerance to fertilizer and resistance to lodging, cold tolerance, strong ratooning ability, no premature senescence of functional leaves at maturity, perennial overwintering roots and wide adaptability. Hong Yahui et al. (1999) introduced the DNA of dense-panicle sorghum into the *japonica* rice variety Eyi 105 using the pollen-tube introduction technology and obtained progenies with a high photosynthetic rate, a maximum of increase of more than 80%. Wan Wenju et al. (1993) used total DNA from maize CYB to immerse the seed embryo of rice varieties XR and 84266, and selected genetically engineered rice 1 (GER－1) with multiple large panicles and a high seed setting rate from the substantially mutated progenies. Zhao Bingran (2003) introduced the total DNA of *Echinochloa crusgalli* into the rice restorer line R207 through stem injection, and selected a new restorer line RB207－1 from the mutant progenies, which showed large panicles with many grains and increased grain weight. The hybrid GDS/RB207－1 has a good plant type and high heterosis. It performs particularly well in mountainous areas at high altitude. In 2005, it was tested at multi-locations and the yield per hectare in a small area reached more than 13,500 kg/ha. 2) Introduce distant favorable genes into hybrid rice parents using genetic engineering techniques. This technique is mainly used to introduce exogenous genes controlling favorable traits of other species into the rice genome by transgenic means to obtain improved rice lines with stable inheritance and expression of exogenous genes. Herbicide resistance genes and insect resistance genes have been successfully introduced into rice using this technology. In collaboration with the HHRRC and the Chinese University of Hong Kong, the genes for the highly photosynthetic C4 photosynthetic enzyme PEPC (phosphoenolpyruvate carboxylase) and PPDK (pyruvate double kinase), which originated from maize, were successfully cloned and introduced into the core parent of super hybrid rice, resulting in a new rice line with 10%－30% higher photosynthetic efficiency than the control (Yuan Longping, 2010). The transformation further polymerized the PPDK regulatory protein gene, and it was found that the phosphorylation level of PPDK was significantly suppressed, photosynthetic carbon sequestration efficiency was enhanced, and biomass and the numbers of panicles and grains were significantly increased in the three-gene polymerized line.

(3) Use of gene editing technology to create new germplasm

The rice genome has been sequenced, and many genes controlling important agronomic traits have been interpreted, providing a solid foundation for the next step of editing and modifying the required rice genetic information. Thus, the use of gene editing technology has become a popular field in biotechnological research worldwide, and in recent years, successful rice breeding using gene editing technology has been reported at home and abroad. In particular, the CRISPR-Cas9 technology has been successfully applied. It does not require exogenous genes and needs only to precisely locate the gene controlling a trait within the rice sequence before one can edit and modify it. Yuan Longping's team in 2017 announced a major scientific achievement – a breakthrough in rice parental cadmium removal technology. This was done using the CRISPR-Cas9 gene editing technology. Specifically, rice materials with low accumulation of cadmium was obtained after knocking out the genes involved in cadmium accumulation in rice, and *indica* hybrid rice parents and combinations with extremely low cadmium content were then produced and planted in highly cadmium-contaminated rice fields. The cadmium content in rice grains of the low-cadmium restorer lines and combinations was about 0.06 mg/kg, which was more than 90% lower than that of the low cadmium control varieties Xiangwanxian 13 and Shenliangyou 5814.

References

[1] DENG HUAFENG, HE QIANG. Study on plant type model of eurytopicity super hybrid rice in Yangtze river basin[M]. Beijing: China Agricultural Press, 2013.

[2] XUE DERONG. Light energy utilization and high yield potential of rice[J]. Guangdong Agricultural Sciences, 1977, 3:27 – 28.

[3] LI MINGQI. Progress in research on photosynthesis[M]. Beijing: Science Press, 1980.

[4] LIU ZHENYE, LIU ZHENQI. Photosynthesis genetics and breeding[M]. Guiyang: Guizhou People's Publishing House, 1984.

[5] ZHANG XIANZHENG. Research methods of crop physiology[M]. Beijing: China Agricultural Press, 1992.

[6] YOSHIO MURATA. Solar energy: utilization and photosynthesis [G]. Recent Advances in Breeding Science, 15, Japanese Compilation of Breeding Science, 1975.

[7] CHEN WENFU, XU ZHENGJIN. Theory and methods of super high yield rice breeding[M]. Beijing: Science Press, 2007.

[8] YUAN LONGPING. Hybrid rice[M]. Beijing: China Agricultural Press, 2002.

[9] QI CHANGHAN, SHI QINGHUA. Study on utilization of light energy and high yield cultivation of rice Ⅰ. Solar radiation resources and rice yield potential in Jiangxi province[J]. Journal of Jiangxi Agricultural University, 1987 (S1): 1 – 5.

[10] LU QIYAO. Potential of light and temperature in rice production in China[J]. Agrometeorology, 1980 (1): 1 – 12.

[11] GAO LIANGZHI, GUO PENG, ZHANG LIZHONG, et al. Light and temperature resources and productivity of rice in China[J]. Chinese Agricultural Sciences, 1984, 17 (1): 17 – 22.

[12] YOSHIDA TAKAICHI. Rice physiology[M]. Beijing: Science Press, 1980.

[13] SUETSUGU I. Records of high rice yield in India. Agric[J]. Technol., 1975,30:212 – 215.

[14] FANG FUPING, CHENG SHIHUA. Rice production capacity in China[J]. Chinese Journal of Rice Science, 2009, 23 (6): 559 – 566.

[15] PARK P K, CHO S Y, MOON H P, et al. Rice varietal improvement in Korea[M]. Suweon: Crop Experiment Station, Rural Development Administration, 1990.

[16] SATO HISAO. Research on super high yield rice breeding[J]. Foreign Agronomy (Rice), 1984 (2): 1-16.

[17] KANEDA TADAYOSHI. Breeding super high yield rice varieties by *indica-japonica* cross[J]. JARQ, 1986, 19 (4): 235-240.

[18] XU ZHENGJIN, CHEN WENFU, ZHANG LONGBU, et al. Status and prospects of rice breeding in Japan [J]. Rice Abstracts, 1990, 9 (5): 1-6.

[19] KUSH G S. Varietal needs for different environments and breeding strategies[M]. In: Muralidharan K, Sid E A. New Frontiers in Rice Research. Hyderabad: Directorate of Rice Research, 1990:68-75.

[20] PENG S, KUSH G S, CASSMAN K G. Evolution of the new plant ideotype for increased for increased yield potential[M]. In: Cassman KG, Breaking the Rice Barrier. Manila: IRRI, 1995:5-12.

[21] YUAN LONGPING. Progress in new plant type breeding[J]. Hybrid Rice, 2011, 26 (4): 72-74.

[22] DENG HUAFENG, HE QIANG, CHEN LIYUN, et al. Study on yield stability of super hybrid rice in Yangtze river basin[J]. Hybrid Rice, 2009, 24 (5): 56-60.

[23] ZHOU KAIDA, MA YUQING, LIU TAIQING, et al. Breeding of heavy-panicle combinations among subspecies of hybrid rice: theory and practice of super high yield breeding of hybrid rice[J]. Journal of Sichuan Agricultural University, 1995, 13 (4): 403-407.

[24] ZHOU KAIDA, WANG XUDONG, LIU TAIQING, et al. Study on interspecific heavy-panicle hybrid rice [J]. Chinese Agricultural Sciences, 1997, 30 (5): 91-93.

[25] YUAN LONGPING. Super high yield hybrid rice breeding[J]. Hybrid Rice, 1997, 12 (6): 1-6.

[26] CHENG SHIHUA, CAO LIYONG, CHEN SHENGUANG, et al. Concept and biological significance of late-stage vigor super hybrid rice[J]. Chinese Journal of Rice Science, 2005, 19 (3): 280-284.

[27] YUAN LONGPING. Further reflections on super high yield hybrid rice breeding[J]. Hybrid Rice, 2012, 27 (6): 1-2.

[28] DONALD C M. The breeding of crop idea-types[J]. Euphytica, 1968, 17:385-403.

[29] ENGLEDOW F L, WADHAM S M. Investigations on yield in the cereals[J]. J Agric. Sci., 1923,13:390-439.

[30] TSUNODA SHIGESABURŌ. Ideal type of rice-regulation of photosynthetic structure [G] //Collection of foreign papers on ideal plant type of rice (2) (translated by Chen Wenfu). Institute of Rice Cultivation, Shenyang Agricultural University, 1985:1-18.

[31] MATSUSHIMA S. A method for maximizing rice yield through "ideal plants" [M]. Yokendo, Tokyo, 1973: 390-393.

[32] YANG SHOUREN. Rice fertilization and irrigation and fertilization management[J]. Liaoning Agricultural Sciences, 1973, 3:19-23.

[33] YANG SHOUREN. Discussion on rice plant type[J]. Journal of Genetics and Genomic, 1977, 4 (2): 109-116.

[34] YANG SHOUREN. Progress in research on rice plant type[J]. Acta Agronomica Sinica, 1982, 8 (3): 205-209.

[35] YANG SHOUREN, ZHANG LONGBU, CHEN WENFU, et al. Verification and evaluation of the "three appropriates" theory in optimizing rice traits combination [J]. Journal of Shenyang Agricultural University, 1994, 25 (1): 1-7.

[36] YANG ZHENYU, LIU WANYOU. Classification of *indica-japonica* F_1 hybrids and its relationship with heterosis

[J]. Chinese Journal of Rice Science, 1991, 5 (4): 151 - 156.

[37] YAN QINQUAN, YANG JUHUA, FU JUN. Relationship between *indica-japonica* degree, combining ability and heterosis of two-line hybrid rice[J]. Journal of Hunan Agricultural University (Natural Science Edition), 2001 (3): 163 - 166.

[38] CHEN LIYUN. Theory and technology of two-line hybrid rice[M]. Shanghai: Shanghai Science and Technology Press, 2001.

[39] AMANTE-BORDEOS A, SITCH. L A, Nelson R, et al. Transfer of bacterial blight and blast resistance from the tetraploid wild rice *Oryza minuta* to cultivated rice, *Oryza sativa* [J]. Theoretical and Applied Genetics, 1992, 84(3 - 4):345 - 354.

[40] MULTANI D S, JENA K K, BRAR D S, et al. Development of monosomic alien addition lines and introgression of genes from *Oryza australiensis* Domin. to cultivated rice *O. sativa* L. [J]. Theoretical and Applied Genetics, 1994, 88(1):102 - 109.

[41] LI DAOYUAN, CHEN CHENGBIN, ZHOU GUANGYU, et al. Study on introduction of wild rice DNA into cultivated rice [G] //Zhou Guangyu, et al. Research Progress of Agricultural Molecular Breeding. Beijing: China Agricultural Science and Technology Press, 1993.

[42] HONG YAHUI, DONG YANYU, ZHAO YAN, et al. Study on introduction of DNA from dense-panicle sorghum into rice[J]. Journal of Hunan Agricultural University, 1999, 25 (2): 87 - 91.

[43] WAN WENJU, PENG KEQIN, ZOU DONGSHENG. Research on genetic engineering rice[J]. Hunan Agricultural Sciences, 1993 (1): 12 - 13.

[44] ZHAO BINGRAN. Study on introducing genomic DNA of distant species into rice[D]. Changsha: Hunan Agricultural University, 2003.

[45] YUAN LONGPING. Research progress in super hybrid rice breeding[G]//Yuan Longping. Collected papers of Yuan Longping. Beijing: Science Press, 2010.

[46] DENG HUABING, DENG QIYUN, CHEN LIYUN, et al. Yield increasing effect of wild rice yield increasing QTL introduced into male parent 9311 of super hybrid mid-season rice[J]. Hybrid Rice, 2007, 2 (4): 49 - 52.

Chapter 4
Molecular Breeding of Hybrid Rice

Mao Bigang / Zhao Bingran

Molecular breeding, i. e., breeding under the guidance of theories of classical genetics, modern molecular biology and molecular genetics, integrates modern biotechnological tools in classical genetic breeding methods and combines phenotype and genotype screening to produce excellent new varieties (Wan, 2007). Rice molecular breeding mainly includes molecular marker-assisted breeding, genome-assisted breeding, transgenic breeding, exogenous DNA introduction breeding and genome-editing breeding. Molecular marker-assisted selection (MAS) in rice breeding has been carried out for many years, and earlier hybrid rice varieties bred through this method with disease and pest resistance have been used in production. With better understanding of gene functions for important agronomic traits of rice and the rapid development of the genome sequencing technology, genomic selection breeding and molecular design breeding have become important directions for hybrid rice breeding. With the support of major national projects, great progress has been made in transgenic hybrid rice breeding, such as in the field of transgenic *Bt* gene for pest resistance. The exogenous DNA introduction technique is a genetic breeding method advocated by Chinese scientists, and many new rice materials have been created using this technique. In recent years, the locus-specific gene editing technique has emerged as an important tool for molecular breeding of rice because it can precisely mutate endogenous genes while eliminating transgenic components.

In 1998, China participated in the International Rice Genome Sequencing Project (IRGSP) as a major initiator and participant, and took the lead in completing the precise sequencing of chromosome 4 of *Oryza sativa L. ssp. japonica cv. Nipponbare* and the whole genome sketch of the super *indica* hybrid rice parent 9311 in 2002. In 2005, IRGSP announced the completion of whole-genome precise sequencing of Nipponbare, followed by rapid development of functional genome research in rice. Related technologies and resource platforms have been continuously improved and expanded, and a large number of important functional genes were isolated and identified. As of 2017, more than 2,300 rice genes had been mapped or cloned (China Rice Data Center, www.ricedata.com and the mechanisms behind some important biological issues in rice breeding have been elucidated. Rice genome sequencing and functional gene research have laid the foundation for technological changes in rice breeding.

Section 1 Techniques and Principles for Hybrid Rice Molecular Breeding

Breeding is a systematic project and even an art, which involves the selection, aggregation and balance of various agronomic traits. Traditional breeding techniques are based on phenotypic selection, and breeders mainly rely on breeding experience to combine superior traits, which involves of a long cycle and great uncertainties. The use of molecular methods can help achieve precise gene transfer, mutation and screening, thus improving breeding efficiency and breaking through some bottlenecks in conventional breeding of hybrid rice. Therefore, the application of molecular technology can lead to the upgrading of hybrid rice breeding technology and promote new development.

Ⅰ. Marker-assisted Selection Breeding Technology

Selection is one of the most important aspects of breeding. Essential selection is to select the genotypes that meet the target requirements in a population. MAS is the breeding technology that links molecular markers closely to target genes to ensure the presence of target traits, which has the advantage of being fast, accurate and not interfered by environmental conditions. MAS can be used as a supplementary means in various breeding processes such as the identification of parental kinship, transfer of quantitative and recessive traits in backcross breeding, selection of hybrid progeny, prediction of hybrid heterosis and verification of variety purity.

1. Types of molecular markers

With the development of molecular biology and related technology, the first-generation molecular markers based on traditional Southern blot, represented by restriction fragment length polymorphism (RFLP), and second-generation molecular markers based on PCR, including simple sequence repeat (SSR), random amplified polymorphism DNA (RAPD), sequence tagged sites (STS), sequence characterized amplified region (SCAR), cleaved amplified polymorphism sequences (CAPS), amplified fragment length polymorphism (AFLP), expressed sequence tag (EST), etc., have emerged. Among these, SSR is the most widely used. The third-generation molecular marker techniques are based on microarrays and high-throughput sequencing, represented by single nucleotide polymorphism (SNP) (Wei Fengjuan et al., 2010). SSR markers are still widely used in many breeding institutions because of the low cost, simple development and design, low requirement for instrumentation and simple operations. SNP markers, on the other hand, are available in huge quantities and can be widely developed within genes, and their selection is highly reliable and can be developed into microarrays with extremely high throughput to achieve fine selection of all genes genome-wide. With the decreasing cost of high-throughput sequencing, SNP markers will be the mainstay in MAS breeding in the future.

2. Genetic basis for marker-assisted selection

The genetic basis of MAS is the close linkage, i.e., co-segregation, of the detected molecular marker and the target gene. Selection for genotypes of target traits with the help of molecular markers is based on the detection of genotypes of molecular markers co-segregation with the target genes to infer and know the genotypes of the target genes. The reliability of selection depends on the recombination frequency between the loci of the target genes and those of the markers, the smaller the genetic distance between the two loci (generally less than 5 cM), the higher the reliability, and vice versa. The above strategy of MAS mainly targets QTLs, and the linkage markers are only closely related to the real target genes. If the molecular marker is within the target gene, the reliability of selection is 100%. At present, many important agronomic traits used in production are only located to QTLs, and before the genes are cloned, we can still use the molecular marker with the closest linkage to the QTLs to complete selection for the agronomic trait.

When MAS focuses on the selection of target genes, it is known as foreground selection, while when it is used for the screening of the genetic background, it is called background selection. Unlike foreground selection, background selection targets the entire genome. In segregated populations, each chromosome may be "reassembled" from bi-parental chromosomes into a heterozygote because of the exchange of homologous chromosomes during the formation of gametes in the previous generation. Therefore, to do selection based on the entire genome, it is necessary to know the composition of each chromosome. It is required that the markers used for selection cover the entire genome, i. e. there must be a complete molecular marker linkage map. When the genotypes of all markers covering the entire genome are known in an individual, it is possible to infer which parent the alleles at each marker locus come from, and thus the composition of all chromosomes in the individual.

While favorable alleles are transferred, unfavorable ones controlling other traits may also be transferred due to the unfavorable linkage with other genes in the donor, a phenomenon called genetic drag or linkage drag. To reduce the effect of linkage drag on the comprehensive traits of breeding materials, reverse screening of genetic background is required to ensure that the genetic background of the original superior materials is maintained to the maximum extent. In MAS, foreground selection is to ensure that the individuals selected from each backcross generation as the next round of backcross parents contain the target genes; while background selection is to accelerate genetic background restoration to the genome of recurrent parents (i. e. recovery rate) in order to shorten the breeding time. Theoretical studies have shown that background selection plays a very significant role.

3. Basic procedure of MAS breeding

The basic process of MAS breeding is similar to that of conventional breeding, except that the detection of molecular markers is added to the conventional phenotypic selection at each breeding generation.

As the first step, the overall protocol is designed based on specific breeding objectives, target gene parental material, target gene loci mapping and molecular marker technology. The second step is to select the target genes that are finely mapped and have clear genetic effects and phenotypic stability. The third step is to select the parents. The parents of a population should be complementary to each other and they need to have the following relationships. The selected parents carry the target allele in the target gene locus; different parents have genotypic differences in the target gene locus; molecular markers closely linked to the target gene must present polymorphism between the gene donor parents and other parents. The fourth step is to construct the breeding population. Generally, if the donor parents show poor traits on the whole, the recipient parents will be used as recurrent parents, and the favorable alleles of the donor parents will be transferred to the background of the recipient after multiple backcrossing. If the donor and recipient parents each have their own strengths, generally only one to three backcrosses will be performed followed by selfing and selection, which may as well start right from F_1. In the last step, the screening of generation materials is carried out and the identification of phenotypes and genotypes is gradually carried out in the process of breeding population generation propagation to select the breeding materials that meet the requirements of the objectives. The above is the core work of molecular breeding (Qian Qian et al., 2007).

4. Advantages of MAS

(1) MAS can be used for traits whose phenotypes are identified with much difficulty. Traits such as fertility restoration, wide compatibility, photo- and thermo-sensitive genic male sterility, disease and pest resistance, drought tolerance, high and low temperature tolerance, etc., which are susceptible to environmental influences, cannot be identified phenotypically, accurately and directly, but are rather time-consuming and laborious to identify. The identification of these phenotypes can be easily achieved using molecular markers.

(2) MAS can be used for multiple (allelic) genes controlling a single trait or for the selection of multiple traits simultaneously. For example, there may be multiple genes affecting the same trait (disease resistance, quality) in different breeding materials. Especially, when different alleles exist at the same locus, it is difficult to identify these alleles by phenotypes. For another example, in order to widen the spectrum and improve the persistence of disease resistance, multiple leaf blight resistance genes can be aggregated, but it is impossible to determine whether multiple genes are introduced from phenotypes alone, and then MAS of different leaf blight resistance genes is the only approach.

(3) MAS can be used to enhance the intensity of selection at the early stage. For example, traits such as PTGMS, high- and low-temperature tolerance, quality and adult plant resistance can be tested in seedlings at an early stage, and individuals with the target genes can be selected for inclusion in the study population as much as possible, which is equivalent to increasing the initial study population and enhancing selection intensity.

(4) MAS can be used for non-destructive trait evaluation and selection. For example, when the plant is evaluated and selected for disease and pest resistance, the number of offspring seeds that may be harvested will be reduced, or even no seeds will be harvested.

(5) MAS can accelerate the breeding process and improve breeding efficiency. Traditional backcrossing requires multiple generations of backcrossing and may also result in linkage drag. In contrast, through MAS, the speed of genetic background recovery of the recurrent parent can be increased through foreground and background screening, while encryption of chromosomal markers of target genes reduces the occurrence of linkage drag (Qian Qian et al., 2007).

II. Genome-assisted Breeding Technology

1. Genome-wide selection

In improving complex traits controlled by polygenes, MAS and marker recurrent selection have two drawbacks. One is that the selection of progeny population is based on QTL mapping. However, QTL mapping results with two parents are sometimes not universal, and QTL mapping results from genetic populations cannot be well applied to breeding populations. The other is that most important agronomic traits are controlled by multiple minor genes, and it is difficult to improve these quantitative traits using individual markers. Meuwissen et al. (2001) first proposed the concept of genomic selection (GS) or genome-wide selection (GWS), which uses marker data or haplotype data across the entire genome and phenotype data of individuals in an initial training population to estimate the genotype-to-phenotype prediction model for each marker in the context of high-density molecular marker genotype identification. In the subsequent breeding population, the estimated effect of each marker and the genotype identification data of individuals are used to predict the phenotype or breeding value (BV) of individuals, and then superior progeny are selected based on the predicted phenotype (Wang Jiankang et al., 2014).

2. Genome-assisted breeding technology

In the past two decades, sequencing technology has been changing rapidly, and the genome sequencing of different rice varieties has accumulated massive data. The development of genomics has promoted the formation and development of genome-assisted breeding (GAB) technology. GAB, as part of genomics, is a new theory and methodology of breeding under genomic assistance or guidance and a genomic sequence-oriented breeding strategy. According to a virtual genome design scheme done in advance, through a series of breeding methods or processes, a superior variety with a large number of favorable genes aggregated, genome coordination, coordinated gene interaction network and optimal genome structure is obtained (Qian Qian et al., 2007).

The difference between conventional breeding and GAB lies in the difference between unknown and known genetic composition of the object of improvement or design. The object of GAB improvement consists of tens of thousands of components (genes), and the process includes gene introduction and elimination, gene expression pattern adjustment, optimization of the core gene composition system and gene structure, etc. GAB requires breeders to start with the design of a variety with optimized genome structure to obtain the parental genome sequence and possible gene interactions of the parental materials. Then, in addition to conventional cross breeding, transgenic and gene modification technologies, as well as gene expression regulation technologies, should be used comprehensively to make the breeding process as precise as a scalpel in surgery. Finally, based on the parental genome information, breeders can simulate breeding, eliminate combinations that do not have advantage or even have disadvantages, thus getting rid of the traditional large-scale testing. Only small-scale precise combining tests are required to select superior combinations. Of course, to achieve such precise breeding, breeders need to establish databases of parental gene information or access public gene information databases of germplasm resources. In addition, a high-throughput testing platform is also required. Major foreign seed companies have achieved GAB for crops such as corn and wheat. With the rapid development of molecular technology in the past decade, some large domestic seed companies and research institutes are also gradually shifting their focus from MAS breeding of a few genes to GAB which will definitely become the mainstream for hybrid rice breeding in the future.

Ⅲ. Transgenic Breeding Technology

Transgenic breeding is the application of DNA recombination technology to introduce exogenous genes into the rice genome by biological, physical or chemical means to produce new rice varieties with stable inheritance and expression of exogenous genes. At present, the main gene transformation techniques used for rice are agrobacterium-mediated transformation and gene gun bombardment (particle bombardment).

1. Agrobacterium-mediated transformation

Agrobacterium-mediated transformation uses the natural plant genetic transformation systems to insert a target gene into a modified T-DNA region, thus achieving the transfer and integration of an exogenous gene into the plant cells by the infection of agrobacterium. Then, the transgenic plants are regenerated with cell and tissue culture techniques. The cells of *Agrobacterium tumefaciens* and *Agrobacterium rhizogenes* contain Ti plasmid and Ri plasmid respectively, with a segment of T-DNA, which can be inserted into the plant genome through *Agrobacterium* infection through plant wound. The Ti plasmid of *Agrobacterium tumefaciens* consists of four regions – the virulence region (Vir region), conjugation region (Con region), origin of replication region (Ori region) and T-DNA region. The 25 bp-long repeats at both ends of the T-DNA region are necessary for the transfer and integration of T-DNA. Generally speaking, T-DNA is preferentially integrated into the transcriptionally active region of plant cells, the highly repetitive region of chromosomes, and the homologous region of T-DNA. The integrated T-DNA also has a certain degree of duplication and deletion. The insertion of T-DNA into a plant genome is random, and it can be inserted into any chromosome during the DNA replication process of plant cells. The different insertion sites can result in different phenotypes and genetic characteristics in the transgenic plants. Although T-DNA can be inserted at multiple physical sites, or several copies can be integrated into the same position in tandem, the possibility of single copy or low copy insertion is still relatively high, which is different from other transformation methods.

Agrobacterium-mediated transformation has the advantages of high transformation frequency, low insertion copy number of exogenous genes, no damage to the recipient, and no need for expensive equipment. Since 1994, Hiei et al. used agrobacterium-mediated transformation of mature embryo callus of *japonica* rice to obtain a large number of transgenic plants, which stimulated a boom of agrobacterium-mediated transformation in rice breeding. Generally, the success rate is higher for *japonica* rice than for *indica* rice, and the transformation frequency for *indica* is not satisfactory, especially for typical *indica* varieties, which brought difficulties to the functional study of *indica*-specific genes and transgenic improvement of *indica* varieties. At present, with a large number of transformation trials and conditions mapped by researchers on *indica* materials, most *indica* varieties can also be successfully transformed with this method.

2. Gene gun bombardment

Gene gun bombardment is a method to transfer the target gene by shooting metal particles (gold or tungsten) absorbed in gene fragments into plant cells at a certain speed using a power system, directly penetrating the cell wall and cell membrane to integrate exogenous gene fragments into the plant genome. It has the advantages of wide application, simple process, short transformation time with high transformation frequency, and low cost. The transformation frequency of a gene gun is directly related to the receptor type, size of the micro-projectiles, bombardment pressure, distance between the stopping disk and the metal particles, receptor pre-treatment, and the culture after receptor bombardment. For plants that cannot be infected by *Agrobacterium tumefaciens*, this method can break the limitation of the vector method (An, Hanbing et al., 1997). Klein et al. (1989) first transformed maize with gene gun bombardment to obtain transgenic plants, and then successfully applied the method to rice, wheat, sorghum, barley and other important cereal crops, which have been widely used.

The success of gene gun bombardment and the frequency of transformation are related not only to the genotype of the test materials, but also to the selection of suitable explants as transformation objects. In the early days, most of the explants used for transformation were embryogenic suspension cells and subcultured embryogenic calli, but the regeneration ability of these explants was reduced after subculture, which would reduce the success rate of gene gun bombardment. Immature embryos, immature embryogenic calli and microspores, which are currently commonly selected as targets for bombardment, have a higher transformation frequency than mature embryos and calli induced. However, different physiological conditions of immature embryos can also affect the transformation frequency of gene gun bombardment, so extensive attempts are often needed in the selection of materials so as to ensure the optimal experimental conditions for different transformation materials.

Ⅳ. Breeding Technology of Exogenous DNA Introduction

Exogenous DNA introduction breeding technology refers to the molecular breeding technology that uses genomic DNA containing the target traits as the donor, and introduces or injects the donor DNA into the recipient plant by asexual reproduction, either directly or indirectly, to screen for mutant progeny containing the target trait and breed new plant varieties (Zhou Guangyu et al., 1988; Dong Yanyu et al., 1994; Zhu Shengwei et al., 2000). It includes methods such as DNA pollen-tube pathway (Zhou Guangyu et al., 1988), spike-stalk injection (Zhou Jianlin et al., 1997) and embryo soaking (Yang Qianjin, 2006).

1. Methods of exogenous DNA introduction

The pollen-tube pathway is a method of injecting a solution of DNA of a distant species containing the target genes into the stigma (or soaking the stigma) after plant pollination, so as to use the pollen-tube pathway naturally formed within the plant during flowering and fertilization to introduce exogenous DNA into fertilized egg cells of the recipient, integrate it into the genome of the recipient, and ultimately develop a new individual or mutant material carrying genetic information of a distant species. It is particularly suitable for crops such as cotton with large floral organs and multiple ovules (Dong Yanyu et al., 1994). The embryo soaking method is derived from the pollen-tube pathway method. It transfers the genetic information of a distant species by soaking germinating seeds or seedlings in exogenous total DNA solution or agrobacterium solution to introduce exogenous DNA into the genome of the recipient (Yang Qianjin, 2006). Spike-stalk injection is a method to obtain mutant material by injecting exogenous DNA solution into the first internode under the neck node of a young panicle at the appropriate developmental stage with a micro-injector so as to introduce and integrate exogenous DNA into the germ cells of the recipient through the processes of vessel transportation (Zhao Bingran et al., 1994). The spike-stalk injection method is particularly suitable for rice, a plant with small spikelets and single ovule. The above methods have been applied to a variety of crops such as cotton, rice, wheat, soybean, watermelon, ramie, and maize to create a large amount of new germplasm by transferring genomic DNA from distant species, and there are also successful examples of exogenous genes used for transferring clones (Pena et al., 1987; Zhou Guangyu, 1993).

2. Basic procedure of breeding

The basic process of breeding hybrid rice by introducing genomic DNA of distant species are as follows. 1) Select appropriate distant species based on the defects of the hybrid rice parents to be improved and extract their genomic DNA. 2) Introduce exogenous DNA into the recipient plants by one of the three methods mentioned above. 3) Select a mutant from the D_1 (or D_2) population. 4) Breed the new parents of hybrid rice by conventional methods. 5) Carry out selection, variety comparison, regional trial and promotion.

3. Characteristics of exogenous DNA introduction

Practice shows that the exogenous DNA introduction technology has four advantages in breeding. 1) It breaks the reproductive barrier between species, and uses distant species to create new germplasm. 2) It avoids the operation process such as tissue culture and does not require a lot of laboratory work. 3) It is easy to stabilize the mutation because only a few donor DNA fragments enter the genome of the recipient. 4) It is easy to integrate with conventional breeding techniques. However, the disadvantages are that the mutation rate is unstable and the occurrence of mutated traits is random.

4. Molecular mechanism of creating germplasm by exogenous DNA introduction

The introduction of genomic DNA from distant species has created a large amount of new crop germplasm with improved traits, but the molecular mechanisms involved have rarely been studied. By applying 3H labeling of large molecules (50 kb) of cotton DNA, the Shanghai Institute of Biochemistry, Chinese Academy of Sciences, and the Institute of Economic Crops, Jiangsu Academy of Agricultural Sciences, demonstrated that the pollen-tube pathway is the only way for exogenous DNA to reach the embryo sac from the micropyle (Huang Junqi et al., 1981). Zhao Bingran et al. (1998) found that, by spike-stalk injection, the exogenous DNA is most likely transported through the vessel before it reaches the end of the vascular bundle and finally enters the cells of the recipient through plasmodesma.

After the exogenous DNA enters the recipient cells, Zhou Guangyu et al. (1979) put forward the hypothesis of DNA fragment hybridization based on the analysis of super-distant (between parents from different genura or at a higher level) hybrid materials in Gramineae plants, suggesting that although the chromosomes of such distant parents are not compatible to each other as a whole, due to the conserved and relatively slow evolution of DNA molecules, some homology may be maintained between the structures of some genes, and thus DNA fragment hybridization, i. e., the replacement of homologous fragments of the recipient genome by exogenous DNA, can occur. Wan Wenju et al. (1992) proposed that genomic DNA from distant species has both gene transfer and bio-mutagenic effects during exogenous

DNA introduction.

Preliminary studies found that the rice mutant lines acquired DNA fragments that were highly homologous to the donor *Oryza. minuta* or *Zizania latifolia* (*Griseb*) Turcz. ex Stapf, but not to the recipient (Zhao et al., 2005; Zhenlan Liu et al., 2000), respectively. A comparative genomic study between donor, recipient and mutant lines by Zhao Bingran et al. revealed that the polymorphism of DNA bands between the mutant lines and the recipient was about 5%, there were DNA fragments in the mutant lines that were highly homologous to the donor but not the recipient, and there were hotspots at the mutant loci (Zhao Bingran et al. 2003; Zhao et al. 2005; Xing et al. 2004). Peng et al. (2016) identified genes associated with variation in rice quality traits in the wild rice DNA introduction variant line YVB and found that allelic variation in the quality key genes *qSW5*, *GS5*, *Wx*, and *GW8* may be due to the introduction of exogenous DNA.

V. Genome-editing Breeding Technology

1. Genome editing

Genome editing is a technique that uses artificially engineered nucleases to precisely edit genes *in vivo*. The most critical step is to use engineered nucleases to generate double-strand breaks (DSBs) at the target site, and to modify the genome through such self-repairing method of homology-directed repair (HDR) or non-homologous end joining (NHEJ). At present, the three most used sequence-specific engineered nucleases are zinc finger nucleases (ZFNs), transcription activator-like effector nucleases (TALENs) and clustered regularly inter-spaced short palindromic repeats (CRISPR/Cas9 system). A common characteristic of these three kinds of enzymes is that they all precisely cleave the DNA double strands at specific sites in the genome, causing DSBs, which can greatly increase the probability of chromosomal recombination events. The repair mechanism of DSBs is highly conserved in eukaryotic cells, and consists of two main repair pathways, HDR and NHEJ. When a homologous sequence donor DNA is present, HDR-based repair can produce a precise site-specific substitution or insertion, while in the absence of a donor DNA, cells are repaired by the NHEJ pathway. Because NHEJ repair is often not precise, a small number of nucleic acids based on insertions or deletions (InDel) are often produced at the DNA strand break site, resulting in mutation (Fig. 4 - 1). Compared with ZFN and TALEN technologies, CRISPR/Cas9 is more efficient in editing (Wang Fujun et al., 2018).

CRISPR/Cas9 is an adaptive immune system of bacteria and archaea against viruses and exogenous DNA invasion. Cas9 protein contains two nuclease structural domains that cleave both of the single strands of the DNA separately. Cas9 first binds to crRNA and tracrRNA to form a complex, then binds and invades DNA through the PAM sequence (5'-NGG-3') and form an RNA-DNA complex structure, which in turn cleaves and breaks the DNA double strand. Due to the simple structure of PAM sequences, a large number of targets can be found in almost all genes. The crRNA and tracrRNA are genetically engineered and ligated together to obtain single-guide RNA (sgRNA). The fused RNAs have similar viability to wild-type RNAs, but are easier for researchers to use because of their simplified structure. By ligating sgRNA-expressing element to the Cas9-expressing element, a plasmid that can express both can be

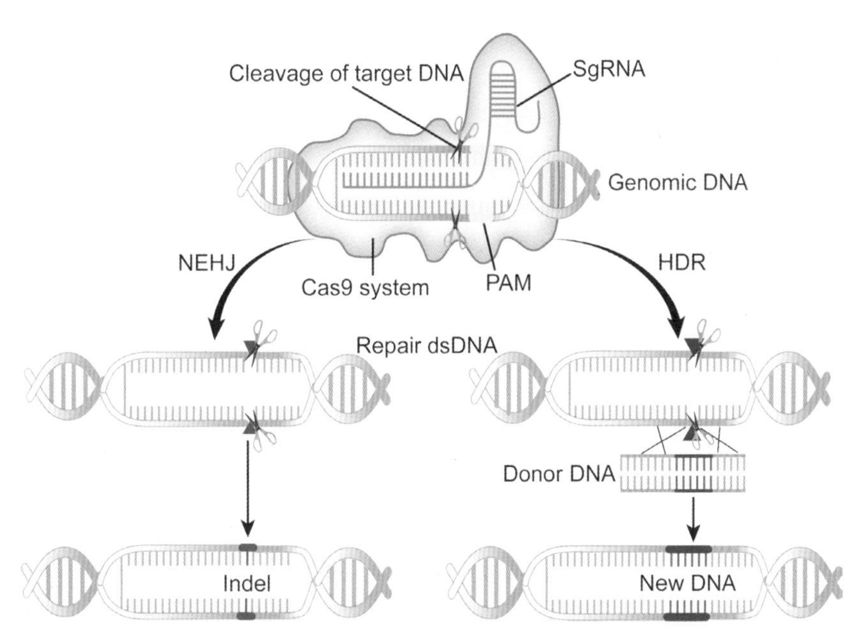

Fig. 4 - 1 Cas9 gene editing and DSB repair in eukaryotic cells

transferred into cells to manipulate the target gene (Ran et al., 2013; Caj et al., 2013). In 2012, Jinek et al. were the first to demonstrate *in vitro* that Cas9 could specifically cleave target DNA sequences under synthetic sgRNA guidance, and in 2013, scientists achieved *in vivo* specific cleavage of a target DNA sequence with a modified CRISPR/Cas9 system. As such operation gets simpler and cheaper, it is rapidly put into application in various fields, making it the mainstream genome editing technology today.

2. Innovation of rice genome-editing technology system

In 2013, three laboratories in China were the first to succeed in using CRISPR/Cas9 technology for targeted knockout studies of rice genes. Gao Caixia's research team at the Institute of Genetics and Developmental Biology (IGDB), Chinese Academy of Sciences (CAS), was the first to use the CRISPR/Cas9 system to mutate four genes in rice in a targeted manner, including *OsPDS*, *OsBADH2*, *Os02g23823* and *OsMPK2*, with a target mutation rate of 4%-9.4% in transgenic rice plants, which was the first application of genome editing technology to plants (Shan et al., 2013). Meanwhile, Zhai Lijia's laboratory at Peking University used the CRISPR/Cas9 system to target mutations in the rice chlorophyll B synthesis gene *CAO1* and the tiller angle control gene *LAZY1* (Miao et al., 2013). In the laboratory of Zhu Jiankang, Shanghai Research Center for Plant Stress Biology, CAS, the rice leaf curl control gene *ROC5*, the chloroplast formation-related gene *SPP*, and the young seedling albino *YSA* were the targeted mutations (Feng et al., 2013). Since then, there have been continuous innovations in the genome editing system, resulting in technologies such as multi-gene targeting and single-base editing.

(1) Multi-genome editing technology in rice

Efficient multi-genome editing is often required when breeding rice varieties with multiple superior trait aggregation. The CRRI and Yangzhou University (YU) collaborated to develop a multi-genome editing system that enables rapid assembly of multiple sgRNAs, which can be used to obtain multiple mutations in the current generation with only one transformation (Wang et al., 2015). Liu Yaoguang's team at South China Agricultural University (SCAU) developed a CRISPR/Cas9 vector system consisting of multiple sgRNA expression cassettes to simultaneously perform targeted editing at 46 target sites in rice and the average effective editing rate reached 85.4%, mostly homozygous mutation and bi-allelic site mutation (Ma et al., 2015). In most cases, CRISPR/Cas9 genome editing produces base substitutions or small fragment insertion deletions (InDels), while large fragment deletions are rare. Using CRISPR/Cas9, Syngenta researchers achieved the deletion of a 10 kb fragment of the *DEP1* gene in *indica* rice, and this efficient large fragment deletion editing will also extend the application of the genome editing technology (Wang et al., 2017).

(2) Efficient base editing technology

The implementation of base editing can further expand the application of genome editing to create germplasm resources with new functions. The effect of base editing can be achieved by changing guanine or cytosine to adenine or uracil (C→T, or G→A) respectively, through deamination. Gao Caixia's team at the IGDB, CAS, has successfully constructed an efficient and precise base editing system. Using Cas protein nCas9 (Cas9-D10 anickase) to fuse two base editing enzymes, cytosine deaminase *APOBEC1* and uracil glucosylase inhibitor (UGI), respectively, the base editing system in plants was constructed and base substitution was performed for rice endogenous genes *OsCDC48*, *OsNRT1.1B* and *OsSPL14*. The results showed that the pnCas9-PBE system was outstanding for base editing, where *OsCDC48* produced 43.48% base mutations and the edited genotypes did not contain unintended InDel mutations (Zong et al., 2017). The nCas9-PBE base editing system can be used for base substitution in many crops, resulting in amino acid substitution or termination mutations, producing more genomic types with unique traits, and providing crop breeders with more base materials or varieties.

3. Principles of genome-editing breeding technology

The genome editing breeding technology is essentially different from the existing transgenic breeding technology in that genome editing adjusts or modifies a specific DNA sequence that has been determined in the genome. It is to modify the genome of an organism by knocking out, inserting or replacing one or several bases, or a DNA sequence, so that the functions of a negatively regulated gene is lost or weakened and the expression of a positively regulated gene is enhanced, allowing the crop to obtain superior traits without introducing exogenous genes. The genome editing breeding technology can break the bottleneck of conventional breeding and rapidly create new germplasm for hybrid rice breeding. Compared with traditional transgenic breeding, genome editing can eliminate transgenic safety concerns by eliminating exogenous genes through selfing or hybridization after targeted modification of specific genes. Therefore, since the successful application of the genome editing technology to plants, the optimization of this technology and its application to crop genetic improvement has become a major focus of investment and R&D in various countries and many international agricultural biotechnology companies.

In the process of genome editing, it is of critical importance to find a pair of "scissors" with its own "navigation system". The CRISPR/Cas9 technology is a new pair of "scissors" that has emerged in recent years. The CRISPR/Cas9 system is a genome editing technique in which the Cas9 protein is directed to cleave genomic DNA under the guidance of sgRNA to change gene function. This technology introduces foreign DNA fragments (such as marker genes, bacterial plasmids, etc.) into rice during specific operations. Since the inserted fragment (Cas9 system) and the edited target gene are generally not on the same chromosome (at least not closely linked); with the exchange and separation of chromosomes, the inserted fragment and the mutant target gene will enter different offspring individuals. Through the screening of transgenic components, individuals having the inserted fragments can be eliminated while the edited individuals without transgenic components will remain.

In addition, backcrossing with donor parents can prevent unintended traits caused by off-target editing systems. Because of its simple structure, high efficiency and ease to use, the CRISPR/Cas9 system has become a very popular tool for genome editing. This tool has great potential in medical science and agricultural breeding because of its ability to make targeted modifications to an organism's own genes, using precision-guided gene scissors that can efficiently and accurately modify the genome according to human will. For example, while traditional methods may take years or even decades to improve a variety, using the CRISPR/Cas9 technology, it may only take a few weeks to improve a single gene in a variety.

Section 2 Gene Cloning for Important Agronomic Traits in Rice

Cloning of important agronomic trait genes and molecular network resolution are the basis of molecular breeding in rice. With the rapid development of sequencing and molecular technology in the past two decades, rice functional genome research has also developed rapidly, and a large number of genes for important agronomic traits such as disease and pest resistance, high yield and efficiency, quality, and stress tolerance have been cloned or found gene markers, laying the foundation for molecular breeding research through transfer, selection or regulation and mutagenesis.

Ⅰ. Genes for Disease and Pest Resistance in Rice

The main diseases and pests that affect rice production include rice blast, bacterial leaf blight, sheath blight, false smut, brown planthopper (BPH), stem borers, and leaf folder. In the past decade, researchers have identified and cloned a series of genes for resistance to rice blast, bacterial leaf blight and brown planthopper and applied them to improve the resistance of rice varieties.

1. Genes for rice blast resistance

Rice blast is an extremely widespread disease in rice producing areas around the world, and it is also the most harmful rice disease in China, where the incidence of rice blast reaches more than 3.8 million hectares of rice fields per year, with losses of hundreds of millions of kilograms of rice, which is about 11%-13% of the annual rice yield, posing a grave threat to China's food security (Ao Junjie et al., 2015). Rice blast prevention and control is generally achieved through chemical control or by growing resistant varieties. The problem with chemical control is that there is no long-term effective fungicide, the cost is high and it can cause environment pollution. In contrast, breeding rice varieties resistant to rice blast is a more economical and effective way to control rice blast.

To date, more than 90 major resistance genes and more than 350 resistance QTLs have been identified and mapped out in different rice germplasms. Most of the mapped genes are located on chromosomes 6, 11 and 12. For example, 14 genes are mapped on chromosome 6 in a cluster near the centromere, including *Pi2*, *Pi9*, *Pi50*, *Pigm*, *Piz* and *Piz-t*, which are all complex alleles at the *Piz* locus. There is a larger cluster of resistance genes containing 22 localized genes at the end of the long arm of chromosome 11, including *Pi34*, *Pb1*, *Pi38*, *Pi44*, *Pikur2*, *Pi7*, *Pilm2*, *Pi18* and *Pif*, all distributed in a cluster near the *Pik* locus. In addition, 14 genes, including *Pita*, *Pita2* and *Pi6*, are also distributed in a cluster near the centromere of chromosome 12 (He Xiuying et al., 2014).

Since the first rice blast resistance gene was cloned in 1999, 24 dominant resistance genes and two recessive resistance genes have been finely mapped and cloned using the gene map-based cloning technology (Table 4 - 1). In terms of the proteins encoded by the genes, *pi21* encodes a proline-rich protein, *Pid2* encodes a receptor-like protein kinase, *Bsr-d1* encodes a C2H2-type transcription factor, *Bsr-k1* encodes a TPR protein, and the other genes encode NBS-LRR protein. The mapping of rice blast resistance laid the foundation for effective MAS. After the accurate mapping of resistance genes, functional markers of resistance genes can be developed, making the resistance improvement of hybrid rice parents more reliable and much more efficient.

Table 4 - 1 Cloned genes for blast resistance

Gene	Symbol	Chromosome	Gene query number	Reference
Rice blast resistance gene	*Pish*; *Pi35*	1	LOC_Os01g57340	Takahashiet al., 2010
Rice blast resistance gene	*Pit*	1	LOC_Os01g05620	Bryanet al., 2000
Rice blast resistance gene	*Pi37*	1	DQ92349. 1	Linet al., 2007
Rice blast resistance gene	*Pib*	2	AB013448	Wanget al., 1999
Rice blast resistance gene	*Bsr-d1*	3	LOC_Os03g32220	Liet al., 2017
Rice blast resistance gene	*pi21*	4	LOC_Os04g32850	Fukuokaet al., 2009
Rice blast resistance gene	*Pi63*	4	AB872124	Xuet al., 2014
Rice blast resistance gene	*Pi-d2*; *Pid2*	6	LOC_Os06g29810	Chenet al., 2006
Rice blast resistance gene	*Pid3*; *Pi25*	6	LOC_Os06g22460	Shanget al., 2009
Rice blast resistance gene	*Pi9*; *Pigm*; *Pi2/Piz-5*; *Pi50*; *Piz*; *Pi*	6	LOC_Os06g17900	Quet al., 2006
Rice blast resistance gene	*Pizt*;	6		Zhouet al., 2006
Rice blast resistance gene	*Pi36*	8	LOC_Os08g05440	Liuet al., 2007
Rice blast resistance gene	*Pi56(t)*	9	LOC_Os09g16000	Liuet al., 2013

Continued

Gene	Symbol	Chromosome	Gene query number	Reference
Rice blast resistance gene	*Pi5*; *Pi5-1*; *Pi3*; *Pi-i*	9	LOC_Os09g15840	Leeet al. , 2009
Rice blast resistance gene	*bsr-k1*	10	Os10g0548200	Zhouet al. , 2018
Rice blast resistance gene	*Pik-m*; *Pik-p*	11	AB462324	Ashikawa et al. , 2008
Rice blast resistance gene	*Pigm*	11	KV904633	Denget al. , 2017
Rice blast resistance gene	*Pb-1*	11	AB570371	Hayashiet al. , 2010
Rice blast resistance gene	*Pi1*	11	HQ606329	Huaet al. , 2012
Rice blast resistance gene	*Pik*	11	HM048900	Zhai et al. , 2011
Rice blast resistance gene	*Pik-h*; *Pi54*; *Pi54rh*	11	LOC_Os11g42010	Sharmaet al. , 2005
Rice blast resistance gene	*Pia*; *PiCO39*	11	LOC_Os11g11790	Zenget al. , 2011
Rice blast resistance gene	*Pita*; *Pi-4a*	12	LOC_Os12g18360	Bryanet al. , 2000

The first cloned rice blast resistance gene, *Pib*, encodes a protein composed of 1,251 amino acids, which is induced and regulated by environmental conditions such as temperature and light (Wang et al., 1999). *Pita* is the second cloned rice blast resistance gene, which encodes a plasma membrane receptor protein 928 amino acids in length. There is only one amino acid difference at the *Pita* locus for resistance or susceptibility, i. e. alanine for resistance, serine for susceptibility at amino acid site 918 (Bryan et al., 2000).

The *Pi9* gene showed high resistance to 43 rice blast strains from 13 countries (Liu et al., 2002), which was constitutively expressed in resistant plants and was not induced by rice blast infection (Qu et al., 2006). *Pi9*, *Pi2/Piz-5*, *Pi50*, *Piz* and *Pi* are alleles, and both *Pi9* and *Pi2* encode a protein composed of 1,032 amino acids. The encoded products of *Piz*-t and *Pi2* differed by eight amino acids only in three LRRs, and these eight mutant amino acids caused the difference in resistance specialization.

The resistance of *Pikm*, *Pikh*, *Pi1*, *Pik* and *Pia* genes cloned on chromosome 11 is jointly affected by two adjacent NBS-LRR resistance proteins. *Pikm* consists of *Pikm1-TS* (1,143aa) and *Pikm2-TS* (1,021aa), and *Pikp* is consists of *KP3* (1,142aa) and *KP4* (1,021aa). *Pikm1-TS* is 95% homology with *KP3* protein and *Pikm2-TS* is 99% homology with *KP4*. While *Pi1* is consisted of *Pi1*-1 (1,143aa) and *Pi1-2* (1,021aa), *Pik* consists of *Pik1*-1 (1,143aa) and *Pik1*-2 (1,052aa), and *Pia* is consisted of *Pia-1* (966aa) and *Pia-2* (1,116aa). Similarly, these alleles encode proteins with high homology (He Xiuying et al., 2014).

The broad-spectrum resistance gene, *Pigm*, is a gene cluster containing multiple NBS-LRR resistance genes and consists of two functional proteins, *PigmR* and *PigmS*. *PigmR* is constitutively expressed in above-ground organs of rice, forming a homodimer that exerts broad-spectrum disease resistance. However, *PigmR* leads to lower 1,000-grain weight and lower yield. *PigmS* is epigenetically regulated and is only highly expressed in rice pollen, with low expression in leaves, stems, and other pathogen-infected tissues. Yet, it can increase rice seed setting rate and counteract the effect of *PigmR* on yield. *PigmS* can compete with *PigmR* to form heterodimers to suppress *PigmR*-mediated broad-spectrum resistance. However, the low level of *PigmS* expression provides a "refuge" for pathogens, and the evolutionary selection pressure of pathogens reduces, slowing down the pathogenic evolution of pathogenicity to *PigmR*, so that *PigmR*-mediated disease resistance is durable (Deng et al., 2017).

Chen Xuewei's team at Sichuan Agricultural University (SAU) discovered a natural variation in the promoter of *Bsr-d1*, a gene encode a C2H2-like transcription factor with broad-spectrum and durable resistance to rice blast, in the broad-spectrum and highly resistant rice Digu. A key base variation at position 618 in the promoter region of the gene *bsr-d1* resulted in enhanced promoter binding of the upstream MYB transcription factor to *bsr-d1*, which inhibits the expression of *bsr-d1* in response to induction by *Magnaporthe grisea* and led to down-regulation of the expression of the H_2O_2 degradase gene directly regulated by *Bsr-d1*, resulting in intracellular H_2O_2 enrichment and improved immune response and disease resistance (Li et al., 2017). In addition, the team screened a disease-resistant mutant *bsr-k1* by artificial chemical mutagenesis, and cloning revealed that the *Bsr-k1* gene encodes a TPR protein with RNA-binding activity that binds to the mRNAs of multiple *OsPALs* gene members (such as *OsPAL1*-7) associated with the immune response, and folds for degradation, ultimately resulting in reduced lignin synthesis and weakened disease resistance. The loss of *Bsr-k1* protein function causes the accumulation of *OsPAL* gene mRNA, conferring resistance to rice blast and bacterial leaf blight (Zhou et al., 2018). The discovery of these two novel broad-spectrum disease resistance mechanisms has greatly enriched the molecular theoretical basis of immune response and disease resistance in rice and provides new insights for breeding hybrid rice parents with durable resistance to rice blast.

The gene *pi21* comes from the upland rice variety *Owari hatamochi* and encodes a 266 amino acid protein with a proline-rich C-terminal and an N-terminal heavy metal-binding and protein-interacting structural domain. *pi21* has 21bp and 48bp deletions in the disease-resistance gene compared to the susceptible variety. *pi21* is a non-small species-specific gene associated with basal resistance and stimulates a slow disease resistance response, and this low-rate induced disease resistance response may be a slow disease resistance response or a novel mechanism for a durable disease resistance response (Fukuoka et al., 2009).

2. Resistant genes to bacterial blight

Rice bacterial blight is a bacterial vascular disease caused by Gram-negative bacterium *Xanthomonas oryzae pv. Oryzae* (*Xoo*). Since its first discovery in 1884 in Fukuoka, Japan, it has become one of the most important diseases in rice production (T. W. Mew, 1987) and can reduce rice yields by 20%-30%, or in severe cases up to 50%, even causing a complete crop failure. The disease is more likely to occur in humid and low-lying areas, and generally more serious in *indica* than in *japonica*, in double-cropping late rice than in early rice, and in single-cropping mid-season rice than in single-cropping late rice (Chen Hesheng et al., 1986). In recent years, the prevalence of bacterial blight in the Yangtze River Basin and South China has been on the rise. With the development of the Belt and Road Initiative, hybrid rice will be gradually promoted to Southeast Asia where bacterial blight is the main rice disease; the promoted varieties must have bacterial blight resistance.

Up to now, 40 bacterial blight resistance genes have been identified from cultivated and wild rice, 27 are dominant (*Xa*) and 13 are recessive (*xa*). Among these, 32 have been mapped and 9 have been isolated and cloned (Table 4 – 2), of which *Xa21*, *Xa23* and *Xa27* are from wild rice.

Table 4 – 2 Cloned Genes with Bacterial Leaf Blight Resistance

Gene	Symbol	Chromosome	Gene query number	Reference
Bacterial blight resistance gene	*Xa1*	4	LOC_Os04g53120	Yoshimuraet al., 1998
Bacterial blight resistance gene; bacterial leaf streak resistance gene	*Xa5*	5	LOC_Os05g01710	Blairet al., 2003
Bacterial blight resistance gene	*Xa27*	6	LOC_Os06g39810	Guet al., 2005
Bacterial blight resistance gene;	*xa13*; *Os8N3*	8	LOC_Os08g42350	Chuet al., 2006
Gene of susceptibility to bacterial blight; bacterial blight resistance gene	*Xa26*; *Xa3*	11	LOC_Os11g47210	Sunet al., 2004
Bacterial blight resistance gene	*Xa23*	11	LOC_Os11g37620	Wanget al., 2014
Bacterial blight resistance gene	*Xa21*; *Xa-21*	11	LOC_Os11g35500	Songet al., 1995
Gene of susceptibility to bacterial blight	*Os11N3*	11	LOC_Os11g31190	Antonyet al., 2010
Bacterial blight resistance gene; TAL effector-mediated resistance gene	*Xa10*	11	JX025645	Tianet al., 2014
Bacterial blight resistance gene	*Xa3*; *Xa4b*; *Xaw*; *Xa6*; *xa9*	11		Xianget al., 2007
Bacterial blight resistance gene; sucrose transporter gene	*xa25*; *OsSWEET13*	12	LOC_Os12g29220	Liuet al., 2011
Bacterial blight resistance gene	*Xa4*		KU761305	Huet al., 2017

Xa21 is the first cloned bacterial blight resistance gene from *Oryza longistaminata*. *Xa21* encodes a receptor-like protein kinase consisting of 1,025 amino acids. It is divided into nine regions, starting from the amino terminus, signal peptide region, unknown functional region, leucine-rich repeats (LRRs), charged region, transmembrane domain (TMD), charged region, juxtamembrane region, serine/threonine kinase (STK) region and carboxyl terminal (CT) region. Among them, LRRs and STK are two important functional domains related to the resistance expression of *Xa21*. The former is composed of 23 incomplete LRRs involved in protein interactions and associated with recognition of pathogens; the latter is a typical signaling molecule containing 11 subdomains and 15 conserved amino acids (Song et al., 1995).

The *Xa23* gene comes from China's common wild rice *Oryza rufipogon*, which shows high resistance to all available domestic and foreign differential lines of bacterial blight, and is completely dominant and resistant throughout the growth period of rice (Wang et al., 2014). The gene is a class of executor R genes and the susceptible *xa23* gene has the same open reading frame (ORF113) as the resistant *Xa23* gene but the TALE effector binding element (EBE) of *AvrXa23* is missing in the promoter region. Under normal conditions, *Xa23* 's ORF113 is transcribed at low levels in both disease-resistant and susceptible varieties, but is highly expressed in resistant plants induced by the pathogen and unchanged in susceptible varieties. Susceptible *xa23* in JG30 and resistant *Xa23* in CBB23 have a 7 bp polymorphism in the promoter region at a site that coincides with *AvrXa23* EBE, and *Xa23* is functions and resists disease by recognizing TALEs in pathogens.

The resistant variety IRBB27 contains the *Xa27* gene. The coding sequence of *Xa27* is identical in the susceptible variety IR24 and the resistant variety IRBB27, except for two differences in the promoter region. The *Xa27* promoter in IR24 has an extra 10 bp sequence about 1.4 kb upstream of the ATG and an extra 25 bp sequence in front of the TA frame compared to IRBB27, which causes the difference in gene expression. The *Xa27* disease resistance allele and the disease susceptibility allele encode the same protein, but only the disease resistance allele is expressed after inoculation with pathogenic bacteria carrying the nuclear-localized Type Ⅲ effector *avrXa27* (Gu et al. 2005).

Recently, researchers have discovered that *Xa4* encodes a cell wall-related kinase that enhances cell wall strength by promoting cellulose synthesis, building a strong fortress for plant cells and defending them against bacterial blight infection. At the same time, the enhanced cell wall greatly improves the mechanical strength of the rice stem, which enhances the lodging resistance of rice to a certain extent. This "strong wall" defense strategy of *Xa4* ensures durable resistance to bacterial blight, while achieving excellent agronomic traits (Hu et al., 2017).

The recessive resistance gene *xa13* is an allele of *Os8N3*, a member of the rhizobia (NODULIN, N3) gene family. *Os8N3* is a host susceptibility gene for bacterial blight and is a member of the *MtN3* gene family, encoding a membrane-intrinsic protein. The expression of *Os8N3* is induced by the bacterial blight *PXO99A* and is dependent on Type-Ⅲ effector gene *PthXo1*. Both transcription activator-like (TAL) effectors *AvrXa7* and *PthXo3* activate the expression of *Os11N3*, another member of the N3 family. *Os11N3* insertional mutation or RNA-mediated silencing results in loss of specific susceptibility to those pathogenic races dependent on the effectors *AvrXa7* and *PthXo3*. *Os8N3* and *Os11N3* encode closely related proteins, which contribute to the role of N3 proteins in promoting the pathogenesis of bacterial blight (Chu et al., 2006; Antony et al., 2010).

The *xa25* gene encodes a protein belonging to the *MtN3*/saliva family, which is commonly found in eukaryotes. The proteins encoded by the recessive *xa25* and the dominant *Xa25* have a difference in eight amino acids (Liu et al., 2011). The sucrose transporter gene *OsSWEET13* can act as a disease susceptibility gene for the TAL effector *PthXo2*. *OsSWEET13* is allelic to *xa25*, and there is a potential susceptibility to *PthXo2*-mediated bacterial blight due to changes in the *OsSWEET13* promoter in *japonica* rice (Zhou et al., 2015).

The *xa5* gene is a recessive gene for resistance to bacterial blight and possibly to resistance and possible bacterial leaf streak. The protein *xa5* encoded is the γ subunit transcription factor Ⅱ A (*TF* Ⅱ Aγ). Unlike previously discovered disease resistance genes. *TF* Ⅱ Aγ is a transcription factor of eukaryotes (Blair et al., 2003). There are two base differences in *xa5* between resistant IRBB5 and susceptible variety Nipponbare as well as IR24, which results in the change of valine at position 39 in IRBB5 into glutamate in Nipponbare and IR24. The amino acid site is located on the surface of three-dimensional protein structure, which may be related to protein interaction.

3. Brown planthopper resistance genes

Brown planthopper (BPH, *nilaparvata lugens*) is a monophagous rice pest, belonging to the genus *Nilaparvata* of *Homoptera Delphacidae*. It is widely distributed in China and is seasonal, migratory, highly reproductive and violent, and is one of the main pests of rice in China. BPH affects more than 13.34 million hectares of land across the country annually, resulting in yield losses up to 2.5 billion kilograms, and the situation is getting worse every year. In addition, BPH is also the vector of the rice grassy stunt virus (RGSV) and rice ragged stunt virus (RRSV) which seriously affects the production and safety of rice. Most rice varieties have poor BPH resistance, and the pest is mainly controlled by using chemical pesticides. However, the use of pesticides often induces BPH pesticide resistance, and some pesticides can even stimulate egg laying by brown planthoppers. The broad-spectrum pesticides used to control BPH also kill the pest's natural predators, which makes BPH more rampant. Therefore, using resistant rice varieties is the safest and most effective way to control BPH with no impact on rice quality and the environment (Wang Hui et al., 2016). Since BPH resistance cannot be accurately identified in the field, it is the most effective to breed new hybrid rice varieties with resistance by using MAS.

Research on genes of BPH resistance began in the 1970s. So far, 34 BPH resistance loci have been reported, including 19 dominant genes and 15 recessive ones. A total of 28 of these have been mapped, and eight have been successfully cloned (Table 4-3). The resistance loci are mainly on chromosomes 2, 3, 4, 6, 8 and 12.

Table 4-3 Genes with BPH resistance

Gene	Symbol	Chromosome	Gene query No.	Reference
BPH resistance gene	*Bph14*; *Qbp1*	3	LOC_Os03g63150	Duet al., 2009
Lectin receptor kinase; BPH resistance gene	*OsLecRK3*; *Bph3*	4	LOC_Os04g12580	Liuet al., 2015
Lectin receptor kinase; BPH resistance gene	*OsLecRK1*; *Bph3*	4	LOC_Os04g12540	Liuet al., 2015
Lectin receptor kinase; BPH resistance gene	*OsLecRK2*; *Bph3*	4	Os04g0202350	Liuet al., 2015
BPH resistance gene	*Bphi008a*	6	LOC_Os06g29730	Huet al., 2011
BPH resistance gene	*Bph32*	6	LOC_Os06g03240	Renet al., 2016
BPH resistance gene	*BPH29*	6	LOC_Os06g01860	Wanget al., 2015

Continued

Gene	Symbol	Chromosome	Gene query No.	Reference
BPH resistance gene	*BPH18*	12	LOC_Os12g37290	Jiet al. , 2016
BPH resistance gene	*BPH1*, *BPH2*, *BPH7*, *BPH9*, *BPH10*, *BPH21*, *BPH26*	12	LOC_Os12g37280	Zhaoet al. , 2016
BPH resistance gene	*BPH6*	4	KX818197	Guoet al. , 2018

Bph14 is the first cloned BPH resistance gene, encoding a protein consisting of 1,323 amino acids and containing a coil domain, a nucleotide-binding domain and a leucine-rich repeat (CC-NB-LRR). *Bph14* activates salicylic acid signaling pathways following BPH infection, inducing callose deposition and trypsin inhibitor production in phloem cells, which reduces the feeding, growth rate and longevity (Du et al. , 2009). The BPH-inducible gene *Bphi008a* enhances resistance to BPH in rice, acting downstream of the ethylene signaling pathway and located in the nucleus (Hu et al. , 2011). *BPH29* encodes a resistant protein containing the B3 structural domain. Its introduction into TN1 can improve the resistance of transgenic plants to BPH. *BPH29* activates the salicylic acid signaling pathway and inhibits the jasmonic acid/ethylene pathway in response to BPH infestation (Wang et al. , 2015). *BPH18* encodes a CC-NBS-NBS-LRR protein that is composed of two genes – *Os12g37290* encoding the NBS structural domain and *Os12g37280* encoding the LRR structural domain (Ji et al. , 2016). *BPH26*, an allele of *Bph2*, encodes a CC-NBS-LRR protein and can inhibit sucking of phloem sieve tubes by the BPH (Tamura et al. , 2014). *BPH18* and *BPH26* are alleles with different functions and the former has the dual functions of antixenosis and antibiosis (Ji et al. , 2016). Wan Jianmin's team at Nanjing Agricultural University (NAU) cloned *Bph3*, a cluster of three genes encoding plasma membrane-localized lectin receptor kinases, namely *OsLecRK1*, *OsLecRK 2* and *OsLecRK 3* (Liu et al. , 2015).

He Guangcun's team at Wuhan University mapped *BPH9* on the long arm of chromosome 12, and found that all the seven BPH resistance genes (*BPH1*, *BPH2*, *BPH7*, *BPH10*, *BPH18*, *BPH21*, *BPH26*) previously located on this chromosome segment were alleles of *BPH9*. The *BPH9* protein activates the salicylic-jasmonic acid signaling pathway and is in the functions of antixenosis and antibiosis. Because allelic variation in the *BPH9* gene confers to rice resistance against different biotypes of BPH, it is an important strategy to cope with variation in BPH population (Zhao et al. , 2016). In 2018, He Guangcun's group cloned another dominant broad-spectrum pest resistance gene *BPH6*, which is a new type of pest resistance gene. *BPH6* protein is located in the exocyst complex and interacts with the exocyst complex subunit *EXO70E1* to regulate rice cell secretion and maintain cell wall integrity, thereby, hindering BPH feeding. *Bph6* regulates various hormonal pathways such as *SA*, *JA*, and *CK*, specially the regulation of *CK*, i. e. cytokinin, which plays an important role in rice pest resistance. The *Bph6* gene is highly resistant to several biotypes of BPH and white-backed planthopper (WBPH), and its resistance mechanisms are antixenosis, antibiosis, and pest tolerance. *Bph6* has no adverse effect on rice growth and yield and is highly resistant in both *indica* and *japonica* backgrounds. Therefore, it is highly valuable to the breeding of hybrid rice with BPH resistance (Guo et al., 2018).

Ⅱ. Genes Related to High Yield

Yield traits are complex quantitative traits, and rice yield is composed of three factors – effective number of panicles per unit area, number of grains per panicle, and grain weight. Grain weight is mainly controlled by four factors – grain length, grain width, grain thickness, and grain fullness. In addition, factors such as plant type, panicle type, and growth duration also affect rice yield per unit area (Zhu Yiwang et al., 2016).

1. Important genes for grain type, panicle type and grains per panicle

In the past decade, many important genes affecting rice yield have been cloned by researchers in China (Table 4 – 4), including *GIF1*, *GW5*, *GW7*, *GW8*, *GS5*, *GS3*, *GSE5*, etc., which control grain type, grain weight and grain number of panicles. In 2015, teams led by Chu Chengcai at IGDB and

Table 4 – 4 Genes related to grain and plant type

Gene locus	Gene query No.	Expressed protein	Trait(s) controlled	Reference
Gn1a	LOC_Os01g10110	Cytokinin-degrading enzymes	Grains per panicle	Ashikari et al., 2005
GIF1	LOC_Os04g33740	Cell wall invertase	Grain fullness	Wanget al., 2008
GW5	ABJ90467	Nuclear localization protein that interacts with polyubiquitin	Major gene controlling grain width and weight	Wenget al., 2008
GW8/ OsSPL16	LOC_Os08g41940	Transcription factors containing SBP domain	Grain size, grain shape and rice quality	Wanget al., 2012
GS5	LOC_Os05g06660	Serine carboxypeptidase	Regulates grain size positively	Liet al., 2011
DEP1	LOC_Os09g26999	G-protein γ subunit	Erect and dense panicle	Huanget al., 2009
DEP2	LOC_Os07g42410	Proteins located in endoplasmic reticulum	Erect and dense panicle; small and round grain	Liet al., 2010
GL2	LOC_Os02g47280	GRF transcription factor	Grain length, width and weight	Cheet al., 2015
GS3	0s03g0407400	Transmembrane proteins with four domains	Major gene controlling grain and grain weight	Maoet al., 2010
IPA1	LOC_Os08g39890	Squamosa promoter binding protein	Plant height, tillers and grains per panicle	Jiaoet al., 2010
OsPPKL1	LOC_Os03g44500	Protein serine/threonine phosphatase	Grain length	Zhsng et al., 2012

Continued

Gene locus	Gene query No.	Expressed protein	Trait(s) controlled	Reference
OsMKK4	LOC_Os02g54600	Mitogen-activated protein kinase	Grain shape, panicle shape and plant height	Dusn et al., 2014
TGW6	LOC_Os06g41850	IAA glucose hydrolase	Grain weight	Idnimsru et al., 2013
Ghd7	LOC_Os07gl5770	CCT structural protein	Heading date, plant height and grain number per panicle	Xue et al., 2006
DTH8	LOC_Os08g07740	Protein containing CBFD-NFY-HMF	Yield, plant height and heading date	Weiet al., 2010
PTB1	LOC_Os05g05280	Protein containing RING-FINGER	Seed setting rate	Liet al., 2013
Bg1	LOC_Os03g07920	Auxin-induced positional functional proteins	Grain shape	Liuet al., 2015
Bg2/GE	LOC_Os07g41240	CYP78M3protein	Grain length, width, thickness, 1,000-grain weight and embryo size	Xuet al., 2015
FUWA	LOC_Os02gl3950	Protein containingNHL domain	Plant height, tillers, panicle number, grain shape and 1,000-grain weight	Chen et al., 2015
OsSPL13/GLW7	LOC_Os07g32170	SBP-type transcription factor	Grain length and weight	Siet al., 2016
NOG1	LOC_OsQlg54860	Enoyl CoA hydratase/isomerase protein	Panicle number	Huo et al., 2017
OsOTUB1	LOC_Os08g42540	Deubiquitinase	Ideal plant type	Wang Set al., 2017

Zhao Mingfu at the Fujian Academy of Agricultural Sciences (FAAS) cloned a dominant gene *Gl2* controlling grain length, from the large-grain rice material *RW11*, which increased grain size without affecting other important yield traits, resulting in a 16.6% increase in yield per plant (Che et al., 2015). In order to identify genes controlling grain size in rice, a GWAS study on grain size was completed in different rice populations by the team of Academician Han Bin at the CAS, and functional analysis of the grain shape-related QTLs was performed by analyzing expression patterns, genetic variation and T-DNA insertional mutations. A major gene site encoding plant specific transcription factor *GLW7* was discovered and confirmed that *GLW7* can positively regulate cell size of rice grain hull, thereby affecting grain length and yield of rice (Si et al., 2016). The key genes controlling rice panicle type are *DEP1* and *DEP2*. The mutation of *DEP1* promotes cell division, thus increasing rice yield by increasing the number of branches per panicle and grains per panicle (Huang et al., 2009), while *DEP2* also controls grain size in addition to regulating panicle type, and the mutant *dep2* shows the phenotype of erect panicle and small round grains (Li et al., 2010). The team of Zhang Qifa from Huazhong Agricultural University (HAU) systematically identified the functions of five subunits of the heterotrimer G protein complex in regulating rice grain length. Among them, G_α protein controls grain size, G_β protein is essential for plant survival and growth, and three G_γ proteins, namely, *DEP1*, *GGC2*, and *GS3*, show antagonistic effects in regulating grain size. When formed in complex with G_β proteins, both *DEP1* and *GGC2* protein can increase grain length individually or in combination. In contrast, *GS3* has no effect on grain size alone, but reduces grain length when interacting competitively with G_β protein. Different genetic manipulation of G protein subunits can artificially increase grain length by 19% or reduce it by 35%, resulting in 28% yield increase or 40% yield reduction (Sun et al., 2018).

The number of grains per panicle is a key factor determining rice yield. Sun Chuanqing and Tan Lubin's team at China Agricultural University (CAU) cloned *NOG1* (number of grains 1), a gene related to the grain number per panicle, which encodes an enoyl-CoA hydratase/isomerase protein. It increases the number of grains per panicle without any negative effect on other yield-related traits such as panicle number, flowering date, seed setting rate and grain weight. Further study showed that in cultivated varieties with more grains per panicle, the promoter region of *NOG1* contained two copies of a 12 bp fragment, while in wild rice with fewer grains per panicle; there is only one copy of this 12 bp fragment. The additional 12 bp fragment enhances the expression of *NOG1*, which eventually leads to the increase of grains per panicle (Huo et al., 2017).

The effect of *NOG1* on yield is affected by promoter regulatory sequences. The QTL *SGDP7*, which controls the number of grains and 1,000-grain weight, cloned by Xing Yongzhong's team at HAU, has a similar regulatory pattern. *SGDP7* is actually frizzy panicle (*FZP*), a cloned panicle development gene which has the function of preventing the formation of axillary bud meristem and establishing floral meristem and is closely related to rice yield. Further study revealed that an 18 bp fragment of transcriptional silencer was replicated at a point 5.3 kb upstream of *FZP* in Chuan 7, forming a copy number variant of *CNV*-18bp. *CNV*-18bp inhibits *FZP* expression, resulting in longer branching time per panicle, thus significantly increasing grain number per panicle, and slightly reducing 1,000-grain weight, but bringing a 15% increase in rice yield. *CNV*-18bp is the silencer of *FZP* because the transcription inhibitor *OsBZR1* can combine with the *CGTG* gene sequence in *CNV*-18bp to inhibit the expression of *FZP*. Studies have shown that the silencer *CNV*-18bp controls the balance between grain number per panicle and 1,000-grain weight by affecting the expression of *FZP*, and ultimately affects the yield (Bai et al., 2017).

The gene *PTB1* (pollen tube blocked 1) was cloned from a female sterile mutant of *indica* rice Shuhui 202 at the Rice Research Institute of Sichuan Agricultural University. The *PTB1* protein contains the C3H2C3 type RINGFINGER domain. *PTB1* positively regulates the seed setting rate by promoting the growth of pollen tubes. *PTB1* gene expression is affected by promoter haplotype and environmental temperature, and has a significant positive correlation with seed setting rate (Li et al., 2013).

2. Genes of ideal plant types

Rice plant type improvement plays an important role in increasing rice yield. So far, rice plant type improvement has mainly gone through two stages, semi-dwarf breeding and ideal plant type breeding. In the 1970s, Japanese scholar S. Matsushima proposed the theory of "ideal plant type" and some specific plant type indicators for rice, which led to the development of ideal plant type breeding for rice based on good plant morphology rather than just yield election. In the 1980s, Yang Shouren put forward and improved the theory of rice breeding based on a combination of ideal plant type with heterosis utilization, and developed a series of rice varieties with high yield and good quality. In the late 1980s, IRRI proposed a new plant type breeding program to achieve a breakthrough in terms of plant type and yield under the premise of fewer tillers, larger panicles and stronger stems.

Academician Li Jiayang and researcher Qian Qian's team cloned *IPA1* (ideal plant architecture 1), a gene for ideal plant type. After mutation, the gene will reduce the number of tillers, increase the number of grains per panicle and 1,000-grain weight, and make the stem stronger, thereby increasing lodging resistance (Jiao et al., 2010). Further studies showed that all the genes for ideal plant types in super high-yielding *indica-japonica* intersubspecific combinations in China have semi-dominant allelic mutations of *IPA1*. *IPA1* is a transcription factor *OsSPL14* encoding SBP-box, which is involved in the regulation of several growth and development processes in rice. Studies of upstream and downstream regulatory networks showed that *IPA1* regulates rice tillering through *TB1* and regulates plant height and panicle length through *DEP1*. The upstream of *IPA1* is regulated by *miR156* and *miR529*. More recent studies have showed that in the nucleus, *IPA1* interacts with an *E3* ligase *IPI1* (IPA1 interacting protein 1), and *IPI1* is able to polyubiquitinate *IPA1* to regulate its protein content. In addition, the type of ubiquitination modification is different in different plant tissues, and the difference in type in turn determines whether the state of *IPA1* protein is degraded or stabilized (Wang J et al., 2017).

Fu Xiangdong's team at CAS successfully cloned *NPT1* (new plant type 1), another regulatory gene for new plant type. The gene encodes a deubiquitinase highly homologous with human protein *OTUB1*. *OsOTUB1* has the depolymerization activity of ubiquitin chain at *K48* and *K63*. Meanwhile, it interacts with *OsSPL14* (*IPA1*) to inhibit the function of *OsSPL14* by depolymerizing its K63 ubiquitin chain. In addition, the study showed that the aggregation of the excellent alleles of *npt1* and *dep1-1* can be a new strategy to increase rice yield. This study not only discovered a new gene for the ideal plant type, but also established the genetic relationship between the three important genes of *NPT*, *IPA1* and *DEP1*, which provided a new strategy for rice yield increase (Wang et al., 2017). Li Jiayang's team also cooperated with He Zuhua's team and cloned a QTL (*qWS8/ipa1-2D*) from high-yielding late *japonica* hybrid rice Yongyou 12. This QTL is a tandem repeat upstream of *IPA1*, a gene for the ideal plant type. This repeat sequence can inhibit DNA methylation modification of *IPA1*, resulting in the loose state of *IPA1* promoter's chromatin structure, thus promoting the expression of *IPA1*, resulting in ideal plant type and yield increase (Zhang et al., 2017).

Plant type is an important factor that determines rice yield. Genetic improvement aiming at breeding the ideal plant type has greatly improved rice yield. *SPL* (squamosapromote binding protein (SBP)-like) proteins are a special class of transcription factors in plants, and they contain highly conserved SBPDNA binding domains. Under the regulation of numerous microRNAs, *SPL* proteins play an important role in the formation of rice plant types. These proteins can inhibit tillering, but only promoting panicle branching under moderate expression. Therefore, fine regulation of *SPL* proteins will contribute to the formation of the ideal plant type and the increase of rice yield (Wang et al., 2017).

3. Genes related to growth period

Zhang Qifai's team at Huazhong Agricultural University cloned *Ghd7* from Minghui 63, which encodes a 257 amino acid nucleoprotein. The product is a *CCT* (*CO*, *CO-like and Timing CAB1*) structural protein, which not only participates in the regulation of flowering, but also has a general promoting effect on plant growth, differentiation and biological yield. Under long daylength conditions, the enhanced expression of *Ghd7* can delay heading, increase plant height and grains per panicle, while the natural mutant with weakened functions can be planted in temperate or even colder regions. Therefore, *Ghd7* plays a very important role in increasing yield potential and improving adaptability of rice across the globe (Xue et al., 2008).

Wan Jianmin's team at Nanjing Agricultural University cloned *DTH8*, a gene that inhibits heading under long daylength conditions, encoding a polypeptide composed of 297 amino acids and containing a CBFD-NFYB-HMF structure domain. *DTH8/Ghd8/LHD1* was confirmed to encode the *HAP3H* subunit of the transcription factor CCAAT box binding protein, which can simultaneously regulate rice yield, plant height and heading date (Wei et al., 2010; Yan et al., 2011). *DTH8*, expressed in many tissues and independent of *Ghd7* and *Hd1*, can down-regulate the transcription of *Ehd1* and *Hd3a* under long-daylength conditions. *Ghd8* can also delay rice flowering by regulating *Ehd1*, *RFT1* and *Hd3a*, but promotes flowering under short daylength conditions. *Ghd8* can up-regulate the expression of *MOC1*, a gene controlling tillering and lateral branching in rice, thereby increasing the number of tillers, primary branches and secondary branches (Yan et al., 2011).

Ⅲ. Genes for Efficient Use of Nutrients

Nitrogen, phosphorus, and potassium are the nutrients with the highest demand throughout the life of rice (*Oryza sativa* L.) and are known as the "three elements of fertilizer". They are not only closely related to rice yield and quality, but also essential to the synthesis and metabolism of physiological substances in rice (Xu Xiaoming et al., 2016). The efficient use of fertilizer is an important direction of modern agricultural development. Cultivating crop varieties with efficient fertilizer use is of great significance for reducing planting cost, improving yield and quality, and reducing environmental pollution.

1. Genes for efficient use of nitrogen

Obara et al. (2011) mapped five QTLs using the degree of root growth at different NH_4^+ concentrations as an indicator. Among them, *qRL1. 1* could significantly increase root length under high NH_4^+ concentration. Further fine mapping showed that *OsAAT2*, encoding aspartate aminotransferase, is a candidate gene for *qRL1. 1*. Bi et al. (2009) found the gene *OsENOD93 - 1* to be expressed at a high level in roots and *OsENOD93 - 1* could improve the nitrogen utilization efficiency of rice while also increasing the dry weight of biomass and yield.

Fu Xiangdong's research team at IGDB found that different allelic variants of the *DEP1* gene have different responses to nitrogen (regarding plant height and tiller number). Rice carrying the *dep1 – 1* allele is not sensitive to nitrogen during vegetative growth, and its ability to absorb and assimilate nitrogen is enhanced, resulting in improved harvest index and yield. *DEP1* encodes the γ-subunit of plant G protein, which, composed of the α, β and γ subunits, is an important signal transduction protein that regulates the growth and development of animals and plants. The *DEP1* protein interacts with G_α(*RGA1*) and G_β(*RGB1*) in vivo. Further studies showed that a decrease in *RGA1* activity or an increase in *RGB1* activity inhibits the response of rice growth to nitrogen. This indicates that the G protein complex participates in the regulation of plant perception and response to nitrogen signals. Therefore, the response of rice to nitrogen can be changed by regulating G protein activity, and high rice yield can be achieved by appropriately reducing the amount of nitrogen applied (Sun et al., 2014).

Chu Chengcai's team at IGDB cloned *OsNRT1. 1B*, a nitrogen-efficiency gene from *indica* rice, which encodes a nitrate transporter. It has the functions of not only nitrate absorption and transport, but also nitrate signal sensing, transmission and amplification, thereby affecting the levels of nitrate absorption, transport and assimilation. *OsNRT1. 1B* has a one-base difference between *japonica* and *indica* rice. A comparison shows that for *indica* rice, *OsNRT1. 1B* has higher activity for nitrate absorption and transportation, while for *indica* lines containing *OsNRT1. 1B*, the number of tillers and yield of near isogenic lines are significantly increased (Hu et al., 2015). Nitrate nitrogen and ammonium nitrogen are the main forms of nitrogen used by plants. As an aquatic plant, rice mainly utilizes ammonium nitrogen. The team also cloned *OsNRT1. 1A*, another nitrogen-efficiency gene located in the vacuole membrane and induced by ammonium salt. It participates in the regulation of nitrate and ammonium salt in rice cells. Over-expression of *OsNRT1. 1A* in different rice varieties and under different nitrogen fertilizer conditions can significantly increase the biomass and yield, and shorten the maturity period (Wang et al., 2018).

The research teams at CAS and CNRRI discovered *ARE1*, a key gene that regulates the efficiency of nitrogen utilization in rice. It encodes a functional conserved protein located in the chloroplast. Its mutation can delay the senescence of rice plants and increase the yield by 10%–20% in the absence of nitrogen. Researchers analyzed 2,155 rice germplasms and found that small insertion in the promoter region of *ARE1* in many germplasms caused decreased *ARE1* expression, which in turn resulted in higher efficiency in nitrogen use in these germplasms (Wang et al., 2018).

2. Genes for efficient use of phosphorus

Phosphorus is an important component of some enzymes in rice, which play an important role in the transportation, transformation and storage of substances. Phosphorus also significantly promotes the growth of rice roots. Wasaki et al. (2003) cloned *OsPI1*, a gene that is very sensitive to phosphorus nutrition and can significantly enhance the tolerance of rice to low phosphorus stress. The transcription of this gene disappears rapidly after the application of phosphorus fertilizer to phosphorus-deficient plants, while the expression level of this gene increased significantly under the condition of phosphorus deficiency, indicating that *OsPI1* can improve the tolerance of plants to low phosphorus. Rico et al. (2012) cloned *PSTOL1*, a phosphorus efficiency gene. Research results showed that over-expression of *PSTOL1* in varieties with phosphorus starvation intolerance can significantly increase their yield in phosphorus deficient soils and that *PSTOL1* is an early root growth enhancer that enhances plant access to phosphorus and other nutrients.

There are 21 *PAPs* genes in rice leaves or roots that are induced to be expressed by low phosphorus stress, and all *PAPs* promoters contain one or two *OsPHR2* binding elements. Phosphorus starvation-induced over-expression of *OsPHR2* can increase the activity of acid phosphatase in the plant and secreted by the root system (Zhang et al., 2011). Jia et al. (2011) found that the phosphate transporter gene *OsPht1;8* (*OsPT8*) in rice regulates the uptake and transport of phosphorus in rice and can increase the absorption and accumulation of phosphorus in the plant. *OsPT8* is also involved in the regulation of the dynamic balance of phosphorus in rice. *OsPT8* has important effects on rice growth and development.

3. Genes for efficient use of potassium

Potassium basically exists in an ionic state in rice, mostly concentrated in young tissues and cells, and playing an important role in the formation of starch and sugar. Potassium can also promote photosynthesis, nitrogen and phosphorus uptake, and root growth, and improve resistance to drought, cold, lodging and pest. By analyzing a whole cDNA expression library of rice, Obata et al. (2007) found that the K^+ channel gene *OsHAK1 – 17* could increase K^+ uptake in rice. Banuelos et al. (2002) cloned and isolated *OsHAK1 – 17*, 17 genes encoding K^+ transporter, indicating that *OsHAK* could increase the absorption and transportation of K^+ in roots. Lan et al. (2010) showed that the highly adsorbed K^+ transporter gene *OsHKT2* is expressed in many tissues, including root hairs and soft tissue tubular cells, and that the encoded protein is present in the plasma membrane, which may represent a novel mechanism for cation uptake and efflux.

Ⅳ. Quality-related Genes

High yield and good quality have always been the main goals of hybrid rice variety improvement. At present, the quality of rice in China is generally low, which affects its market competitiveness to a certain extent. Rice quality is a comprehensive trait, including the characteristics of rice or rice-related products to meet the needs of consumers, production and processing. In China, quality standards covered in rice variety approval mainly include head rice yield of milling quality, length/width ratio in appearance, and chalkiness grain rate, chalkiness degree, gel consistency and amylose content of cooking and eating quality. Therefore, strengthening genetic research on rice quality traits is an important research direction, as it can help clarify the molecular mechanism of quality formation. It is important to combine molecular and conventional breeding so as to develop new hybrid rice varieties with high quality.

The early studied genes of rice quality are the genes starch synthesis-related genes in endosperm, including granule bound starch synthase (GBSS), adenosine diphosphate glucose (ADPG), pyrophosphorylase, starch branching enzyme and starch debranching enzyme (SDBE). The combination of these genes and their alleles directly affects the amylose content of rice endosperm, and in turn, affects the quality of rice. Fragrance is one of the important criteria for evaluating rice quality. The mutation of *OsBADH2*, a gene controlling fragrance, can lead to the constant accumulation of 2-acetyl-1-pyrroline, which can form fragrant rice leaves and grains (Chen et al., 2008).

Rice contains a lot of storage proteins, which is the second largest substance in rice after starch. Among them, glutenin has the highest content in rice seeds, accounting for more than 60% of the total proteins, and it is the primary target of rice protein improvement. By screening a large number of mutagenic materials, Wan Jianmin's team at Nanjing Agricultural University obtained a series of mutants with abnormal accumulation of rice glutenin precursor, and successively cloned *OsVPE1* (Wang et al., 2009), *GPA1/Rab5a* (Wang et al., 2010), *GPA2/VPS9a* (Liu et al., 2013), *GPA3* (Ren et al., 2014), and *GPA4* (Wang et al., 2016), which are involved in the accumulation of rice grain proteins. These genes participate in the regulation of rice glutenin shear maturation, post-Golgi sorting and glutenin endoplasmic reticulum output, respectively, building our understanding of the molecular network pathways of glutenin synthesis, sorting and deposition, and laying a theoretical foundation for regulating the content and composition of glutenin so as to improve rice quality.

Chalkiness is the white opaque part formed by the loose arrangement of endosperm starch grains and protein particles during the filling period. It greatly affects the edible yield of rice (whole grain rice rate) and has a great impact on the appearance (transparency), cooking taste and nutritional quality (straight chain starch content, gel consistency and protein content) of rice. Therefore, chalkiness is one of the most important indicators of rice quality and an important factor for affecting the quality and yield of rice. He Yuqing's research team at Huazhong Agricultural University cloned *Chalk5* (Li et al., 2014), the first major gene cloned for chalkiness grain rate in rice. More than 90% of the dry weight of rice grains is composed of storage starch and proteins. Protein content is not only a key indicator used to determine nutritional quality, but it also exerts a great impact on the appearance and eating quality of rice. Therefore, controlling the protein content of rice not only is important for the nutritional value, but also has great impact on the economic value. In the same year, the research team published a paper on the cloning of *OsAAP6*, another gene affecting rice quality and an amino acid transporter. It regulates the nutritional quality and cooking and eating quality of rice by regulating the synthesis and accumulation of starch and the storage of proteins in rice seeds. *OsAAP6* is a constitutively expressed gene with relatively high expression in microtubule tissues, and it is a positive regulator of protein content in rice seeds. The gene can promote the absorption and transport of amino acids in rice roots and plays an important role in regulating the distribution of free amino acids *in vivo*. Researchers analyzed 197 mini-core collections, and found that two common polymorphic loci in the promoter region of *OsAAP6* gene are closely related to the protein content in the seeds of *indica* varieties (Peng et al., 2014).

The continuous pursuit of synergistic improvement of high yield and quality is a major goal and challenge for rice breeders. The length/width ratio of rice grain is an important factor affecting rice quality. In 2012, Fu Xiangdong's research team from IGDB and Zhang Guiquan's research team from South China Agricultural University successfully cloned *GW8*, a key gene encoding a transcription factor containing the SBP domain (OsSPL16), from a high-quality Basmati variety in Pakistan, which could help improve rice quality and yield. In Basmati rice, a variation in the promoter of the *GW8* gene leads to a decrease in gene expression, which can make the grains more elongated and also affect the starch grain arrangement and chalkiness, improving the quality of rice in terms of appearance and taste. *GW8* can affect the quality and yield of rice by regulating the grain width (Wang S et al., 2012). In 2015, Fu Xiangdong's research team identified *GW7*, another important gene controlling grain shape, from high-quality hybrid rice maintainer line Taifeng B (TFB). Research results showed that there is an *OsSPL16* binding site in the *GW7* promoter sequence. *OsSPL16* can control rice quality by directly binding to *GW7* promoter and negatively regulating *GW7* expression. Combining the allelic variants of the *OsSPL16* and *GW7* genes in high-yielding rice in China can significantly improve rice quality and yield (Wang S et al., 2015). The cloning of *OsSPL16-GW7* in rice revealed the molecular mystery of the synergistic improvement of rice quality and yield, and provided a new gene with important application for high yield and quality molecular breeding in rice. Subsequently, the research team successfully cloned *LGY3*, an important gene that controls rice yield and improves rice quality. *LGY3* encodes *OsMADS1*, a member of the MADS-box family of proteins. The β and γ subunit dimers of the G protein are co-factors of *OsMADS1* and regulate the transcriptional activity of *OsMADS1* by directly interacting with *OsMADS1* and further affecting the rice grain type regulatory pathway genes. In addition, the gene has a natural variant, *OsMADS1lgy3*, which encodes a truncated C-terminal *OsMADS1* protein. *OsMADS1lgy3* can increase grain length, reduce chalky grain rate and chalky area, which in turn affects rice yield and appearance quality. Combination of the three allelic variants, *OsMADS1lgy3*, *DEP1* and *GS3* can improve rice quality and yield simultaneously (Liu et al., 2018).

Ⅴ. Genes for Tolerance to Abiotic Stress

In addition to the biotic stresses such as pests, diseases and weeds, rice is also subject to the abiotic stresses such as unfavorable climate, poor soil and water conditions. In recent years, extreme and persistent high temperature in summer, frequently drought in some areas and "cold dew wind" during flowering in southern double-season rice have caused significant losses to rice production in southern China, especially in the middle and lower reaches of the Yangtze River. For this reason, it is also important to discover genes for abiotic stress tolerance in rice.

1. Genes for heat tolerance

In recent years, high temperature and heat damage to rice in the Yangtze River Basin has occurred frequently, often resulting in large-scale reduction of rice yield. Therefore, it is of great significance to study the mechanism of the damage caused by high temperature to rice and explore the genetic resources of rice with tolerance to high temperature so as to breed new rice varieties with tolerance to high temperature for rice production.

Using African rice grown in the tropics to construct a genetic population with Asian cultivated rice, Lin Hongxuan's team at Shanghai Institute of Plant Physiology and Ecology (SIPPE), CAS, successfully cloned QTL thermo-tolerance 1 (*OgTT1*), a major gene controlling the high temperature tolerance of African rice. *OgTT1* encodes a α2 subunit of the *26S* proteasome and the allele in African rice not only responds more effectively to high temperature at the transcriptional level, but also encode a protein that allows the proteasome in the cell to degrade ubiquitinated substrates at a higher rate when temperature is high. Proteomic analysis shows that this faster degradation could result in a significant reduction in both the type and amount of toxic denatured proteins accumulated in rice cells, which in turn protects the plant cells. This study revealed a new mechanism for plant cells to respond to high temperature, i. e. timely and effective removal of denatured proteins is essential to maintain intracellular protein homeostasis at high temperature (Li et al., 2015). The *OgTT1* from African rice can be directly applied to the breeding of rice with tolerance to high temperature through molecular marker breeding based on conventional hybridization, which provides valuable gene resources for crop improvement.

Xue Yongbiao and Cheng Zhukuan's research teams at IGDB jointly cloned a new heat tolerance gene, *TOGR1* (thermotolerant growth required 1). *TOGR1* acts as a nuclear-located DEAD-boxRNA helicase, protects rice from high temperature damage in the form of pre-rRNA chaperone. Further studies showed that *TOGR1* aggregates to the small subunit (SSU) of the ribosome, which ensures the unfolding of misfolded pre-rRNA precursor into the correct conformation and ensures the effective processing of rRNA required for cell division under high temperature (Wang et al., 2016). This study interprets a new molecular mechanism of regulating the high-temperature tolerance of rice, which provides a theoretical basis for breeding new rice varieties with high-temperature tolerance.

2. Genes for drought and salinity tolerance

Lin Hongxuan's research team at SIPPE, CAS, obtained a stable inheritable mutant with higher drought and salinity tolerance, *dst* (drought and salt tolerance), and cloned the gene through large-scale screening of the mutant library of rice EMS mutagenesis. *DST* encodes a protein containing only a C2H2-type zinc finger structural domain and is a novel nuclear transcription factor. In the *dst* mutant, two amino acid variants of this protein significantly reduced the transcriptional activation activity of *DST*. *DST*, as a negative regulator of stress resistance, directly down-regulates the expression of genes related to hydrogen peroxide metabolism when its function is absent, decreasing the ability to remove hydrogen peroxide, thus increasing the accumulation of hydrogen peroxide in guard cells, promoting the closure of leaf stomata, reducing water evaporation, and ultimately improving the drought and salt tolerance of rice (Huang et al., 2009). *DCA1* is a *DST*-interacting protein that functions as a *DST* transcriptional co-activator, and down-regulation of *DCA1* significantly enhances drought and salt tolerance in rice, while over-expression of *DCA1* increases susceptibility to stress treatments (Cui et al., 2015). A collaborative study of IGDB and HHRRC also found that the drought and salt tolerance gene *DST* directly regulates the expression of *Gn1a* (*OsCKX2*) in the reproductive meristem. *DSTreg1*, a semi-dominant allele of *DST*, can disrupt the *DST*-induced regulation of *OsCKX2* expression in the reproductive apical meristem and increase cytokinin, resulting in enhanced meristem vigor, accelerated panicle branching, increased grain number per panicle, and higher yield per plant (Li et al., 2013). The *LP2* gene, cloned by Wan Jianmin's team at CAAS, encodes a leucine-rich receptor kinase that is down-regulated by drought and ABA-induced expression. The accumulation of H_2O_2 decreases in *LP2* over-expressing plants, while open stomata on leaves are increased and hyper-sensitive to drought. *LP2* transcription is directly regulated by *DST* and interacts with the drought responsive aquaporins *OsPIP1. 1*, *OsPIP1. 3* and *OsPIP2. 3* to play the role of kinase in the plasm membrane (Wu et al., 2015).

Xiong Lizhong's team at Huazhong Agricultural University discovered a gene *DWA1* (drought-induced wax accumulation 1) in rice that specifically controls epidermal wax synthesis under drought stress. This gene is highly conserved in vascular plants and encodes an unreported giant protein composed of 2,391 amino acids. *DWA1* is specifically expressed in vascular tissues and epidermis, and is strongly induced by drought and other stresses. Under normal growth conditions, there is no significant difference between rice with this gene deletion and wild-type rice, but under drought stress, rice with this gene deletion is extremely sensitive to drought due to leaf epidermal wax defects and is therefore more likely to suffer severe yield loss. Further studies revealed that this gene encodes a new key enzyme in the wax synthesis pathway, which regulates epidermal wax synthesis by controlling the synthesis and accumulation of extra-long-chain fatty acids (VLCFAs) under drought stress, thereby controlling the plant's ability to adapt to drought (Zhu et al., 2013).

Lan et al. (2015) fine-mapped the seedling salt tolerance mutant gene *SST* to a 17 kb interval on rice chromosome 6 BAC clone B1047G05, where only one predicted gene, encoding the *OsSPL10* (squamosa promoter-binding-like protein 10), is present. Compared to the wild type, the *sst* mutant has a deletion at position 232 of the ORF of this gene, resulting in a shift mutation that leads to premature termination of protein translation. Ogawa et al. (2011) and Toda et al. (2013) cloned salt tolerance genes *RSS1* and *RSS3* from salt-sensitive mutants *rss1* and *rss3* respectively. *RSS1* participates in the regulation of the cell cycle and it is an important factor to maintain the activity and vitality of meristematic cells under salt stress. *RSS3* regulates the expression of jasmonate-responsive genes and plays an important role in maintaining root cell elongation at an appropriate rate under salt stress. Takagi et al. (2015) used MutMap, a new gene mapping technology, to quickly identify *OsRR22*, the gene site controlling salt tolerance enhancement of mutant *hst1*, which encodes a Type-B response regulator protein. The results showed that *hst1* can tolerate 0.75% salinity, which is of great value for breeding salt-tolerant rice.

3. Genes for cold tolerance

Rice is a thermophilic plant originated in the tropics and cold damage is one of the main disasters in the production of the early and late rice in Northeast China, Southwest China and South China. The annual yield loss in the country caused by cold damage can be up to 3 – 5 million tons. Therefore, improving the cold tolerance of rice varieties is of great significance for expanding the growing area of rice and improving the quality of rice in areas of high latitudes and altitudes. Zhong Kang's team at CAS found that both *indica* near-isogenic lines with the *COLD1* gene derived from *japonica* materials and with over-expression of *COLD1* in *japonica* showed significantly enhanced cold tolerance, while the loss-of-function mutant *cold1 – 1* or antisense lines show strong sensitivity to cold. *COLD1* encodes a G-protein signal regulator located in the plasma membrane and endoplasmic reticulum. Seven *SNPs* were identified after analyzing the *COLD1* of 127 different cultivated and wild rice varieties, among which, *japonica*-specific *SNP2* influenced *COLD1* activity and conferred cold tolerance in *japonica* rice. The study revealed a new mechanism by which *COLD1* alleles and specific *SNP* s obtained through domestication conferred cold tolerance in rice (Ma et al., 2015).

Li Zichao's team at China Agricultural University cloned *CTB4a* (*LOC_Os04g04330*), an important heading stage cold-tolerance gene which encodes a conserved leucine-rich repeat-receptor like kinase

(LRR-RLK) that can interact with AtpB, the β subunit of ATP synthase, and affects the activity of ATP synthase to ensure energy supply during grain filling under cold conditions. Haplotype analysis of 119 rice varieties showed that polymorphism in the promoter region of *CTB4a* determined the degree of cold response in different rice varieties and also showed the influence of the domestication process of cold tolerance in *japonica* rice on this gene locus. Improving the cold tolerance of rice at the booting stage helps increase the seed setting rate and avoid cold damage. Therefore, the cloning of *CTB4a* is of great value for breeding rice varieties at can tolerate cold at the booting stage (Zhang et al., 2017).

Although a large number of genes for important agronomic traits have been cloned, only a few of them have been used in hybrid rice breeding so far. The application value of the cloned genes needs to be further evaluated and it is necessary to further explore the genes for favorable traits in hybrid rice varieties based on modern cultivated rice, farm rice varieties, wild rice or other gramineous plant resources and improve the gene reciprocal regulatory network so as to lay a solid foundation of molecular breeding.

Section 3 Practice of Hybrid Rice Molecular Breeding

Molecular breeding of hybrid rice shows strong development momentum. In the past decade or so, MAS has produced a series of parents and combinations that are resistant to diseases and pests, and with high quality and high yield. Genome-wide molecular design breeding with the help of the high-throughput SNP molecular marker technology has become more and more common. Significant progress has been made in transgenic breeding, represented by pest resistance and herbicide tolerance. In recent years, the application of genome editing in breeding has been in full swing. The joining of various molecular breeding technologies contributes to the upgrading of hybrid rice breeding technology.

I. Marker-assisted Breeding of Hybrid Rice

1. Marker-assisted breeding for disease and pest resistance

The occurrence of diseases and pests can cause serious yield reduction in rice, and the extensive use of pesticides brings problems in regard to rice quality and ecological security. The most effective and economical way to control diseases and pests is to cultivate resistant varieties. However, there are many types of diseases and pests, and the genes controlling the occurrence of diseases and pests are complex and changeable. Moreover, the occurrence of diseases and pests often changes with the environment and climate, and phenotypic selection requires certain environmental conditions and is not accurate. Therefore, molecular breeding for disease and pest resistance based on genotype selection has its advantages. In recent years, it has been applied to hybrid rice breeding and played an important role in rice production.

(1) Marker-assisted breeding of blast resistance genes in rice

With the interpretation of the functions of a large number of blast resistance genes, MAS technology has been widely used in breeding for rice blast resistance, with remarkable success. A large number of rice blast resistant restorer lines and sterile lines have been bred. Wang Jun et al. (2011) transferred rice blast resistance genes *Pita* and *Pib*, and the rice stripe resistance gene *Stv-bi* into high-yielding varieties, and bred 74121, a rice variety with high yield, high quality and multi-resistance. Yin Desuo (2011) and Wen Shaoshan (2012) introduced the *Pi9* gene into Yangdao 6, R6547 and Luhui 17, and the recipient lines showed a higher blast resistance identified in the disease nursery. Liu Wuge et al. (2012) used the MAS technology to breed male sterile lines Jifeng A and Anfeng A carrying disease resistance genes *Pi1* and *Pi2*. Yu Shouwu et al. (2013) used *Si13070D*, the tightly linked *STS* marker of *Pi25*, to detect the target genes and obtained five two-line male sterile lines 16S, 38S, 39s, 61S and 73s with good overall performance. Tu Shihang et al. (2015) used BL47 which contained *Pi25* as the donor parent and Fudao B as the recipient, and used the molecular marker *Si13070C* to detect the *Pi25* gene, combined with conventional breeding methods and seedling blast identification in disease nursery, and bred CP4A, a CMS line with excellent overall traits. Yang Ping et al. (2015) obtained three improved homozygous restorer lines with the target genes by using Gumei 4 carrying the rice blast resistance gene *Pigm* as antigen and Chunhui 350 as the recurrent recipient parent and the resistance frequency to 20 representative pathogen strains in Jiangxi province in recent years has been between 85% and 100%. Xing Xuan et al. (2016) developed R288 - Pi9, a new rice line with high blast resistance, by using the functional marker *Clon2 - 1* of *Pi9*, and selecting 75 - 1 - 127 as the donor parent and R288 as the recipient parent. The selection efficiency of *clon2 - 1* was 100%. Dong Ruixia et al. (2017) used Bl27 carrying the rice blast resistance gene *Pi25* as the antigen donor, and Zhenda B, a rice maintainer line with high quality, strong combining ability and susceptibility to rice blast, as the recipient parent to breed through crossing and backcrossing new germplasms of rice blast resistant maintainer lines, then transferred them to the sterile line Zhenda A by test crossing and backcrossing and obtained 157A, a sterile line with high blast resistance.

The current research and application of rice blast resistance genes are still faced with the following issues. 1) The long-term use of the same or similar rice blast antigens and resistance genes has promoted the formation of new dominant pathogen races, and the resistance of newly bred varieties is decreasing, while there is a shortage of broad-spectrum or durable antigen materials and resistance genes. 2) The overall agronomic traits of antigen materials carrying disease resistance genes are often unsatisfactory for they tend to have poor quality, low yield and/or high plant height. When they are used as antigens for variety breeding, the phenomenon of "linkage drag" tends to occur, that is, while the disease resistance gene is introduced into the improved material, other undesirable traits are also introduced, which increases the time and difficulty of breeding material improvement. 3) Although many genes have been mapped, due to differences in the mapping population and identification of strains, coupled with the fact that rice blast resistance genes are mostly distributed in clusters, it is possible that more of these genes are alleles to the genes that are currently clearly mapped (He Xiuying et al., 2014).

In response to the above problems, geneticists and breeders are working on the following six aspects to further alleviate rice blast infestation. 1) Mining and identifying new rice blast antigens and resistance genes from local germplasm, wild rice and cultivars, and paying attention to the rational layout and rotation of varieties with different resistance gene types in production. 2) Applying traditional breeding methods, supplemented by MAS, artificial inoculation and multi-location disease nursery identification, to broaden the resistance spectrum and the persistence of resistant varieties. 3) Clarifying the relationship between genes in the gene cluster and developing functional markers through fine gene mapping, cloning and resistance spectrum analysis. 4) Analyzing the major genes and background minor genes of the resistant materials before using them, and introducing the major genes and minor genes into the improved lines simultaneously. 5) Developing functional markers of disease resistance genes or using two pairs of markers closest to the target gene, and combining background and foreground selection requirements to densify the markers near the target chromosome and target genes to prevent "linkage drag". 6) Using genome editing techniques to replace the susceptible gene fragments of susceptible varieties with resistant gene fragments or knocking out the susceptible gene to make them resistant (recessive blast resistant gene) and avoid the introduction of undesirable traits during cross breeding.

(2) Marker-assisted breeding of BPH resistance genes

Unlike rice blast resistant materials, which can be screened in disease nurseries, BPH resistant materials have no stable screening environment and large scale screening is thus difficult to conduct. The prevalence of BPH varies with the years and climatic conditions, and the phenotypic identification of plant field resistance is inaccurate. However, the selection of resistant parents can be accelerated by foreground and background screening in early generations. When the resistant strains are genetically stabilized, it is necessary to further identify the resistance by artificial inoculation at the seedling stage and natural induction without pesticides.

In August 2010, BPH68S developed by Zhu Yingying's team at Wuhan University was recognized by the Department of Science and Technology, Hubei province. It is a newly bred male sterile line with BPH resistance for two-line hybrid rice. BPH68S was bred by crossing and backcrossing with the genes *Bph14* and *Bph15* based on MAS. Hybrid rice Liangyou 234 derived from BPH68S is the first rice variety with BPH resistance and sets a good example of combining MAS with conventional breeding techniques.

Subsequently, a number of entities in China have extensively carried out MAS breeding of hybrid rice with BPH resistance, especially for the improvement of resistance in restorer lines. For example, Liu Kaiyu et al. (2011) introduced *Bph3* and *Bph24* (t) into Guanghui 998, Minghui 63, R15, R29 and 9311, and obtained 32 *Bph3* introgression lines, 22 *Bph24* (T) introgression lines, and 13 *Bph3* and *Bph24* (t) superior polymeric lines. Artificial inoculation shows that the resistance of *Bph3* and *Bph24* (t) introgression lines to BPH are moderate or high, and that of *Bph3* and *Bph24* (t) introgression lines is the highest. Zhao Peng et al. (2013) successfully polymerized *Bph20* (t), *Bph21* (t) and the rice blast resistance gene *Pi9* into the maintainer line Bo-Ⅲ B, and bred five materials with resistance to both BPH and rice blast. Yan Chengye et al. (2014) introduced *Bph14* and *Bph15* into the restorer line R1005 simultaneously through MAS, hybridization and backcrossing, and bred homozygous lines CY11711 - 14, CY11712 - 5 and CY11714 - 100, all of which are highly resistant to BPH at the seedling stage. Hu Wei et al. (2015) introduced *Bph3*, *Bph14* and *Bph15* into Guinongzhanzhong, a high-yielding rice variety in South China, and significantly improved its BPH resistance.

From 2015 to 2017, Zhao Bingran's team at HHRRC used the BPH resistant restorer line Luoyang 69 provided by Professor He Guangcun of Wuhan University, and introduced *Bph6* and *Bph9* into the strong restorer line R8117 through hybridization, backcrossing, and foreground and background selection. The identification of resistance through artificial inoculation at the seedling stage in laboratory showed that the recipient parent R8117 is susceptible, while the new restorer line is resistant to BPH, as much as its donor parent Luoyang 69, and the accuracy of marker selection is above 95%.

The result above showed that MAS is very effective for the breeding of rice varieties with BPH resistance. Different from conventional breeding, phenotype identification at an early generation is not needed, and rapid improvement of parents can be achieved just through MAS to ensure that the resistance genes are not lost, and the original parents have superior traits.

(3) Marker-assisted breeding of bacterial blight resistant genes

At present, *Xa4*, *Xa7*, *Xa21* and *Xa23* are the most used genes for leaf blight resistance in production, and one or more of these genes can be aggregated through MAS to basically solve the resistance problem of the varieties. Xue Qingzhong (1998) introduced *Xa21* from IRBB21 into susceptible restorer lines Minghui 63 and Milyang 46, and bred improved restorer lines and new hybrid rice combinations with resistance to bacterial blight. Deng Qiming et al. (2005) carried out an analysis of the aggregation and effect of *Xa21*, *Xa4* and *Xa23*, genes with resistance to bacterial blight. The resistance of the three-gene cumulative lines was significantly stronger than that of the two-gene cumulative lines and single-gene varieties, indicating that the use of the MAS technology to aggregate multiple resistance genes into the same rice variety can significantly improve the resistance and expand the resistance spectrum. Luo Yanchang et al. (2005) developed a sterile line R106A that aggregates *Xa21* and *Xa23* genes and has high resistance to leaf blight throughout the growth period. Zheng Jiatuan et al. (2009) bred a series of bacterial blight resistant lines by using resistant materials containing *Xa23* through MAS. Lan Yanrong et al. (2011) obtained four lines carrying *Xa21* or *Xa7*, by conventional backcross and MAS, and improved the resistance of Hua 201S to bacterial blight. Luo et al. (2012) polymerized *Xa4*, *Xa21* and *Xa27* into the restorer line XH2431 by MAS, and obtained materials with significantly enhanced resistance and broader resistance spectrum. Huang et al. (2012) successfully polymerized *Xa7*, *Xa21*, *Xa22* and *Xa23* into the excellent hybrid rice restorer line Huahui 1035 using the MAS method. The progeny materials showed different degrees of resistance to 11 representative strains in China.

In the past decade, a number of excellent male sterile lines and restorer lines with disease and pest resistance have been bred through MAS, and applied in rice production. Practice also showed that because of the diversity of rice blast in physiological races, resistance works better when MAS and disease nursery identification are carried out simultaneously. It is relatively easy to breed varieties with resistance to bacterial blight and BPH. For example, the problem of bacterial blight or BPH can be basically solved by aggregating genes with resistance to bacterial blight such as *Xa21* and *Xa23* or introducing genes with resistance to BPH such as *Bph3*, *Bph6* and *Bph9*. Besides these three major diseases and pests, rice false smut and sheath blight have also become more frequent over the past years, which seriously affects rice yield, quality and food safety. However, due to the lack of resources for resistance, the breeding of varieties with resistance to these diseases is underdeveloped and more needs to be done in basic and applied research at the molecular level.

2. Molecular maker-assisted breeding of high-yield and ideal plant types

Yang Yishan et al. (2006) bred Yuanhui 611, a new late-season *indica* restorer line, by using test cross materials containing the high-yield QTLs *yld1. 1* and *yld2. 1* (Xiao et al., 1996) of wild rice from Malaysia as gene donor. Wu Jun et al. (2010) bred the restorer line R163 with super rice parent 9311 as the recipient and recurrent parent, and they also bred the two-line hybrid rice Y-Liangyou 7, which was applied in wide scale of production, by crossing R163 with Y58S.

Academicians Li Jiayang and researcher Qian Qian's teams found that using different alleles of the *IPA1* gene to achieve moderate expression of *IPA1* (*OsSPL14*) is the key to forming ideal plants with large panicles, appropriate tillering and thick stems for lodging resistance. They succeeded in breeding the new Jiayouzhongke series of varieties using MAS to aggregate high-yielding *IPA1* alleles. For two consecutive years, the average yield of these varieties planted on 666. 7-ha demonstration plots increased by more than 20% over that of local varieties, and they were suitable for mechanized or direct seeding cultivation.

IRRI researchers introduced the alleles *Gn1a*-type3 and *OsSPL14 WFP* into local main *indica* varieties by MAS, and compared the number of grains per panicle of BC_3F_2 and BC_3F_3 populations with their donor and recipient parents in a multi-location experiment. With an *indica* background, the *Gn1a*-type3 locus had no significant effect on the number of grains per panicle, while the *OsSPL14 WFP* locus had a significant effect on yield and could increase the grain number per panicle by 10. 6%-59. 3% in different backgrounds. Subsequently, they bred five high-yielding varieties using *OsSPL14 WFP* and MAS and achieved a yield increase between 28. 4% and 83. 5% compared with that of the recipient parent and a yield increase of 64. 7% compared with IRRI156, the high-yielding control variety (Sung et al., 2018).

Molecular biologists are striving to use the high-throughput sequencing technology to analyze the high-yielding and high-quality genotypes and genotypic combinations of hybrid rice and parents currently in wide use in production, expecting to continuously improve the yield of rice varieties and achieve the balanced improvement of other overall agronomic traits through genome-wide MAS.

3. Molecular breeding of quality-related genes

Zhang Shilu et al. (2005) used four low amylose content (AC) *indica* varieties (R367, 91499, Yanhui 559, and Hui 527) as donors of high-quality genes, and a three-line *indica* rice restorer line 057 which has a high combining ability for yield, as the recurrent recipient parent. Molecular markers were used to select the genotypes controlling the AC and improve 057 with high AC through backcrossing. The AC values of three expression types (*GG*, *TT*, *GT*) of the gene *Wx* identified by molecular markers were determined and analyzed, and the AC of 057 was effectively reduced through MAS.

Chen Sheng et al. (2008) used *PCR-Acc I* to improve the quality traits of the parents of Xieyou 57 by MAS, and successfully reduced the AC of Xieqingzao to a moderately low level (12. 5%), resulting in softer gel consistency and greatly improved AC homogeneity. Wang Yan et al. (2009) introduced the *alk* and *fgr* allele fragments of Chinese fragrant rice into Minghui 63, and significantly improved the appearance quality, cooking quality and eating quality of the recipient lines.

Ren Sanjuan et al. (2011) bred an *indica* CMS line with good quality and fragrance with the MAS technology. Thirteen *indica* maintainer lines were tested for fragrance, and the functional molecular marker *1F/1R* for fragrance gene (*fgr*) was used for PCR molecular detection. Yixiang B was selected as the Type-I band (*fgr/fgr*), and the rest of the materials were used as the Type - Ⅱ band (*Fgr/Fgr*). The improved fragrant type Ⅱ - 32B was bred using Ⅱ - 32B/Yixiang B hybrid progenies with MAS. The selected plants were then backcrossed with Ⅱ - 32A for five generations to obtain a number of good-quality fragrant male sterile lines (maintainers) with stable traits, such as Zhenongxiang A (B).

With a careful design, the joint team of Li Jiayang at the IGRD, CAS, and Qian Qian at the CRRI used Teqing as the recipient and Nipponbare and 9311, which have good cooking and appearance quality, as the donors to optimize the combinations of 28 target genes related to rice yield, appearance quality, cooking and eating quality and ecological adaptability. After more than eight years of efforts, they successfully polymerized the superior alleles of high-quality target genes into the recipient materials through hybridization, backcrossing and MAS. The high-yielding traits of Teqing were fully retained, and the qualities of appearance, cooking and eating were significantly improved, with more hybrids derived from the improved parents (Zeng et al., 2017).

Ⅱ. Transgenic Breeding for Pest Resistance and Herbicide Tolerance

1. Transgenic breeding for pest resistance

Diseases and pests accompany the whole process of rice production. The use of chemicals is not only costly, but will also cause problems of serious pollution. Rice itself lacks efficient pest resistant genes, so the target genes for pest resistant transgenic rice are mainly exogenous ones, including the genes of *bacillus thuringiensis*, pest protease inhibitor, exogenous lectin, chitinase, nutritional pesticidal protein, pest hormone, etc. (Wang Feng et al., 2000). The gene *Bt* from *bacillus thuringiensis* is the most widely used and efficient pest resistance gene in the world. It confers high resistance to *lepidoptera*, *diptera* and *coleoptera*, and is safe to human, animals and the environment. In addition to *Bt*, some protease inhibitors, plant lectin, ribosome inactivating protein and plant secondary metabolite genes also have good pest resistance and have been widely used in the research and development of transgenic pest-resistant rice (Xu Xiuxiu et al., 2013).

(1) Development of pest-resistant transgenic rice

In 1981, the first *Bt* insecticidal gene was cloned. In 1993, the first *Bt* rice line was successfully developed. In 2000, *Bt* rice lines entered field trial one after another. Huahui 1, *Bt*-Shanyou 63 (with genetic transformation of *cry1Ab/ry1Ac*) developed by Huazhong Agricultural University and Kemingdao (trans-*cry1Ab* gene) developed by Zhejiang University, showed high resistance to Chilo suppressalis, Tryporyza incertulas, and Cnaphalocrocis medinalis throughout the whole growth stages (Xu Xiuxiu et al., 2013).

As of April 2011, 701 *Bt* insecticidal genes have been cloned and named worldwide. These genes come from more than 30 countries and regions, and China contributed 259, the most (Zhang Jie et al., 2011). Although *Bt* genes has been the most successful and widely used pest resistance gene in transgenic plants, lasting pest resistance can be achieved only through the combination of *Bt* genes and other pest resistance genes. The transgenic rice with *cry1Ac+CpTI* in China and the transgenic rice with *cry1Ab+Xa21 +GNA* in India are relatively successful examples. From the perspective of food safety, Ye et al. (2009) introduced the green tissue-specific expression of the *rbcS* promoter-driven *cry1C* gene into the *japonica* rice variety Zhonghua 11 and obtained a transgenic line with high pest resistance and *Bt* toxin expression only in the stem and leaf parts of rice susceptible to insect attack. The expression of *Bt* toxin in the leaves of this line was three-fold higher than that of the transgenic line using the *Ubiquitin* promoter, but the content of *Bt* toxin in the endosperm was extremely low.

(2) Development of pest resistant transgenic rice in China

China is one of the largest rice producing countries in the world. *Bt* pest resistant rice has great economic, ecological and social value as its planting in China could increase the yield by 8%, reduce the use of pesticides by 80% and bring an annual income of about four billion US dollars (Huang et al., 2005). The research teams at IGDB, Huazhong Agricultural University, Fujian Academy of Agricultural Sciences, etc. mainly conducted research on bivalent transgenic insect resistant rice and adopted technologies such as selection marker removal transformation, intracellular localization and efficient and stable expression to obtain transgenic rice lines with high resistance to *lepidopteran* without selection markers. With joint efforts from Sichuan Agricultural University, Hubei Academy of Agricultural Sciences, Guangdong Academy of Agricultural Sciences, Jiangxi Academy of Agricultural Sciences and other outstanding breeding units, the insect-resistance genes were transferred into main rice cultivars and hybrid rice parents suitable for the upper Yangtze River rice region, the middle and lower Yangtze River rice region and the South China rice region, and formed a large number of insect-resistant hybrid rice combinations. Multi-site field experiments showed that transgenic rice parents and their combinations showed high resistance to *lepidopteran* pests such as *Cchilo suppressalis*, *Tryporyza incertulas* and *Cnaphalocrocis medinalis*. When no pesticide is applied, the transgenic rice outperforms the control varieties in terms of growth and damage, and shows obvious yield increase (Zhu Zhen et al., 2010).

2. Molecular breeding for herbicide tolerance

In recent years, weeds have caused annual economic losses of 10%−20% of the total crop output, and a variety of herbicides have been widely developed and used to reduce losses (Lou Shilin et al., 2002). Herbicide-resistant crops must be selected and bred to give full play to the effects of herbicides. However, among the available rice germplasm resources, rice with natural resistance to herbicides is almost nonexistent, and conventional breeding is greatly limited. The introduction of herbicide-resistance genes into rice using genetic engineering techniques or the mutation of endogenous herbicide-sensitive genes in rice by chemical mutagenesis to breed new herbicide-resistant rice varieties provide new ways to prevent and control weeds.

(1) Mechanisms of herbicides and strategies of creating herbicide tolerance

Inhibition of key enzymes in plant physiological metabolic processes, which can cause weed death, is the main mechanism by which chemical herbicides kill weeds in agricultural fields. These metabolic processes include photosynthesis, amino acid metabolism and others. The herbicides of glyphosate and glufosinate inhibit the key enzymes of plant fragrant amino acid synthesis, namely 5-enolpyruvy-shikimate-

3-phosphate synthase (EPSPS) and glutamine synthetase (GS), which plays an important role in the regulation of ammonia assimilation and nitrogen metabolism, respectively. Glyphosate molecules entering the plant when glyphosate is applied to weeds, compete with phosphoenolpyruvate (PEP) to bind to the active site of EPSPS, which terminates the synthesis pathway of fragrant amino acids, causing a deficiency of amino acids such as phenylalanine, tyrosine and tryptophan, and ultimately leading to plant death (Wang Xiujun et al., 2008).

There are usually three strategies to create herbicide tolerant transgenic crops. First, to overexpress herbicide-acting target protein so that plants can still carry out normal physiological metabolism after herbicide uptake; second, to modify the target protein to make them less efficient in binding with herbicides, thereby improving plant tolerance; third, to degrade or detoxify herbicides before they take effect by introducing herbicide-degrading enzymes or enzyme systems. Currently, commercial glyphosate tolerance is mostly based on the introduction of genes with non-sensitive target enzymes (such as EPSPs) (Qiu Long et al., 2012).

(2) Research and development of herbicide resistant rice abroad

The main herbicide-resistant genes widely used in rice are 5-enolpyruvy-shikimate-3-phosphate synthase (EPSPS) gene, acetolactate synthase (ALS) gene, glutamine synthetase (GS) gene, etc. (Wu Faqiang et al., 2009). Glyphosate is a widely used herbicide with the advantages of being cheap, having no toxicity, easy decomposition and causing no environmental pollution. EPSPS genes with high resistance to glyphosate were obtained by screening with glyphosate added to *E. coli* medium or isolated from soil heavily contaminated with glyphosate, and glyphosate-resistant rice was obtained by transgenesis, such as Monsanto's Nunda-resistant rice. Through transgenic breeding, we can obtain glyphosate resistant rice varieties, such as the Roundup Ready rice developed by Monsanto. Imidazolinone resistant rice can be obtained without transgenic breeding by screening the endogenous ALS gene of mutant rice, such as Clear-field rice jointly developed by BASF and Rice Biotechnology, Inc. USA. Glutamine synthetase is the target of glufosinate, and glufosinate resistant rice can be bred by bar gene transformation. In 1999, the United States approved the commercial production of bar gene transgenic herbicide tolerant rice LL-RICE06 and LLRICE62 of Sanofi-Aventis, and, in 2000, the rice was approved to be used as food.

(3) Application of herbicide resistance genes in hybrid rice breeding

In 1996, CRRI introduced herbicide-resistance genes *bar* and *cp4-EPSPS* into rice by gene gun bombardment for the first time, and successfully developed glufosinate and glyphosate resistant direct-seeding transgenic rice line Jiahe 98 and hybrid rice combination Liaoyou 1046, respectively. In the same period, the Institute of Subtropical Agroecology (ISA), CAS, bred a new herbicide-resistant rice line Bar 68 − 1 and its combinations. The success of the trans-herbicide resistance genes not only solved the problem of chemical weed control in direct-seeded rice, but also solved the key technical problem of seed purity in hybrid rice production. *Bar*-transformed rice lines, such as the restorer line T2070 and direct-seeded rice TR3 and T-Xiushui 11, are more resistant to herbicides and can also be used to maintain seed purity for hybrid rice. South China Botanical Garden, CAS, bred new combinations Ⅱ − You 86B and Teyou 86B by using Minghui 86B (containing the *bar* gene), which is resistant to the herbicide Liberty (Wu

Faqiang et al. , 2009). The team of Zhen Zhu at the Institute of Genetic Development, Chinese Academy of Sciences, randomly mutated the EPSPS gene of rice by error-prone PCR and introduced the EPSPS-deficient *E. coli* strain AB2829. After screening for glyphosate-resistant strains, the mutant EPSPS gene was isolated as a change from a proline to a leucine at position 106 of the polypeptide (base 317 changed from C to T). The mutant strain showed a 70-fold decrease in affinity for glyphosate and a three-fold increase in resistance to glyphosate (Zhou et al. , 2006). This gene has been used by Fujian Academy of Agricultural Sciences to breed new herbicide-resistant hybrid rice varieties.

Ⅲ. Breeding by Introducing Exogenous DNA

The main methods applied to rice to transfer genetic materials from distant species to create new germplasm are the pollen tube pathway method and the panicle-stalk injection method. The introduction of genomic DNA from distant species has proved in the past decades that new materials with improved agronomic traits can be created, thus creating an alternative way to exploit the advantages of distant hybrids.

Zhao Bingran's research team at HHRRC created rich variation materials of rice by means of panicle-stalk injection. For example, R254, a variety with a 43% increase in grain number per panicle and 13. 9% increase in 1,000-grain weight, was obtained by introducing maize genomic DNA into the restorer line R644. RB207, a large-panicle and large-grain restorer line, was created by introducing the DNA of barnyardgrass into the restorer line Xianhui 207, bringing an increase of 50% in both grain numbers per panicle and 1,000-grain weight compared with Xianhui 207, as well as good grain quality. 330, a variety with high resistance to rice blast at the seedling stage, was created by introducing the genomic DNA of *Oryzaminuta* (4N=48, BBCC) into the restorer line Minghui 63. The mutant line YVB with improved rice quality was bred when the same genomic DNA was introduced into V20B. The ERV1 variant line was obtained by introducing the genomic DNA of *Oryzaeichingeri* (2n=24, CC) with tall plant and long panicles into the restorer line RH78 and its number of grains per panicle increased from 202 to 325, and the plant height increased from 99. 8 cm to 131. 4 cm. The female sterile mutant *fsv1* was created by transferring the genomic DNA of apomictic *Panicummaximum* into the restorer line Gui 99.

In recent years, the genomic DNA of sequenced sorghum (BTx623) has been introduced into *indica* rice 9311 and a mutant material S931 was created with strong stem, resistance to lodging, large panicles with significantly increased grain density but fast filling at the base of the panicle. A new restorer line 2017C105 with excellent overall agronomic traits was bred in 2017 by crossing S931 with R94, an intermediate material for restorer line breeding.

Ⅳ. Genome-editing Breeding

Genome-editing breeding is to realize the expression regulation of endogenous genes through genome editing, so as to achieve the precise improvement of key traits or multiple traits such as yield, quality and resistance of varieties. Genome editing can avoid the linkage drag in the process of hybridization and backcrossing, making the molecular design of varieties truly possible and solving the bottleneck that cannot be broken by conventional means.

1. Breeding of new photo-thermo sensitive male sterile lines

The fertility of two-line hybrid rice is controlled by nuclear gene, and there is no requirement for the restorer and maintainer lines, resulting in free combination, simplified seed production, low cost, high use of rice germplasm resources and high probability of breeding excellent combinations. The current two-line sterile line is bred by conventional crossing methods, and each generation involves low-temperature reproduction of intermediate materials and slow stabilization with linkage drag. Zhuang Chuxiong's team at South China Agricultural University first used the Cas9 technology to knock out the PTGMS gene *TMS5* and create a new two-line male sterile line (Zhou et al., 2016). This technology can accelerate the breeding of two-line male sterile lines, and it only takes two generations to obtain genetically stable results without transgenic components. Theoretically, any fertile material can be developed into a two-line male sterile line by knocking out the PTGMS gene. In September 2017, a series of new PTGMS materials developed with CRISPR/Cas9-directed knockout of *TMS5* were displayed at Fuyang experimental station of CRRI. Eleven *tms5* PTGMS lines, including *japonica* Chunjiang 119 and Chunjiang 23, high grain quality *indica* Wushansimiao and Yuejingsimiao have uniform growth, excellent morphology and complete sterility. The yields of Chunjiang 119S/CH87 and Wushansimiao S/6089 - 100 were significantly higher than that of the control variety Fengliangyou 4. The application of the genome-editing technology will broaden the genetic background of two-line male sterile lines and promote the utilization of heterosis.

The MYB transcription factor *CSA* (carbon starved anther) participates in regulating sucrose distribution during anther development in rice, and its mutants causes photo-sensitive male sterility (Zhang et al., 2013). Li et al. (2016) used the CRISPR/Cas9 technology for targeted editing of *CSA* genes, in which some of the mutant materials showed male sterility under short daylength conditions and fertile under long daylength conditions, which provided a new way to breed PTGMS lines. In 2017, Zhang Dabing's team cloned a new rice PTGMS gene, *TMS10*, which encodes a leucine receptor kinase that plays an important regulatory role in anther development. They used the CRISPR/Cas9 genome-editing technology and obtained *TMS10* homozygous mutants in *japonica* and *indica* rice. All male sterile lines showed phenotypes of sterility under high temperature conditions and fertility under low temperature conditions, indicating that *TMS10* is functionally conserved in *japonica* and *indica* rice, and can be used to develop new PTGMS lines (Yu et al., 2017).

2. Improvement of disease resistance, quality, yield and wide compatibility

Wang et al. (2016) used the CRISPR/Cas9 system to mutate *OsERF922*, a negative regulatory gene of rice blast, and found that six homozygous mutant lines of the T_2 generation showed stronger resistance to rice blast than wild types at the seedling and tillering stages, while there are no significant changes in other traits. Zhang Huijun et al. (2016) used CRISPR/Cas9 to edit the *Pi21* and *OsBadh2* of Kongyu 131 and improved its blast resistance and fragrance quality.

Amylose content is closely related to rice quality. Ma et al. (2015) used the CRISPR/Cas9 technology to carry out targeted mutation of *OsWaxy*, the amylose synthase gene of T65, and the amylose content of the mutant decreased from 14.6% to 2.6%, thus acquiring waxy quality. Sun et al. (2017)

carried out defined-loci editing of *SBE* Ⅱ *b*, the starch branching enzyme gene. The amylose content of the *SBE* Ⅱ *b* mutant increased from 15% to 25% compared with that of the wild type.

Li et al. (2016) conducted defined-loci editing of four yield-related genes in *japonica* rice Zhonghua 11, including *Gn1a*, *DEP1*, *GS3* and *IPA1*. In the T_2 generation, *Gn1a*, *DEP1* and *GS3* mutants showed dense and erect panicle type with increased grain number per panicle and 1,000-grain weight, but semi dwarf and long awn appeared in *DEP1* and *GS3* mutants, and multi tillers and few tillers appeared in *IPA1* mutant. Shen Lan et al. (2017) knocked out eight agronomic traits-related genes (*DEP1*, *EP3*, *Gn1a*, *GS3*, *GW2*, *IPA1*, *OsBADH2*, *Hd1*) in rice, and the mutation frequencies of these genes were 50%, 100%, 67%, 81%, 83%, 97%, 67% and 78%, respectively. They obtained 25 multi-gene knockout mutants with different gene combination patterns, which greatly enriched the types of germplasm resources.

The erect-panicle gene *DEP1* is an important gene related to the yield trait. There is *DEP1* in *japonica* rice, a mutant gene that can promote cell division, make semi-dwarf plants, and increase panicle density and the number of branches and grains per panicle, so as to increase the yield of rice. However, the mutated gene *dep1* does not exist in *indica*. It will be time-consuming and labor-intensive to transfer *DEP1* from *japonica* rice to *indica* rice by conventional breeding methods. Syngenta researchers used the CRISPR/Cas9 technology to directly knock out a 10-kb fragment of the *DEP1* gene region of *indica* rice, and obtained new rice materials with yield increasing potential (Wang et al., 2017).

OsPDCD5, the programmed cell death gene, is a gene that negatively regulates rice yield (Su Wei, 2006). After it is knocked out with the defined-loci gene-editing technique, the growth duration of the mutant line is delayed, and all kinds of plant traits are strengthened, with a significant increase in final yield compared with wild types. In 2017, new high-yielding breeding materials with *OsPDCD5* knockout were displayed in the experimental base of Fudan University in Taicang city of Jiangsu province. The trial showed that the yield of Changhui T025 and Huazhan improved lines with *OsPDCD5* knockout increased by 15%–30% compared with the control lines, and the corresponding hybrid rice yield also had significantly higher biomass and grain yield than the control combination.

The progenies produced by *indica-japonica* cross usually have good agronomic traits but sterile, which greatly limits the utilization of heterosis in rice. Chen Letian's team of South China Agricultural University used the CRISPR/Cas9 technology to knock out the genes of *SaF* or *SaM* to obtain wide-compatibility rice materials (Xie et al., 2017).

3. Development of low-cadmium hybrid rice

In recent years, the excessive cadmium content of rice has become a serious food safety concern due to the contamination of the soil with cadmium and other heavy metals. The so-called low cadmium rice bred by conventional breeding techniques for emergency use still fails to meet the requirements for cadmium content when cultivated in heavily cadmium-contaminated fields.

Zhao Bingran's team at HHRRC used Huazhan and Longke 638S, the core parents of hybrid rice widely used in production, as materials, mutated *OsNramp5*, the major cadmium uptake gene of the two parents with the defined-loci genome-editing technology, and created the new restorer line Dige 1 and the

PTGMS line Dige 1S with good agronomic traits, stable low cadmium content and no exogenous genes. Then, they combined Dige 1 and Dige 1S to breed the low-cadmium hybrid combination Liangyoudige 1 (Tang et al., 2017).

In 2017, Dige 1 and Liangyoudige 1 were planted in trial fields heavily cadmium contaminated (whole cadmium content 1.5 mg/kg, pH 6.1), and the average cadmium content of these two varieties were 0.065 mg/kg and 0.056 mg/kg, respectively, a decrease of more than 90% compared with that of the control varieties Xiangwanxian 13 (1.48 mg/kg) and Shenliangyou 5814 (0.65 mg/kg) as well as the original varieties of Huazhan (1.31 mg/kg) and Longliangyou-Huazhan (0.84 mg/kg). This technology is expected to fundamentally solve the problem of cadmium-contaminated rice in China, and has the advantages of being economical, practical, safe which ensures broad application prospects.

4. Development of herbicide resistant rice through genome editing

Xu et al. (2014) used genome editing to knock out the bentazone sensitive lethal gene in rice, and obtained bentazone-sensitive mutation. This two-line male sterile line with *BEL* deletion can be used to solve the problem of hybrid seed impurity caused by selfing of the male sterile line. Almost all of the glyphosate-resistant crops currently in production were obtained by introducing the EPSPS gene of the Agrobacterium CP4 strain. The commercialization of transgenic varieties is greatly hampered by the biosafety concern of transgenic breeding. Gao Caixia and Li Jiayang's research team established collaboratively a CRISPR/Cas9-based gene substitution and defined-loci gene insertion system in rice using the non-homologous end-joining (NHEJ) method. A defined-loci substitution of two amino acids (*T102i* and *P106S*, TIPS) in the conserved region of the rice endogenous gene *OsEPSPS* was realized, and the heterozygote of TIPS defined-loci substitution was obtained in the T_0 generation. The results of the transmission analysis showed that the TIPS mutation in the EPSPS gene was resistant to glyphosate and stably inherited by the next generation (Li et al., 2016).

Currently, the Swedish Agricultural Commission has determined that being genome-edited is not transgenic. The U.S. agricultural regulatory authorities do not consider mutants produced by the plant self-repair mechanisms to be genetically modified, and genome-edited plants are not defined as genetically modified organisms (GMO). A clear line should be drawn between GMO and genome-edited crop (GEC) and the difference between them should be explained to the public. GMO is the product in which an exogenous DNA sequence is introduced with transgenic technology, while GEC is the product of editing and modifying genes held by the organism itself. Harvard geneticist George Church also said that CRISPR would be the end of "transgenesis". Whether the CRISPR/Cas9 technology is transgenic or not remains controversial in China and effective regulation of products produced using this technology is still needed by authority. To provide guidance for the application of genome editing to crop breeding, Academician Li Jiayang of the CAS and others have proposed a regulatory framework for genome-edited crops: First, the risk of uncontrolled dissemination of GEC should be minimized during the research phase. Second, it should be ensured that exogenous DNA is completely removed from GEC. Third, changes in the target DNA should be accurately recorded, and if a new sequence is introduced by homologous recombination, the relationship between the donor and the recipient must be clarified. Fourth, detection and determination is needed to confirm the absence of unexpected secondary editing events at major targets and consider the consequences of potential off-target events based on reference genome information and whole genome sequencing technologies. Finally, the above four points should be provided in the records of new varieties. With the five points mentioned above, GEC can be sufficiently regulated with simply the standards for conventional crop varieties (Huang et al., 2016).

Genome editing, represented by CRISPR/Cas9, is highly efficient, low-cost, and applicable to all species, simple to operate, and fast. In the post-genome era, as a new technology, genome editing will have a broad application prospect.

Hybrid rice technology has made great contributions to food safety in China and even in the world, but at the same time, its development also faces many challenges. For example, the conflict between the change of rice production methods and the high price of hybrid rice seeds, as well as the market demand for further improvement of hybrid rice grain quality. In recent years, molecular technology has been developing rapidly, and molecular breeding of hybrid rice is beginning to deliver. Through the combination of various molecular technologies, it is expected to promote the further development of the hybrid rice industry and science. Hybrid rice molecular breeding in the future will focus on breeding varieties with good eating quality, less pesticide, less chemical fertilizer, early maturity, good adaptability to climate change, especially suitability for direct seeding, mechanization, and other less laborious and simple cultivation methods without sacrificing the yield. In addition, the establishment and use of molecular technology systems to breed parents and combinations that can be mixed in sowing and harvesting and suitable for mechanized seed production is also an important direction for hybrid rice research.

References

[1] WAN JIANMIN. Status quo and prospects of molecular breeding of rice in China[J]. Journal of Agricultural Science and Technology, 2007, 9 (2): 1-9.

[2] WEI FENGJUAN, CHEN XIUCHEN. Molecular marker technology and its application in rice breeding[J]. Guangdong Agricultural Sciences, 2010, 37 (8): 185-187.

[3] QIAN QIAN. Gene-design breeding of rice[M]. Beijing: Science Press, 2007.

[4] MEUWISSEN T H, HAYES B J, GODDARD M E. Prediction of total genetic value using genome-wide dense marker maps[J]. Genetics, 2001, 157 (4): 1819-1829.

[5] WANG JIANKANG, LI HUIHUI, ZHANG LUYAN. Gene mapping and breeding design[M]. Beijing: Science Press, 2014.

[6] HIEI Y, OHTA S, KOMARI T, et al. Efficient transformation of rice (*Oryza sativa* L.) mediated by Agrobacterium and sequence analysis of the boundaries of the T-DNA[J]. The Plant Journal, 1994, 6 (2): 271-282.

[7] AN HANBING, ZHU ZHEN. Application of gene gun in plant genetic transformation[J]. Progress in Biotechnology, 1997 (1): 18-26.

[8] KLEIN T, SANFORD J, FROMM M. Genetic transformation of maize cells by particle bombardment[J]. Plant Physiology, 1989, 91 (1): 440-444.

[9] ZHOU GUANGYU, WENG JIAN, GONG ZHENZHEN, et al. Introduction of exogenous DNA into plants after pollination in agricultural molecular breeding[J]. Chinese Agricultural Sciences, 1988, 21 (3): 1-6.

[10] DONG YANYU, HONG YAHUI, REN CHUNMEI, et al. Application of exogenous DNA introduction technology in plant molecular breeding[J]. Journal of Hunan Agricultural University, 1994, 20 (6): 513-521.

[11] ZHU SHENGWEI, HUANG GUOCUN, SUN JINGSAN. Research progress on direct introduction of exogenous DNA into recipient plants[J]. Chinese Bulletin of Botany, 2000, 17 (1): 11-16.

[12] ZHOU JIANLIN, LI YANGSHENG, JIA LINGHUI, et al. A preliminary report on molecular breeding technology of introducing *echinochloa crusgalli* DNA into rice by panicle-stalk injection[J]. Research of Agricultural Modernization, 1997 (4): 44-45.

[13] YANG QIANJIN. Research progress in introducing exogenous DNA into rice by embryo soaking[J]. Journal of Anhui Agricultural Sciences, 2006 (24): 6452-6454.

[14] ZHAO BINGRAN, WU JINGHUA, WANG GUIYUAN. A preliminary study on introduction of exogenous DNA into rice by spike-stalk injection[J]. Hybrid Rice, 1994 (2): 37-38.

[15] PEÑA A, LÖRZ H, SCHELL J. Transgenic rye plants obtained by injecting DNA into young floral tillers [J]. Nature, 1987, 325(6101):274-276.

[16] ZHOU GUANGYU, CHEN SHANBAO, HUANG JUNQI. Research progress of agricultural molecular breeding[M]. Beijing: China Agricultural Science and Technology Press, 1993.

[17] HUANG JUNQI, QIAN SIYING, LIU GUILING, et al. Variation of upland cotton traits induced by exogenous DNA from gossypium barbadense[J]. Journal of Genetics and Genomics, 1981,8 (1): 56-62.

[18] ZHAO BINGRAN, HUANG JIANLIANG, LIU CHUNLIN, et al. A study on the *in vivo* transport of exogenous DNA injected into stem and the female sterile mutant[J]. Journal of Hunan Agricultural University, 1998 (6): 436-441.

[19] ZHOU GUANGYU, GONG ZHENZHEN, WANG ZIFEN. Molecular basis of distant hybridization-a demonstration of DNA fragment hybridization hypothesis[J]. Journal of Genetics and Genomics, 1979 (4): 405-413.

[20] WAN WENJU, ZOU DONGSHENG, PENG KEQIN. On the dual functions of biological mutation and exogenous DNA introduction[J]. Journal of Hunan Agricultural University, 1992, 18 (4): 886-891.

[21] ZHAO B, XING Q, XIA H, et al. DNA polymorphism among Yewei B, V20B, and *Oryza minuta* JS presl. ex CB presl[J]. Journal of integrative plant biology, 2005, 47(12):1485-1492.

[22] LIU ZHENLAN, DONG YUZHU, LIU BAO. Cloning of species DNA sequence of zizania latifolia and its application in detection of introduction of zizania latifolia DNA into rice[J]. Chinese Bulletin of Botany, 2000, 42 (3): 324-326.

[23] ZHAO BINGRAN. A study on introducing genomic DNA of distant species into rice[D]. Changsha: Hunan Agricultural University Press, 2003.

[24] XING Q, ZHAO B, XU K, et al. Test of agronomic characteristics and amplified fragment length polymorphism analysis of new rice germplasm developed from transformation of genomic DNA of distant relatives[J]. Plant Molecular Biology Reporter, 2004, 22(2): 155-164.

[25] PENG Y, HU Y, MAO B, et al. Genetic analysis for rice grain quality traits in the YVB stable variant line using RAD-seq[J]. Molecular Genetics and Genomics, 2016, 291 (1): 297-307.

[26] WANG FUJUN, ZHAO KAIJUN. Progress and challenges of genome-editing technology in crop genetic improvement[J]. Scientia Agricultura Sinica, 2018, 51 (1): 1-16.

[27] RAN F A, HSU P D, WRIGHT J, et al. Genome engineering using the CRISPR/Cas9 system[J]. Nature Protocol, 2013, 32 (12): 815.

[28] GAJ T, GERSBACH C, BARBAS R. ZFN, TALEN, and CRISPR/Cas-based methods for genome engineering[J]. Trends in Biotechnology, 2013, 31 (7): 397 - 405.

[29] SHAN Q, WANG Y, Li J, et al. Targeted genome modification of crop plants using a CRISPR/Cas system [J]. Nature Biotechnology, 2013, 31 (8): 686 - 688.

[30] MIAO J, GUO D, ZHANG J, et al. Targeted mutagenesis in rice using CRISPR/Cas system[J]. Cell Research, 2013, 23 (10): 1233 - 1236.

[31] FENG Z, ZHANG B, DING W, et al. Efficient genome editing in plants using a CRISPR/Cas system[J]. Cell Research, 2013, 23 (10): 1229 - 1232.

[32] WANG C, SHEN L, FU Y, et al. A simple CRISPR/Cas9 system for multiplex genome editing in rice[J]. Journal of Genetics and Genomics, 2015, 42 (12): 703 - 706.

[33] MA X, ZHANG Q, ZHU Q, et al. A robust CRISPR/Cas9 system for convenient, high-efficiency multiplex genome editing in monocot and dicot plants[J]. Molecular Plant, 2015, 8 (8): 1274 - 1284.

[34] ZONG Y, WANG Y, LI C, et al. Precise base editing in rice, wheat and maize with a Cas9-cytidine deaminase fusion[J]. Nature Biotechnology, 2017, 35 (5): 438.

[35] WANG Y, GENG L, YUAN M, et al. Deletion of a target gene in indica rice via CRISPR/Cas9[J]. Plant Cell Reports, 2017, 36 (8): 1 - 11.

[36] AO JUNJIE, HU HUI, LI JUNKAI, et al. Research progress on inheritance and gene cloning of rice blast resistance[J]. Journal of Yangtze University (Natural Science Edition), 2015, 12 (33): 32 - 35.

[37] HE XIUYING, WANG LING, WU WEIHUAI, et al. Research progress on mapping and cloning rice blast resistant genes and application in breeding[J]. Chinese Agricultural Science Bulletin, 2014, 30 (06): 1 - 12.

[38] TAKAHASHI A, HAYASHI N, MIYAO A, et al. Unique features of the rice blast resistance *Pish* locus revealed by large scale retrotransposon-tagging[J]. BMC Plant Biology, 2010, 10: 175.

[39] BRYAN G, WU K, FARRALL L, et al. A single amino acid difference distinguishes resistant and susceptible alleles of the rice blast resistance gene *Pita* [J]. The Plant Cell, 2000, 12 (11): 2033 - 2046.

[40] LIN F, CHEN S, QUE Z, et al. The blast pesistance gene *Pi37* encodes a nucleotide binding site leucine-rich repeat protein and is a member of a resistance gene cluster on rice chromosome 1[J]. Genetics, 2007, 177 (3): 1871 - 1880.

[41] WANG Z, YANO M, YAMANOUCHI U, et al. The Pib gene for rice blast resistance belongs to the nucleotide binding and leucine-rich repeat class of plant disease resistance genes[J]. The Plant Journal, 1999, 19 (1): 55 - 64.

[42] FUKUOKA S, SAKA N, KOGA H, et al. Loss of function of a proline-containing protein confers durable disease resistance in rice[J]. Science, 2009, 325 (5943): 998 - 1001.

[43] XU X, HAYASHI N, WANG C, et al. Rice blast resistance gene *Pikahei-1(t)*, a member of a resistance gene cluster on chromosome 4, encodes a nucleotide-binding site and leucine-rich repeat protein[J]. Molecular Breeding, 2014, 34 (2): 691 - 700.

[44] CHEN X, SHANG J, CHEN D, et al. A B-lectin receptor kinase gene conferring rice blast resistance[J]. The Plant Journal, 2006, 46 (5): 794 - 804.

[45] QU S, LIU G, ZHOU B, et al. The broad-spectrum blast resistance gene *Pi9* encodes a nucleotide-binding

site-leucine-rich repeat protein and is a member of a multi-gene family in rice[J]. Genetics, 2006, 172 (3): 1901 - 1914.

[46] LIU G, LU G, ZENG L, et al. Two broad-spectrum blast resistance genes, *Pi9* (t) and *Pi2* (t), are physically linked on rice chromosome 6[J]. Molecular Genetics and Genomics, 2002, 267 (4): 472 - 480.

[47] ZHOU B, QU S, LIU G, et al. The eight amino-acid differences within three leucine-rich repeats between *Pi2* and *Pizt* resistance proteins determine the resistance specificity to magnaporthe grisea[J]. Molecular Plant-Microbe Interactions, 2006, 19 (11): 1216 - 1228.

[48] LIU X, LIN F, WANG L, et al. The in silico map-based cloning of *Pi36*, a rice coiled-coil-nucleotide-binding site-leucine-rich repeat gene that confers race-specific resistance to the blast[J]. Fungus Genetics, 2007, 176 (4): 2541 - 2549.

[49] LEE S, SONG M, SEO Y, et al. Rice *Pi5*-mediated resistance to magnaportheoryzae requires the presence of two coiled-coil-nucleotide-binding-leucine-rich repeat[J]. Genes Genetics, 2009, 181 (4): 1627 - 1638.

[50] ASHIKAWA I, HAYASHI N, YAMANE H, et al. Two adjacent nucleotide-binding site-leucine-rich repeat class genes are required to confer *Pikm*-specific rice blast resistance[J]. Genetics, 2008, 180 (4): 2267 - 2276.

[51] DENG Y, ZHAI K, XIE Z, et al. Epigenetic regulation of antagonistic receptors confers rice blast resistance with yield balance[J]. Science, 2017, 355 (6328): 962.

[52] HAYASHI N, INOUE H, KATO T, et al. Durable panicle blast-resistance gene *Pb1* encodes an atypical CC-NBS-LRR protein and was generated by acquiring a promoter through local genome duplication[J]. The Plant Journal, 2010, 64 (3): 498 - 510.

[53] HUA L, WU J, CHEN C, et al. The isolation of *Pi1*, an allele at the *Pik* locus which confers broad spectrum resistance to rice blast[J]. Theoretical and Applied Genetics, 2012, 125 (5): 1047 - 1055.

[54] ZHAI C, LIN F, DONG Z, et al. The isolation and characterization of *Pik*, a rice blast resistance gene which emerged after rice domestication[J]. New Phytologist, 2011, 189 (1): 321 - 334.

[55] SHARMA T, MADHAV M, SINGH B, et al. High-resolution mapping, cloning and molecular characterization of the *Pi-kh* gene of rice, which confers resistance to Magnaporthe grisea[J]. Molecular Genetics and Genomics, 2005, 274 (6): 569 - 578.

[56] ZENG X, YANG X, ZHAO Z, et al. Characterization and fine mapping of the rice blast resistance gene *Pia* [J]. Science China Life Sciences, 2011, 54 (4): 372 - 378.

[57] LI W, ZHU Z, CHERN M, et al. A natural allele of a transcription factor in rice confers broad-spectrum blast resistance[J]. Cell, 2017, 170 (1): 114 - 126.

[58] ZHOU X, LIAO H, CHERNET M, et al. Loss of function of a rice TPR-domain RNA-binding protein confers broad-spectrum disease resistance[J]. Proceedings of the National Academy of Sciences of the United States of America, 2018, 115 (12): 3174 - 3179.

[59] MEW T W. Current status and future prospects of research on bacterial blight of rice[J]. Annual Review of Phytopathology, 1987, 25 (1): 359 - 382.

[60] CHEN HESHENG, MAO FUTING, REN JIANHUA. A study on the overwintering bacteriàl sources of rice bacterial blight[J]. Journal of Zhejiang Agricultural University, 1986 (1): 77 - 82.

[61] SONG W, WANG G, CHEN L, et al. A receptor kinase-like protein encoded by the rice disease resistance gene, *Xa21* [J]. Science, 1995, 270: 1804 - 180.

[62] WANG C, FAN Y, ZHENG C, et al. High-resolution genetic mapping of rice bacterial blight resistance gene

Xa23 [J]. Molecular Genetics and Genomics, 2014, 289 (5): 745 - 753.

[63] GU K, YANG B, TIAN D, et al. R gene expression induced by a type - Ⅲ effector triggers disease resistance in rice[J]. Nature, 2005, 435: 1122 - 1125.

[64] XIANG Y, CAO Y, XU C, et al. *Xa3*, conferring resistance for rice bacterial blight and encoding a receptor kinase-like protein, is the same as *Xa26* [J]. Theoretical and Applied Genetics, 2006, 113 (7): 1347 - 1355.

[65] HU K, CAO J, ZHANG J et al. Improvement of multiple agronomic traits by a disease resistance gene via cell wall reinforcement[J]. Nature Plants, 2017, 3: 17009.

[66] CHU Z, FU B, YANG H, et al. Targeting *xa13*, a recessive gene for bacterial blight resistance in rice[J]. Theoretical and Applied Genetics, 2006, 112 (3): 455 - 461.

[67] ANTONY G, ZHOU J, HUANG S, et al. Rice *xa13* recessive resistance to bacterial blight is defeated by induction of the disease susceptibility gene *Os11N3* [J]. The Plant Cell, 2010, 22 (11): 3864 - 3876.

[68] LIU Q, YUAN M, ZHOU Y, et al. A paralog of the MtN3/saliva family recessively confers race-specific resistance to *Xanthomonas oryzae* in rice[J]. Plant, Cell & Environment, 2011, 34 (11): 1958 - 1969.

[69] ZHOU J, PENG Z, LONG J, et al. Gene targeting by the TAL effector *PthXo2* reveals cryptic resistance gene for bacterial blight of rice[J]. Plant Journal, 2015, 82 (4): 632 - 643.

[70] MATTHEW W B, AMANDA J G, ANJALI S I, et al. High resolution genetic mapping and candidate gene identification at the *xa5* locus for bacterial blight resistance in rice (*Oryza sativa* L.) [J]. Theoretical and Applied Genetics, 2003, 107 (1): 62 - 73.

[71] WANG HUI, YAN ZHI, CHEN JINJIE, et al. Research progress and prospects of *Bph* resistant genes in rice [J]. Hybrid Rice, 2016, 31 (04): 1 - 5.

[72] DU B, ZHANG W, LIU B, et al. Identification and characterization of *Bph14*, a gene conferring resistance to brown planthopper in rice[J]. Proceedings of the National Academy of Sciences, 2009, 106 (52): 22163 - 22168.

[73] LIU Y, WU H, CHEN H, et al. A gene cluster encoding lectin receptor kinases confers broad-spectrum and durable pest resistance in rice[J]. Nature Biotechnology, 2015, 33 (3): 301 - 305.

[74] HU J, ZHOU J, PENG X, et al. The *Bphi008a* gene interacts with the ethylene pathway and transcriptionally regulates MAPK genes in the response of rice to brown planthopper feeding[J]. Plant Physiology, 2011, 156 (2): 856 - 872.

[75] REN J, GAO F, WU X, et al. *Bph32*, a novel gene encoding an unknown SCR domain-containing protein, confers resistance against the brown planthopper in rice[J]. Scientific Reports, 2016, 6: 37645.

[76] WANG Y, CAO L, ZHANG Y, et al. Map-based cloning and characterization of *Bph29*, a B3 domain-containing recessive gene conferring brown planthopper resistance in rice[J]. Journal of Experimental Botany, 2015, 66 (19): 6035 - 6045.

[77] JI H, KIM SR, KIM YH, et al. Map-based cloning and characterization of the *Bph18* gene from wild rice conferring resistance to brown planthopper pest[J]. Scientific Reports, 2016, 6: 34376.

[78] ZHAO Y, HUANG J, WANG Z, et al. Allelic diversity in an NLR gene *BPH9* enables rice to combat planthopper variation[J]. Proceedings of the National Academy of Sciences of the United States of America, 2016, 113 (45): 12850 - 12855.

[79] GUO J, XU C, WU D, et al. *Bph6* encodesan exocyst-localized protein and confers broad resistance to planthoppers in rice[J]. Nature Genetics, 2018, 50 (2): 297 - 306.

[80] ZHU YIWANG, LIN YARONG, CHEN LIANG. Research progress in rice molecular breeding in china[J]. Journal of Xiamen University (Natural Science Edition), 2016, 55 (05): 661-671.

[81] ASHIKARI M, SAKAKIBARA H, LIN S, et al. Cytokinin oxidase regulates rice grain production[J]. Science, 2005, 309 (5735): 741.

[82] WANG E, WANG J, ZHU X, et al. Control of rice grain-filling and yield by a gene with a potential signature of domestication[J]. Nature Genetics, 2008, 40 (11): 1370-1374.

[83] WENG J, GU S, WAN X, et al. Isolation and initial characterization of *GW5*, a major QTL associated with rice kernel width and weight[J]. Cell Research, 2008, 18 (12): 1199-1209.

[84] WANG S, WU K, YUAN Q, et al. Control of grain size, shape and quality by *OsSPL16* in rice[J]. Nature Genetics, 2012, 44 (8): 950-954.

[85] LI Y, FAN C, XING Y, et al. Natural variation in *GS5* plays an important role in regulating grain size and yield in rice[J]. Nature Genetics, 2011, 43 (12): 1266-1269.

[86] DUAN P, XU J, ZENG D, et al. Natural variation in the promoter of *GSE5* contributes to grain size diversity in rice[J]. Molecular Plant, 2017, 10 (5): 685.

[87] HUANG X, QIAN Q, LIU Z, et al. Natural variation at the *DEP1* locus enhances grain yield in rice[J]. Nature Genetics, 2009, 41 (4): 494-497.

[88] LI F, LIU W, TANG J, et al. Rice dense and erect panicle 2 is essential for determining panicle outgrowth and elongation[J]. Cell Research, 2010, 20 (7): 838.

[89] CHE R, TONG H, SHI B, et al. Control of grain size and rice yield by *GL2*-mediated brassinosteroid responses [J]. Nature Plants, 2015, 2: 15195.

[90] MAO H, SUN S, YAO J, et al. Linking differential domain functions of the *GS3* protein to natural variation of grain size in rice[J]. Proceedings of the National Academy of Sciences of the United States of America, 2010, 107 (45): 19579-19584.

[91] JIAO Y, WANG Y, XUE D, et al. Regulation of *OsSPL14* by *OsmiR156* defines ideal plant architecture in rice [J]. Nature Genetics, 2010, 42 (6): 541-544.

[92] ZHANG X, WANG J, HUANG J, et al. Rare allele of *OsPPKL1* associated with grain length causes extra-large grain and a significant yield increase in rice[J]. Proceedings of the National Academy of Sciences, 2012, 109 (52): 21534-21539.

[93] DUAN P, RAO Y, ZENG D, et al. *Small Grain1*, which encodes a mitogen-activated protein kinase kinase 4, influences grain size in rice[J]. The Plant Journal, 2014, 77 (4): 547-557.

[94] ISHIMARU K, HIROTSU N, MADOKA Y, et al. Loss of function of the IAA-glucose hydrolase gene *TGW6* enhances rice grain weight and increases yield[J]. Nature Genetics, 2013, 45 (6): 707-711.

[95] LIU L, TONG H, XIAO Y, et al. Activation of *BigGrain1* significantly improves grain size by regulating auxin transport in rice[J]. Proceedings of the National Academy of Sciences of the United States of America, 2015, 112 (35): 11102-11107.

[96] XU F, FANG J, OU S, et al. Variations in *CYP78A13* coding region influence grain size and yield in rice [J]. Plant Cell and Environment, 2015, 38 (4): 800-811.

[97] CHEN J, GAO H, ZHENG X, et al. An evolutionarily conserved gene, *FUWA*, plays a role in determining panicle architecture, grain shape and grain weight in rice[J]. The Plant Journal, 2015, 83 (3): 427-438.

[98] SI L, CHEN J, HUANG X, et al. *OsSPL13* controls grain size in cultivated rice[J]. Nature Genetics, 2016, 48 (4): 447-456.

[99] SUN S, WANG L, MAO H, et al. A G-protein pathway determines grain size in rice[J]. Nature Communications, 2018, 9 (1): 851.

[100] HUO X, WU S, ZHU Z, et al. *NOG1* increases grain production in rice[J]. Nature Communications, 2017, 8 (1): 1497.

[101] BAI X, HUANG Y, HU Y, et al. Duplication of an upstream silencer of *FZP* increases grain yield in rice [J]. Nature Plants, 2017, 3 (11): 885 - 893.

[102] LI S, LI W, HUANG B, et al. Natural variation in *PTB1* regulates rice seed setting rate by controlling pollen tube growth[J]. Nature Communications, 2013, 4 (7): 2793.

[103] WANG J, YU H, XIONG G, et al. Tissue-Specific Ubiquitination by *IPA1* interacting protein 1 Modulates *IPA1* protein levels to regulate plant architecture in rice[J]. Plant Cell, 2017, 29 (4): 697 - 707.

[104] WANG S, WU K, QIAN Q, et al. Non-canonical regulation of SPL transcription factors by a human OTUB1-like deubiquitinase defines a new plant type rice associated with higher grain yield[J]. Cell Research, 2017, 27 (9): 1142 - 1156.

[105] WANG L, ZHANG Q. Boosting rice yield by fine - tuning SPL gene expression[J]. Trends in Plant Science, 2017, 22 (8): 643 - 644.

[106] ZHANG L, YU H, MA B, et al. Anaturaltandem array alleviates epigenetic repression of *IPA1* and leads to superior yielding rice[J]. Nature Communications, 2017, 8: 14789.

[107] XUE W, XING Y, WENG X, et al. Natural variation in *Ghd7* is an important regulator of heading date and yield potential in rice[J]. Nature Genetics, 2009, 40 (6): 761 - 767.

[108] WEI X, XU J, GUO H, et al. *DTH8* suppresses flowering in rice, influencing plant height and yield potential simultaneously[J]. Plant Physiology, 2010, 153 (4): 1747 - 1758.

[109] YAN W, WANG P, CHEN H, et al. A major QTL, *Ghd8*, plays pleiotropic roles in regulating grain productivity, plant height, and heading date in rice[J]. Molecular Plant, 2011, 4 (2): 319 - 330.

[110] XU XIAOMING, ZHANG YINGXIN, WANG HUIMIN, et al. Research progress in genetic traits of nitrogen, phosphorus and potassium uptake and utilization in rice[J]. Journal of Nuclear Agricultural Sciences, 2016, 30 (04): 685 - 694.

[111] OBARA M, TAKEDA T, HAYAKAWA T, et al. Mapping quantitative trait loci controlling root length in rice seedlings grown with low or sufficient, supply using backcross recombinant lines derived from a cross between *Oryza sativa* L. and *Oryza glaberrima* Steud[J]. Soil Science & Plant Nutrition, 2011, 57 (1): 80 - 92.

[112] BI Y, KANT S, CLARKE J, et al. Increased nitrogen-use efficiency in transgenic rice plants over-expressing a nitrogen-responsive early nodul in gene identified from rice expression profiling[J]. Plant Cell & Environment, 2009, 32 (12): 1749.

[113] SUN H, QIAN Q, WU K, et al. Heterotrimeric G proteins regulate nitrogen-use efficiency in rice[J]. Nature Genetics, 2014, 46 (6): 652 - 656.

[114] HU B, WANG W, OU S, et al. Variation in *NRT1. 1B* contributes to nitrate-use divergence between rice subspecies[J]. Nature Genetics, 2015, 47 (7): 834 - 838.

[115] WANG W, HU B, YUAN D, et al. Expression of the nitrate transporter gene *OsNRT1. 1A / OsNPF6. 3* confers high yield and early maturation in rice[J]. The Plant Cell, 2018, 30 (3): 638 - 651.

[116] WANG Q, NIAN J, XIE X, et al. Genetic variations in *ARE1* mediate grain yield by modulating nitrogen utilization in rice[J]. Nature Communications, 2018, 9 (1): 735.

[117] WASAKI J, YONETANI R, SHINANO T, et al. Expression of the *OsPI1* gene, cloned from rice roots using cDNA microarray, rapidly responds to phosphorus status[J]. New Phytologist, 2003, 158 (2): 239 - 248.

[118] GAMUYAO R, CHIN JH, PARIASCA-TANAKA J, et al. The protein kinase Pstol1 from traditional rice confers tolerance of phosphorus deficiency[J]. Nature, 2012, 488 (7412): 535 - 539.

[119] ZHANG Q, WANG C, TIAN J, et al. Identification of rice purple acid phosphatases related to posphate starvation signalling[J]. Plant Biology, 2011, 13 (1): 7 - 15.

[120] JIA H, REN H, GU M, et al. The phosphate transporter gene *OsPht1* ; 8 is involved in phosphate homeostasis in rice[J]. Plant Physiology, 2011, 156 (3): 1164 - 1175.

[121] OBATA T, KITAMOTO H, NAKAMURA A, et al. Rice shaker potassium channel *OsKAT1* confers tolerance to salinity stress on yeast and rice cells[J]. Plant Physiology, 2007, 144 (4): 1978 - 1985.

[122] LAN W, WEI W, WANG S, et al. A rice high-affinity potassium transporter (HKT) conceals a calcium-permeable cation channel[J]. Proceedings of the National Academy of Sciences of the United States of America, 2010, 107 (15): 7089 - 7094.

[123] CHEN S, YANG Y, SHI W, et al. *Badh2*, encoding betaine aldehyde dehydrogenase, inhibits the biosynthesis of 2-acetyl-1-pyrroline, a major component in rice fragrance[J]. Plant Cell, 2008, 20 (7): 1850 - 1861.

[124] WANG Y, ZHU S, LIU S, et al. The vacuolar processing enzyme *OsVPE1* is required for efficient glutenin processing in rice[J]. Plant Journal, 2009, 58 (4): 606 - 617.

[125] WANG Y, REN Y, LIU X, et al. *OsRab5a* regulates endomembrane organization and storage protein trafficking in rice endosperm cells[J]. Plant Journal, 2010, 64 (5): 812 - 824.

[126] LIU F, REN Y, WANG Y, et al. *OsVPS9A* functions cooperatively with *OsRAB5A* to regulate post-golgi dense vesicle-mediated storage protein trafficking to the protein storage vacuole in rice endosperm cells[J]. Molecular Plant, 2013, 6 (6): 1918 - 1932.

[127] REN Y, WANG Y, LIU F, et al. Glutelin precursor accumulation 3 encodes a regulator of post-Golgi vesicular traffic essential for vacuolar protein sorting in rice endosperm[J]. Plant Cell, 2014, 26 (1): 410.

[128] WANG Y, LIU F, REN Y, et al. Golgi transport 1B regulates protein export from endoplasmic reticulum in rice endosperm cells[J]. Plant Cell, 2016, 28 (11): 2850.

[129] LI Y, FAN C, XING Y, et al. *Chalk5* encodes a vacuolar H(+)-translocating pyrophosphatase influencing grain chalkiness in rice[J]. Nature Genetics, 2014, 46 (6): 398 - 404.

[130] PENG B, KONG H, LI Y, et al. *OsAAP6* functions as an important regulator of grain protein content and nutritional quality in rice[J]. Nature Communications, 2014, 5 (1): 4847.

[131] WANG S, WU K, YUAN Q, et al. Control of grain size, shape and quality by *OsSPL16* in rice[J]. Nature Genetics, 2012, 44 (8): 950 - 954.

[132] WANG S, LI S, LIU Q, et al. The *OsSPL16-GW7* regulatory module determines grain shape and simultaneously improves rice yield and grain quality[J]. Nature Genetics, 2015, 47 (8): 949 - 954.

[133] LIU Q, HAN R, WU K, et al. G-protein βγ subunits determine grain size through interaction with MADS-domain transcription factors in rice[J]. Nature Communications, 2018, 9: 852.

[134] LI X M, CHAO D Y, WU Y, et al. Natural alleles of a proteasome α2 subunit gene contribute to thermo tolerance and adaptation of African rice[J]. Nature Genetics, 2015,47(7): 827 - 833.

[135] WANG D, QIN B, LI X, et al. Nucleolar DEAD-Box RNA helicase *TOGR1* regulates thermo-tolerant

growth as a pre-rRNA chaperone in rice[J]. Plos Genetics, 2016, 12(2): e1005844.

[136] HUANG X, CHAO D, GAO J, et al. A previously unknown zinc finger protein, DST, regulates drought and salt tolerance in rice via stomatal aperture control[J]. Genes and development, 2009, 23(15):1805.

[137] CUI L, SHAN J, SHI M, et al. *DCA1* acts as a transcriptional co-activator of DST and contributes to drought and salt tolerance in rice[J]. Plos Genetics, 2015, 11 (10): e1005617.

[138] WU F, SHENG P, TAN J, et al. Plasma membrane receptor-like kinase leaf panicle 2 acts downstream of the drought and salt tolerance transcription factor to regulate drought sensitivity in rice[J]. Journal of Experimental Botany, 2015, 66 (1): 271-281.

[139] LI S, ZHAO B, YUAN D, et al. Rice zinc finger protein DST enhances grain production through controlling *Gn1a/OsCKX2* expression[J]. Proceedings of the National Academy of Sciences of the United States of America, 2013, 110 (8)3167-3172.

[140] ZHU X, XIONG L. Putative megaenzyme *DWA1* plays essential roles in drought resistance by regulating stress-induced wax deposition in rice[J]. Proceedings of the National Academy of Sciences of the United States of America, 2013, 110 (44): 17790-17795.

[141] LAN T, ZHANG S, LIU T, et al. Fine mapping and candidate identification of SST, a gene controlling seedling salt tolerance in rice (*Oryza sativa* L.)[J]. Euphytica, 2015, 205 (1): 269-274.

[142] OGAWA D, ABE K, MIYAO A, et al. *RSS1* regulates the cell cycle and maintains meristematic activity under stress conditions in rice[J]. Nature Communications, 2011, 2 (1): 121-132.

[143] TODA Y, TANAKA M, OGAWA D, et al. Rice salt sensitive 3 forms a ternary complex with JAZ and Class-C bHLH factors and regulates jasmonate-induced gene expression and root cell elongation[J]. Plant Cell, 2013, 25 (5): 1709-1725.

[144] TAKAGI H, TAMIRU M, ABE A, et al. MutMap accelerates breeding of a salt-tolerant rice cultivar[J]. Nature Biotechnology, 2015, 33 (5): 445-449.

[145] MA Y, DAI X, XU Y, et al. *COLD1* confers chilling tolerance in rice[J]. Cell, 2015, 160 (6): 1209-1221.

[146] ZHANG Z, LI J, PAN Y, et al. Natural variation in *CTB4a* enhances rice adaptation to cold habitats[J]. Nature Communications, 2017, 8: 14788.

[147] WANG JUN, YANG JIE, CHEN ZHIDE, et al. Aggregation of Rice Resistant Genes *Pi-ta*, *Pi-b* and *Stv-b-i* by Molecular Marker-Assisted Selection[J]. Acta Agronomica Sinica, 2011, 37 (6): 975-981.

[148] YIN DESUO, XIA MINGYUAN, LI JINBO, et al. Development of STS linked marker of rice blast resistant gene *Pi9* and its application in marker assisted breeding[J]. Chinese Journal of Rice Science, 2011, 25 (1): 25-30.

[149] WEN SHAOSHAN, GAO BIJUN. Introduction of rice blast resistant gene *Pi-9* (T) into rice restorer line Luhui 17 by marker-assisted selection[J]. Molecular Plant Breeding, 2012, 10 (1): 42-47.

[150] XING XUAN, LIU XIONGLUN, CHEN HAILONG, et al. Improvement of rice blast resistance of *R288* by marker assisted selection of *Pi9* [J]. Crop Research, 2016, 30 (5): 487-491.

[151] LIU WUGE, WANG FENG, LIU ZHENRONG, et al. Improving blast resistance of three-line male sterile line rongfeng a by polymerizing *Pi-1* and *Pi-2* by molecular marker technology[J]. Molecular Plant Breeding, 2012, 10 (5): 575-582.

[152] YU SHOUWU, ZHENG XUEQIANG, FAN TIANYUN, et al. Molecular marker assisted selection of PTGMS line with blast resistant gene *Pi25* [J]. China Rice, 2013, 19 (3): 15-17.

[153] TU SHIHANG, ZHOU PENG, ZHENG YI, et al. Breeding of rice blast resistant three-line male sterile line by molecular marker assisted selection of *Pi25* [J]. Molecular Plant Breeding, 2015, 13 (9): 1911 - 1917.

[154] YANG PING, ZOU GUOXING, CHEN CHUNLIAN, et al. Improvement of blast resistance in Chunhui 350 by molecular marker assisted selection[J]. Molecular Plant Breeding, 2015, 13 (4): 741 - 747.

[155] LIU KAIYU, LU SHUANGNAN, QIU JUNLI, et al. Breeding of *Bph* resistant rice restorer lines[J]. Molecular Plant Breeding, 2011, 9 (4): 410 - 417.

[156] ZHAO PENG, FENG RANRAN, XIAO QIAOZHEN, et al. Screening of rice lines with *bph20* (*T*) and *bph21* (*T*) and rice blast resistant gene *Pi9* [J]. Journal of Southern Agriculture, 2013, 44 (6): 885 - 892.

[157] YAN CHENGYE, MAMADOU GANDEKA, ZHU ZIJIAN, et al. Improvement of *Bph* resistance in rice restorer line *R1005* by molecular marker assisted selection[J]. Journal of Huazhong Agricultural University, 2014, 33 (5): 8 - 14.

[158] HU WEI, LI YANFANG, HU KAN, et al. Improvement of *Bph* resistance in *Guinongzhan* by marker assisted selection of *Bph* resistance genes[J]. Molecular Plant Breeding, 2015, 13 (5): 951 - 960.

[159] DONG RUIXIA, WANG HONGFEI, DONG LIANFEI, et al. Improvement of rice blast resistance of CMS line Zhenda A and its hybrids by molecular marker assisted selection [J]. Journal of Plant Genetic Resources, 2017, 18 (3): 573 - 586.

[160] XUE QINGZHONG, ZHANG NENGYI, XIONG ZHAOFEI, et al. Breeding restorer lines resistant to bacterial blight by molecular marker-assisted selection [J]. Journal of Zhejiang Agricultural University, 1998 (6): 19 - 20.

[161] DENG QIMING, ZHOU YUZHEN, JIANG ZHAOXUE, et al. Aggregation and effect analysis of resistant genes *Xa21*, *Xa4* and *Xa23* to bacterial blight[J]. Acta Agronomica Sinica, 2005 (9): 1241 - 1246.

[162] LUO YANCHANG, WU SHUANG, WANG SHOUHAI, et al. Breeding of three-line CMS line *R106A* with polygene resistance to rice bacterial blight[J]. Scientia Agricultura Sinica, 2005 (11): 14 - 21.

[163] ZHENG JIATUAN, TU SHIHANG, ZHANG JIANFU, et al. Marker assisted selection of restorer lines containing xanthomonas oryzae pv. oryzae resistant gene *Xa23* [J]. Chinese Journal of Rice Science, 2009, 23 (4): 437 - 439.

[164] LAN YANRONG, WANG JUNYI, WANG YI, et al. Improvement of bacterial blight resistance of photo-thermo sensitive genic male sterile line *Hua201s* by molecular marker-assisted selection[J]. Chinese Journal of Rice Science, 2011, 25 (2): 169 - 174.

[165] LUO Y, SANGHA J S, WANG S, et al. Marker-assisted breeding of *Xa4*, *Xa21* and *Xa27* in the restorer lines of hybrid rice for broad-spectrum and enhanced disease resistance to bacterial blight[J]. Molecular Breeding, 2012, 30 (4): 1601 - 1610.

[166] HUANG B, XU J, HOU M, et al. Introgression of bacterial blight resistance genes *Xa7*, *Xa21*, *Xa22*, and *Xa23*, into hybrid rice restorer lines by molecular marker-assisted selection[J]. Euphytica, 2012, 187 (3): 449 - 459.

[167] XIAO J, GRANDILLO S, SANG N A, et al. Genes from wild rice improve yield[J]. Nature, 1996, 384 (6606): 223 - 224.

[168] YANG YISHAN, DENG QIYUN, CHEN LIYUN, et al. Yield increasing effect of introducing high-yield QTL from wild rice into restorer line of late rice[J]. Molecular Plant Breeding, 2006 (1): 59 - 64.

[169] WU JUN, ZHUANG WEN, XIONG YUEDONG, et al. Breeding a new hybrid rice combination Y-Liangyou 7 with good quality and high yield by introducing yield-increasing QTL of wild rice[J]. Hybrid Rice, 2010, 25 (4): 20-22.

[170] KIM S, RAMOS J, HIZON R, et al. Introgression of a functional epigenetic *OsSPL14* ^{WFP} allele into elite indica rice genomes greatly improved panicle traits and grain yield[J]. Scientific Report, 2018, 8: 3833.

[171] ZHANG SHILU, NI DAHU, YI CHENGXIN, et al. Reducing amylose content of *indica* rice *057* by molecular marker-assisted selection[J]. Chinese Journal of Rice Science 2005 (5): 467-470.

[172] CHEN SHENG, NI DAHU, LU XUZHONG, et al. Reducing amylose content of Xieyou 57 by molecular marker technology[J]. Chinese Journal of Rice Science 2008 (6): 597-602.

[173] WANG YAN, FU XINMIN, GAO JIEJUN, et al. Molecular marker assisted selection for improving rice quality of restorer line Minghui 63[J]. Molecular Plant Breeding, 2009, 7 (4): 661-665.

[174] REN SANJUAN, ZHOU YIFENG, SUN CHU, et al. Efficient breeding of *indica* fragrant male sterile lines using functional molecular marker *1F/1R* of fragrance gene[J]. Journal of Agricultural Biotechnology, 2011, 19 (4): 589-596.

[175] ZENG D, TIAN Z, RAO Y, et al. Rational design of high-yield and superior-quality rice[J]. Nature plants, 2017, 3:17031.

[176] WANG FENG. Status Quo, Problems and development strategies of transgenic rice breeding[J]. Fujian Journal of Agricultural Sciences, 2000 (S1): 141-144.

[177] XU XIUXIU, HAN LANZHI, PENG YUFA, et al. R&D and application of transgenic pest resistant rice and its development strategy in China[J]. Journal of Environmental Entomology, 2013, 35 (2): 242-252.

[178] ZHANG JIE, SHU CHANGLONG, ZHANG CHUNGE. Patent protection and trend of *Bt* pesticidal genes [J]. Plant Protection, 2011, 37 (3): 1-6, 11.

[179] YE G, SHU Q, YAO H, et al. Field evaluation of resistance of transgenic rice containing a synthetic *cry1Ab* gene from bacillus thuringiensis berliner to two stem borers[J]. Journal of Economic Entomology, 2001, 94 (1): 271-276.

[180] HUANG J, HU R, ROZELLE S, et al. Pest-resistant GM rice in farmers' fields-assessing productivity and health effects in China[J]. Science, 2005, 308 (5722): 688.

[181] ZHU ZHEN, QU LEQING, ZHANG LEI. Transgenic rice research and new variety breeding[J]. Biotechnology, 2010 (3): 27-34.

[182] LOU SHILIN, YANG SHENGCHANG, LONG MINNAN. Genetic engineering[M]. Beijing: Science Press, 2002.

[183] WU FAQIANG, WANG SHIQUAN, LI SHUANGCHENG, et al. Research progress in and safety of herbicide resistant transgenic rice[J]. Molecular Plant Breeding, 2006, 4 (6): 846-852.

[184] WANG XIUJUN, LANG ZHIHONG, DAN ANSHAN, et al. Mechanism of herbicides as inhibitors of amino acid biosynthesis and research progress in herbicide tolerant transgenic plants[J]. China Biotechnology, 2008, 28 (2): 110-116.

[185] QIU LONG, MA CHONGLIE, LIU BOLIN, et al. Research progress in and development prospects of herbicide tolerant transgenic crops[J]. Scientia Agricultura Sinica, 2012, 45 (12): 2357-2363.

[186] ZHOU M, XU H, WEI X, et al. Identification of a glyphosate-resistant mutant of rice 5-enolpyruvylshikimate 3-phosphate synthase using a directed evolution strategy[J]. Plant Physiology, 2006, 140 (1): 184-

195.

[187] ZHOU H, HE M, LI J, et al. Development of commercial thermo-sensitive genic male sterile rice accelerates hybrid rice breeding using the CRISPR/Cas9-mediated *TMS5* editing system[J]. Scientific Reports. 2016, 22, 6: 37395.

[188] HUI Z, LIANG W, YANG X, et al. Carbon starved anther encodes a MYB domain protein that regulates sugar partitioning required for rice pollen development[J]. Plant Cell, 2010, 22 (3): 672-689.

[189] ZHANG H, XU C, HE Y, et al. Mutation in CSA creates a new photoperiod-sensitive genic male sterile line applicable for hybrid rice seed production[J]. Proceedings of the National Academy of Sciences of the United States of America, 2013, 110 (1): 76-81.

[190] LI Q, ZHANG D, CHEN M, et al. Development of *japonica*, photo-sensitive genic male sterile rice lines by editing carbon starved anther, using CRISPR/Cas9[J]. Journal of Genetics and Genomics, 2016, 43 (6): 415-419.

[191] YU J, HAN J, KIM Y, et al. Two rice receptor-like kinases maintain male fertility under changing temperatures[J]. Proceedings of the National Academy of Sciences of the United States of America, 2017, 114 (46): 12327-12332.

[192] WANG F, WANG C, LIU P, et al. Enhanced rice blast resistance by CRISPR/Cas9-targeted mutagenesis of the ERF transcription factor gene *OsERF922* [J]. Plos One, 2016, 11 (4): e0154027.

[193] ZHANG HUIJUN. Rice blast resistance and fragrance quality improved by *pi21* and *osbadh2* editing[D]. Wuhan: Huazhong Agricultural University Press, 2016.

[194] MA X, ZHANG Q, ZHU Q, et al. A robust CRISPR/Cas9 system for convenient, high-efficiency multiplex genome editing in monocot and dicot plants[J]. Molecular Plant, 2015, 8: 1274-1284.

[195] SUN Y, JIAO C, LIU Z, et al. Generation of high-amylose rice through crispr/cas9-mediated targeted mutagenesis of starch branching enzymes[J]. Frontiers in Plant Science, 2017, 8 (223): 298.

[196] LI M, LI X, ZHOU Z, et al. Reassessment of the four yield-related genes *Gn1a*, *DEP1*, *GS3*, and *IPA1* in rice using a CRISPR/Cas9 system[J]. Frontiers in Plant Science, 2016, 7 (12217): 377.

[197] SHEN LAN. Rice polygene editing based on CRISPR/Cas9 and its application in breeding[D]. Yangzhou: Yangzhou University Press, 2017.

[198] SU WEI. Cloning and functional analysis of genes related to programmed cell death in rice[D]. Shanghai: Fudan University Press, 2006.

[199] XIE Y, NIU B, LONG Y, et al. Suppression or knockout of *SaF/SaM* overcomes the Sa-mediated hybrid male sterility in rice[J]. Journal of Integrative Plant Biology, 2017, 59 (9): 669-679.

[200] TANG L, MAO B, LI Y, et al. Knockout of *OsNramp5* using the CRISPR/Cas9 system produces low Cd-accumulating indica rice without compromising yield[J]. Scientific Reports, 2017, 7: 14438.

[201] XU R, LI H, QIN R, et al. Gene targeting using the *Agrobacterium tumefaciens*-mediated CRISPR-Cas system in rice[J]. Rice, 2014, 7: 5.

[202] LI J, MENG X, ZONG Y, et al. Gene replacements and insertions in rice by intron targeting using CRISPR/Cas9[J]. Nature Plants, 2016, 2 (10): 16139.

[203] HUANG S, WEIGEL D, BEACHY R, et al. A proposed regulatory framework for genome-edited crops[J]. Nature Genetics, 2016, 48 (2): 109-111.

Part 2

Breeding

Chapter 5 Breeding of Super Hybrid Rice Male Sterile Line

Deng Qiyun / Bai Bin / Yao Dongping

Section 1 Sources of Cytoplasmic Male Sterile Lines and Their Main Types

I. Sources of Male Sterile Lines

The male sterile lines used in three-line hybrid rice breeding are all of nucleocytoplasmic male sterility, usually called cytoplasmic male sterility (CMS), or simply cytoplasmic sterility. The basic characteristics of CMS lines include abnormal development of male organs, absence of anthers or thin, withered and indehiscent anthers containing aborted pollen, and no self-seeding. Obtaining original CMS materials is a prerequisite for breeding CMS lines through natural mutations, interspecific hybridization and intraspecific hybridization.

1. Natural mutation

Stamen development is sensitive to external environmental changes, and natural mutant single plant with male sterility can often be found in nature. On November 23, 1970, Li Bihu, an assistant of Yuan Longping at Anjiang Agricultural School in Hunan province, and Feng Keshan, a technician at Nanhong Farm in Yaxian county, Hainan, found a pollen-deficient natural mutant plant (Fig. 5 - 1) in a common wild rice population in Yaxian county, Hainan, which was of a creeping plant type with strong tillering ability, narrow leaves, thin stems, thin grains,

Fig. 5 - 1 The scene of wild pollen-abortive plants was found in Sanya, Hainan (HHRRC)

long and red awn, extremely easy shattering, purple leaf sheaths and lemma tips, exserted stigma, thin, indehiscent and light-yellow anthers containing abortive pollen, sensitive to the daylength, typical short-daylength plants. The sterile cytoplasm of the wild abortive (WA) sterile lines currently in wide use in China came from this sterile mutant of the common wild rice.

2. Interspecific hybridization

Interspecific hybridization refers to the distant hybridization between different species. In 1958, Katsuo Kiyoshi of Tohoku University, Japan, obtained a male sterile line in the hybrid progeny of Chinese red awn wild rice and Fujisaka 5 (*japonica* rice), and then bred the Fujisaka 5 male sterile line with the cytoplasm of Chinese red awn wild rice. Chinese rice scientists have produced male sterile lines with various wild rice cytoplasms using crosses between various ecotypes of common wild rice and common cultivated rice (Table 5 - 1). Generally speaking, it is easier to obtain CMS materials by using common wild rice as the female parent and cultivated rice as the male parent, while it is more difficult to obtain CMS by reciprocal crossing, in which case even if some male sterile plants could be obtained, it would be difficult to stabilize the sterility. For example, the Institute of Agricultural Science in Pingxiang of Jiangxi province (1978) obtained pollen-free sterile plants from the cross of Ping Dwarf 58 x South China wild rice, but it has not been possible to breed a stable sterile line.

Due to some special biological traits of wild rice, the following three points should be considered when crossing it with cultivated rice. First, wild rice is very daylength-sensitive and required corresponding shortdaylength to enter the reproductive growth stage. Therefore, wild rice and their low generations of hybrid progenies must be short daylength treated after the four-leaf stage when planted in the Yangtze River Basin and further north, or planted in the early season in Southern China. Otherwise, the plants will not head normally or the heading will be delayed. Second, wild rice shows extremely easy shattering, so the crossing bags should remain over the crossed panicles before harvest. Finally, the seed dormancy of wild rice and its low generation materials is long and very strong, so if the seeds need to be sown after harvest, they must be repeatedly turned over when sun-dried before soaking, or the dried seeds must be broken dormancy in adrying cabinet with 59 ℃ for 72 hours. The glumes can be peeled off during germination to improve the germination rate.

Table 5 - 1 Main CMS materials obtained from common wild rice crossed with cultivated rice in China

Material	Combination	Institution	Year
Guangxun 3A	Yacheng wild rice/Guangxuan 3	Guangxi Academy of Agricultural Sciences	1975
62A	Yanglan wild rice/62	Zhangqing Agricultural School	1975
Jingyu 1A	Sanya Hongye/Jingyu 1	Institute of Crop Sciences, Chinese Academy of Agricultural Sciences	1975
Liantangzao A	Red awn wild rice/Liantangzao	Wuhan University	1975
Erjiuqing A	Tengqiao wild rice/Erjiuqing	Hubei Academy of Agricultural Sciences	1975

Continued

Material	Combination	Institution	Year
Jinnante 43A	Liuzhou red awn wild rice/Jinnante 43	Guangxi Academy of Agricultural Sciences	1976
Liuyezhenshan 97A	Liuzhou red or white awn wild rice/ Zhenshan 97	Hunan Academy of Agricultural Sciences	1974
Guangxuanzao A	Hepu wild rice/Guangxuanzao	Hunan Academy of Agricultural Sciences	1975
IR28A	Tiandong wild rice/IR28	Hunan Academy of Agricultural Sciences	1978

3. Intraspecific hybridization

Intraspecific hybridization includes *indica-japonica* intersubspecific hybridization and hybridization between different varieties within the same subspecies.

In 1966, Japanese scholar Shinjyo used Indian spring *indica* rice Chinsurah-Boro Ⅱ as the female parent and crossed it with *japonica* rice variety Taichung 65 from Taiwan of China to obtain sterile plants, and bred the BT-type sterile line Taichung 65A. After that, Chinese scholars also bred a number of new male sterile lines through *indica-japonica* hybridization. The key to obtaining new sterile lines from *indica-japonica* crosses is the selection of parents. Past experience shows that it is easier to breed CMS lines using low-latitude spring *indica* varieties from India, *indica* varieties from South Asia, late-season *indica* rice from South China and *indica* rice from the Yunnan-Guizhou Plateau as the female parent and cross them with *japonica* varieties widely planted in Japan and China. Examples of such combinations include IR24 x Xiuling, Tianjidu x Fujisaka 5, Jingquannuo x Nantaijing, and Eshandabaigu x Hongmaoying.

In 1972, IRRI used Taichung Native 1 (TN1), an *indica* rice variety from Taiwan of China, as the female parent and crossed it with Pankhari 203, an *indica* rice variety from India, to breed the Pankhari male sterile line. The seed setting rate of the second-generation backcross was less than 3.4%. The institute also bred the D388 male sterile line by crossing Peta with D388.

Chinese scholars have done a lot of work on breeding male sterile lines through intraspecific hybridization. Sichuan Agricultural University (SAU) crossed *Gambiaka kokum*, an *indica* rice variety from West Africa, with early-season *indica* rice from China, and succeeded in obtaining male sterile plants from their progenies and breeding the GA-type male sterile lines. Hunan Academy of Agricultural Sciences (HAAS) crossed geographically distant *indica* rice varieties to obtain sterile plants in the hybrids of Gu-Y12 and Zhenshan 97, Indonesia Paddy 6 and Pyongyang 9, IR665 and Guiluai 8, Qiugu'ai 2 and Pyongyang 9, Qiutangzao 1 and Bonian'ai, Shaxianfengmenbai and Zhenshan 97, and bred sterile lines. Yunnan Agricultural University obtained *japonica* male sterile lines by crossing and reciprocal crossing local high-plateau *japonica* Zhaotongbeizigu and modern *japonica* Keqing 3.

For intraspecific crosses, it is more difficult to obtain sterile lines due to the close genetic relationship between the parents, so two points need to be noted in the selection process. 1) First, the parents must be geographically distant or of different ecotypes. For example, it can be crosses between a foreign variety and a Chinese variety, a Southeast Asian variety and a Yangtze River variety, a photo-sensitive South China late-season *indica* and a thermo-sensitive Yangtze River Basin early-season *indica*, a low-latitude high-altitude *japonica* from the Yunnan-Guizhou Plateau and a high-latitude low-altitude *japonica* from northern China, etc. 2) Generally speaking, there will be no sterile plants in the F_1 intervarietal hybrids, so the original male parent is used for one or two backcrossing, and then allow progeny selfing and segregating with 300 – 500 or more plants in each generation, from which male sterile plants can be selected for further backcrossing.

Ⅱ. Main Types of CMS Lines

Based on the cytoplasmic source, current CMS lines can be classified as follows.

1. WA CMS lines

WA CMS lines are the most widely used cytoplasmic male sterile lines in China and the sterile cytoplasm is derived from the natural pollen-abortive plant of common wild rice. The main characters of this type of sterile lines are strong tillering, thin stems, narrow leaves, long and slender grains, well-developed and exserted stigma, thin and yellowish anthers without dehiscence, and no selfed seed setting, and the pollen abortion is the typical abortive type, belonging to sporophyte male sterility. The representative WA-type CMS lines include Erjiunan 1A, V20A, Zhenshan 97A, Jin 23A, Tianfeng A, etc. Among them, the combination of Shanyou 63, which used Zhenshan 97A as the female parent, is the most widely used hybrid rice combination with the largest planting area in China, accumulatively 170 million hectares since 1983. Jin 23A is a CMS line with good rice quality, and its hybrid combinations such as Jinyou 299 and Jinyou 527 are all super rice varieties recognized by China's Ministry of Agriculture (MOA). The following is the breeding process of Jin 23A (Fig. 5 – 2).

2. ID-type CMS lines

The cytoplasm of Indonesia Paddy (ID-type) CMS lines is derived from Indonesia Paddy, which is of the sporophyte male sterility type. The pollen development of ID-type CMS lines shows abnormality after the tetrad stage of the pollen mother cells, specifically, during the period of cell disintegration in the tapetum, there is still a thicker layer of cells in the tapetum until the late mononuclear stage, and then the tapetum disintegrates rapidly and the microspores gradually become deformed and completely abortive. The main representatives of ID-type CMS lines include Ⅱ – 32A, You-I A and T98A etc. Ⅱ – 32A is a stable CMS line developed by HHRRC by crossing Zhenshan 97B with IR665 and then backcrossing the progeny with Indonesia Paddy rice Zhending A (Glutinous rice). Ⅱ – 32A has the characteristics of long growth duration, compact plant type, good flowering habit and high percentage of stigma exsertion. Eight super rice varieties recognized by China's MOA, including Ⅱ – Youming 86, Ⅱ – You084, Ⅱ – You602 and Ⅱ – Youhang 2, have been bred with Ⅱ – 32A. HHRRC also bred You-I A, an early *indica* CMS line with good quality and a high outcrossing rate, by crossing and backcrossing Ⅱ – 32A, the female parent, with Xieqingzao B's small-grain mutant, the male parent. ID-type and WA-type CMS lines are the two main types of CMS lines used in *indica* hybrid rice production. The breeding process of You-I A male sterile line is shown in Fig. 5 – 3.

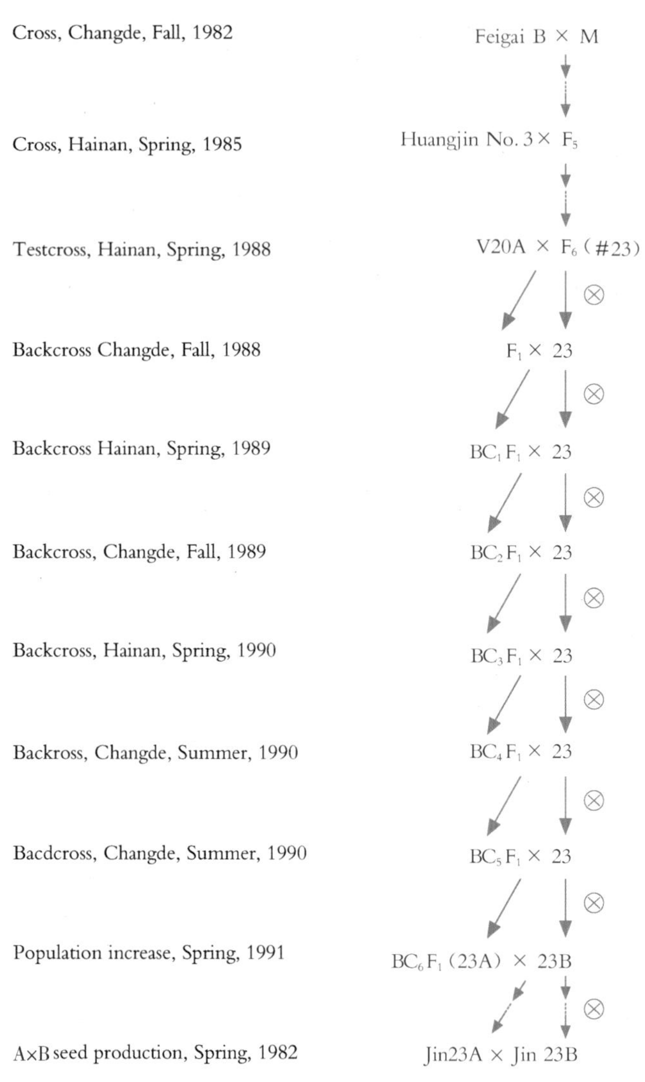

Fig. 5 - 2 Breeding process of Jin 23A male sterile line (Xia Shengping, 1992)

3. D-Type CMS lines

The cytoplasm of D-type CMS lines is derived from Dissi D52/37 which belongs to the sporophyte male sterility type. In 1972, Sichuan Agricultural College discovered a male sterile early maturing plant in the F_7 population of Dissi D52/37//Aijiaonante. After crossing it with the *indica* rice variety Italia B, it was found that Italia B was able to maintain the sterility. Then, Zhenshan 97 was used to cross and back-cross with the sterile plant, and obtained the D-type D-Shan A. The pollen abortion period and characteristics of D-type CMS are similar to those of the WA type. The representative D-type CMS lines include D-Shan A, D297A, Yixiang 1A, etc. Among these, Yixiang 1A, in particular, has good agronomic traits, good outcrossing habits and high grain quality. The breeding process of Yixiang 1A is shown in Fig. 5 - 4.

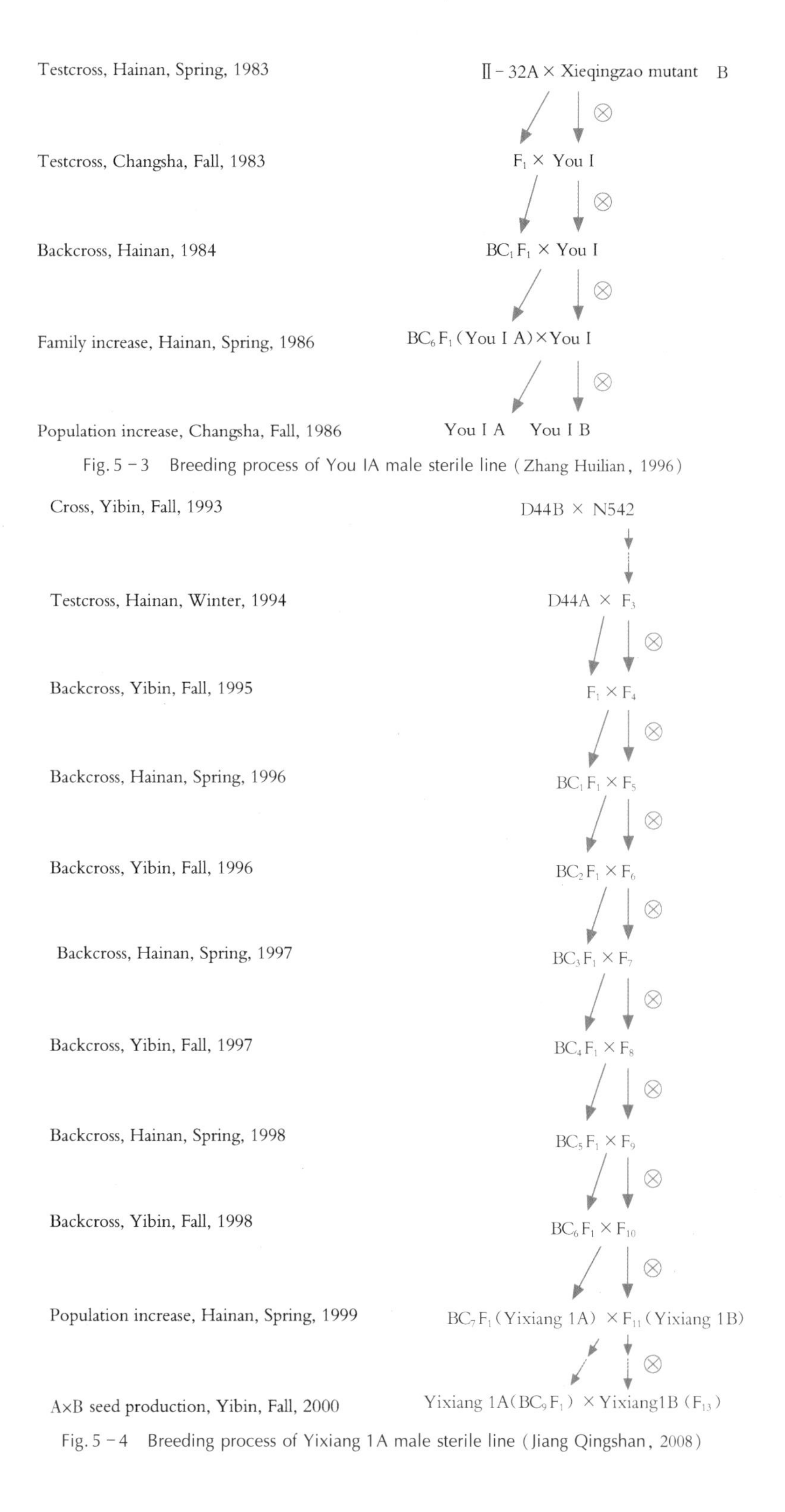

Fig. 5 - 3 Breeding process of You IA male sterile line (Zhang Huilian, 1996)

Fig. 5 - 4 Breeding process of Yixiang 1A male sterile line (Jiang Qingshan, 2008)

4. GA-type CMS line

GA-type CMS lines are a series of CMS lines bred from sterile plants segregated from the progenies of a hybrid between Gambiaka, an *indica* rice variety from West Africa and Aijiaonante. In the process of stabilizing the sterility, some used geographically distant intrasubspecific crosses (*indica-indica* crosses) and some used *indica-japonica* intersubspecific crosses, and the selected sterile lines differed in pollen abortion due to the different stabilization and maintenance lines used. For example, Chaoyang 1A belongs to the typical abortion type and Qingxiaojinzao A belongs to the stained abortion type (Fig. 5 - 5).

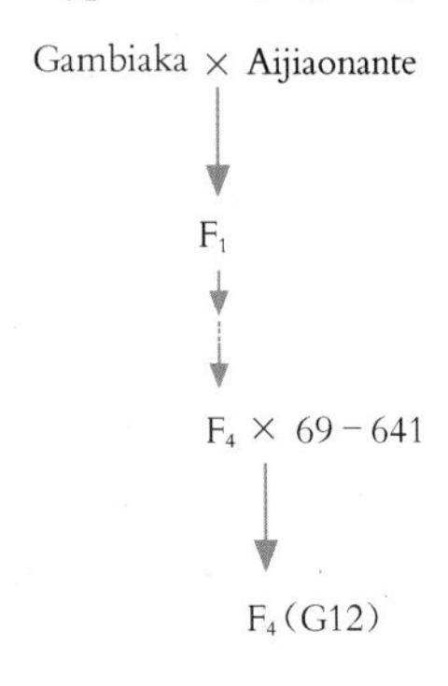

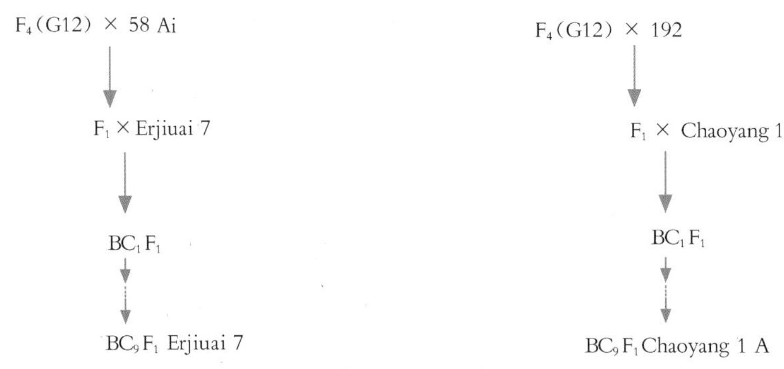

Fig. 5 - 5 Breeding process of GA-type CMS lines (Li Shifen, 1997)

5. DA-type CMS line

The cytoplasm of the DA-type CMS lines is derived from a dwarf wild rice in Jiangxi province, and belongs to the sporophyte male sterile type. Xieqingzao A is a representative of the DA-type CMS lines. It was bred by using a dwarf male sterile plant/Zhujun//Xiezhen 1, which served as the female parent to cross with a progeny of Junxie/Wenxuanqing//Aitangzao 5. The CMS line showed high percentage of stigma double exsertion, good flowering habit, moderate plant height, delicate growth, and disease resistance. After continuous selective backcrossing, BC_4 became basically stable in the summer of 1982, and was named Xieqingzao A.

6. Honglian-type CMS line

The cytoplasm of Honglian (HL)-type CMS lines is derived from Hongmang wild rice, which is a type of gametophyte sterility. Most of the pollen of HL-type CMS lines is aborted at the binuclear stage, with mainly spherical abortion, and a small amount of stained pollen with iodine-potassium iodide staining. The restorer spectrum of HL-type CMS lines is wider than that of WA-type CMS lines, and most varieties of early-season and mid-season rice growing in the Yangtze River Basin can restore its fertility. The representatives of the HL-type CMS lines are Honglian A, Yuetai A, Luohong 3A, etc.

7. BT-type CMS lines

The cytoplasm of BT-type CMS lines in China is derived from the *japonica* male sterile line Taichung 65A introduced from Japan, which is of the gametophyte male sterile type with different characters of pollen abortion from those of WA and HL types. The pollen of BT CMS lines is of the stained abortion type and the pollen is aborted at the trinuclear stage. Since 1973, many research institutes in China have bred a large number of BT-type *japonica* CMS lines by using Taichung 65A through crosses and backcrosses. The representatives of BT-type CMS lines include Liming A, Jingyin 66A, Liuqianxin A, etc.

8. Dian-type CMS lines

The cytoplasm of the Dian-type CMS lines is derived from the male sterile plant discovered in the *japonica* rice variety Taipei 8. Some were derived from DT2- and DT4-type male sterile lines or from D9-type male sterile lines of common wild rice. They are of the gametophyte sterility type. The Dian-type CMS Lines and the BT-type CMS lines are similar in pollen abortion stages and characteristics. The representatives of Dian-type CMS lines include Fengjin A and Hexi 42 – 7A.

Section 2 Breeding of CMS Lines

Ⅰ. Criteria of Excellent CMS Lines

A good CMS line should have the following traits.

1) Stable sterility. The sterility of CMS lines should not be recovered by multiple backcrossing of a maintainer line or fluctuate with environmental conditions (such as the rise and fall of temperature).

2) High ability for the sterility to be restored by a restorer line. High compatibility with restorer lines, wide restoring spectrum, more restorer lines and high seed setting rate in the hybrids, and non-susceptibility to changes of environment conditions.

3) Good flowering habit, well-developed floral organs, and a high outcrossing seed setting rate. The good flowering habit means that the CMS line flowers early in the day, with a concentrated flowering period, large angle and long time of glume opening, no or less glume closure during flowering. Well-developed floral organs mean that the stigma has a high percentage of exsertion and high vitality. Good flowering habit and well-developed floral organs are the basis of high yield in seed production for CMS lines.

4) Good combining ability, easy to be used to breed for high-heterosis and superior hybrids. This requires the CMS line to have a high-yielding plant type and the corresponding physiological basis, and is complementary to the restorer line in some major economic traits. The level of heterosis is related to the genetic distance and genetic relationship between the parents, so for an excellent CMS line to have good combining ability, the genetic difference in major traits between the CMS line and the restorer line should

be appropriately increased, and the introduction of the restorer line into the CMS line should be avoided.

5) Good grain quality. A good CMS line must have good grain transparency with low chalkiness, high milling and heading rice yield, and good cooking and eating quality.

6) High resistance to diseases and insects. It should be resist to at least local major diseases or insects.

Ⅱ. Transformation Breeding of CMS Lines

In order to continuously increase hybrid yield, quality and seed production, the rice male sterile lines already used in the production must be continuously improved and enhanced. At the same time, the wide range of rice cultivation also necessitates the breeding of a wide variety of sterile lines adapted to various ecological environments and farming systems. The most efficient and economical way to breed new sterile lines is through the transfer of sterile lines that have been bred. At present, many new isocytoplasmic CMS lines have been bred through transformation breeding with those widely used CMS lines in China such as the WA type, ID type and Gang type.

The process of transformation breeding of a CMS line can be divided into two steps – test crossing and selective backcrossing.

In test crossing, a CMS line, as the female parent, is crossed with a selected maintainer line, as the male parent. The fertility performance of the F_1, BC_1 and BC_2 populations should be observed. The F_1 population must be completely sterile, and the spikelets in the upper, middle and lower parts of the panicles should all be carefully examined. Attention should be paid to the appearance of the plants when using *indica* CMS lines to trans-breed *indica* male sterile lines. There is a great possibility that a new male sterile line can be bred if the panicles of F_1 plants are partially enclosed in the flag leaf sheath, the anthers in spikelets are degenerated, the morphology is similar to that of the original female parent, and no pollen staining is found in microscopic examination. It is very unlikely to breed any new sterile line if the panicles of F_1 plants are less enclosed or completely exserted, with fat anthers in the lower glumes and some stained pollen grains, and generally, the seed setting rate will increase after one or two backcrosses or always have a few selfed seeds. In some F_1 plants trans-bred using *indica* CMS lines to *indica* sterile lines, the panicles are completely exserted, and all anthers are light yellow, but very small and slender, rod-shaped and indehiscent, that is, completely sterile. In this case, more observations should be made on the BC_1 or even BC_2 populations. If there are some dehiscent pollen and selfed seeds in these two generations, the transformation breeding is not going to succeed. If a progeny of *japonica* or *indica /japonica* is used as the maintainer line for trans-breeding, it is possible to develop a new gametophyte male sterile line if the F_1, BC_1 and even later generations show thin rod-shaped anthers and complete sterility. For the trans-breeding of sterile *japonica* lines with gametophyte male sterility of the stained abortion type, the sterility examination should be carried out with the main purpose of observing the selfed seeds of sterile plants.

The second step is selective backcrossing. Combinations that are proved to be successful in the test crossing should be continuously backcrossed using the male parent for nuclear substitution with the purpose of transferring the plant type of the female parent to one similar to that of the male parent with stable sterility as soon as possible. Selective backcrossing is to select single plants with many excellent traits and good flowering habits from progenies with a high percentage of sterile plants and high sterility to be pair-backcrossed. The first step is to select a combination, then select breeding line families within the combination, and finally select the best single plant within the family. In the process of backcrossing, if the sterile plants gradually show some poor traits, such as serious glume closure, untimely and unconcentrated flowering, and small angle of glume opening, it indicates that this material has no value in production and should be abandoned.

Theoretically, the probability of a complete substitution of the parental nucleus is 0. 0156 at BC_2, which means that if there are 300 – 400 plants in the BC_2 population, there would be five to six backcross progenies that are completely homotypic with the male parent. By backcrossing these single plants, the trans-breeding process can be completed at BC_3, forming a stable new male sterile line. If the BC_2 population cannot reach the size of 300 – 400 plants, the nucleus substitution process may be completed at BC_3 with 5 – 10 single plants within a BC_2 population of 50 – 100 plants (probability 0. 1250). Because of the high probability of complete nuclear substitution by BC_3, experienced breeders can generally breed new male sterile lines when it proceeds to BC_4. In order to accelerate the screening process and obtain good plants more accurately, 5 – 10 backcross families should be maintained for each combination from BC_2 to BC_3. The traits of all the lines in BC_3 should be comprehensively examined. In order to expand the BC_4 population, more than 1,000 superior plants should be selected, and their fertility and degree of nuclear substitution should be examined. The male sterile lines that meet the requirements can be used in production.

For some good materials of crucial importance, which is very likely to breed new sterile lines, in order to shorten the breeding time, test crossing and trans-breeding can be carried out at the same time when they are still in the stage of segregation at an low generation and then the tested progenies and male parent can be selected and stabilized synchronously. However, this requires a large population of both the male parent and the test crossing progenies, and the number of male parent lines for backcrossing should be increased; otherwise, it will be difficult to achieve the expected results. The population size of a low generation of male parent is determined by segregation. Where there is greater segregation, the population size should be appropriately larger. As the generations increased, there will be more plants to meet the breeding objectives, and the population size can be reduced rapidly discarding lines with undesirable traits. In this synchronous stabilization, because the morphology of the progenies changes with the backcrossing generation of the male parent, there is no need to maintain a large population of the low-generation progenies. Once the male parent becomes basically stable, the female parent population should be expanded so as to examine whether the female parent lines are basically homotypic with the male parents in terms of fertility and other traits. If the breeding objectives have been met, the new sterile line and maintainer line are obtained simultaneously.

It is difficult to carry out synchronous stabilization of male parent and trans-breeding backcross, because it requires a great foresight, careful planning and right working method for the selection of both parents and backcross progenies in each generation. If anything goes wrong, it would be possible to get half the results with double the effort or, in worse cases, the selection efforts would come to nothing at all.

Generally, synchronous stabilization of male parent and trans-breeding backcross should not be carried out on those low-generation materials developed by crossing parents with widely different traits, because segregation will remain serious over many generations.

Ⅲ. Breeding of CMS Maintainer Lines

A CMS line is bred by nuclear substitution with a maintainer line and the CMS agronomic traits are basically determined by its maintainer line. Therefore, the breeding of a good CMS line should start with the selection of an excellent CMS maintainer line.

1. Parent selection

In the parental selection, in addition to genetic distance with the restorer lines, good combining ability, high resistance and good rice quality, other traits specific to a maintainer line should also be examined, such as high outcrossing and high ability to maintain sterility. By pedigree analysis, the main parental sources of the superior *indica* maintainer lines bred in China include Zhenshan 97B, V20B and Xieqingzao B. The improvement of overall traits can be conducted by introducing exotic high-quality rice germplasm and local resistant varieties. Sichuan Agricultural University (1995) used the strong maintainer lines V41B and Zhenshan 97B to make single cross and multiple crosses with Erjiuai and Yaaizao, respectively, and breed Gang 46B, an early maturing maintainer line with good plant and leaf morphology, large panicles, many grains and good outcrossing habit, to breed Gang 46A.

2. Breeding methods

(1) Cross breeding

Cross breeding uses existing maintainer lines to cross with one or several good parents to select new maintainer lines from their offspring. According to the way of hybridization, it can be divided into single cross and multiple cross. Wufeng B was derived from a single cross of You-IB and G9248. G9248 has the traits of early maturity and high grain quality, while You-IB has a good maintenance performance. Wufeng B has both good quality and good maintenance performance (Fig. 5-6). Shen 95B, developed at Shenzhen Graduate School of Tsinghua University, was derived from multi-crossing of Boro-Ⅱ, Cypress and Bengal wild rice with the strong maintainer lines Zhenshan 97B, V20B and Fengyuan B and backcrossing to Jin 23A, finally obtaining Shen 95A.

(2) Backcross breeding

Backcross breeding refers to the breeding method of using a non-recurrent parent with a certain good trait to cross and backcross with a good maintainer line (recurrent parent) so as to improve a certain trait of the recurrent parent. Backcross breeding is mostly used to improve the resistance of maintainer lines. For example, Guangdong Academy of Agricultural Sciences (2014) crossed Rongfeng B with Bl122, a material carrying the broad-spectrum rice blast resistance gene *pi-1*, and the F_1 plants obtained were backcrossed with Rongfeng B continuously. Combining agronomic traits selection and molecular marker tracking, the maintainer line Jifeng B with strong resistance to rice blast and the corresponding male sterile line Jifeng A were obtained. Most of the combinations bred with this male sterile line showed moderate or high resistance to rice blast, which was a significant improvement (Table 5-2).

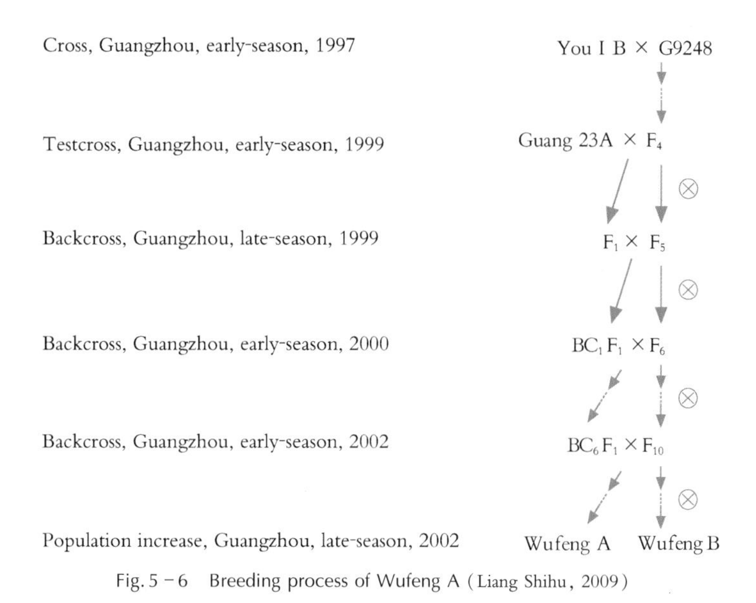

Fig. 5 - 6 Breeding process of Wufeng A (Liang Shihu, 2009)

Table 5 - 2 Resistance to rice blast of new CMS line Jifeng A

Combination	Year	Frequency of resistance	Resistance grading	Yield (t/ha)	Yield increase over control variety (%)	Control variety
Jigengyou 512	2011	97.30	Resistance	7.47	10.28	Tianyou 122
	2012	97.90	High	6.55	3.12	
Jigengyou 1008	2011	100.00	High	6.83	-0.13	Youyou 122
	2012	97.87	Medium	6.91	5.55	
Jigengyou 1002	2011	100.00	High	7.42	14.30	Bo-Ⅲ You 273
	2012	100.00	High	7.59	8.16	
Jigengyou 3550	2011	100.00	Medium	7.37	14.18	Bo-Ⅲ You 273
	2012	91.49	High	7.74	10.39	

Section 3 PTGMS Resources and Their Fertility Transition

Ⅰ. Methods to Obtain PTGMS Resources

There are three possible ways to obtain PTGMS resources, natural mutation, distant hybridization and artificial mutation.

1. Natural mutation

Nongken 58S, Annong S-1 and 5460S discovered in China are all PTGMS mutations under natural conditions. In early October 1973, Shi Mingsong discovered three male sterile plants in a land plot growing Nongken 58, a late-season single-cropping *japonica* rice variety. Research showed that this material is sterile under the condition of long daylength and high temperature, but is fertile under the condition of short daylength and low temperature, so it can be used as both the maintainer line and the sterile line. In 1985, the material was officially named Hubei photoperiod sensitive genic male sterile rice (HPGMR). In 1987, a scholar at Anjiang Agricultural School, Hunan province, discovered a natural thermo-sensitive male sterile mutant in the F_5 population of Chao 40B/H285//6209 - 3, and named it Annong S - 1, which shows male sterility at high temperature and male fertility at low temperature. In addition to Nongken 58S and Annong S - 1, Fujian Agricultural University also discovered a natural thermo-sensitive male sterile mutant in the restorer line 5460, and named it 5460S.

2. Distant hybridization

Wild segregation occurs in the progenies of distant hybridization, and it also occurs to male fertility. For example, Zhou Tingbo et al. bred two PTGMS materials, Hengnong S - 1 and 87N - 123 - R26, by crossing long-awn wild rice with R0183 and then crossing the resulted plants with Ce64. The two materials show completely different reactions to daylength and temperature. Hengnong S - 1 shows male sterility under high temperature and long daylength, and male fertility under low temperature and short daylength, while 87N - 123 - R26 was fertile under high temperature and long daylength, but sterile under low temperature and short daylength. In addition, PTGMS materials may also appear in the hybridization between geographically distant varieties. For example, X88, which was bred by crossing Egyptian rice with Japanese rice in Japan, at about 10 - 25 days before heading, i. e. from spikelet differentiation to pollen mother cell formation, a daylength longer than 13. 75 h induces sterility, while a daylength shorter than 13. 5 h induces fertility.

3. Artificial mutagenesis

Radiation-induced mutation is one of the ways to induce PTGMS. For example, the Agricultural Research Center of Japan found H89 - 1 in the progenies of the Liming variety treated with 20,000R γ-rays. According to observations by researchers from Japan and IRRI, H89 - 1 showed total sterility at 31 ℃/24 ℃, semi-sterility at 28 ℃/21 ℃, and fertility at 25 ℃/18 ℃. S. S. Virmani et al. also obtained the thermo-sensitive male sterile mutant IR32464 - 20 - 1 - 3 - 2B by radiation, which shows male sterility at 32 ℃/24 ℃, semi-sterility at 27 ℃/21 ℃, and fertility at 24 ℃/18 ℃. Besides radiation-induced mutation, chemical mutagenesis can also create PTGMS mutants. For example, MT, discovered by N. J. Rutgar of the United States, was obtained from M201 treated with ethyl methane sulfonic acid. Its fertility is controlled by the photoperiod, but it might also be affected by temperature.

Ⅱ. PTGMS Fertility Transition and Photoperiod and Temperature

1. Effects of photoperiod and temperature on the fertility of PTGMS lines

The understanding of the mechanism of fertility transition in photosensitive sterile rice was initially focused on the effect of photoperiod, and then the temperature-sensitive male sterile lines were discovered. Due to the large fluctuation of temperature in nature, researchers prefer to select typical PTGMS lines that are insensitive or less sensitive to temperature. As research progressed, it was found that the fertility of PTGMS lines is affected by both the daylength and the temperature. No absolute daylength sensitivity has been found for either *japonica* or *indica* rice, that is, they are not purely daylength-sensitive, but have an obvious daylength-temperature compensation mechanism. In addition, sunlight itself is a kind of thermal radiation, and light and temperature are inseparable. Therefore, in most cases, it is collectively called "photo-thermo sensitive male sterility". According to the dominant factor of the fertility transition, it can be divided into two categories, photosensitive nuclear sterility (PGMS) with daylength as the dominant factor and temperature-sensitive nuclear sterility (TGMS) with temperature as the dominant factor.

(1) Effects of daylength on fertility

The sensitive developmental period for daylength-induced fertility transition in PGMS rice is the second branching peduncle and spikelet primordium differentiation of young panicles to pollen mother cell formation, in which, the pistil formation to pollen mother cell formation is the most sensitive period. Under natural conditions, the critical daylength for fertility transition in PGMS lines is 13.5 – 14 h. The daylength-induced fertility transition of PGMS is not an intermittent leap, i. e., it is not completely sterile when the daylength is longer than a certain critical level, or completely fertile when the daylength is shorter than a certain critical level; rather, there is a continuous transition process, i. e., within a certain daylength range, as the daylength extends, it gradually becomes sterile, showing certain characteristics of a quantitative change. Accordingly, Xue Guangxing et al. (1990) proposed the concepts of induction critical daylength and abortion critical daylength. The former refers to the daylength at which a PGMS line starts to become abortive while the latter refers to the daylength for complete abortion in a PGMS line. Under the natural conditions in Beijing, the induction critical daylength of the PGMS line Eyi 105S is 13 h 25 min, while the abortion critical daylength is 14 h 20 min. Researchers nowadays have recognized that the fertility transition in PGMS rice is controlled not only by daylength, but also by temperature. Deng Qiyun et al. (1996) separated each PGMS population into four parts and observed them under four different conditions, long daylength and low temperature, short daylength and low temperature, long daylength and high temperature, and short daylength and high temperature. It was found that under a low temperature, a long daylength could not induce complete sterility in 7001S and other PGMS rice lines, while under high temperature conditions, a short daylength could not induce a high selfed seed setting rate either (Table 5 – 3).

(2) Effect of temperature on fertility transition

With limited research conditions and experience in the early days, researchers did not know enough about the effect of temperature on the fertility transition of PTGMS lines. With the emergence of a number of PTGMS rice lines such as Annong S – 1, Hengnong S – 1 and 5460S, researchers paid more attention to the role of temperature on fertility transition.

Chen Liangbi et al. (1993) treated the PTGMS lines Annong S – 1, Hengnong S – 1, Hengnong S – 2 and W7415S with low temperature (24 ℃/22 ℃) and long daylength, and found that Hengnong S – 1 and Hengnong S – 2 only needed three days of low temperature treatment during meiosis to prevent sterility gene expression, while Annong S – 1 and W7415S needed more than seven days of continuous

Table 5 – 3 Fertility performance of PGMS lines under different daylength and temperature conditions

Material	Deep staining rate of pollen (%)				Selfed seed setting rate (%)			
	Ⅰ	Ⅱ	Ⅲ	Ⅳ	Ⅰ	Ⅱ	Ⅲ	Ⅳ
7001S	1.3	35.8	0.2	14.4	0.0	6.4	0.1	4.7
8902S	9.8	10.2	0.1	0.7	2.9	10.2	3.3	4.7
1147S	20.2	20.4	2.4	2.1	7.5	0.1	0.6	1.2
Peiai 64S	11.2	11.5	0.1	0.0	0.0	0.0	0.0	0.0

Note. Ⅰ, Ⅱ, Ⅲ and Ⅳ are respectively long daylength and low temperature (daily average temperature at 23.8 ℃, and temperature range 19 ℃–28 ℃), short daylength and low temperature (daily average temperature at 23.3 ℃ and temperature range from 19 ℃–28 ℃), long daylength and high temperature (daily average temperature at 30.0 ℃ and natural long daylength and high temperature in mid- and late July in Changsha city), and short daylength and high temperature (daily average temperature at 31.0 ℃ and natural conditions in mid- and late July in Changsha plus dark room treatment).

low temperature treatment to prevent their sterility gene expression (Table 5 – 4). Zeng Hanlai et al. (1993) treated W6154S with high temperature and low temperature at different stages of young panicle development and found that W6154S is sensitive to temperature from pistil formation to the mononuclear pollen stage, and it is most sensitive to temperature at the meiosis stage, during which three days of low temperature can induce fertility.

Table 5 – 4 Development periods when PTGMS lines are sensitive to low temperature

Sterile line	Control (average temperature 29 ℃)		Artificial low temperature (24 ℃/22 ℃)																			
	10		1		2		3		4		5		6		7		8		9		10	
	P	S	P	S	P	S	P	S	P	S	P	S	P	S	P	S	P	S	P	S	P	S
Annong S – 1	0	0	0	0	0	0	0	0	0	0	0	0	0	0	38.8	19.3	0	0	58.1	24.0	57.2	26.3
Hengnong S – 1	0	0	0	0	0	0	0	0	37.1	53	0	0	0	0	67.2	32.8	0	0	72.4	39.6	71.0	37.3
Hengnong S – 2	0	0	0	0	0	0	0	0	31.5	3.8	0	0	0	0	58.1	18.7	0	0	64.5	37.7	79.3	45.2
W7415S	0	0	0	0	0	0	0	0	0	0	0	0	0	0	15.3	1.8	0	0	36.5	7.7	48.7	12.5
Xiangzaoshan 3	98.8	96.5	/	/	/	/	/	/	/	/	/	/	/	/	/	/	/	/	/	/	96.1	92.3

Note. *P* stands for pollen fertility (%), *S* stands for selfing seed setting rate (%); 1 – 10 refers to – 1: Spikelet primordium differentiation stage; 2: Pistil and stamen differentiation stage; 3: Pollen mother cell formation stage; 4: Meiosis stage; 5: Stage from spikelet primordium to pistil and stamen differentiation; 6: Stage from stamen and pistil differentiation to pollen mother cell formation; 7: Stage from pollen mother cell formation to meiosis; 8: Stage from spikelet primordium to pollen mother cell formation; 9: Stage from stamen and pistil differentiation to meiosis; 10: Stage from spikelet primordium differentiation to meiosis.

Different PTGMS lines have different thermo-sensitive periods for fertility transition. Deng Qiyun et al. (1997) used a phytotron to study the relationship between fertility transition and temperature in different PTGMS lines by giving long daylength and low temperature treatments for four consecutive days at the third, fourth, fifth and sixth stages of young spike differentiation and 15, 11 and 7 days at the third, fourth and fifth stages of young panicle differentiation, respectively. The treatment period with significant pollen fluctuations was used as the sensitive developmental period.

The results showed that low temperature from the secondary branching and spikelet primordium differentiation stage of young panicle development to the pollen content filling stage had a certain effect on the fertility transition of all participating PTGMS lines, i. e. there was a common developmental period of fertility transition sensitivity (common sensitivity period). At the same time, the most sensitive period of fertility transition to temperature varied among the 26 lines examined. Specifically, 73% of the lines examined were most sensitive from pollen mother cell formation to meiosis, i. e., 10 - 14 d before flowering, such as Annong S - 1, etc.; 19% of the lines were most sensitive from pistil and stamen formation to pollen mother cell formation, i. e., 12 - 17 d before flowering, such as Pei Dwarf 64S, etc.; and 8% of the sterile lines were most sensitive from 3 - 8 d before flowering, i. e., pollen content filling, such as 870S, etc. (Table 5 - 5).

Table 5 - 5 Analysis of the most temperature-sensitive period of fertility transition in different PTGMS lines

Material	Most sensitive period (Days before flowering)	Material	Most sensitive period (Days before flowering)	Material	Most sensitive period (Days before flowering)
Annong S	8 - 13	Lunhui 22S	9 - 13	Xiang 125S	8 - 12
644S	8 - 13	1147S	12 - 16	867S	12 - 18
8421S	8 - 11	Peiai 64S - 05	11 - 16	Ce49S	9 - 13
861S	8 - 14	Peiai 64S - 25	11 - 17	Ce64S	10 - 14
N8S	7 - 11	Peiai 64S - 35	11 - 16	26S	8 - 11
338S	7 - 13	Anxiang S	8 - 13	133S	7 - 12
LS2	5 - 8	G10S	7 - 11	870S	3 - 7
100S	8 - 13	A113S	10 - 14	92 - 40S	9 - 13
545S	8 - 12	CIS28-10	10 - 14	/	/

Based on the results of their own study and previous work, Wu Xiaojin et al. (1992) proposed the hypothesis of three sensitive periods, i. e., there may be three sensitive periods for temperature-induced fertility transition in PTGMS rice - strong sensitive period (P_1), weak sensitive period (P_2) and micro sensitive period (P_3) (Fig. 5 - 7). The temperature conditions in the strongly sensitive period P_1 are decisive for the fertility transition. In the strong sensitive period, as long as a certain intensity of temperature below or equal to the critical low temperature is encountered, the sterile lines will be fertile; if a temperature above or equal to the critical high temperature is encountered, the sterile lines will be sterile. This sensitive developmental period is also the most temperature-sensitive period for PTGMS fertility transition as mentioned by Chen Liangbi et al. (1993) and Deng Qiyun et al. (1996).

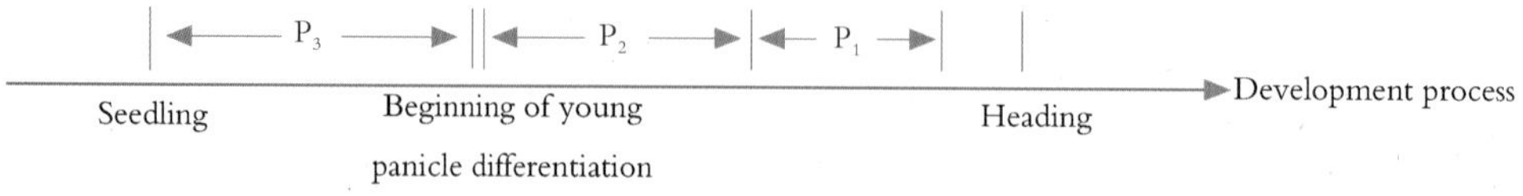

|◄— P_3 —►||◄— P_2 —►|◄— P_1 —►| |

Seedling — Beginning of young panicle differentiation — Heading —► Development process

P_1: Period of strong sensitivity (meiosis stage); P_2: Period of weak sensitivity (from the 3rd stage of young panicle differentiation to the stage of pollen mother cell formation); P_3: Period of slight sensitivity (vegetative growth stage)

Fig. 5-7 Three sensitive periods of PTGMS lines for temperature-induced fertility transition

The temperature during P_2 (weak sensitivity) is not decisive for fertility transition in PTGMS rice, but it can influence the temperature threshold required for P_1 to induce fertility transition. If P_2 is a period of lasting high temperature, the critical temperature for the required fertility transition during P_1 will be lower; if P_2 is a period of low or a moderate temperature, the critical temperature for the fertility transition will be higher. P_3 is a period of growth and development in which PTGMS rice has a weak response to temperature for fertility transition. Zhang Ziguo et al. (1993) studied the effect of daylength and temperature conditions during the vegetative growth period on the fertility of the PTGMS line W6154S and found that the daylength and temperature conditions during the vegetative growth period had a certain effect on the critical temperature and pollen fertility of the PTGMS line W6154S after the differentiation of young spikelets, and it can be concluded that the vegetative growth period is the micro-sensitive period for temperature-induced fertility transition.

2. Function modes of daylength and temperature on fertility transition in PTGMS rice

Although it has been shown that all currently utilized PTGMS line are subject to the effect of both daylength and temperature, they can be broadly classified into three types, photo-sensitive, temperature-sensitive, and photo-temperature sensitive, depending on whether daylength or temperature sensitivity is the main factor affecting the sterile line. For a PTGMS line, fertility transition is dominated by daylength, with temperature playing a supplementary role. In a certain temperature range, the photo-sensitivity of fertility can be clearly expressed, when temperature is higher than a critical point, high temperature masked the role of daylength, then any daylength results in sterility. This critical point is called the upper limit critical temperature of photo-sensitive sterility (critical high temperature). When it is below a critical temperature, low temperature would also mask the role of daylength, then any daylength is associated with stable fertility. This critical point is called the lower limit critical temperature of photo-sensitive fertility (critical low temperature). The range between these two limits is the photosensitive temperature range. Within this range, long daylength induces sterility while short daylength induces fertility, and there is a complementary effect between daylength and temperature, i. e., the critical daylength shortens as the temperature increases; conversely, the critical daylength lengthens when the temperature decreases. Yuan Longping (1992), Zhang Ziguo (1992) and Liu Yibai (1991) proposed a similar model of photo-thermo interaction for fertility transition in PTGMS lines (Fig. 5-8). It should be noted that when the temperature is higher than the biological upper limit or lower than the biological lower limit, rice will suffer from physiological damage and cannot undergo normal development or form pollen normally. Representative sterile lines of the long daylength sterile type include Nongken 58S, N5088S, 7001S, ZNU 11S, etc. In addition to the common long daylength sterile type, there are also a few reports of the short daylength sterile type, in which case short daylength induces sterility while long daylength induces fertile within the photosensitive temperature range (e. g. Yi D1 - S).

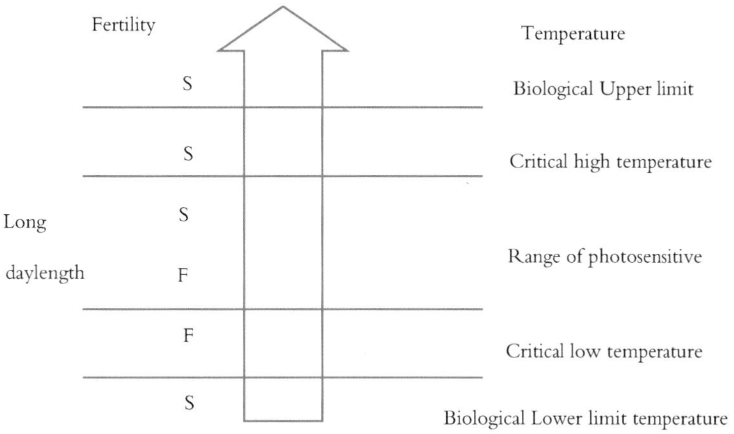

Fertility
Temperature
S
Biological Upper limit
S
Critical high temperature
Long daylength
S
F
Range of photosensitive
F
Critical low temperature
S
Biological Lower limit temperature

S: Sterile F: Fertile

Fig. 5 - 8 Function mode of PTGMS fertility transition

Temperature-sensitive sterile lines are temperature-dependent, and they are sterile when the temperature is higher than the critical temperature and fertile when it is lower than the critical temperature or fertile when the temperature is higher than the critical temperature and sterile when it is lower than the critical temperature, and daylength has less effect on the fertility transition. Wu Xiaojin et al. (1992) observed that Annong S - 1 was sterile from March 17 to 22, 1990, when the average daily temperature of Sanya was between 23. 8 ℃ and 25. 1 ℃, but it was fertile from February 11 to 19, 1991, when the average daily temperature there ranged from 24. 1 ℃ to 25. 3 ℃, indicating that there is a transition temperature between the critical temperature of sterility and that of fertility. Within this transition temperature range, the PTGMS lines can be either fertile or sterile. Therefore, the fertility transition of PTGMS rice can be illustrated by Fig. 5 - 9. When the temperature is lower than the critical low temperature of physiological sterility or higher than the critical high temperature of fertility transition, the PTGMS rice is sterile under any conditions. When the temperature is higher than the critical low temperature of physiological sterility but lower than the critical low temperature of fertility transition, the PTGMS rice is fertile. Within the range of transition temperature between the critical low temperature and the critical high temperature, the fertility of PTGMS lines is determined by the following three factors. The first is temperature at the stage before transition, if the temperature at the prior stage is higher than the transition temperature, the PTGMS lines may be sterile and if it is lower than the transition temperature, the PTGMS may be fertile. The second is the duration of the transition temperature. If the transition temperature persists for a long time, the PTGMS line may be sterile, if it persists only for a short period, the PTGMS may be fertile. The third is daylength and other conditions, it may be sterile under long daylength and fertile under short daylength conditions. The representative male sterile lines with sterility under high temperature include Annong S－1, Hengnong S－1, Peiai 64S, Y58S, etc. In addition to the common high temperature male sterile type, there are also a small number of materials reported as low-temperature male sterile. The sensitive period of this type of male sterile lines is below the critical temperature, and they are fertile when the temperature is above the critical temperature. This type of PTGMS lines undergoes fertility transition under certain conditions of photo-thermo interaction. For example, long daylength and high temperature types are similar to long daylength sterile lines in terms of fertility transition, but the difference is that the primary and secondary roles of daylength and temperature cannot be distinguished, such as Luguang S, 3418S, W9593S, etc. Lu Xinggui et al. (2001) found that when the temperature is low, such as when the average daily temperature is below 26 ℃, daylength plays a major role; however, when the temperature is high, temperature is the main determinant.

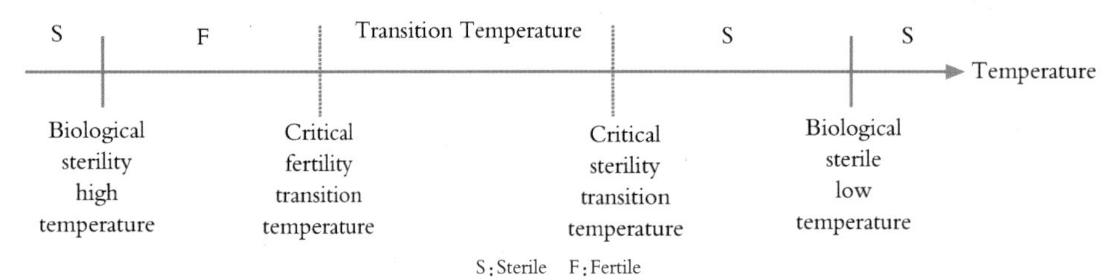

Fig. 5－9 Fertility transition of PTGMS lines

Ⅲ. Genetic Basis of PTGMS

1. Inheritance of PTGMS male sterility

The F_1 hybrids of Nongken 58S and conventional rice varieties have normal fertility, which shows that the photo-sensitive male sterility of Nongken 58S is controlled by recessive genes, while the conventional varieties have dominant restoring genes. Based on the normal pollen fertility in F_1 hybrids and the fertility segregation in the F_2 population in the cross and reciprocal cross, it can be inferred that the photo-sensitive sterility of Nongken 58S is sporophyte sterility. The insignificant difference in fertility for the cross and reciprocal cross infers that the fertility restoration of the F_1 population is controlled by nuclear genes and the cytoplasm plays no role in the fertility. Many scholars in China have studied the genetic pattern of PTGMS represented by Nongken 58S, and most of them believe that the sterility of Nongken 58S and its derived male sterile lines is controlled by a pair of recessive genes. For example, Shi Mingsong (1987), Lu Xinggui (1986) and Zhu Yingying (1987) think that the fertility of Nongken 58S is controlled by a pair of recessive genes, and that under long daylength condition, the F_2 population shows 3 : 1 fertility segregation, and the BC population shows fertility segregation at 1 : 1.

However, there are some inconsistent results regarding the genetic model of male sterility. In a study by Lei Jianxun et al. (1989), the difference between Nongken 58S and Nongken 58 involves a pair of major recessive genes, but the difference between Nongken 58S and other *japonica* rice varieties involves two pairs of major recessive genes, and the fertility segregation ratios of the F_2 and BC_1 populations are

15 : 1 and 3 : 1 respectively. Mei et al. (1990) suggested that the inheritance of photosensitive male sterility in Nongken 58S is characterized by the inheritance of qualitative-quantitative traits and the shape of the curve of F_2 seed setting rate distribution, which is related to the artificial grouping method.

Sheng Xiaobang (1992) showed that the two pairs of genes controlling photosensitive male sterility in Nongken 58S differed in their mode of interactions in different types of *japonica* rice varieties. In early- and mid-season *japonica* varieties, the F_2 segregation ratio was 9 : 6 : 1 in an additive action, in late-season *japonica* varieties, the F_2 segregation ratio was 9 : 3 : 3 : 1, in an independent segregation, and in the background of Nongken 58, the F_2 segregation ratio was 9 : 3 : 4 in recessive epistasis action. 32001S is an *indica*-type photosensitive sterile line that was bred with Nongken 58S as the sterility gene donor. Zhang et al. (1994) evaluated the fertility of 650 plants in the F_2 population of 32001S x Minghui 63 under natural conditions in Wuhan in 1991. The result showed that the photo-sensitive male sterility of 32001S might be determined by the complementary effect of two pairs of recessive genes.

In contrast to the single source of PTGMS genes, which were mostly derived from Nongken 58S's nuclear sterility genes, there were more abundant PTGMS materials with a wide range of sterility genes, and the PTGMS loci were not fully allelic among the sterile lines. Although the fertility gene of the PTGMS line Peiai 64S is the same as that of Nongken 58S, there is no allelic sterility gene between Annong S and its derived sterile lines and Nongken 58S. Li Bihu et al. (1990) crossed Annong S with 13 different varieties and found all of the 195 F_1 plants and 3,818 out of the 4,887 F_2 plants to be fertile while the remaining 1,069 of F_2 plants to be sterile. The ratio of fertile plants to sterile plants was 3.571 : 1, basically in line with the theoretical ratio of 3 : 1, indicating that the male sterility of Annong S is probably controlled by a pair of recessive genes. Wu Xiaojin et al. (1992) carried out a study on Hengnong S-1 and four early *indica* rice varieties. By the fertility segregation ratio of the F_2 and BC populations, it was inferred that the fertility of Hengnong S-1 was controlled by a pair of recessive genes. At present, the research results are basically consistent - the male sterility of Annong S is controlled by a pair of recessive genes yet is also influenced by minor genes.

2. Genetic heterogeneity of PTGMS rice

There is genetic heterogeneity in the male sterility of PTGMS rice, which is characterized by - 1) segregation in high generations, specifically up to F_5 or F_6 populations, which may be weakened under high-pressure selection, but is difficult to eliminate (Deng Qiyun, 1998); 2) changes of the critical temperature for fertility transition as the generations increases, which leads to the proposal of the concept of "genetic drift" (Yuan Longping, 1994); 3) inconsistency in the daylength for abortion in single plants in the same sterile line; Xue Guangxing (1996) observed that the critical daylength for abortion in single plants in a Nongken 58S population ranges from 13.8 h to 14.3 h, and can be even longer in some cases); 4) the PTGMS genes are not allelic. Lu Xinggui et al. (1994) reported that the male sterility genes of Nongken 58S and its derived male sterile lines are not allelic. After comparing the alleles of PTGMS genes, Sun Zongxiu et al. (1994) listed out three types of non-allelic male sterile genes, i.e., non-allelic between the genes of Nongken 58S and its derived *indica* male sterile lines, non-allelic between the genes of *indica* lines derived from Nongken 58S, and non-allelic among different sources of *indica* male sterile lines.

Selection of practical PTGMS lines requires consideration of difference in the daylength and temperature response of PTGMS genes placed in different backgrounds. Sun Zongxiu et al. (1991) used Nongken 58S and the sterile lines N5047S, WD－1S and Zhongming 2－S, all derived from Nongken 58S, as materials and treated them with different daylength (12 h and 15 h) and temperatures (23.6 ℃ and 29.6 ℃) under artificially controlled conditions, and found that they responded differently to daylength and temperature.

The fertility performance of Zhongming 2－S was similar to that of Nongken 58S. It is sterile under long daylength and high temperature conditions, and showed low seed setting under long daylength and low temperature conditions; although it was fertile under short daylength, but the selfed seeding rate was obviously lower than under short daylength and low temperature conditions. N5047S showed sterility under both long daylength high temperature and long daylength and low temperature conditions, the seeding setting rate was relatively low under short daylength and low temperature conditions, but even lower under short daylength and high temperature conditions. WD－1S set no seeds under all conditions (Table 5－6).

Table 5－6 Fertility performance of Nongken 58S and its derived PTGMS lines under artificially controlled daylength and temperature conditions

Sterile line	Combination of daylength and temperature treatment			
	23.6 ℃/12 h	23.6 ℃/15 h	29.6 ℃/12 h	29.6 ℃/15 h
Nongken 58S	26.0 ± 14.3	0.2 ± 0.7	7.5 ± 7.2	0
N5047S	7.7 ± 9.6	0	0.9 ± 1.3	0
WD－1S	2.3 ± 5.1	0	0	0
Zhongming 2－S	31.7 ± 20.7	1.8 ± 1.8	2.8 ± 3.3	0.2 ± 0.5

In 1993, Deng Qiyun examined the fertility of Annong S－1 and some of its derived male sterile lines such as 545S, 1356－1S, A113S, Ce49－32S and Ce64S under the condition of artificially controlled temperature during sensitive periods. The results showed that the fertility performance of the sterile lines differed after treatment with an average daily temperature of 24 ℃ (day/night temperature: 27 ℃/19 ℃) for four days during the sensitive period (Table 5－7). The fertility transition temperature of the sterile lines 545S, 1356－1S, Ce49－32S and Ce64S was lower than 24 ℃; that for 168－95S was also significantly lower than that of Annong S－1, but that for A113S was similar to that for Annong S－1. Difference in the fertility performance of Annong S－1 PTGMS genes placed in different genetic backgrounds is mainly reflected in the difference in the temperature at the starting point of the induced fertility transition. Wu Xiaojin et al. (1991) concluded, based on the analysis of available experiments and observations, that the changes in temperature at the starting point of the induced fertility transition may be continuous when Annong S－1's PTGMS is placed in different genetic backgrounds (Fig. 5－10). Evidence to support this argument includes－1) When using Annong S－1 as a nuclear sterility gene donor, lifelong sterility types (always sterile regardless of temperature and daylength, probably because the starting temperature for inducing fertility transition is below or close to the lower limit for biological sterility), extreme low temperature-sensitive types (starting temperature ≤22 ℃), low temperature-sensitive types (starting temperature 22 ℃–24 ℃), and high temperature sensitive type (starting temperature >26 ℃), have all been found; 2) The fertility expression of different sterile lines of the same type is also slightly different as shown in Table 5-7. In the sterile lines whose starting point temperature for inducing fertility transition is lower than 24 ℃, the degree of abortion of Ce 64S is higher than that of 545S, and that of 545S is higher than that of 1356-1S. In sterile lines whose starting point temperature for inducing fertility transition is higher than 24 ℃, the abortion degree of 168-95S is higher than that of Annong S-1, and the abortion degree of Annong S-1 is higher than that of A113S; 3) In the offspring that use Annong S-1 as a nuclear sterility gene donor, the starting temperature of fertility transition of most sterile plants or sterile lines is close to that of Annong S-1, with only a few showing low temperature sensitivity or high temperature sensitivity, and even fewer showing life-long sterility or extremely low temperature sensitivity.

Table 5-7 Fertility performance of Annong S-1 and some derived male sterile lines under controlled temperature (24 ℃) (Deng Qiyun, 1993)

Sterile line	Pollen abortion rate (%)	Selfed seed setting rate when bagged (%)
545S	99.8	0.00
1356-1S	98.3	0.00
168-95S	94.6	7.18
A113S	69.5	18.40
Ce 49-32S	100.0	0.00
Ce 64S	100.0	0.00
Annong S-1	88.3	18.00

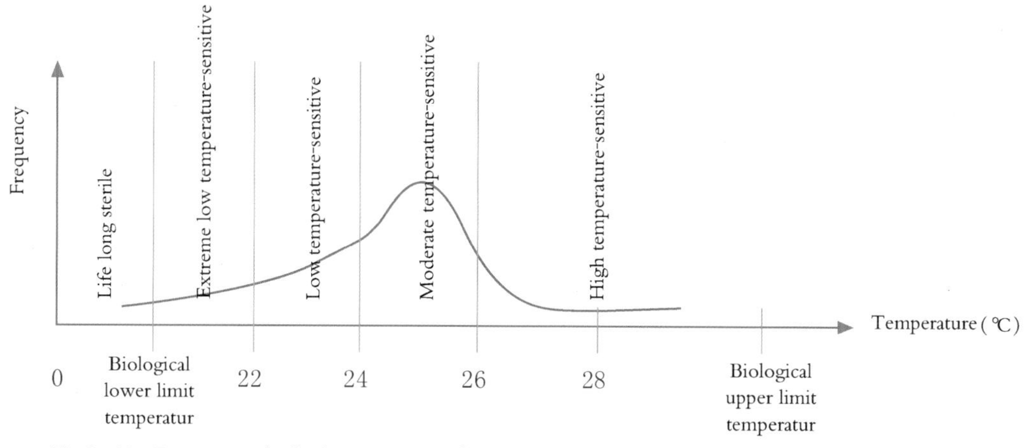

Fig. 5-10 Temperature for fertility transition with Annong S-1's PTGMS gene in different genetic backgrounds

Regarding the genetic mechanism and causes of sterility instability in PTGMS lines, most scholars have attributed it to the role of genetic background. Deng Qiyun et al. (1998, 2003) suggested that the genetic behavior of PTGMS in rice is controlled by a few pairs of major genes and is also influenced by minor genes sensitive to ecological factors such as temperature and daylength, so PTGMS is expressed as a qualitative-quantitative trait. Therefore, it is inevitable that under natural propagation conditions, the exchange and recombination of PTGMS lines selected through conventional methods lead to a slight variation in the starting point temperature of sterility, and the selection effect of the natural climate contributes to the accumulation of these slight variations, resulting in a generation-by-generation increase in the sterility starting temperature and a more pronounced drift after multiple generations. This is the essence of genetic drift. He Yuqing et al. (1998) suggested that minor-effective genes of different genetic backgrounds are the main cause of sterility instability and fertility convertibility in PTGMS lines, while mutation and recombination of minor-effective genes may be the main cause of critical temperature change and genetic drift. Liao Fuming (1996, 2000, 2001, 2003) suggested that the control of minor-effective polygenes in regard to the starting temperature of sterility is the genetic mechanism of unstable expression of sterility in PTGMS lines, and the impure genetic basis or genetic heterozygosity at the starting temperature of sterility is the intrinsic cause of unstable expression of sterility. It is proposed that the quantitative traits of sterility expression should be fully considered in breeding, and the progeny selection method combining pollen culture and pedigree selection method should be used in the breeding to achieve the objective of selecting PTGMS lines with stable sterility expression.

Section 4 Breeding of PTGMS Lines

I . Selection Objectives for the Breeding of Practically Valuable PTGMS Lines

For two-line hybrid rice to be applied to large-scale production, there must not only be strong hybrid heterosis, high resistance and good grain quality, but also low risk in seed production and high yield, of which the key point is the practicality of the PTGMS line. At present, many approved PTGMS lines in China have been through testing and screening for many years, but only some of the hybrids derived from these sterile lines have passed validation at the provincial level, indicating the importance of breeding and selecting practical PTGMS lines (Chen Liyun, 2010).

The main criteria for the selection of PTGMS lines of high practical value are as follows.

1) Stable sterility – both the biological lower limit and critical sterility temperature are low, stable sterility for safe seed production and high and stable yield.

2) Excellent overall agronomic traits – relatively short plant height, strong stems, moderately compact plant type, good early-growing and strong tillering, large biomass.

3) Excellent outcrossing traits – uniform heading, concentrated flowering, high stigma exsertion rate (>70%), high stigma vitality and good outcrossing compatibility.

4) Good rice quality – head rice rate above 50%, chalky grain rate below 20%, amylose content

16%−24% , gel consistency above 60 mm, Grade V alkali spreading value, and good tasting quality.

5) High resistance − moderate resistance to rice blast, tolerance to both high- and low-temperature stress, and lower susceptibility to bacterial blight, rice kernel smut and sheath blight.

6) Good combining ability − easy fertility restoration, good combining ability, easy breeding of heterotic hybrids, and stable hybrid yield.

Ⅱ. Methods to Breed PTGMS Lines

1. Trans-breeding

The main method to breed PTGMS lines is to use existing PTGMS materials for trans-breeding, which is to cross an existing PTGMS line with one or more excellent parents, and then select new PTGMS lines from the progenies. According to the crosses made, there are single cross and multiple cross trans-breeding.

(1) Single-cross trans-breeding

Since PTGMS lines are featured by fertility transition, nuclear sterile lines can be used as either the female or the male parent in a cross. Single-cross trans-breeding is fast with a small F_2 population. The PTGMS lines bred through single cross trans-breeding include C815S, HD9802S, Xin'an S and Xiangling 628S. The process of single-cross trans-breeding is similar to that of common hybridization. Table 5 − 8 shows the breeding process of C815S.

Table 5 − 8 Breeding process of C815S (Chen Liyun, 2012)

Season/Year	Location	Generation	Remarks
Winter 1996	Sanya	F_1	A sterile plant with good overall traits selected from 5SH038 (Anxiang S/Xiandang//02428) F_6 and crossed with Peiai 64S, 36 F_1 seeds harvested
Summer and autumn 1997	Changsha	F_2	25 single plants planted, plants with good traits selected and ratooned, bagged self-pollinated and 12 single plants harvested
Winter 1997	Sanya	F_3	8 rows planted, 7SH05S showed an ideal plant type with a low fertility transition temperature, selected and bagged self-pollinated, 21 single plants harvested
Summer 1998	Changsha	F_4	15 rows planted and 8 plants with good plant type selected and ratooned, 23 plants with better fertility than Peiai 64S (8S019) selected and bagged self-pollinated and seeds harvested
Winter 1998	Sanya	F_5	18 rows planted. The fertility of 8SH015 was better than that of Peiai 64S with other traits consistent meeting the breeding objectives and relatively stable agronomic traits, 10 plants from the selected rows bagged for reproduction
Spring, summer and autumn 1999	Changsha	F_6	10 rows planted, and one of them, 9S02, was selected and isolated for reproduction, further purified before participating in the ecological practicality validation of Hunan province.

Continued

Season/Year	Location	Generation	Remarks
Winter 1999	Sanya	F_7	Further purified, isolated reproduction, some for test crossing and evaluated on the breeding site.
Spring, summer and autumn 2000	Changsha	F_8	Continued the ecological practicality validation by Hunan province, traits of outcrossing, combining ability, hybrid performance, fertility, seed production evaluated on-site
2000 – 2003			Pressurized selection carried out, further studies on response to daylength and temperature, fertility transition, and breeding and seed production techniques carried out.
2004			Validated and approved for production by Hunan Crop Variety Approval Committee.

The key to single-cross breeding lies in the F_2 population. The population size, which is generally grown under long daylength and high temperature conditions, depends on the number of pairs of genes controlling the PTGMS trait and the difference between the two parents' traits. If the genes controlling the PTGMS are simple, the F_2 population can be smaller; conversely, if the genes controlling the PTGMS are complex, the F_2 population should be larger. For example, the frequency of sterile plants in the F_2 generation in Liaoning was less than 1% when crossing Nongken 58S to an early *japonica* variety with weak photo receptivity; the frequency of sterile plants in the F_2 generation in Wuhan ranged from 1% to 7% when crossing Nongken 58S-derived sterile lines as nuclear sterility gene donors with different types of varieties or breeding lines. The frequency of sterile strains in the F_2 population in Changsha is generally around 25% when crossing Annong S – 1 with different types of varieties or breeding lines. Therefore, the F_2 population should be larger when using Nongken 58S and its derived PTGMS lines as the nuclear sterility gene donor, while the F_2 population can be smaller when using Annong S – 1 and its derived temperature sensitive sterile lines as the nuclear sterility gene donor. In addition, if the genetic differences between the parents are large and the traits that need to be recombined are many and complex, the F_2 population should be larger; conversely, if the genetic differences between the parents are little and simple, the F_2 population can be smaller.

(2) Multiple-cross trans-breeding

The purpose of multiple-cross trans-breeding is to combine the superior traits of multiple parents. Technically, there are two ways to do multiple-cross breeding – two or more rounds of single-crossing or three-way crossing.

1) Two or more rounds of single-cross breeding. This method is to select good single sterile plants or lines from the progeny of the first single-cross and cross them with another parent, and then carry out a second single-cross breeding. The procedure is actually two or more single-cross breeding. For example, Y58S was selected through multiple single-crosses (Fig. 5 – 11). 454S, a hybrid progeny of Annong S – 1 and Changfei 22B, as the female parent, was crossed with 168S as the male parent, which was a hybrid

progeny of Annong S-1 and Lemont, an American glabrous hull rice variety. PTGMS material was then selected from the F_2 population and bred into Guangye 0058S after it was stabilized over several generations. Then, it was crossed with Peiai 64S and an excellent single plant was selected from the F_2 population. Y58S was the result of this process after it gets stable over multiple generations.

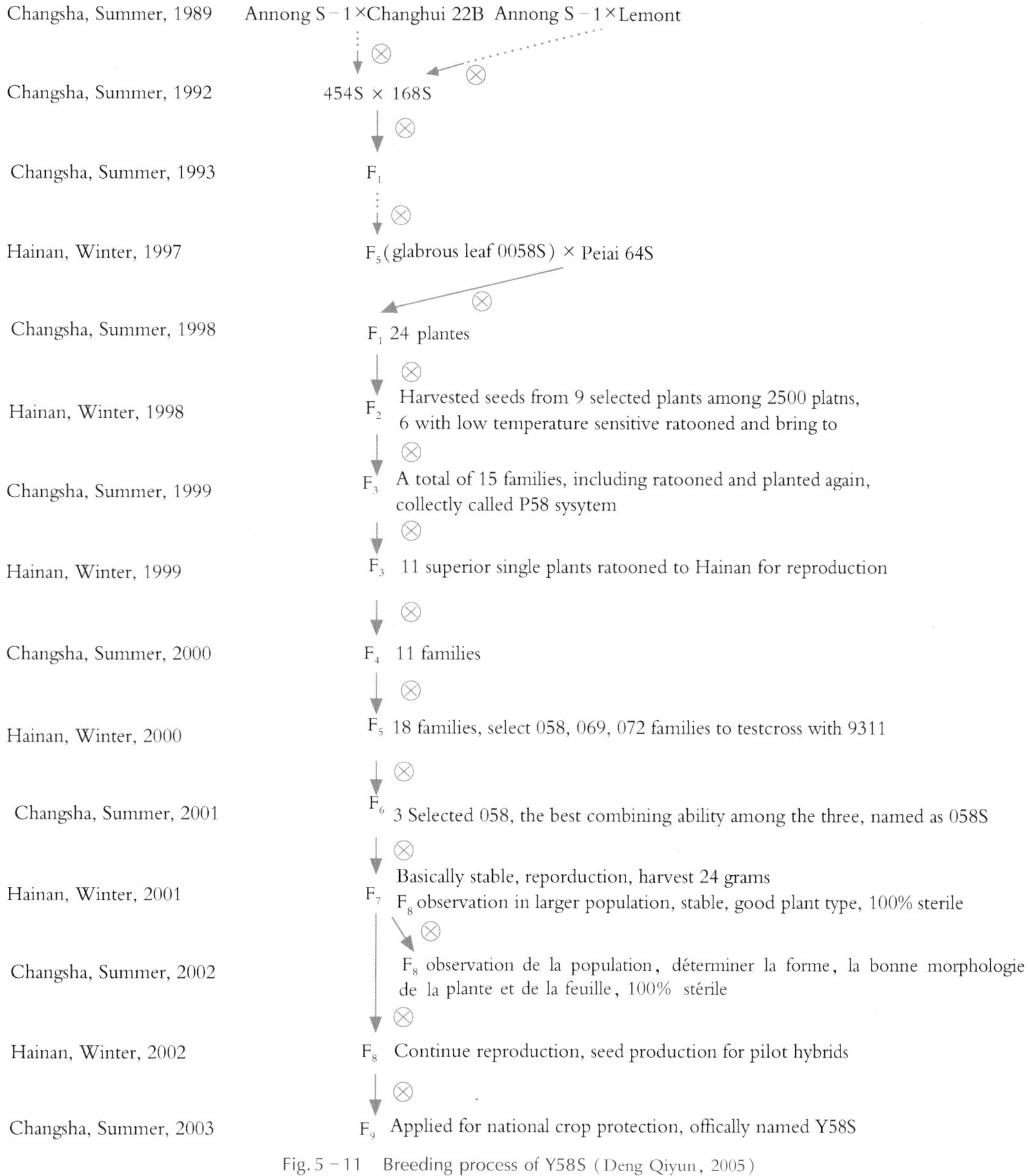

Fig. 5-11 Breeding process of Y58S (Deng Qiyun, 2005)

2) Three-way crossing. In this case, a PTGMS gene donor is crossed with the first parent, and then an F_1 plant is crossed with a second parent.

2. Backcrossing

Backcrossing is a method to introduce PTGMS genes into a good recurrent parent. The purpose is to breed sterile lines with traits other than PTGMS and similar to those of the recurrent parent. A recurrent parent is generally parent with excellent overall traits and good combining ability. Technically, alternate-generation backcrossing is more commonly used, and the premise is finding sterile plants. First, sterile plants with traits similar to the recurrent parents are selected from the generation of fertility segregation, and then crossed with the recurrent parent. Peiai 64S, the first practical low-temperature PTGMS line in China, was bred in this way (Fig. 5 – 12).

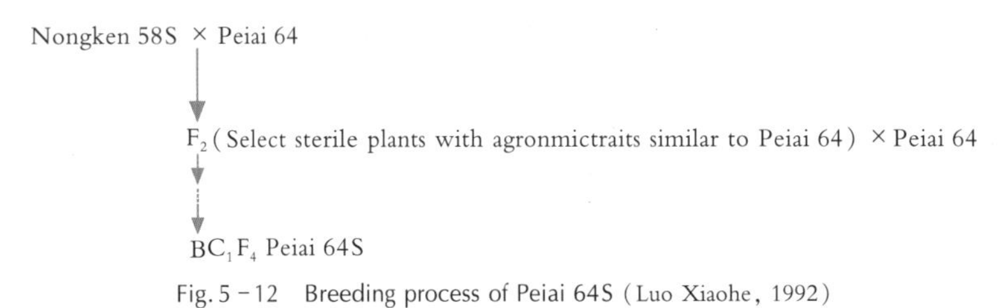

Fig. 5 – 12 Breeding process of Peiai 64S (Luo Xiaohe, 1992)

3. Population improvement

Population improvement is to cross a PTGMS gene donor with multiple parents with complementary traits, mix the F_2 seeds and bulkily plant the mixed F_2 seeds (population) under the condition of long daylength and high temperature, conduct artificial pollination to make the population fully heterozygous, and then mix the seeds of good sterile and fertile plants for the next round of multi-crossing and selection. The procedure takes two forms – alternate-generation random multi-crossing and continuous random multi-crossing. Zhun S bred by HHRRC was selected through two rounds of random multi-crossing and pedigree selection.

4. Anther culture

Anther culture can stabilize PTGMS lines quickly and accelerate the breeding progress. In general, the technical operation is to cultivate the anthers of S (PTGMS gene donor)/preferred variety (line) F_1 or S/another PTGMS line F_1. The seeds are regenerated under short daylength and low temperature to obtain self-pollinated seeds, and then after two or three generations of pedigree selection, stable sterile lines are obtained.

Ⅲ. Principles of Parent Selection for the Breeding of Practically Valuable PTGMS Lines

1. Selection of PGMS gene donors

So far, among the original PTGMS lines found in China, only Nongken 58S has strong PGMS, and the fertility transition of Annong S – 1, Hengnong S – 1 and 5460S is less associated with daylength. Therefore, for the selection of a practical photosensitive line, it is more appropriate to use Nongken 58S or its derived PGMS lines as the nuclear sterility gene donor, which requires a larger F_2 population for fertility segregation. The results of existing studies show that the frequency of complete male sterility in the F_2 population is very low, generally less than 8%, or even less than 1% in some combinations, and can vary with the combination and location. If the frequency of good sterile plants is calculated as 1% of the total number of sterile plants, the F_2 population should have more than 10,000 plants.

2. Selection of TGMS gene donors

Through an analysis of the pedigrees of 130 TGMS lines, covering two-line hybrids that have been recognized and protected by new variety rights in China since 1994, it was found that there are mainly two sources of the TGMS lines. The first type, and also the majority, comes from the TGMS lines selected in the early years such as Nongken 58S, Annong S－1, and their derived sterile lines. The second type is the newly discovered male sterile lines, which are mostly natural mutants or results of cross breeding (Si Huamin, 2012). The second type covers only a small number of lines, but is highly regarded because it may differ from the sterility genes of Nongken 58S and Annong S－1. One example of this is Yannong S, a naturally mutated male sterile line discovered in the late-season *indica* rice variety 3714. HD9802S is bred through systematic breeding by selecting sterile plant from the F_2 population of Huda 51 (female) and Hongfuzao (male).

3. Selection of recipient parents

For the selection of recipient parents of PTGMS lines, in addition to the parental selection principles required for general breeding, such as high combining ability, good adaptability (including pest and disease resistance), overall agronomic traits, and avoidance of the same poor traits as the donor parent (trait complementarity), special attention should also be paid to the following two points. 1) Recipient parents with photo-thermo-insensitivity or weak photo-sensitivity are preferred, because if the new sterile line is too photo-sensitive, the plants will have no heading at the appropriate time for safe seed production, making it no good for seed production; 2) High outcrossing. The yield of seed production depends, to a large extent, on the outcrossing of the sterile parent, and the outcrossing rate, to a great extent, depends on the outcrossing habits of the male sterile line. Therefore, outcrossing traits should be considered when selecting recipient parents. Parents with early and concentrated flowering, high stigma exsertion rate, well-developed stigma and large glume opening angle are preferred so as to cultivate male sterile lines with high outcrossing rates.

Ⅳ. Techniques for Selecting PTGMS Lines with High Practical Value

There are mainly two methods to select PTGMS lines with high practical value－high-pressure screening in low generations and high-pressure screening in high generations (Fig. 5－13 and Fig. 5－14). The low-generation high-pressure screening method use natural variability and controlled temperature conditions to first screen for single sterile plants or lines with low fertility transition temperatures in low generations under certain selection pressure, and then select for agronomic and quality traits, adaptability and outcrossing habits of the candidates. After selecting sterile lines with low fertility transition temperature and stable agronomic traits, testcrosses are made to select sterile lines with good combining ability. In high-generation high-pressure screening, the agronomic traits, quality traits, adaptability and outcrossing habits of the low generations are first checked, then, the sterile lines with good combining ability are tested through test crossing in the F_5 and F_6 populations. After the sterile lines with stable morphological traits, excellent comprehensive traits and strong combining ability are obtained, the sterile plants or lines with a low threshold temperature for fertility transition are selected under certain pressure using natural and artificially controlled temperatures. High-pressure screening in high generations is used to select for agronomic and quality traits, adaptability and outcrossing habits in the low generations, and then screen for sterile lines with good combining ability in the F_5 and F_6 populations. Once the selected lines are stable in terms of morphological and overall traits and show good combining ability, single sterile plants or lines with low fertility transition temperature shall be selected under natural or controlled temperature conditions with a certain selection pressure. The high-generation high-pressure screening is feasible because in a high-generation population of a sterile line, there will be a certain frequency of fertility variation, producing single plants which have the same traits and combining ability with the original sterile line, with difference only in fertility performance.

Selection of practical PTGMS lines with low critical fertility transition temperature can be divided into three steps.

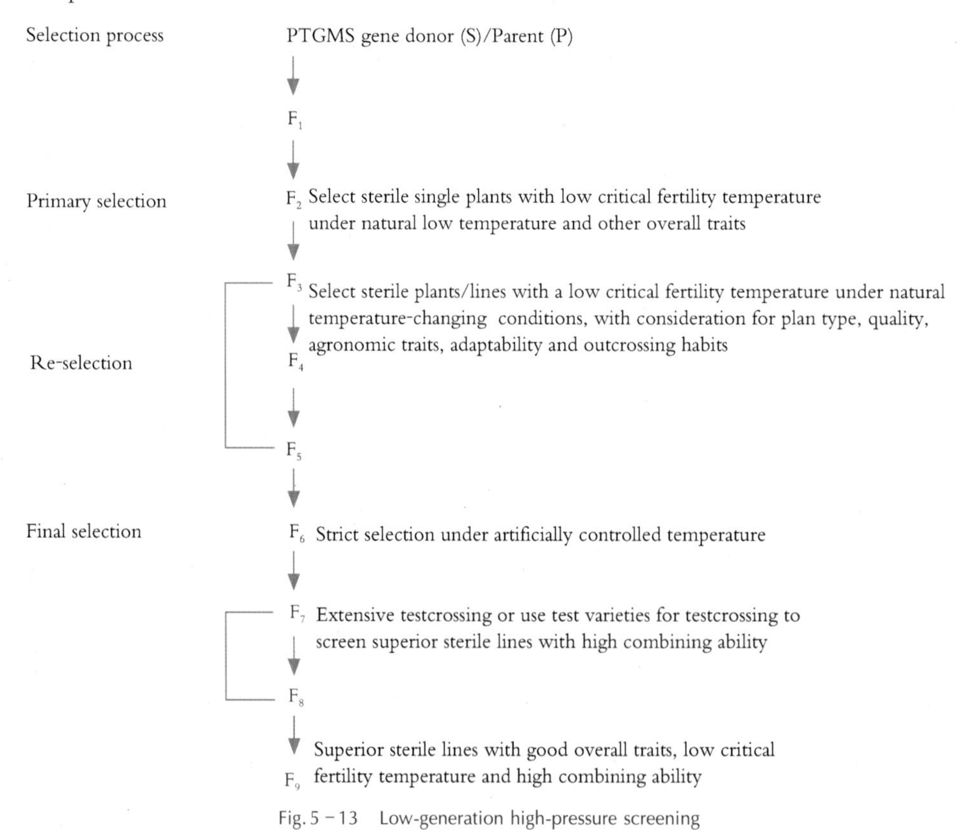

Fig. 5 - 13 Low-generation high-pressure screening

(1) Primary selection

The F_2 population (Fig. 5 - 13) or segregation population (Fig. 5 - 14) of a medium or high generation of male sterile lines are planted under natural temperature conditions, and sterile single plants with a low critical fertility transition temperature are selected under certain pressure based on the breeding objectives. Generally speaking, the temperature fluctuations in the Yangtze River Basin in mid-late June or mid-late September, and in Sanya from mid-late February to early March, is a good time to screen for sterility or lines with a low critical fertility transition temperature under natural conditions. In addition, screening at high altitudes is also feasible. The technical operation is to arrange the sensitive period at the time of frequent natural temperature changes, and then select sterile single plants with the required temperature for the critical fertility transition temperature according to the requirements of the breeding objectives and breeding performance.

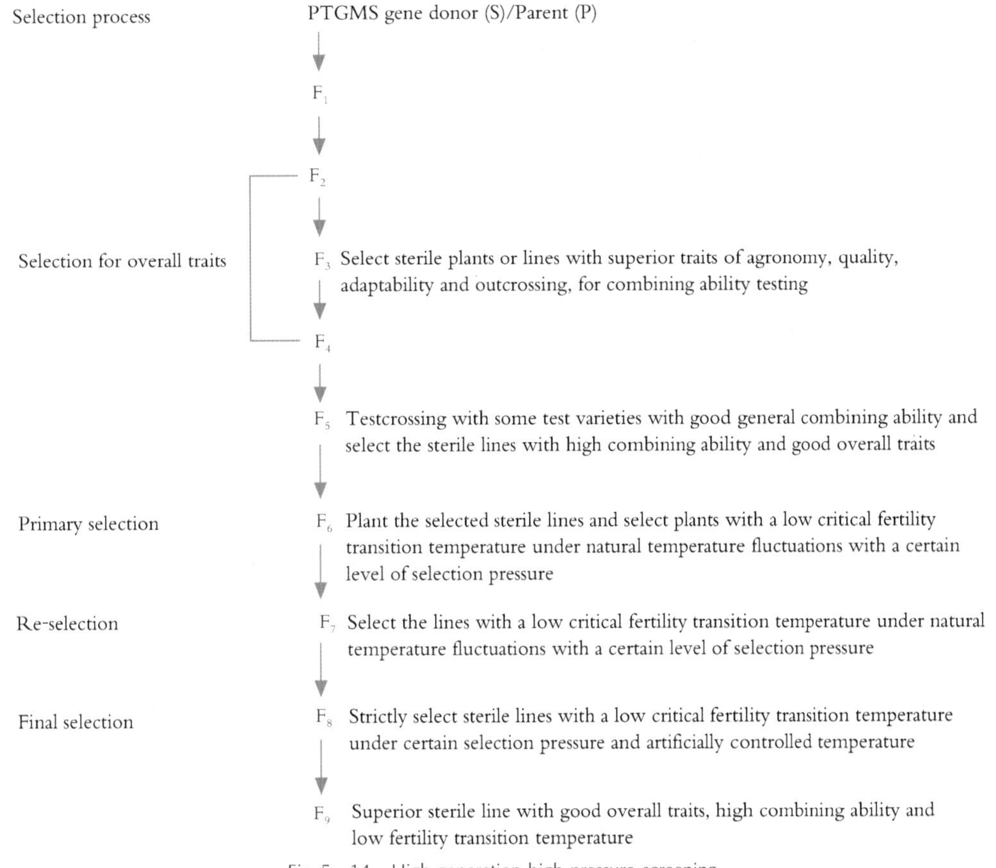

Fig. 5 - 14 High-generation high-pressure screening

(2) Re-selection

The sterile plants obtained though the primary selection are re-selected at low pressure for regeneration. After the seeds are obtained, continue with the selection process for one generation or more under the condition of natural temperature change.

(3) Final selection

The sterile lines obtained from the re-selection are strictly selected for low fertility transition temperature using artificially controlled temperature conditions under selection pressure set according to the breeding objectives.

Ⅴ. Stability Verification of PTGMS Fertility Transition Temperature

1. Principles for the stability verification of PTGMS fertility transition temperature

The stability of fertility transition of PTGMS lines is an important basis for ensuring the safety of two-line hybrid rice seed production. In order to ensure that the verification is accurate, reliable and practical, the following principles should be followed in an artificially controlled climate chamber.

(1) Long daylength and low temperature

PTGMS lines, whether photosensitive or thermo-sensitive, are only of practical value if they have a low critical fertility transition temperature, and require a longer period of time to restore fertility at temperatures below the critical fertility transition temperature. Considering the compensation effect of long daylength on low temperature, the verification of practical sterile lines should be carried out under long daylength and low temperature conditions to ensure fertility stability.

(2) Accuracy and reliability

The most temperature-sensitive period of different gene sources and different types of sterile lines are different, and there is always a slight difference in the developmental progress between individuals of the same sterile line, so it is difficult to ensure the reliability of the verification if the candidate is only treated in one growth stage. Therefore, multiple sets of treatments for different developmental stages will be necessary to ensure accurate and reliable verification results.

(3) Natural simulation

The specific daylength and temperature levels and the duration of the treatment must be similar to the low temperature spells of a certain region in mid-summer, including the intensity of low temperature, the pattern of temperature change, etc.

(4) Hierarchical treatment

The intensity of stable sterility despite low temperature may vary across different lines. Therefore, low temperature treatments should have different intensities to identify sterile lines with different levels of fertility stability.

2. Techniques for verification of PTGMS fertility transition stability

(1) Temperature and daylength

1) Temperature – According to the principle of natural simulation, the specific temperature for the verification of practical sterile lines for fertility stability must be determined according to the frequency and intensity of low temperature spells that may occur during the seed production season in different regions. The appropriate conditions for the central China rice region are 4 consecutive days with a daily average temperature of 23.5 ℃, daily maximum temperature 27 ℃, and daily minimum temperature 19 ℃. The temperature change should also be similar to the diurnal temperature variation pattern (Deng Qiyun, 1996).

2) Daylength – Under natural conditions, low temperature in mid-summer is often accompanied by rain and high humidity, weak radiation, and leaf temperature close to air temperature. According to the detailed analysis of low temperature data in mid-summer of 1989 in Changsha and observation, Deng Qiyun et al. (1996) concluded that the average light intensity during periods of continuous abnormal low

temperature and rainy weather in summer was about 8,000 lx. Therefore, when setting the artificial climate conditions, the suitable light intensity should be 8,000 – 10,000 lx to simulate low temperature and rainy spells in mid-summer. If the light intensity is too high, radiation will be strong and affect the relative humidity, resulting in a great difference between leaf temperature and air temperature, thus affecting the accuracy of fertility stability verification. If the light intensity is too low, it is not conducive to the growth and development of plants. In Central China, the daylength should be set at 13.5 h.

(2) Practical techniques for verifying fertility stability of practically valuable male sterile lines

According to the principles of the above verification method, long daylength and low temperature treatment should be done at four growth stages for eight groups of plants (Table 5 – 9) for the verification of the fertility stability of practical PTGMS lines under artificial climate conditions.

1) Long daylength and low temperature – the daylength is 13.5 h with a light intensity of 8,000 – 10,000 lx; average daily temperature 23.5 ℃ (19 ℃/27 ℃).

2) Intensity levels of low temperature – four intensity levels of low temperature should be used, respectively 4, 7, 11 and 15 consecutive days.

3) Eight groups – According to the development stages, each population of a male sterile line is divided into eight groups to be treated in chambers in succession. Groups *d*, *e*, *f*, *g* and *h* (Table 5 – 9) are treated for four days at the third, fourth, fifth, sixth and seventh stages of young panicle differentiation to ensure that at least one group can experience the low temperature of 23.5 ℃ (Level-2 low temperature) for four days during the most sensitive period. If a male sterile line is still stable and completely sterile after two levels of low-temperature treatment, the probability of safe seed production in Changsha in July and August will exceed 95%. Group *c* was treated with low temperature for seven days (Level-1 low temperature). If all the male sterile lines remain sterile, they can tolerate low temperature similar to that of mid-summer 1989. Groups *b* and *a* are treated for 11 days and 15 days of low temperature respectively. Under these two levels of low temperature, the male sterile lines generally show different degrees of instability. The reproduction difficulty of male sterile lines can be evaluated according to their fertility performance, so as to take corresponding measures, such as cold water irrigation and high-altitude growing to ensure safe parental seed production.

Table 5 – 9 Treatment of four intensity levels on eight groups under long daylength and low temperature (Deng Qiyun et al., 1996)

Group	Initial treatment stage *	Days before anthesis (days)	Treatment duration (days)	Minimum number of treated plants
a	End of Stage 3	16 – 22	15	10
b	End of Stage 4	13 – 18	11	10
c	End of Stage 5	10 – 15	7	10
d	Middle of Stage 3	18 – 24	4	10
e	Middle of Stage 4	15 – 20	4	10

Continued

Group	Initial treatment stage *	Days before anthesis (days)	Treatment duration (days)	Minimum number of treated plants
f	Middle of Stage 5	12 - 16	4	10
g	Middle of Stage 6	9 - 13	4	10
h	Start of Stage 7	4 - 9	4	10

Note. * the development period of young panicles.

References

[1] CHEN LIYUN, XIAO YINGHUI. Mechanism of PTGMS and breeding strategy of PTGMS lines[J]. Chinese Journal of Rice Science, 2010, 24 (2): 103 - 107.

[2] CHEN LIYUN. Research on two-line hybrid rice[M]. Shanghai: Shanghai Scientific & Technical Publishers, 2012.

[3] CHEN LIANGBI, LI XUNZHEN, ZHOU GUANGQIA. Effect of temperature on gene expression of photo-sensitive and thermo-sensitive genic male sterile rice[J]. Acta Agronomica Sinica, 1993, 19 (1): 47 - 54.

[4] CHENG SHIHUA, SUN ZONGXIU, MIN SHAOKAI, et al. Studies on responses to daylength and temperature of PTGMS rice Ⅰ. Fertility performance of PTGMS rice in Hangzhou (30°05'N) under natural conditions[J]. Chinese Journal of Rice Science, 1990, 4 (4): 157 - 163.

[5] CHENG SHIHUA, SUN ZONGXIU, SI HUAMIN, et al. Classification of photo-thermo responsive fertility transition of dual-purpose genic male sterile lines of rice[J]. Scientia Agricultura Sinica, 1996, 29 (4): 11 - 16.

[6] DENG HUAFENG, SHU FUBEI, YUAN DINGYANG. Research on and itilization of Annong S - 1 [J]. Hybrid Rice, 1999, 14 (3): 1 - 3.

[7] DENG QIYUN, FU XIQIN. Study on fertility stability of PTGMS rice Ⅲ. Threshold temperature drift of sterility and its control technology[J]. Journal of Hunan Agricultural University (Natural Science Edition), 1998, 24 (1): 8 - 13.

[8] DENG QIYUN, OU AIHUI, FU XIQIN, et al. Discussions on methods of fertility stability verification for practically valuable PTGMS lines[J]. Journal of Hunan Agricultural University, 1996, 22 (3): 217 - 221.

[9] DENG QIYUN, OU AIHUI, FU XIQIN. Studies on fertility stability of PTGMS rice Ⅰ. Analysis of daylength and temperature response of PTGMS lines of rice[J]. Hybrid Rice, 1996, 11 (2): 23 - 27.

[10] DENG QIYUN, SHENG XIAOBANG, LI XINQI. Study on the inheritance of PTGMS of *indica* rice[J]. Chinese Journal of Applied Ecology, 2002 (3): 376 - 378.

[11] DENG QIYUN. Genetic study on male sterility of PTGMS *indica* rice[D]. Changsha: Hunan Agricultural University Press, 1997.

[12] DENG QIYUN, YUAN LONGPING. Studies on the fertility stability of PTGMS rice and verification techniques[J]. Chinese Journal of Rice Science, 1998, 12 (4): 200 - 206.

[13] DENG QIYUN. Breeding of PTGMS line Y58S with wide compatibility in rice[J]. Hybrid Rice, 2005, 20 (2): 15 - 18.

[14] DUAN MEIJUAN, YUAN DINGYANG, DENG QIYUN, et al. Studies on fertility stability of PTGMS rice Ⅳ. Law of threshold temperature drift of sterility[J]. Hybrid Rice, 2003, 18 (2): 62 - 64.

[15] FAN YOURONG, CAO XIAOFENG, ZHANG QIFA. Research progress of PTGMS rice[J]. Chinese Science Bulletin, 2016, 61 (35): 3822 - 3832.

[16] HE YUQING, YANG JING, XU CAIGUO, et al. Genetic study on fertility instability and conversion of PTGMS *indica* rice[J]. Journal of Huazhong Agricultural University, 1998, 17 (4): 305 - 311.

[17] JIANG QINGSHAN, LIN GANG, ZHAO DEMING, et al. Breeding and utilization of good-quality fragrant male sterile line Yixiang 1A[J]. Hybrid Rice, 2008, 23 (2): 11 - 14.

[18] JIANG DAGANG, LU SEN, ZHOU HAI, et al. Mapping of thermo-sensitive male sterile gene *tms5* in rice using EST and SSR markers[J]. Chinese Science Bulletin, 2006, 51 (2): 148 - 151.

[19] LEI JIANXUN, LI ZEBING. Study on genetic law of Hubei PTGMS rice Ⅰ. Analysis of fertility of progenies from the cross between original PTGMs rice and middle *japonica* rice[J]. Hybrid Rice, 1989 (2): 39 - 43.

[20] LI BIHU, DENG HUAFENG. Discovery and preliminary study of annong S - 1[C]//Selected papers on utilization of photo-thermo sensitive genic male sterility and interspecific heterosis in Rice, 1990:87.

[21] LI SHIFEN. Breeding, utilization and genetic study of Gang-type and DA-type hybrid rice[J]. Hybrid Rice, 1997 (S1): 1.

[22] LIAO FUMING, YUAN LONGPING. Strategies for genetic purification of PTGMS lines of rice[J]. Hybrid Rice, 1996 (6): 1 - 4.

[23] LIAO FUMING, YUAN LONGPING. Study on fertility expression of the PTGMS line Peiai 64S under low temperature[J]. Scientia Agricultura Sinica, 2000, 33 (1): 1 - 9.

[24] LIAO FUMING, YUAN LONGPING, YANG YISHAN. Study on male sterility stability of the practically valuable PTGMS line Peiai 64S [J]. Chinese Journal of Rice Science, 2001, 15 (1): 1 - 6.

[25] LIAO FUMING, YUAN LONGPING. Review on the genetic mechanism and causes of unstable expression of PTGMS in rice[J]. Hybrid Rice, 2003, 18 (2): 1 - 6.

[26] LIANG SHIHU, LI CHUANGUO, LI SHUGUANG, et al. Breeding of *indica* hybrid rice Wufengyou 2168 with good quality, high yield and disease resistance[J]. Bulletin of Agricultural Science and Technology, 2009 (7): 132 - 134.

[27] LIU YIBAI, HE HAOHUA, RAO ZHIXIANG, et al. Study on the function mechanism of daylength and temperature on the fertility of dual-purpose genic male sterile lines of rice[J]. Journal of Jiangxi Agricultural University, 1991, 13 (1): 1 - 7.

[28] LIU WUGE, WANG FENG, LIU ZHENRONG, et al. Breeding and application of early maturing blast resistant CMS line Jifeng A[J]. Hybrid Rice, 2014, 29 (6): 16 - 18.

[29] LU XINGGUI, GU MINGHONG, LI CHENGQUAN. Theory and technology of two-line hybrid rice[M]. Beijing: Science Press, 2001.

[30] LU XINGGUI, WANG JILIN. Research and utilization of PTGMS rice in Hubei province Ⅰ. Observation of fertility stability[J]. Hybrid Rice, 1986, 1:004.

[31] LU XINGGUI, YUAN QIANHUA, YAO KEMIN, et al. Climatic adaptability of main types of PTGMS lines in China[J]. Chinese Journal of Rice Science, 2001, 15 (2): 81 - 87.

[32] LUO XIAOHE, QIU ZHIZHONG, LI RENHUA. Peiai 64S, a dual - purpose male sterile line with low threshold temperature for male sterility[J]. Hybrid Rice, 1992, 7 (1): 27 - 29.

[33] MEI GUOZHI, WANG XIANGMING, WANG MINGQUAN. Genetic analysis of Nongken 58S's photo-sensi-

tive male sterility[J]. Journal of Huazhong Agricultural University, 1990, 9 (4): 400 – 406.

[34] SHENG XIAOBANG. Genetic study on the male sterility of the PTGMS line Nongken 58S[J]. Journal of Hunan Agricultural University, 1992, 6 (1): 5 – 14.

[35] SHI MINGSONG, SHI XINHUA, WANG GENHUA. Discovery and utilization of PTGMS rice in Hubei province[J]. Journal of Wuhan University (HPGMR special issue), 1987:2 – 6.

[36] SI HUAMIN, FU YAPING, LIU WENZHEN, et al. Pedigree analysis of PTGMS lines of rice[J]. Acta Agronomica Sinica, 2012, 38 (3): 394 – 407.

[37] SUN ZONGXIU, CHENG SHIHUA, MIN SHAOKAI, et al. Studies on responses of PTGMS rice to daylength and temperature Ⅱ. Fertility verification for PTGMS *japonica* line under artificial control[J]. Chinese Journal of Rice Science, 1991, 5 (2): 56 – 60.

[38] SUN ZONGXIU, CHENG SHIHUA. Hybrid rice breeding: from three line, two line to one line[M]. Beijing: China Agricultural Science and Technology Press, 1994.

[39] WU XIAOJIN, YIN HUAQI, SUN MEIYUAN, et al. Discussion on breeding and utilization of PTGMS rice [J]. Hybrid Rice, 1992, 6:19.

[40] WU XIAOJIN, YIN HUAQI. Breeding of PTGMS rice-selection of male sterile gene donor and ways to accelerate breeding[J]. Hunan Agricultural Sciences, 1992 (3): 15 – 16, 25.

[41] WU XIAOJIN, YIN HUAQI, YIN HUAJUE. Preliminary study on comprehensive effects of temperature on annong S – 1 and W6154S[J]. Crop Research, 1991, 5 (2): 4 – 6.

[42] WU XIAOJIN, YIN HUAQI. Inheritance and stability of PTGMS rice[J]. Chinese Journal of Rice Science, 1992, 6 (2): 63 – 69.

[43] XIA SHENGPING, LI YILIANG, JIA XIANYONG, et al. Breeding of CMS *indica* line good-quality Jin 23A [J]. Hybrid Rice, 1992, 5:29 – 31.

[44] XUE GUANGXING, CHEN CHANGLI, CHEN PING. Analysis of photoperiod effect (PE) of PTMGS *japonica* rice and its hybrids[J]. Acta Agronomica Sinica, 1996, 22 (3): 271 – 278.

[45] XUE GUANGXING, ZHAO JIANZONG. Preliminary study on critical daylength of photo-sensitive male sterility in rice and its response to environmental factors[J]. Acta Agronomica Sinica, 1990, 16 (2): 112 – 122.

[46] YUAN LONGPING. Research progress of two-line hybrid rice[J]. Scientia Agricultura Sinica, 1990, 23 (03): 1 – 6.

[47] YUAN LONGPING. Purification and seed production of PTGMS lines of rice[J]. Hybrid Rice, 2000 (S2): 37.

[48] YUAN LONGPING. Technical strategies for breeding PTGMS lines of rice[J]. Hybrid Rice, 1992 (1): 1 – 4.

[49] YUAN LONGPING. Hybrid rice breeding strategies[J]. Hybrid rice, 1987, 1 (1): 3.

[50] YUAN LONGPING. Hybrid rice[M]. Beijing: China Agricultural Publishing House, 2002.

[51] ZENG HANLAI, ZHANG ZIGUO, YUAN SHENGCHAO, et al. Study on the thermo-sensitive period for fertility transition in PTGMS rice[J]. Journal of Huazhong Agricultural University, 1993, 12 (5): 401 – 406.

[52] ZHANG HUILIAN, DENG YINGDE. Breeding and application of You Ⅰ A, a CMS line with high outcrossing rate and good quality[J]. Hybrid Rice, 1996 (2): 4 – 6.

[53] ZHANG XIAOGUO, ZHU YINGGUO. Inheritance of male sterility in PTGMS rice in Hubei province[J]. Genetics, 1991, 13 (3): 1 – 3.

[54] ZHANG ZIGUO, ZENG HANLAI, LI YUZHEN, et al. Effect of daylength and temperature on the fertility

transition of PTGMS *indica* rice[J]. Hybrid Rice, 1992, 5:34-36.

[55] ZHANG ZIGUO, LU KAIYANG, ZENG HANLAI, et al. Study on stability of photoperiod and temperature for fertility transition of PTGMS lines of rice[J]. Hybrid Rice, 1994 (1): 4-8.

[56] ZHANG ZIGUO, LU XINGGUI, YUAN LONGPING. Selection and identification of critical temperature for fertility transition of PTGMS rice[J]. Hybrid Rice, 1992, 6:29-32.

[57] ZHANG ZIGUO, YUAN SHENGCHAO, ZENG HANLAI, et al. Genetic study on two photoperiod responses in PTGMS rice[J]. Journal of Huazhong Agricultural University, 1992, 11 (1): 7-14.

[58] ZHOU HAI, ZHOU MING, YANG YUANZHU, et al. *RNase ZS1* processes *Ubl40* mRNA to control PTGMS in rice[J]. Genetics, 2014, 36 (12): 1274.

[59] ZHU YINGGUO, YANG DAICHANG. Research and utilization of PTGMS rice[M]. Wuhan: Wuhan University Press, 1992.

[60] ZHU YINGGUO, YU JINHONG. Study on fertility stability and genetic behavior of PTGMS rice in Hubei province[J]. Journal of Wuhan University (HPGMR special issue), 1987:61-67.

[61] VIRMANI S S. Heterosis and hybrid rice breeding[M]. Springer Science & Business Media, 2012.

[62] ZHANG Q, SHEN B Z, DAI X K, et al. Using bulked extremes and recessive class to map genes for photoperiod-sensitive genic male sterility in rice[J]. Proceedings of the National Academy of Sciences, 1994, 91 (18):8675-8679.

Chapter 6 Breeding of Super Hybrid Rice Restorer Lines

Deng Qiyun / Wu Jun / Zhuang Wen

Section 1 Inheritance of Restorer Genes

The restorer gene of two-line hybrid rice is actually an allele of the male sterility gene of the PTGMS line. Theoretically, all the existing rice varieties with normal fertility are the restorer lines of PTGMS lines, but there are no maintainer lines. However, in breeding practice, there are a few varieties that do not have full ability to restore fertility and some varieties that can only fully restore fertility in some PTGMS lines. This inheritance of the sterility of PTGMS lines may be related to the genetic background or compatibility of their parents.

Cytoplasmic male sterility is utilized in breeding three-line hybrid rice. According to the relationship between the restorer line and the maintainer line, CMS lines mainly include the WA type, HL type and BT type. The first two types are quite different in terms of the relationship between the restorer line and the maintainer line.

Ⅰ. Genetic Analysis and Gene Cloning for Fertility Restoration in WA-type Male Sterile Lines

WA-type male sterility can be restored by two pairs of restorer genes, Rf_3 and Rf_4, which were initially mapped on chromosome 1 and chromosome 10, respectively. More specifically, Rf_3 is mapped at 6 cM from the RG532 marker on chromosome 1 and Rf_4 is mapped at 3.3 cM from the G4003 marker on chromosome 10 (Yao et al., 1997). Zhang Qunyu et al. (2002) further mapped Rf_4 on chromosome 10 at 0.9 cM from the Y3-8 marker using a segregated population of near-isogenic lines (NIL). Progress has been made in the map-based cloning of Rf_4, but not for Rf_3. Wang et al. (2006) and Hu et al. (2014) finely mapped Rf_4 within a 137 kb interval on chromosome 10 where there is a cluster of 10 to 11 pentatricopeptide repeat (PPR) genes adjacent to the previously cloned CMS-BT and CMS-HL restorer gene Rf_{1a}(Rf_5). Verified through genetic transformation, *PPR9-782-M* was confirmed to have the function of restoring fertility in CMS-WA lines, which is the function of the Rf_4 gene. This gene encodes 782 amino acids. Both the proteins encoded by Rf_4 and the *PPR3-791-M* encoded by Rf_{1a} contain 18 PPR motifs with 86% amino acid sequence similarity, but they can only specifically restore fertility to WA and BT/HL types respectively. Sequence analysis showed that functional (dominant) Rf_4 has multiple allelic variations, while non-functional (recessive) mutations can be divided into the *japonica* Rf_4-*j* (a large number of base mutations) and the *indica* Rf_4-*i* (containing two fragment insertions producing premature termination codon). It was showed that Rf_4 restored fertility by degrading the *WA352c* transcripts at the post-transcription level, while Rf_3 has no effect on the *WA352c* transcripts, but suppressed the accumulation of *WA352c* protein production to achieve the biological function of fertility restoration.

Classical genetics defines male sterility caused by genetic interactions between recessive nuclear restorer genes and cytoplasmic sterility genes as nucleo-plasmic interactive sterility. In contrast, recent cloning of restorer genes has shown that dominant nuclear restorer genes restore fertility by suppressing CMS gene expression. In particular, *PPR3 – 791* and *PPR2 – 506* encoded by the CMS-BT restorer genes Rf_{1a} and Rf_{1b} located on chromosome 10 enter the mitochondria, and specifically cleave and degrade *B-atp6/orfH79*, which is the transcript of sterility genes. The CMS-HL restorer gene Rf_5 is actually the CMS-BT restorer gene Rf_{1a} whose encoded protein forms a complex with another nuclear gene encoded protein *GRP*, which leaves the transcript of CMS-HL sterile gene *orfH79* to achieve fertility restoration. The *PPR9 – 782 – M* protein, encoded by the CMS-WA restorer gene Rf_4, enters the mitochondria, and restores fertility by degrading the transcripts of *WA352* through an unknown mechanism. In addition, the interaction of the *WA352c* protein with the nuclear-encoded mitochondrial protein *COX11* is the molecular basis of the onset of male sterility, while the specific accumulation of the *WA352c* protein in the tapetum at the pollen mother cell stage may also be controlled by nuclear genes. Thus, plant CMS and its restorability involves nucleo-plasmic gene interactions at different levels.

Ⅱ. Genetic Analysis and Gene Cloning for Fertility Restoration in HL-type Male Sterility

The HL-type CMS lines have two pairs of restorer genes, Rf_5 and Rf_6. Rf_5 was first discovered in the restorer line Milyang 23. The Rf_5 NILs was obtained through cross and backcross, and a backcross population was constructed to map Rf_5 between the SSR markers RM6469 and RM25659 on chromosome 10. Hu et al. obtained candidate clones by screening the BAC library of Milyang 23 and sequenced the subclones and transgenic complementation was performed for each possible candidate gene. The results showed that only PPR791 could restore the fertility of YTA, and the T_1 population showed a 1 : 1 gametophyte segregation genetically. Rf_5, a PPR gene encoding 791 amino acids, is expressed in all tissues and cytological mapping shows that the protein is localized in the mitochondria, and is the same gene as Rf_1 (Rf_{1a}, PPR791) of BT-type sterile lines. In hybrid F_1, the sterility gene transcripts, either the 2.0 kb *atp6-orfH79* or the 0.5 kb *orfH79* (s), are cut into smaller fragments, thus failing to translate and restore fertility in HL-type hybrid rice.

Numerous experiments have demonstrated that Rf_5 does not directly interact with *atp6-orfH79*, so how this restorer gene processes the sterility gene transcript is an important scientific question to elucidate the mechanism of fertility restoration. Several Rf_5 interacting proteins were obtained by such biochemical techniques as yeast two-hybrid, BiFC, Pull-down and co-immunoprecipitation. It was found that the glycine rich protein (*GRP162*) can specifically bind to the sterility gene transcript *atp6-orfH79* through its RNA binding domain. *GRP162* can form a dimer, which is consistent with the result that *GRP162* has two binding sites to sterile transcript. The 400 – 500 kDa protein complex of *Rf5* and *GRP162* is named the restoration fertility complex (RFC). Recent studies have discovered a new subunit (RFC subunit 3, RFC3), with a transmembrane structure that interacts with *Rf5* at the C-terminus and *GRP162* at the N-terminus, which specifically produces gametophytic male sterility in HL-type hybrid rice in transgenically interfered materials, and further mechanistic studies showed changes in the size of the RFC. Therefore, it is inferred that the fertility restoration of HL-type hybrid rice is completed through a protein complex, in which Rf_5 functions as recruiter, *GRP162* forms a dimer to bind the sterile gene transcript, and *RFC3* is responsible for proper assembly of the subunits of the protein complex. Since the size of the RFC is between 400 kDa and 500 kDa, there are still other protein subunits to be discovered, and how these subunits are involved in fertility restoration needs to be further investigated.

9311 and its derivatives are among the most widely used restorer lines in China. Genetic studies show that 9311 has two non-allelic restorer genes for the HL-type CMS, and the restoration rate is 50% when Rf_5 or Rf_6 is present alone in HL-CMS hybrid rice, and 75% when both Rf_5 and Rf_6 are present, with more stable fertility. In addition to Rf_5 on chromosome 10, another restorer gene with comparable restoring ability on chromosome 8 was found and was named Rf_6(Huang et al., 2015). Research revealed that Rf_6 can restore fertility in both HL-type and the BT-type CMS. Rf_6 is finely mapped between RM3710 and RM22242 markers on chromosome 8 by constructing 19,355 F_2 plants and 554 BC_1F_1 plants. The co-disjunction molecular marker *ID200 – 1* was developed based on the sequence of a repeat missing from a sterile line within a PPR gene in this region between the sterile line and the restorer line. The gene in 9311 is 2,685 bp long, encoding 894 amino acids, and is named PPR894, while the rf_6 in YTA has only 786 amino acids. The transgenic complementation experiment showed that PPR894 could restore the fertility of the HL-type CMS line YTA, and was inherited in the gametophytic pattern in transgenic progeny. The Rf_6 protein is also located in the mitochondria, which is consistent with the presence of the sterile gene product in in the mitochondria. Although Rf_6 also belongs to the PPR gene family, this gene is a very special PPR novel gene because the third, fourth and fifth PPR tandem structural units of Rf_6 have a duplication and thus have a restoring function, if the structural units connected by these three PPR tandems do not have a duplication, it will not have a restoring function. Research on the mechanism of Rf_6 also shows that Rf_6 could not directly interact with sterility gene transcripts, and by yeast two-hybrid library and Pull-down validation, the specific interaction protein, hexokinase 6 (*HXK6*), of Rf_6 was obtained, and transgenic interference plants of *HXK6* also show gametophytic male sterility, while sterility gene transcripts *atp6-orfH79* processing was also disrupted (Huang et al., 2015). The results showed that there was no interaction between Rf_6 and Rf_5 interactive proteins, so, it was inferred that Rf_6 processes sterility gene transcripts as another protein complex to achieve fertility restoration, and in-depth molecular mechanism research is still in progress.

Section 2 Criteria of Super Hybrid Rice Restorer Lines

A good restorer line of super hybrid rice should meet the following criteria.

1) Good plant and leaf morphology – appropriate plant height, medium tillering ability, large panicles and more grains, high seed setting rate, good grain filling, high yield potential, and good grain quality;

2) Strong restoring ability – stable seed setting rate in F_1 hybrids with only small fluctuation when planted in different years and seasons;

3) Good flowering habits – a long flowering period, early and concentrated flowering, well-developed anthers and sufficient pollen;

4) Wide adaptability – insensitivity or low sensitivity to daylength and temperature, with only a small difference in growth duration when planted in the same season of different years;

5) Good general combining ability – significant hybrid heterosis when combined with multiple sterile lines;

6) Resistance to diseases and lodging – tolerance to fertilizer and lodging, resistance or moderate resistance to major rice diseases and pests such as blast, bacterial blight and planthopper.

Section 3 Breeding Methods of Restorer Lines

At present, the most common and effective methods to breed restorer lines are testcross, hybridization selection, backcross and mutation breeding.

I. Testcross

This means to cross sterile lines with existing rice varieties (lines), and according to the performance of the hybrids (F_1) to select the best varieties (lines) with strong restoration, good combining ability with obvious heterosis, to be the restorer lines.

1. Principles for selecting parents for testcross

At present, there are two main types of male sterility used in rice production in China – PTGMS and CMS. There are large differences in the selection of parents for testcross due to their different genetic mechanisms of fertility.

(1) Testcrossing parent selection for PTGMS lines

Most conventional rice varieties are restorer lines of PTGMS lines with a wide restorer spectrum and free-mating. However, in breeding practice, not all varieties with restoring ability can be used to breed hybrid rice combinations with strong heterosis. Only a few of them can actually serve as restorer lines. According to the genetic mechanism of heterosis in rice and years of breeding experience, there is a certain correlation between the geographic distribution of the restorer parents of good hybrid rice combinations and the genetic composition of the PTGMS lines. The general trend is – for sterile lines with early- and mid-season *indica* varieties in China's Yangtze River basin as the genetic background, the selection of test-crossing parents should target Southeast Asian *indica* varieties and South China late-season *indica* varieties; for sterile lines with Southeast Asian *indica* varieties as the genetic background, the testcrossing parents should be selected from early- and mid-season *indica* varieties in China's Yangtze River basin; for sterile lines with a complex genetic background as the female parent, the selection of testcrossing parents is generally not limited by geographical distribution.

(2) Testcrossing parent selection for CMS lines

At present, major CMS lines used in China include the WA, BT, and HL types, and restricted by the restorer-maintainer relationship, there is no free-mating. At the same time, due to the different sterile cytoplasm sources of the WA, BT and HL types, the distribution of maintainer and restorer lines is also somewhat geographically diverse. Generally speaking, restorer varieties for the WA type, relatively rare, are mainly *indica* varieties distributed in low-latitude and low-altitude tropical and subtropical regions.

Hunan Academy of Agricultural Sciences (1975) used WA-type male sterile lines to testcross with Southeast Asian varieties and late-season *indica* varieties from South China, and found only 4% of the 375 test-crossed varieties to have restoring ability. Further pedigree analysis of these restorer varieties showed that most of the restorer varieties from Southeast Asia, such as IR24 and IR26, are related to Peta. Most of the varieties with restoring ability from South China such as Qiugu'ai and Qiutang'ai are related to Indonesian Paddy. It can be concluded that restorer genes of the WA type mainly come from several original rice varieties in Southeast Asia. Therefore, *indica* rice varieties from Southeast Asia and late-season *indica* rice varieties from South China, which are related to Peta or Indonesian Paddy, should be selected as parents for testcrossing.

According to the relationship between rice varieties, the evolution process and the differentiation of fertility genes, it is generally believed that *indica* rice evolved from wild rice, while *japonica* evolved from *indica* ; and that with the evolution process, cytoplasmic sterility genes gradually converted to fertility genes, while the nuclear restoring genes gradually converted to sterility genes. Research shows that no cultivated *japonica* rice varieties were found to have the restoring ability when BT CMS lines were testcrossed with existing cultivars. Hong Delin et al. (1985) used eight *japonica* CMS lines, including BT, Dian, L, ID and WA types, to testcross with 706 *japonica* rice varieties from the Taihu Lake region of China, 111 *japonica* varieties from Yunnan province and 187 *japonica* varieties from abroad. The results showed that most of the *japonica* varieties from China have no restoring ability, while only a few have weak restoring ability or partial restoring ability. A few high-stalk original varieties can restore Dian- and L-type male sterile lines with the seed setting rates higher than 70% . He concluded that there was no BT restoring gene in the existing cultivated *japonica* varieties. Meanwhile, in the process of testcrossing for restorer varieties, some Southeast Asian *indica* rice varieties such as IR8 and IR24 were found to have restoring ability to BT-type male sterile lines. This suggests that BT-type restorer genes are mainly distributed in *indica* varieties in low-latitude and low-altitude tropical and subtropical regions. However, because *indica* and *japonica* are two different subspecies, progenies of *indica-japonica* crossing cannot be used directly as restorer lines because of the large genetic difference between their parents, physiological incompatibility, and low hybrid fertility. Although very few original *indica* varieties and individual *javanica* varieties were found to have direct restoring ability to BT-type sterile lines during testcrossing, *indica* and *javanica* varieties flower early, while the *japonica* sterile lines flower late, and the flowering time was seriously asynchronous, so the seed production yield is low, and it is also difficult to apply to production. Therefore, the selection of testcrossing parents for BT CMS lines should focus on *japonica*-like varieties derived from *indica* and *japonica* crosses and related to IR8 and IR24.

Most restorer varieties for the HL type are distributed in temperate and subtropical regions. Generally, *indica* varieties in the Yangtze River Basin and South China have restoring ability for it. Therefore, the parents selected for testcrossing for the HL type should be *indica* varieties in these regions.

2. Methods for selecting parents for testcrossing

(1) Primary test

Typical individual plants are selected from the varieties (lines) that meet the breeding objectives and crossed with representative sterile lines. Generally, there should be more than 30 seeds for each crossing. F_1 hybrids and their parents should be planted next to each other, and dozens of hybrids and their male parents are planted as single plants. Major economic traits such as growth period are to be recorded, and the anther dehiscence and pollen plumpness of the hybrids examined at the heading stage. If the anther dehiscence is normal, the pollen is full, and the seed setting is good after maturity, it indicates that the variety has the ability to restore fertility. If the fertility of the hybrids or other traits such as growth period show segregation, it indicates that the variety is not genetically pure. As for whether such varieties should go into further test or be abandoned depends on the performance of the hybrids, if the hybrid vigor is obvious, and other economic traits are in line with the breeding objectives, more than one single plant can be selected to continue crossing in pairs until full stabilization. For example, researchers at Anjiang Agricultural School in Hunan province crossed male sterile lines with IR9761 - 19 - 1 introduced from IRRI. The hybrids showed segregation in growth period. Therefore, individual plants with different maturity stages were selected for further crossing and a number of early-maturing restorer lines, such as Ce 64 - 7, Ce 49 and Ce 48, were bred in succession.

(2) Re-testcross

The varieties found to have restoring ability in the primary test can be retested. More than 150 hybrid seeds and more than 100 plants are needed for retest, and control varieties need to be chosen. The growth period and other economic traits are recorded in detail, and a normal seed setting rate proves the restoring ability. The hybrids with good seed setting and strong heterosis should go into test for yield. After overall evaluation of the growth period, yield and other economic traits, varieties with no obvious heterosis, significantly lower yield than the control varieties and poor resistance should be abandoned. The varieties selected through re-testcross can be used for small-scale seed production before they are used for heterosis validation or plot yield trials in the next season.

(3) Effect of testcross screening and its evaluation

Testcrossing is one of the main methods to select restorer lines for hybrid rice from existing rice resources. In the early 1970s, after the successful breeding of the WA CMS lines, a number of varieties with restoring ability were selected from Southeast Asian rice varieties, such as IR24, IR26, IR661, Taiyin 1 and Gu 223, which soon completed the three-line sets. A number of hybrid rice combinations with strong heterosis, such as Nanyou 2, Nanyou 3, Shanyou 2, Shanyou 6 and Weiyou 6, were developed and widely used in production. In the mid-1980s, the restorer line Ce 64 -7 was bred through testcrossing, giving rise to a number of mid-maturing hybrid rice combinations with strong heterosis, such as Weiyou 64 and Shanyou 64. This enriched the portfolio of hybrid rice combinations and formed sets of mid- and late-maturing rice combinations among late-season hybrid rice in the Yangtze River Basin, which contributed to the overall development of hybrid rice. Then, early-maturing restorer lines such as Ce 49 and Ce 48 were bred by test crossing with IR9761 -19 -1 and developed a number of early-maturing hybrid rice lines such as Weiyou 49 and Weiyou 48, thereby expanding the planting of the double-cropping early-season hybrid rice in China from 25°N to the south of 30°N. In the middle and late 1980s, the restorer line Milyang 46 was selected by testcrossing, and a batch of double-cropping late-season hybrid rice combinations with a moderate maturity date, strong resistance, wide adaptability and obvious heterosis were bred and they rapidly replaced a number of conventional combinations such as Shanyou 6 and Weiyou 6 which had been in use for many years and had weak resistance to diseases and pests. In the 1990s, two-line hybrid rice was successfully cultivated in China, and the restorer lines of a group of hybrid rice combinations, such as Liangyoupeite (Peiai 64S/Teqing), Peizaishanqing (Peiai 64S/Shanqing 11), Xiangliangyou 68 (Xiang 125S/D 68), Liangyoupeijiu (Peiai 64S/9311), Y-Liangyou 1 (Y58S/9311), Fengliangyou 1 (Guangzhan 63S/9311) and Yangliangyou 6 (Guangzhan 63 - 4S/9311), were bred by testcrossing and widely used in production. Among these combinations, Liangyoupeijiu and Y-Liangyou 1 have become the main hybrid rice varieties with the largest annual growing area in China. The breeding of a number of mid-maturing early-season hybrid rice combinations with good quality and high yield, such as Xiangliangyou 68, has basically solved the long-standing problem of early maturity without high quality or high quality without early maturity during the breeding of double-cropping early-season *indica* hybrid combinations in the Yangtze River Basin. It is simple, efficient and remarkably effective to breed good restorer lines using existing rice resources. It will be one of the main methods to breed hybrid rice, especially the restorer lines of two-line hybrid rice.

II. Hybridization Breeding

1. Principle for selecting hybrid parents

Based on years of breeding practice and experience, the following principles should be followed in the selection of hybrid parents.

1) Appropriate plant and leaf morphology - high canopy, low panicle layer, medium and large panicles and strong lodging resistance with the upper three leaves long, erect, narrow, concave and thick to increase the population's efficiency of solar energy utilization to effectively increase the "source";

2) Big genetic difference and complementary traits - high yield, strong disease and pest resistance and good rice quality; or more superior traits than inferior ones, with complementarity; parents with distant genetic relationship preferred over those with close genetic relationship;

3) Intersubspecific heterosis utilization – varieties with wide compatibility, such as 02428, and Lunhui 422, are preferred as one of the parents in order to use the high heterosis of intersubspecific crosses.

4) High general combining ability – research show a higher frequency of superior individual plants in the offspring of hybrid combinations derived from parents with good general combining ability. Most of the good WA-type restorer lines currently in use in China are selected from combinations of Minghui 63, Ce 64 – 7, Milyang 46 and other varieties with good general combining ability.

5) High restoring ability – Genetic studies show that strong restorer lines are generally selected from crosses with a strong restorer line as at least one of the parents, and it is difficult to obtain a strong restorer line from a cross of weak restorer parents.

2. Methods of breeding restorer lines for two-line hybrid rice

Hybridization is one of the main methods for breeding restorer lines. PTGMS lines have a wide restoring spectrum and free-mating in terms of combinations. The method of breeding PTGMS restorer lines is the same as that of breeding conventional variety.

(1) Selection methods for hybridization breeding

At present, there are mainly two selection methods for hybridization breeding – system breeding (pedigree selection) and bulk-population selection.

1) In pedigree selection, F_2 is the generation of gene segregation and recombination. Generally, more than 5,000 plants are required to be planted with appropriately wider spacing than usual, and sound fertilizer and irrigation management. The selection criteria for F_2 single plants should not be too strict. The number of single plants to be selected should be based on the frequency of good single plants in the combination, more for combinations with a higher frequency and fewer, or even none for populations with a lower frequency. Generally, 30 – 50 individual plants should be selected from each combination and the selected F_2 plants will proceed to produce an F_3 family population composed of 50 – 100 single plants. The traits of F_3 plants are not yet stable so only qualitative traits with high heritability, such as growth duration and plant height, should be considered in the selection process, while quantitative traits controlled by multiple pairs of genes should be examined with appropriately relaxed criteria. Generally, 3 – 5 plants should be selected from each family and more can be selected from those families with outstanding performance. Selected F_3 plants will proceed to produce an F_4 population with each single F_3 plant forming a sub-family of 50 – 100 plants. In the F_4 population, qualitative traits controlled by a few genes tend to stabilize and plants should be selected according to the breeding objectives. Families or subfamilies with poor performance should be abandoned. Selected F_4 plants will continue to produce an F_5 population, composed of 100 plants in each family. Up to F_5, most traits have basically stabilized, so the restoring ability and heterosis can be evaluated. Individual plants should be selected from each family in strict accordance with the breeding objectives and testcrossed with sterile lines. Superior individual plants meeting the requirements of the breeding objectives are selected from the F_6 population according to the performance of F_1 single-crossing progenies, and those plants with weak heterosis or poor disease resistance shall be abandoned.

2) Bulk-population selection is based on the genetic laws of gene segregation, recombination and ho-

mozygosity in hybrid progenies. In bulk-population selection, mixed sowing, planting and harvesting are adopted, and selection is not made in the low generations, but only in higher generations when the traits are basically stable, i. e. when the genes controlling the traits of hybrid progenies are basically homozygous. The occurrence probability of homozygous genotypes is related to the generation and the number of gene pairs that control the traits. Therefore, the generation at which selection is to be carried out should be determined according to the number of genes that control the main traits involved in the breeding objectives. Yang Jike (1980) proposed that selection should only be carried out in the F_6 population when more than 80% homozygous genes for most traits appear in the population. In order to improve the selection efficiency of bulk-population selection, attention should be paid to the following points from F_2 to F_6 -1) evaluating the combining ability of the original parents as early as possible to ensure that good combinations are selected; 2) growing the population in a special cultivation environment, so as to eliminate naturally those individual plants that cannot adapt to the environment; 3) for those qualitative traits with high heritability, such as heading date and plant height, selection should be done early so as to eliminate the small number of individual plants that fail to meet the requirements of the breeding objectives; 4) selecting those single plants with particularly excellent traits in any generation. Bulk-population selection can greatly reduce the workload in the field compared to other methods that require selection work from a low generation, but it requires appropriately increased breeding population size and area, as well as a long breeding period.

(2) Comprehensive utilization of breeding technologies

Comprehensive use of a wide range of breeding methods on the basis of hybridization breeding is the main approach to breeding restorer lines with strong heterosis. HHRRC has bred a great number of restorer lines with excellent comprehensive agronomic traits and outstanding other traits by such methods as hybridization of geographically distant varieties, intervarietal and intersubspecific hybridization, backcrossing, distant hybridization assisted by molecular technology, total DNA introduction and testcrossing. For example, the restorer lines Xianhui 207, Xianghui 111, Xianghui 227 and Xianghui 299 were bred by *indica-japonica*, *indica-javanica* and *japonica-javanica* intersubspecific hybridization; 0293 and 0389 were bred by intervarietal hybridization and testcrossing; and Yuanhui 2, Yuanhui 611 and R163 were bred by distant hybridization plus molecular marker-assisted breeding. These strong restorer lines have been used to form hybrid rice combinations with high quality and multiple resistance, super hybrid rice combinations or potential high-yielding hybrids, and some of them have been widely used in production.

R163, a strong restorer line, was bred by using the super hybrid rice restorer line 9311 as the recipient and recurrent parent and introducing yield-increasing QTLs from Malaysian common wild rice through molecular marker-assisted selection (MAS). On the basis of previous selection and experience of traditional breeding, single plants with excellent agronomic traits in comparison with the recurrent parent were selected for backcrossing, and genetic background comparison was not carried out until BC_4F_4 and BC_6F_3 to analyze how much the genetic background of the introgression lines have restored to recurrent parents. BC_6F_3 lines having the high-yielding QTLs *yld1. 1* and *yld2. 1* and a high degree of genetic background restoration were selected. Analysis of the yield structure showed that the yield of the BC_6F_3 lines was higher than that of the recipient 9311 due to increased number of effective panicles per plant, a higher seed setting rate and increased 1,000-grain weight. Lines with uniform performance and excellent overall agronomic traits were selected to testcross with Y58S. R163, a new restorer line with significant yield increase and high combining ability, was selected based on evaluation of hybrids to identify those showing high heterosis and excellent overall traits. Y-Liangyou 7, a combination derived from R163, was approved by Hunan provincial authorities in 2008 and recognized as a mid-season super hybrid rice variety. Yuanhui 2, a strong restorer line bred by using a good individual plants selected from the early-generation (BC_3F_1) progenies of R163 as female and Shuhui 527 as the male parent, then selecting excellent individuals from the F_2 population to go through five generations of selfing. It was combined with Y58S to breed Y-Liangyou 2, a representative hybrid variety for attainment of the third phase goal of China's super hybrid rice breeding project. On-site yield evaluation by the MOA showed that, in 2011, the average yield per hectare of Y-Liangyou 2 reached 13.9 t/ha, and achieving the objective of the third phase of China's super rice breeding project ahead of schedule.

One of the most effective ways to breed strong super hybrid rice restorer lines is to use lines with wide compatibility as a bridge to introduce some *japonica* relatives into the genetic background of *indica* rice and construct an *indica-japonica* intermediate material bank, so as to realize high-level utilization of *indica-japonica* intersubspecific heterosis. In August 2004, a *japonica* variety with wide compatibility, 02428, was crossed with E32 and the F_1 plants were then crossed with Xianhui 207 and Huanhui 422 successively. In 2007, excellent individual plants were selected from the F_3 population of the three-way crossing and crossed again with Yangdao 6. After four years and eight generations of pedigree selection, an excellent and stable line was obtained. This stable line was used as the male parent for testcrossing with Peiai 64S, Y58S and other good male sterile lines. The combinations showed great performance, and the male parent, named R900 (Fig. 6 - 1), was combined with Y58S to breed Y-Liangyou 900, a pioneer super hybrid rice variety of the fourth phase of China's super rice breeding project. It was also crossed with Guangxiang 24S to breed Xiangliangyou 900 (Chaoyouqian), a super hybrid rice combination of the fifth phase of the project, with a yield of more than 16 t/ha.

3. Methods to breed restorer lines for three-line hybrid rice

CMS lines are restricted by the relationship between the restorer and the maintainer, so the CMS parents do not have much freedom in combination, and restorer line breeding is quite complex. There are the one-time hybridization methods and the composite hybridization method.

(1) One-time hybridization

1) Sterile line (A) and restorer line (R)

This is a simple method to breed restorer lines from A/R combinations. Starting from F_2, fertile plants with excellent agronomic traits are selected, and most plants in the F_4 or F_5 population should have stable fertility and normal seed setting. Individual plants with homozygous restorer genotype can be selected through testcrossing and used to breed new restorer lines, which are known as iso-plasmic restorer lines because their cytoplasm comes from the same male sterile line. By adopting this method, Guangxi Academy of Agricultural Sciences (GAAC) succeeded in breeding Tonghui 601, 616, 621 and 613, and

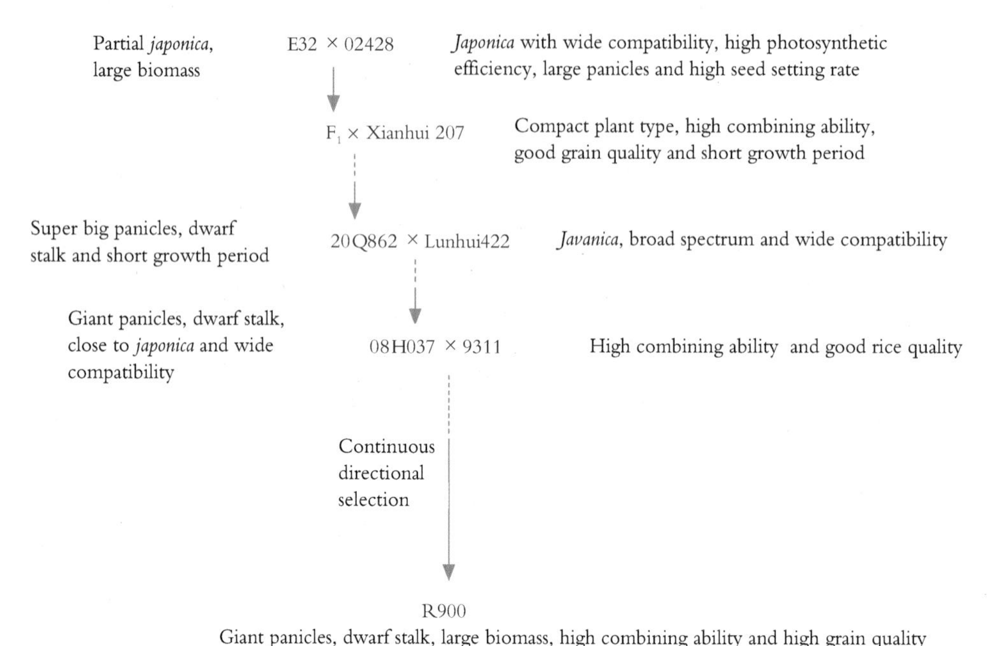

Fig. 6 - 1 Pedigree of *indica-japonica* intermediate type restorer line R900

HHRRC succeeded in breeding long-grain Tonghui and short-grain Tonghui lines. However, it should be noted that, according to years of breeding practice, when the iso-plasmic restorer lines are crossed with sterile lines, their progenies would have no obvious heterosis because of the small genetic difference between the parents. Generally, the above methods are no longer in use today for restorer line breeding.

2) Restorer line (R) and restorer line (R)

R/R crosses aim to combine good traits of two restorer lines, or to improve a certain trait of one parent, such as growth period or resistance. Both of the two parents in this combination have restorer genes. Although there is gene recombination and segregation, the genotype of the plants does not change in terms of fertility restoration. That is to say, in this hybrid combination, every single plant in the population of each generation has the restoring ability. Therefore, when a hybrid combination is used to breed restorer lines, there is no need to carry out testcrossing in early generations. After the main traits of each individual plant are basically stable, the primary test and retest can be carried out, so as to select single plants with strong restoring ability, superior traits and significant heterosis for the breeding of new restorer lines. The restorer lines bred by this method mainly include Minghui 63 and Minghui 77 bred by Sanming Institute of Agricultural Sciences (SIAS), Fujian province, Gui 33 bred by Guangxi Academy of Agricultural Sciences, Wan 3 bred by HHRRC, and Changhui 121 bred by Jiangxi Agricultural University.

3) Restorer line (R) and maintainer line (B) or maintainer line (B) and restorer line (R)

R/B or B/R is one of the most commonly used methods to breed restorer lines. The restorer bred lines through this method mainly include Xianhui 207 bred by HHRRC, R198 bred by Hunan Agricultural University, and Zhenhui 129 bred by Zhenjiang Institute of Agricultural Sciences, Jiangsu province. In the selection of R/B or B/R hybridization populations, F_2 and later-generations will have plants with different restorer genotypes because only one parent has the restorer gene. The number of plants with homozygous restorer genotypes will increase in later generations as selfing continues. For example, when a WA-type restorer line is crossed with a maintainer line, individual plants with homozygous restorer genotypes will account for about 6.25% of the F_2 population, 14.06% of the F_3 population and 19.14% of the F_4 population. However, these individual plants cannot be distinguished from individual plants with other genotypes in morphology and genotypes of fertility can only be identified through testcrossing.

In order to breed individual plants with homozygous restorer genes through this method as soon as possible, Wang Sanliang (1981) proposed an early-generation testcrossing method on the basis of the inheritance laws of restorer genes in rice. The details are as follows.

① Determination of the generation for testcrossing. There will be individual plants with homozygous restorer genotypes in the F_2 population no matter how many gene pairs control this trait, and the more gene pairs involved, the lower the probability of getting individual plants with homozygous restorer genotypes. If we are to ensure a 99% or 95% probability of getting at least one individual plant with homozygous restorer genotypes in each generation, the number of individual plants to be tested in each generation should be determined using the formula $n \geqslant \lg \alpha / \lg P$, where n stands for the number of individual plants to be testcrossed, P refers to the probability of getting plants with other genotypes, and α denotes the probability of an allowable miss. For example, supposing the trait of fertility restoration is controlled by two pairs of genes and we want a 99% probability of getting at least one individual plant with homozygous restoring genotype, with testcrossing starting from F_2. Then, according to Table 6-1, the probability of getting other genotypes is 93.75% (P=0.9375), and the allowable miss (α) is 1%. Using the above formula, we know that n=71, i.e. we need to testcross 71 plants. Similarly, the number of individual plants, systems or system groups to be testcrossed in F_3 and later generations can be calculated. It can also be seen from the table that the earlier the testcrossing is done, the heavier the workload of testcrossing, but the lighter the workload of field planting. Therefore, early-generation testcrossing is easier.

Table 6-1 Probability (%) of homozygous restorer genotype and other genotypes in F_1 selfing progenies of individual plants with n pairs of heterozygous genotypes

Generation	n=1		n=2		n=3	
	Homozygous restorer genotype	Other genotypes	Homozygous restorer genotype	Other genotypes	Homozygous restorer genotype	Other genotypes
F_1	0	100	0	100	0	100
F_2	25.00	75.00	6.25	93.75	1.56	98.44
F_3	37.50	62.50	14.06	85.94	5.27	94.73
F_4	43.75	56.25	19.14	80.86	8.37	91.63
F_5	46.88	43.12	21.97	78.03	10.30	89.70
F_6	48.44	51.56	23.46	76.54	11.36	88.64

② Determination of the minimum number of F_2 plants to be testcrossed. According to the genetic studies on rice restorer genes, fertility restoration is controlled by one (*japonica*) or two (*indica*) pairs of genes, which is a quality trait, and testcrossing from F_2 is more suitable. A total of 16 single plants should be testcrossed for the selection of a *japonica* restorer line, while 71 is needed for the selection of an *indica* restorer line.

③ Determination of the number of each plant's progenies to be planted. In order to determine the genotypes of the testcrossed plants, according to the formula $n \geq \lg \alpha / \lg P$, where n stands for the number of plants to be planted, P is the probability of R_1r_1 or R_1R_2 heterozygous genotype, and α denotes the allowable miss, for F_2, $P=0.5$, the intended probability of getting a plant with a homozygous genotype is 99.9%, and $\alpha=0.001$. Putting these into the formula, we obtain $n=10$, i. e. at least 10 plants should be planted in the progenies of each tested plant. If fertility is restored in all 10 plants, the genotype of the tested plant is homozygous, i. e. $F(R_1Rr_1)$ (*japonica*) or $F(R_1R_1R_2R_2)$ (*indica*). If some of the 10 plants are fertile, some partial fertile, and some sterile, the genotype of the tested plant is $F(R_1r_1)$ or $F(R_1{-}R_2{-})$, i. e. heterozygous. If some of the 10 plants are partially fertile and some sterile, the genotype of the tested plant is $S(R_1{-}r_2r_2)$ or $S(r_1r_1R_2{-})$, i. e. heterozygous. If all the 10 plants are sterile, the genotype of the tested plant is $F(r_1r_1)$ or $F(r_1r_1r_2r_2)$, i. e. homozygous maintainer genotype.

④ Treatment of hybrid progenies. F_2 is the generation of gene recombination and segregation. Therefore, it is necessary to expand the F_2 population as much as possible according to breeding conditions and select superior individual plants for testcrossing. Each F_2 testcrossed individual will proceed to produce an F_3 population, forming a family of 50 - 100 plants each. Generally, one to four or more families will be found to have homozygous restorer genotypes and individual plants can be selected from these families. Because each family has a very small population, the selection criteria should not be too strict, especially for quantitative traits controlled by multiple gene pairs. It is even possible that none of the plants is selected. Selected F_3 plants will proceed to F_4, and each single plant will form a sub-family, which must have more than 100 plants, from which superior plants meeting the requirements of the breeding objectives are to be selected to produce F_5. When most traits become basically stable in F_5, retesting and combining ability evaluation can be carried out.

In addition, for the B/R or R/B populations, both pedigree selection and bulk-population selection can be used. However, if pedigree selection is used, the selected F_2 plants have to be more than enough, and also, more plants need to be selected from F_3 and F_4, in multiple families, in case of a loss of the fertility restoring genes in a small selection population which may result in failure to achieve the breeding objectives.

(2) Composite crossing (multiple crossing)

Composite crossing is usually adopted to transfer the superior traits from more than two parents into one variety. Adopting this method, Sichuan Agricultural University used the single recessive nuclear sterile material MS and Minghui 63, Milyang 46 and other restorer lines as the base material to establish a recurrent population, from which good single plants were selected and pedigree-selected for four generations to produce Shuhui 498, which was combined with Jiangyu F32A, to produce super hybrid rice F-You 498. Hunan Agriculture University bred the late-maturing restorer line R518 through pedigree selection for multiple generations in 2005 from the progenies of 9113/Minghui 63//Shuhui 527, and then crossed it with H28A to develop H-You 518. After four years and eight generations of systematic selection, resistance and quality verification, and testing for restoring ability and superiority, Guanghui 308 was bred in 2001 by the Rice Research Institute of Guangdong Academy of Agricultural Sciences, using Guanghui 122, a widely used restorer line, as the male parent, and Chaoliuzhan/Sanhezhan, a high-quality disease-resistant intermediate material, as the female parent, and the super rice Wuyou 308 was bred with Wufeng A in 2001. CRRI used C57 (Liaoning BT *japonica* restorer line) as the female parent and (300 × IR26) F_1 as the male parent, and then bred the restorer line Zhonghui 9308 by pedigree selection for four generations, and paired it with Xieqingzao A to produce the super hybrid rice Xieyou 9308. The dried seeds of Mingchuai 86 after space-borne by a returnable satellite were selected by Fujian Academy of Agricultural Sciences to plant, and single superior plants was used as the mother plant and crossed with Tainong 67, the F_2 *indica*-typical single plant was then used as the female parent and crossed with N175. After five generations of selfing and selection, the restorer line Fuhui 673 was developed (Fig. 6-2), which was combined with Yixiang 1A to develop the super rice Yiyou 673.

The number of parents used in composite crossing is large, and there are both restorer and maintainer lines in the parents, which results in the diversity of mating methods and complexity of the relationships of restoring genes. The following is a brief description of the genetic behavior and selection methods of some common combinations, $(R/R)F_1xR$, $(R/B)F_1xR$, $(R/B)F_1x(R/B)F_1$, $(R/B)F_1xB$.

1) $(R/R)F_1 \times R$ or $(R/R)F_1 \times (R/R)F_1$

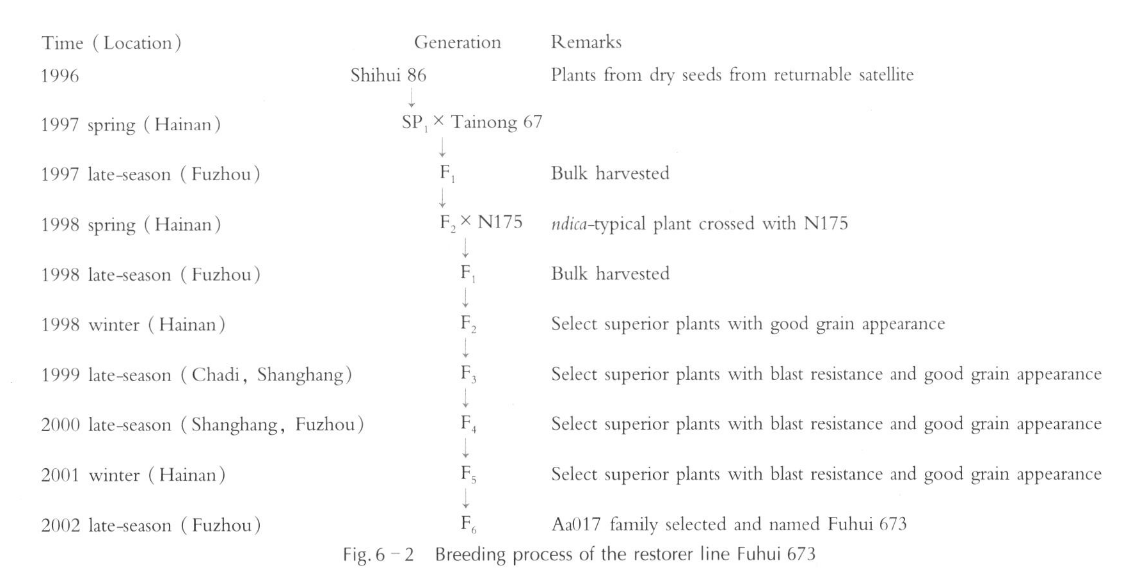

Fig. 6-2 Breeding process of the restorer line Fuhui 673

In such a combination, since each parent involved in the cross is a restorer line with the same genotype, i. e. the $F(R_1R_1R_2R_2)$ homozygous restorer genotype, the hybrid of the first cross (F_1) and the hybrid of the second cross (F_1) contain the same homozygous restorer genotype $F(R_1R_1R_2R_2)$. After the F_1 self-crossing of the second cross, no further segregation occurred in the F_2 population with respect to restorer genes, and all single plants in the F_2 and subsequent generations are of the $F(R_1R_1R_2R_2)$ homozygous restorer genotype. Therefore, in the selection of restorer lines for such combinations, it is not necessary to consider the restoring ability of each single plant, instead, attention should be paid to other traits.

2) (R/B) F_1 ×R

When selecting restorer lines with this combination, since one of the parents is a restorer line with the genotype $F(R_1R_1R_2R_2)$ and the other parent is a maintainer lines with the genotype $F(r_1r_1r_2r_2)$ in the first cross, the hybrid (F_1) genotype is $F(R_1r_1R_2r_2)$, a heterozygous genotype.

This heterozygous genotype produces four male and female gametes, R_1R_2, R_1r_2, r_1R_2 and r_1r_2, respectively, and a second cross is made with this genotype as the female and the restorer line as the male parent. The restorer line produces only one male gamete, R_1R_2. If all four gametes have an equal probability of receiving pollen and the same fertilizing ability, then four genotypic single plants, $F(R_1R_1R_2R_2)$, $F(R_1R_1R_2r_2)$, $F(R_1r_1R_2R_2)$, and $F(R_1r_1R_2r_2)$, will appear in the hybrids (F_1) of the second cross, among which the plants with the $F(R_1R_1R_2R_2)$ homozygous restorer genotype account for 39.06% of the total population. Then, they will account for 47.26% of the F_3, 51.66% of the F_4, and 53.93% of the F_5 population, gradually approaching 56.25% with more generations of selfing. It can be seen that in this type of combination, the frequency of homozygous restorer genotypes in all generations of hybrids is relatively high, and no matter at which generation is the selection done for testcrossing with sterile lines, homozygous restorer genotypes can be selected to breed new restorer lines. However, in breeding practice, the situation is much more complicated, mainly in the second cross, which involves artificial emasculation, and the number of hybrid seeds produced by artificial emasculation is limited. In this case, the four female gametes are unlikely to have equal chances of receiving pollen and getting fertilized, and there will not be an equal number of single plants with the four genotypes in the F_1 population. Thus, it is difficult to calculate the probability of occurrence of the homozygous restorer genotypes in F_2 and subsequent generations. However, one thing is for sure, when the r_1r_2 gametes in the female parent are combined with the R_1R_2 gametes in the male parent, the genotype of the hybrid F_1 is $F(R_1r_1R_2r_2)$. This genotypic monoculture still segregated 6.25% of the homozygous restorer genotypic single plants in F_2, not to mention that the r_1r_2 gametes accounted for only 25% of the total number of gametes, while the other three gametes also have a 25% chance of receiving pollen. Therefore, it is unlikely that the hybrids (F_1) will be exclusively the $F(R_1r_1R_2r_2)$ genotype, and the probability of homozygous restorer genotypes in F_2 must be greater than 6.25%. The probability of having a homozygous restorer genotype in F_2 must be greater than 6.25%. As long as the cross is tested with a sterile line, a homozygous restorer genotype can be selected from it.

3) (R/B) F_1 ×(R/B) F_1

In this hybrid combination, after the first cross, which is always R/B, F_1 plants are of the $F(R_1r_1R_2r_2)$ heterozygous genotype, and they produce four kinds of gametes, R_2R_2, R_1r_2, r_1R_2 and

r_1r_2. When two F_1 plants are crossed and if all gametes have an equal chance of pollination and fertilization, there will be individual plants with multiple genotypes after the second cross, with 6.25% of the population having the F($R_1R_1R_2R_2$) homozygous restorer genotype. The F_2 population will segregate and the plants with the F($R_1R_1R_2R_2$) restorer genotype will account for 14.06% of the total. The frequency will be 19.14% in F_3 and 21.97% in F_4. As selfing continues to later generations, the individual plants of homozygous restorer genotype will gradually account for 25% of the total population.

These are the oretical data obtained under the condition that both male and female gametes have an equal chance of pollination and fertilization. However, in breeding practice, due to the influence of artificial emasculation, the number of hybrid seeds produced is small, and the probability of pollination and fertilization of male and female gametes is unequal. The F_1 of the second cross may have one or more of multiple genotypes, while the segregation at F_2 is determined by the genotype of single F_1 plants. Therefore, the selection of restorer lines in this type of crosses is complicated, and the best way is to start testcrossing from F_2. If all the progenies of the testcrosses are restorer lines, it indicates that the tested single plant has the F($R_1R_1R_2R_2$) homozygous restorer genotype. If the fertility of the progenies is segregated and there are fertile, partial fertile and sterile ones, the tested single plant has the S(R_1-R_2-) heterozygous genotype. The restorer genes of these genotypes are not homozygous, and it is necessary to continue to select single plants for testcrossing until there is no fertility segregation in the progenies. If the progenies of the testcross are partially or completely sterile, it means that the tested single plant may have the S($R_1-r_1r_2$), S($r_1r_1R_2-$), or S($r_1r_1r_2r_2$) genotype and these individual plants have weak or no restoring ability and should be abandoned as soon as possible.

4) (R/B)F_1 × B

Breeding *indica* restorer lines through this combination involves more work more difficulties and some risks. The F_1 of the first cross has the F($R_1r_1R_2r_2$) heterozygous genotype, and four kinds of gametes will be produced, while the male parent in the second cross is a maintainer line and only one kind of gamete will be produced. Under the condition that gametes have equal ability to receive pollen and get fertilized, the F_1 of the second cross will have four genotypes, only one ($R_1r_1R_2r_2$) of which can segregate the genotype F($R_1R_1R_2R_2$), the homozygous restorer genotype in the F_2 population, while the other three genotypes are not restorer genotypes. According to the genetic law of restorer genes in rice, plants with the homozygous restorer genotype only account for 1.56% of the total F_2, 3.51% of the F_3 and 4.7% of the F_4 population, gradually approaching 6.25% as selfing continues. It can be seen that in this combination, the frequency of the homozygous restorer genotype in each generation of hybrids is very low, which makes the selection difficult to complete, and it is even more risky when artificial emasculation is carried out for the second cross, which cannot ensure equal opportunity for the female gamete to receive pollen and get fertilized. It is thus difficult to assess whether the genotype F($R_1r_1R_2r_2$) can appear in the hybrids (F_1). Therefore, to select restorer lines in this way, it is necessary to start testcrossing in F_2 or F_3, and to test as many single plants as possible so as to determine whether there is a single plant of the genotype F($R_1r_1R_2r_2$) in F_2 or F_3 based on the fertility performance of the progenies. If all the test progenies are partially sterile or sterile, it indicates that the gametes R_1R_2 is not combined in the second cross and there is no $F(R_1R_1R_2\ R_2)$ homozygous restoring genotype, so all the plants should be abandoned. If several progenies have fully restored fertility, or some progenies are fertile while others are partially fertile or sterile, it means that there are $F(R_1r_1R_2\ r_2)$ plants in the F_2 or F_3 population. In this case, parents of the fully restored plants or plants with segregated fertility should be selected for further testcrossing until full fertility restoration in the progenies.

However, in the selection of *japonica* restorer lines, because the restorer genes of *japonica* rice originate from *indica*, a different subspecies from *japonica*, resulting in incompatibility, *indica* rice with restorer genes cannot be directly used as restorer lines. In order to introduce the *indica* restorer genes into *japonica* varieties, solving the problem of incompatibility and accelerating the stability of hybrid progenies, the restorer line selection is generally carried out by the mating method of (R/B) F_1/B. At the same time, there is only one pair of restorer genes in *japonica* restorer lines. According to the inheritance pattern of restorer genes, the frequency of single plants with a homozygous restorer genotype at each generation of this combination is relatively high, 12.5% for F_2, 16.75% for F_3, 22.88% for F_4, and gradually approaching 25.0% with the increase of selfing generations. Therefore, the use of this mating method is not only conductive to the stability of each trait in the hybrids, but also easy to screen out homozygous restorer genotypes by testcrossing and breed new restorer lines. For example, the C57 restorer line developed by the Institute of Rice Crop Science, Liaoning Academy of Agricultural Sciences, is a cross of IR8, which has good yield, restorer genes and semi-dwarf genes, and Keqing 3, which was used as the male parent, and then the F_1 was crossed with Jingyin 35.

Ⅲ. Backcross Breeding (Directional Breeding)

In the process of breeding restorer lines by testcrossing, some varieties with several good traits in plant and leaf morphology, resistance, grain quality and high yield potential, but no restoring ability are usually found to be used as restorer lines in three-line hybrid. In order to keep the good traits of these varieties and give them restoring ability, multiple backcrossing is generally adopted to transfer the restorer genes into these varieties. The procedures are as follows – using the F_1 of a cross of A/R as the female parent and the maintainer line (hereinafter referred to as "Variety A") as the male parent for crossing. The genotype of the female parent is $S(R_1r_1R_2\ r_2)$, producing four kinds of female gametes; while the genotype of the male parent is $S(r_1r_{1r2}r_2)$, producing only one kind of male gamete. After crossing, there will be individual plants with four genotypes in the F_1 population. Among them, only individual plants with the $S(R_1r_1R_2\ r_2)$ genotype will show normal fertility, while those with other genotypes will all show partial or complete sterility. That is to say, all plants with restorer genes are fertile, while those with one or no restorer gene are partially fertile or completely sterile. Therefore, the plants with normal fertility in F_1 are selected to backcross with Variety A for the first time. Similarly, in BC_1, only those plants with restorer genes are fertile, which are selected to backcross with Variety A for the second time. In this way, individual plants with stable traits and fertility as well as normal seed setting are selected after backcrossing for three or four times and selfing for once or twice and used for testcrossing with the sterile lines. The progenies that have completely restored fertility are selected for the breeding of a homotypic restorer line of

Variety A.

The restorer line of multiple-backcrosses, except for a few traits that the restorer gene is linked to its restoring ability, the rest of the traits are from Variety A. There is considerable difficulty in selecting *indica*-type three-line restorer lines by using this method.

1) In each cross or backcross, as many hybrid seeds as possible should be produced to expand the population size of F_1 and its backcross generations (BC_1, BC_2,...).

2) When finding no plants with normal fertility in the population of F_1 or its backcross generations (BC_1, BC_2, ...), the breeding work should be immediately terminated and restart hybridization.

Molecular marker assisted selection can improve the efficiency of backcrossing.

By combining conventional backcross breeding with MAS techniques, several identified restorer genes, such as *Rf3*, *Rf4*, *Rf5*, *Rf6*, etc., can be tracked for introduction to significantly improve the degree of restoration. At the same time, the resistance of restorer lines to diseases and pests can be improved by tracking resistance genes. Zhang Honggen et al. (2018) crossed R1093, a line carrying *Rf6*, with C418, a BT-type restorer line carrying *Rf1*, thereby introducing *Rf6* into C418 for integration breeding of *Rf6* and *Rf1*, and obtained six improved lines whose agronomic traits were basically close to those of C418. The testcross results showed that the restoration degree of HL-type CMS lines by the improved line carrying *Rf6* was more than 85%, good for rice production. Therefore, the introduction of *Rf6* is an important way to breed HL-type restorer lines because it can effectively improve the restoring ability of BT-type restorer lines to HL-type CMS lines.

Ⅳ. Mutation Breeding

Mutation breeding makes use of physical or chemical factors to artificially induce genetic variation in crops, obtaining usable mutants within a short period of time, and then selects mutants according to the requirements of the breeding objectives and breed directly or indirectly new varieties to be used in production. Mutation breeding has played an important role in breeding new varieties and creating new germplasm, especially in rice breeding where mutagenesis has been most prominent. Mutation breeding includes physical and chemical mutagenesis. At present, physical mutagenesis is more frequently used in breeding, among which radiation mutagenesis breeding and space-induced mutation breeding is the most fruitful. Radiation mutation uses χ, γ, α and β rays, as well as neutron and ultraviolet beams to treat organisms so as to create new variants. Space mutation breeding uses the space environment (high vacuum, microgravity, strong radiation, etc.) that can be accessed by space vehicles (returnable satellites, spaceships, high-altitude balloons, etc.) to induce genetic variation in plant seeds, and the mutagenic offspring is will then go through ground screening to breed new germplasm, new materials and new varieties. Using a comprehensive breeding method combining radiation mutagenesis, hybridization breeding, and adverse temperature screening, the Sichuan Institute of Atomic Energy successfully bred 12 restorer lines with strong restoring ability, combining ability and resistance, including Fuhui 838 and its derivative restorer lines Fuhui 718, Fuhui 305, Zhonghui 218, Mianhui 3728 and Nuohui 1. A total of 43 hybrid combinations where these restorer lines are combined with WA, Gang, D and ID CMS lines, passed national or provincial crop variety evaluation. These varieties show high seed setting rates, tolerance to low temperature and heat damage, wide adaptability, and stable high yield in large-scale planting, and their planting area has totaled over 40 million hectares. In particular, Ⅱ-You 838 is a well-known hybrid variety which has been in use in production for a long time with a large total planting area second only to Shanyou 63. In 2005, it became a control combination used in China's regional yield trials for *indica* varieties, and it is also a hybrid rice variety exported to Vietnam and other Southeast Asian countries several years ago. An analysis of 245 new rice varieties produced through direct or indirect radiation in Zhejiang province, showed that 89.9% of them originated from Funong 709 and Zhefu 802 and 81.8% of the Yongyou series of *indica-japonica* hybrid male sterile lines originated from Funong 709.

The Rice Research Institute of Fujian Academy of Agricultural Sciences used space breeding technology to treat the dry seeds of the restorer line Minghui 86 with high-altitude radiation, and succeeded in breeding the restorer line Hang 1 through selection over multiple generations and multi-location resistance evaluation under different ecological conditions, and combined it with Ⅱ-32A and Longpute A to develop super hybrid rice Ⅱ-Youhang 1 and Teyouhang 1. After that, the dry seeds of Minghui 86, which had undergone high-altitude radiation mutation aboard a satellite, were planted under different ecological conditions in Fujian province, Sanya city of Hainan province and other locations, and, by adopting shuttle breeding and directional breeding, the restorer line Hang 2 was developed, whose overall traits were better than those of Minghui 86, and was combined with Ⅱ-32A to develop super hybrid rice Ⅱ-Youhang 2. The Super Rice Research and Development Center of Jiangxi province and other institutions used the SP3 space mutants of large-panicle and large-grain strong restorer line Kehui 752 as the male parent and the high-quality restorer line R225 as the female parent to breed the high-quality strong restorer line Yuehui 1573 in 2010 after four years and eight generations of conventional breeding and space mutation. It was then combined with Wufeng A to breed super hybrid rice Wuyouhang 1573. There are three methods to breed restorer lines by radiation-induced mutation.

1) Existing good restorer lines are treated with radiation to induce mutation and select variants for the breeding of new restorer lines. For example, through radiation treatment of the restorer line IR36, Wenzhou Agricultural Science Research Institute of Zhejiang province succeeded in breeding 36-Fu, a restorer line that matures earlier than IR36. It was then used to develop the mid-maturing late-season hybrid Shanyou 36 Fu for the Yangtze River Basin. The Sichuan Institute of Atomic Nuclear Application Technology used ^{60}Co-γ to treat Taiyin 1 and succeeded in breeding Fu 06, a new restorer line maturing 20 days earlier.

2) Breed new restorer lines through radiation treatment of hybrid progenies to induce mutations and select mutants from their progenies. For example, HHRRC treated the F_1 hybrids of Minghui 63 and 26 Zhaizao with ^{60}Co-γ and bred the restorer line Wan 3, and used it to develop a batch of mid-maturing late-season *indica* hybrids, including Shanyouwan 3 and Weiyouwan 3 for planting on a large scale in the Yangtze River Basin. The inoculation of anthers with ^{60}Co-γ rays after cold treatment of the main panicles and the panicles from high-internode tillers of the F_1 hybrids of Mingchuai 63/Zigui was carried out by Zhang Zhixiong et al. (1995) to obtain a certain number of diploid plants, among which, a new re-

storer line, Chuanhui 802, was bred and it has a good plant type, strong tilling ability, larger panicles and more grains, high restoring ability and strong combining ability, and was used to develop Ⅱ – You 802 and other hybrids.

3) Radiation treatment of existing rice varieties or hybrid progenies, from which mutants are selected as hybrid parents. Wu Maoli et al. (2000) selected the partial *japonica* D091 restorer line from the *indica-japonica* cross 02428///(Gui 630/Guichao 2)//IR8γ/IR1529 – 680 – 3γ with good plant type, large panicles, strong restoring ability, good combining ability, moderate resistance to diseases and pests, and sufficient pollen, and developed the hybrid combination Nuoyou 2. (Gui630/Guichao 2)γ, IR8γ and IR1529 – 680 – 3γ are all mutants selected by ^{60}Co – γ radiation. In addition, Xu Yungui from the Department of Biology, Wuhan University, used laser treatment on Guangluai 4 to select the Jiguang 4 restorer line from the mutants and from a hybrid combination for trial planting.

References

[1] CHEN LETIAN, LIU YAOGUANG. Discovery, utilization and molecular mechanism of WA cytoplasmic male sterility in rice[J]. Chinese Science Bulletin, 2016 (35): 3804 – 3812.

[2] DENG DASHENG, CHEN HAO, DENG WENMIN, et al. Breeding and application of rice restorer line Fuhui 838 and its derivative lines[J]. Journal of Nuclear Agricultural Sciences, 2009, 23 (2): 175 – 179.

[3] HUANG WENCHAO, HU JUN, ZHU RENSHAN, et al. Research and development of HL type hybrid rice[J]. Science China: Life Science, 2012 (9): 689 – 698.

[4] LU YANTING, CHEN JINYUE, ZHANG XIAOMING, et al. Research progress of radiation rice breeding in Zhejiang province[J]. Journal of Nuclear Agricultural Sciences, 2017, 31 (8): 1500 – 1508.

[5] MAO XINYU, WANG SHUSEN, ZHANG HONGHUA, et al. Breeding and application of Shanyou 36 Fu[J]. Hybrid Rice, 1989 (5): 33 – 35.

[6] REN GUANGJUN, YAN LONGAN, XIE HUAAN. Review and prospect of three-line hybrid rice breeding[J]. Chinese Science Bulletin, 2016 (35): 3748 – 3760.

[7] WU JUN, DENG QIYUN, YUAN DINGYANG, et al. Research progress of super hybrid rice[J]. Chinese Science Bulletin, 2016 (35): 65 – 7.

[8] WU JUN, ZHUANG WEN, XIONG YUEDONG, et al. Breeding new hybrid rice combination Y Liangyou 7 with good quality and high yield by introducing yield-increasing QTLs of wild rice[J]. Hybrid Rice, 2010, 25 (4).

[9] WU JUN, DENG QIYUN, ZHUANG WEN, et al. Breeding and application of super hybrid rice pioneer combination Y Liangyou 2 for the third phase[J]. Hybrid Rice, 2015, 30 (2): 14 – 16.

[10] WU MAOLI, LIU YUSHENG, YANG CHENGMING, et al. Breeding and application of *indica-japonica* line D091 [J]. Hybrid Rice, 2000 (S1): 3, 21.

[11] YOU QINGRU, ZHENG JIATUAN, YANG DONG. Breeding and application of new mid-season rice combination Chuanyou 673[J]. Hybrid Rice, 2011, 26 (5): 18 – 21.

[12] YUAN LONGPING. Hybrid rice[M]. Beijing: China Agricultural Press, 2002.

[13] ZHANG HONGGEN, ZHONG CHONGYUAN, SI HUA, et al. Restoring ability of MAS-improved C418 to HL CMS *japonica* rice lines[J]. Chinese Journal of Rice Science, 2018, 32 (5): 445 – 452.

[14] ZHANG QUNYU, LIU YAOGUANG, ZHANG GUIQUAN, et al. Molecular marker mapping of WA CMS restoring gene *Rf4*. Journal of Genetics and Genomics, 2002, 29:1001 – 1004.

[15] HONG DELIN, TANG YUGENG. Studies on *japonica* rice male sterility restoring genes Ⅰ. Geographical distribution of *japonica* rice male sterility restoring genes[J]. Jiangsu Journal of Agricultural Sciences, 1985, 1 (4): 1 – 5.

[16] YANG JIKE. Theoretical basis of quantitative heredity of rice group breeding[J]. Heredity, 1980, 2 (4): 38 – 41.

[17] ZHANG ZHIXIONG, ZHANG ANZHONG, XIANG YUEWU, et al. New hybrid rice combination Ⅱ – You 802 developed by anther culture of pure lines[J]. Hybrid Rice, 1995 (6): 36.

[18] XU YUNGUI. V20A × Jiguang 4: a new hybrid rice combination[J]. Hubei Agricultural Sciences, 1983 (2): 11 – 13.

[19] HU J, HUANG W C, HUANG Q, et al. Mechanism of *OrfH79* inhibition with artificial restorer gene *Mt-GRP162* [J]. New Phytol, 2013, 199: 52 – 58.

[20] HU J, WANG K, HUANG W, et al. The rice pentatricopeptide repeat protein *Rf5* restores fertility in HL CMS lines via a complex with the glycine-rich protein *GRP162*[J]. Plant Cell, 2012, 24: 109 – 122.

[21] HU JUN, HUANG WENCHAO, HUANG QI, et al. Mitochondria and cytoplasmic male sterility in plants [J]. Mitochondrion, 2014, 19 Pt B: 282 – 288.

[22] HUANG W, YU C, HU J, et al. Pentatricopeptide-repeat family protein *RF6* functions with hexokinase 6 to rescue rice cytoplasmic male sterility[J]. Proc Natl Acad Sci U S A, 2015, 112(48): 14984 – 14989.

[23] HUIWU TANG, XINGMEI ZHENG, CHULIANG LI, et al. Multi-step formation, evolution, and functionalization of new cytoplasmic male sterility genes in the plant mitochondrial genomes[J]. Cell Research, 2017, 32(1): 130.

[24] LUO D, XU H, LIU Z, et al. A detrimental mitochondrial-nuclear interaction causes cytoplasmic male sterility in rice[J]. Nat Genet, 2013, 45: 573 – 577.

[25] QIN X, HUANG Q, XIAO H, et al. The rice *DUF1620*-containing and *WD40*-like repeat protein is required for the assembly of the restoration of fertility complex[J]. New Phytol, 2016, 210(3): 934 – 945.

[26] TANG H, LUO D, ZHOU D, et al. The rice restorer *Rf4* for wild-abortive cytoplasmic male sterility encodes a mitochondrial-localized PPR protein that functions in reduction of *WA352* transcripts[J]. Mol Plant, 2014, 7: 1497 – 1500.

[27] WANG Z, ZOU Y, LI X, et al. Cytoplasmic male sterility of rice with boro Ⅱ cytoplasm is caused by a cytotoxic peptide and is restored by two related PPR motif genes via distinct modes of mRNA silencing[J]. The Plant Cell, 2006, 18: 676 – 687.

[28] YAO F Y, XU C G, YU S B, et al. Mapping and genetic analysis of two fertility restorer loci in the wild-abortive cytoplasmic male sterility system of rice (*Oryza Sativa* L.)[J]. Euphytica, 1997, 98: 183 – 187.

[29] ZHANG G, LU Y, BHARAJ T S, et al. Mapping of the *Rf3* nuclear fertility-restoring gene for WA cytoplasmic male sterility in rice using RAPD and RFLP markers[J]. Theor Appl Genet, 1997, 94: 27 – 33.

Chapter 7

Breeding of Super Hybrid Rice Combinations

Yang Yuanzhu / Wang Kai / Fu Chenjian / Xie Zhimei / Liu Shanshan / Qin Peng

Section 1 Process of Breeding Super Hybrid Rice Combinations

Ⅰ. Objectives of Breeding Super Hybrid Rice Combinations

In 1981, Japan proposed the idea of breeding super high-yielding rice and established a large-scale national cooperative research project "Development of Super-high Yielding Rice and Establishment of Cultivation Techniques", the so called "Ni 753 Plan", with the aim of breeding varieties with high yield potential, complemented by corresponding cultivation technology, to achieve a yield per unit area 50% higher than that of the control variety in 15 years with a brown rice yield of 7.5 – 9.8 t/ha in low-yielding areas and over 10 t/ha in high-yielding areas. In 1989, IRRI launched the "New Plant Type (NPT)" rice breeding project with the goal of producing rice varieties with new plant types and a yield potential 20%–30% higher than that of varieties commonly in production at that time, and a yield of 13 – 15 t/ha by 2005. In 1994, IRRI reported its results of NPT research at a meeting held by the Consultative Group for International Agricultural Research (CGIAR). The term "super rice" appeared in media reports and has become frequently used ever since (Yuan Longping, 2008; Fei Zhenjiang, 2014).

In 1996, China's MOA launched China's Super Rice Research Project and formed the Super Rice Collaborative Group with 11 major domestic rice research institutes in China, including China Rice Research Institute (CRRI) and Hunan Hybrid Rice Research Center (HHRRC). The main objective of this project is to breed high-yielding rice varieties with a yield increase of 15% and 30% over that of high-yielding varieties by 2000 and 2005, respectively (Table 7 – 1), to achieve a stable rice yield of 9.0 – 10.5 t/ha in a large area (6.67 ha, or 100 mu) by 2000, to exceed 12.0 t/ha by 2005, to reach 13.5 t/ha by 2015, and to form a technological system for super rice cultivation. Yuan Longping suggested that the yield target of super hybrid rice should vary with the times, ecological regions and planting seasons, and it is more reasonable to use daily yield per unit area instead of the absolute yield as the indicator in the breeding program. In 1997, it was recommended that the target for super-high yielding hybrid rice breeding during the Ninth Five-Year Plan period was 100 kg/ha.

Table 7 - 1 Target yield (t/ha) of super rice varieties in the first and second phases (Yuan Longping, 1997)

Variety/Phase	Super conventional rice				Super hybrid rice			Yield increase (%)
	Early-season *indica*	Early-, mid- and late-season *indica*	Southern China Single-cropping *japonica*	Northern China *japonica*	Early-season *indica*	Single-cropping *indica* and *japonica*	Late-season *indica*	
Current high yield	6.75	7.50	7.50	8.25	7.50	8.25	7.50	
1996 - 2000 (Phase 1)	9.00	9.75	9.75	10.50	9.75	10.50	9.75	>15
2001 - 2005 (Phase 2)	10.50	11.25	11.25	12.00	11.25	12.00	11.25	>30

Note. Phenotypes on two plots of 6.67 ha in the same ecological area for two consecutive years.

In 2005, the No. 1 Document of the State Council proposed the establishment of a super rice promotion project. In the same year, the MOA issued Measures for Confirmation of Super Rice Variety (Trial) (NBK [2005] No. 39), and defined super rice varieties (hybrids) as new rice varieties with significantly increased yield potential, high quality and strong resistance, bred by combining ideal plant types, through heterosis utilization, and with super high yielding rice cultivation techniques. The Science and Education Department of the MOA organized experts to formulate China Super Rice Research and Extension Plan (2005 -2010)" (referred to as the "Plan"). The Plan points out that the basic principles of China's super rice research and promotion is "high yield, high quality, wide adaptation, good seeds combined with good cultivation methods, integration of research, demonstration and promotion". While improving yields per unit area, the types of super rice in different ecological zones were also enriched, taking into account yield, grain quality and resistance. Based on the traits of super rice varieties, more efforts were done for the integration of technologies that were practical, simple cost-saving and efficient to form a better technical system, improve farmers' enthusiasm and increase economic benefits. The government was to take lead efforts to strengthen the training, demonstration and promotion of new super rice varieties so as to increase the total planting area. By 2010, 20 leading, super rice varieties should be developed and promoted to 30% of the national total rice planting area (about 8 million hectares), with an average yield increase of 900 kg/ha, leading to a significant increase in the national total rice yield and maintaining China's leading position in the world in rice breeding. The yield target was adjusted according to the results of the previous phases of the project and the estimation of rice yield potential (Table 7 -2).

The breeding objectives of super hybrid rice are not static. They change with different ecological conditions, social development, production methods, and ecological environment. In 2008, the MOA revised the Super Rice Variety Confirmation Measures (Trial) issued in 2005 to established various indicators for super rice varieties, forming Super Rice Variety Confirmation Measures (NBK [2008] No. 38) as shown in Table 7 -3.

Table 7 - 2 Yield, quality and resistance of super rice (Cheng Shihua, 2010)

Area		Early-season ricein the Yangtze River Basin	Northeastern China early-maturing *japonica*, Yangtze River Basin middle maturing late-season rice	Southern China early- and late-season rice and Yangtze River Basin late-season rice	Yangtze River Basin single-cropping rice and Northeastern mid-maturing *japonica*	Upper reaches of Yangtze River late-maturing single-cropping rice and Northeastern late-maturing *japonica*
Growth period (d)		102 - 112	121 - 130	121 - 130	135 - 150	150 - 170
Yield	Fertilizer tolerance	9.00 t/ha	10.20 t/ha	10.80 t/ha	11.70 t/ha	12.75 t/ha
	Wide adaptability	Yield increases more than 8% or yield at 1/3 testing locations increases more than 15% with a similar growth duration of check variety in regional yield trials above the provincial level.				
Quality	Northern China *japonica* meets Grade 2, Southern China late-season *indica* meets Grade 3, and Southern China early-season and single-season *indica* meets Grade 4 quality standard issued by MOA.					
Resistance	Resistant to 1 - 2 local main diseases and pests.					

Notes. 1. Yield of Northern China *japonica* rice is 300 kg/ha lower than that of Southern China *indica* rice with the same growth duration.

2. The maximum yield target of 13.5 t/ha (upper Yangtze late-maturing single-cropping rice and Northeastern China late-maturing *japonica* rice) is the target for 2015, the Plan's maximum yield target is 12.75 t/ha for 2010.

Table 7 - 3 Main indicators of super rice varieties

Area	Yangtze River Basin early-maturing early-season rice	Yangtze River Basin mid- and late-maturing early-season rice	Yangtze River Basin Mid-maturing late-season rice; Southern China photo-sensitive late-season rice	South China early- and late-season rice; Southern China late-maturing late-season rice; Northern China early-maturing *japonica*	Yangtze River Basin single-cropping rice; Northeast China mid-maturing *japonica*	Upper reaches of the Yangtze River late-maturing singlecropping rice; Northeastern China latematuring *japonica*
Growth duration (d)	≤105	≤115	≤125	≤132	≤158	≤170
Yield in 6.67 ha continuous plot (kg/ha)	≥8,250	≥9,000	≥9,900	≥10,800	≥11,700	≥12,750
Quality	Northern China *japonica* meets or exceeded Grade 2, Southern China late-season *indica* meets or exceeded Grade 3, Southern China early-season *indica* and single-season rice meet or exceeded Grade 4 quality standards issued by the MOA.					
Resistance	Resistant to 1 - 2 local main diseases and pests.					
Planting area	More than 3,333.33 ha per year in two years after the variety is confirmed.					

Ⅱ. Procedure of Breeding Super Hybrid Rice Combinations

1. Parent selection for super hybrid rice breeding

(1) Selecting parents with ideal plant types

Excellent plant morphology is the basis of super high yield, and Yuan Longping proposed that super high yielding hybrid rice breeding should make full use of the complementary effect of the excellent traits of both parents and make more improvement in morphology (Yuan, 1997). The ideal plant type enables rice to adapt to environmental conditions, strike the best balance between the "source", "flow" and "sink", and maximize the efficiency of solar energy utilization, so as to achieve the goal of super-high yield. China has vast rice growing areas with different ecological conditions. Since dwarf rice breeding, breeders have been working on and improving ideal plant types for different ecological areas.

Based on the semi-dwarf and early-clustering plant type, Huang Yaoxiang (1983, 2001) suggested the early-clustering semi-dwarf model as the ideal plant type for South China super rice from the perspective of ecological breeding. Under the climate conditions in South China, early- and late-season rice varieties have relatively a short growth duration and should grow faster to make full use of the thermal and solar conditions in the early growth stage for high yield. The designed traits of the plant type for both early- and late-season super rice varieties are 105 - 115 cM plant height, 9 - 18 panicles per hill, 150 - 250 grains per panicle, strong root viability, a growth duration of 115 - 140 days, a harvest index of 0.60, and a yield potential up to 13 - 15 t/ha. Representative varieties include Guichao 2, Tesan'ai 2 and other super high yielding conventional early- and late-season rice varieties.

Yang Shouren, with a focus on the influence of the panicle shape of *japonica* rice on population structure and the status of receiving solar energy, proposed the "upright and large panicles" model for Northern China *japonica* rice (Xu Zhengjin, 1996), and held that the upright panicle shape would be an important breakthrough in the pursuit of ideal rice plant types. Based on this, Chen Wenfu (2003) designed a quantitative model for the plant type of super high yielding of *japonica* rice - 105 cM of plant height, upright and large panicles, medium to high tillering ability, 15 - 18 panicles per hill, 150 - 200 grains per panicle, high biological yield and strong overall resistance, a growth duration of 155 - 160 days, a harvest index of 0.55 - 0.60 and a yield potential up to 12 - 15 t/ha. Representative varieties include conventional super high yielding *japonica* varieties Shennong 265 and Liaojing 263.

Zhou Kaida (1995) proposed a "subspecific heavy panicle type" model for super rice in the Sichuan Basin based on the ecological conditions of single-cropping rice in Sichuan, where there is less wind, high humidity, high temperature and more clouds and fog. Under such ecological conditions, moderately increased plant height, reduced panicle number and increased panicle weight are more conducive to the population's photosynthesis and dry material production and can reduce the harm of diseases and pests so as to achieve super high yield. The plant type parameters are as follows - 120 - 125 cm in plant height, 26 - 30 cm in panicle length, 200 grains per panicle, and above 5 g per panicle in weight. Representative varieties include super high yielding three-line hybrids Ⅱ - You 6078 and Gangyou 188.

Based on the ecological conditions in the middle and lower reaches of the Yangtze River, Yuan Longping (1997) proposed the structure of an ideal super-high yielding rice with "high canopy, short panicle layer, and medium-large panicles", which unites the "three high" traits, i. e. high biological yield, high harvest index and high resistance to lodging, to fully utilize solar energy and improve yield. "High canopy" means the upper three leaves are long, straight, narrow, concave and thick with a large leaf area index for strong photosynthesis, resulting in a sufficient "source". A short panicle layer means 100 cm of plant height, drooping panicles, and the top of the panicles only 60 – 70 cm above the ground, so as to lower the gravity center to ensure strong lodging resistance, a smooth "flow"; medium to large panicles with a single panicle weight of 5 – 6 g, 2.7 – 3 million effective panicles per hectare and 200 grains per panicle, leading to a large "sink". The plants should also have a high economic coefficient, specifically higher than 0.55, mainly relying on increasing the biological yield so as to further increase rice yield; and a moderate maturity period with a daily yield of 100 kg/ha. The representative combinations include Peiai 64S/E32 and 29S/510 (Fig. 7 – 1).

Fig. 7 – 1 Plant type of Peiai 64S/E32

The growing season of early-season rice in the double-cropping rice areas of the Yangtze River Basin is characterized by frequent cold spells in late spring, continuous rains, and low temperatures and less sunlight in early summer, heavy rain in mid- and late June, and high temperatures, hot and dry winds in July.

Therefore, Yang Yuanzhu suggested the model of ideal plant type for super early-season rice in the double-cropping rice growing area in the middle and lower reaches of the Yangtze River. 1) Medium to strong culms – 100 cm in height, high biological yield, a harvest index above 0.55, and thick, short, tough and wrapped culms with high resistance to lodging; 2) Early and quick growth – early tillering at low tiller nodes, moderately compact, strong tillering ability, panicle-bearing tillers above 75%, effective panicles up to 3.75 million per hectare to ensure a large "sink"; 3) Slanting early leaves and upright late leaves – uncoiled leaves at the early leaf stage, light green in color and slightly thin so as to make full use of direct light and improve the light interception rate – the upper three leaves erect at the late leaf stage, thicker and slightly concave, enabling the plant to make full use of scattered light and improve the photosynthetic efficiency, thereby ensuring sufficient "source"; 4) Medium-sized panicles and a high seed setting rate – long panicles with a low grain setting density especially at the panicle bottom, more primary branches and light two-phase grain filling, around 130 grains per panicle, a seed setting rate of 85%, and 25 –28 g of 1,000-grain weight; 5) Strong roots without early senescence – well-developed and vigorous roots, strong rooting ability and fast expansion with no senescence in the late stage, ensuring smooth "flow" (Yang Yuanzhu, 2010).

Development tends to follow a spiraling path, and the same is true for the plant type of super hybrid rice. Based on the general rule that biological yield increases with plant height while the harvest index remains at around 0.5, Yuan Longping proposed that the super hybrid rice plant type should change from tall to short and then to semi-dwarf, semi-tall, new tall and super-tall (Fig. 7 – 2) (Yuan, 2012)

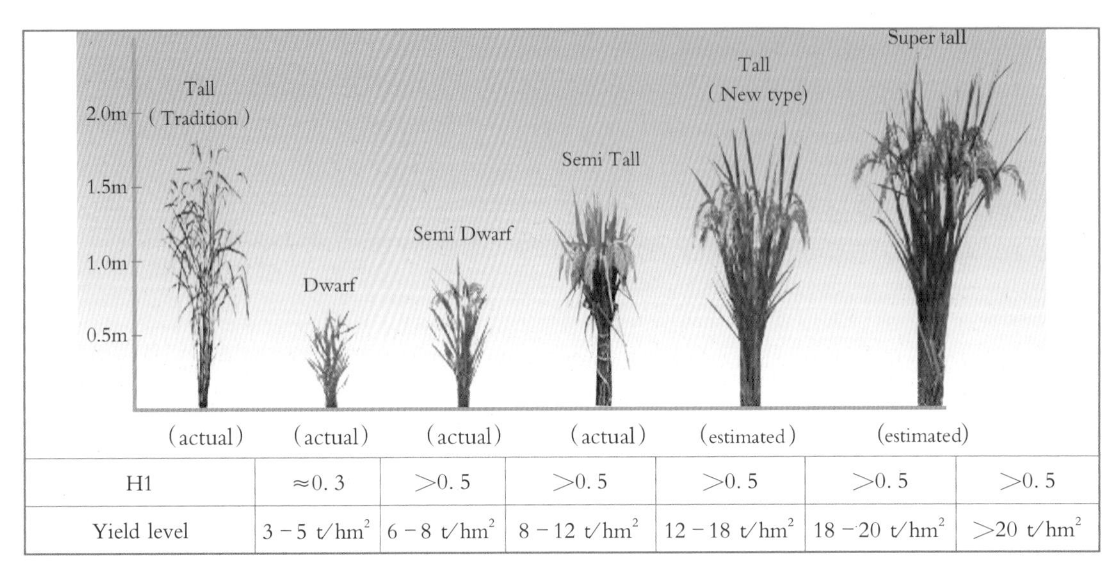

H1	≈0. 3	>0. 5	>0. 5	>0. 5	>0. 5	>0. 5
Yield level	3 – 5 t/hm^2	6 – 8 t/hm^2	8 – 12 t/hm^2	12 – 18 t/hm^2	18 – 20 t/hm^2	>20 t/hm^2

Fig. 7 – 2 Rice plant type development trend

The ideal plant type is the ideal leaf and culm shapes that can balance the "source, sink and flow" to the maximum extent and achieve the maximum economic yield in a specific rice growing environment. The breeding history and experience of super hybrid rice show that the ideal plant type of super hybrid rice is not invariable, and there are specific ideal plant type patterns suitable for different specific rice growing environments under different ecological conditions.

The ideal plant type of super hybrid rice is the result of inheritance from parents. When selecting parents to produce different combinations, we should pay attention to the following points. 1) According to the requirements of different ecological conditions for the ideal plant type, parents should have as many traits of the intended ideal plant type as possible; 2) When the parents cannot both have the same good traits, they should be complementary in such traits so as to eliminate disadvantages; 3) When selecting parents, priority should be given to sterile lines of the ideal plant type. Hybrid rice breeding experience shows that the traits controlling the plant type of hybrid rice combinations are partly influenced by cytoplasmic inheritance, and the plant types of hybrid rice combinations are mainly determined by the female parent. Therefore, hybrid rice combinations are more similar to their female parents in plant types.

(2) Selection of parents with certain genetic distance

Yuan Longping proposed the combination of morphological improvement and heterozygous advantage as the technical route of hybrid rice breeding for super-high yield according to the characteristics of hybrid rice breeding (Yuan Longping, 1997). Genetic diversity is the genetic basis of hybrid heterosis, and the selection of two parents with a big genetic distance maximizes the utilization of heterosis between the two parents. Yuan Longping concluded that the level of rice heterosis exists as follows, *indica-japonica* > *indica-javanica* > *japonica-javanica* > *indica-indica* > *japonica-japonica* crosses (Yuan Longping, 2008), and proposed the technical route of three development stages from low to high level of heterosis utilization, i. e., from intervarietal to intersubspecies to distant heterosis utilization (Yuan Longping, 1987). *Japonica-*

japonica, *indica-indica* and *japonica-javanica* crosses are al intervarietal (or inter-ecotype) hybrid heterosis, which is the first stage of hybrid heterosis utilization and the main form of heterosis utilization at present. The *indica-japonica* and *indica-javanica* crosses are the second stage, i. e. intersubspecific heterosis utilization, but with problems such as low and unstable seed setting rates and poor grain filling. Yuan Longping proposed eight strategies for the breeding of *indica-japonica* intersubspecific hybrid combinations. 1) Seeking height in dwarfism, using dwarf genes to solve the problem of excessive height of hybrid plants, and increasing the plant height appropriately to improve biological yield without lodging; 2) Seeking nearness in distance, partially using intersubspecific hybrid heterosis to overcome the physiological dysfunctions and unfavorable traits arising from the excessive genetic differences in typical intersubspecific hybrid rice; 3) Using dominant complementary effect of parents' good traits while maintaining a relatively large genetic distance between the parents to avoid overlap in order to fully play the role of overdominance; 4) Seeking medium to large panicles, not oversize or huge panicles, to help coordinate between the "source" and the "sink", so that it has a higher seed setting rate and better grain filling, increase panicle length, the number of primary branches, and the numbers of panicles and grains; 5) Selecting hybrids with a high grain-to-leaf ratio; 6) Selecting varieties and lines with good and extraordinarily good grains, not a high 1,000-grain weight but a high unit weight; 7) Combining *javanica-indica* intermediate materials with high quality and long grains with *indica* rice to produce *indica*-like quality rice. Combine *javanica* rice or *javanica-japonica* intermediate materials with short grains with *japonica* rice to produce *japonica*-like quality rice; 8) For ecological adaptation, use mainly *indica-javanica* crosses in *indica* rice areas, and *japonica-javanica* crosses in *japonica* rice areas, supplemented by *indica-japonica* crosses in both regions (Yuan Longping, 1996).

Under the premise that fertility incompatibility can be eliminated, heterosis is closely related to the genetic difference between the parents. Heterotic groups can be formed according to the genetic relationship between the parents (genetic diversity). Two heterotic groups have been formed for the inbred lines of modern maize in the United States, the stiff-stalk (SS) group and the non-stiff stalk (NS) group (Mikel, 2006). Hybrid heterosis is high for parents from different heterotic groups, while hybrid heterosis is low for parents from the same heterotic group. The creation of maize heterotic groups and hybrid breeding on this basis has greatly improved the efficiency of heterosis utilization, and provided good reference for heterosis utilization in rice. The breeding experience of hybrid rice shows that there are also heterotic groups in rice. For example, CMS lines (maintainer lines) and restorer lines are from two different major heterotic groups, namely the early-season *indica* ecotype in the Yangtze River Basin and the mid- and late-season *indica* ecotype in South Asia and Southeast Asia. PTGMS two-line hybrid rice form a new heterotic group different from the two three-line *indica* groups, but the male parents of two-line hybrids belong to the same heterotic group of three-line restorer line parents (Wang Kai, 2014; Wang, 2006). Before forming a combination, the parents should be checked for their heterotic groups according to their genetic distance using molecular markers, and parents from different heterotic groups and with thus a high genetic distance should be selected for crossing whenever possible.

(3) Selection of parents with high combining ability and strong heterosis

The combining ability of the parents is closely related to the hybrid's traits (Yuan Longping, 2002).

General combining ability (GCA) is mainly based on the additive effect of genes and the characteristic showed in phenotypic sense is that the effects of genes are cumulative. (Arne, 2010) and inherited stably. Typical examples of such parents include Quan 9311A and Ⅱ-32A of CMS lines, and Guangzhan 63S, P88S, C815S and Longke 638S of PTGMS, and Minghui 63, 9311, R1128, Minghui 86 and Lehui 188 of hybrid male parents (Zhong Richao, 2015). Special combining ability (SCA) is mainly determined by the non-additive effects of genes, mainly the dominance, overdominance and epistatic effects (Arne, 2010). The unfavorable recessive gene effect of the parents is concealed by the favorable dominant gene effect (dominance effect), or the interaction among alleles (overdominance effect) caused by the heterozygous combination of parents' genotypes, resulting in obvious heterosis in the hybrid F_1 population over the parents. Three-line hybrid rice developed in the early days mainly used the dominance and overdominance effects. The parents, such as V20A, Jin 23A, Yue 4A, R402, Yuehui 9113, Ce 64, etc., have no obvious advantage; however, the expansion of the genetic distance between the parents resulted in significant heterosis in the progenies (Weiyou 64, Jinyou 402, Yueyou 9113, etc.) over the parents. Chen Liyun proposed that when it proceeded into the stage of super hybrid rice breeding from general breeding, it became difficult to meet the breeding requirements of super high yield by relying solely on heterosis. Super hybrid rice breeding today must be based on the overall improvement of parental traits achieved through the selection and breeding of super parents. Only by scientifically using heterosis can further breakthroughs be achieved in super hybrid rice breeding (Chen Liyun, 2007).

Since the success of three-line and two-line hybrid rice breeding, a large number of three-line and two-line hybrid rice parents have been bred. However, not all parents with large genetic differences and excellent phenotypic traits can be used to breed super hybrid rice combinations. It has been proved that only those three- and two-line parents with good combining ability and strong heterosis can breed super hybrid rice combinations suitable for production. Representative male sterile lines include Ⅱ-32A, Jin 23A, Tianfeng A, Wufeng A, Y58S, Guangzhan 63-4S, Peiai 64S, Zhun S, Zhu 1S, Lu 18S, Xiangling 628S, Longke 638S and representative restorer lines are Yuanhui 2, R900, Shuhui 527, Zhonghui 8006, Huazhan, 9311, F5032, Hua 819, Hua 268 and other male sterile lines and restorer lines. In sum, super hybrid rice combinations can only be bred using lines with high combining ability as the parents.

2. Testcrossing

After the hybrid parents are selected, testcross and hybrid selection are to be carried out. The selection and breeding of superior varieties is a low probability event. For a better combination of super hybrid rice, the best way is to use multiple sterile lines with high combining ability, strong heterosis and ideal plant type as the female parent to do an incomplete diallel cross in a large-scale test with strong dominant male parents, Yuan Longping Agricultural High-tech Co., Ltd. (LPHT) carries out as many as 30,000 to 40,000 testcrosses annually. Under the premise of reasonable selections of parents for super hybrid rice breeding, only an increase in the scale of testcrossing can increase the opportunity of getting super hybrid rice combinations. According to the required amount of seeds, the testcrossing methods can be artificial glume cutting, paired bagging, direct planting together, cloth-covered isolation, etc. for super hybrid rice testing.

The incomplete diallel cross of sterile lines (female) and male parents, while producing yield test results, can also be used to calculate the GCA of the parents and the SCA of the hybrids for an evaluation of the combining ability of the super hybrid rice parents. Moreover, it can help construct a super rice heterosis prediction model and realize super rice hybrid heterosis prediction based on the results of whole-genome genotype analysis of both parents (e. g., resequencing) and the analysis of hybrid F_1 phenotype, thus realizing accurate design-based breeding of super hybrid rice (Zhen, 2017).

3. Combination evaluation

The general process of a combination evaluation includes heterosis validation, single-site, multiple-site, and regional and production yield trials.

(1) Heterosis validation

The purpose of heterosis validation is to carry out preliminary evaluation of heterosis in yield and other major agronomic traits of the tested combinations. Superior combinations with super rice potential are selected for single-site comparison test, while the inferior hybrids will be abandoned. It is recommended to plant five rows per combination and 20 plants per row, making 100 plants in total. The method of interval comparison is adopted and a super hybrid rice control variety is set for every 10 combinations.

(2) Single-site yield trial

Combinations that have strong heterosis and have passed the heterosis validation shall proceed to single-site yield trial. The trial goes for randomized blocks with three replications on a rectangular plot of 13.33 m^2 with a length-width ratio of (2 – 3) : 1, and a super hybrid rice control variety is set for every 13 – 15 combinations.

(3) Multiple-site yield trial

Combinations with strong heterosis that selected out from single-site evaluation will move on to multiple-site yield trial. In natural agricultural regions, sites with representative natural and cultivation conditions are selected and the test goes for randomized blocks with three replications on rectangular plots of 13.33 m^2 with a length-width ratio of (2 – 3) : 1, and a super hybrid rice control variety is set for every 13 – 14 combinations.

(4) Regional and production yield trials

In different natural areas within the same ecological region, selected trial sites should be those that can represent the soil characteristics, climatic conditions, cropping system and production level of the areas. Agricultural authorities are responsible for the organization of regional and joint tests, and breeding, cultivation and promotion enterprises are to carry out green-channel tests for home-bred varieties to find out the yield, yield stability, adaptability, rice quality, resistance and other important traits of the combinations according to a unified test plan and operation procedures, so as to determine the value of the combination for production and the suitable areas. For varieties delivering outstanding performance in regional tests, production test shall be carried out along with the regional test in the next year under conditions close to those for field production, so as to speed up progress.

4. Combination approval

New hybrid rice combinations that meet the provincial or national standard for new rice variety approval shall be submitted to the provincial or national variety examination and approval committee for examination and approval.

5. Yield estimation and variety acceptance in 6.67-ha demonstration land plots

In regional tests of rice varieties at or above the provincial level, varieties with a growth period similar to that of the control variety and an average yield increase of more than 8% over a two-year period should be cultivated in 6.67-ha high-yielding demonstration plots for one year. Varieties with yield less than 8% higher than that of the control variety need to be cultivated in different places for two years in 6.67-ha land plots. Then, these varieties will be examined by an expert panel organized by the MOA according to the Super Rice Variety Confirmation Method (NBK [2008] No. 38), and the varieties that meet the requirements of super rice confirmation (Table 7 – 1) will be approved as super rice varieties.

Section 2 Principles for Breeding Super Hybrid Rice Combinations

It is a long-term goal of hybrid rice breeding to breed new super hybrid rice varieties with significantly increased yield as well as resistance and rice quality similar to those of the control varieties by the three-line or two-line method. In order to achieve the goal of super hybrid rice breeding, parent lines must be selected based on the principle of crossing rice varieties with strong heterosis (Yuan Longping, 1987). In 1997, Yuan Longping proposed to take the route of morphology improvement through heterosis utilization for super high yield hybrid rice breeding (Yuan Longping, 1997), which has become the guiding principle for super hybrid rice breeding in China. He also proposed a three-stage framework for heterosis utilization, going from low heterosis combinations to high heterosis ones, or more specifically, from intervarietal heterosis utilization to intersubspecific heterosis utilization and then to distant heterosis utilization (Yuan Longping, 1987). Based on the breeding experience of super hybrid rice at home and abroad, the following principles should be followed in practice.

Ⅰ. Principles for Intervarietal (Inter-Ecotype) Heterosis Utilization

Intervarietal hybrid heterosis utilization is the first stage of hybrid rice breeding proposed by Yuan Longping, which involves mainly the utilization of the hybrid heterosis between varieties within a subspecies (Yuan Longping, 1987). Due to the close genetic relationship between varieties and the limited heterosis, the heterosis between ecotypes should be fully utilized when selecting super hybrid rice combinations. The first generation of high-yielding three-line hybrid *indica* rice, represented by Shanyou 63, has sterile lines (maintainers) and restorer lines that belong to two major ecotypes, namely the early-season *indica* ecotype in the Yangtze River Basin and the mid- and late-*indica* ecotype (and its derivatives) in South and Southeast Asia, so it can also be assumed that Chinese early-season *indica* sterile lines (maintainers) and foreign restorer lines (and their derivatives) should belong to two heterotic groups, and their intergroup crosses fit a strong heterosis utilization pattern (Wang Kai, 2014). When using intervarietal hybrid heterosis for the selection of super hybrid rice combinations, the genetic distance between the two parents should be as large as possible in order to make full use of the hybrid heterosis between varieties or ecotypes.

Further breakthroughs in yield and other aspects (e. g., rice quality and resistance) using inter-ecotype hybrid heterosis require innovations in materials and methods, such as more and better use of foreign *indica* ecotype resources and widened genetic distance between *indica* heterotic groups in China. The *Aus* rice group is an independent sub-group of *indica* rice (Fig. 7 - 3). It mainly consists of early-maturing, drought- and waterlogging-resistant varieties (Glaszmann, 1987) planted in the *Aus* season (March to July in Bangladesh and West Bengal of India). There are a large number of superior genes in this group, up to 80%-90% of the germplasm with the gene *Pup1* (Chin, 2010), high-temperature tolerance (Ye, 2012), submergence tolerance (Xu, 2012, 2006) and wide compatibility (Kumar, 1992). As a group related to *indica* rice, it has no reproductive isolation from other *indica* subgroups, and contains a large number of superior genes. However, the *Aus* rice group has not received as much attention from breeders as *indica* and *japonica* in breeding. In the process of classification and utilization of heterotic groups, the *Aus* group can be treated as an *indica* ecological heterosis group (3000 Rice Genomes Project, 2014), and introduce the superior genes from the *Aus* group through hybridization breeding into existing core parents of super rice to improve them toward a (new) hybrid *indica* parental group, eventually becoming a new heterotic group. At the same time, the superior genes in the *Aus* rice group are transferred to the male parents of core super rice varieties, so as to further improve the utilization of intersubspecific heterosis.

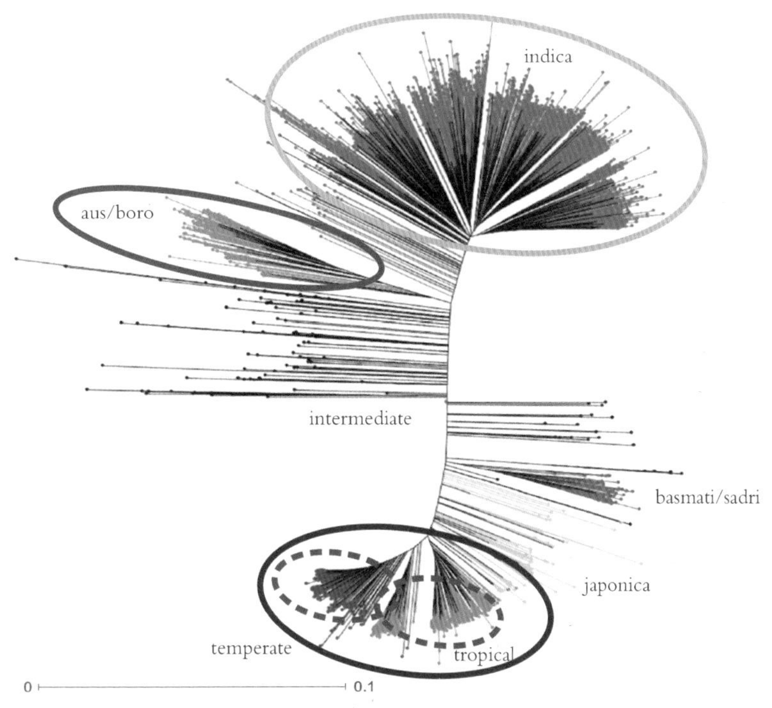

Fig. 7 - 3 3000 genetic clustering map of rice resources (3000 Rice Genomes Project, 2014)

Ⅱ. Principles for Intersubspecific Heterosis Utilization

Intersubspecific heterosis utilization refers to the utilization of heterosis between *indica* and *japonica* rice, the second developmental stage of hybrid rice breeding as proposed by Yuan Longping (1987). There is a wide genetic distance between the *indica* and *japonica* subspecies, so *indica-japonica* hybrids have strong heterosis, such as tall plants, thick stems for lodging resistance, well-developed root systems, large panicles with more grains, strong germination and tillering ability, strong regeneration ability and high stress resistance. However, the genetic incompatibility and hybrid sterility in typical *indica-japonica* hybrids result in abnormal fertilization and poor seed setting rates (generally only about 30%). There are also some other problems, such as late maturity, poor grain filling and excessive plant height. In the 1970s, Yang Shouren (1973) put forward the idea of partial utilization of *indica-japonica* hybrid heterosis. Yang Zhenyu (1994) also proposed the ideas of "*indica-japonica* bridging" to improve compatibility, facilitate favorable gene exchange and maintain moderate difference. By bridging *indica* and *japonica* rice, C57, a *japonica* restorer line partially related to *indica*, was developed, thereby realizing partial utilization of *indica-japonica* intersubspecific heterosis. The incompatibility between *indica* and *japonica* varieties is the biggest obstacle to the utilization of *indica-japonica* intersubspecific heterosis. The discovery of wide compatibility varieties or genes and the study of intersubspecific sterile genes have made it possible to overcome hybrid sterility and utilize intersubspecific heterosis (Morinaga, 1985; Chen, 2008; Guo, 2016).

1. Partial utilization of heterosis among subspecies

The utilization of wide compatibility germplasm and the strategy of partial utilization of heterosis in *indica-japonica* hybrids have accelerated the realization of the objectives of the super hybrid rice breeding from Phase 1 to Phase 4. The successful breeding of the *indica-japonica* intermediate PTGMS line Peiai 64S [Nongken 58S/(Peidi/Aihuangmi//Ce 64)] with wide compatibility provided an opportunity for the direct utilization of the intersubspecific hybrid heterosis. Peiai 64S was then combined with the *indica* parent 9311 to produce the Phase-1 super hybrid rice combination Liangyoupeijiu and was widely used in production. The Phase-2, Phase-3 and Phase-4 super hybrid rice varieties Y Liangyou 1, Y Liangyou 2 and Y Liangyou 900 selected by Deng Qiyun et al. were all produced by pairing Y58S as the female parent with *indica* parents 9311, Yuanhui 2 and R900, respectively (Deng Qiyun, 2005; Li Jianwu, 2013; Li Jianwu, 2014; Wu Jun, 2016). Y58S was developed by selecting sterile single plants with desirable plant traits from the progenies of a multi-parental cross (Annong S-1/Changfei 22B//Annong S-1/Lemont)/Peiai 64S (Deng Qiyun, 2005), which combined the excellent traits of high quality, high solar energy efficiency, disease and stress resistance of *javanica* Lemont (tropical *japonica* rice) and the wide compatibility, high combining ability and excellent leaf morphology of Peiai 64S, realizing the integration of favorable polygenes. Y Liangyou 900, the representative achievement of super hybrid rice breeding in the fourth phase, was bred using Y58S as the female parent and R900, a weak photosensitive *indica-japonica* intermediate restorer line with strong heterosis, as the male parent. It is the hybrid between a PTGMS wide-compatibility *indica-japonica* intermediate male sterile line and an *indica-japonica* intermediate restorer line, and thus has strong heterosis. In 2014, it yielded over 15 t/ha in 6.67-ha demonstration plots for the first time (Wu Jun et al, 2016) (Fig. 7-4). When using the two-line method to partially exploit intersubspecific hybrid heterosis in breeding super hybrid rice, one of the two parents should have a certain degree of *japonica* kinship and wide compatibility.

Fig. 7 - 4 Y Liangyou 900 during maturity period (Wu Jun et al., 2016)

Comparing the representative super hybrid rice varieties of different phases with Shanyou 63 in terms of panicle structure and yield composition, it was found that the effective panicles of super hybrid rice from phase 1 to phase 4 showed a decreasing trend, but the number of grains per panicle increased significantly, resulting in a 10% increase in the total number of spikelets per unit area in each phase (Table 7 - 4). Although the ratios of "farmers' yield" to "experts' yield" (i. e., actual yield or yield potential for large area extension) decreased from 90.7% for Phase 1 to 78.7% for Phase 4, the absolute yield per unit area always maintained an increase of 8%-10%, indicating that super hybrid rice not only has higher yield potential under super-high yielding conditions, but also has significant yield increase under general ecological and common cultivation conditions (Wu Jun, 2016).

Table 7 - 4 Yield performance of representative super hybrid rice in different phases under common cultivation conditions

Combination	Parents	Effective panicles (10,000/ha)	Grains per panicle	Total spikelets (10,000/ha)	Actual yield (kg/ha)	Yield potential (kg/ha)	Discount rate (%)	Relative yield increase (%)
Shanyou 63 (CK)	Zhenshan 97A/ Minghui 63	259.5	146.1	37,912.5	8,599.5	9,000	95.6	0
Liangyoupeijiu	Peiai 64S/9311	225.0	179.0	45,645.0	9,522.0	10,500	90.7	10.7
Y Liangyou 1	Y58S/9311	273.0	182.8	49,905.0	10,209.0	12,000	85.1	18.7
Y Liangyou 2	Y58S/Yuanhui 2	231.0	237.5	54,862.5	11,100.0	13,500	82.2	29.1
Y Liangyou 900	Y58S/R900	211.5	288.7	61,060.5	11,800.5	15,000	78.7	37.2

Note. Discount rate = Actual yield of large-scale planting/yield potential × 100%.

Based on the above achievements and progress in super hybrid rice breeding, Yuan Longping launched the fifth phase of the super hybrid rice breeding project, aiming at a yield of 16 t/ha by 2020. In 2015, the yield of Chaoyouqian, the pioneer combination of the fifth phase, which was bred through morphological improvement and partial utilization of *indica-japonica* intersubspecific heterosis, delivered yields exceeding 15 t/ha in five demonstration plots, including a 16. 01 kg/ha in the 6. 67-ha demonstration plot in Gejiu city, Yunnan province, attaining the goal of 16 t/ha set for the fifth phase. This combination set new local high rice yield records in 11 provinces (autonomous regions and municipalities), including Hainan, Guangdong, and Guangxi, and proceeded to be planted in more than 80 demonstration sites in more than 10 provinces (autonomous regions and municipalities) nationwide in 2016. Chaoyouqian set three world records – the average yield of 16. 32 t/ha was a new rice yield record in the world for large-scale rice planting, the average yield of 18. 80 t/ha for single-cropping rice plus ratooned rice in Qichun, Hubei province, set the record for single-cropping rice plus ratooned rice in the middle and lower reaches of the Yangtze River, and the average combined yield of double-cropping rice in Xingning, Guangdong province, 23. 07 t/ha, set the world record for double-cropping rice yield. In 2017, 31 contiguous high-yielding super rice demonstration sites, 6. 67-ha each, were established in 13 municipalities, provinces and autonomous regions. Chaoyouqian, which got national approved in 2012 under the name Xiangliangyou 900, was planted in the 7. 7-ha high-yielding research plot in Yongnian district of Handan city, Hebei province, passed the test by an expert group organized by the Hebei Provincial Department of Science and Technology with an average yield of 17. 24 t/ha, hitting the mark of 17 t/ha for the first time and setting a new world record for large-scale planting.

Before the bottleneck of typical *indica-japonica* intersubspecific heterosis utilization is solved, maximizing the utilization of inter-ecotype heterosis and partial intersubspecific heterosis is still one of the directions for super hybrid rice breeding. It is an important strategy to effectively tap into the inter-ecotype and intersubspecific heterosis by introducing *japonica* components into two-line male sterile lines while retaining modern early-season rice components.

In super hybrid rice development, early-season super hybrid rice has long suffered from long growth duration and insufficient yield heterosis. Eight early-season super hybrid rice varieties have been approved by the MOA for the middle and lower reaches of the Yangtze River, but three were canceled due to limited planting area. Among the remaining five hybrids, three use PTGMS lines Zhu 1S, Lu 18S and the Zhu 1S derived Xiangling 628S bred by Yang Yuanzhu as the female parent. Zhu 1S and Lu 18S, which have low critical sterility and low critical fertility temperatures and excellent overall agronomic traits, were bred with a sterile plant found in the *indica-japonica* hybrid cross (Kangluozao//Kefuhong 2/Xiangzaoxian 3//02428), after selection under the stress of double pressure from natural and artificial low temperatures in multiple generations, a selection method created by Yang Yuanzhu (Yang Yuanzhu, 2007). The nuclear sterility gene of Zhu 1S was then used to breed the super hybrid rice sterile line Xiangling 628S. Zhu 1S and Lu 18S, both have good GCA and strong heterosis, respectively contain 9. 1% and 9. 0% of the genetic components of 02428, a *japonica* rice variety with wide compatibility (Lu Jingjiao, 2014). The cytoplasmic DNA of Zhu 1S and Lu 18S is typically *indica*, which is different from that of Peiai 64S and other *japonica* varieties (Liu Ping, 2008). Therefore, the hybrids derived from Zhu 1S and Lu 18S have wide adaptability to the ecological conditions of early-season *indica* rice in South China, with cold tolerance at the seedling stage, early growth, rapid development at the early stage, no leaf senescence at the late stage, quick grain filling and high seed setting rates. The yield of early-season hybrid rice in the Yangtze River Basin reached a new level due to the partial utilization of *indica-japonica* intersubspecific heterosis. A total of 33 early- and mid-maturing hybrids of Zhu 1S and Lu 18S with a growth duration of 108 days or less have been approved for production. The average yields in regional yield trials reached 7.47 t/ha, 5.4% higher than that of the control variety. The maturity date is one day earlier than the control, which has solved the problem of "early maturity without heterosis or heterosis without early maturity" of early-season hybrid rice in the Yangtze River Basin.

The tall plants, poor fertilizer tolerance and weak lodging resistance are important factors that affect the high and stable yield of super hybrid rice. Yang Yuanzhu obtained the SV14S dwarf mutant from Zhu 1S through somaclone variation technique (Liu Xuanming, 2002) and the length of SV14S basal internodes 1 to 3 was only 1/3 to 1/2 that Zhu 1S, and the culm wall becomes thicker with reduced length of parenchyma cell. For example, the culm wall at the fourth internode is 1/3 thicker than that of Zhu 1S and the length of parenchyma cells is only 1/3 of that of Zhu 1S's (Fig. 7-5). Genetic analysis showed that the shortening of basal internodes is due to a semi-dominant gene named shortened basal internodes (*SBI*). With map-based cloning combined with genomic techniques, it was found that *SBI* encodes *GA2*, a previously unreported oxidase highly expressed in the basal internodes. Enzyme function analysis showed that this oxidase can render active gibberellin inactive. There are two alleles of *SBI* in rice, which results in significant differences in the catalytic activity of the *SBI* enzyme. High-activity *SBI* locus significantly reduces the content of active gibberellin in the basal internodes of rice stalks, thereby inhibiting the

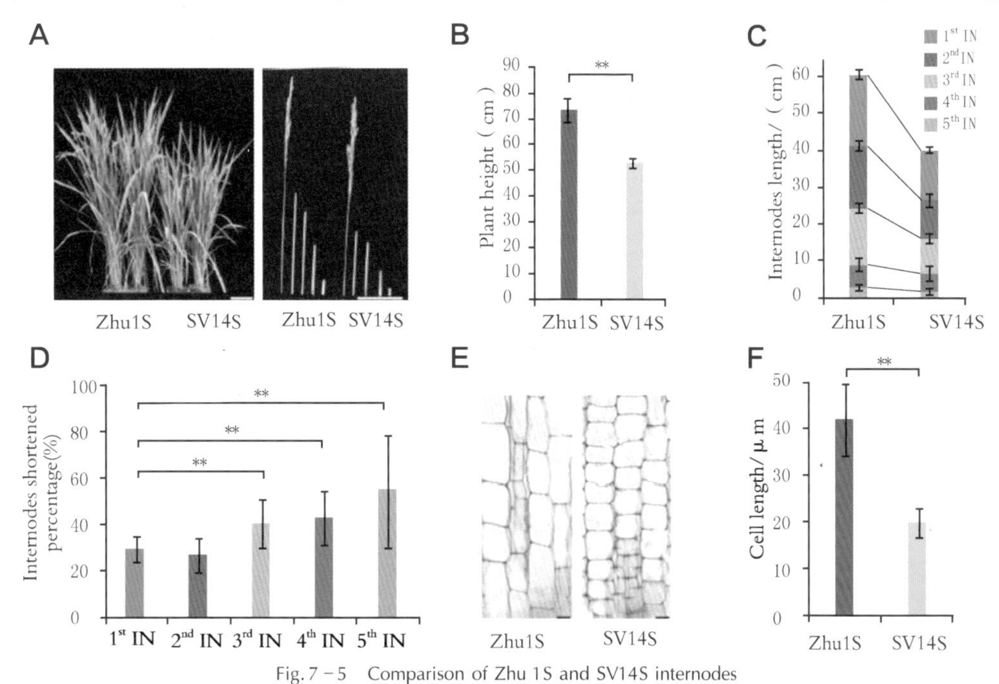

Fig. 7-5 Comparison of Zhu 1S and SV14S internodes

A. Plant type (14 weeks) and stalk length (16 weeks) between Zhu 1S and SV14S (Scale: 10 cm); B. Plant height of Zhu 1S and SV14S at maturity; C. Internode length between Zhu 1S and SV14S; D. Percentage of internode length of SV14S to that of Zhu 1S; E. Longitudinal section of the fourth stem node of Zhu 1S and SV14S; F. Cell length of Zhu 1S and SV14S

elongation of basal internodes (Fig. 7 - 6 and Fig. 7 - 7) (Liu, 2017). The mutant SV14S was crossed with excellent blast-resistant male parent ZR02 and the hybrid progenies were screened under stress and

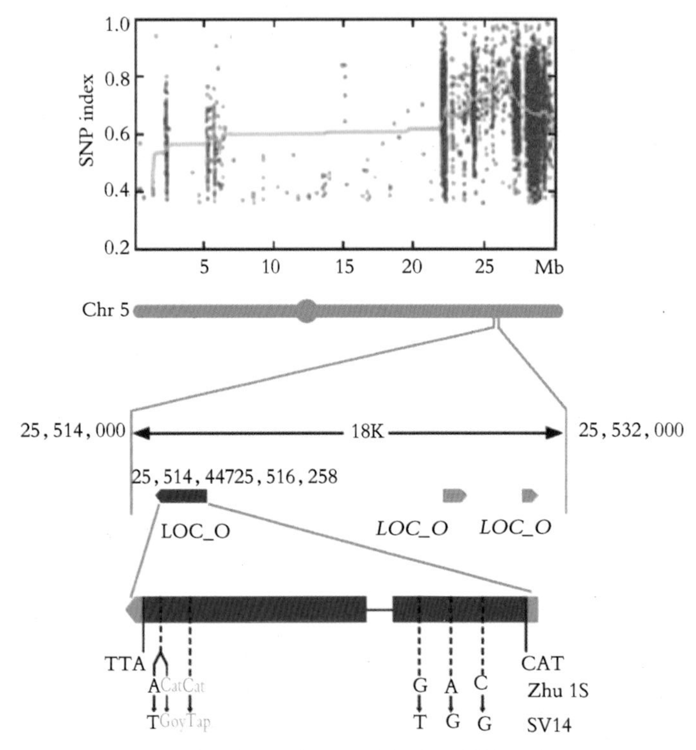

Fig. 7 - 6 Map-based cloning of *SBI*

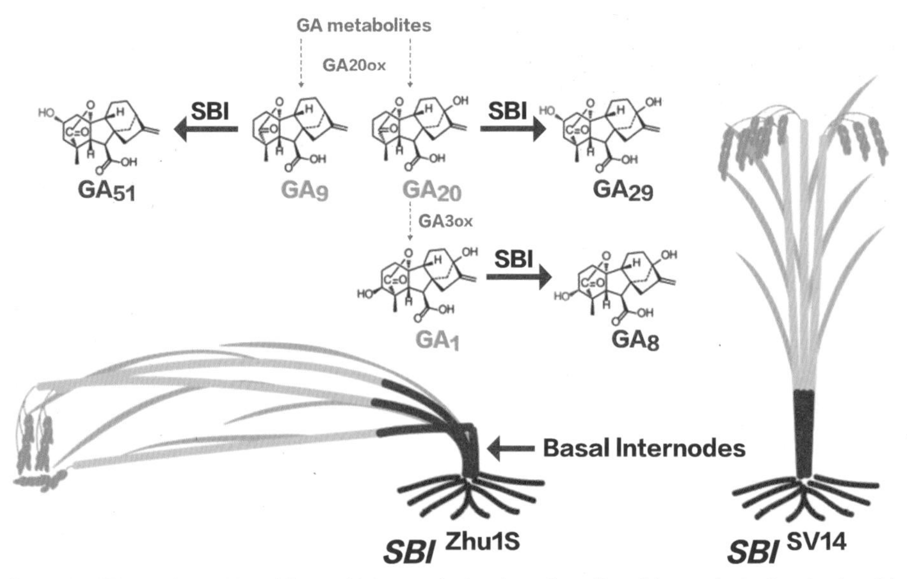

Fig. 7 - 7 *SBI* encodes a *GA2* oxidase, which controls the elongation of basal internodes by inactivating *GA*

cultivated in a targeted manner by adopting the natural and artificial low temperature dual-stress selection method (Yang Yuanzhu, 2007) to breed the excellent lodging-resistant male sterile line Xiangling 628S (Fu Chenjian, 2010). Lingliangyou 268, a combination bred using Xiangling 628S, was approved as super rice by the MOA in 2011.

Longke 638S and Jing 4155S, both mid-season *indica* PTGMS lines with *japonica* genetic kinship, were developed by crossing Zhu 1S, an early-season *indica* wide-compatibility two-line sterile line containing 9.1% of *japonica* genetic components (Lu Jingjiao, 2014) with a mid-season *indica* two-line sterile line with *japonica* components by LPHT. Analysis using 2,120 *indica-japonica* specific SNP molecular markers showed that Longke 638S and Jing 4155S had 8.7% and 11.7% of the genetic components of *japonica* rice, respectively. Whole-genome re-sequencing showed that Longke 638S and Jing 4155S had 56.1% and 51.0% of genetic components of early-season *indica* rice respectively. Longke 638S and Jing 4155S have more than half of the genetic background of early-season *indica* plus some *japonica* genetic background, and they are genetically distant from the restorer lines of the three-line system, male parents of the two-line system and conventional rice varieties in South China, so they have high GCA to produce strong heterosis and a total of 82 combinations bred using these two PTGMS lines have been approved at or above the provincial level. Among them, Longliangyou 1988 achieved a yield increase of 12.2% in the yield trial in the late-season *indica* region of South China (Table 7-5 and Fig. 7-8). Longliangyou-Huazhan was approved by the MOA in 2017. Longliangyou 1988, Longliangyou 1308, Jingliangyou 1377 and Jingliangyou-Huazhan have been accepted by experts of the MOA in 6.67-ha demonstrations. In the super hybrid rice yield test on 6.67-ha demonstration plots in Villager Group 4 of Liujia village, Jiuxiang township, Hanyuan county, Yaan city, Sichuan province, in 2017, Jingliangyou 1377 achieved an average yield of 15.60 t/ha (Table 7-6).

2. Utilization of *indica-japonica* intersubspecific heterosis

Although the use of wide compatibility genes provides an opportunity for the utilization of *indica-japonica* intersubspecific heterosis, there has been no typical case of successful direct utilization so far due to such problems as poor compatibility, tall plants and heterosis over parents in growth duration. At present, the main method to exploit intersubspecific hybrid heterosis is the mating of typical *japonica* rice with *indica*-leaning material or typical *indica* rice with *japonica*-leaning material (collectively referred to as *indica-japonica* hybrid heterosis utilization). Typical intersubspecific heterosis utilization currently mainly adopt the three-line method, and there are two ways - *indica* sterile line × *japonica*-leaning restorer line and *japonica* sterile line × *indi* ca-leaning restorer line. However, due to the difficulty in breeding the restorer lines and the limited genetic diversity of *japonica* rice resources, breeders failed to breed excellent hybrid rice combinations that can be widely used in production through this method. Compared with the three-line method of crossing an *indica* sterile line with a *japonica* restorer line, the method of crossing a *japonica* sterile line with an *indica* restorer line is better for the utilization of intersubspecific heterosis thanks to the rich *indica* rice resources and the easy breeding of *indica* restorer lines. Examples include the Yongyou hybrid series bred by Ningbo Academy of Agricultural Sciences, the Chunyou hybrid series bred by CRRI, the Zheyou hybrid series bred by Zhejiang Academy of Agricultural Sciences, and the Jiayouzhongke hybrid series

Table 7 – 5 Performance of some Longliangyou and Jingliangyou hybrids with strong heterosis in regional yield trials

Variety	Combination	Maturity group	Approval No.	Yield (t/ha)	Yield increase over the control (%)	Growth period (d)	Growth period over the control (d)	Resistance to rice blast (Grade)		Rice quality (grade by national standards)
								Two-year average	Highest loss of panicle to blast	
Longliangyou-Huazhan	Longke 638S× Huazhan	Mid-season *indica* in the middle and lower reaches of the Yangtze River	GS2015026	9. 70	8. 40	140. 1	2. 0	22	5	
		Mid-season *indica* in Wuling Mountain Area	GS2016045	9. 20	6. 59	149. 3	1. 5	1. 85	3	3
		Single-cropping mid-season in the upper reaches of the Yangtze River	GS20170008	9. 39	3. 60	157. 9	3. 6	2. 8	3	2
		Late-season *indica* in South China	GS20170008	7. 66	8. 20	115	1. 5	3. 7	3	
Jingliangyou -Huazhan	Jing 4155S× Huazhan	Late-season *indica* in South China	GS2016602	7. 79	1244	117. 4	22	3. 4	5	3
		Mid-season *indica* in the middle and lower reaches of the Yangtze River	GS20176071	10. 70	5. 90	138. 5	1. 2	24	3	
		Mid-season rice in Wuling Mountain Area	GS20176071	9. 29	5. 56	150	0. 2	1. 7	1	3
		Single-cropping mid-season rice in the upper reaches of the Yangtze River	GS2016022	9. 18	5. 30	157. 6	2. 2	3. 0	3	3
Longliangyou 1968	Longke 638S× R1968	Late-season *indica* in South China	GS20176010	7. 63	12. 20	118	2. 8	3. 9	7	
		Mid-season *indica* in the middle and lower reaches of the Yangtze River	GS2016609	9. 85	7. 99	138. 6	2. 4	3. 6	5	
Jingliangyou 1212	Jing 4155S× R1212	Late-season *indica* in South China	GS2016601	7. 66	10. 57	116. 9	1. 8	40	5	3
Longliangyou 534	Longke 638S× R534	Single-cropping mid-season rice in the middle and lower reaches of the Yangtze River	GS20170001	9. 80	7. 70	1425	3. 0	3. 1	5	3
		Late-season *indica* in South China	GS2016603	7. 66	10. 57	118. 7	3. 5	3. 1	3	3
Longliangyou-Huanglizhan	Longke 638S× Huanglizhan	Late-maturing mid-season *indica* in the middle and lower reaches of the Yangtze River	GS20176002	9. 66	6. 00	139. 2	3. 0	29	5	3
		Late-season *indica* in South China	GS2016604	7. 62	9. 99	116. 9	1. 8	40	5	2
Longliangyou 1377	Longke 638S×R1377	Late-season *indica* in South China	GS20176007	7. 37	8. 30	119. 5	43	29	3	
		Single-cropping mid-season rice in the middle and lower reaches of the Yangtze River	GS20176007	9. 86	6. 10	142. 1	46	2. 8	5	3
Longliangyou 1308	Longke 638S× Huahui 1308	Late maturing mid-season *indica* in the middle and lower reaches of the Yangtze River	GS20176065	9. 80	8. 30	137. 2	3. 9	2. 4	3	

Continued

Variety	Combination	Maturity group	Approval No.	Yield (t/ha)	Yield increase over the control (%)	Growth period (d)	Growth period over the control (d)	Resistance to rice blast (Grade)		Rice quality (grade by national standards)
								Two-year average	Highest loss of panicle to blast	
Longliangyou 1206	Longke 638S×R1206	Single-cropping mid-season rice in the middle and lower reaches of the Yangtze River	GS20176009	9.90	6.50	141.1	1.4	29	5	
		Late-season *indica* in South China	GS20176009	7.36	8.20	117.5	23	3.8	7	
Longliangyou 1813	Longke 638S×R1813	Single-cropping mid-season rice in the middle and lower reaches of the Yangtze River	GS20176024	9.81	8.20	143.2	3.6	46	5	
Longliangyou 836	Longke 638S×R336	Single-cropping mid-season rice in the middle and lower reaches of the Yangtze River	GS20170044	9.73	7.30	142.7	3.4	3.1	5	2
Longliangyou 2010	Longke 638S×R2010	Late-maturing mid-season *indica* in the middle and lower reaches of the Yangtze River	GS20176073	10.70	7.00	137.4	0.1	5.6	9	
Longliangyou 837	Longke 638S×Huahui1337	Late-maturing mid-season *indica* in the middle and lower reaches of the Yangtze River	GS20176064	9.71	6.70	137.2	3.5	3.4	3	
Longliangyou 987	Longke 638S×R967	Late-maturing mid-season *indica* in the middle and lower reaches of the Yangtze River	GS20176059	9.69	6.30	137.7	3.4	5.0	9	
Longliangyou 1212	Longke 638S×R1212	Mid-season rice in Wuling Mountain Area	GS20170022	9.83	6.29	149.1	1.0	1.8	1	3
		Single-cropping mid-season rice in the middle and lower reaches of the Yangtze River	GS20170022	9.91	6.10	140.4	3.2	3.0	5	3

Table 7-6 Yield per unit area of Longliangyou and Jingliangyou series super rice at acceptance (t/ha)

	2016		2017		
	Longhui, Hunan	Xinjin, Chengdu City, Sichuan	Taoyuan, Hunan	Yaan, Sichuan	Mianyang, Sichuan
Longliangyou 1988	11.91		11.85		
Longliangyou 1308	14.19				
Longliangyou 1377				15.59	
Jingliangyou-Huazhan		11.74			11.83

Fig. 7 – 8 Field performance of some Longliangyou and Jingliangyou combinations with strong heterosis

bred by Jiaxing Academy of Agricultural Sciences and the Institute of Genetics and Developmental Biology, CAS. All of these *indica-japonica* hybrids show strong heterosis and super-high yield potential. Among these hybrids, Yongyou 12, Yongyou 15, Yongyou 538, Yongyou 2640, Chunyou 84 and Zheyou 18 have been approved as *indica-japonica* super hybrid rice by the MOA. The yields of these super hybrid rice varieties increased by 7.8%–26.3%, and the highest yield of Yongyou 538, in particular with obvious strong heterosis, increased 26.3% in regional yield trials. Yongyou 4543, Quanjingyou 1, Jiayouzhongke 1, Jiayouzhongke 6 and other *indica-japonica* hybrids showed significant heterosis, and their yield reached more than 11.25 t/ha when planted as single-cropping late-season rice in regional yield trials (Table 7 – 7). Jiayouzhongke 6, a hybrid approved at the national level, achieved an average yield of 11.35 t/ha in the national regional yield trials of late-season *japonica* rice in the middle and lower reaches of the Yangtze River from 2014 to 2016, setting the record of national regional yield trials in South China. Jiayouzhongke 1, a hybrid approved by Shanghai municipal authorities, achieved an average yield of 11.84 t/ha in 2014 and 2015, the highest in Shanghai.

However, there are also some disadvantages for the hybrid model of *japonica* sterile lines and *indica* restorer lines. The main issue is the low seed production yield due to the poor flowering habits such as low stigma exsertion and late flowering of *japonica* male sterile lines (Lin Jianrong, 2006), which seriously affects the commercialization of *indica-japonica* super hybrid rice varieties. The breeding of *japonica* male sterile

Table 7 - 7 Performance of hybrid rice combinations of *japonica* male sterile lines and *indica* restorer lines in regional tests

Variety	Combination	Approval No.	Year of test	Control variety	Yield (kg/ha)	Compared with control variety (%)	Growth period (d)	Plant height (cm)	Effective panicles (10,000/ha)	Grains per panicle	Seed setting rate (%)	1,000-grain weight (g)
Yongyou 12	Yongjing 2A/F5032	ZJA2010015	2007—2008	Xiushui 09	8. 48	16. 20	154. 1	120. 9	184. 5	327. 0	72. 4	22. 5
Yongyou 15	Jingshuang A/F5032	ZJA2012017	2008—2009	Liangyoupeijiu	8. 96	8. 60	138. 7	127. 9	178. 5	235. 1	78. 5	28. 9
Yongyou 538	Yongjing 3A/F7538	ZJA2013022	2011—2012	Jiayou 2	10. 78	26. 30	153. 5	114. 0	210. 0	289. 2	84. 9	22. 5
Yongyou 2640	Yongjing 26A/F7540	ZJA2013024	2010—2011	Xiushui 417	7. 76	10. 90	125. 7	96. 0	286. 5	189. 4	75. 9	24. 4
Yongyou 84	Chunjiang 16A/C84	ZJA2013020	2010—2011	Jiayou 2	10. 29	22. 90	156. 7	120. 0	210. 0	244. 9	83. 6	25. 2
Yongyou 18	Zhe 04A/Zhehui 818	ZJA2012020	2010—2011	Yongyou 9	9. 93	7. 80	153. 6	122. 0	195. 0	306. 1	76. 3	23. 2
Yongyou 4543	Yongjing 45A/F7543	ZJA20170017	2014—2015	Yongyou 8	11. 60	7. 21	170. 4	112. 0	225. 0	—	85. 9	22. 7
Yongyou 1540	Yongjing 15A/F7540	ZJA2017014	2015—2016	Ning 81	10. 27	21. 4	144. 7	99. 9	256. 5	223. 5	80. 9	23. 2
Yongyou 7850	Yongjing 78A/F9250	CNA20170065 (single-cropping late-season *japonica* in the middle and lower reaches of the Yangtze River)	2014—2015	Jiayou 5	10. 85	8. 6	154. 7	116. 6	225. 0	254. 0	88. 6	23. 5
Quanjingyou 1	Quanjing 1A/Quanguanghui 1	CNA20170066 (single-cropping late-season *japonica* in the middle and lower reaches of the Yangtze River)	2015—2016	Jiayou 5	11. 02	9. 5	146. 3	125. 7	243. 0	291. 2	83. 4	22. 5
Jiayouzhongke 1	Jia 66A/Zhongkejiahui 1	HUA2016004	2014—2015	Huayou 14	11. 84	16. 40	157. 5	110. 3	220. 5	234. 0	87. 1	28. 4
Jiayouzhongke 6	Jia 66A/Zhongke 6	CNA20170063 (single-cropping late-season *japonica* in the middle and lower reaches of the Yangtze River)	2014—2015	Jiayou 5	11. 35	13. 60	152. 8	117. 8	214. 5	256. 1	83. 9	29. 4

lines with early flowering and high outcrossing is an important direction to solve the problem of low seed production yield of *indica-japonica* hybrid rice.

Two-line hybrid rice is not restricted by the relationship between the restorer line and the maintainer line, and thus has the advantage of free mating. Yuan Longping proposed the utilization of *indica-japonica* intersubspecific heterosis by adopting the two-line method as the focus of the second strategic development stage of hybrid rice breeding (Yuan Longping, 2006). Although there is currently no typical two-line *indica-japonica* hybrid rice used in production, it is expected that the two-line method will become one of the important means to utilize typical *indica-japonica* intersubspecific heterosis. The following principles can help achieve the goal. 1) Two-line *indica* male sterile lines can be crossed with existing *japonica*-leaning restorer lines (*indica* male sterile lines crossed with *japonica* male parents); 2) Two-line intermediate *japonica* male sterile lines with early flowering and high stigma exsertion can be crossed with *japonica*-leaning restorer lines (*indica* as female and *japonica* as male parent); 3) At least one of the parents should have wide compatibility, least hybrid sterility loci (five sterility loci and two regulation genes in rice have been map-base cloned, Ouyang Yidan, 2016; Shen, 2017). By following the above principles, we will hopefully breed typical *indica-japonica* intersubspecific hybrid rice varieties with high seed production yield, strong heterosis and moderate growth duration.

Although typical intersubspecific hybrid rice has strong heterosis and super high yield potential, it also has disadvantages, such as high plants, long growth duration, unstable seed setting rate, susceptibility to false smut, low seed production yield, and easy deterioration of rice quality (Lin Jianrong, 2012). Traditional genetic improvement techniques combined with modern molecular biological technologies such as molecular design for breeding and genome editing should be adopted to provide technical support and solve the problems for further development of *indica-japonica* intersubspecific hybrid rice.

3. Utilization of interspecific heterosis

Cultivated rice consists of two species – *Oryza sativa* L. and *Oryza glaberrima* Steud. The former covers the three subspecies of *indica*, *japonica* and *javanica*. *Indica* is mainly planted in tropical and subtropical regions, and *japonica* and *javanica* are planted in temperate and tropical regions respectively. Yuan Longping (1990) proposed the general degrees of heterosis level of intersubspecific heterosis among the three rice subspecies as follows, *indica* × *japonica* > *indica* × *javanica* > *japonica* × *javanica* > *indica* × *indica* > *japonica* × *japonica*, and the utilization of *indica-japonica* intersubspecific heterosis has always been a strategic focus of super hybrid rice breeding. As a species related to *Oryza sativa* L., *Oryza glaberrima* Steud has interspecific heterosis with *Oryza sativa* L (Nevame, 2012). However, just like in the case of *indica-japonica* hybridization, hybrid sterility is a major barrier in the utilization of interspecific crosses, which leads to decreased seed setting rates and a lack of significant heterosis in yield, greatly restricting the utilization of distant heterosis (Heuer, 2003). Xie et al. (2017) conducted in-depth research on the *S1* locus controlling the sterility of combinations of the two species, and cloned the key gene *OgTPR1*. It was found that the gene function for the development of male and female gametes was not affected by the knockout of *OgTPR1*, but can specifically eliminate the *S1* locus-mediated sterility of this hybrid rice combination. With research on the mechanism of sterility in the hybrids of this combination, the problem of sterility can be solved to a certain extent by searching for and using wide compatibility genes in the hybrids and knocking out the sterility genes by backcross substitution and gene editing (Sarla, 2005), thereby achieving partial utilization of interspecific heterosis.

Ⅲ. Breeding Principles Based on Heterosis Groups and Heterosis Utilization Models

Heterosis utilization is the basis of the successful breeding of hybrid rice, and its genetic basis is the utilization of genetic diversity. Past experience of hybrid breeding shows that the parents of combinations with strong heterosis often come from different heterotic groups (Reif, 2005; Wang Shengjun, 2007; Wang, 2015). A heterotic group consists of a group of germplasms from different or the same sources. The hybrids of these germplasms show similar combining ability when crossed with those from other heterotic groups, and their hybrids show relatively strong heterosis. However, when germplasms from the same heterotic group are crossed, their hybrid progenies show low heterosis. Group-based heterosis utilization refers to crossing materials from two different heterotic groups to combine their germplasms and breed progenies with strong heterosis (Melchinger, 1998). The theory of heterotic group is based on the long-term practice of hybrid maize breeding, which is still in progress. Although its genetic basis remains to be explained, there is no doubt that this theory plays a great role in guiding modern hybrid maize breeding. The theories of heterotic group and heterosis utilization models can help breeders select hybrid parents, simplify breeding procedures, reduce workload, and improve breeding efficiency. Although there is no systematic report on rice heterotic group research, general conclusions can be drawn from hybrid rice research over the past three decades and existing reports on the study of rice heterotic groups. 1) The three-line *indica* hybrids are coming from two main heterotic groups, the early-season *indica* ecotype from the Yangtze River Basin and the mid- and late-season *indica* ecotype from South Asia and Southeast Asia. Therefore, it can be concluded that the CMS lines (maintainer lines) in China and foreign restorer lines belong to two different heterotic groups, and the cross between them is a model of strong heterosis utilization; 2) The female parent group can be divided into two subgroups, Xieqingzao and Zhenshan 97; while the male parent group can also be divided into two subgroups, the lines or varieties derived from IR24 and IR26 and the lines or varieties derived from IRRI varieties, such as Minghui 63 (its restorer line is derived from IR30) and its derivative lines; 3) The male sterile lines of two-line hybrid rice are a separate heterotic group independent of the two groups for three-line hybrid *indica* rice, but the two-line male parent and three-line restorer line belong to the same heterotic group (Wang Kai, 2014). Based on simplified re-sequencing and pedigree analysis, LPHT classified 190 major hybrid rice parents into four heterotic groups, the group of restorer lines (Ⅰ), the group of two-line male sterile lines (Ⅱ), the group of three-line CMS lines (Ⅲ), and the group of modern early-season *indica* rice (Ⅳ). Among them, Group I can be further divided into the three subgroups of two-line restorers (Ⅰ-1), three-line restorers (Ⅰ-2), and Guangdong conventional rice subgroup (Ⅰ-3). Subgroup Ⅰ-1 mainly consists of Huazhan and its related lines, the derivative lines of Minghui 63 and the derivative lines of 9311. Subgroup Ⅰ-2 mainly consists of three-line restorer lines, while Subgroup Ⅰ-3 mainly consists of conventional rice varieties in South China. The modern early-season *indica* rice group can be further divided into the subgroups of two-line early-season restorer lines (Ⅳ－1) and the subgroup of two-line male sterile lines derived from Zhu 1S (Ⅳ－2) (Fig. 7－9).

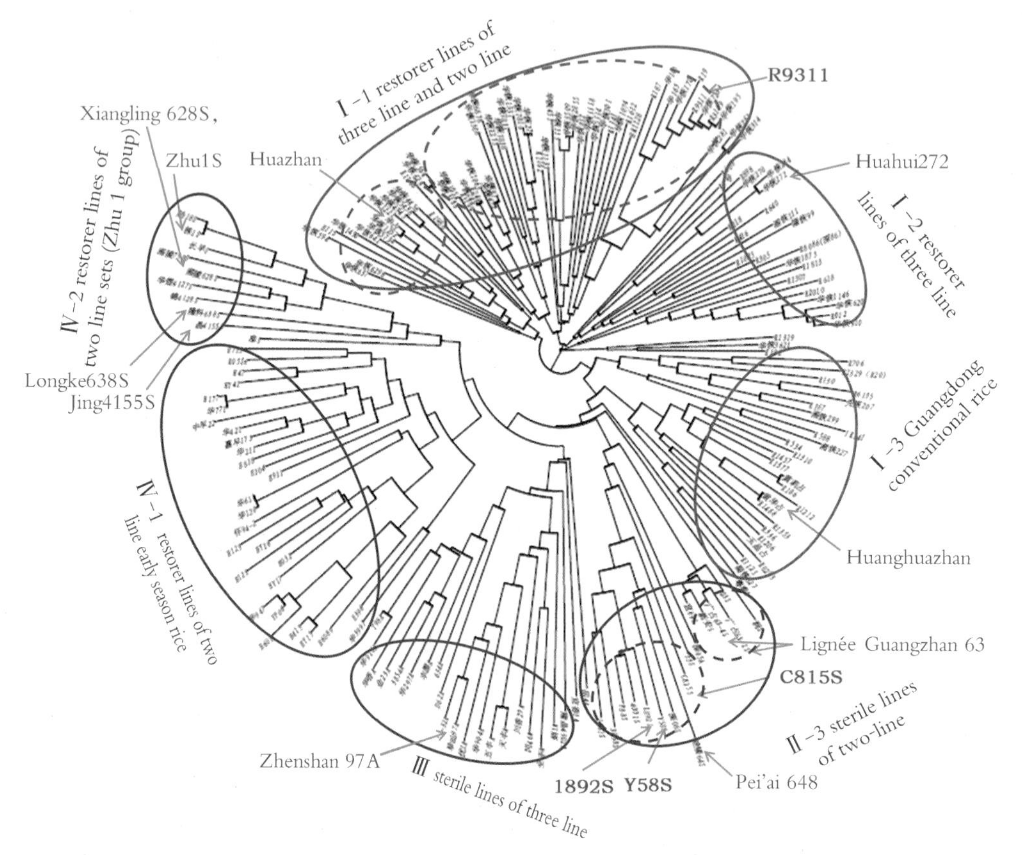

Fig. 7－9 Genetic grouping of major parent lines of hybrid rice

The group of the PTGMS Zhu 1S and its derived lines is a separate group with close genetic distance to the modern early-season *indica* rice in the Yangtze River Basin, but it is distant from the three-line restorer line, the male parent of two-line mid-season rice, and conventional rice in South China. The representative male sterile lines, Longke 638S and Jing 4155S, have high combining ability and yields when combined with three-line restorer lines, mid-season male parents of two-line hybrid rice and South China conventional rice. A total of 82 hybrids derived from these two females have been approved since 2014.

Ⅳ. Principles for Selecting for Cooking and Eating Quality

The endosperm is the edible part of rice and its quality determines the quality of rice. The endosperm is triploid (3n) formed by the combination of two female polar nuclei and one male sperm nucleus. Different from conventional rice varieties, hybrid rice uses the F_1 hybrids of two genetically different parents. F_2 grains are harvested from the F_1 plants, and the heterozygous genes in F_1 undergo triploid segregation in F_2 grains. Take the genotype Aa as an example. The endosperm genotypes of the seeds from the inbred lines of this plant are segregated as AAA : AAa : Aaa : aaa = 1 : 1 : 1 : 1 (Sano, 1984). If the gene is a major gene, the endosperm traits of F_2 grains controlled by this gene will be significantly segregated. Amylose content, gel consistency and gelatinization temperature, as the most important indicators determining the cooking and eating quality of rice, belong to endosperm triploid inheritance. If the differences of amylose content, gel consistency and gelatinization temperature between the parents of hybrid rice are too distant, the cooking quality of harvested hybrid rice grains will show obvious segregation. Different cooking qualities require different cooking conditions; therefore, the overall eating quality of rice is seriously affected by the mixture of rice with different cooking qualities. Take rice amylose content as an example. The waxy (*Wx*) gene is the main gene controlling rice amylose content and plays a decisive role in rice cooking and eating quality (Preiss, 1991; Smith, 1997; Wang, 2017).

The *Wx* gene has abundant variants, and at least six functional alleles have been discovered. Wx^a, as one of the six alleles, mainly exists in *indica* varieties, and controls the formation of high amylose content. The amylose content of rice is generally more than 25%, and the highest is more than 30%. For example, the male sterile lines Zhenshan 97A, Longtefu A, Tianfeng A and their respective maintainer lines are of the Wx^a allelic type, and their amylose content is higher than 25%, putting them into the category of hard rice. Wx^b mainly exists in *japonica* varieties whose amylose content ranges from 15% to 18%, so they belong to the category of sticky rice. Most varieties in the middle and lower reaches of the Yangtze River and rice growing areas in northern China contain this allele (Zhu Jihui, 2015). If a parent with Wx^a (high amylose content) is combined with a parent with Wx^b (low amylose content), its progenies will generally show intermediate amylose content, and they can meet the standard for international Grade C or even Grade A quality in terms of amylose content. However, due to the segregation mode of $Wx^aWx^aWx^a : Wx^aWx^aWx^b : Wx^aWx^bWx^b : Wx^bWx^bWx^b = 1 : 1 : 1 : 1$ in endosperm genotypes, 1/4 of these rice grains are of the $Wx^aWx^aWx^a$ genotype with high amylose content (25%-30%), and 1/4 of them are of the $Wx^bWx^bWx^b$ genotype with low amylose content (15%-18%), and the rest of these rice grains are of heterozygous genotypes ($Wx^aWx^aWx^b$ and $Wx^aWx^bWx^b$) with intermediate amylose content (18%-25%), so the grains harvested from a rice hybrid is actually mixed types of rice (Wang, 2017). The average quality index of the mixed rice may be very good, but it has poor taste and low market popularity.

In the breeding of super hybrid rice, the male sterile lines and restorer lines with similar cooking and eating quality genes (or alleles) should be selected so as to reduce the segregation of endosperm genes between hybrid rice grains. Rice aroma is mainly controlled by the recessive gene *Badh2* (Bradbury, 2015), so the optimal option for the breeding of aromatic super rice combinations should be the selection of both parents with aroma. If only one of the parents has aroma, only 1/4 of the hybrid rice grains will have aroma.

Section 3 Super Hybrid Rice Combinations

According to the requirements of the Super Rice Variety Confirmation Method (Trial) (NBK [2005] No. 39) or Super Rice Variety Confirmation Method (NBK [2005] No. 38), 166 super rice varieties (combinations) have been approved by the MOA since 2005, including 108 super hybrid rice varieties, accounting for 65.1%. After the approval, 36 varieties (hybrids) were later canceled because their planting areas failed to meet the requirements. As of March 2017, 130 varieties (hybrids) had been approved and were still in production, of which 94 were super hybrid rice varieties, accounting for 72.3%.

Ⅰ. Three-line Super Hybrid Rice

1. Three-line *indica* super hybrid rice

As of March 2017, the MOA had approved 62 three-line *indica-indica* super hybrid rice varieties, of which nine were later canceled due to insufficient planting area (Table 7-8). A total of 81 super hybrid parents, including 29 CMS lines and 52 restorer lines for three-line sets, were used to breed 62 three-line *indica-indica* super hybrid rice varieties (Table 7-9). Among the sterile lines, Ⅱ-32A, Tianfeng A and Wufeng A were used to breed the largest number of super hybrid rice varieties, specifically eight, seven and seven, respectively. Among the restorer lines, Shuhui 527 and Zhonghui 8006 were used to breed the largest number of super hybrid rice varieties, specifically four and three, respectively.

2. Three-line *indica-japonica* super hybrid rice

As of March 2017, the MOA had approved 10 three-line *indica-japonica* super hybrid rice varieties, of which three were later canceled due to insufficient planting area (Table 7-10). Five of them were bred by Ningbo Academy of Agricultural Sciences or Ningbo Seed Industry Co., Ltd.

Ⅱ. Two-line Super Hybrid Rice

As of March 2017, the MOA had approved 36 two-line super hybrid rice varieties, of which two were canceled due to insufficient planting area (Table 7-11). A total of 50 super hybrid parents were used to breed 36 two-line *indica-indica* super hybrid rice varieties, including 18 male sterile lines and 32 male parents (Table 7-12). Among the male sterile lines, Y58S and Guangzhan 63-4S were used to breed the largest number of super hybrid rice varieties (combinations), specifically seven and four, respectively. Among the male parents, 9311, Hua 819 and Huazhan were used to breed the largest number of super hybrid rice varieties, specifically three, two and two respectively.

Table 7 - 8 Three-line *indica-indica* hybrid rice

Variety	Combination	Leading institution	Year of approval	Approval No.	Year of cancellation
Tianyou 998	Tianfeng A×Guanghui 998	Rice Research Institute, Guangdong Academy of Agricultural Science (GAAS)	2005	GS2006052, YS2004008, JXS2005041	
Ⅱ - You 1	Ⅱ - 32A×Hang 1	Rice Research Institute, Fujian Academy of Agricultural Science (FAAS)	2005	GS2005023, MS2004003	
Teyouhang 1	Longtepu A×Hang 1	Rice Research Institute, FAAS	2005	GS2005007, MS2003002, ZS2004015, YS2008020	
Zhongzheyou 1	Zhongzhe A×Hanghui 570	China National Rice Research Institute (CNRRI)	2005	ZS2004009, XS2008026, QS2011005, HBS2012004	
Ⅱ - You 7	Ⅱ - 32A×Luhui 17	Rice and Sorghum Research Institute, Sichuan Academy of Agricultural Science (SAAS)	2005	CS82, YNS [2001]369	
Ⅱ - You 602	Ⅱ - 32A×Luhui 602	Rice and Sorghum Research Institute, SAAS	2005	GS2004004, CS2002030	
Ⅱ - Youming 86	Ⅱ - 32A×Minghui 86	Sanming Institute of Agricultural Sciences	2005	GS2001012, QS228, MS2001009	
Ⅱ - You 162	Ⅱ - 32A×Shuhui 162	Rice Research Institute, Sichuan Agricultural University (SAU)	2005	GS20000003, CS(97) 64, ZS195, HBS008—2001	
D-You 527	D62A×Shuhui 527	Rice Research Institute, SAU	2005	GS2003005, QS242, CS135, MS2002002	
Xieyou 527	Xieqingzao A×Shuhui 527	Rice Research Institute, SAU	2005	GS2004008, CS2003003, HBS2004007	
Fengyou 299	Fengyuan A×Xianghui 299	Hunan Hybrid Rice Research Center (HHRRC)	2005	XS2004011	
Jinyou 299	Jin 23A×Xianghui 299	HHRRC	2005	JXS2005091, GXS2005002, SS2009005	
Ⅱ - You 7954	Ⅱ - 32A×Xianghui 7954	Institute of Crop and Nuclear Technology Utilization, Zhejiang Academy of Agricultural Sciences	2005	GS2004019, ZS378	
Ⅱ - You 084	Ⅱ - 32A×Zhenhui 084	Zhenjiang Institute of Agricultural Sciences in Hilly Areas of Jiangsu province	2005	GS2003054, SS200103	
Guodao 3	Zhong 8A×Zhonghui 8006	CNRRI	2005	ZS2004011, JXS2004027	2017

Continued 1

Variety	Combination	Leading institution	Year of approval	Approval No.	Year of cancellation
Guodao 1	Zhong 9A×Zhonghui 8006	CNRRI	2005	GS2004032, JXS2004009, YS2006050	
Xieyou 9308	Xieqingzao A × Zhonghui 9308	CNRRI	2005	ZS194	2014
Qiannanyou 2058	K22A×QN2058	Qiannan Academy of Agricultural Sciences	2006	QS2005009	2008
Q-You 6	Q2A×R1005	Chongqing Seed Company	2006	GS2006028, QS2005014, YS2005001, XS2006032, HS2006008	
Tianyou 122	Tianfeng A×Guanghui 122	Rice Research Institute, GAAS	2006	GS2009029, YS2005022	
D-You 202	D62A×Shuhui 202	Rice Research Institute, SAU	2006	GS2007007, CS2004010, ZS2005001, GXS2005010, WS06010503, HBS2007010	
Yifeng 8	K22A×Shuhui 527	Sorghum Research Institute, Sichuan Academy of Agricultural Science (SAAS)	2006	GS2006020	2017
Jinyou 527	Jin 23A×Shuhui 527	Rice Research Institute, SAU	2006	GS2004012, CS2002002	
Ganxin 688	Tianfeng A×Chnaghui 121	Jiangxi Agricultural University (JAU)	2007	JXS2006032	
Ⅱ-Youhang 2	Ⅱ-32A× Hang 2	Rice Research Institute, FAAS	2007	GS2007020, WS06010497	
Guodao 6	Neixiang 2A×Zhonghui 8006	CNRRI	2007	GS2007011, GS2006034, YS2007007	
Rongyou 3	Rongfeng A×R3	Rice Research Institute, GAAS	2009	GS2009009, JXS2006062	
Jinyou 458	Jin 23A×R458	Rice Research Institute, Jiangxi Academy of Agricultural Science (JAAS)	2009	GS2008007, JXS2003005	2017
Luoyou 8	Luohong 3A×R8108	College of Life Sciences, Wuhan University	2009	GS2007023, HBS2006005	
Chunguang 1	G4A×Chunhui 350	Rice Research Institute, JAAS	2009	JXS2006055	2013
Wufengyou T025	Wufeng A×Changhui T025	JAU	2010	GS2010024, JXS2008013	
Wuyou 308	Wufeng A×Guanghui 308	Rice Research Institute, GAAS	2010	GS2008014, YS2006059	
Tianyou 3301	Tianfeng A×Minhui 3301	Biotechnology Research Institute, FAAS	2010	GS2010016, MS2008023, QS2011015	
Xinfengyou 22	Xinfeng A×Zhehui 22	Jiangxi Dazhong Seed Industry Co., Ltd.	2010	JXS2007034	2014

Continued 2

Variety	Combination	Leading institution	Year of approval	Approval No.	Year of cancellation
Teyou 582	Longtepu A×Gui 582	Rice Research Institute, JAAS	2011	GXS2009010	
03-You 66	03A×Zaohui 66	Rice Research Institute, JAAS	2011	JXS2007025	2015
Shenyou 9516	Shen 95A×R7116	Tsinghua Shenzhen International Graduate School	2012	YS2010042	
Q-You 8	Q3A×R78	Chongqing Zongyi Seed Industry Co., Ltd.	2012	CQS2008007	2016
Yiyou 673	Yixiang 1A×Fuhui 673	Rice Research Institute, FAAS	2012	DS201,0005	
Tianyou-Huazhan	Tianfeng A×Huazhan	CNRRI	2012	GS2012001, GS2011008, GS2008020, QS2012009, YS2011036, HBS2011006	
Dexiang 4103	Dexiang 074A×Luhui H103	Rice and Sorghum Research Institute, SAAS	2012	GS2012024, CS2008001	
Jinyou 785	Jin 23A×Qianhui 785	Guizhou Rice Research Institute	2012	QS01,0002	
H-You 518	H28A×51084	Hunan Agricultural University	2013	GS011020, XS2010032	
Tianyou 3618	Tianfeng A×Guanghui 3618	Rice Research Institute, GAAS	2013	YS2009004	
Tianyou-Huazhan	Tianfeng A×Huazhan	CNRRI	2013	GS2012001, GS2011008, GS2008020, QS2012009, YS2011036, HBS2011006	
Zhong-9-You 8012	Zhong 9A×Zhonghui 8012	CNRRI	2013	GS2009019	
Rongyou 225	Rongfeng A×R225	Rice Research Institute, JAAS	2014	GS2012029, JXS2009017	
Wufengyou 615	Wufeng A×Guanghui 615	Rice Research Institute, GAAS	2014	YS2012011	
F-You 498	FS3A×Shuhui 498	Rice Research Institute, SAU	2014	GS2011006, XS2009019	
Shengtaiyou 722	Shengtai A×Yuehui 9722	Hunan Dongting High-tech Seed Industry Co., Ltd.	2014	XS2012016	
Nei 5-You 8015	Neixiang 5A×Zhonghui 8015	CNRRI	2014	GS2010020	
Shenyou 1029	Shen 95A×R1029	Jiangxi Modern Seed Industry Co., Ltd.	2015	GS2013031	

Continued 3

Variety	Combination	Leading institution	Year of approval	Approval No.	Year of cancellation
Yixiangyou 2115	Yixiang 1A×Yahui 2115	College of Agriculture, Sichuan Agricultural University	2015	GS2012003, CS2011001	
Jiyou 225	Jifeng A×R225	Rice Research Institute, JAAS	2016	JXS2014013	
Wuyou 662	Wufeng A×R662	Jiangxi Huinong Seed Industry Co., Ltd.	2016	JXS2012010	
Deyou 4727	Dexiang 074A×Chenghui 727	Rice and Sorghum Research Institute, SAAS	2016	GS2014019, CS2014004, DS2013007	
Fengtianyou 553	Fengtian 1A×Guihui 553	Rice Research Institute, Guangxi Academy of Agricultural Siences	2016	GXS2013027, YS2016052	
Wuyouhang 1573	Wufeng A×Yuehui1573	Jiangxi Super Rice Research and Development Center	2016	JXS2014019	
Wufengyou 286	Wufeng A×Zhonghui 286	Jiangxi Modern Seed Industry Co., Ltd.	2016	GS2015002, JXS2014005	
Wuyou 116	Wufeng A×R7116	Guangdong Modern Agriculture Group Co., Ltd.	2017	YS2015045	
Jifengyou 1002	Jifeng A×Guanghui 1002	Rice Research Institute, GAAS	2017	YS2013040	
Yixiang 4245	Yixiang 1A×Yihui 4245	Yibin Academy of Agricultural Sciences	2017	GS2012008, CS009004	

Table 7 - 9 Parents of three-line *indica-indica* hybrid rice and the number of super hybrid rice varieties bred

Sterile line	No. of super hybrid rice varieties bred	Restorer line	No. of super hybrid rice varieties bred	Restorer line	No. of super hybrid rice varieties bred
Ⅱ-32A	8	Shuhui 527	4	Hang 2	1
Tianfeng A	7	Zhonghui 8006	3	Hanghui 570	1
Wufeng A	7	R225	2	Luhui 17	1
Jin 23A	4	R7116	2	Luhui 602	1
Yixiang 1A	3	Hang 1	2	Luhui H103	1
D62A	2	Huazhan	2	Minhui 3301	1

Continued

Sterile line	No. of super hybrid rice varieties bred	Restorer line	No. of super hybrid rice varieties bred	Restorer line	No. of super hybrid rice varieties bred
K22A	2	Xianghui 299	2	Minghui 86	1
Dexiang 074A	2	51084	1	Qianhui 785	1
Jifeng A	2	QN2058	1	Shuhui 162	1
Longtepu A	2	R1005	1	Shuhui 202	1
Rongfeng A	2	R1029	1	Shuhui 498	1
Shen 95A	2	R3	1	Yahui 2115	1
Xieqingzao A	2	R458	1	Yihui 4245	1
Zhong 9A	2	R662	1	Yuehui 9722	1
03A	1	R78	1	Yuehui 1573	1
FS3A	1	R8108	1	Zohui 66	1
G4A	1	Changhui 121	1	Zhehui 22	1
H28A	1	Changhui T025	1	Zhehui 7954	1
Q2A	1	Chenghui 727	1	Zhenhui 084	1
Q3A	1	Chunhui 350	1	Zhonghui 286	1
Fengtian 1A	1	Fuhui 673	1	Zhonghui 8012	1
Fengyuan A	1	Guanghui 1002	1	Zhonghui 8015	1
Luohong 3A	1	Guanghui 122	1	Zhonghui 9308	1
Neixiang 2A	1	Guanghui 308	1		
Neixiang 5A	1	Guanghui 3618	1		
Shengtai A	1	Guanghui 615	1		
Xinfeng A	1	Guanghui 998	1		
Zhong 8A	1	Gui 582	1		
Zhongzhe A	1	Guihui 553	1		

Table 7 - 10 Three-line *indica-japonica* super hybrid rice

Variety	Combination	Leading institution	Year of approval	Approval No.	Year of cancellation
Liaoyou 1052	105A×C52	Rice Research Institute, Liaoning Academy of Agricultural Sciences	2005	LS2009125	2010
Ⅲ-You 98	MH2003A×R18	Rice Research Institute, Anhui Academy of Agricultural Sciences	2005	WS02010333	2011
Liaoyou 5218	Liao 5216A×C418	Rice Research Institute, Liaoning Academy of Agricultural Sciences	2005	LS [2001] 89	2011
Yongyou 6	Yongjing 2A×K4806	Rice Research Institute, Ningbo Academy of Agricultural Sciences (NAAS)	2006	ZS2005020 MS2007020	
Yongyou 12	Yongjing 2A×F5032	NAAS	2011	ZS2010015	
Yongyou 15	JIngshuang A×F5032	NAAS	2013	ZS2012017 MS2013006	
Yongyou 84	Chunjiang 16A×C84	CNRRI	2015	ZS2013020	
Yongyou 538	Yongjing 3A×F7538	Ningbo Seed Industry Co., Ltd.	2015	ZS2013022	
Zheyou 18	Zhe 04A×Zhehui818	Institute of Crop and Nuclear Technology Utilization, Zhejiang Academy of Agricultural Sciences	2015	ZS2012020	
Yongyou 2640	Yongjing 26A×F7540	Ningbo Seed Industry Co., Ltd.	2017	MS2016022 SS201507 ZS2013024	

Table 7 - 11 Two-line super hybrid rice

Variety	Combination	Leading institution	Year of approval	Approval No.	Year of cancellation
Liangyoupeijiu	Peiai 64S×9311	Institute of Food Crops, Jiangsu Academy of Agricultural Sciences (JAAS)	2005	GS2001001, SS313, XS300, MS2001007, GXS2001117, HBS006—2001	
Zhunliangyou 527	Zhun S×Shuhui 527	HHRRC	2005	GS2005026, GS2006004, XS006—2003, MS2006024	
Y-You 1	Y58S×9311	HHRRC	2006	GS2008001, GS2013008, XS2006036, YS2015047	
Liangyou 287	HD9802S×R287	College of Life Sciences, Hubei University	2006	HBS2005001, GXS2006003	

Continued 1

Variety	Combination	Leading institution	Year of approval	Approval No.	Year of cancellation
Xinliangyou 6	Xin'an S×Anxuan 6	Anhui Quanyin Agricultural High-tech Research Institute	2006	GS2007016, WS0501M60, SS200602	
Zhuliangyou 819	Zhu 1S×Hua 819	Hunan AVA Seed Industry Research Institute	2006	JXS2006004, XS2005010	
Peizataifeng	Peiai64S×Taifengzhan	College of Agriculture, South China Agricultural University	2006	GS2005002, YS20M013, JXS2006014	
Xinliangyou 6380	03S×D208	Rice Research Institute, Nanjing Agricultural University	2007	GS2008012, SS200703	
Fengliangyou 4	Feng 39S×Yandao 4 – Xuan	Hefei Fengle Seed Industry Co., Ltd.	2007	GS2009012, WS06010501	
Yangliangyou 6	Guangzhan 63 – 4S×9311	Lixiahe Region Agricultural Research Institute of Jiangsu province	2009	GS2005024, SS200302, QS2003002, SHS2005003, HBS2005005	
Fengliangyouxiang 1	Guangzhan 63S × Fengxianghui 1	Institute of Food Crops, Hubei Academy of Agricultural Sciences	2009	GS2007017, XS2006037, JXS2006022, WS07010622	
Luliangyou 819	Lu 18S×Hua 819	Hunan AVA Seed Industry Research Institute	2009	GS2008005, XS2008002	
Peiliangyou 3076	Peiai 64S×R3076	Institute of Food Crops, Hubei Academy of Agricultural Sciences	2010	HBS20060M	2014
Guiliangyou 2	Guike – 2S×Guihui 582	Nanning Branch, China National Rice Improvement Center	2010	GXS2008006	
Zhunliangyou 1141	Zhun S×R1141	Hunan Longping Seed Industry Co., Ltd.	2011	CQS20100010005, XS2008021	2014
Lingliangyou 268	Xiangling 628S×Hua 268	Hunan AVA Seed Industry Research Institute	2011	GS2008008	
Huiliangyou 6	1892S×Yangdao 6 – Xuan	Rice Research Institute, Anhui Academy of Agricultural Sciences	2011	GS2012019, WS2008003	
Zhunliangyou 608	Zhun S×R608	Hunan Longping Seed Industry Co., Ltd.	2012	GS2009032, XS2010018, XS2010027, 2015005	

Continued 2

Variety	Combination	Leading institution	Year of approval	Approval No.	Year of cancellation
Shenliangyou 5814	Y58S×Bing 4114	China National Hybrid Rice R&D Center	2012	GS2009016, GS20170013, YS2008023, QS2013001	
Guangliangyouxiang 66	Guangzhan 63 - 4S×Xianghui 66	Hubei Agricultural Technology Promotion Station	2012	GS2012028, HBS2009005, YUS2011004	
Y Liangyou 087	Y58S×R087	Wode Crop Research Institute, Nanning city	2013	GXS2010014, YS2015049	
Y Liangyou 5867	Y58S×R674	Keyuan Seed Industry Co., Ltd., Jiangxi province	2014	GS2012027, JXS2010002, ZS2011016	
Guangliangyou 272	Guangzhan 63 - 4S×R7272	Institute of Food Crops, Hubei Academy of Agricultural Sciences	2014	HBS2012003	
Liangyou 038	03S×R828	Jiangxi Tianya Seed Industry Co., Ltd.	2014	JXS2010006	
Liangyou 616	Guangzhan 63 - 4S×Fuhui616	Fujian Nongjia Seed Industry Co., Ltd., China National Seed Group Co., Ltd.	2014	MS2012003	
C - Liangyou - Huazhan	C815S Huazhan	Beijing Golden Nonghua Seed Technology Co., Ltd.	2014	GS2013003, GS2015022, GS2016002, 2016008, JXS2015008, HBS2013008	
Y Liangyou 2	Y58S×Yuanhui 2	HHRRC	2014	GS2013027	
Liangyou 6	HD9602S×Zaohui 6	Hubei Jingchu Seed Industry Co., Ltd.	2014	GS2011003	
N Liangyou 2	N118S×R302	Changsha Nianfeng Seed Industry Co., Ltd.	2015	XS2013010	
H Liangyou 991	HD90O2S× R991	Guangxi Zhaohe Seed Industry Co., Ltd.	2015	GXS2011017	
Shenliangyou 870	Shen 08S×P5470	Guangdong Zhaohe Seed Industry Co., Ltd.	2016	YS2014037	
Huiliangyou 996	1892S×R996	Hefei Keyuan Institute of Agricultural Sciences	2016	GS2012021	

Continued 3

Variety	Combination	Leading institution	Year of approval	Approval No.	Year of cancellation
Shenliangyou 8386	Shen 08S×R1386	Guangxi Zhaohe Seed Industry Co., Ltd.	2017	GXS2015007	
Y Liangyou 900	Y58S×R900	Hunan Yuanchuang Super Rice Technology Co., Ltd.	2017	GS2016044, GS2015034, YS2016021	
Y Liangyou 1173	Y583×Hanghui 1173	National Plant Space Breeding Engineering Technology Research Center, South China Agricultural University	2017	YS2015016	
Longliangyou-Huazhan	Longke 638S×Huazhan	Hunan Longping Agricultural Hi-Tech Co., Ltd.	2017	GS2015026, GS2016045, GS20170008, XS2015014, JXS2015003, MS2016028	

Table 7 - 12 Parents of two-line super hybrid rice and number of super hybrid rice varieties (combinations) bred

Female parent	Number of super hybrid rice varieties bred	Male parent	Number of super hybrid rice varieties bred
Y58S	7	9311	3
Guangzhan 63 - 4S	4	Hua 819	2
HD9802S	3	Huazhan	2
Peiai 64S	3	D208	1
Zhun S	3	P5470	1
03S	2	R0S7	1
1892S	2	R1141	1
Shen 08S	2	R1386	1
C815S	1	R287	1
N118S	1	R302	1
Feng 39S	1	R3076	1

Continued

Female parent	Number of super hybrid rice varieties bred	Male parent	Number of super hybrid rice varieties bred
Guangzhan 63S	1	R606	1
Guike - 2S	1	R674	1
Longke 638S	1	R7272	1
Lu 18S	1	R828	1
Xiangling 628S	1	R900	1
Xin'an S	1	R991	1
Zhu 1S	1	R996	1
		Anxuan 6	1
		Bing 4114	1
		Fengxianghui 1	1
		Fuhui 616	1
		Guihui 582	1
		Hanghui 1173	1
		Hua 268	1
		Shuhui 527	1
		Taifengzhan	1
		Xianghui 66	1
		Yandao 4 - Xuan	1
		Yangdao 6 - Xuan	1
		Yuanhui 2	1
		Zaohui 6	1

References

[1] CHEN LIYUN, XIAO YINGHUI, TANG WENBANG, et al. Three-step design and practice of super hybrid rice breeding[J]. Chinese Journal of Rice Science, 2007, 21 (1): 90-94.

[2] CHEN WENFU, XU ZHENGJIN, ZHANG LONGBU. Super high yield rice breeding: from theory to practice[J]. Journal of Shenyang Agricultural University, 2003, 34 (5): 324-327.

[3] CHENG SHIHUA. Chinese super rice breeding in China[M]. Beijing: Science Press, 2010.

[4] DENG QIYUN. Breeding of Y58S, a photo-thermo sensitive rice male sterile line with wide adaptability[J]. Hybrid Rice, 2005, 20:15-18.

[5] FEI ZHENJIANG, DONG HUALIN, WU XIAOZHI, et al. Theory and practice of super rice breeding[J]. Hubei Agricultural Sciences, 2014, 53 (23): 5633-5637.

[6] FU CHENJIAN, QIN PENG, HU XIAOCHUN, et al. Breeding of Xiangling 628S, a dwarf thermo-sensitive genic rice male sterile line with lodging resistance[J]. Hybrid Rice, 2010 (S1): 177-181.

[7] HUANG YAOXIANG, CHEN SHUNJIA, CHEN JINCAN, et al. Rice cluster breeding[J]. Guangdong Agricultural Sciences, 1983 (01): 1-6.

[8] HUANG YAOXIANG. Ecological breeding project of chinese semi-dwarf super rice with early growth, deep root, super high yield and superb quality[J]. Guangdong Agricultural Sciences, 2001 (3): 2-6.

[9] LI JIANWU, ZHANG YUZHU, WU JUN, et al. Study on high-yielding cultivation techniques of new super high yield rice combination Y Liangyou 900 with yield of 15.40 t/ha[J]. China Rice, 2014, 20 (6): 1-4.

[10] LI JIANWU, DENG QIYUN, WU JUN, et al. Traits and high-yielding cultivation techniques of new super hybrid rice combination Y Liangyou 2[J]. Hybrid Rice, 2013, 28 (01): 49-51.

[11] LIN JIANRONG, SONG XINWEI, WU MINGGUO. Biological traits and heterosis utilization of four *indica-japonica* intermediate restorer lines with wide compatibility[J]. Chinese Journal of Rice Science, 2012, 26 (6): 656-662.

[12] LIN JIANRONG, WU MINGGUO, SONG XINWEI. Relationship between flowering habit and outcrossing seed setting rate of CMS lines in three-line *japonica* rice[J]. Hybrid Rice, 2006 (5): 69-72.

[13] LIU PING, DAI XIAOJUN, YANG YUANZHU, et al. Study on the *indica* and *japonica* properties of the PTGMS line Zhu 1S[J]. Acta Agronomica Sinica, 2008, 34 (12): 2112-2120.

[14] LIU XUANMING, YANG YUANZHU, CHEN CAIYAN, et al. Screening of dwarf mutants of the PTGMS line Zhu 1S in rice by somaclonal variation[J]. Chinese Journal of Rice Science, 2002, 16 (4): 321-325.

[15] LU JINGJIAO. Analysis of *indica* and *japonica* components of main hybrid rice parents in south China[D]. Changsha: Hunan Normal University Press, 2014.

[16] OUYANG YIDAN. Sterility and wide compatibility of *indica-japonica* hybrid rice[J]. Chinese Science Bulletin, 2016 (35): 3833-3841.

[17] WANG KAI. Analysis of heterosis groups and fine mapping of QTL *qhus6. 1* for hull silicon content in *indica* rice[D]. Beijing: Chinese Academy of Agricultural Sciences, 2014.

[18] WANG SHENGJUN, LU ZUOMEI. A preliminary study on the heterosis of parents of *indica* rice in China[J]. Journal of Nanjing Agricultural University, 2007, 30 (1): 14-18.

[19] WU JUN, DENG QIYUN, YUAN DINGYANG, et al. Progress in research on super hybrid rice[J]. Chinese Science Bulletin, 2016 (35): 3787-3796.

[20] XU ZHENGJIN, CHEN WENFU, ZHOU HONGFEI, et al. Physiological and ecological traits and utilization prospects of erect-panicle rice population[J]. Chinese Science Bulletin, 1996 (12): 1122-1126.

[21] YANG SHOUREN. Studies on the breeding of *indica-japonica* hybrids[J]. Hereditas, 1973 (2): 34-38.

[22] YANG YUANZHU, FU CHENJIAN, HU XIAOCHUN, et al. Progress, problems and countermeasures of early-season hybrid rice breeding in the Yangtze river basin[J]. Hybrid Rice, 2010 (S1): 68-74.

[23] YANG YUANZHU, FU CHENJIAN, HU XIAOCHUN, et al. Discovery of Zhu 1S's PTGMS gene and breeding of super hybrid early-season rice[J]. China Rice, 2007 (6): 17-22.

[24] YANG ZHENYU. Progress in *japonica* hybrid rice breeding[J]. Hybrid Rice, 1994, 19:46-49.

[25] YU CAN, YANG YUANZHU, QIN PENG, et al. Comparison of hybrid rice matching methods[J]. Crop Research, 2014 (4): 416-418.

[26] YUAN LONGPING. Hybrid rice[M]. Beijing: China Agricultural Publishing House, 2002.

[27] YUAN LONGPING. Research on super hybrid rice[M]. Shanghai: Shanghai Science and Technology Press, 2006.

[28] YUAN LONGPING. Progress of super hybrid rice breeding[J]. China Rice, 2008, 6 (1): 1-3.

[29] YUAN LONGPING. Research progress of two-line hybrid rice[J]. Scientia Agricultura Sinica, 1990, 23 (3): 1-6.

[30] YUAN LONGPING. Further ideas for breeding super high yielding hybrid rice[J]. Hybrid Rice, 2012, 27 (6): 1-2.

[31] YUAN LONGPING. Strategy for breeding intersubspecific hybrid rice combinations[J]. Hybrid Rice, 1996 (2): 1-3.

[32] YUAN LONGPING. Super high yielding hybrid rice breeding[J]. Hybrid Rice, 1997, 12 (6): 1-6.

[33] YUAN LONGPING. Breeding strategy of hybrid rice[J]. Hybrid Rice, 1987 (1): 1-2.

[34] ZHONG RICHAO, CHEN YUEJIN, YANG YUANZHU. Study on combining ability of nine hybrid rice parents[J]. Journal of Shaoyang University (Natural Science Edition), 2015, 12 (4): 36-42.

[35] ZHOU KAIDA, MA YUQING, LIU TAIQING, et al. Breeding of heavy-panicle combinations among subspecies of hybrid rice-theory and practice of super high yielding hybrid rice breeding[J]. Journal of Sichuan Agricultural University, 1995 (4): 403-407.

[36] ZHU JIHUI, ZHANG CHANGQUAN, GU MINGHONG, et al. Research progress in allelic variation and breeding utilization of *Wx* gene in rice[J]. Chinese Rice Science, 2015, 29 (4): 431-438.

[37] HALLAUER, ARNEL R, MARCELO J, et al. Quantitative genetics in maize breeding[M]. Springer Science & Business Media, 2001.

[38] BRADBURY L M T, FITZGERALD T L, HENRY R J, et al. The gene for fragrance in rice[J]. Plant Biotechnology Journal, 2005, 3(3):363-370.

[39] CHEN J, DING J, OUYANG Y, et al. A triallelic system of *S5* is a major regulator of the reproductive barrier and compatibility of *indica-japonica* hybrids in rice[J]. Proceedings of the National Academy of Sciences, 2008, 105(32):11436-11441.

[40] CHIN J H, LU X, HAEFELE S M, et al. Development and application of gene-based markers for the major rice QTL phosphorus uptake 1[J]. Theoretical & Applied Genetics, 2010, 120(6):1087-1088.

[41] GLASZMANN J C. Isozymes and classification of Asian rice varieties[J]. Theoretical and Applied Genetics, 1987, 74(1):21-30.

[42] GUO J, XU X, LI W, et al. Overcoming intersubspecific hybrid sterility in rice by developing *indica*-compatible *japonica* lines[J]. Scientific Reports, 2016, 6:26878.

[43] HEUER S, MIéZAN K M. Assessing hybrid sterility in *Oryza glaberrima* × *O. sativa* hybrid progenies by

PCR marker analysis and crossing with wide compatibility varieties[J]. Theoretical and Applied Genetics, 2003, 107(5):902 - 909.

[44] KUMAR R V, VIRMANI S S. Wide compatibility in rice (*Oryza sativa* L.)[J]. Euphytica, 1992, 64:71 - 80.

[45] LIU C, ZHENG S, GUI J, et al. Shortened basal internodes encode a gibberellin 2-oxidase and contribute to lodging resistance in rice[J]. Molecular Plant, 2018, 11(2):288 - 299.

[46] MELCHINGER A E, GUMBER R K. Overview of heterosis and heterotic groups in agronomic crops[J]. Concepts and Breeding of Heterosis in Crop Plants, 1998, 1:29 - 44.

[47] MIKEL M A, DUDLEY J W. Evolution of North American dent corn from public to proprietary germplasm [J]. Crop Science, 2006, 46:1193 - 1205.

[48] MORINAGA T, KURIYAMA H. Intermediate type of rice in the subcontinent of India and Java[J]. Japanese Journal of Breeding, 1958, 7(4):253 - 259.

[49] NEVAME A Y M, ANDREW E, SISONG Z, et al. Identification of interspecific grain yield heterosis between two cultivated rice species *Oryza sativa* L. and *Oryza glaberrima* steud[J]. Australian Journal of Crop Science, 2012, 6(11):1558.

[50] PREISS J. Biology and molecular biology of starch synthesis and its regulation[J]. Plant Mol Cell Biol, 1991, 7(20):5880 - 5883.

[51] REIF J C, HALLAUER A R, MELCHINGER A E. Heterosis and heterotic patterns in maize[J]. Maydica, 2005, 50: 215 - 223.

[52] SARLA N, SWAMY B P M. *Oryza glaberrima* : a source for the improvement of *Oryza sativa* [J]. Current Science, 2005: 955 - 963.

[53] SHEN R, WANG L, LIU X, et al. Genomic structural variation-mediated allelic suppression causes hybrid male sterility in rice[J]. Nature Communications, 2017, 8(1):1310.

[54] SANO Y. Differential regulation of waxy gene expression in rice endosperm[J]. Theoretical and Applied Genetics, 1984, 68(5):467 - 473.

[55] SMITH A M, DENYER K, MARTIN C. The synthesis of the starch granule[J]. Annual Review of Plant Biology, 1997, 48(1):67 - 87.

[56] WANG K, QIU F, LARAZO W, et al. Heterotic groups of tropical *indica* rice germplasm[J]. Theoretical and Applied Genetics, 2015, 128(3):421 - 430.

[57] WANG K, ZHOU Q, LIU J, et al. Genetic effects of *Wx* allele combinations on apparent amylose content in tropical hybrid rice[J]. Cereal Chemistry, 2017, 94(5):887 - 891.

[58] WANG S, WAN J, LU Z. Parental cluster analysis in *indica* hybrid rice (*Oryza sativa* L.) by SSR analysis [J]. Zuowuxuebao, 2006, 32(10):1437 - 1443.

[59] XIE Y, XU P, HUANG J, et al. Interspecific hybrid sterility in rice is mediated by *OgTPR1* at the *S1* locus encoding a peptidase-like protein[J]. Molecular Plant, 2017, 10(8):1137 - 1140.

[60] XU K, XU X, FUKAO T, et al. Sub1A is an ethylene-response-factor-like gene that confers submergence tolerance to rice[J]. Nature (London), 2006, 442(7103):705 - 708.

[61] YE C, ARGAYOSO M A, REDOÑA E D, et al. Mapping QTL for heat tolerance at flowering stage in rice using SNP markers[J]. Plant Breeding, 2012, 131(1):33 - 41.

[62] ZHEN G, QIN P, LIU K Y, et al. Genome-wide dissection of heterosis for yield traits in two-line hybrid rice populations[J]. Scientific Reports, 2017, 7(1): 7635.

Chapter 8

Prospects of Third-generation Hybrid Rice Breeding

Li Xinqi / Li Yali

The development direction of crop heterosis utilization is to integrate superior traits, widen genetic differences between parents, enhance free-mating, reduce the costs of hybrid seed production, and continuously improve the level of heterosis. At present, there are many factors hampering crop heterosis utilization, and it is difficult to breed superior combinations, which affects the potential of crop heterosis utilization. Although the research and utilization of hybrid rice in China is a great achievement, in breeding science, hybrid rice is still in the early stages of development with a relatively small planting area as a share of the world's total. The breeding of rice male sterile lines is restricted by specific genetic background, and it is a long cycle with slow progress. The male sterile lines cannot meet the needs of hybrid rice breeding due to their unstable fertility, limited pool of sterility genes and less free mating. At present, the rapid development of biological science leads to rapid progress in hybrid rice breeding towards simpler procedures and higher efficiency. History shows that each new stage of hybrid rice breeding is a new breakthrough that pushes the yield of rice to a higher level brings a leap to agricultural production.

Section 1 Concept of Third-generation Hybrid Rice

Ⅰ. First-generation Hybrid Rice

The first-generation hybrid rice is the three-line hybrid rice system with CMS lines, maintainer lines and restorer lines as the genetic tools. It is a great achievement in the history of rice breeding and promotion, making China the first country in the world to successfully apply hybrid rice to large-scale production, and so far, it has been promoted to more than 330 million hectares of land in China.

Ⅱ. Second-generation Hybrid Rice

The second-generation hybrid rice is the two-line hybrid rice system with PTGMS lines and male parents as the genetic tools. Research on the second-generation hybrid rice began in 1973 and succeeded in 1995. It is now the most important means of rice production and heterosis utilization.

Ⅲ. Third-generation Hybrid Rice

The third-generation hybrid rice is a new type of hybrid rice system with common recessive genic male sterile (RGMS) line as the female parent and a conventional rice variety or line as the male parent.

Common RGMS is more frequent. Compared with PTGMS lines, common RGMS is completely sterile and not affected by the environment. It is an ideal genetic tool for the utilization of crop heterosis due to its simple sterility inheritance and can meet the requirements for the breeding of superior male sterile lines for agricultural crops. Its great potential is reflected in – 1) Sterility is generally controlled by only one pair of recessive genes, and sterility expression is not restricted by genetic background, so it is possible to transfer sterility genes to any rice variety to make it completely male sterile; 2) Any conventional rice has a dominant fertile gene that is allelic to the sterility gene, and can be used as a restorer line for a common RGMS line; 3) The common RGMS line is not as susceptible to environmental effects as the PTGMS line, and is stably sterile during a normal growing season in rice growing regions, and can be used for seed production. Currently, many dominant fertility genes have been cloned in maize, rice and *Arabidopsis* (Table 8 – 1) and they all determine the normal development of male gametophytes, such as *MSCA1* and *MS45* in maize; *SPL/NZZ*, *AMS*, *MS*1, *MS*2, *NEF*1 and *AtGPAT1* in *Arabidopsis*; and *MSP1*, *EAT1*, *TDR*, *CYP703A3* and *CYP704B2* in rice. Abnormalities in any of the relevant fertile genes at each stage from rice stamen primordium differentiation to the formation and release of mature pollen grains may result in failure to form viable pollen and produce male sterility. Here, the *TDR* gene and several related fertility genes are used as examples to briefly describe the principles of their control of male fertility.

The *TDR* (tapetum degeneration retardation) encodes a *bHLH* transcription factor that binds directly to the promoter regions of the programmed cell death (PCD) genes *OsCP1* and *OsC6* and positively regulates the PCD process in the tapetum and pollen wall cell formation. The *tdr* mutant has delayed degradation of the tapetum and rapid degradation after microspore release resulting in complete male sterility.

The expression of *CYP703A3* is directly regulated by *TDR*, which regulates PCD of the tapetum and the pollen wall cell formation. *CYP703A3* is a cytochrome P450 hydroxylase that acts as a chain hydroxylase on the specific substrate lauric acid to form heptahydroxylauric acid. Due to the insertion of a single base, the *cyp703a3 – 2* mutant has defects in the development of the cuticle and the outer wall of the pollen on the surface of the anther. The content of cuticle monomer and wax components is significantly reduced, resulting in abnormal anther development, and the anther becomes smaller and white-yellow, unable to form mature pollen grains, leading to male sterility and failure to form mature seeds.

The *bHLH* transcription factor encoded by *EAT1* gene acts downstream of *TDR* and cooperates with *TDR*. It is specifically expressed in the tapetum and positively regulates the PCD of another tapetum cells. It can directly regulate the expression of *OsAP25* and *OsAP37*, two genes encoding aspartic protease which induces PCD in yeast and plant cells and plays an important role in the development and maturity of pollen grains. The abortion characteristics of the rice *eat1* mutant are similar to those of *tdr* in terms of sterility. Although microspores can be formed through meiosis, due to the delayed PCD of the tapetum, the energy supply of the microspores is insufficient, and the microspores are degraded after being released from the tetrad, and the anthers are dry and shriveled. The pollen grains are aborted, and ultimately cannot form functional pollen grains, resulting in male sterility.

Previously, rice RGMS materials could not be propagated by self-reproduction, nor could the seeds of sterile lines be mass-produced by crosses, etc. It was difficult to obtain 100% sterile plants of rice (GMS) lines for commercial production of hybrid rice seeds. The use of sterile plant (msms) crossed with heterozygous fertile plants (MSms) can only obtain 50% sterile plants.

Table 8 - 1 Dominant fertility genes of rice RGMS

Nuclear fertility gene	Protein encoded by the fertility gene	Gene function (s)
PAIR1	Coiled-coil domain protein	Homologous chromosomes synapsis
PAIR2	HORMA domain protein	Homologous chromosomes synapsis
ZEP1	Coiled-coil domain protein	Synaptonemal complex formation in meiosis
MEL1	ARGONAUTE family protein	Cell division before meiosis of germ cells
PSS1	Kinessin family protein	Meiotic dynamics of male gametes
UDT1	bHLH transcription factor	Tapetum degradation
GAMYB4	MYB transcription factor	Aleurone layer and pollen sac development
PTC1	PHD-finger transcription factor	Tapetum and pollen development
API5	Anti-apoptotic protein 5	Delayed degradation of tapetum
WDA1	Carbon lyase	Lipid synthesis and pollen exine formation
DPW	Fatty acid reductase	Pollen sac and exine development
MADS3	Homeomorphic Class-C transcription factors	Pollen sac development at the late stage and pollen development
OSC6	Lipid transfer family proteins	Liposomes and pollen exine development
RIP1	WD40 domain protein	Pollen maturation and germination
CSA	MYB transcription factor	Pollen and pollen sac distribution
AID1	MYB transcription factor	Anther dehiscence

Modern genetic engineering techniques provide an effective way to solve the reproduction problem of common RGMS materials. China National Hybrid Rice R&D Center (CNHRRDC) used *EAT1*, a rice pollen fertility gene, and its corresponding sterile mutant *eat1* to construct the linkage-expression vectors of the fertility gene, fluorescent protein gene and pollen lethal gene and transform common RGMS mutants, so as to obtain transgenic plants with normal fertility. In 2015, CNHRRDC succeeded in creating the genetically engineered GMS line Gt1s, the reproductive line Gt1S and the third-generation hybrid rice.

The third-generation hybrid rice technology not only has the advantages of stable fertility of CMS lines and free mating of PTGMS lines, but also overcomes the disadvantages of limited mating of CMS lines and possible fertility fluctuation of PTGMS lines. Moreover, it shows great application potential as seed production is very simple and easy.

A genetically engineered GMS line is a common RGMS line obtained by means of genetic engineering and it could be applied on a large scale to commercial reproduction. The fertility of Gt1S, the reproductive line carrying *eat1*, is restored. After selfing, half of the seeds on each panicle are colored while the other half is colorless (Fig. 8 – 1). Seeds with different colors can be separated with a color sorter. Those without red fluorescence are genetically engineered GMS lines with complete sterility and 100% sterile plants, which can be used in hybrid rice seed production. Because they do not contain transgenic components, the hybrid rice seeds produced using them are also non-transgenic. The seeds with red fluorescence are fertile and can be used to reproduce male sterile lines. The colored and colorless seeds account for 50% respectively in the next generation. This kind of transgenic line can bring fertility to common RGMS mutants, so it is called the reproductive line of genetically engineered GMS rice. The method of creating genetically engineered GMS lines using the linkage-expression vectors of the above three genes is

①Seeds with red fluorescence of the common RGMS line Gt1S and seeds without red fluorescence under color sorter

② Seeds of the common RGMS line Gt1s and the common RGMS breeding line Gt1S (red) with glume removed under natural light

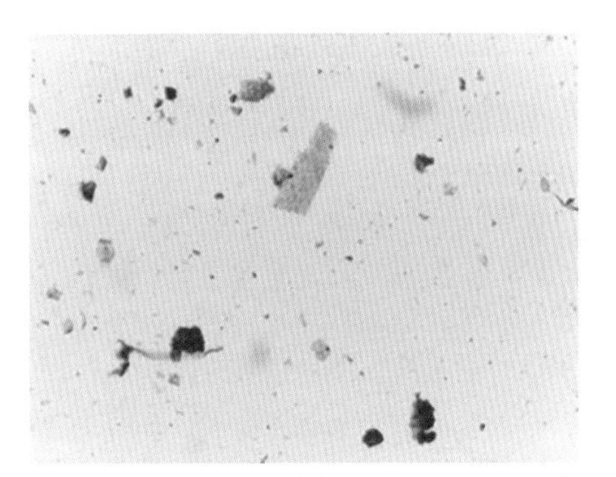

③Complete pollen abortion of the common RGMS line Gt1s under microscope

④Common RGMS line Gt1s

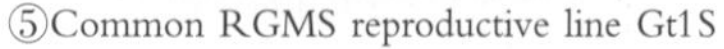

⑤Common RGMS reproductive line Gt1S　　⑥Third-generation hybrid rice combination Gt1s/L180

Fig. 8 - 1　Gt1s, Gt1S and their hybrid combination Gt1s/L880

called the third-generation hybrid rice SPT technology. The third-generation hybrid rice is obtained by combining genetically engineered GMS lines as female parents with conventional varieties (lines) as the male parents (Fig. 8 - 2, Fig. 8 - 3).

Fig. 8 - 2　Third-generation *japonica* hybrid rice combination (Gt1s/H33)

Fig. 8 - 3　Third-generation *indica-japonica* hybrid rice combination (Gt3s/E889)

Section 2　Principles for the Breeding of Third-generation Hybrid Rice

Ⅰ. Fluorescent Protein and Fluorescent Color Separation

As a molecular tag, fluorescent protein is used to label individuals, tissues, cells, subcells, virus particles and protein mapping. It has been widely used in analytical biotechnology and intracellular molecular tracing.

Fluorescence is a substance that is excited after absorbing electromagnetic radiation, and the excited atoms or molecules emit radiation with the same or different wavelength as the excitation radiation during the de-excitation process. When the excitation source stops irradiating the specimen, the re-emission process stops immediately. The object is exposed to shorter wavelengths of light, stores the energy, and then slowly emits longer wavelengths of light, which is called fluorescence. The first green fluorescent protein (GFP) was discovered by Osuma Shimomu et al. in 1962 in a jellyfish with the scientific name *Aequorea victoria*. Since most organisms have weak spontaneous green fluorescence, the high background in intracellular imaging affects the sensitivity of GFP detection. At the same time, GFP may be involved in the apoptotic process, making it difficult to establish a GFP-stable line.

In 1999, Matz et al. reported *drFP583*, the first red fluorescent protein from coral. Its commercial name is *DsRed*, which is composed of 225 amino acid residues with a maximum absorption wavelength of 558 nm and a maximum emission wavelength of 583 nm. The fluorescent protein has high quantum yield and photo-stability, and is less affected by pH, with no significant change in absorption and emission light intensity within the pH range of 5 – 12. So far, all red fluorescent proteins have been isolated and evolved from different species of *Corallimorpharia* or *Actiniaria* under *Anthozo* a. Compared with GFP expression in seeds, *DsRed* expression is more obvious and red fluorescence can be detected even under white light. Moreover, the protein composition of transgenic seeds is not affected by *DsRed*, so the *DsRed* gene is more suitable as a reporter gene for transgenic crops than GFP.

The barley-derived *Ltp2* promoter in monocotyledonous transgenic plants regulates the expression of exogenous genes specifically in the aleurone layer without affecting normal seed development. The cloned endosperm promoters include *2S*, *VP1*, *mZE40 – 2* and *Nam – 1* in maize, *napB*, *Bn-FAE1. 1*, *Napin*, *BcNA* and *FAD2* in rape, and *γTMT*, *Wsi18* and glutenin promoter in rice. The aleurone layer-specific expression of promoter-linked red fluorescent protein gene enables specific expression of red fluorescence in the aleurone layer.

The male sterile reproductive line can be obtained by introducing two sets of elements, a linkage-expressed fertility restoration gene and a red fluorescent protein gene regulated by a aleurone layer-specific expression promoter, into homozygous RGMS plants (Table 8 – 2).

Self-crossing of the reproductive line results in 1/4 of the colorless sterile seeds and 3/4 of the red fluorescent seeds (Table 8 – 3), of which the red fluorescent seeds could be separated by a fluorescent color sorter (Fig. 8 – 4).

Fig. 8 – 4 Brown rice of colored (fertile) and colorless (sterile) seeds of genetically engineered GMS reproductive line separated by color sorter

Table 8 - 2 Trans-linked fluorescent and fertile genes to obtain fluorescent genetically engineered nuclear sterile lines

♀ Gamete genotype	♂ Gamete genotypes
	ms / *MS* + *DsRed*
ms	*msms* / *MS* + *DsRed* (red)

Note. *ms* is a sterility gene; *MS* is a fertility gene; *DsRed* is a red fluorescent gene.

Table 8 - 3 Selfing of genetically engineered GMS lines by fluorescence sorting

♀ Gamete genotype (s)	♂ Gamete genotypes	
	ms	*ms* / *MS* + *DsRed*
ms	*msms* (sterile gene type, colorless)	*msms* / *MS* + *DsRed* (red)
ms / *MS* + *DsRed*	*msms* / *MS* + *DsRed* (red)	*msms* / *MSMS* + *DsRed* (red)

Note. *ms* is a sterility gene; *MS* is a fertility gene; *DsRed* is a red fluorescent gene.

A color sorter automatically sorts out granular materials by color according to the difference of the optical characteristics of the materials, using the electronic eye of its ultra-high speed sensor. It can identify heterochromatic areas as small as 0.08 mm^2, and is widely used in the selection of rice and other food crops, and reject all chalky rice grains at a processing rate of 3 - 10 t/h. A color sorter is mainly composed of the feeding system, optical detection system, signal processing system and separation execution system (Fig. 8 - 5).

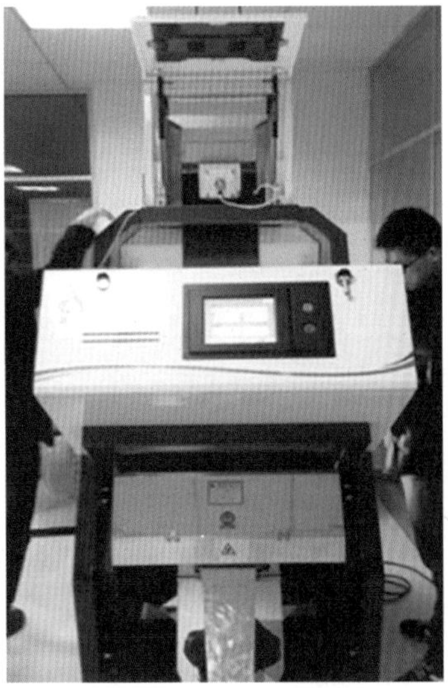

Fig. 8 - 5 Color sorter

When a color sorter is working, the materials to be sorted enter the machine from the hopper at the top, and go through the vibration produced by the vibrator device, the materials then slide along the channel accelerates down into the observation area inside the sorting chamber, and passes between the sensor and the background plate. Under the light, according to the light intensity and color change, the system generates output signal to drive the solenoid valve to work and blow out the different color particles into the waste chamber of the receiving hopper, while the good materials continue into the chamber of the receiving hopper for finished products, thereby achieving the separation.

Anhui Meyer Optoelectronic Technology Co., Ltd. and CNHRRDC joined hands to develop a fluorescent color sorter for genetically engineered GMS rice lines using green light. When the light is irradiated to the rice seeds, the seeds will turn red; the fluorescent grains will reflect the green light as well as the red light, while the non-fluorescent rice grains will only reflect the green light. A band-pass filter is mounted in front of the camera to filter out the green light and let the red light pass. At this time, the fluorescent rice captured by the camera is red, while the green light reflected by the non-fluorescent rice is filtered out, and a black image is formed. Therefore, there is clear difference in the color of the rice seeds with and without red fluorescence (Fig. 8 - 6).

Fig. 8 - 6 Red fluorescent seeds of genetically engineered GMS line are excited to show red fluorescence, while the seeds of non-fluorescent sterile lines do not show the color.

Endosperm xenia refers to the phenomenon in which the endosperm of seeds produced by contemporary sexual hybrids exhibit pollen donor traits. For example, the yellow endosperm of maize is a dominant trait. If the pollen of yellow maize is used to pollinate the PTGMS lines of white maize, the seeds with yellow endosperm will be produced on the male sterile lines, and show the dominant traits of the male parents. In this case, it can be used as a direct method to identify true and false hybrids.

Crossing a common RGMS plant with a fertility gene donor and a red fluorescent gene-linked reproductive line is also a way to propagate a common RGMS line. The number of chromosomes in the endosperm is 3 *n*, in which 2 *n* is from the polar nucleus of the common RGMS line and 1 *n* is from the sperm nucleus of the red fluorescent parent. The 3 *n* endosperm directly shows red fluorescence, a trait of the male parent, under the influence of sperm nucleus (Table 8 - 4).

Table 8 - 4 Genetically engineered GMS lines by endosperm xenia induction and fluorescence sorting

♀ Gamete genotype	♂ Gamete genotype(s)	
	ms	*ms* / *MS* + *DsRed*
ms	*msmsms* (endosperm genotypes of male sterile lines; colorless)	*msmsms* / *MS* + *DsRed* (endosperm genotypes of reproductive lines; red)

Ⅱ. Pollen Lethal Gene and Pollen/Anther Specific Expression Promoters

The pollen lethal genes in plants include *ZM-AA1*, *pep1* and *SGB6* in maize, *argE* and *dam* in *Escherichia coli*, *Osg1* in rice, *Barnase* in *Bacillus amyloliquefaciens*, *rolB* and *rolC* in *Agrobacterium rhizogenes*, *CytA* in *Bacillus thuringiensis*, *DTAβ* in *Corynebacteriophage*, *CHS* in *Paeonia lactiflora*, *Wun1* in potato and *pelE* in *Erwinia chrysanthemi*. For example, the *RNase* (*Barnase*), *RnaseT1* and *DTA* genes of *Bacillus* are cytotoxic genes, whose expression by the tapetum-specific promoter *TA29* will lead to pollen abortion of such transgenic crops as tobacco, rape and maize. The product of *Escherichia coli*'s *argE* gene can remove non-toxic substances and induce the production of L-foscarnosin, a toxic substance. Kriete et al. linked the *argE* gene with the homolog of the tapetum-specific promoter *TA29* to construct a chimeric gene, and then introduced it into tobacco plants. The toxin was released after *N-ac-Ptis* was applied at the pollen development stage, resulting in empty anther and male sterility. β-1, and the 3-glucanase (*Osg1*) gene can cause male sterility through early degradation of the callose wall. Besides, the *pelE* gene as well as the *rolB* and *rolC* genes of *Agrobacterium rhizogenes* can cause pollen abnormality and male sterility by fusing with an anther-specific promoter.

The tissue-specific expression of pollen development-related genes in pollen or anther requires correct regulation by regulatory factors related to pollen or anther-specific expression at a specific time and location. The main pollen- or anther-specific promoters in plants include *PG47*, *5126*, *ZmC5*, *ZmC13*, *Zmabp1*, *ZmPSK1*, *Mpcbp*, *Zmpro1* and *AC444* in maize, *TA13*, *TA26*, *TA29* and *NTP303* in tobacco, *OsSCP1*, *OsSCP2*, *OsSCP3*, *OsRTS*, *OsIPA* and *OsIPK* in rice, *A3*, *A6* and *A9* in *Arabidopsis thaliana*, *tap1* in *Antirrhnum*, and *Bp10* and *Bp19* in European rape. In-depth studies have been carried out on *Zmc13*, *5126* and *PG47* in maize. In modern biotech breeding, these promoters can be linked with pollen lethal genes to construct expression vectors, and be specifically expressed in the pollen of transformed plants to cause pollen abortion.

The fertility restoration gene of a RGMS mutant is linked to its specific promoters, along with the pollen lethal gene and its specific promoter and then transfer into the male nuclear sterility mutant of the corresponding crop. The fertility restoration gene in plants can restore fertility to pollen microspores containing the sterility genes, while pollen lethal genes can degrade transgenic pollen containing the fertility restoration genes by late pollen development, leaving only non-transgenic pollen containing the RGMS gene. However, there are two types of female gametes, and after pollination, two genotypes of seeds will be produced, one is heterozygous seeds containing fertility restoration genes and sterile mutation genes, i. e., the RGMS reproductive line, and the other is the non-transgenic seeds (*ms/ms*) containing only the male sterility mutant gene, namely the RGMS line, thus achieving the goal of integrating the functions of restoration and the maintenance (Table 8 - 5).

Table 8 - 5 Reproductive lines of genetically engineered GMS lines obtained by self-crossing sterile mutants with trans-linked lethal and fertile genes

♀ Gamete genotype(s)	♂ Gamete genotype(s)	
	ms (fertile gamete)	*ms* / *MS*+*ZmAA1* (Abortive gametes)
ms	*msms* (Sterile line genotype)	—
ms / *MS*+*ZmAA1*	*msms* / *MS* + *ZmAA1* (Breeding line genotypes)	—

Note. *msms* is a homozygous sterility gene; *MS* is a fertility gene; *ZmAA1* is a pollen lethal gene.

The pollen of the reproductive line of genetically engineered GMS lines does not contain any transformed gene, while the female gamete has two genotypes, *ms* and *MS*. Half of the seeds obtained through selfing are sterile and the other half of them are fertile and can be used as the reproductive line. In breeding, it is necessary to increase the number of the male parent plants because only half of the reproductive line will have fertile pollen (Table 8 - 6).

Table 8 - 6 Reproduction of genetically engineered GMS lines obtained by trans-linked pollen lethal and fertility genes

♀ Gamete genotype	♂ Gamete genotype(s)	
	ms (fertile gamete)	*ms* / *MS* + *ZmAA1* (abortive gametes)
ms	*msms* (sterile line genotype)	—

Note. *msms* is a homozygous sterile gene; *MS* is a fertile gene; *ZmAA1* is a pollen lethal gene.

Ⅲ. SPT Technology

A PCT patent, the SPT technology, was designed in 1993 by Plant Genetic System, USA. Three sets of gene-linked expression elements, including fertility restoration gene, screening reporter gene and pollen lethal gene, are introduced into a RGMS line plants to obtain the reproductive line of the male nuclear sterile line plants, and then the reproductive line can be self-fertilized to reproduce the male sterile lines and the reproductive line (Fig. 8 - 7). It is a collection of technologies from the aforementioned red fluorescence pathway and pollen lethal pathway.

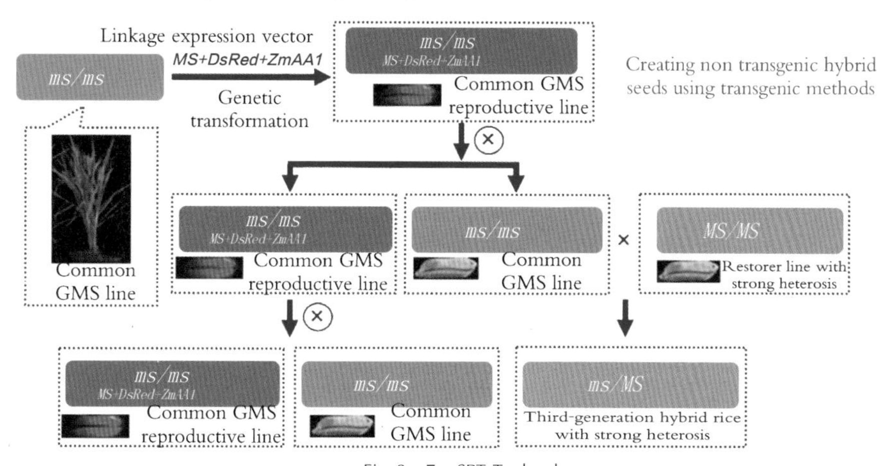

Fig. 8 - 7 SPT Technology

Note. *ms* is a sterile gene; *MS* is a fertile gene; *DsRed* is a red fluorescent protein gene; *ZmAA1* is a pollen lethal gene

The third-generation hybrid rice breeding technology developed by CNHRRDC is an application of the SPT technology in rice breeding. A plant expression vector is constructed and introduced into rice RGMS mutants through *Agrobacterium*-mediated transformation by tightly linking the genes controlling rice pollen fertility restoration (*EAT1*), the pollen lethal gene (*ZmAA1*) and the red fluorescent protein marker gene (*DsRed*). The fertility gene is transferred into the male sterile mutants to restore fertility. The pollen lethal gene causes the male gamete containing transgenic fragments to lose reproductive viability and the pollen to wither before maturing, while pollen without transgenic components matures. The female gametes consisted of two types, one identical to the female gametes of the sterile mutant *eat1*, and the other containing three sets of gene-linked expression vectors. Selfing of the reproductive lines enables the propagation of the RGMS line and the reproductive line, achieving dual use for one line. The red fluorescent gene is used as a reporter gene for screening, and the offspring seeds with red fluorescence are seeds of the reproductive line, while those without red fluorescence are seeds of the sterile line. Thus, the reproductive line and the sterile line can be separated completely with a color sorter to obtain the homozygous sterile line needed for hybrid seed production. Half of the seeds on each panicle of the genetically engineered GMS line Gt3s are red and the other half are colorless. Those colorless seeds are male sterile, therefore, can be used for hybrid rice seed production. The red seeds are fertile and can be used to reproduce the RGMS line, each accounting for half of total F_1 population (because the introduced *ZmAA1* can make the male gamete sterile). The red seeds and colorless ones can be completely separated by a color sorter. It can be seen that the seed production and breeding of genetically engineered GMS lines are very simple and easy (Table 8 - 7 and Table 8 - 8).

Table 8 - 7 Genetic analysis of genetically engineered GMS line with the SPT technology

♀ Gamete genotype(s)	♂ Gamete genotype(s)	
	ms (fertile gamete)	*ms* / *MS* + *ZmAA1* (abortive gametes)
ms	*msms* (sterile line genotype)	—
ms / *MS* + *DsRed* + *ZmAA1*	*msms* / *MS* + *DsRed* + *ZmAA1* (breeding line genotype)	—

Note. *ms* is a sterile gene; *MS* is a fertile gene; *DsRed* is a gene encoding red fluorescent protein; *ZmAA1* is a pollen lethal gene.

Table 8 - 8 Genetic analysis of F_1 hybrids of an SPT genetically engineered GMS line and its reproductive line

♀ Gamete genotype	♂ Gamete genotype(s)	
	ms (fertile gamete)	*ms* / *MS*+*DsRed*+*ZmAA1* (abortive gametes)
ms	*msms* (sterile line genotype)	—

Note. *ms* is a sterile gene; *MS* is a fertile gene; *DsRed* is a gene encoding red fluorescent protein; *ZmAA1* is a pollen lethal gene.

Ⅳ. Other Utilization Methods

It is possible to develop new ways to exploit common RGMS based on rice recessive nuclear sterility gene function and its expression using regulatory elements of fertility genes. For example, if conditions (such as inducibility) are used to control the promoter of genetic expression of a fertility restoration gene, and the gene is transferred normally into the male sterile plant as a complementary gene, then if suitable conditions for the expression of the specific promoter are not provided, the fertility restoration gene of the plant cannot be expressed and a male sterile line can be obtained; on the contrary, when suitable conditions for the expression of the specific promoter are provided, such as by spraying the inducer, the fertility restoration gene of the plant is expressed normally and thus the male sterile line can be successfully reproduced. In addition, the promoter can be used to drive the action of some suppressors of the endogenous fertility genes in the plant itself to achieve the same purpose. Using plant male sterility closely linked to herbicide resistance traits, sterile lines are propagated by killing fertile seeds with herbicides. That is, relying on the constitutive expression of herbicide resistance genes, half of the fertile plants are killed by herbicide application and the other half are retained. However, it is difficult to completely and accurately control the sterility and fertility of male sterile lines in actual production, and it also involves the safety issue of genetically modified organisms because there are genetically modified components in the male sterile lines. As a result, the above methods have not been promoted in production.

Cytoplasmic fertility by chloroplast transformation may be another route for the use of common RGMS. Introduce a fertile gene into the chloroplast genome of a sterile plant by homologous recombination by linking a common GMS fertility gene to a promoter from the chloroplast (or another suitable promoter) and part of the chloroplast homologous sequence to get it expressed; and the transgenic plants with fertile cytoplasm and sterile nuclei are likely to be fertile and can be used as parents to breed sterile lines because the F_1 nuclei and cytoplasm are identical in composition to those of the sterile plants and are likely to be 100% sterile. The sterile lines are then mated with good parents to produce hybrid seeds for field use. The success of this route depends on the effective expression of the fertility genes in the chloroplast and the efficiency of the chloroplast transformation technology.

Section 3 Breeding of Third-generation Hybrid Rice

Ⅰ. Genetically Engineered GMS Lines and Its Breeding Line Cultivation

Crossing and backcrossing are the main ways to breed a genetically engineered RGMS line. The basic requirements for a rice common RGMS line are complete function of the sterility gene, stable sterility, complete abortion, insusceptibility to environment factors, ability of heading at any time and place, more than 99.5% pollen sterility, good flowering habits, large and exserted stigma, large flowering angle, long flowering duration, early and concentrated flowering, good plant and leaf morphology, high GCA, strong resistance and good grain quality. In previous studies, a genetically engineered GMS line is usually given a name or code starting with an upper case G and ending with a lower case s, such as in Gt1s, so as to distinguish them from PTGMS lines. The name of its reproductive line also begins with the capital letter G, but it ends with the capital letter S, such as in Gt1S.

1. Cross breeding of genetically engineered GMS lines

The technical route of obtaining genetically engineered GMS lines through the efficient SPT technology, transforming the linked expression elements of the fertility restoration gene, screening reporter gene and pollen lethal gene, has been working well. Here, we will only introduce the breeding of SPT genetically engineered GMS lines, not the reproductive lines. Because half of the seeds on each panicle of genetically engineered GMS lines are seeds of the sterile line while the other half are seeds of the reproductive line, the breeding of genetically engineered GMS lines is actually the breeding of genetically engineered GMS reproductive line. The red fluorescent gene is linked with the fertility gene and pollen lethal gene.

In the cultivation of each generation of the breeding lines, seeds with red fluorescence should be selected because no fertility restoration gene, screening reporter gene or pollen lethal gene of the reproductive line exists in the colorless seeds.

(1) Single-cross breeding

A genetically engineered RGSM breeding line is used as the female parent to cross with a male parent that has certain required traits. A genetically engineered RGMS reproductive line to be improved cannot be used as the male parent, but they can be used as the female parent because there is no fertility gene, fluorescence gene or pollen lethal gene of these breeding lines in mature pollen.

An existing genetically engineered RGMS line is used as the female parent to cross with a male parent with target traits. Half of the seeds on the panicles of the female parent are colored, while the other half are colorless. The colorless seeds are removed, while the colored seeds are to be planted to produce an F_1 population. Half of the seeds on the panicles of F_1 plants are colorless (non-transgenic seeds), and the other half are colored (transgenic seeds) and the colored seeds are kept and planted to form an F_2 population. All plants in the F_2 population are fertile, and half of the seeds on the panicles are colorless, while the other half are colored. The individual plants with the target traits are selected according to the breeding objectives, and their colored seeds are kept to be planted to form F_3 families. Among the F_3 families, 1/4 will have an equal number of colorless (sterile without transgenic gene) and colored (fertile with transgenic genes) seeds. The colored seeds in the fertile F_3 plants are kept to form an F_4 population and the screening and selection for the target traits will continue in the F_4 families; 2/4 of the F_4 families have heterozygous sterility genes, 1/8 of them are colorless sterile plants (genotype is *msms*, Table 8 – 9), 3/8 are colorless fertile plants, and 4/8 are colored fertile plants. Starting from the F_3 population, the morphology, quality traits, adaptability and outcrossing habits should be examined in the selection process, and the sterile plants can be used for testcrossing with other male parents, and be evaluated at F_4. F_5 will be suitable for further trait evaluation and combining ability test according to the selected F_4 families or plants. Seed production research can start from F_6 and purification and selection should continue into F_7 for seed production and yield trials. Except those seeds for testcross purpose, all other progenies from the breeding crosses can be filtered with a color sorter to discard those colorless seeds to increase the efficiency and accuracy of selection.

(2) Multiple-cross breeding

Multiple-cross breeding refers to the breeding of new genetically engineered RGMS lines and the corresponding reproductive lines using three or more parents with the purpose of integrating superior traits. In practice, there are two forms of multiple crossing, two or more rounds of single-crossing and three-way crossing.

1) Two or more rounds of single-crossing.

Good plants are selected from the reproductive line of the first single-cross, and crossed with another parent. The breeding process is actually composed of two or more rounds of single crossing. The selection methods used in single-cross breeding can be used here for the integration of superior traits according to the breeding objectives.

Table 8 – 9 Genotypes and selection methods for single-cross breeding of genetically engineered GMS line

Generation	Genotypes of the female parent		Genotype (s) of the male parent	Breeding method	Selection method
Current generation	Plants before fertilization	Seeds after fertilization *	*MSMS*	Crossing	Keep colored seeds
	msns + *MRZ*	1 *MSms* : 1 *MSms* + *MRZ*			
F_1	*MStns* + *MRZ*	1 *MSMS* : 2 *MSms* : 1 *msms* : 1 *MSMS* + *MRZ* : 2 *MSms* + *MRZ* : 1 *msms* + *MRZ*	*MSms* + *MRZ*	Selfing	Keep colored seeds
F_2	*MSMS* + *MRZ*	1 *MSMS* : 1 *MSMS* + *MRZ*	*MSMS* + *MRZ*	Selfing	Select plants with targeted traits, keepcolored seeds
	MStns + *MRZ*	1 *MSMS* : 2 *MSms* : 1 *msms* : 1 *MSMS* + *MRZ* : 2 *MSms* + *MRZ* : 1 *msms* + *MRZ*	*MSms* + *MRZ*		
	msns + *MRZ*	1 *msms* : 1 *msms* + *MRZ*	*msms* + *MRZ*		
F_3	*MSMS* + *MRZ*	1 *MSMS* : 1 *MSMS* + *MRZ*	*MSMS* + *MRZ*	Selfing	Discard
	MSns + *MRZ*	1 *MSMS* : 2 *MSms* : 1 *msms* : 1 *MSMS* + *MRZ* : 2 *MSms* + *MRZ* : 1 *msms* + *MRZ*	*MSms* + *MRZ*		Select plants with targeted traits, keep colored seeds
	msms + *MRZ*	1 *msms* : 1 *msms* + *MRZ*	*msms* + *MRZ*		Select plants with targeted traits, keep colored seeds
F_4	*MSMS* + *MRZ*	1*MSMS* : 1*MSMS* + *MRZ*	*MSMS* + *MRZ*	Selfing	Discard
	MSns + *MRZ*	1 *MSMS* : 2 *MSms* : 1 *msms* : 1 *MSMS* + *MRZ* : 2 *MSms* + *MRZ* : 1 *msms* + *MRZ*	*MSms* + *MRZ*		Select plants with targeted traits, keep colored seeds
	msms + *MRZ*	1 *msms* : 1 *msms* + *MRZ*	*msms* + *MRZ*		Select plants with targeted traits, keep colored seeds, sterile plants for test-crossing

Note. *ms* stands for sterility gene; *MS* stands for fertile gene; *MRZ* stands for the ternary complex of introduced fertility restoration gene (MS), screening reporter gene (*DsRed*) and pollen lethal gene (*ZmAA1*) * The number in front of each genotype indicates its share in the segregated population.

2) Three-way crossing.

In three-way crossing, a genetically engineered RGMS reproductive line is crossed with the first parent, and its F_1 is then used as the female parent to cross with a second parent. Three-way crossing is faster than two rounds of single crossing. It is based on the fact that 1/8 of the red fluorescent seeds of the F_2 population will be sterile because sterility is controlled by a pair of recessive genes. Gt5s is a line bred through three-way crossing of Gt1s/R12//Lunhui 422 (Fig. 8 - 8).

Common GMS breeding line Gt5S(left) and common GMS line Gt5s (right)

Performance of the third-generation hybrid rice combination Gt5s/R900

Fig. 8 - 8 Third-generation hybrid rice Gt5s, its breeding line and hybrid combination

2. Backcross breeding of genetically engineered RGMS line

Backcrossing is a method of transferring the sterility gene and three-sets of gene linkage expression vectors from a genetically engineered RGMS line into a superior recurrent parent through backcrossing so as to breed genetically engineered RGMS line and its reproductive line. Its purpose is to breed a sterile line with traits other than male sterility, which are similar to those of the recurrent parents. Recurrent parents are generally a parent with superior overall traits and strong combining ability. In practice, there are two kinds of backcrossing - alternate-generation backcrossing and direct backcrossing.

(1) Alternate-generation backcrossing

The typical feature of this method is that the backcross is performed only if a sterile plant of red fluorescent seeds is found in the progenies of the backcross. The general operation is to select single plants of genetically engineered RGMS reproductive lines with traits similar to those of the recurrent parents at the fertility segregation generation and backcross them with the recurrent parent. Alternate-generation backcrossing is used to improve the traits of the recurrent parent, selected by selfing and backcrossing segregation.

(2) Direct backcrossing

Direct backcrossing is to use genetically engineered RGMS reproductive lines to cross the F_1 generation directly with the recurrent parent, without waiting till the segregation generation. The advantage of direct backcrossing is that it can accelerate breeding. However, if some traits of the recurrent parent also need to be improved, it may be more difficult to achieve the goal with direct backcrossing.

The inheritance of the male sterility gene in genetically engineered RGMS lines is simple, and plants of red fluorescent seeds can be selected randomly for continuous backcrossing as long as there are enough plants from red fluorescent seeds in each backcross generation. Generally, there should be about 1.5% of male sterile plants in B_3F_2 after three consecutive rounds of backcrossing, the plant and leaf morphology as well as growth duration of the B_3F_2 population are basically stable, and the traits of the population are basically the same as those of the recurrent parent. It shows that continuous backcrossing is a fast and effective method for directional breeding of genetically engineered RGMS lines and their reproductive lines.

3. Technology of treating hybrid progenies in genetic engineering for GMS line breeding

The selection effect of genetically engineered RGMS lines and its reproductive lines not only is closely related to the selection of parents and mating methods, but also depends to a large extent on the breeder's handling techniques for hybrid progenies. Practically valuable male sterile lines with good traits can only be bred with proper handling methods. After genetically engineered RGMS plants are selected, their other traits should be examined so as to ensure that the plants with superior traits are selected while those with poor traits are abandoned. It is very easy to obtain plants of the reproductive line of a genetically engineered RGMS line, so the plants of the reproductive line should be selected from the F_2 or F_3 population according to the breeding objectives. In theory, the mechanism of inheritance of many important traits is relatively simple. Major genes also play a very significant role in determining the quantitative traits, many of which can be stabilized quickly. In selection, those plants without desirable main traits should be abandoned as early as possible because they will consume time, energy and financial resources that will not pay off if we continue with them through seed preparation, planting and breeding. In general, the lines planted in F_5 and F_6 should be materials with superior traits, great practical value and high combining ability, but without major defects, except for those that can be used as intermediate materials due to their special traits. The most promising materials should be the focus in the treatment of hybrid progenies, and there should be priorities in breeding.

In the first round of selection, the overall trait performance of the segregation population should be taken into consideration. For example, an F_2 population should not be selected if the overall traits are poor and there are no outstanding sterile plants to be eliminated. Male sterility results in poor flowering traits, delayed and prolonged flowering, lower glume opening rate and lower stigma activity. The traits of the floral organs and flowering habits after heading should be closely observed, such as flowering time, flowering period, glume closure rate, stigma exsertion rate and post-flowering closure. Lines with good floral traits, good flowering habits, strong stigma vitality and high outcrossing rate should be selected, and those with serious glume closure, low stigma vitality and low outcrossing rate should be abandoned. The sterile lines and their reproductive lines that can be used to breed hybrids with superior traits, significant heterosis and high yield potential should be selected. The combining ability and heterosis should be stabilized and determined, and the reproductive lines of sterile lines with poor combining ability should be abandoned.

MAS can effectively accelerate the breeding process of genetically engineered RGMS lines. The linkage of molecular marker to the common RGMS gene can be used as the selective marker of male sterility gene-assisted breeding. Red fluorescence can be used as a selective marker for the red fluorescent fertility gene and pollen lethal gene.

In practice, group selection in the intermediate phase of generation advancement is also a simple and effective method to speed up the breeding process. The good sterile plants obtained from the hybrid progenies will undergo focused group selection. A large number of seeds are obtained from plants selected from the F_2 population, and more than 3,000 plants will be planted as the F_3 population. Superior plants are then selected and seeds collected from each selected plant. The F_4 population should be planted in 1,000 rows, with 14 – 20 plants in each row when there are a few lines with stable traits. The theoretical basis is that there are 12 pairs of homologous chromosomes in rice, so the frequency of plants with completely homozygous chromosomes is 1/4,096 in the F_2 population if chromosome exchange and error are not considered. Most of the plants have 5 – 7 pairs of homozygous chromosomes. If the plants of the reproductive line selected from the F_2 population have six pairs of homozygous chromosomes, and the share of plants with 12 pairs of homozygous chromosomes in the F_3 population will be 1/64. In practice, 5 – 10 plants with highly homozygous chromosomes will be found in every 1,000 F_2 plants if chromosome exchange and random errors are taken into consideration, and these plants will show uniform traits in the F_3 population. Group selection is usually effective only when F_2 plants and the F_3 population have outstanding traits that meet the breeding objectives. In F_4 lines, the superior plants with less stable traits can also be treated by group selection. F_5 population can be moderately smaller, and F_6 lines can be reduced slightly.

Haploid breeding, including anther culture and parthenogenesis, cannot be directly used to accelerate the breeding of SPT genetically engineered GMS lines, because the pollen of sterile breeding lines does not contain fertility genes, pollen lethal genes or red fluorescent genes. Besides, the reproductive line of genetically engineered RGMS line can produce fertile pollen for seed production only when the pollen lethal gene in them is heterozygous.

4. Directional genetic transformation of superior common RGMS materials

After obtaining common RGMS plants by means of crossing, backcrossing and gene knockout, three-sets of gene linkage expression vectors including the fertility gene, pollen lethal gene and red fluorescent gene will be used for genetic transformation to carry out directional breeding of genetically engineered GMS lines and their reproductive lines.

(1) Inoculation and induction of callus

Callus is induced from young panicles of male sterile lines. When a target male sterile plant is at the stage of young panicle differentiation, the young panicles are taken out for the induction of callus when these panicles are 0.5 – 4.0 cm in length.

(2) Subculture of callus

The light yellow, dense and relatively dry calli are peeled off and inoculated into the fresh subculture medium for cultivation, and they are transferred into a new subculture medium every 10 days.

(3) Infection of calli by *agrobacterium tumefaciens*

Agrobacterium tumefaciens carrying the fertility gene, pollen lethal gene and red fluorescent gene are used to infect the calli. Vigorous calli are selected from those that have been subcultured for about one month, and then they are transferred into the culture medium containing *agrobacterium tumefaciens* for 30 minutes.

(4) Co-culture

The calli of rice infected with *agrobacterium tumefaciens* are air dried before they are spread on the sterile filter paper in the co-culture medium to be cultured in a dark room at 28 ℃ for 2 – 3 days. After they are washed with sterile ultra-pure water, the *agrobacterium tumefaciens* residues on the surface of these calli will be washed off with carboxybenzyl and cephalosporin.

(5) Selection of callus

The cleaned calli are inoculated into the selection medium to be cultured in a dark room at 28 ℃ for 10 days, and then they are transferred into a new selection medium for secondary selection every 15 days.

(6) Callus differentiation

The selected fresh calli are transferred and inoculated into the pre-differentiation medium to be cultured in a dark room at 28 ℃ for two weeks, and then they are transferred and inoculated into the differentiation medium to be cultured in a light room at 28 ℃ for about 20 days.

(7) Rooting of calli

The calli will be transferred to aseptic rooting medium for rooting culture when their green buds grow to about 1 cm during differentiation.

(8) Acclimatization of seedlings

When the seedlings in the rooting medium grow to the top of the culture bottle, and strike thick and strong adventitious roots (a fully developed root system), they will be transplanted into basin soil with moderate hardness, sufficient moisture and moderate fertilizer for cultivation.

(9) Detection of transgenic plants

PCR is carried out using the fragment detection primer to detect whether the rice plants are the intended transgenic plants. The correct PCR products are sequenced for final verification.

(10) Fertility restoration and selfing of genetically engineered GMS lines

The original plants of a genetically engineered RGMS line obtained from tissue culture are transplanted into transgenic test fields under normal field management. If the plants are fertile, half of their seeds will have no fluorescence and be completely sterile, while the other half of the seeds will be red and fertile, which will be used as the target genetically engineered RGMS line and its reproductive line.

Ⅱ. Progress in Third-generation Hybrid Rice Breeding

The third-generation hybrid rice is bred by combining a genetically engineered RGMS line with a conventional restorer line. It combined the advantages of stable male sterility of first-generation hybrid rice and free-mating of second-generation hybrid rice. Since the research team led by Academician Yuan Longping started research on third-generation hybrid rice in 2011, a mature breeding technology system has been established, a number of third-generation hybrid RGMS lines such as G3 – 1s and restorer lines such as Qin 89 have been bred, and a group of pilot combinations with strong heterosis, such as Sanyou

1, have been cultivated for small-scale tests and demonstration.

The third-generation hybrid rice was planted in four different double-cropping late-season rice ecological areas in Hengnan county and Taoyuan county of Hunan province in 2019 and showed strong heterosis. The average yield per unit area of Sanyou 1 in Hengnan county was more than 15.69 t/ha, attaining the 15 t/ha target. The average yield of G3 - 1s/Qin 19 in the demonstration plots in Taoyuan county was 12.57 t/ha, exceeding the 12 t/ha target.

Based on the preliminary success of the demonstration in 2019, third-generation hybrid rice was included in the "Three-One Project" (a scientific and technological innovation project aiming to feed one person with rice harvested from a 200 m^2 land plot) of Hunan province in 2020 as a double-cropping late-season rice for high-yielding demonstration. Among them, the demonstration plot in Hengnan county where the double-cropping late-season rice Sanyou 1 was trial planted, covered an area of two hectares. On November 2, 2020, Hunan province's Association of Agricultural Science Societies organized experts from Fujian Academy of Agricultural Sciences, CRRI, Jiangxi Academy of Agricultural Sciences, Wuhan University, South China Agricultural University, Guangxi Academy of Agricultural Sciences, Southwest University, the Department of Agriculture and Rural Affairs of Hunan province, Hunan Agricultural University and Hunan Normal University to carry out yield tests on the demonstration plots. The expert group inspected all demonstration plots, and randomly selected three hilly land plots for machine harvest and evaluated the yields according to relevant standards stipulated by the Ministry of Agriculture and Rural Affairs, and the average yield of these three fields reached 13.68 t/ha. In addition, the average yield of double-cropping early-season rice (early-season hybrid rice combination Zhuliangyou 168), measured in July 2020, reached 9.29 t/ha, and the average annual yield of double-cropping rice reached 22.96 t/ha, achieving the goal proposed by Academician Yuan Longping, i.e. an annual yield above 22.5 t/ha for double-cropping rice, and setting a new record for the annual yield of double-cropping rice in the middle and lower reaches of the Yangtze River. It is worth noting that in 2020, the average daily temperature at Hengnan demonstration plots during the month after heading was only 20.65 ℃, which was 2.95 ℃ lower than the average daily temperature of 23.60 ℃ in the same period of 2019, and the total sunlight hours were only 43.85 h, down by 143.95 h from the 187.80 h of 2019. Under such extremely unfavorable solar and temperature conditions, the average yield of third-generation hybrid rice was still as high as 13.68 t/ha, pushing the annual yield of double-cropping rice beyond 22.50 t/ha, which is a major breakthrough in double-cropping rice regions.

Ⅲ. Prospects of Third-generation Hybrid Rice Breeding and Application

1. Defects of the three-line and two-line systems

The major defects of the three-line hybrid system are as follows. 1) There are but a few excellent parent lines available and there is not much freedom for mating. The probability of finding a maintainer line is lower than 0.1% and that of finding a restorer line is lower than 5% in conventional *indica* varieties. The genetic difference between sterile lines is small. 2) The restorer gene is not fully dominant, so it is difficult for the fertility of the hybrids to reach the level of conventional varieties. 3) The fertility of the hybrids is not stable enough, and resistance to extreme weather conditions is poor. 4) Low cytoplasm diversity may lead to risks. Due to the limitation of the genetic background, it is difficult to breed CMS lines and integrate superior traits. The breeding efficiency is also very low. Therefore, there are only a few CMS lines with good overall traits.

The issue of minor restorer genes is a major problem for the selection of CMS lines. Rice varieties generally contain recessive or partially recessive minor restorer genes. In the breeding of maintainer lines through hybridization, these genes are segregated. As a result, it is difficult to determine a true maintainer. Only when the nucleus of a stable line is completely substituted into the sterile cytoplasm can we determine the true maintainer or real performance of good flowering habits. Therefore, in breeding practice, it is very likely that a sterile line obtained at the end is not completely sterile or does not have good flowering habits. According to Virmani (1994), about 17% of the hybrid progenies of two maintainer lines are not maintainers. It is difficult to find a maintainer line with complete sterility, so the sterile line with many sterile cytoplasm are not completely sterile. As a result, most of them cannot be used in production.

The two-line hybrid system based on PTGMS breaks the limitation of maintainer-restorer relationship in the three-line hybrid system, greatly improves free mating, and makes it easier to obtain hybrid combinations with great yield potential. However, the fertility of a PTGMS line is unstable due to its sensitivity to the environment and also induced the drift of the fertility transition temperature. This may cause self-pollination and seed setting in a PTGMS line, thus reducing the purity of hybrid seeds. The risk is high because weather conditions may change quite significantly.

Male sterility leads to poor flowering habits, delayed and non-concentrated flowering, lower spikelet opening rate and weak stigma vigor. The higher the degree of abortion, the worse the flowering habits. In hybrid breeding, the fertility sensitive period of the PTGMS lines is usually arranged in a time of high temperature, resulting in poor outcrossing habits and low yield of seed production. Therefore, regions with relatively low temperature in the booting stage are chosen for seed production with the two-line method. However, it may cause fertility fluctuation in a PTGMS line, and lead to some degree of self-pollination and seed setting, resulting in seed production failure.

2. Breeding of third-generation hybrid rice combinations

Although existing superior hybrid rice varieties have high yield, they cannot meet the breeding requirements regarding adaptability, disease and pest resistance, grain quality and other traits in different regions due to the limitations of the three-line and two-line hybrid rice systems. The three-line maintainer lines and PTGMS lines are often inferior to conventional rice varieties in agronomic traits, resistance, quality, yield potential, etc.

The greater the genetic difference between the parents is, the stronger the heterosis will be. It is more likely for parents with superior traits to produce hybrids with over-parent yield and standard heterosis. High yield, high quality, strong resistance and ecological adaptation are the common goals of crop breeding. It can be predicted that the third-generation hybrid rice can fully tap the potential of heterosis and significantly improve the heterosis of hybrids by improving free mating, GCA and SCA.

The genetically engineered RGMS lines and the male parent of third-generation hybrid rice can bring together any superior genes without being affected by the genetic background. Various good traits can be integrated through the recombination of various good agronomic traits. The yield of seed production can be close to that of conventional rice varieties, and the cost can be close to that of conventional variety seeds. However, rice quality can be greatly improved through the recombination of good traits and free mating. Different consumption needs can be satisfied by the diversity of rice products. The introduction of various resistance genes can realize the integration of multiple resistances or similar resistance genes, and reduce losses caused by diseases and pests to the largest possible extent. The reproductive lines of sterile lines are capable of seed setting through selfing, contributing to stable yield.

At present, hybrid rice from China is planted in such countries as the United States, India, Vietnam, Indonesia, Bangladesh and Pakistan, and its annual global planting area outside China reaches more than 5.2 million hectares, accounting for only less than 5% of the total rice planting area in the world, but with an average yield increase about 2 t/ha. The existing hybrid rice varieties lack high adaptability, and are susceptible to diseases and pests. For example, some varieties are not resistant to tropical diseases and pests or have poor rice quality; some are easily affected by changes in ecological conditions and have limited scopes of application; some do not have the cooking and eating qualities that fit the preference of the local people. There are many cytoplasmic minor restorer genes in tropical rice varieties, so the CMS lines bred with these varieties are not completely sterile. It is difficult to breed PTGMS lines because it is difficult to propagate and advance the generations under high temperature, which results in poor traits of the parents, little freedom for mating and poor heterosis. However, the above obstacles can be overcome by the third-generation hybrid rice breeding technology, so the third-generation hybrid rice breeding using genetically engineered RGMS lines as the genetic tool not only has great strategic significance for China's food security, but it also will bring about great changes to rice planting across the world.

Third-generation hybrid rice is also a solution to the problem of heterosis utilization of *japonica* rice. The annual planting area of *japonica* rice in the world is about 15 million hectares with an annual yield of 110 million tons. The planting area of *japonica* rice in China accounts for 56.1% of the total, and the yield accounts for 58.5% of the world's total. Most *japonica* hybrid rice varieties used in production do not have strong competitive edges over conventional *japonica* rice, and there are few excellent *japonica* hybrid rice combinations that organically combine high quality, high yield and strong stress resistance. On the contrary, most of them have poor grain quality and show no significant yield heterosis. The yield and purity in *japonica* hybrid rice seed production also need to be improved. At present, there is but a small genetic difference between the parents of *japonica* hybrid rice used in large-scale production, resulting in poor heterosis. Most male sterile lines are of the gametophyte sterile type, so they are not completely sterile, which seriously affects the purity of the hybrid progenies in terms of fertility. Worse still, the outcrossing performance of *japonica* male sterile lines is not as good as that of *indica* male sterile lines, which results in low yield, poor purity and high costs of seed production, leading to reduction in yield and deterioration in rice quality.

3. Technical breakthroughs to be made regarding third-generation hybrid rice

(1) Breed genetically engineered RGMS lines with high heterosis, high combining ability, high outcrossing ability, strong resistance and good quality by integration so as to continuously improve the heterosis level of rice and develop new super hybrid rice varieties than can adapt to the conditions in different regions.

(2) Understand the characteristics of the growth and development of and the genetics underlying third-generation hybrid rice, as well as the differences between third-generation and the first- and second-generation hybrid rice, and carry out research on the cultivation, demonstration and promotion of third-generation hybrid rice so as to provide theoretical guidance for promotion.

(3) Studies on the flowering and outcrossing habits of genetically engineered RGMS lines and clarify the genetics of high outcrossing rate, flowering time, glume opening rate, stigma vitality and exsertion rate, so as to provide theoretical guidance for the production and application of genetically engineered RGMS lines.

(4) Explore the optimal conditions for fluorescent protein expression and high yield of male sterile lines, as well as the purity and quality of male sterile lines in different environment and sorting conditions, set up a stable fluorescent marker screening system and establish a platform based on mechanized precision color sorting technology for large-scale seed preparation of genetically engineered RGMS lines.

(5) Establish a safety standard for the evaluation system for genetically engineered RGMS lines and their reproductive lines so as to provide theoretical basis and scientific basis for safe production of genetically engineered RGMS lines.

(6) Based on the functional and expression research of rice RGMS genes, conduct research on the regulation and stability of RGMS mutations in rice, obtain candidate genes and regulatory elements with independent intellectual property rights that can be used in third-generation hybrid rice breeding, upgrade core regulatory elements and tagging elements, and provide technical support for the upgrading and application of genetically engineered RGMS lines.

References

[1] FAN JINYU, CUI ZONGQIANG, ZHANG XIAN'EN. Spectral diversity and *in vitro* molecular evolution of red fluorescent protein[J]. Progress in Biochemistry and Biophysics, 2008,35 (10): 1112 – 1120.

[2] HU ZHONGXIAO, TIAN YAN, XU QIUSHENG. Analysis on the promotion process and status quo of hybrid rice in China[J]. Hybrid Rice, 2016 (2): 1 – 8.

[3] KUANG FEITING, YUAN DINGYANG, LI LI, et al. A novel method for vector making: recombinant-fusion PCR[J]. Genomics and Applied Biology, 2012, 31 (6): 634 – 639.

[4] KUANG FEITING. Preliminary study on engineered GMS system in *oryza sativa*[C]. Changsha: Central South University, 2013.

[5] LEI YONGQUN, SONG SHUFENG, LI XINQI. Development of rice heterosis utilization technology[J]. Hybrid Rice, 2017, 32 (3): 1 – 4.

[6] LEI YONGQUN. Transformation and expression of fertility genes in rice genetically engineered genic male ster-

ile lines[D]. Changsha: Central South University, 2017.

[7] LI XINQI, ZHAO CHANGPING, XIAO JINHUA, et al. Technical analysis of utilization of spontaneous and artificial genic male sterility in molecular breeding of hybrid crops[J]. Science and Technology Review, 2006, 24 (11): 39 - 44.

[8] LI XINQI, ZHAO CHANGPING, YUAN LONGPING. A method of hybrid crop breeding using the secondary restoring genes of nucleocytoplasmic male sterile lines. China: 200610072717.2 [P], 2007 - 10 - 10.

[9] LI XINQI, KUANG FEITING, YUAN DINGYANG, et al. A method of hybrid crop breeding. China: 201210513350.9 [P], 2013 - 03 - 20.

[10] LIAO FUMING, YUAN LONGPING. Discussion on genetically purifying PTGMS lines at critical sterility-inducing temperature[J]. Hybrid Rice, 1996 (6): 1 - 4.

[11] MA XIQING, FANG CAICHEN, DENG LIANWU, et al. Research progress and breeding application of RGMS gene in rice[J]. Chinese Journal of Rice Science, 2012, 26 (5): 511 - 520.

[12] TAN HEXIN, WEN TIEQIAO, ZHANG DABING. Molecular mechanism of rice pollen development[J]. Chinese Bulletin of Botany, 2007, 24 (3): 330 - 339.

[13] WANG CHAO, AN XUELI, ZHANG ZENGWEI, et al. Research progress and prospects of plant RGMS gene breeding technology system[J]. China Biotechnology, 2013, 33 (10): 124 - 130.

[14] WU SUOWEI, WAN XIANGYUAN. Establishment of technology system for male sterility hybrid breeding and seed production of main crops by biotechnology[J]. China Biotechnology, 2018, 38 (1): 78 - 87.

[15] YUAN LONGPING. Preliminary research success of the third-generation hybrid rice[J]. Chinese Science Bulletin, 2016, 61 (31): 3404.

[16] ALBERTSEN M C, FOX T W, HERSHEY H P, et al. Nucleotide sequences mediating plant male fertility and method of using same. Patent No. WO2007002267, 2006.

[17] TAYLOR L, MO Y. Methods for the regulation of plant fertility. Patent No. WO93/18142 [P], 1993.

[18] JI C H, LI H Y, CHEN L B, et al. A novel rice *bHLH* transcription factor, *DTD*, acts coordinately with *TDR* in controlling tapetum function and pollen development[J]. Molecular Plant, 2013, 6 (5): 1715 - 1718.

[19] DIRKS R, TRINKS K, UIJTEWAAL B, et al. Process for generating male sterile plants. Patent No. WO94/29465 [P], 1994.

[20] HAN M J, JUNG K H, YI G W, et al. Rice immature pollen (*RIP1*) is a regulator of late pollen development[J]. Plant Cell Physiol, 2006, 47 (11): 1457 - 1472.

[21] LI H, PINOT F, SAUVEPLANE V, et al. Cytochrome P450 family member *CYP704B 2* catalyzes the ω-hydroxylation of fatty acids and is required for anther cutin biosynthesis and pollen exine formation in rice[J]. Plant Cell, 2010, 22 (1): 173 - 190.

[22] LI N, ZHANG D S, LIU H S, et al. The rice tapetum degeneration retardation gene is required for tapetum degradation and anther development[J]. Plant Cell, 2006, 18 (11): 2999 - 3014.

[23] MATZ M V, FRADKOV A F, LABAS Y A, et al. Fluorescent proteins from non-bioluminescent Anthozoa species [J]. Nat Biotechnol, 1999, 17 (10): 969 - 973.

[24] NIU N N, LIANG W Q, YANG X J, et al. *EAT1* promotes tapetal cell death by regulating aspartic proteases during male reproductive development in rice[J]. Nature Communications, 2013, 4: 1445.

[25] PEREZ-PRAT E, VAN LOOKEREN CAMPAGNE M M. Hybrid seed production and the challenge of propagating male sterile plants[J]. Trends Plant Sci, 2002, 7: 199 - 203.

[26] VIRMANI S S. Heterosis and hybrid rice breeding[M]. Berlin: Springer-Verlag, 1994.
[27] WILLIAMS M, LEEMANS J. Maintenance of male sterile plants [P]. Patent No. WO93/25695 [P], 1993.
[28] YANG X J, WU D, SHI J X, et al. Rice *CYP703A3*, a cytochrome P450 hydroxylase, is essential for development of anther cuticle and pollen exine[J]. Journal of Integrative Plant Biology, 2014, 56 (10): 979 - 994.

Part 3

Cultivation

Chapter 9 Ecological Adaptability of Super Hybrid Rice

Ma Guohui / Wei Zhongwei

Section 1 Ecological Conditions for Super Hybrid Rice

The high yield potential of super hybrid rice depends not only on its own genetics, but also on certain ecological conditions. Suitable ecological conditions are conducive to the high yield potential. The high yield potential of super hybrid rice can only be fully exploited by adopting a scientific and reasonable planting layout, making full use of ecological advantages, and ensuring sound cultivation management that coordinates between varieties and ecological conditions. The ecological conditions required by super hybrid rice mainly involve temperature, sunlight, soil fertility, soil moisture and various other ecological factors.

Ⅰ. Temperature

A suitable temperature is one of the environmental conditions necessary for rice growth. There are the "triple base point" temperatures, i.e. the optimum, minimum and maximum temperatures for crop's life activities. At the optimum temperature, the crop grows and develops rapidly and well; at the maximum and minimum temperatures, the crop stops growing, but can stay alive. If the temperature goes above the maximum or below the minimum, different degrees of damage will be caused to the crop till death. The "three base point" temperatures are the most basic temperature indicators widely used to determine the effectiveness of temperature and the season and area to plant a crop, and to calculate crop growth rates, photosynthetic potential and yield potential.

Effective cumulative temperature is a basic requirement for rice growth for a season. Generally, the cumulative temperature needs to reach 2,000 ℃–3,700 ℃ over 110 – 200 d. With sufficient sunlight, precipitation, soil fertility and other environmental conditions, a double-cropping system can be adopted if the cumulative temperature reaches 5,800 ℃–9,300 ℃ over 260 d or more.

Rice damaged by heat will show a range of symptoms, the stems are prone to dryness and crack, the leaves show dead spots, browning and yellowing, sunburn, and in severe cases, the whole plant dies; and male sterility, inflorescence or ovary shedding and other abnormal phenomena occur. The damage of high temperature to plants can be divided into two parts. Direct damage means that high temperature directly affects the structure of the cytoplasm, and symptoms appear in a short period of time, and can spread from the heated parts to the non-heated parts; indirect damage means that high temperature leads to abnormal metabolism, gradually causing plant damage over a longer process. High temperatures often cause excessive transpiration and water loss in rice, which is similar to drought damage, causing a series of metabolic disorders due to cellular water loss, resulting in poor growth.

The heading and flowering period is the most sensitive period of rice to high temperature, and pollen is affected by high temperature and loses its activity, resulting in a significant decrease in fertilization rate, accelerated filling rate and shortened filling time (Wang Jialong et al., 2006). Meanwhile, pollen quality and quantity are reduced; anther dehiscence is prevented, and even if pollen falls on the stigma, it does not germinate normally, which eventually leads to more empty grains.

It has been shown that the degree of pollen sterility is affected by the dehiscence of pollen sacs under high temperature conditions and varies significantly among rice varieties; 1 – 4 d from the day of flowering is the period when rice is most susceptible to heat damage, and the rate of fertility reduction increases as heat damage increases. In terms of physiology, high temperature stress causes damage to the ultrastructure of rice chloroplasts, which starts to degrade the chloroplasts, decreases the sunlight trapping ability, and reduces the activity of enzymes involved in dark reactions, ultimately reducing photosynthesis (Ai Qing et al., 2008).

According to analysis of the ecological and climatic conditions at the super hybrid rice super-high yielding base in Longhui, Hunan, where yields exceed 15.0 t/ha, the climatic and ecological conditions suitable for rice super-high yielding cultivation are an average daily canopy temperature of 25 ℃– 28 ℃ from stem elongation to maturity, with a diurnal temperature difference greater than 10 ℃; an active cumulative temperature of more than 3,700 ℃ and more than 1,200 h of sunshine during the whole reproductive period (Li Jianwu et al., 2015).

II. Sunlight

Sunlight is one of the fundamental factor affecting plant growth and development. The combination of daylength, quality, and intensity affects yield formation in super hybrid rice. Among them, daylength mainly affects the process of vegetative and reproductive growth and development of rice. Light quality can regulate the growth, morphogenesis, photosynthesis, material metabolism and gene expression of the crop to some extent. Specifically, blue light promotes the growth and development of rice roots, increases the number of roots and root vigor, and increases the total and active absorption area of seedlings (Pu Gaobin et al., 2005), while blue-violet light inhibits the growth of rice by increasing the activity of indoleacetic acid (IAA) oxidase and decreasing the level of IAA; UV light has the effect of increasing the activity of IAA oxidase and inhibiting the activity of amylase, thus hindering the synthesis and utilization of starch.

Carbohydrate redistribution in the stem sheath during rice grain filling is significantly influenced by light intensity, and the transfer of accumulated assimilates in the stem sheath increases significantly under low light conditions. Rice shading significantly increases the transport of nonstructural carbohydrates from the stem sheath to reproductive organs, and this fraction is mainly transported to the strong spikelets. Since the activity of soluble starch synthase, starch granule-bound starch synthase and starch branching enzymes is reduced under shaded growth conditions, starch accumulation in seeds is also affected by light intensity during the grain filling period, mainly in the form of reduced amylose and sucrose content.

Super hybrid rice is generally characterized by photosynthetic production with high efficiency of solar energy utilization, high photosynthetic rate, and resistance to photo-oxidation. Compared with Shanyou 63, the photosynthetic characteristics of two super hybrid rice varieties, Liangyoupeijiu and Peiai 64S/E32, are significantly higher, indicating that these two super rice hybrids have obvious advantages in solar energy utilization and good ecological adaptability to different light conditions. The leaf photosynthesis rate of super hybrid rice Xieyou 9308 is significantly higher than that of Xieyou 63 during the full-heading and yellow-ripening periods. The chlorophyll content of Liangyoupeijiu has an average decline period of 20 days, which is three days longer than Ⅱ - You 58, and the average half-life of chlorophyll content of leaves is 25 days, which is four days longer than Ⅱ - You 58, indicating that super hybrid rice has better light capture capacity (Cheng Shihua et al., 2005).

Ⅲ. Fertilization Management

Rice is more sensitive to nitrogen and potassium fertilizer, and reasonable application of nitrogen and potassium can improve rice yield to a certain extent. However, in actual production, farmers generally have the habit of applying excessive nitrogen but seriously insufficient phosphorus and potassium, and the ratio and period of nitrogen, phosphorus and potassium application is out of balance, leaving super-high yielding rice varieties unable to unleash their high yield potential. In addition, a survey found that farmers usually apply compound fertilizer once and for all, which severely restricted the high yield potential of super hybrid rice. How to improve rice yields through reasonable allocation of nitrogen, phosphorus and potassium fertilizer is one of the important issues facing super-high yielding rice production at present. Some studies have shown that high-yielding rice has less nutrient uptake in the early growth stages and increases the uptake in the middle and late stages. Conventional fertilization is heavy on nitrogen and light on potassium, heavy on the early stages and light on the later stages, which is not conducive to the growth and high yielding super hybrid rice (Pan Shenggang et al., 2011)

In the cultivation of rice, appropriate late application of nitrogen and potassium fertilizer facilitates healthy rice growth. By appropriately reducing the amount of basal tillering fertilizer and increasing the amount of panicle fertilizer at the later stage, the total number of grains per panicle, the number of filled grains per panicle, the seed setting rate and the grain weight can be significantly increased to finally achieve high yield (Zeng Yongjun et al., 2008). On the basis of the total amount of nitrogen applied in the whole period, a reasonable combination of basal fertilizer, tiller fertilizer, panicle fertilizer and grain fertilizer is important for the growth and later seed setting of rice. The yield with panicle fertilizer is significantly higher than without. With an equal total amount of nitrogen fertilizer, the nitrogen uptake and yield of rice are proportional to how many times the fertilizer is applied. The optimal ratio of basal fertilizer : tiller fertilizer : panicle fertilizer : grain fertilizer is 3 : 3 : 1. The delayed application of a high proportion of nitrogen fertilizer also shows no benefit to the growth and yield of rice. The share of nitrogen fertilizer for delayed application shall generally be 20%-40%.

Nitrogen status has an obvious influence on tillering and panicle formation, and increased nitrogen fertilization is beneficial to tillering and the number of effective panicles. The increase in plant nitrogen content promotes secondary branching and increases the number of spikelets (Kazuhiro et al., 1994). The amount of panicle fertilizer applied has an effect on the number of primary, secondary and even tertiary branches. The appropriate application of panicle fertilizer promotes the formation of spikelets, but excessive use can also reduce the number of spikelets. It was found that nitrogen accumulation during panicle differentiation of rice showed a quadratic curve relationship with the numbers of panicles and grains. This shows that rice plants have suitable nitrogen levels during the spikelet differentiation period, and too high or too low nitrogen fertilizer is not conducive to the formation of large panicles. Appropriate nitrogen application can increase grain weight, but seed setting decreases with the increase of nitrogen application.

In terms of metabolic physiology, carbon and nitrogen metabolism are important metabolic processes in rice, affecting the synthesis and transport of photosynthetic products, the uptake and utilization of mineral nutrients, and protein synthesis. Carbon metabolism provides carbon and energy for nitrogen metabolism, while nitrogen metabolism provides enzymes and photosynthetic pigments for carbon metabolism, which are interrelated and require common reducing power, ATP and carbon skeleton, and are closely related to plant growth and development, as well as yield and quality formation. In the absence of nitrogen, the normal growth of rice plants will be seriously affected, resulting in small biomass, fewer tillers, and shorter plants, promoting the synthesized sucrose to synthesize fructosans in the parenchyma cells of the stem and sheath phloem, so as to maintain the concentration difference between the "source" and the "sink", and then deliver the accumulated matters to the seeds at the time of grain filling, compensating for the insufficient accumulation of assimilates due to reduced photosynthesis.

Nitrogen level can also affect the storage of starch and soluble carbohydrates in rice. When nitrogen fertilizer is insufficient during the cultivation process, more starch will accumulate in stems, while more nitrogen fertilizer will help with the movement of dry matter. When the nitrogen level reaches 300 kg/ha, the vegetative organs will remain green and the leaves cannot senesce normally, which is not conducive to the transfer of the soluble carbohydrates accumulated in the stem sheath to the panicles.

Ⅳ. Soil Moisture

Super hybrid rice requires large amounts of water, so severely insufficient soil moisture (drought) clearly can cause serious harm to the growth of rice plants. Research found that the drought resistance of rice is a resistance to and delay of the impact of a dry environment through its own physiological mechanism. Delay here means to delay the effect of drought on plant through the plant's own water storage and the water stored in the soil; resistance is a protective mechanism to reduce the damage of drought on the plant through regulation of the plant's internal physiological environment (e. g. increase in body fluid concentration to reduce the rate of water dispersal, physiological regulation of leaf curling, closing stomata to reduce transpiration water consumption). Excessive soil moisture is also not conducive to the normal growth of rice, and the demand for soil moisture is different across different periods of super hybrid rice development. If a plant is submerged in the tiller stage, the plant height will significantly increase with the internodes significantly elongated, thus causing the stalk to be weak and bent, and the matter accumulation insufficient. In the early booting period, the rice plant is in the stage of vegetative and reproductive growth and transformation and excessive soil moisture will affect the normal growth of young panicles and reduce lodging resistance. In the late booting period, it shifts from vegetative growth to reproductive growth, and the rice plant is metabolically active and particularly sensitive to soil moisture. Excessive water will reduce the leaf area and the photosynthetic capacity. In addition, flooding can inhibit spikelet differentiation and reduce the seed setting set rate and grain weight (Ling Qihong et al., 2008; Zhang Hongcheng et al., 2010).

Section 2 Ecological Adaptability of Super Hybrid Rice to Soil Conditions

High crop yield is closely related to soil quality. Yuan Longping proposed the "four good" concept for high-yielding rice, in which "good field" refers torice fields with high soil fertility and excellent soil structure, as well as good physical and chemical properties. With "good seeds, good cultivation methods, and good climatic conditions" in place, "good soil conditions" (good field) are decisive for high rice yield (Zou Yingbin et al., 2006), the realization the yield potential of the good seeds, and the yield differences of super hybrid rice in the same ecological region in the same year.

Ⅰ. Influence of Soil Fertility on the Yield of Super Hybrid Rice

Soil fertility is usually a major indicator of a good field. The contribution of soil fertility to crop yield is generally measured by the soil fertility contribution ratio (the ratio of crop yield without fertilizer to crop yield with appropriate fertilization), the level of which depends on the crop type, climate, and soil properties. The results of statistical analysis on the soil fertility contribution of major grain crops and influencing factors in China showed that rice had the highest soil fertility contribution among the three major crops (60.2±12.5%), and the main factor determining the soil fertility contribution in southern China rice fields was the phosphorus supply capacity of the soil (Tang Yonghua et al., 2008).

Rice yield in China increased from 4,324 kg/ha in 1981 to about 6,000 kg/ha in 1999 and then stagnated for over a decade from 1999 to 2010 (Grassini et al., 2013; Xiong et al., 2014). This stagnation was not compensated by the relative gains from the optimization of management practices and improvement of varieties (Xiong et al., 2014). In addition to yield reduction caused by climate extremes and changes in the cropping system (double-cropping to single-cropping or paddy field to upland field), the large area of low- and medium-yielding fields with poor soil fertility is also an important factor for the stagnation of rice yield in China.

Super hybrid rice has achieved a yield up to 15 t/ha in some high-altitude single-cropping regions and

12 t/ha in ecologically suitable single-cropping regions, showing great yield potential. However, the yield advantage of high-yielding rice has not fundamentally improved China's overall yield level, and the main challenge is the huge gap between the yield potential of high-yielding hybrid rice varieties and the actual yield. Ma Guohui et al. compared the actual yield of early-, mid- and late-season rice in Hunan province with those in regional trials from 2003 to 2012 and found that the difference between the actual yield of improved varieties and the yield of regional trials was about 1,650 kg/ha, a gap of 24%–30%, while the yield difference between super-high yielding fields and general fields was even greater, reaching 38.6%. A comparative study found that the yield of super hybrid rice is more than twice higher than the national average. The main reason for the failure to fully tap into the yield potential is the relatively low quality of farmland, which limits the yield potential of high-yielding varieties, and the yield level of high-yielding varieties grown on low- and medium-yielding fields can even be lower than that of conventional rice varieties (Xu Minggang et al., 2016).

A study on the super hybrid rice variety Xudao 3 reveals that the yield increased with soil fertility despite the different nitrogen levels and the same was true for the yields of brown, milled and head rice. The milled rice yield of the high soil fertility field also showed the same trend under the same level of nitrogen application (Zhang Jun et al., 2011). Studies on super hybrid rice Liangyoupejiu found that its slow leaf senescence rate, despite the advantages of a longer chlorophyll functional period and a more stable photosynthetic rate at the later stage, led to a low seed setting rate due to insufficient nutrient supply and poor soil physical properties in the field, which limited its yield increase potential (Xiong Xurang et al., 2005).

In the same ecological region, rice paddy with high soil fertility may achieve super-high yield potential with a small amount of fertilizer, while rice paddy with medium to low soil fertility will not achieve super-high yield potential even if fertilizer application is increased. An investigation in a demonstration area shows that where the soil layer is thick with sufficient basal fertilizer, the seedlings are robust and vigorous with bigger grains and no sign of early senescence; while where the soil layer is thinner and less fertile, seedlings are sparse with low tiller nests or low vigor. Excessive fertilization may produce a large quantity of upper roots, increase ineffective tillering, degrade the physical and chemical properties of the soil, aggravate pests and diseases, increase early senescence and lodging, and ultimately result in a gap of 1,500–3,000 kg/ha in yield. Thus, soil environment or soil fertility is a major restriction on the realization of the super high yield potential of super hybrid rice (Xiong Xurang et al., 2005).

Ⅱ. Feedback Effect of Super Hybrid Rice Yield on Soil Fertility

Soil quality improvement is the basis for high crop yield, and organic matter in rice fields plays a central role in soil fertility improvement, i.e., soil organic carbon is the core indicator of soil quality. Soil organic carbon circulation and accumulation has an important impact on various physical, chemical, and biological properties and processes of the soil, and is the material basis and key mechanism for the formation of systematic processes and productivity. Xu Minggang et al. (2016) analyzed the results of multi-site experiments across the country and found that there was a positive correlation between crop yield, stability and soil organic matter, e. g. , 0. 1% of organic matter in southern China rice areas can translate into 600 – 900 kg/ha of yield. On average, a 0. 1% increase in soil organic matter increases the stability of yield by about 10% . Organic matter is an important source of plant nutrients, which can effectively improve soil nutrient retention and buffering. Soil organic matter exists in the form of organic colloids, and colloidal particles have a large negative charge, which can adsorb cations and water as well as phosphorus, iron and aluminum ions to form complexes or chelates, avoiding the precipitation of insoluble phosphate, thus improving soil fertilizer retention. It can also improve the buffering effect of soil against acid and alkali, and improve the resistance of plants to acid and alkali stress. Organic matter can improve the physical properties of soil and is an indispensable cementing substance for the formation of water-stable agglomerate structure, so it helps clayey soil to form a good soil structure.

In agricultural soils, fresh organic matter entering the soil includes naturally returned plant residues and root secretions, artificially returned organic fertilizers, etc. The results of numerous regional surveys and long-term site-specific experiments show that paddy ecosystems can significantly increase soil organic carbon content, more than other land use in the same ecological region (Huang & Sun, 2006; Sun et al. , 2009) , mainly due to the high organic carbon input from paddy fields. The biomass of super hybrid rice is relatively large, and likewise the amount of organic carbon naturally returned through the root system, stubbles, and root secretion is larger. In addition, the alternately dry and wet water management pattern creates special oxidation and reduction conditions in the rhizospheric soil. The stability of soil aggregates is thus enhanced and soil structure tends to be better after continuous super rice cultivation (Meng Yuanduo et al. , 2011). A comparison between the Longhui base of super hybrid rice cultivation in Hunan province and the Changsha headquarters experimental base of HHRRC revealed that the content of large agglomerates in the Longhui base was much higher than that in the Changsha base. Specifically, there are 3. 6 times more agglomerates >2 mm and 7. 9 times more agglomerates 0. 2 – 2. 0 mm in the Longhui base than in the Changsha base, while the latter has 1. 2 times more particles <0. 02 mm than the former. The increase of agglomerates is beneficial to the accumulation of soil nutrients. For example, the organic matter and total nitrogen in the Longhuai base are 4. 8% and 17. 1% higher than in the Changsha base, respectively, which is conducive to the realization of rice yield potential, so that the actual yield in Longhuai can reach 15. 1 t/ha, 27. 9% higher than in the Changsha base (Li Jianwu et al. , 2015).

Ⅲ. Effect of Soil Moisture Management on Hybrid Rice Yield

Rice is the most water-intensive irrigated crop in the world, and soil moisture plays an important role in nutrient utilization efficiency and rice yield formation. With increasingly tight water supply for irrigation, effective soil moisture management is of great practical importance to fully utilize the limited water resources to ensure high and stable yield of super hybrid rice, enhance water conservation in rice fields, and mitigate the environmental impacts (water pollution and greenhouse gas emissions) of rice production.

Cheng Jianping et al. (2008) studied the effects of drought stress (different soil water potentials) and nitrogen nutrition on the physiological properties, yield and nitrogen fertilizer utilization efficiency of the super hybrid rice variety Liangyoupeijiu. The results are as follows. 1) At the same nitrogen fertilizer application level, the net photosynthetic rate of leaves, the contents of chlorophyll a, chlorophyll b, and the two combined, the SPAD value and the leaf water potential decreased as the soil water potential declines, while the activity of chlorophyll a, chlorophyll b, malondialdehyde, and peroxidase increases. 2) At the same nitrogen fertilizer level, rice yield decreases with the decrease of soil water potential; when the soil was slightly dry, the yield shows the pattern high nitrogen > medium nitrogen > low nitrogen; while when the soil moisture is sufficient or the soil is very dry, the yield order is medium nitrogen > high nitrogen > low nitrogen.

The high matter production capacity of superhybrid rice is closely related to its large root volume and high root vigor. However, root morphology and function are limited by the soil environment, which can be influenced with soil water management. A comparative analysis shows that the root density, root vigor, population growth rate and relative growth rate of hybrid rice were significantly higher when the wet irrigation mode is adopted than when submerging irrigation is used; hybrid rice roots form aerenchyma the latest when submerged, while root vigor is higher with the alternating wet and dry or controlled water irrigation mode (Liu Famou et al., 2011). Water management affects the rhizospheric oxygen supply as it regulates the proportion of solids, gas, and liquid in the soil and hybrid rice is characterized by high root vigor at high dissolved oxygen levels. He Shengde et al. (2006) showed that compared with submerging, rhizospheric oxygen supply can increase the redox potential of the soil, enhance the dry mass weight of rice organs, promote tillering, and increase the numbers of effective panicles, primary branches, and filled grains on each panicle, thus significantly boosting the yield.

Ⅳ. Paddy Land Fertility Problems and Good Field Cultivation Measures

Current, arable land in China generally suffer from the following problems. The overall quality of arable land in China is low, with a large proportion of low- and medium-yielding fields (71.0% of the total arable land area) and many obstacles (89% have obstacles); soil degradation is serious, manifested in a thinning arable layer and the accelerated acidification of red soils (Xu Minggang et al., 2016). Huang Guoqin (2009) summarized the problems facing the sustainable development of rice farming systems in the south China and proposed the weakening intensity of land nourishment and the deterioration of farmland environment as the major problems limiting rice yield. First of all, long-term double-cropping has had negative impacts on soil fertility, which is mainly reflected in the deterioration of soil physical properties, lopsided soil nutrient consumption, the accumulation of toxic substances in the soil and the spread of diseases, pests, and weeds as a result of long-term multi-cropping solely with rice. Secondly, the intensity of land nourishment in southern China has reduced significantly. The traditional "land cultivation" activities (e.g. water, fertilizer, and weed management in rice fields) have disappeared in the tillage system; field management measures have become much less frequent; and old land nourishment measures such as the agricultural green manure, farm manure have become less used or disappeared. Reduced land nourishment has led to decline in soil fertility and increased dependence on chemical fertilizer, resulting in the waste of crop straw resources and further declining land fertility.

Soil organic carbon is an important indicator of soil fertility. Many past studies have revealed that the key to improving farmland quality is to increase organic matter in the soil. An effective way to do this is to recycle and make efficient use of all kinds of fertilizer resources, including farm manure, straws, green manure, and agricultural waste (Xu Minggang et al., 2016). It has been proved by both short-term and long-term experiments that the combination of chemical fertilizer and organic fertilizer is the best way to improve soil fertility (Huang et al., 2006; Liu et al., 2014; Xu Minggang et al., 2016), significantly improving the physical properties of the soil, especially in South China where the soil is relatively sticky and heavy (Liu Lisheng, et al., 2015; Chen et al., 2016).

For the improvement of low- and medium-yielding fields and soil fertility enhancement, the first step is to identify the obstacle factors and gradually restore the basic soil fertility by developing various environment-friendly soil conditioners to eliminate these obstacle factors (e.g. acidified soil, secondary salinized soil, alkaline soil, submerged fields, etc.) (Xu Minggang et al., 2016). In addition, southern China rice fields face a greater risk of land use change, and the conversion of irrigated rice field to upland field, double-cropping to single-cropping field, fallow, and abandonment may have an impact on soil fertility, making it more important to protect existing high-yielding fields from nutrient loss, soil erosion, and soil fertility degradation caused by improper management. For high-yielding fields, efforts can be made to maintain good soil structure and biological functions through the recycling of organic resources within the system, conservation tillage, and intercropping rotation.

Section 3 Super Hybrid Rice Cultivation Zoning

Ⅰ. Yield Performance of Super Hybrid Rice in Different Ecological Regions

Super high yield and ecological suitability experiments carried out on 6.67-ha fieldsfor the new combination of super hybrid rice Xiangliangyou 900 in Haitang Bay of Sanya; Xingyi of Guizhou; Lechang of Guangdong; Nandan and Guanyang of Guangxi; Guidong, Qidong and Xupu of Hunan; Jinhua of Zhejiang; Nanchuan of Chongqing; Yongding district of Zhangjiajie, Taoyuan and Cili of Hunan; Tongcheng of Anhui; Chongzhou of Sichuan; Baihu Farm in Lujiang of Anhui; Suixian county of Suizhou, Hubei; Hanzhong of Shaanxi; and Juxian county of Rizhao, Shandong in 2015; as well as in Gejiu of Yunnan and Yongnian of Hebei in 2016. The geological locations of these experiment sites range from 15.8 m (Sanya) to 1,287 m (Gejiu of Yunnan) in altitude, from E103°17' (Gejiu of Yunnan) to E119°23' (Jinhua of Zhejiang) in longitude, and from N18°15' (Sanya) to N36°33' (Yongnian of Hebei) in latitude. In total, the experiments spanned two years and data were collected from 21 ecological sites in 14 provinces, municipalities, and autonomous regions, with basically all types of ecological areas for the growing of super hybrid rice covered. The difference in ecological area had a significant impact on the yield performance of the same super hybrid rice variety. The yield variation of Xiangliangyou 900 ranged from 11.18 t/ha to 16.32 t/ha, with a large yield variation. Further analysis of yield components in each ecological are shows that yield components such as effective panicles, number of grains per panicle, and seed setting rate varied widely, indicating that ecological are influences yield by affecting effective panicles, number of grains per panicle, and seed setting rate. The common features of the super-high yield obtained are the synergy of large panicles and multiple spikes and the achievement of high "sink" capacity and high seed setting rate (Wei Zhongwei. et al., unpublished). In 2015, an experiment was carried out in Longhui, Hunan, to examine the ecological suitability of the three super hybrid rice varieties Y Liangyou 900, Xiangliangyou 2, and Shenliangyou 1813 at the four different altitudes of 300 m, 450 m, 600 m and 750 m. The results showed that altitude had a significant influence on the growth duration, plant height, panicle length, and yield of all three varieties. Their growth duration ranged from 150 days to 175 days, the higher the altitude, the longer the duration. At altitudes above 450 m, the plant height, panicle length, and yield all dropped significantly as the altitude increases. In particular, at 750 m, all the figures were the lowest. The yield declines mainly because the total number of grains and the seed setting rate decreased. All varieties had the highest yield at 450 m, which, therefore, can be regarded as the best in the region (Wei Zhongwei et al., unpublished).

The above two experiments show that although super hybrid rice has super-high yield potential, the ecological environment plays a major role in the realization of such potential, so ecological suitability is clearly an important consideration for super hybrid rice.

Similar results have been obtained previously in other studies. Deng Huafeng et al. (2009) examined the yield stability of 12 super hybrid rice combinations at seven ecological experiment sites in the Yangtze River Basin. The result show that the yield of all 12 varieties varied significantly or very significantly across the years and sites, and the interaction was also either significant or very significant, meaning that different rice combinations show vastly different suitability in the various ecological areas. Therefore, when promoting super high yielding rice, it is necessary to consider not only its yield potential but also its ecological suitability. Ao Hejun et al. (2008) examined the yield stability of super hybrid rice at different sites in Hunan province and found the difference in yield to be extremely significant and the yield between years to be also significant, indicating that super hybrid rice has its suitable planting areas. Li Ganghua (2010) carried out density experiments in eight typical ecological areas in China to examine the ecological difference in rice yield across different ecological areas. The results show great differences in growth duration, yield composition, "source" and "sink" size, grain-leaf ratio, plant architecture and dry matter accumulation and distribution across different ecological areas. Xiao Wei (2008) concluded that Liangyou 293 and Liangyoupeijiu both had their suitable planting dates in Changsha. If planted earlier, it is likely to be hurt by high temperature during the heading and seed setting stages; if sowed later, it is likely to meet cold front in the later stages, which is also detrimental to the yield.

A lot of studies have been done on the mechanism of the impact of ecological areas on rice yield. Yang Huijie et al. (2001) conducted a comparison between Longhai of Fujian, a general ecological area, and Taoyuan of Yunnan, a suitable ecological area, and found that super high yielding rice accumulated huge biomass so the yield increased together with the accumulated dry matter. As a result, in a suitable ecological area, the physiological and ecological traits were all given full play and the yield potentials were more likely to be fully realized. Yuan Xiaole, et al. (2009) found that rice with well-developed roots, high "sink" capacity, strong capability to produce and accumulate matter, and high photosynthetic efficiency depends heavily on ecological suitability for high yield. In a suitable environment, the rice plants develop more roots, absorb more nutrition, and deliver a higher yield. Li Jianwu et al. (2015) analyzed the difference in yield of the super hybrid rice variety Y Liangyou 900 in Longhui, a suitable ecological area, and in Changsha, a general ecological area, and found Longhui to provide clearly more suitable ecological conditions from initial heading to maturity, with a more suitable temperature (around 28 ℃ on average with no high temperature above 37 ℃) to prevent harm; in addition, the temperature varied more between day and night with a difference 3. 7 ℃ greater than in Changsha, boosting grain filling and the grain weight.

Therefore, when the mechanism of ecological factors' effect on yield is clarified, the key ecological environment and cultivation measures that affect the number of effective panicles and seed setting rate should be the focus of our attention. Liu Jun et al. (1996) concluded that large-panicle super hybrid rice varieties with suitable climatic and ecological conditions for high panicle rate and high seed setting rate are a reliable way to achieve super high yield for hybrid rice. Yang Huijie et al. (2000) suggested that the yield of super hybrid rice requires larger panicles in suitable ecological conditions so as to increase the number of spikelets per unit area and ensure a high "sink" capacity. Ai Zhiyong et al. (2010) showed that the main reason for the poor yield stability and poor high yield repeatability of super hybrid rice was the poor ecological adaptability and the strict requirements for a suitable environment. In an unsuitable ecological area, the seed setting rate is unstable and there are fewer effective panicles, Therefore, increasing the effective panicles and improving the stability of seed setting are the keys to high and stable yield of super hybrid rice. In summary, in the process of promoting super hybrid rice to more general ecological areas, cultivation and control techniques should be developed to increase panicle number and stabilize seed setting rate, so as to increase panicles, stabilize seed setting and expanding "sink", realizing high and stable yield of super hybrid rice in general ecological areas.

Ⅱ. Ecological Adaptability of Super Hybrid Rice

Since the ecological factors in the natural environment are generally fixed, it is particularly important to clarify the ecological adaptation of rice in order to maximize itsgrowth advantage. The concept of ecological adaptability is the ability of a species to achieve equilibrium in an ecological environment, and the ecological adaptability of rice emphasizes the ability to adapt to different ecological conditions and resist adversity, and ultimately deliver stable high yields. The ecological adaptability of rice involves the adaptability to location, temperature, water, light, and soil fertility.

In recent years, with the wide application of super hybrid rice varieties, some of the varieties have shown some problems. For example, Liangyoupeijiu and some other super hybrid rice varieties are not tolerant of high temperature during the grain filling and seed setting stages, resulting in a significant reduction in the seed setting rate under high temperature. In some areas, this led to a significant reduction in yield. Therefore, varieties with poor heat tolerance can easily suffer losses. In addition, some varieties are also difficult to unleash their super-high yield potential in large-scale production due to susceptibility to rice blast, false smut, and other diseases, or intolerance to low temperature at late stages.

In order to meet this challenge, Chinese scientists have adhered to the breeding goal of high yield, high quality, and wide adaptation, and actively taken on research related to the wide adaptation of super hybrid rice and made promising progress. For example, HHRRC developed the widely adaptive PTGMS line Y58S, which has many excellent traits such as high compatibility, strong disease and stress resistance, high temperature tolerance during grain filling and seed setting, and low temperature tolerance during the late reproductive period. Therefore, it is suitable for the breeding of widely adaptive super hybrid rice. The Y Liangyou super hybrid rice varieties derived from the PTGMS line have good resistance to biotic and abiotic stresses and strong ecological adaptability. Indoor evaluation and large-scale production have shown that several representative varieties, such as Y Liangyou 1, Y Liangyou 2 and Y Liangyou 900 have good resistance to high temperature, low temperature and drought.

Y Liangyou 1 has strong resistance to abiotic stresses such as high and low temperatures, drought, and also has strong resistance to biotic stresses such as rice blast, false smut and bacterial leaf blight. In 2006, 2008 and 2013, it passed the national evaluation of southern *indica* rice in three ecological areas, namely the middle and lower reaches of the Yangtze River, the early-season rice area in southern China and the upper reaches of the Yangtze River. Indoor experiments under controlled conditions showed that the resistance to high and low temperatures, and drought tolerance of Y Liangyou l was significantly better than that of Liangyoupeijiu. In 2004, its yield ranked first in both the Hunan super rice group (high fertility group) and the mid-season hilly area late-maturing group (medium and late maturing group), with a yield increase of 11.2% and 9.04% over the control variety Liangyoupeijiu and Ⅱ-You 58, respectively, showing its strong adaptability to soil fertility conditions. Statistics show that Y Liangyou l has been the most widely promoted hybrid rice variety in China since 2010, and has been recognized by the MOA as a major promoted rice variety in the Yangtze River basin for six consecutive years, and the total planting area has reached four million hectares and is still growing.

Deng Huafeng et al. (2009) classified super hybrid rice into two types based on the results of yield performance, stability of yield and seed setting rate in various ecological sites. One is the super hybrid rice with wide adaptability that can adapt to different ecological environments. This is the solution to an important issue, difficult point, and hot topic facing super hybrid rice breeding at present. This type of super hybrid rice, on the basis of an 8% yield increase over the control variety in regional tests, has shown high and stable seed setting rates in different ecological areas and cropping systems, and the yield potential can be fully and steadily tapped. Among all the varieties tested, Zhunliangyou 527, Honglianyou 6, C Liangyou 87 and Y Liangyou 1 belong to this category. The other type of super hybrid rice adapts to specific ecological areas, tentatively called the regional type of super hybrid rice. This type of super hybrid rice can show high yields and high seed setting rates in certain ecological areas, but its super high yield potential cannot be fully tapped in other areas. These varieties include Ⅱ-Youming 86, Liangyou 293 and Liangyoupeijiu.

Super hybrid rice has strict ecological suitability and their suitable planting areas have specific characteristics. Super hybrid rice should be planted and promoted according to these. Before the promotion and application of super hybrid rice, a specific super hybrid rice combination is used to carry out comprehensive trials in multiple ecological areas, to initially clarify its suitable planting areas or suitable soil, meteorology, altitude and other ecological environmental conditions, to determine suitable planting zoning. For locations where it has not been planted before, it is necessary to conduct more rigorous introduction and ecological suitability tests than usual to clarify the suitability.

Any variety can only unleash its full high yield potential in the most suitable cultivation environment. All cultivation techniques are aimed at creating a suitable cultivation environment in terms of irrigation, insulated nursery for the cultivation of seedlings, fertilization, etc. However, some natural ecological factors are uncontrollable, such as temperature, light and rainfall, so it is required to develop some cultivation measures (such as sowing period) and try everything possible to avoid the influence from those unfavorable factors, in order to strive for the best ecological environment conducive to high yield. Therefore, the establishment of cultivation and control technology for stable high yields is the main direction in the research of the ecological suitability of super hybrid rice.

Ⅲ. Climatic-ecological Zoning of Super Hybrid Rice Cultivation in Southern China

Super rice is divided into super conventional rice and super hybrid rice. There are *indica* and *japonica* varieties in both types. *Indica* varieties are mostly hybrid rice, while *japonica* varieties are mainly conventional rice. Super hybrid rice is mainly planted in the south, while conventional super *japonica* rice is mainly planted in the north. So far, whether it is double-cropping rice or single-cropping rice in southern China, large-scale production of super rice involves mainly super hybrid rice. Both super hybrid rice and common hybrid rice prefer high temperature, high humidity and short daylength, and their planting areas are basically the same. The dominant areas for super hybrid rice cultivation are mainly in the south. According to the hybrid rice zoning of the National Hybrid Rice Meteorological Research Collaborative Group (1980) (Table 9 - 1), and the safe growing season, accumulated temperature and the date of low temperatures in autumn in various locations, the super hybrid rice area in the southern rice area can be divided into three categories and six zones.

Table 9 - 1 Hybrid rice zoning

Maturity type	Safe growing season (d)	Accumulated temperature (℃)	Low temperature date
Single-cropping hybrid rice	110 - 160	2,400 - 3,800	Early September
Double-cropping hybrid rice minor-planting	160 - 180	3,800 - 4,300	Mid-September
Double-cropping hybrid rice main-planting	180 - 200	4,300 - 4,800	Late September
Double-cropping hybrid rice	>200	>4,800	Early and mid-October

Region Ⅰ - 1 - Single-cropping hybrid rice zone at the northern edge of the southern rice region. This zone is located north of Nanjing and Hankou, and south of Zhengzhou and Xuzhou, with a safe growing season of 150 - 160 d, accumulated temperature of 3,500 ℃- 3,800 ℃, low temperatures appearing in late August to early September, enough days for single-cropping hybrid rice cultivation, but not for double cropping. When planted on wheat stubble in mid to late April, it has a total of growth duration of 135 – 145 d and 95 – 105 d from sowing to heading. Rice can head in early to mid-August and mature in mid- to late September, and the heading period is not likely to meet low temperature days in autumn.

Region Ⅰ – 2 – The Guizhou low plateau and western Sichuan mountainous single-cropping hybrid rice area. The heat conditions in different parts of the region vary greatly. Most areas have a cumulative temperature above 2,400 ℃ and a safe growing season of 110 d or more, and thus can grow single-cropping mid-season season hybrid rice. In spring, it warms up early, so seeds can be sown in late March to early April. Low temperature damage in autumn mostly occurs in late August to early September. Due to the absence of high temperature in summer, the total growth duration is exceptionally long, generally 160 – 170 d, with a long period of 110 – 120 d from sowing to heading and 50 d for grain filling and maturing. Hybrid rice in this area tends to be well developed.

Region Ⅱ – 1 – The main planting area of double-cropping late-season rice. The area is to the south of Nanchang and Huaihua and north of Fuzhou, Chenzhou and Guilin. The elevation is below 400 m, and areas 400 – 600 min elevation is the transitional zone. Conventional early rice in the early season plus hybrid rice in the late season are planted thanks to the good heat conditions. Early-season rice is sown in mid- to late March, and autumn low temperature starts in late September. Thus, the safe growing season lasts more than 180 d. The cumulative temperature exceeds 4,300 ℃. The early-season conventional rice is mainly medium- to late-maturing varieties, and the late-season hybrid rice has abundant time. Since the 1980s, the growth duration of early-season hybrid rice is similar to that of local medium- to late-maturing conventional rice varieties, and the area planted with double-cropping hybrid rice has been expanding year by year.

Region Ⅱ – 2 – Yangtze River basin double-cropping late-season hybrid rice collocation area. The growing area has a poor heat condition for one early-season of conventional rice and one late-season of hybrid late rice. Early-season rice is sown in late March to early April; autumn low temperature starts in mid-September, safe growing season 160 – 180 d, cumulative temperature 3,800 ℃– 4,300 ℃. It is the northern border areas for double-cropping rice. Early rice is mainly medium-maturing varieties as time is not quite sufficient for late-season rice. Hybrid late-season rice is suitable for only some small parts of the area. Since the 1990s, as the growth duration of early-season hybrid rice is similar to that of medium- and late-maturing conventional rice, double-cropping hybrid rice began to be planted in demonstration.

Region Ⅱ – 3 – The Sichuan Basin double-cropping late-season hybrid rice area. Although the total heat is similar to that in Region Ⅱ – 1, the heat distribution is different from what it is like in the middle and lower reaches of the Yangtze River, with good heat conditions in the early season and relatively poor heat conditions in autumn, and more rains and earlier low temperature occurrence in autumn. Double-cropping late-season hybrid rice cultivation is mainly arranged in the central southeastern part of the basin. The growth duration of hybrid rice in this area encounters high temperatures of about 27 ℃, and the average temperature of July to August can reach 30 ℃. The high temperature accelerates the growth and development of individuals and restricts the development of high-yielding populations, which is unfavorable to high-yielding cultivation.

Region Ⅲ – Southern China double-cropping hybrid rice area with the best heat condition. Here, early-season rice is generally sown in late February to early March and harvested in mid-July; typical late-season rice varieties are sown in late June, safely reach full heading in early to mid-October and are harvested in mid-November, with a safe growing season of more than 200 d and a cumulative temperature of more than 4,800 ℃. Double-cropping hybrid rice can be planted. In Shaoguan, Guangdong, the northern boundary of this area, early-season hybrid rice is sown in early March, with an early-season growing period of 140 d. Late-season hybrid rice is planted from late July to early August, and full heading occurs in late September to early October. In South China's Pearl River Delta, sowing occurs in mid- to late February, the early-season rice has a growing period of 160 – 170 d, good not only for early-season rice, but also for some extra-late conventional rice varieties with a longer reproductive period. However, the early-season rice in this area has "dragon-boat water" from late May to early June and typhoon damage during maturity. The average temperature from May to September is higher than 27 ℃ and individual development is faster, with a shorter growth duration. Dry matter accumulation is thus insufficient, so the yield is generally more stable but not high.

Ⅳ. Climatic Zoning of Hybrid Rice Cultivation in Yunnan Province

Yunnan, China, has a unique low-latitude plateau climate with a distinctive three-dimensional climatic structure. The zoning of hybrid rice cultivation is more complex here. Zhu Yong et al. (1999) converted the agro-meteorological indicators required for hybrid rice growth and development and yield formation into the following commonly used agro-climatic zoning indicators. ① The upper limit of safe cultivation of hybrid rice is about 1,400 m above sea level east of the Ailao Mountains and about 1,450 m above sea level west of the Ailao Mountains; ② average annual temperature above 17 ℃; ③ average temperature above 21 ℃ in June, July and August, and above 22 ℃ in July; ④ ≥10 ℃ active cumulative temperature above 5,500 ℃. According to the different climatic characteristics, the hybrid rice growing areas in Yunnan are divided as follows.

1) Tropical, low-heat river valley double-cropping hybrid rice early- and late-season succession area. This area includes Jinghong and Mengla in Xishuangbanna; Yuanjiang, Honghe and Hekou in the Yuanjiang River Valley; Mengding in Lincang; Ruili in Dehong; Lujiangba in Baoshan; Liuku in Nujiang; and Yuanmou and Qiaojia in the Jinsha River Valley. The annual average temperature is higher than 20 ℃, the average temperature in July is higher than 24 ℃; the average temperature in October is higher than 20 ℃, and the ≥ 10 ℃ active cumulative temperature is 7,300 ℃ or more. The characteristics are adequate heat , safe double-cropping system for early- and late-season succession of hybrid rice, and high seed setting rates. However, because the average temperature from June to August is above 24 ℃, the growth duration is short. Therefore, consideration should be given to the increase of the number of effective panicles and grains, focusing on grain weight. From the perspective of single-cropping yield, this area is a stable production area for hybrid rice, but not a high-yielding area. Under current production condi-

tions, with a little effort, it will not be difficult to achieve a yield of 15 t/ha for the early and late seasons.

2) Single-cropping hybrid rice area in southeastern Yunnan. This area includes Xinping, Guangnan, Mile, Jianshui, Shiping, Mengzi, Kaiyuan, Pingbian, Wenshan, Maguan, Malipo, Xichu, Funing and some other counties and cities and the low-altitude areas of Qubei and Yanshan. The average annual temperature in this area is 17 ℃-20 ℃; the average temperature in July is 22 ℃-24 ℃; and the sunshine hours from June to August is more than 400 h, which is the best combination of light and temperature in the hybrid rice growing areas in Yunnan. From the perspective of climatic productivity, it is a high yielding area for hybrid rice. The yield of large-area planting here is above 9 t/ha, and it is entirely possible to push it further up to 10.5 t/ha with improved cultivation techniques. In this area, due to the slow warming in spring and the "spring cold", it is important to use plastic film to cover seedling nurseries so as to ensure safe heading. Efforts need to be made to make full use of the sufficient light and heat resources before the start of the rainy season so that the high yield potential can be fully tapped. In production, attention should be paid to having more panicles and more grains, as well as a higher seed setting rate.

3) Double-cropping rice, conventional rice and hybrid rice early- and late-season succession area in southern Yunnan. This area includes Jinping, Jiangcheng, Simao, Puer, Mojiang, Jingdong, Nanjian, Yunxian, Yongde, Zhenkang, Gengma, Lincang, Shuangjiang, Canyuan, Lancang, Menglian, Jinggu, Menghai and other counties and cities. The average annual temperature is 17 ℃-20 ℃, the sunshine hours in June to August, except in Nanjian, is 240 - 400 h. It has the least sunshine in all hybrid rice growing areas. The climatic conditions in the western part of this area are better than those in the eastern part, with the largest area suitable for growing hybrid rice. However, with less sunshine, more precipitation, more pests and diseases, the seed setting rate is low and the yield level is between that of Region I and Region Ⅱ. In production, attention should be paid to the coordinated population development, especially disease and pest prevention and control, so as to improve the seed setting rate for a higher and more stable yield.

4) Conventional rice and hybrid rice succession area in southwestern Yunnan. This area includes Shidian, Yingjiang, Lianghe, Luxi, Longchuan and other counties and cities. The average annual temperature is 17 ℃-20 ℃ and the average temperature in July is 21 ℃-24 ℃; the sunshine hours from June to August total 350 - 400 h. The climatic productivity here is second only to that of the single-cropping hybrid rice area in southeastern Yunnan. It is characterized by a relatively small difference in sunshine duration, with fewer hours of sunshine in July. The main consideration in production should be to increase the number of panicles and grains, as well as to improve the seed setting rate.

5) Northern single-cropping hybrid rice area. This area includes the valley areas in Fugong, Huaping, Yongren, Yongshan, Suijiang, Yanjin, and Weixin counties. The total area is small and geographically dispersed, with an annual average temperature of 17 ℃-21 ℃; an average temperature of 23 ℃-27 ℃ in July, and 400 - 600 h of sunshine from June to August, which is the most abundant in the hybrid rice growing areas in Yunnan. The climate of the area is characterized by late but fast warming in spring, early cooling in autumn, and relatively obvious continental temperatures. If irrigation can be guaranteed, plastic film can be used to cover seedling nurseries, and timely sowing and early planting can be

ensured to make full use of the light and heat resources, this area can also be a high yielding area for hybrid rice.

References

[1] WANG JIALONG, CHEN XINBO. Research progress on rice heat tolerance. Hunan Agricultural Sciences, 2006, (6): 23 - 26.

[2] AI QING, MU TONGMIN. Research progress on rice heat tolerance. Hubei Agricultural Sciences, 2008, 47 (1): 107 - 111.

[3] LI JIANWU, ZHANG YUZHU, WU JUN, et al. Study on the "four food" supporting technology system for super hybrid rice with 15.0 t/ha yield. China Rice, 2015, 21 (4): 1 - 6.

[4] PU GAOBIN, LIU SHIQI, LIU LEI, et al. Effects of different light quality on the growth and physiological characteristics of tomato seedlings. Acta Horticulturae Sinica, 2005, 32 (3): 420 - 425.

[5] CHENG SHIHUA, CAO LIYONG, CHEN SHENGUANG, et al. Concept and biological significance of late functional super hybrid rice. Chinese Journal of Rice Science, 2005, 19 (3): 280 - 284.

[6] PAN SHENGGANG, HUANG SHENGQI, ZHANG FAN, et al. Growth and development characteristics of mid-season *indica* hybrid rice for super high yield cultivation. Acta Agronomica Sinica, 2011, 37 (3): 537 - 544.

[7] ZENG YONGJUN, SHI QINGHUA, PAN XIAOHUA, et al. Effects of nitrogen application on nitrogen utilization characteristics and yield formation of high yield early-season rice. Acta Agronomica Sinica, 2008, 34(8): 1409 - 1416.

[8] KAZUHIRO, KOBAYASI, TAKESHI H. The effect of plant nitrogen condition during reproductive stage on the differentiation of spikelets and rachis-branches in rice. JPN. J. Crop. Sci., 1994, 63(2):193 - 199.

[9] LING QIHONG. Formation and development of rice cultivation theory and technology system with Chinese characteristics: commemoration of the 100th anniversary of Chen Yongkang. Jiangsu Journal of Agricultural Sciences, 2008, 24 (2): 101 - 113.

[10] ZHANG HONGCHENG, WU GUICHENG, WU WENGE, et al. Quantitative rice cultivation model for super high yield featuring carefully cultivated seedlings, stable early period, controlled tillering, excellent middle period, big panicles, and strong late period. Scientia Agricultura Sinica, 2010, 43(13):2645 - 2660.

[11] ZOU YINGBIN, AO HEJUN, WANG SHUHONG, et al. Studies on the "three determination" cultivation of super hybrid rice Ⅰ. Concepts and theoretical basis. Chinese Agricultural Science Bulletin, 2006, 22 (5): 158 - 162.

[12] TANG YONGHUA, HUANG YAO. Statistical analysis on soil fertility contribution to major grain crops andits influencing factors in the Chinese mainland. Journal of Agro-Environment Science, 2008, 27(4): 21 - 27.

[13] GRASSINI P, ESKRIDGE K M, CASSMAN K G. Distinguishing between yield advances and yield plateaus in historical crop production trends. Nature Communications[J]. 2013, 4, 2918.

[14] XIONG W, VELDE M V D, HOLMAN I P, et al. Can climate-smart agriculture reverse the recent slowing of rice yield growth in China? [J]. Agriculture Ecosystems Environment, 2014, 196: 125 - 136.

[15] XU MINGGANG, LU CHANG'AI, ZHANG WENJU, et al. Cultivated land quality and improvement strategy in China. Chinese Journal of Agricultural Resources and Regional Planning, 2016, 37 (7): 8 - 14.

[16] ZHANG JUN, ZHANG HONGCHENG, DUAN XIANGMAO, et al. Effects of soil fertility and nitrogen application on yield, quality and nitrogen utilization efficiency of super hybrid rice. Acta Agrinomica Sinica, 2011, 37(11): 2020 - 2029.

[17] XIONG XURANG, PEI YOULIANG AND MA GUOHUI. On the main restrictive factors and countermeasures for the super high yield cultivation of super hybrid rice in hunan province Ⅱ. Restrictive factors for super high yield cultivation. Hunan Agricultural Sciences, 2005, (2): 21 - 22.

[18] HUANG Y, SUN W J. Changes in topsoil organic carbon of croplands in mainland China over the last two decades[J]. Chinese Science Bulletin. 2006, 51:1785 - 1803.

[19] SUN W J, HUANG Y, ZHANG W, et al. Estimating topsoil SOC sequestration in croplands of eastern China from 1980 to 2000[J]. Australian Journal of Soil Research, 2009, 47: 261 - 272.

[20] MENG YUANDUO, PAN GENXING. Effects of continuous cultivation of super hybrid rice on the stability of soil organic carbon and aggregates. Journal of Agro-Environment Science, 2011, 30(9): 1822 - 1829.

[21] CHENG JIANPING, CAO COUGUI, CAI MINGLI, et al. Effects of different soil water potential and nitrogen nutrition on physiological characteristics and yield of hybrid rice. Journal of Plant Nutrition and Fertilizers, 2008, 14(2): 199 - 206.

[22] LIU FAMOU, ZHU LIANFENG, XU JIAYING, et al. Hybrid rice root growth advantages and response and regulation of environmental factors. China Rice, 2011, 17 (4): 6 - 10.

[23] HE SHENGDE, LIN XIANQING, ZHU DEFENG, et al. Effects of rhizospheric oxygen supply on soil redox potential and yield in hybrid rice. Hybrid Rice, 2006, 21(3): 78 - 80.

[24] HUANG GUOQIN. Ten problems in sustainable development of paddy fields in south China. Tillage and Cultivation, 2009, (3): 1 - 2.

[25] LIU S L, HUANG D Y, CHEN A L, et al. Differential responses of crop yields and soil organic carbon stock to fertilization and rice straw incorporation in three cropping systems in the subtropics[J]. Agriculture Ecosystems Environment, 2014, 184: 51 - 58.

[26] LIU LISHENG, XU MINGGANG, ZHANG LU, et al. Evolution of soil particulate organic carbon in long-term green manure rice paddy. Journal of Plant Nutrition and Fertilizers, 2015, 21 (6): 1439 - 1446.

[27] CHEN A L, XIE X L, DORODNIKOV M, et al. Response of paddy soilorganic carbon accumulation to changes in long-term yield-driven carbon inputs in subtropical China[J]. Agriculture Ecosystems Environment, 2016, 232: 302 - 311.

[28] DENG HUAFENG, HE QIANG, CHEN LIYUN, et al. Study on yield stability of super hybrid rice in the Yangtze river basin. Hybrid Rice, 2009, 24 (5): 56 - 60.

[29] AO HEJUN, WANG SHUHONG, ZOU YINGBIN, et al. Study on dry matter production characteristics and yield stability of super hybrid rice. Scientia Agricultura Sinica, 2008, 41 (7): 1927 - 1936.

[30] LI GANGHUA. Study on yield formation mechanism and quantitative cultivation techniques of ultra-high-yielding rice. Nanjing: Nanjing Agricultural University, 2010.

[31] XIAO WEI. Effects of sowing period on yield formation and grain quality of super hybrid rice. China Rice, 2008, (5): 41 - 43.

[32] YANG HUIJIE, LI YIZHEN, YANG RENCUI, et al. Study on dry matter production characteristics of super high yielding rice. Chinese Journal of Rice Science, 2001, 15(4): 265 - 270.

[33] YUAN XIAOLE, PAN XIAOHUA, SHI QINGHUA, et al. Source-sink coordination of early- and mid-season super rice varieties. Acta Agronomica Sinica, 2009, 35 (9): 1744 - 1748.

[34] LIU JUN, YU TIEQIAO, HE HANLIN. Study on the climatic and ecological characteristics of super high yield rice yield formation. Journal of Hunan Agricultural University, 1996, 22 (4): 326 - 332.

[35] YANG HUIJIE, YANG RENCUI, LI YIZHEN et al. Analysis of yield potential and yield composition of super high yield rice varieties. Fujian Journal of Agricultural Sciences, 2000, 15 (3): 1 - 8.

[36] AI ZHIYONG, QING XIANGUO, PENG JIMING. Study on ecological adaptability of double-cropping super hybrid rice varieties in Hunan province. Proceedings of the First Chinese Congress of Hybrid Rice, 2010.

Chapter 10

Growth and Development of Super Hybrid Rice

Zhang Yuzhu / Guo Xiayu / Wei Zhongwei

Section 1 Organ Formation of Super Hybrid Rice

The life of rice can be divided into three stages according to the different organs developed – vegetative growth, vegetative and reproductive growth, and reproductive growth (Table 10 – 1). The vegetative growth stage is the period from sowing to panicle differentiation, during which vegetative organs including roots, stems, leaves and tillers are formed. The reproductive and vegetative growth stage refers to the period from panicle initiation to heading, during which vegetative organs such as roots, stems and leaves continue to grow but, more prominently, stems elongate and young panicles are formed. The reproductive growth stage refers to the period from heading to new seed maturity, characterized by heading, flowering and seed setting, producing mature new seeds.

Ⅰ. Seed Germination and Seedling Growth

Three basic conditions need to be met for rice seeds to germinate – moisture, temperature and oxygen. When these conditions are met at the same time, the rice seeds start to germinate. Germination can be divided into three stages – swelling, sprouting and germinating. When seeds are placed in water, the cellular protoplasm in the seeds is in a hydrophilic gel state, so it rapidly absorbs water and swells until the water inside the cells reaches saturation. As the water absorption of the seeds increases, the enzyme activity increases in the scutellum absorption layer and the endosperm aleurone layer of the seed embryo and respiration becomes intensified. Meanwhile, the stored substances in the endosperm are constantly converted into soluble substances such as sugars and amino acids, and transported to the embryo cells, leading to cell division and rapid elongation. When the size of the embryo increases to a certain extent, it will burst out of the seed coat, which is called "breaking the chest" or "revealing the white". Normally, the radicle breaks through the seed coat and then grows the embryo, but under flooded conditions, the embryo grows first and then the root. After the seed sprouts, the embryo continues to grow. When the radicle length is equal to the grain length and the germ length reaches half of the grain length, it is called "germination". When the young shoot grows, the original three leaves (including incomplete leaf) and the growth point in the embryo grow and differentiate at the same time, but the coleoptile appears first. The coleoptile has two veins (vascular bundles), but does not contain chlorophyll, which means that it does not perform photosynthesis. When the incomplete leaf (the first true leaf) exserts from the coleoptile of the embryo, chlorophyll is created and photosynthesis starts. This process is called "emergence". After the incomplete leaf, leaves that emerge have blades and sheaths, and hence are called "complete leaves". Each leaf is named after its order of emergence (second, third... *N*-th).

Table 10 – 1 Rice growth stages

<table>
<tr><th colspan="5">Vegetative growth</th><th colspan="3">Vegetative and reproductive growth</th><th colspan="3">Reproductive growth</th></tr>
<tr><td>Seedling</td><td>Seedling and tillering</td><td colspan="3">Tillering</td><td colspan="3">Young panicle development</td><td colspan="3">Flowering and seed setting</td></tr>
<tr><td colspan="2">Seedling</td><td>Seedling reestablishment</td><td>Effective tillering</td><td>Ineffective tillering</td><td>Differentiation</td><td>Formation</td><td>Completion</td><td>Milky ripe</td><td>Dough</td><td>Full ripe</td></tr>
<tr><td colspan="2">Foundation stage for panicle quantity</td><td colspan="3">Crucial stage for panicle quantity, foundationstage for grain quantity</td><td colspan="3">Consolidating stage for panicle quantity, crucial stage for grain quantity, foundation stage for grain weight</td><td colspan="3">Crucial stage for grain weight</td></tr>
</table>

When the first leaf exserts, two adventitious roots begin to grow on the coleoptile nodes, and three more adventitious roots grow during the exsertion of the first leaf, which forms the seedling. The seedling mainly relies on the nutrients stored in the endosperm before the three-leaf stage, after which it absorbs the inorganic nutrients, water from the soil, and organic nutrients produced by the leaves. Therefore, the stage before the three-leaf stage of seedlings is referred to as the weaning stage, which is a transition period from absorbing endosperm nutrition to living on its own nutrition, i. e., from heterotrophy to autotrophy (Fig. 10 – 1).

Seed germination and seedling growth of super hybrid rice are somehow different from those of conventional rice, which is manifested in seed viability and the basic conditionsrequired.

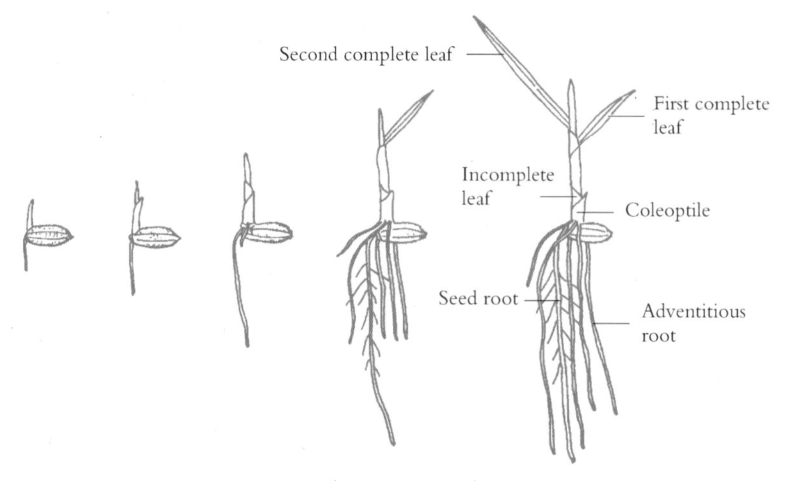

Fig. 10 – 1 Seed germination and seedling growth

1. Seed viability

Seed viability refers to the potential ability of a seed to germinate or the viability of the seed embryo, measured by germinating rate. Super hybrid rice seeds integrate favorable genes from both parents and have an over-parent heterosis in viability with enhanced respiration, and more vigorous initial metabolism, and much lower storability than conventional rice seeds under normal storage conditions. Therefore, to maintain the viability of super hybrid rice seeds, the requirements for storage conditions must be more stringent than for conventional rice seeds. Firstly, moisture must be strictly controlled when seeds go into storage. Seed moisture content is a key factor affecting the safe storage of seeds. It must be 13% or lower when the seeds go into storage. Secondly, the relative humidity in the bin during the storage of seeds should be controlled at 65% or lower. Meanwhile, attention should be paid to the temperature in the storehouse, as a higher temperature will enhance seed respiration and lead to pests and molds. Seeds are most likely to be damaged and deteriorated during late spring, summer and early autumn. Low-temperature preservation is the best.

2. Moisture

Moisture plays two main roles. First, it is the only source of hydrogen nutrient; and second, it is indispensable for metabolism as a carrier of nutrients and excretions. The moisture content of rice seeds to start germination is about 40% of seed weight, and the moisture content to start sprouting is about 24% of seed weight. Low moisture content results in slow germination. The time for the rice grain to reach saturation water absorption, on the one hand, is affected by the temperature of the soaking water. In a certain temperature range, the higher the temperature, the faster the seeds absorb water, the shorter the time to reach saturation water absorption. On the other hand, it is influenced by the rice variety and combination type. Since super hybrid rice seeds have stronger enzyme activity after water absorption, their soaking time is shorter than that of conventional rice seeds.

3. Temperature

Seed germination is a physiological and biochemical process that takes place with the participation of a series of enzymes. Enzyme catalysis is closely related to temperature. If the temperature is too low, even if the rice seeds absorb enough water and oxygen, it is difficult to germinate. The minimum temperature for germination is 8 ℃–10 ℃ for *japonica* rice and 12 ℃ for *indica* rice; while the maximum temperature is 44 ℃ and the optimum temperature is 28 ℃–32 ℃. The temperature requirement for rice germination varies by variety, especially the minimum temperature, which varies greatly among varieties. The minimum temperature for germination of rice from cold regions is low, and the earlier-maturing varieties have relatively a higher germination rate under low temperature conditions. The minimum temperature requirement for seedling growth is 10 ℃ for *japonica* rice and 12 ℃ for *indica* rice. The optimum temperature for *japonica* rice is 18.5 ℃–33.5 ℃, and for *indica* rice is 25 ℃–35 ℃, when the temperature is lower than 15 ℃, rice grows and develops slowly. Tillering stops when the average daily temperature is below 15 ℃–17 ℃. When the temperature drops below 8 ℃ or rises above 35 ℃, rice growth and development stops. Super hybrid rice is mostly *indica*, and the origin of the restorer lines is mostly Southeast Asia such as the Philippines, so the germination starting temperature is higher, generally 12 ℃–13 ℃, and the optimum temperature is about 32 ℃.

4. Oxygen

All the energy required for the germination of rice seeds is converted to energy by respiration. During the lifetime of rice, the respiration per unit area of the plant body is the largest during the germination period. At the same time, rice can only undergo cell division and organ differentiation under aerobic conditions to maintain normal growth and development. The growth and differentiation rate of rice seed shoots and seedlings is related to the oxygen content in the air. When the oxygen content is lower than 21%, the higher the oxygen concentration, the better the growth, but when it is above 21%, the growth is inhibited. The life activities of super hybrid rice are vigorous during the germination period and seedling growth period, requiring sufficient oxygen to maintain good growth potential. In actual production, the improvement of super hybrid rice seed soaking and the change of seedling cultivation (water seedling to wet seedling or dry seedling), the fundamental purpose is to fully meet the oxygen demand of super hybrid rice seed germination and seedling growth.

Ⅱ. Leaf Growth

1. The process of leaf growth

Rice leaves grow from the apical meristem of the growth point of the stem, which differentiates into leaf primordia in an alternate order during seed germination, followed by gradual elongation of the leaf primordia. First is the elongation of the blade, then the elongation of the leaf sheath. When the blade elongation reaches 8 – 10 mm, at the base of the roll-shape blade appears a gap, and then at this gap differentiates the ligule and auricle. The whole process of rice leaf differentiation from leaf primordia to leaf blade and leaf sheath formation until death after completing its function is a continuous process, but it can be roughly divided into five stages according to the main growth and functional status at different times.

(1) Leaf primordium differentiation

The tunica and corpus cells at the base of the growing point at the stem end begin to divide and proliferate, and leaf primordia appear, and the meristematic tissue of the leaf primordia divides continuously, growing toward the stem growing point in the upper part, and differentiating laterally toward its surrounding the stem growing point. When the left is entangled with the right, the height of the upper part has exceeded the stem growing point, forming a snow-cap (also known as root cap-like) encircling the stem growing point. At this time, the main veins begin to differentiate at the tip of the young leaves followed by large vascular bundles on both sides and small vascular bundles within large ones. When the length of young leaves is close to 1 cm, the ligule and auricles are differentiated in the lower part, followed by leaf sheath differentiation underneath. At this point, the differentiation of leaf primordia is largely completed. After some time, when the young leaf primordium is snow cap-shaped, the leaf blade prototype is largely formed. When the leaf blade length reaches more than 1 mm, the differentiation is in the period of determining the number of large vascular bundles and small vascular bundles. The advantage of super hybrid rice in leaf growth is that the length and width of the leaves are larger than those of conventional rice, and there are also significantly more vascular bundles.

(2) Elongation and growth

Leaf elongation is caused by cell division and proliferation of phloem tissue and cell elongation. After leaf differentiation is completed, the growth is shifted to elongation-based growth. At the same time, the cellular structures of various tissues on the epidermis of the leaf flesh tissue gradually differentiate and form, and the stomata gradually differentiate downward from the apical part of the leaf. After leaf blade elongation, followed by leaf sheath elongation, when the tip of the leaf is drawn out in the leaf sheath of the previous leaf, the differentiation of all the tissues at the base of the leaf blade is completed. With the continued growth of the leaf, the leaf sheath is also elongated rapidly, until the leaf blade is all out and unfolded. Soon, leaf sheath elongation stops.

(3) Protoplast enrichment

The protoplasm filling period begins when the leaf tip exposes the collar of the previous leaf and ends when the leaf sheath reaches full length. After the leaf blade protrudes from the sheath of the next leaf, chloroplasts are formed and photosynthesis and transpiration begin. During this phase, the cell wall constituents of leaf cells increase, making the tissue stronger, while protein synthesis is accelerated and protoplasm concentration is approximately doubled. When the leaf blade is fully unfolded, the leaf sheath reaches its full length, the auricles and ligule are drawn out, and thus leaf blade growth is completed.

(4) Functioning

After the protoplasm enrichment period, the leaf area increases to the maximum, the leaf photosynthesis has the maximum intensity and lasts the longest. This is the primetime of leaf function. The duration of the functional period is related to the location and order of leaves, and also affected by the population structure and environmental conditions. The upper leaves have a longer functional period.

(5) Aging

Protoplasm in leaf cells is gradually destroyed and cellular function declines until death.

2. Growth relationship between leaves

In the rice seed embryo, two young leaves (including incomplete leaves) and a leaf primordium are formed at maturity, and after sowing, as the leaves emerge, new leaves continue to differentiate. Therefore, there are more than one young leaves in the newly emerged leaf at different stages. In the three-leaf stage during weaning, young leaves experience the slowest differentiation and growth, and there is only one young leaf and one leaf primordium in the newly emerged leaf within a short period of time. From the six-leaf stage to panicle differentiation, there are three young leaves and one leaf primordium in the newly emerged leaf.

When the newly emerged leaf blade emerges from the collar of the previous leaf, it is also the time of leaf sheath elongation of the same leaf and leaf blade elongation of the latter leaf, and the leaf primordia of the posterior trifoliate (before five leaves) or posterior quadrufoliate (after six leaves to panicle differentiation) also differentiate. The relationship between the four is as follows. N-th leaf emerges ≈ N-th sheath elongation ≈ $(N+1)$-th blade elongation ≈ $(N+3)$-th (prior to five-leaf) or $(N+4)$-th (from six-leaf to panicle differentiation) leaf primordium differentiation.

3. Number of leaves on the main stem and leaf length

The number of leaves on a rice plant is counted from the total number of leaves (referring to complete leaves) on the main stem. Most rice varieties are 11 – 19 leaves. The number of leaves on a rice plant is directly related to the growth duration of the variety, and varieties with short growth durations have fewer leaves. The leaf emergence rate of the main stem is about 3 d before the weaning stage for the first three leaves. The leaf emergence interval at the tillering stage is 5 – 6 d, and at the stalk elongation stage is 7 – 9 d. The number of main stem leaves varies with the growth duration of the variety, generally 10 – 13 leaves for early-season varieties whose growth duration is 95 – 120 d, 10 – 14 leaves for late-season varieties whose growth duration is 105 – 125 d, 14 – 15 leaves for medium-season varieties whose growth duration is 125 – 150 d, and more than 16 leaves for single-cropping late-season varieties whose growth duration is above 150 d. With a set sowing season in a given ecological area, the development of the same variety is relatively stable at different leaf development stages, but when cultivated under different conditions, the number of leaves may increase with a prolonged growth duration, and decrease with a shortened growth duration. According to observation data of HHRRC, under the conditions of Changsha, Hunan, super hybrid rice cultivated as single-cropping late-season rice (sowed on May 13), the leaf numbers for the varieties are Liangyoupeijiu 13.5, Y Liangyou 1 14.6, Y Liangyou 900 15.3, and Xiangliangyou 900 15.8. However, under the conditions in Longhui (relatively high altitude), Hunan, the leaf numbers of the super hybrid rice (sowed on May 10) are 14.6 for Liangyoupeijiu, 15.5 for Y Liangyou 1, 16.2 for Y Lingyou 900 and 16.4 for Xiangliangyou 900.

There is a relatively stable pattern of variation in the length of leaves at each leaf position of the rice plant. Starting from the first leaf upward, the leaf lengths change from short to long, and then from long to short by the antepenult leaf. The first leaf of super hybrid rice is longer than that of conventional rice, generally 1 – 2 cm, with the longest leaf appearing in the antepenult leaf (long growth duration) or the penultimate leaf (short growth duration), and the flag leaf is short and wide.

4. Leaf blade lifespan

The lifespan of leaves at different positions is different. Normally, it increases with the rise of the phyllotaxis. The life span of 1 – 3 leaves in the seedling stage is generally 10 – 15 d, those in the mid-stage are 30 – 50 d and the last three leaves are over 50 d. Leaf longevity is related to the variety, and also affected by environmental factors, including abnormal temperature, nitrogen imbalance and water deficiency.

Ⅲ. Tillering

1. The process of tiller primordium differentiation

The tiller primordium meristem evolves from the basal meristem of the stem growing point. When the stem growth point differentiates into leaf primordium which further differentiates into a snow-cap shape, the tiller bud protrusions of the inferior leaves differentiate below the base of the leaf primordium (the side where the leaf edge is enclosed) (Fig. 10 – 2). As it continues to expand and differentiate, the tiller sheath (front leaf) is formed, followed by the first leaf primordium and others. When the differentiation of the tiller primordium is completed, its mother leaf is also exserted.

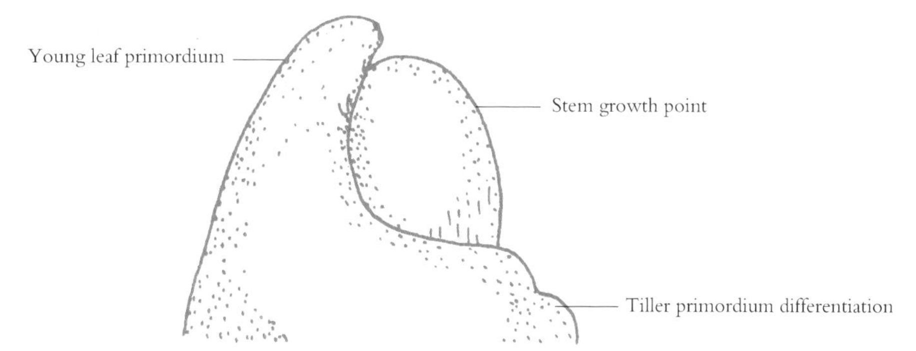

Fig 10 - 2 Tiller primordiumdifferentiation (Yuan Longping, 2002)

2. Tiller bud insertion node and differentiation substitute

Each stem node of the rice plant has a tiller bud except for the spike neck node. As the seed matures, there are three tiller primordia in the embryo. The tiller primordium of each coleoptile degenerates at the late stage of embryo development, and the tiller primordia of the incomplete leaf node also degrade gradually at the seed germination stage. Except for the tiller primordia of the coleoptile and incomplete leaf node, the differentiation of tiller buds and tiller leaves increases with their corresponding patterns.

The differentiation of tiller buds keeps a certain interval with the differentiation of mother stem phyllopodium, and accordingly keeps differentiating upward. Generally, when the n-th leaf of the mother stem are exserted, the tiller primordia of the $n+4$-th leaf node start to differentiate, the tiller primordia of the $n+2$-th and $n+1$-th leaf nodes differentiate and expand, and the tiller primordia of the n-th leaf node have differentiated into the first leaf primordia to form a complete tiller bud. The differentiation of tiller buds is not related to the elongation of the tiller and has little to do with external environmental conditions. The relative differentiation between the tiller primordia and the parent stem leaves at each leaf position is as described above, except for the shoot sheath nodes and incomplete leaf nodes.

After the tiller buds are formed, regardless of whether the tiller elongates or not, the tiller buds keep differentiating into leaves, and the differentiation of the leaf primordia of the mother stem is synchronized, and each time the mother stem increases a leaf primordia, the tiller buds also increase a leaf primordia. If the tiller bud does not elongate, the leaf primordia will be divided into multiple layers in the form of cabbage-shaped and wrapped in the "dormant bud".

3. Tillering

Tiller buds start to elongate under suitable conditions, with tiller sheath (front leaf) being the first. The tiller sheath has two longitudinal prismatic protrusions without a blade, holding the mother stem between the two ridges, and enclosing the tiller with a jig-like portion beyond the two prongs. The tillering sheath is wrapped in the leaf sheath of the mother stem, which is not easily visible when the tiller emerges, and there is no chlorophyll, thus no photosynthesis. Most of the leaves seen at the time of tillering are the first leaves, which are backed by its mother leaf toward the same direction. The emergence period when the first leaf emerges is referred to as tillering, after which leaves emerge as fast as those on the main stem under normal conditions.

The lowest node is the first leaf node (the axillary buds of the shoot sheath node and incomplete leaf node have long been degraded and are difficult to sprout as tillers) , and the uppermost node is where the flag leaf grows. However, it is difficult for the axillary buds on the elongated stem to sprout (flag leaves in particular). They only sprout in the cases of lodging, broken panicles or excessive nutrition in the later stage. Therefore, the highest tiller node is generally derived by deducting the number of elongation nodes from the number of stem nodes. For example, if there are a total of 16 stem nodes and five elongation nodes, the highest tiller node is the 11th node (16−5 =11). It is easy for the axillary buds on the elongation nodes of some varieties to sprout, so they can be used to produce ratooning rice. The order of tiller is from bottom to top with the increase of main stem leaves. Tillers can produce tillers. A tiller produced by the main stem is a primary tiller, one produced by a primary tiller is a secondary tiller, and one produced by a secondary tiller is a tertiary tiller. Hybrid rice has larger numbers of primary, secondary and tertiary tillers. Although there is an inherent possibility of tiller emergence at each node, whether it elongates into a tiller depends on the prevailing conditions. When the conditions are not suitable, the tiller buds will remain dormant and only leaf differentiation takes place without tillering.

To describe the position of the tillering nodes, each tiller is represented by a number. Primary tillers are represented by the numbers of their node on the main stem. For example, if primary tillering occurs on node 6, the tiller is called the sixth tiller; and if it occurs on node 7, the tiller is the seventh tiller. Secondary tillering is represented by two numbers connected with a hyphen. For example, secondary tillering at the first node of the 6th tiller is "6−1" (the former number indicates the position of the primary tiller, while the latter refers to the position of the secondary tillering on the primary tiller). The same applies to tertiary tillering. For example, in "6 − 1 − 1", the first digit indicates the position of the primary tiller, the middle digit indicates the position of the secondary tiller on the primary tiller, and the last digit indicates the position of the tertiary tiller on the secondary tiller. The number of leaves on the main stem or a tiller can be described with the relationship between the tiller node (as the denominator) and the phyllotaxis (as the numerator). For example, in "8/0", "0" refers to the main stem, while "8" indicates the eighth leaf. In "4/6 − 1 − 1", "6 − 1 − 1" indicates the tillering position and "4" indicates the fourth leaf on the indicated tiller.

4. Concurrent leaf-tiller emergence

When a new leaf emerges from the mother stem of a rice plant, a first leaf also emerges at the third tiller node under the new leaf. This happens to the main stem and primary tillers, as well as to the primary and secondary and tertiary tillers. Such a relationship between the leaves on the tillers and those on the mother stem is referred to as concurrent leaf-tiller emergence.

Tillering sheath nodes can also produce tillers, but only in a small quantity. The tiller sheath is represented with the letter *P*. The tiller sheath is one node lower than the first leaf tiller. Therefore, when the third leaf exserts from the tiller, the tiller sheath of that tiller grows the first leaf.

The phenomenon of leaf and tiller extending together only shows the corresponding relationship between tiller buds and mother stem leaves in general, but it does not indicate that the corresponding tiller will definitely extend when the new leaves of the mother stem extend because the extension depends on various internal and external factors. Internal factors such as the carbon and nitrogen content of the plant and carbon to nitrogen ratio, especially nitrogen content, are closely related to the occurrence of tillering, while external factors such as temperature, light and water also play a role. For example, if the leaf area coefficient is too large in the seedling and production field, tillering will stop.

The cultivation of super hybrid rice requires sparse sowing for strong seedlings, but its leaf area grows rapidly and tillering stopsearlier. Tillering generally does not occur in the middle and late stages of the seedling field and the late growth stage in the production field. When the conventional water-cultivated seedlings are used for transplantation, tillers are not able to sprout for a while after transplanting due to the effect of planting injury. Normally, when three new leaves grow after transplantation, tillers sprout out of the axillary buds from the last leaf in the field. For example, if the seedlings with six full leaves are transplanted, when the eighth leaf emerges in the field, the fifth axillary bud emerges as well. When transplantation with soil (tray, including cast planting), tillering occurs earlier before the third leaf grows in the field. For example, when the fourth leaf gives birth to the fifth leaf, the axillary buds of the second leaf extend at the same time.

There is not a universal pattern for concurrent emergence. For example, before the ineffective tiller dies, its leaf emergence rate gradually slows down and the birth of tiller leaves lags behind the leaf emergence rate of the mother stem. Another situation is the dormant axillary buds that have not sprouted after the coextension period. When the field conditions improve, these dormant axillary buds sprout into tillers again, and then the birth of tiller leaves has lagged behind the corresponding coextension leaves of the mother stem, and this phenomenon is more common in super hybrid rice. For example, when the mother stem is producing the ninth leaf, the 6th tiller position does not have the same extension due to the plant's nitrogen deficiency; when the mother stem is producing the tenth leaf, the nitrogen condition has improved and the axillary buds of the seventh tiller position sprout tillers according to the same extension rule, and the axillary buds of the sixth tiller position also sprout and pull out at the same time. The sixth and seventh tillers are drawn at the same time, except that the 6th tillers lost the advantage of lower position, and its economic coefficient is similar to that of the seventh tillers.

5. Effective and ineffective tillering

The effectiveness and ineffectiveness of tiller is based on the ability of the tiller to produce seeds. Generally, a tiller with five or more seeds is considered an effective tiller; otherwise, it is considered an ineffective tiller. The length of effective tillering period varies with variety, generally 7 – 12 d for early-maturing varieties, 14 – 18 d for medium-maturing varieties and 20 d for late-maturing varieties. In production, it is called the initial tillering stage when 10% of the plants in the field start to tiller, tillering stage when 50% of the plants start tillering, and tilling peak stage when 80% of the plants start tillering. In the process of tiller increase, the effective tiller termination period is when the number of tillers equals to the number of final tillers that become panicles. In fact, not all tillers formed before the effective tillering period ends are effective. Tillers formed after that can still be effective. According to Japan's S. Matsushima, the true effective tillering terminates at the peak tillering stage. Practice shows that the termination of effective tillering actually depends on factors such as the degree of population shade, nutritional conditions and harvesting period.

However, in terms of growth transition, there is still a period dominated by effective tillering. The tiller itself does not grow from the first node until the fourth leaf, when the root system can grow autotrophically. Therefore, the tiller must have more than three leaves to have a high probability of becoming a panicle. After the stem starts to elongate, the center of growth shifts to a period of mainly reproductive growth. If the tillers still have fewer than three or four leaves, the possibility of them becoming ineffective tillers increases. It takes five days for a tiller to grow a leaf, and 15 days to grow three. Therefore, tillering that happens 15 days before stem elongation is more likely to produce effective tillers and the earlier the tillering happens, the more likely it will be effective. Based on this, it is believed that tillering is effective if there are four leaves or more (three leaves plus one newly emerged leaf) when elongation starts and ineffective if there are fewer than three leaves (two leaves and one newly emerged leaf).

Ⅳ. Root Growth

1. Growth of adventitious roots

There are two types of rice roots, i. e. seed roots and adventitious roots (Fig. 10 - 3). After sowing, seed roots grow downward. When the first complete-leaf emerges, adventitious roots, also known as nodal roots, begin to grow. The first five adventitious roots that grow from the coleoptile nodes are called coleoptile roots, or "chicken feet roots". Afterwards, in below the surface of the ground, adventitious roots (nodal roots) grow from bottom to top along tillering nodes (root nodes).

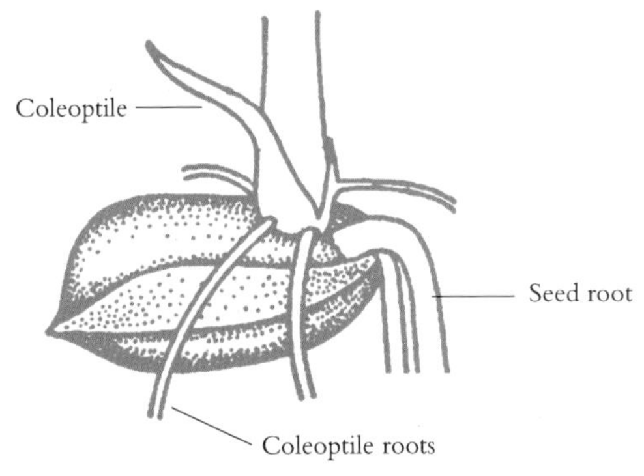

Fig. 10 - 3 Rice roots (Yuan Longping, 2002)

A rice root system has two parts according to the position of the root nodes – upper nodal roots and lower nodal roots (Fig. 10 - 4). The lower nodal roots are the functional roots at the tillering stage, whose number and length increase with the number of tillers, and they grow downward. The upper nodal roots are the main functional roots for the last three leaves which determine yields before and after stem elongation. The distribution of the whole root system in the soil is related to the growth duration. Roots are sparse in the vegetative growth period, and they grow further in quantity and depth during the heading stage before reaching the peak.

Compared with conventional rice, super hybrid rice has a better root system in terms of quantity, depth and robustness. The HHRRC studied the root traits of super hybrid rice Xiangliangyou 900 and Y Liangyou 1 in different growth stages and found that the dry weight, volume and density of the roots of each individual stem or the whole population decreased as they grew deeper into the soil in different growth periods. Roots were mostly distributed in the upper layer (0 - 10 cm) of the soil, while fewer roots could be found in the lower layer (below 10 cm). The upper layer contained over 75% of the dry weight and volume of the root in different growth periods. The dry weight, volume and density of roots

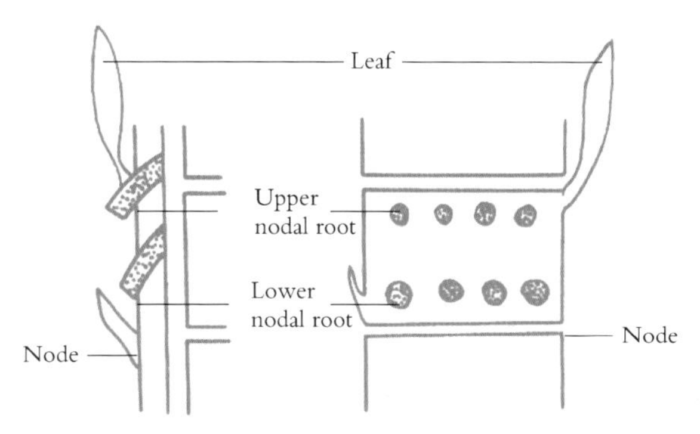

Fig. 10 - 4 Rice root nodes (Yuan Longping, 2002)

were high in the upper layer, lower layer and all layers of soil; but the dry weight and volume were much higher in the lower layer (below 10 cm), indicating that the roots are inclined to grow deeper (Table 10 - 2). Therefore, it is necessary to regulate the super hybrid population by deep ploughing and deep fertilization for cultivation, and intermittent irrigation at the later growth period, so as to have a deeper and better root system.

Table 10 - 2 Dry weight, volume and density of roots of super hybrid rice in soil in different growth periods

Trait	Soil layer (cm)	Xiangliangyou 900					Y Liangyou 1				
		Peak tillering stage	Full heading stage	18 days after full heading	35 days after full heading	Maturity stage	Peak tillering stage	Full heading stage	18 days after full heading	35 days after full heading	Maturity stage
Individual stem dry weight (g)	0 - 10	0. 178	0. 348	0. 343	0. 329	0. 282	0. 131	0. 219	0. 205	0. 189	0. 166
	10 - 30	0. 039	0. 119	0. 113	0. 109	0. 081	0. 029	0. 071	0. 065	0. 051	0. 035
	Total	0. 217	0. 467	0. 456	0. 438	0. 363	0. 160	0. 290	0. 270	0. 240	0. 201
Individual stem root volume (cm^3)	0 - 10	1. 744	2. 288	2. 253	2. 172	2. 002	1. 307	1. 734	1. 707	1. 668	1. 596
	10 - 30	0. 417	0. 834	0. 828	0. 823	0. 641	0. 325	0. 579	0. 563	0. 497	0. 388
	Total	2. 161	3. 122	3. 081	2. 995	2. 643	1. 632	2. 313	2. 270	2. 165	1. 984
Root dry weight of the population (t/ha)	0 - 10	0. 513	0. 586	0. 578	0. 554	0. 475	0. 496	0. 475	0. 444	0. 410	0. 360
	10 - 30	0. 113	0. 200	0. 190	0. 184	0. 136	0. 110	0. 154	0. 141	0. 111	0. 076
	Total	0. 626	0. 787	0. 768	0. 738	0. 611	0. 606	0. 629	0. 585	0. 520	0. 436
Root volume of the population (m^3/ha)	0 - 10	5. 031	3. 853	3. 794	3. 658	3. 372	4947	3. 759	3. 700	3. 616	3. 460
	10 - 30	1. 203	1. 405	1. 395	1. 386	1. 060	1. 230	1. 255	1. 220	1. 077	0. 841
	Total	6. 234	5. 258	5. 189	5. 044	4451	6. 178	5. 014	4921	4693	4. 301
Root dry weight ratio (%)	0 - 10	82. 03	7452	75. 22	75. 11	77. 69	81. 88	75. 52	75. 93	78. 75	82. 59
	10 - 30	17. 97	25. 48	24. 78	2489	2231	18. 13	2448	24. 07	21. 25	17. 41
Root volume ratio (%)	0 - 10	80. 70	73. 29	73. 13	7252	75. 75	80. 09	7497	75. 20	77. 04	80. 44
	10 - 30	19. 30	26. 71	26. 87	27. 48	24. 25	19. 91	25. 03	2480	2296	19. 56

Continued

Trait	Soil layer (cm)	Xiangliangyou 900					Y Liangyou 1				
		Peak tillering stage	Full heading stage	18 days after full heading	35 days after full heading	Maturity stage	Peak tillering stage	Full heading stage	18 days after full heading	35 days after full heading	Maturity stage
Root density (g/cm^3)	0 - 10	0. 102	0. 152	0. 152	0. 151	0. 141	0. 100	0. 126	0. 120	0. 113	0. 104
	10 - 30	0. 094	0. 143	0. 136	0. 132	0. 126	0. 069	0. 123	0. 115	0. 103	0. 090
	0 - 30	0. 100	0. 150	0. 148	0. 146	0. 137	0. 098	0. 125	0. 119	0. 111	0. 101

2. Growth sequence of roots

The growth of rice roots follows the rule of $N-3$. When the N-th leaf is exserted, it is the growth period of the ($N-3$)-th node. The growth sequence of the lower nodal roots can be represented by as follows.

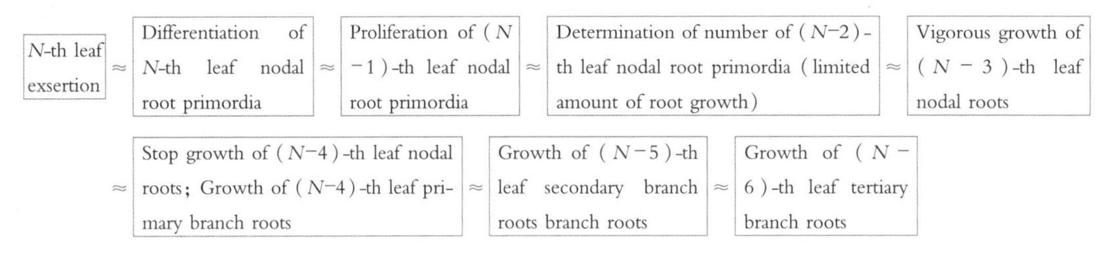

The growth sequence of the upper nodal roots can be represented by the following figure.

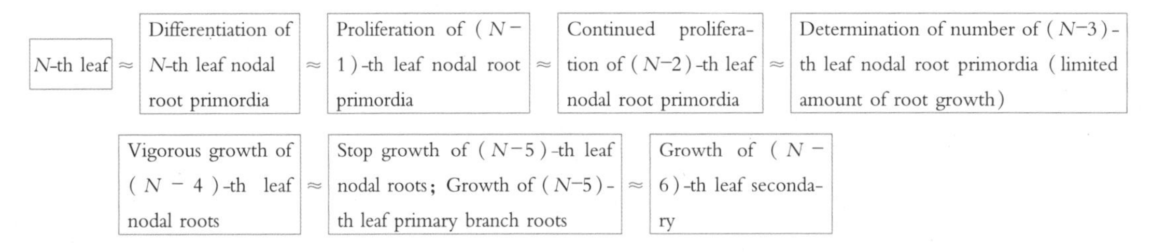

V. Stem Growth

1. The process of stem growth

The early stage of stem growth is characterized by the emergence of new stem nodes and leaves through the activity of the apical meristem. Throughout panicle differentiation, the meristem at the top of the stem degenerates. Stem growth in the late stage depends on the intercalary meristem. The period when the stem nodes undergo intercalary growth and begin to elongate to 1 - 2 cm is referred to as the stem elongation stage. Therefore, rice stem grows from the top and ends with intercalary growth. The process is divided into four stages.

(1) Tissue differentiation

First, the growth cone protomeristem of a rice plant differentiates into various primary meristems, which further differentiate into stem nodes and tissues on the internodes, such as the transfusion tissue, mechanical tissue and parenchymatous tissue. It takes about 15 days for an internode to differentiate. The stage of tissue differentiation is the basis for the robustness of the stem, and therefore has an impact on the quality of tillers and the size of panicles.

(2) Internode elongation and thickening

On the basis of the completion of tissue differentiation in the previous stage, intercalary meristemson the internode base undergo vigorous division and elongation; so do the cortex meristems and the ancillary meristems of the small vascular bundles, increasing the thickness of the stem (Fig. 10 - 5). Cell division occurs in the dividing area of the internode base; the differentiation of internode tissues occurs in the differentiating area; and only longitudinal elongation occurs in the elongating area. It is only a few centimeters long from the dividing area to the elongation zone, above which are mature tissues that do not further elongate. In the whole rice stem, the upper internodes have more active intercalary meristems, whose cell division and elongation are more vigorous. Therefore, upper internodes are generally longer.

The elongation and thickening period for each internode lasts generally seven days, which is the key period to determine the length and thickness of the stem. Although the lower internodes do not elongate, the thickness is determined at this period and the robustness of the lower stem nodes has a direct bearing on the robustness of the upper internodes. The thickness of the stem determines the size of the panicles. It is now known that the number of primary branches in a panicle is 1/4 - 1/3 that of large vascular bundles on the first elongation internode, and also that of the large vascular bundles on the neck-panicle internodes (or fewer than it by one or two).

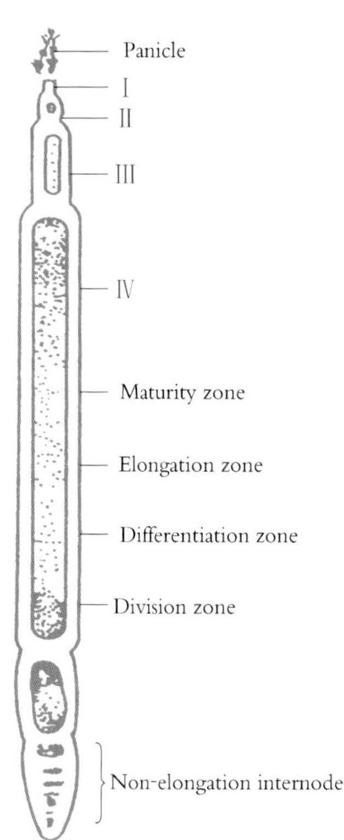

Fig. 10 - 5 Meristem of rice stems
(Yuan Longping, 2002)

(3) Substance enrichment

After the elongation period, substances on the nodes and internodes are gradually enriched, adding to their hardness and maximizing unit volume and weight. Growth at this stage determines the robustness and lodging resistance of the plant, while the quantity of substances stored determines the fullness of the panicles. The substances during the stem substance enrichment period come from the leaves at the lower part of the internodes and the photosynthetic products in the leaves on the nodes underneath. Therefore, it is important to maintain vigorous growth of the leaves for the enrichment of the stem.

(4) Substance export

After heading, starch stored in the stem is hydrolyzed and transferred to the grains. Generally, about three weeks after heading, the weight of the stem drops to the lowest, which is only 1/3 - 1/2 the weight before heading. During nutrient transfer, the main influencing factors are moisture and fertilization. A lack of moisture affects the normal physiological activities of rice plants and hinders nutrient transfer. Nitrogen content should be maintained at a moderate level. If it is too high, starch transfer will slow down; while if it is too low, the leaves will have early senescence, thereby reducing photosynthesis.

2. Internode elongation

Elongation starts with the lower internodes first and then extends upward. However, within the same period, there are three internodes elongating simultaneously, generally the late stage of the first internode is the end of the elongation peak period of the second internode and the beginning of the elongation period of the third internode. The neck-panicle internode (the top internode) begins elongation slowly over 10 days before heading, and the elongation turns most rapidly 1 – 2 days before heading.

3. Correspondence between the elongation of internodes and that of other organs

The elongation of internodes is closely associated to the growth of other organs. In terms of nodal differences, the vigorous growth positions of leaves, leaf sheaths, internodes, tillers, and roots all had relatively constant differences. For example, the elongation of internodes occurs two or three nodes lower than the elongating leaf and one node lower than the elongating leaf sheath. Root development and tillering occur three nodes lower than leaf emergence. The relationship is showed below.

N-th leaf elongation	≈	(*N*−1)-th leaf sheath elongation	≈	(*N*−2)-th and (*N*−2)-th leaf internodal elongation	≈	(*N*−3)-th leaf node growth	≈	(*N*−4)-th leaf node tillering and elongation

The relationship between internode elongation and panicle differentiation mainly depends on the number of elongation nodes of the variety. There are three circumstances. Firstly, there are only four elongation nodes, and panicle differentiation occurs before the first internode elongates. For some extreme early-maturing dwarf varieties the differentiation may be even earlier. Secondly, there are five elongation nodes, and panicle differentiation overlaps with the elongation of the first internode. Thirdly, there are six elongation nodes, and panicle differentiation occurs after elongation. Most super hybrid rice varieties are semi-high varieties with a long growth period, and five or six elongation internodes.

Ⅵ. Panicle Growth

1. Process of young panicle differentiation and growth

After completion of the photoperiod required for developmental stage transition, panicle differentiation begins when the flag leaf differentiation is completed and the stem growth cone differentiates into the first bract primordium. The differentiation and development of the young panicle to the morphology formation of the panicle and the completion of all internal reproductive cells is a continuous process. Prof. Ding Ying divides panicle development into eight stages. The first four stages are young panicle formation stages (when organs are formed), while the last four stages are booting stages (when reproductive cells develop).

(1) First bract differentiation

When a young panicle starts differentiation, a ring-shaped protrusion emerges at the growth cone base opposite the flag leaf primordium. This is the first bract primordium (Fig. 10 – 6). The first bract differentiates on the neck panicle node, with the rhachis in the upper part. Therefore, the period when the first bract differentiates is also referred to as the panicle-neck differentiation period, which is the start of reproductive growth. Previous research suggested two distinctive features of the differentiation of the first bract. First, when the leaf primordium differentiates into a protrusion, the previous leaf grows to the peridium growth point; yet, the first bract primordium is already differentiated before the flag leaf primordium covers the growth point. Second, the angle between the protrusion and the growth cone is an acute angle during early leaf primordium differentiation, while that between the protrusion and the growth cone is an obtuse angle later. However, it has also been found with observation through scanning electron microscope that there is no morphological difference between the first bract primordium and the flag leaf primordium, but only internal physiological differences. Therefore, it is suggested that the bract proliferation period be regarded as the first phase of panicle differentiation based on the morphological observation of the differentiation. After the first bract differentiates, new bracts are differentiated in a spiral shape with an opening of 2/5 along the growth cone. In sequence, these new bracts are called the second bract, the third bracts, etc. , and this is the bract proliferation stage. Super hybrid rice is mostly with large panicles and more than 10 bracts.

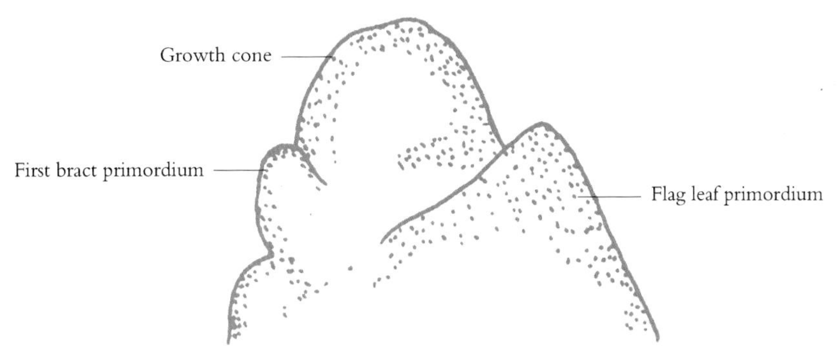

Fig. 10 - 6 Differentiation pattern of the first bract primordium and the flag leaf primordium (Yuan Longping, 2002)

(2) Primary branch differentiation

When the first bract primordia enlarge, new transverse lines continue to differentiate at the base of the growth cone, which are the second and third bract primordia. The appearance of these bracts marks the beginning of the differentiation of the primary branch primordia (Fig. 10 - 7, I). Soon after the bracts proliferate, a protrusion is formed in the axillary part of the first bract, which is the primary branch primordium. The order of differentiation of the primary branch primordia is from bottom to top, gradually proceeding toward the tip of the growth cone. The end of primary branch differentiation is marked by the cessation of differentiated growth at the growing point of the stem and the beginning of white bract hairs at the bract bearing site (Fig. 10 - 7, Ⅱ)

(3) Secondary branch primordium and spikelet primordium differentiation

At the end of the differentiation of the primary branch, the most recently differentiated primary branch primordium at the top of the growth cone grows the fastest and differentiates into new bracts at its basewith small protrusions in the bract axis, which are the secondary branch primordium. A young panicle is 0.5 - 1.0 mm long, all covered with bract hair (Fig. 10 - 8). The growth rate of the secondary branch

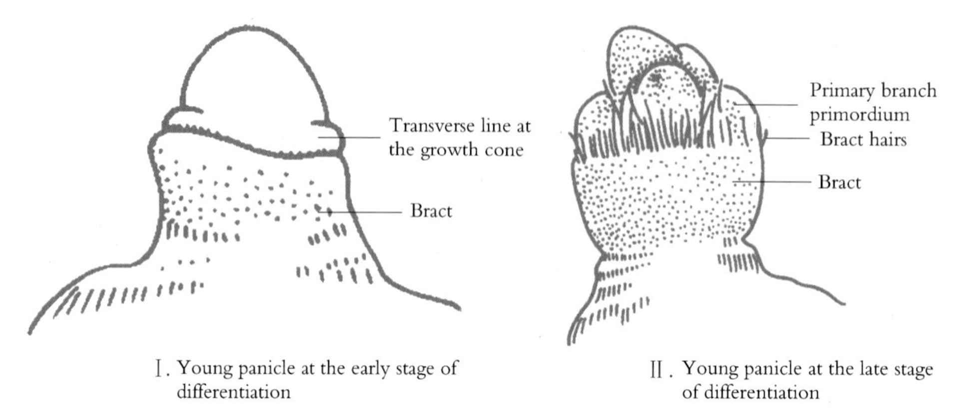

Ⅰ. Young panicle at the early stage of differentiation

Ⅱ. Young panicle at the late stage of differentiation

Fig. 10 – 7 Primary branch differentiation (Yuan Longping, 2002)

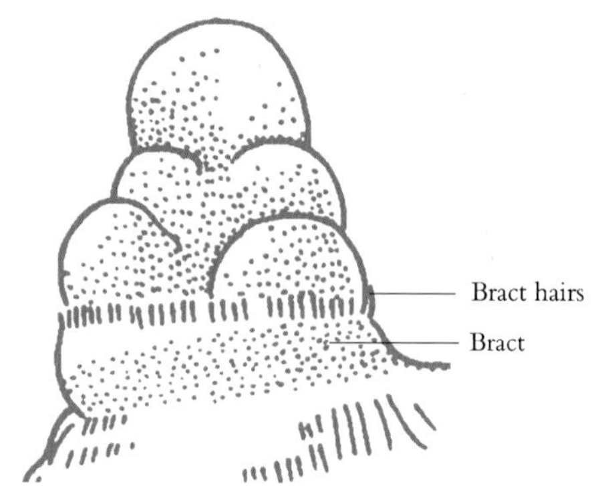

Fig. 10 – 8 Secondary branch differentiation (Yuan Longping, 2002)

primordium is opposite to the order of differentiation, with the upper positions differentiate faster than the lower ones. On a whole panicle, the secondary branches on the primary branch at the top of the panicle develop faster than those at the base of panicle axis, i. e. the off-top development. The number of secondary branches is most closely related to the total number of spikelets on a panicle. One of the main advantages of super hybrid rice is that it has larger panicles and more secondary branches. Therefore, it is important to ensure good conditions for the growth of hybrid rice during secondary branch differentiation.

After secondary branch differentiation, a protrusion of the degenerated glume primordia begins to appear at the tip of thefirst primary branch at the top of the rachis, followed by two rows of re-protrusions of the glume primordia also on the secondary branch. After the emergence of the first and second degenerated glume primordia and the sterile glume primordia, the inner and outer glume primordia are differentiated. The differentiation of the glume primordia is early in the top branching peduncle of the rachis and late in the lower part as far as a whole spike is concerned. For a branching peduncle, the apical grain is the earliest to differentiate, followed by the basal first grain, and then upwards in order. Therefore, the penultimate grain of each branchial peduncle differentiates the latest. When the differentiation of the glumes on the primary pedicel is completed, before the differentiation of pistils and stamens, and soon after the differentiation of the glumes on the lower part of the spike, the secondary branch and glume differentiation period is over.

(4) Pistil and stamen formation

In the spikelet primordium which develops fastest on the upper part of the panicle, some small protrusions emerge in the inner and outer glumes as pistil and stamen primordia, which are crowded together and surrounded by the inner and outer glumes, resembling a nest of eggs when viewed microscopically (Fig. 10 – 9). This differentiation advances from the glumes in the upper part of the spike to the glumes in the lower part of the spike. When the glumes on the lowermost secondary pedicels of the spike are differentiated one after another, the highest number of glumes of the whole spike is fixed subsequently, the rachis and branches begin to elongate rapidly, the inner and outer glumes also elongate and close to each other, the stamens differentiate into anthers and filaments, and the pistils differentiate into stigmas, styles and ovaries. At this point, the panicle organs are all differentiated, the young panicle prototype has been formed with a full length of 5 – 10 mm. Thereafter, young panicle development shifts from the differentiation stage to the reproductive cell formation stage, i. e. the booting stage.

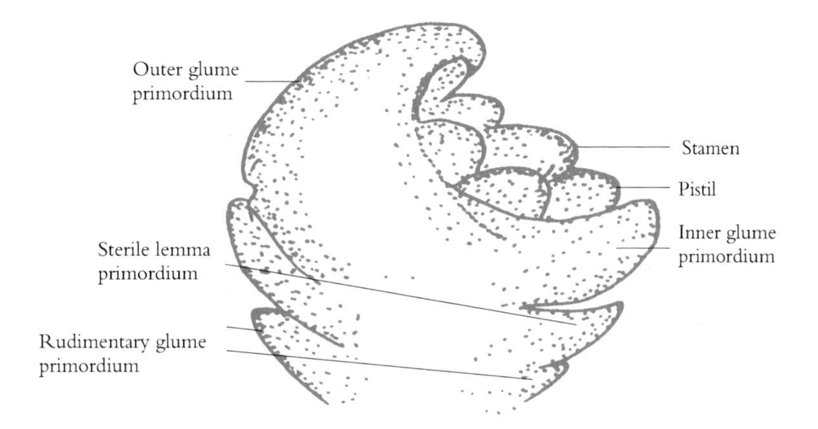

Fig. 10 – 9 Pistil and stamen primordium differentiation (Yuan Longping, 2002)

(5) Pollen mother cell formation

When the inner and outer glumes close, the anthers of the stamens differentiate into four locules, at which time large and irregular pollen mother cells can be seen in the anthers (Fig. 10 – 10), while a stigma protrudes from the tip of the pistil primordia. At this time, the flag leaf is in the process of exsertion, the glumes are nearly 2 mm long, about 1/4 of their final length, and the young spikelets are 1.5 – 4.0 cm long.

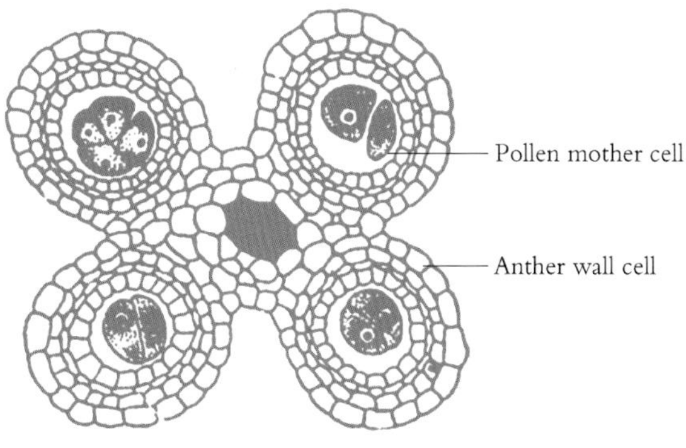

Fig. 10 – 10 Pollen mother cell formation (Yuan Longping, 2002)

(6) Pollen mother cell meiosis

A pollen mother cell undergoes two consecutive cell divisions (meiosis and mitosis) to form four daughter cells with 12 chromosomes, called tetrads (Fig. 10 – 11), which soon afterward disperse into four mononuclear pollen. During this period, the young panicle elongates the fastest, normally from 3 – 4 cm to over 10 cm. The spikelet elongates to half its final length, and the anthers turn yellowish green. From the appearance of the morphology, it is the meiotic peak period when the flag leaf collar is in the process of extending flush with its next leaf collar. Pollen mother cells undergo meiosis for 24 – 48 h and this period is important in the development process which requires stringent external conditions because unfavorable conditions may cause the branch spikelet to degenerate and the glume volume to become smaller.

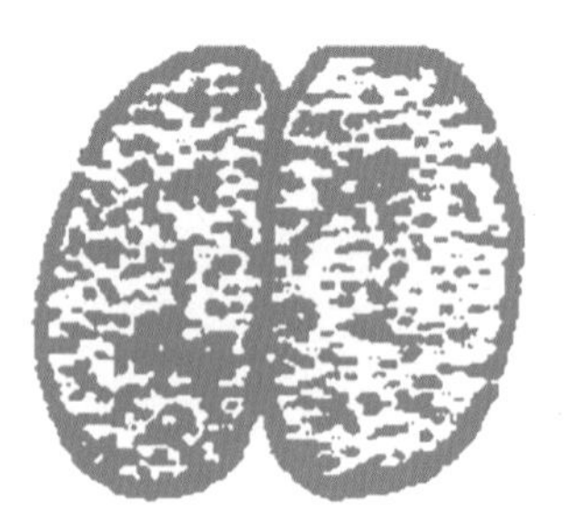

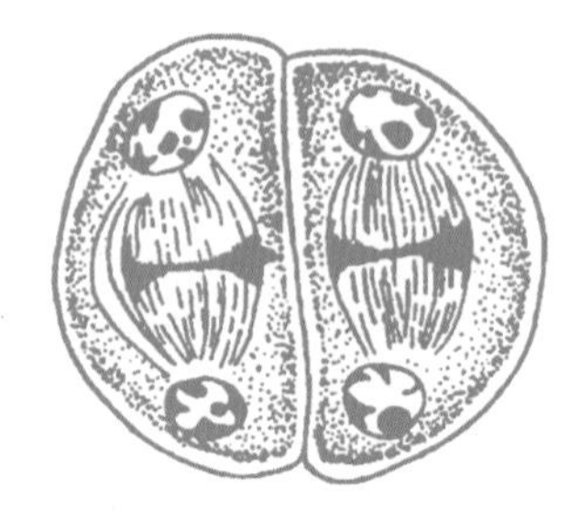

Ⅰ. Beginning of secondmeiosis　Ⅱ. Pollen mother cell meiosis　Ⅲ. Formation of tetrads

Fig. 10 – 11　Pollen mother cell meiosis (Yuan Longping, 2002)

(7) Pollen content enrichment

After the tetrad produced by meiosis disperses into mononuclear pollen, the size increases and the pollen shell is formed. The germination aperture appears, the pollen is continuously filled, and the pollen nucleus divides to form a reproductive nucleus and a vegetative nucleus, becoming a binuclear pollen grain. At this time, the longitudinal elongation of the outer glume basically stops, the length of the glume reaches about 85% of its full length, the chlorophyll of the glumes begins to increase, the volume of the pistils and stamens and the transverse width of the glume increase rapidly, feather-like protrusions appear on the stigma, and the pollen is filled (Fig. 10 – 12).

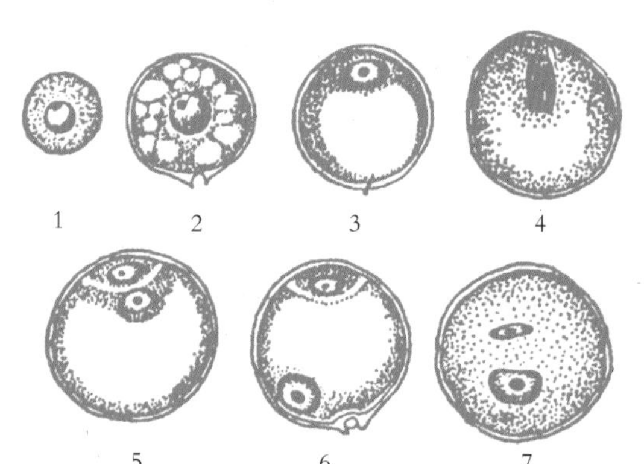

1. Formation of mononuclear pollen
2. Formation of mononuclear pollen shell
3. Formation of pollen germination aperture
4. Beginning of pollen nucleus division
5. Formation of reproductive nucleus and vegetative nucleus
6. Formation of binuclear pollen
7. Pollen content enrichment

Fig. 10 – 12　Pollen filling (Yuan Longping, 2002)

(8) Pollen formation

One or two days before heading, the pollen is filled and the reproductive nucleus further splits into two sperm nuclei, plus one vegetative nucleus, forming the trinuclear pollen. This marks the end of the entire process of pollen development. Next comes heading and flowering. This stage is also characterized by the increase in chlorophyll content in the inner and outer glumes, rapid elongation of the filaments, growing starch content in the pollen, and yellow anthers (Fig. 10 – 13).

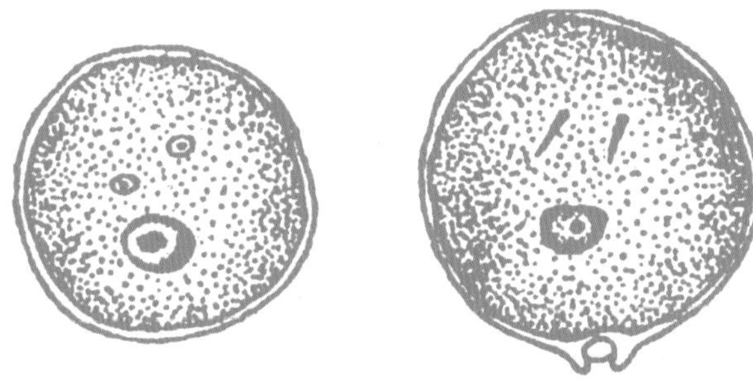

Fig. 10 - 13 Mature pollen (Yuan Longping, 2002)

2. Young panicle differentiation and development

The duration of young panicle differentiation and development differs with variety, growth duration, temperature, and nutrition status. The whole process lasts 25 - 35 days. Normally, there is a drastic difference in the duration of reproductive organ formation (from bract to stamen differentiation) before panicle differentiation, but only a slight difference in the duration of reproductive cell formation. Panicle differentiation lasts for different time periods at different stages. Bract differentiation needs normally 2 - 3 days, primary branch differentiation takes 4 - 5 days, secondary branch differentiation and spikelet differentiation need 6 - 7 days, and pistil and stamen formation requires 4 - 5 days. All are measured on a whole panicle basis. If measured by spikelet in reproductive cell differentiation, pollen mother cell formation needs 2 - 3 days, pollen mother cell meiosis takes two days, pollen filling spans 7 - 8 days, and the pollen formation takes 3 - 4 days.

Since super hybrid rice combinations mostly have alonger growth period and their panicle differentiation lasts longer than that of conventional rice varieties, generally 30 - 35 days. From the duration of panicle differentiation of several super hybrid rice combinations (Table 10 - 3), it can be seen that there are differences in the entire panicle differentiation period of different combinations, and also in the duration of the various differentiation stages.

Table 10 - 3 Duration from panicle differentiation to initial heading of super hybrid rice (day)

Stage	YLiangyou 1	Y Liangyou 900	Xiangliangyou 900
From first bract differentiation to initial heading	32. 5	33. 5	34
From primary branch differentiation to initial heading	27. 5	28. 5	29
From secondary branch differentiation to initial heading	22. 5	22	22. 5
From pistil and stamen formation to initial heading	17	17. 5	17. 5
From pollen mother cell formation to initial heading	12	13. 5	14
From pollen mother cell meiosis to initial heading	9. 5	10	10. 5
From pollen content enrichment to initial heading	6	6. 5	7
From pollen completion to initial heading	2	3	3

3. Identification of panicle differentiation and development stages

The identification of the period of spike differentiation and development should generally be done through anatomical examination with the aid of a microscope, but usually we can use organ development relationships to make projections. For example, there are the leaf age index method, the leaf age residual method, and the young spike length method (Table 10 − 4).

Table 10 − 4 Identification of panicle development stages

Method	Stage 1	Stage 2	Stage 3	Stage 4	Stage 5	Stage 6	Stage 7	Stage 8
Leaf age index (%)	76±	82±	85±	92±	95±	97±	100±	
Remaining leaf primordium number (%)	3.0±	2.5±	2.0±	1.2±	0.6±	0.5±	0	
Young panicle length (mm)	< 0.1	0.1 − 1	1 − 2	4	25	60	90	
Duration (d)	2 − 3	4 − 5	6 − 7	4 − 5	2 − 3	2	7 − 8	2 − 3
Observation	N/A	Hair	Hair tuft	Grains	Glumes	Half length	Green	Panicle emergence

Ⅶ. Heading, Flowering, Pollination and Seed Setting

After panicle differentiation, it comes the stages of heading, flowering, fertilization, grain filling and maturity. This is the seed setting stage.

1. Heading

One or two days after the maturity of the pollen and embryo sac of spikelets in the upper part of the plant, the flag leaf sheath will emerge at the tip of the panicle. This is the heading stage. The initial heading stage is when heading occurs to 10% of the plants in a field, and the heading stage is when 50% of the plants head, and the full heading stage is when 80% of them head. It takes about five days for a panicle to emerge completely and 7 − 10 days for a bunch of panicles in a plant, and one to two weeks for a whole population. The optimum temperature for heading is 25 ℃− 35 ℃. If it exceeds 40 ℃, or drop below 20 ℃, heading will not happen normally, or the panicles will be enclosed in the sheaths. In practice, the end of the safe period for full heading of *japonica* is the last day with an average daily temperature above 20 ℃, while it is above 22 ℃− 23 ℃ for *indica* rice.

2. Flowering, pollination and fertilization

Under normal circumstances, flowering starts on the day or the next day of panicle exsertion. Early-season rice starts to flower around 7 am, and is in full bloom from 11 am − 12 pm; while late-season rice starts to flower from 8 − 9 am, and is in full bloom from 11 am − 1 pm, with fewer flowers after 2 pm. Flowering lasts 5 − 6 days for a single panicle of early- and mid-season rice and 7 − 8 days for late-season rice. In a rice plant, the main stem usually flowers first, followed by the lower tillers and then the higher tillers. Rice is a hermaphrodite crop. When spikelets are flowering, anthers are dehisced and pollen are scattered on stigmas for fertilization. Fertilization normally lasts five to seven hours after flowering. The optimum temperature for rice flowering is about 30 ℃, with the minimum temperature of 13 ℃-15 ℃, and the maximum of 40 ℃-45 ℃.

3. Grain filling and seed setting

After fertilization, the embryo and endosperm of the zygote enter the developmental stage, which-spans about 10 days. Meanwhile, the rice grains continue to grow, first longitudinally and then by length and width followed by thickening. Normally, their length, width and thickness all reach their inherent level over 10 days after flowering. There is a difference in the growth potential of the spikelets in one single panicle. Normally, early-flowering spikelets and those on primary branches are stronger, while late-flowering spikelets and those on secondary branches are mostly weaker. Strong flowers develop so fast that it takes about seven days to fill the spikelet, while weak flowers develop slowly. Super hybrid rice generally have large panicles, with more weak flowers on secondary branches, and some require more than 40 days to grow to their inherent size.

The maturity of rice grains come after four periods. 1) The milky ripe period which is 3 - 10 days after flowering. The rice grains are filled with white starch milk and the moisture content is about 86% ; then the milk becomes thicker and the glume turns green; 2) The dough period - The endosperm turns from a milky state to a hard state, which can be deformed by manual pressure; the glume turns from green to yellow; 3) The full ripe stage - It is the most suitable period for harvesting when all spike-stalks and husks turn yellow and the rice grains become hard; 4) The dead ripe stage - With most of the glumes and branches dying, they turn gray; the grains are prone to shedding, and the panicles and stalks to breakage.

Section 2 Growth and Development of Super Hybrid Rice

Ⅰ. Growth Duration of Different Combinations

The growth duration of rice varieties ranges from below 100 days to more than 180 days. Within the growth duration, reproductive growth generally lasts 60 - 70 days, and the rest are for vegetative growth. Therefore, the difference in the growth duration among different varieties lies mainly in their vegetative growth period. According to the conditions of different rice varieties, their growth and development can usually be divided into three stages - vegetative growth, vegetative growth and reproductive growth, and reproductive growth. Vegetative growth can be further divided into basic vegetative growth (short day-length high-temperature growth), and variable vegetative growth, which is affected by the photoperiod and the temperature.

Super hybrid rice generally has a longer growth duration, but it varies with varieties. According to statistics from super hybrid rice demonstration and promotion bases approved by the MOA in the past five years (2014—2018) (Table 10 - 5), single-cropping rice varieties, including two-line *indica* hybrid rice, three-line *indica* hybrid rice and *indica-japonica* hybrid rice, have a growth and development period of 132.4 -

Table 10 - 5 Full growth duration of single-cropping super hybrid rice approved by the MOA (2014—2018)

Type	Variety	Growth duration (d)	Approval No.	Year of approval
Two-line *indica*	Longliangyou 1988	138. 6	GS2016609	2018
	Shenliangyou 136	138. 5	GS2016030	2018
	Jingliangyou-Huazhan	157. 6	GS2016022	2018
	Y Liangyou 900	140. 7	GS2015034	2017
	Longliangyou-Huazhan	140. 1	GS2015026	2017
	Weiliangyou 996	132. 4	GS2012021	2016
	N Liangyou 2	141. 8	XS2013010	2015
	Y Liangyou 2	139	GS2013027	2014
	Y Liangyou 5867	138	GS2012027	2014
	C-Liangyou-Huazhan	134	HBS2013010	2014
	Guangliangyou 272	140	HBS2012003	2014
	Liangyou 616	143	MS2012003	2014
Three-line *indica*	Neixiang 6 You 9	155. 9	GS2015007	2018
	Shuyou 217	155. 4	GS2015013	2018
	Huyou 772	157. 6	GS2016024	2018
	Yixiang 4245	159. 2	GS2012008	2017
	Deyou 4727	158. 4	GS2014019	2016
	Yixiangyou 2115	156. 7	GS2012003	2015
	Nei 5 You 8015	133	GS2010020	2014
	F-You 498	157	GS2011006	2014
	Tianyou-Huazhan	133	HBS2011006	2013
	Zhong 9 You 8012	133	GS2009019	2013
Indica-japonica hybrid rice	Yongyou 1504	151	GS2015040	2018
	Yongyou 2640	149	SS201507	2017
	Yongyou 538	153. 5	ZS2013022	2015
	Chunyou 84	156. 7	ZS2013020	2015
	Zheyou 18	153. 6	ZS2012020	2015
	Yongyou 15	139	ZS2012017	2013

159.2 days. The full growth duration is long in general, but there is a big difference among different varieties. For example, the full growth duration of Huiliangyou 996 is 26.8 days less than that of Yixiang 4245.

Even for the same variety of super hybrid rice, there is still a difference in the length of the growth duration when the variety is planted in different areas mainly due to the influence of photosensitivity and temperature sensitivity or both. The growth duration changes greatly when rice is planted in different regions (the variable vegetative growth lasts longer) and more so for varieties with high photosensitivity and temperature sensitivity. No variety has the same growth duration across different regions. In recent years, HHRRC has cultivated a new super high yielding hybrid rice variety, Xiangliangyou 900 (Chaoyouqian) and investigated its super high yield and ecological suitability in different rice-growing regions across the country at 80 plots, 6.67 ha each. Analysis of its performance in different regions at similar altitudes (Table 10-6) revealed that the growth duration prolongs as the latitude increases. For example, the growth duration of Chaoyouqian in Xinyu Demonstration Base in Jiangxi province was 140 days, yet moving northward, the growth duration of the same variety in Yongnian Demonstration Base in Handan, Hebei province, was 183 days, which was 43 days longer. Since the growth duration changes with location, when introducing a new variety to a certain area, attention must be paid to the temperature and light sensitivity and the length of its variable vegetative growth duration. Normally, when a combination with high sensitivity to light and temperature is introduced to somewhere in the north, its full growth duration will be significantly prolonged, which may undermine its maturity. The introduction of the same variety to some place in the south may result in a shorter growth duration or early maturity and reduce the yield. To ensure stable yield, it is a better choice to cultivate a combination with a certain level of photosensitivity in a suitable area for its reproductive growth, thereby leading to successful full heading, although the appropriate range of latitude may be limited.

Table 10-6 Growth duration of Xiangliangyou 900 in different areas (2016)

Location	Altitude (m)	Latitude	Sowing date (m/d)	Transplanting date (m/d)	Maturity date (m/d)	Full Growth period (d)
Fengxi Village, Zhushan Town, Yushui District, Xinyu City, Jiangxi Province	41.0	N27°47′17.05″	5/23	6/16	10/10	140
Wangjia Village, Wannian County, Shangrao City, Jiangxi Province	30.0	N28°45′52.25″	5/9	6/5	10/1	145
Tangxi Town, Wucheng District, Jinhua City, Zhejiang Province	60.2	N29°03′20.81″	5/18	6/12	10/13	148
Xindu Town, Tongcheng City, Anqing City, Anhui Province	28.0	N30°51′5.66″	4/28	5/20	9/28	153
Houbai Town, Jurong County, Zhenjiang City, Jiangsu Province	29.1	N31°48′26.71″	5/15	6/24	10/17	155
Zhuanqiao Town, Guangshan County, Xinyang City, Henan Province	46.0	N31°50′48.24″	4/15	5/9	9/17	156

Continued

Location	Altitude (m)	Latitude	Sowing date (m/d)	Transplanting date (m/d)	Maturity date (m/d)	Full Growth period (d)
Daludong Village, Yanzhuang Town, Juxian County, Rizhao City, Shandong Province	22.8	N35°39′53.32″	4/1	5/5	9/26	178
Guangfu Town, Yongnian District, Handan City, Hebei Province	41.0	N36°42′3.03″	4/10	5/15	10/10	183

There can also be differences in the growth duration of the same variety of rice planted in the same area because of different climate conditions and different sowing dates. For example, an investigation in the influence of sowing dates on the growth duration of Liangyoupeijiu was done in Gaoyou city, Jiangsu province (Table 10 – 7), and it was found that delayed sowing shortened the growth duration by 4 – 5 days. Further analysis shows that the difference was not in the period from the initial heading to maturity but in the period from sowing to young panicle differentiation, i. e., the vegetative growth period. Therefore, early sowing can help ensure a stable yield for Liangyoupeijiu. In comparison, Shanyou 63 matured around eight days earlier than Liangyoupeijiu when the two were sowed at the same time and the difference also was in the vegetative growth period.

Table 10 – 7 Growth duration of varieties with different sowing dates (Jiang Wenchao et al., 2001)

Combination	Sowing date (m/d)	From sowing to heading (d)	From sowing to maturity (d)	Full growth duration (d)
Liangyoupeijiu	4/29	111	47	158
	5/5	106	47	153
	5/11	102	46	148
	5/17	98	46	144
Shanyou 63	5/5	98	47	145

Different temperatures at different altitudes in the same latitude also have an obvious effect on the growth duration. For example, an ecological suitability test at different altitudes was carried out in Longhui, Hunan province, with four testing sites 300 – 750 m in elevation, 150 m between every two adjacent sites (Table 10 – 8). It is showed that there is positive correlation between altitude and the growth duration of varieties, the higher the altitude, the longer the growth duration. Therefore, it is necessary to follow this rule of varying growth duration based on altitude and introduce hybrids suitable for local seasonal conditions in order to be successful.

Ⅱ. Overlap and Change of Vegetative Growth Period and Reproductive Growth Period

The whole life of a riceplant can be divided into the vegetative growth period, vegetative and reproductive growth period, and reproductive growth period. An overlap means that part of the vegetative

Table 10 - 8 Growth duration at different altitudes

Altitude (m)	Y Liangyou 900	Xiangliangyou 2	Shenliangyou 1813
300	150 d	150 d	157 d
450	153 d	153 d	160 d
600	164 d	164 d	169 d
750	170 d	170 d	175 d

growth period is pushed back to the period in parallel with the reproductive period, which is mainly marked by part or all of the vigorous tiller stage taking place within the panicle differentiation period.

There are two reasons for such an overlap. Firstly, a variety with a short growth duration moves fast into the next stage and does not have enough time for vegetative growth. Hence, tillers grow vigorously during panicle differentiation. It is also possible that a low temperature in early spring leads to untimely sprouting of the tiller buds and then when the temperature rises to an appropriate level, the tillers sprout during the panicle differentiation stage.

Secondly, an overlap can even occur in varieties with a long growth duration when the seedlings are aged and have limited vegetative growth. For example, Weiyou 46 and Peiliangyou-Teqing in the Yangtze River Basin have a relatively long growth duration when they are cultivated as double-cropping late-season rice, while the period with optimum temperature for rice growth is short. Therefore, early sowing is usually practiced with aged seedlings to ensure safe full heading. The tillering stage in the field is normally in line with the growth of panicles on the main stem, with overlapping stages 1 and 2.

Measures should be taken in cultivation management to tackle such an overlap. In some rice growing areas, mid-growth seedling control and limited after-tillering irrigation and fertilization are adopted for this purpose. These measures are not good for rice with overlapping growth periods, because early control may result in insufficient seedlings, and late control may affect normal panicle differentiation. It could even do harm to super hybrid rice, because these controls not only affect the large panicle heterosis of super hybrid rice, but also results in insufficient panicles per unit area.

The overlap of growth stages is not only related to the length of the growth duration of the hybrid, but also affected by ecological and cultivation conditions. For example, if a hybrid with a short growth duration is planted in spring when the temperature is low, its active tillering period will probably be delayed and an overlap of growth stages will occur. If the temperature is high at the early stage of growth, it would be less likely to have an overlap. For double-cropping late-season rice, if the seedlings are too aged, there will be an increase in the overlap of growth stages, and if the seedlings are young, there is a lower possibility of an overlap. If the seeding density is high and the seedlings are aged, panicle differentiation will occur earlier in the field and the tillering period will start in the later period of panicle differentiation. This is the most serious overlap.

Ⅲ. Division of Main Periods of Carbon and Nitrogen Metabolism

Carbon and nitrogen metabolism are closely related. Carbon metabolism provides carbon and energy for nitrogen metabolism, while nitrogen metabolism provides enzymes and photosynthetic pigments for carbon metabolism. Therefore, a sound regulation of carbon and nitrogen and coordination between carbon and nitrogen metabolism are of great significance for high and stable yields.

The nitrogen content of super hybrid rice is the highest at the tillering stage, while the starch-based carbohydrate content is the lowest at the tillering stage and the highest atthe maturity stage. The carbon to nitrogen ratio is the lowest at the tillering stage and the highest at the maturity stage. According to the changes of the carbon to nitrogen ratio during the lifetime of a rice plant, it can be roughly divided into three different periods – in the first period, nitrogen metabolism is dominant, in the middle period, both carbon and nitrogen metabolism is important, and in the later period, carbon metabolism is dominant. Super hybrid rice is generally in a nitrogen metabolism-dominated stage until the end of May (post-planting to the end of tillering), in a carbon-nitrogen metabolism-dominated stage from June to early July (end of tillering to heading), and in a carbon-metabolism-dominated stage from heading to harvesting in early July.

The nitrogen content of the rice plant is higher at the first stage (after transplanting to the end of tillering), while the starch content and carbon to nitrogen ratio are lower. The plant building materials in this period mainly form nitrogenous substances such as amino acids and proteins, which are used for the rapid growth and building of leaf tillers and other organs.

The middle stage is from the end of tillering to the full heading stage, during which the contents of nitrogen and starch, and the carbon to nitrogen ratio of the rice plant are at a moderate level, while the phosphorus and potassium content (closely related to the vigorous physiological activities) and chlorophyll content are generally at a high level, which is the period of both carbon and nitrogen metabolism (or the transition period of carbon and nitrogen metabolism). This period is both a period of vigorous physiological activities and a period of extremely complex activities. There is both the growth of seedling leaves and the extinction of tillers and both the development of panicles and the storage of carbohydrates; it may form the basis of a high yield or lay the hidden danger of yield reduction. If the level of nitrogen metabolism is raised too much at this time, it will cause overgrowth and lead to "green madness"; if the shift to high carbon metabolism is promoted too early, the plant and the population may suffer undergrowth.

The later stage is from heading to maturity. The rice plant at this stage has the lowest nitrogen content and the highest starch content and carbon to nitrogen ratio. The plant in this period mainly synthesizes soluble sugars, starch and other sugars to form rice grain yield, indicating that carbon metabolism is absolutely dominant.

The key to a high yield of super hybrid rice is large panicles and large grains, and the potential for further yield increase lies in increasing the number of panicles and improving the seed setting rate, both of which are related to the period of concurrent carbon and nitrogen metabolism. Therefore, it is more important for super hybrid rice to have a sound balance between carbon and nitrogen metabolism in the middle stage. The formation of large panicles and large grains in super hybrid rice occurs mainly in this period, and the number of panicles is also related to the tiller formation in this period, while the seed setting rate is closely related to the plant population size and starch reserves formed in this period. In conclusion, a correct balance of the metabolic levels of carbon and nitrogen in the middle stage, ensured through reasonable cultivation management, is the key to high yield of super hybrid rice.

Section 3 Yield Formation of Super Hybrid Rice

There are two key concepts when it comes to rice yield-biological yield and economic yield. Biological yield refers to the total amount of organic matter produced and accumulated by rice during the growth period, i. e. , the total dry matter harvested from the entire plant (excluding the root system), among which organic matter accounts for 90%–95%, and mineral matter accounts for 5%–10%, so organic matter is the main material basis for yield formation. Economic yield, on the other hand, is the amount of the product, i. e. , the rice grains harvested, which we rely on to achieve the purpose of cultivation. Economic yield is part of the biological yield and its formation is based on the biological yield. Without high biological yield, there can be no high economic yield. The efficiency of converting biological yield into economic yield is the economic coefficient, which is generally around 50%. Biological yield, economic yield and economic coefficient are very close related. Under normal growth conditions, the economic coefficient of rice is relatively stable, thus a high biological yield is generally associated with a high economic yield, and improving biological yield is the basis for obtaining high rice yield.

Ⅰ. Formation of Biological Yield

According to the formula "rice yield = biological yield × harvest index (HI)", rice yield is determined by the biological yield and the HI, and increasing either one or both can increase rice yield. At present, the HI of semi-dwarf rice varieties is already very high, and further improvement of HI is quite unlikely. Therefore, to increase rice yield further, we need to depend mainly on improving the biological yield. There are basically two ways to do this. One is to increase plant height and the other is to increase stalk wall thickness, with the latter more difficult than the former. From a morphological point of view, raising plant height is the most effective and feasible way to increase biological yield. Based on the history of high yielding rice breeding, a general trend or rule can be tentatively drawn, that is, the biological yield increases with the increase of plant height when the harvest index remain at around 0. 5. That is, the yield of rice increases with the increase of plant height.

A high biological yield is a major part of the heterosis of super hybrid rice, and the growth per unit time is significantly greater than that of conventional rice, thus laying the foundation for the increase in effective yield. Studies have shown that in the process of biological yield formation, the advantage of dry matter production in super hybrid rice lies mainly in the middle and late stages, with more than 80% of the yield coming from the photosynthetic products after heading (Zhai Huqu et al. , 2002), while this rate is only around 60% for common hybrid rice. Further analysis concluded that the material output,

output rate and conversion rate of the stem sheaths of super hybrid rice were actually not high, and grain filling relied mainly on materials accumulated after heading, not the materials stored in the stem sheaths and other vegetative organs before heading (Wu Wenge et al., 2007).

HHRRC examined the biological yield formation of the super high yielding hybrid rice variety Xiangliangyou 900 in comparison with the control varieties super hybrid rice Y Liangyou 1 and conventional rice 9311 (Table 10 - 9). The result shows that the total biological yield of Xiangliangyou 900 at maturity is significantly higher than that of the control varieties or more specifically, higher by 14.3% and 22.6% respectively than Y Liangyou 1 and 9311, though its HI is slightly lower than that of Y Liangyou 1. This indicates that the yield increase of Xiangliangyou 1 is mainly the result of the significant increase in its biological yield.

Table 10 - 9 Dry matter production and HI (Wei Zhongwei et al., 2015)

Variety	Total dry weight (g/m^2)			Dry weight of panicles at maturity (g/m^2)	Dry matter accumulated after heading (g/m^2)	HI
	Active tillering	Heading	Maturity			
Xiangliangyou 900	346.4a	1,272.1a	2,296.8a	1,293.6a	1,024.7A	0.515a
YLiangyou 1	354.0a	1,214.1a	2,008.9b	1,073.8b	794.8B	0.527a
9311	331.7a	1,336.8a	1,873.7c	958.0c	536.9C	0.480b

During biological yield formation, the total dry weight of Xiangliangyou 900 at the active tillering and heading stages was basically the same asthat of the control varieties, but was higher at maturity. The dry matter accumulation of Xiangliangyou 900 after heading was significantly greater than that of the control varieties, more specifically 28.9% and 90.9% more than that of Y Liangyou 1 and 9311 respectively. There was also a significant difference among the three varieties in the proportion of biological yield accumulated before and after heading at maturity. The biological yield of Xiangliangyou 900 at the heading stage accounted for 55.4% of that at maturity, with the 44.6% accumulated after heading. In comparison, Y Liangyou 1 accumulated 60.4% of dry matter before and 39.6% after heading, and 9311 accumulated respectively 71.3% and 28.7% before and after heading. It suggests that the super high yielding hybrid rice Xiangliangyou 900 accumulates a reasonable amount of dry matter in the early growth stages, and its advantage in producing more dry matter becomes obvious mainly in the middle and late stages of growth.

In addition to photosynthesis after heading, the grain filling materials also come from materials stored in the vegetative organs before heading. The stem and sheath output, output rate and conversion rate of Xiangliangyou 900 were all lower than those of the control varieties (Table 10 - 10), indicating that the amount of dry matter accumulated in the late growth period can meet the need of grain filling needs, and hence there is less dependence on the non-structural substances in the stem sheaths. This is consistent with the large green leaf area and strong stalk vigor in the late growth period of Xiangliangyou 900, which is conducive to maintaining the strong vigor and supporting function of the stalk in the late growth period.

Table 10 - 10 Dry matter transfer (Wei Zhongwei et al., 2015)

Variety	Stem and sheath dry matter transfer		
	Output (g/m^2)	Output rate (%)	Conversion rate (%)
Xiangliangyou 900	71.6Bb	10.18Bb	5.83Bc
YLiangyou 1	220.2Aa	31.18Aa	22.11Aa
9311	79.1Bb	11.32Bb	8.82Bb

Ⅱ. Formation of Rice Yield

The economic yield ofrice is the amount of rice grains produced, which is measured by the number of effective panicles per unit area, grains per panicle, seed setting rate and grain weight (1,000-grain weight). These are the four key components of yield, and the relationship among them can be used as the basis for yield design. The relationship between yield and the four components can be described as

Yield (kg/ha) = number of panicles per unit area (ha) × number of grains per panicle × seed setting rate (%) × grain weight (g) $\times 10^{-6}$

The formation process of each component of rice yield is also the process of organ building during the growth and development of rice, and the formation of each component is somewhat time-bound during the development of rice. The four components of yield are interrelated. For example, the number of effective panicles per unit area, within a certain range, increases with the number of basic seedlings, but when the number of effective panicles per unit area increases beyond a certain range, the contradiction between the number of panicles and the number of grains will intensify, that is, the increase in the number of effective panicles per unit area will cause a decrease in the number of grains per panicle. If the loss in the number of grains per panicle due to an increase in the number of effective panicles cannot be compensated by the increase in the number of panicles, it will lead to a reduction in yield. Grain weight is a relatively stable factor, but a small grain weight can also have a serious impact on yield if the climatic conditions are poor and cultivation is not properly managed. It can be seen that high yield can be obtained only by reasonably selecting varieties, strengthening cultivation management, coordinating between individuals and population, and balancing between the yield factors.

1. Number of panicles

Super hybrid rice is characterized by large panicles in terms of yield components, but due to the limitation of seed quantity, there are normally a small number of panicles. With agiven number of seedlings, the number of panicles depends on the total number of tillers and the tillers' panicle-bearing rate. Their relationship can be described as number of panicles per unit area = number of plants × number of tillers per plant × tiller panicle-bearing rate. The dynamics of stem-tiller development in a rice population is a visual representation of tillering and panicle formation, and ultimately affects yield significantly. Reasonable dynamics of stem-tiller development and high panicle formation rate are one of the basic characteristics of a high-yielding population. For example, the super high yielding hybrid rice Xiangliangyou 900 has early and rapid onset of tiller growth 5 – 25 days after transplantation and it tends to have more stems and tillers than Y Liangyou 1; the number of panicles reaches the expected level 20 days after transplantation and the number of tillers reaches the maximum 35 days after transplantation, both earlier than Y Liangyou 1. The number of panicles reaching and even going beyond the expected level allows more time for the plant to develop its roots and grow large panicles (Wei Zhongwei et al., 2015). The reasonable dynamic pattern of super hybrid rice stem-tiller development should be as follows. 1) the number of basic seedlings is determined during the transplanting period; 2) at the critical leaf age ($N-n$), effective tillering leads to an appropriate number of panicles; 3) at the stem elongation leaf age ($N-n+3$), the number of tillers reaches its peak, with 1.2 – 1.3 folds of the expected number of panicles; and 4) at the heading stage, an appropriate number of panicles are formed.

The period that affects the number of panicles generally starts from the initial tillering stage and ends 7 – 10 days after the active tillering period. The active tillering period is decisive for the number of panicles. Early growth also plays a role, especially the quality of the seedlings and growth during the seedling re-establishment period. As there are less basic seedlings due to limited seed quantity of super hybrid rice, coupled with the thick stalks and large leaves, there is much shading in the late growth period, thus affecting the panicle-bearing rate. Therefore, increasing the number of panicles becomes a key for high yield. To increase the number of panicles, the quantity of basic seedlings should be increased and good field conditions should be created. In addition, field management activities should be carried out to increase effective tillers and reduce ineffective tillers in production. Two measures can be taken for this purpose. One mainly involves management in the tillering stage is to do everything possible to promote the early growth of tillers and to increase the panicle-bearing rate and the number of panicles. The other is to properly control tiller growth at the late stage to prevent ineffective tillering which consumes too much nutrients and affect the efficiency of light energy utilization.

2. Number of spikelets

In terms of the number of grains per panicle, super hybrid rice varieties generally have an advantage over normal hybrid rice varieties (Huang et al., 2011), and the advantage of large panicles of super hybrid rice varieties is associated with their higher number of secondary branching stalks (Huang et al., 2012). The number of spikelets per panicle depends on the amount of differentiation and degradation. Number of spikelets per panicle = number of differentiated spikelets-number of degenerated spikelets. Therefore, on the one hand, the number of differentiated spikelets is increased by applying flower-promoting fertilizers during the panicle differentiation period (generally in the middle and late second stage, mainly nitrogen fertilizer), and on the other hand, flower-preserving fertilizer (applied in the middle and late fifth stage, a combination of nitrogen, phosphorus and potassium fertilizer with nitrogen as the dominant type) are applied in time to reduce degeneration. The number of spikelets per panicle is determined by the number of primary and secondary branches, and the period that determines spikelet differentiation is the branch differentiation and spikelet differentiation periods, of which the more important one is the secondary branch differentiation period, and the main period for spikelet degeneration is the meiosis period. Therefore, it is of great importance to ensure good field conditions for spikelet growth during these two periods. Sufficient nitrogen and water are needed to ensure successful differentiation of branches and spikelets.

In cultivation, if the field is excessively sun-dried or the nitrogen content in the soil and plants is reduced, the differentiation of branches and glumes will inevitably be hampered. Insufficient supply of nitrogen during meiosis, or drought, or insufficient light, will lead to an increase in the number of degenerated spikelets. The amount of spikelet degradation in super hybrid *indica* rice is generally about 30% . In cultivation, reducing the rate of spikelet degradation is important to fully tap the advantage of having large panicles.

3. Seed setting rate

Seed setting rate is the result of the impact of the "sink", "source" and "flow" combined, and it sees the largest variation among the four components of yield, so a high seed setting rate is a prerequisite for super hybrid rice to obtain super high yield. Seed setting rate is highly sensitive to growth conditions and can change dramatically. A minor change in the ecological and cultivation conditions may reduce the seed setting rate significantly. Therefore, care must be taken in all aspects.

The time period affecting the seed setting rate is long, starting from seedling planting and lasting till maturity. The most sensitive periods are meiosis, flowering and grain filling of the primary branch, i. e., from 20 days before to 20 days after heading, totaling 40 days. There are many factors that may affect the seed setting rate, such as poor seedling quality, excessive population density, premature row closure, unsatisfactory climate, poor sun-drying at the middle stage, lack of fertilizer and water during meiosis, bad weather and insufficient nutrition during flowering, and insufficient water and fertilizer during grain filling, which all directly affect seed setting.

There are various reasons for a low seed setting rate. First, the rice embryo may encounter adverse conditions before development, and flowering and fertilization may be disturbed and making it impossible to get pollinated or fertilized resulting in empty grains. For example, when the temperature is too high, some of the spikelets can be prevented from fertilization (such as pollen tube metamorphosis), forming unfertilized empty glumes. When the temperature is above 35 ℃, the flowering and fertilization process is significantly affected. The empty and shriveled grains formed under low temperature conditions are mainly unfertilized grains, and the first stage affected is the pollen microspore development stage (7 – 10 d before flowering); the second is when it affects normal flowering and fertilization (but there are also a few half-filled grains due to low temperature). According to Zhou Guangqia (1984), the seed setting rate decreases to below 50% under natural conditions when the temperature is below 19 ℃ for 3 – 5 days. In addition, there is a difference in the temperature required for flowering and fertilization. There is no serious impact on flowering when the average daily temperature is less than 17 ℃, but the number of flowers tends to be reduced when the temperature is below 20 ℃. Rice flowers in a large amount when the temperature is higher than 20 ℃. However, the temperature required for fertilization is significantly higher, generally a daily average temperature of 22 ℃ and a daily maximum temperature of 25 ℃ or higher on sunny days for normal fertilization. In weather with insufficient sunshine, a temperature of 23 ℃ or higher is required for proper fertilization.

Second, although the rice embryo develops normally, it may not be sufficiently filled since the carbohydrates produced and stored by the plant cannot fill the "sink" (total spikelet capacity), and the "source-sink" balance is broken, forming half-filled grains, which is mainly caused by an inappropriate population structure, such as what multiple rounds of fertilization and dense planting may result. The lower the shade level of the rice plant population, the higher the seed setting rate, but the total number of spikelets per unit area decreases, and the yield will drop at a certain level, so a certain population level should be maintained. Under high temperature conditions, half-filled grains can also be formed and they are actually glumes whose development stalled after fertilization. The reason is either an increase in the respiration intensity which results in excessive consumption of photosynthetic products under high temperature or, the synthesis ability of leaves is weakened under high temperature. Studies have shown that under high temperature, the photosynthetic rate of super rice varieties decreases, the photo-reductive activity of chloroplasts decreases, and the granum lamella in chloroplasts are disorganized, so the production of photosynthetic products is affected.

The main measures to improve the seed setting rate in super hybrid rice cultivation are as follows. First, do research to ensure the most suitable vegetative organ structure and panicle structure of a population according to the different geographical conditions. The specific methods include cultivating strong seedlings with tillers and well-developed root systems in the early stage, planting young and thin seedlings at the right time, field-drying in the middle stage to increase starch accumulation, applying fertilizer to promote meiosis, spraying foliar fertilizer, nourishing the roots and protecting the leaves, etc. Second, make sure tillering occurs in the most suitable season to prevent the influence of high and low temperatures.

4. Grain weight

Super hybrid rice generally has a larger grain size, and the effect of grain weight on yield is greater than that of conventional rice. This is especially true for some early-maturing combinations, where the advantage of large panicles is weaker and high yields often come from the advantage of large grains. Therefore, attention should be paid to the increase of grain weight in cultivation.

Grain weight is determined by both glume volume and endosperm development. Therefore, there are two periods that affect grain weight. The first is the pollen mother cell meiosis period, which determines the size of the glumes (glume capacity) and the second is the post-flowering period till full grain maturity, during which the availability of carbohydrates directly affects the size of the rice grains.

The development of the glumes during meiosis will be affected under the conditions of insufficient nutrients and water and poor weather. The time of filling one grain during the grain filling period is inversely correlated to the final grain weight, i. e., the shorter the time from flowering to fullness, the larger the grain weight; and the slower filling, the lower the grain weight.

With large panicles, many secondary branches and weak flowers, super hybrid rice has rather uneven grain weight. Besides the spikelet position, the main factors affecting the grain filling rate include the population structure, the decay rate of roots and leaves, and the moisture and climate conditions. The most important factors are the population structure and the climate conditions. When the temperature is too high or too low, the synthesis of photosynthetic products of the plant decreases, the filling rate lowers, and the grain weight decreases. When the population is over-shaded, there tend to be insufficient carbohydrates stored before flowering, and also an excessive number of spikelets per unit area, both of which are not conducive to the increase of grain weight.

References

[1] WEI ZHONGWEI, MA GUOHUI. Studies on characteristics of root system of super high yielding hybrid rice combination Chaoyouqian[J]. Hybrid Rice, 2016, 31(5): 51-55.

[2] JIANG WENCHAO, SUN LONGQUAN, XIAO BOQUN, et al. Effects of sowing stage on the growth characteristics and yield of Liangyoupeijiu[J]. Hybrid Rice, 2001, 16(1): 38-40.

[3] ZHAI HUQU, CAO SHUQING, WAN JIANMIN, et al. Relationship between the photosynthetic function and yield of super-high-yielding hybrid rice at the grain filling stage[J]. Science in China (Series C: Life Sciences), 2002, 32(3): 211-217.

[4] WU WENGE, ZHANG HONGCHENG, WU GUICHENG, et al. Preliminary study on super rice population sink characters[J]. Scientia Agricultura Sinica, 2007, 40(2): 250-257.

[5] WEI ZHONGWEI, MA GUOHUI. Biological characteristics and lodging resistance of super high yielding hybrid rice Chaoyouqian[J]. Hybrid Rice, 2015, 30(1): 58-63.

[6] HUANG M, ZOU Y B, JIANG P, et al. Relationship between grain yield and yield components in super hybrid rice[J]. Agricultural Sciences in China, 2011, 10 (10): 1537-1544.

[7] HUANG M, XIA B, ZOU Y B, et al. Improvement in super hybrid rice: a comparative study between super hybrid and inbred varieties[J]. Research on Crops, 2012, 13 (1): 1-10.

[8] ZHOU GUANGQIA, TAN ZHOUZI, LI XUNZHEN. Research on the disorder of hybrid rice seed setting caused by low temperature[J]. Hunan Agricultural Sciences, 1984(4):8-12.

Chapter 11
Physiology for Super Hybrid Rice Cultivation

Huang Min / Chang Shuoqi / Zhu Xinguang / Zou Yingbin

Section 1 Mineral Nutrition Physiology of Super Hybrid Rice

Ⅰ. Essential Mineral Elements for Super Hybrid Rice

Super hybrid rice, like ordinary hybrid rice and conventional rice, requires 13 essential mineral elements. Among them, six are macroelements, namely, nitrogen, phosphorus, potassium, calcium, magnesium and sulfur, and seven are microelements, namely, iron, boron, manganese, zinc, copper, molybdenum and chlorine. In addition, since silicon is absorbed by rice in large amounts and plays an important role in rice yield formation and resistance, silicon is usually referred to as an agronomically essential mineral element for rice.

1. Functions of macroelements

(1) Nitrogen

Nitrogen is the primary mineral element that affects the growth and yield formation of super hybrid rice. It is an essential component of many important organic compounds and genetic materials (e. g. chlorophyll, amino acids, nucleic acids and nucleosides) in the rice plant. It can affect all parameters related to rice yield, and also plays a significant role in regulating rice quality. In addition, nitrogen can also affect the uptake of other macroelements (e. g. phosphorus and potassium).

(2) Phosphorus

Phosphorus is an essential nutrient element for the growth and development of super hybrid rice. It is an important component of organic substances such as adenosine triphosphate, nucleoside, nucleic acid and phospholipid in the rice plant. It participates in many metabolic activities in various ways and plays a significant role in storing and converting energy and maintaining the integrity of cell membranes.

(3) Potassium

Potassium is also an essential nutrient element for the growth and development of super hybrid rice. The amount of potassium in rice is second only to nitrogen. It is involved in physiological processes such as osmoregulation, enzyme activation, cellular pH regulation, anion and cation balance, stomatal respiration regulation and photosynthetic assimilate transport in the rice plant, and plays an important role in increasing rice yield and improving rice quality.

(4) Calcium

Calcium is an important component of the cell walls of super hybrid rice. It bridges the phosphate and phosphate ester on the biomembrane surface with protein carboxyl, which plays an important role in stabilizing biomembrane structure and maintaining cell integrity. In addition, calcium is also closely related to cell elongation, osmoregulation, ionic equilibrium, as well as resistance.

(5) Magnesium

Magnesium is not only an important component of chlorophyll in super hybrid rice, but also involved in the activation of nearly 20 enzymes in the rice plant, including the malic enzyme and glutathione synthase. Magnesium is an element connecting ribose subunits, playing an important role in maintaining the stability of ribosome structures. It is also involved in cell pH regulation and ionic equilibrium.

(6) Sulfur

Sulfur is an important constituent of amino acids (e. g. cysteine, methionine and cystine) required for chlorophyll synthesis in super hybrid rice. Also, it is a constituent of coenzymes in protein synthesis and participates in some redox reactions in rice.

2. Functional properties of trace mineral elements

(1) Iron

Iron is an essential mineral element in the photoreaction stage of photosynthesis in super hybrid rice. It not only plays a crucial role in the photosynthetic electron transport, but also is an important constituent of porphyrin iron andferredoxin. In addition, iron is an important electron acceptor and a catalyst for several enzymes (e. g. catalase, succinate dehydrogenase) in redox reactions.

(2) Boron

Boron plays a significant role in cell wall biosynthesis and structure and biomembrane integrity in super hybrid rice. In addition, it is an essential mineral element for rice carbon metabolism, sugar transport, lignification, nucleic acid synthesis, cell elongation and division, respiration and pollen development.

(3) Manganese

Manganese not only participates in the redox reaction featuring oxygen release and electron transport in the photosynthesis of super hybrid rice, but also plays an important role in activating and regulating various enzymes (e. g. oxidase, peroxidase, dehydrogenase, decarboxylase and kinase). In addition, manganese is an essential mineral element for chloroplast formation and stability, protein synthesis, the reduction of nitrate ions and the tricarboxylic acid cycle.

(4) Zinc

Zinc is an essential mineral element forcytochrome and nucleic acid synthesis, auxin metabolism, enzyme activation, and cell membrane integrity in super hybrid rice. In addition, zinc is a constituent of enzymes for early protein synthesis in rice plants.

(5) Copper

Copper is an essential mineral element for lignin synthesis in super hybrid rice, as well as a constituent of ascorbic acid, oxidase, phenolase and plastocyanin. In addition, copper is also a regulator (e. g. effector, stabilizer and inhibitor) of enzyme reactions in rice plants and also a catalyst for oxidation reactions. It plays a significant role in nitrogen metabolism, hormone metabolism, photosynthesis, respiration, pollen formation and fertilization.

(6) Molybdenum

Molybdenum is an important constituent of the nitrate reductase in super hybrid rice. It not only plays an important role in the nitrogen metabolism of rice plants, but also has certain effects on phosphorus metabolism, photosynthesis and respiration.

(7) Chlorine

As a cofactor of manganese, chlorine participates in water photolysis in super hybrid rice, acting on photosystem Ⅱ. It is not only a mineral element necessary for oxygen release in the Hill reaction, but also promotes photophosphorylation. Chlorine can also regulate the opening and closing of rice stomata and the activation of H^+-pump ATPase.

3. Functional properties of silicon

Silicon is an essential mineral element for forming the cutin and siliceous layer on the surface of the stems and leaves of super hybrid rice. The formation of the siliceous layer on the epidermis plays an important role in improving the resistance of rice plants (e. g. reducing diseases caused by bacteria and fungi and damage by pests including stem borers and planthoppers, and improving lodging resistance) and reducing leaf transpiration. In addition, silicon exerts a significant effect on the plant type of super hybrid rice. Rice plants with sufficient silicon supply have erect leaves and sound growth conditions, which is conducive to better utilization of light energy and nitrogen. Silicon can also help rice plants absorb less manganese and iron and alleviate the toxicity of low-valency iron and manganese under reductive conditions.

Ⅱ. Mineral Element Absorption and Transport of Super Hybrid Rice

1. Absorption

(1) Existence of mineral elements

Mineral elements that super hybrid rice plants absorb exist in three states, 1) dissolved in the soil solution; 2) attached on soil colloids; 3) existing as insoluble salts. The first of the three is the most common.

(2) Characteristics of mineral element absorption

The uptake of mineral elements in soil solution by the root system of super hybrid rice has the following two characteristics. 1) The root system uptakes mineral elements and water separately. Although the root system absorbs mineral elements and water mainly with the non-suberized epidermal cells of the roots, the absorbed amounts of mineral elements and water are out of proportion. Thus, it can be seen that the root uptake of mineral elements and water uptake are two independent processes. However, the independence of the two processes is relative because there is interaction between them. 2) The root uptake of mineral elements is selective. The quantity of mineral element ions absorbed by the root system is not proportional to the their content in the soil environment; it is more in line with the physiological demand of the plant. That is, root uptake of mineral elements is highly selective.

(3) Ways of mineral element absorption

The uptake of mineral elements in soil solution by the root cells of super hybrid rice is carried out in three ways. 1) Active uptake, i. e. , carrier molecules on the plasma membrane of living root cells combine with mineral ions to form a complex, and the complex releases the absorbed ions after reaching the inner side from the outer side of the membrane. This mode is related to respiration and is a process that consumes metabolic energy. Firstly, because carrier molecules are present in a fairly stable plasma membrane structure, energy must be consumed for the maintenance of the membrane structure. Secondly, carrier molecules are complex organic substances and their synthesis and movement must consume energy. Thirdly, it also requires energy to form the complex of the carrier molecule and the absorbed ion and to take the ion from the outer side of the membrane and release it when reaching the inner side. 2) Passive uptake which refers to the uptake of mineral elements by root cells without consuming any energy from respiration, which is independent of other root cell metabolism, and the entry of mineral elements into the root cells is governed only by physical laws (diffusion). 3) Pinocytosis. When the plasma membrane of root living cells folds inward, the mineral elements adsorbed on the plasma membrane can be wrapped in, forming water vesicles, and gradually moved into the cell.

2. Mineral element transport

A small portion of the mineral elements absorbed by the root system of super hybrid rice remains in the root system to participate metabolism of root cells, and most of them are transported to other parts of the rice plant. Mineral ions are transported through protoplasmic streaming in cells and with the plasmodesma between cells. During ion transport, each cell consumes some mineral elements in its metabolism, while actively discharging the rest into the ducts or sieve tubes, which are then transported in upward liquid flows to the plant parts above the ground, mainly to the respiratory growth sites. Studies have shown that most of the inorganic nitrogen absorbed by the roots of a rice plant is transformed into organic nitrogen compounds (e. g. amino acids) in the root system for transport; absorbed phosphorus is mainly transported in the form of orthophosphate; and absorbed potassium is mainly transported in the form of inorganic ions. In addition, some of the mineral elements (e. g. , calcium, iron, etc.) that are transported to the above-ground parts cannot be reused, i. e. , they accumulate after entering young tissues, while others (e. g. , nitrogen, phosphorus, potassium) can be re-transported, i. e. , after being used by young tissues, they can be transported to other parts as the tissues age.

Ⅲ. Absorption and Use of Nitrogen, Phosphorus and Potassium by Super Hybrid Rice

1. Absorption, accumulation and distribution of nitrogen, phosphorus and potassium

(1) Nitrogen

There were significant genotypic differences in the nitrogen uptake of the parts of super hybrid rice above the ground (Table 11 - 1). It can be observed that among the 10 varieties tested, Zhunliangyou 527 has the highest uptake amount (189.09 kg/ha), followed by Zhongzheyou 1 (184.26 kg/ha), Liangyoupeijiu (183.65 kg/ha) and Neiliangyou 6 (182.92 kg/ha). They had significantly higher uptake amounts than the other four varieties (Ⅱ - You 084, Ⅱ - Youhang 1, D-You 527 and Y Liangyou 1) and the conventional hybrid rice variety Shanyou 63. The conventional rice variety Shengtai 1 had the lowest uptake of nitrogen (170.50 kg/ha), which was significantly lower than that of the super and conventional hybrid rice varieties. There is only a slight difference in nitrogen accumulation in the straw of super hybrid rice varieties of different genotypes, but the difference is significant in nitrogen accumulation in rice grains across different genotypes. Table 11 - 1 also shows significant differences among locations and years in terms of the nitrogen accumulation in rice straw, grains or empty grains. In terms of the differences among locations, Guidong had the highest nitrogen accumulation, with an average of 214.90 kg/ha over three years. In terms of the differences among the years, 2009 witnessed the highest amount and the average was 206.30 kg/ha across the three locations. As for the nitrogen distribution in various organs, the average in rice grains was 62.56% (61.2%-65.3%), that in rice straw was 32.8% (31.8%-33.6%), and that in empty grains was 4.6% (2.8%-5.5%).

Table 11 - 1 Absorption, accumulation and distribution of nitrogen in super hybrid rice plants (Zou Yingbinet al., 2011)

Year/location/variety		Straw (kg/ha)		Grains (kg/ha)		Empty grains (kg/ha)		Entire plant (kg/ha)	
		Average	Significance	Average	Significance	Average	Significance	Average	Significance
Year	2007	54.34	b	91.04	c	4.30	c	149.69	c
	2008	61.52	a	118.37	b	6.21	b	186.09	b
	2009	62.17	a	129.69	a	14.45	a	206.30	a
Location	Changsha	51.76	c	99.86	c	7.21	b	158.83	c
	Guidong	69.61	a	134.81	a	10.57	a	214.90	a
	Nanxian	56.65	b	104.43	b	7.17	b	168.25	b
Variety	Ⅱ - You 084	59.16	ab	112.30	bc	8.08	c	179.54	b
	Ⅱ - Youhang 1	59.63	ab	108.81	cd	9.25	b	177.69	b
	D-You 527	60.01	ab	111.28	c	9.97	a	181.26	b
	Y Liangyou 1	59.11	ab	111.79	c	8.49	c	179.39	b
	Liangyoupeijiu	59.27	ab	117.03	b	7.34	d	183.65	ab
	Neiliangyou 6	60.49	a	112.69	bc	9.74	ab	182.92	ab
	Zhongzheyou 1	60.13	ab	116.69	b	7.44	d	184.26	ab
	Zhunliangyou 527	60.23	a	123.44	a	5.42	e	189.09	a
	Shanyou 63	59.58	ab	110.94	c	8.11	c	178.64	b
	Shengtai 1	55.80	b	105.34	d	9.35	ab	170.50	c

Note. The same significance level in one column indicates asignificance level below 5%.

(2) Phosphorus

There were significant genotypic differences in the phosphorus uptake of the parts of super hybrid rice above the ground (Table 11 - 2). It can be observed that among the 10 varieties tested, Ⅱ - Youhang 1 has the highest uptake amount (39.80 kg/ha), followed by Liangyoupeijiu (38.59 kg/ha) and D-You

527 (38.25 kg/ha). Their uptake amounts are significantly higher than the other five varieties (Ⅱ-You 084, Ⅱ-Youhang 1, Neiliangyou 6, Zhongzheyou 1 and Zhunliangyou 527) as well as the conventional hybrid rice variety Shanyou 63 and the conventional rice variety Shengtai 1 which has the lowest uptake (35.72 kg/ha), significantly lower than that of all other varieties covered. There were significant differences in phosphorus accumulation in the straw and grains of super hybrid rice varieties of different genotypes. Table 11-2 also shows that there were significant differences among locations and years in terms of the phosphorus accumulation in rice straw, grains or empty grains. In terms of differences among locations, Guidong had the highest phosphorus accumulation, with an average of 44.56 kg/ha over three years. In terms of the differences across the years, 2008 witnessed the highest uptake and the average was 43.91 kg/ha across the three locations. As for the phosphorus distribution in various organs of the rice plant, the average in rice grains was 71.0% (67.6%-74.4%), that in rice straw was 24.3% (21.7%-27.4%), and that in empty grains was 4.7% (2.8%-5.5%).

Table 11-2 Absorption, accumulation and distribution of phosphorus in super hybrid rice plants
(Zou Yingbin et al., 2011)

Year/location/variety		Straw (kg/ha)		Grains (kg/ha)		Empty grains (kg/ha)		Entire plant (kg/ha)	
		Average	Significance	Average	Significance	Average	Significance	Average	Significance
Year	2007	7.42	b	24.75	c	1.19	c	33.36	c
	2008	12.45	a	29.84	a	1.62	b	43.91	a
	2009	7.67	b	25.66	b	2.47	a	35.80	b
Location	Changsha	7.77	c	25.88	b	1.66	b	35.32	b
	Guidong	11.07	a	31.33	a	2.16	a	44.56	a
	Nanxian	8.69	b	23.05	c	1.46	c	33.20	c
Variety	Ⅱ-You 084	9.35	bc	26.30	abc	1.85	b	37.50	b
	Ⅱ-Youhang 1	10.90	a	26.92	ab	1.98	ab	39.80	a
	D-You 527	9.30	bc	26.84	ab	2.11	a	38.25	ab
	Y Liangyou 1	8.52	cd	27.27	ab	1.89	b	37.68	b
	Liangyoupeijiu	10.10	ab	27.05	ab	1.44	c	38.59	ab
	Neiliangyou 6	9.20	bc	26.01	bc	1.95	ab	37.16	bc
	Zhongzheyou 1	8.10	d	27.83	a	1.48	c	37.40	bc
	Zhunliangyou 527	8.57	cd	27.32	ab	1.05	d	36.94	bc
	Shanyou 63	9.13	c	26.85	ab	1.88	b	37.85	b
	Shengtai 1	8.62	cd	25.14	c	1.97	ab	35.72	c

Note. The same significance level in one column indicates a significance level below 5%.

(3) Potassium

There were significant genotypic differences in the potassium uptake of the parts of super hybrid rice above the ground (Table 11 – 3). It canbe observed that among the 10 varieties tested, Zhunliangyou 527 has the highest amount (165.39 kg/ha), followed by Liangyoupeijiu (161.05 kg/ha), D-You 527 (160.59 kg/ha), Shanyou 63 (160.02 kg/ha) and Ⅱ – Youhang 1 (158.75 kg/ha). They have significantly higher amounts than the other four varieties (Ⅱ – You 084, Y Liangyou 1, Neiliangyou 6 and Zhongzheyou 1). The conventional rice variety Shengtai 1 has the lowest uptake amount (144.98 kg/ha), which was significantly lower than that of all other varieties covered. There are significant differences in potassium accumulation in the straw of super hybrid rice varieties of different genotypes. Yet in terms of the amount accumulated, Zhongzheyou 1 has the highest amount (20.51 kg/ha), while Shanyou 63 (16.68 kg/ha) and Shengtai 1 (16.49 kg/ha) have the lowest. No significant difference was found among other varieties covered (Ⅱ – You 084, Ⅱ – Youhang 1, D-You 527, Y Liangyou 1, Liangyoupeijiu, Neiliangyou 6 and Zhunliangyou 527). Table 11 – 1 also shows that there were significant differences among locations and years in terms of the potassium accumulation in rice straw, grains and empty grains. In terms of differences among locations, Guidong had the highest potassium accumulation, with an average of 179.72 kg/ha over three years. In terms of differences among years, 2009 witnessed the highest

Table 11 – 3 Absorption, accumulation and distribution of potassium in super hybrid rice plants (Zou Yingbin et al., 2011)

Year/location/variety		Straw (kg/ha)		Grains (kg/ha)		Empty grains (kg/ha)		Entire plant (kg/ha)	
		Average	Significance	Average	Significance	Average	Significance	Average	Significance
Year	2007	133.16	b	18.74	a	1.29	c	153.19	b
	2008	132.07	b	17.22	b	1.39	b	150.67	b
	2009	146.34	a	18.83	a	27.3	a	167.89	a
Location	Changsha	126.77	b	15.09	c	1.56	b	143.42	c
	Guidong	154.25	a	23.25	a	2.21	a	179.72	a
	Nanxian	130.54	b	16.45	b	1.63	b	148.62	b
Variety	Ⅱ – Youna	133.55	bc	17.91	b	1.91	b	153.38	b
	Ⅱ – Youhang 1	137.87	ab	18.68	b	2.20	a	158.75	ab
	D-You 527	140.20	ab	18.17	b	2.23	a	160.59	ab
	Y Liangyou 1	135.47	b	19.15	b	1.81	bc	156.44	b
	Liangyoupeijiu	141.34	ab	18.09	b	1.63	d	161.05	ab
	Neiliangyou 6	135.21	b	18.29	b	2.36	a	155.86	b
	Zhongzheyou 1	134.00	bc	20.51	a	1.55	d	156.06	b
	Zhunliangyou 527	145.63	a	18.67	b	1.09	e	165.39	a
	Shanyou 63	141.76	ab	16.68	c	1.57	d	160.02	ab
	Shengtai 1	126.83	c	16.49	c	1.67	cd	144.98	c

Note. The same significance level in one column indicates a significance level below 5%.

amount and the average was 167.89 kg/ha across the three locations. As for the potassium distribution in various organs, the average in straw was 87.2% (85.8%–88.6%), that in rice grains was 11.6% (10.4%–3.1%) and that in empty grains was 1.2% (0.7%–1.5%).

2. Absorption and accumulation of nitrogen, phosphorus and potassium

According to Fig. 11–1, the accumulation of nitrogen, phosphorus and potassium in super hybrid rice reaches 33.8%, 20.2% and 21.0% respectively at the mid-tillering stage (MT); 50.0%, 44.3% and 49.2% at the young panicle differentiation stage (PI); 65.6%, 62.2% and 72.7% at the booting stage (BT); 85.7%, 81.0% and 85.0% at the heading stage (HD); 97.6%, 95.1% and 98.1% at the milk-ripe stage (MK), and close to the total amount at the maturity stage (MA).

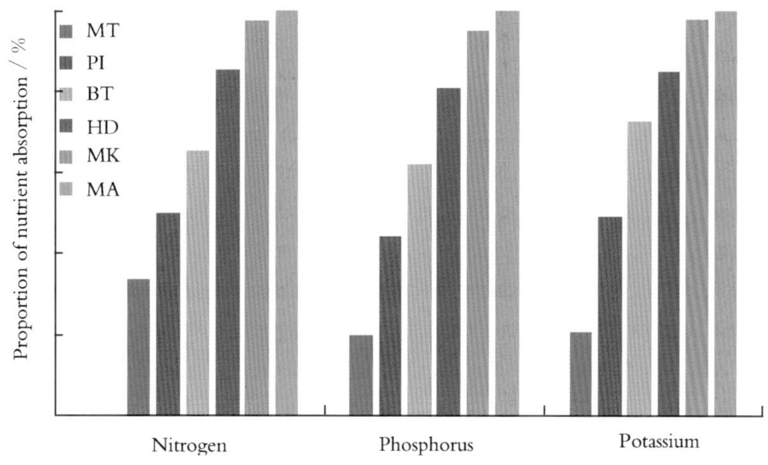

Fig. 11–1 Accumulation of nitrogen, phosphorus and potassium in super hybrid rice (ZouYingbin and Xia Shengping, 2011)

3. Required amount of nitrogen, phosphorus and potassium

According to Zou Yingbin et al. (2011), the nitrogen requirement of super hybrid rice ranged from 17.99 kg to 19.27 kg per 1,000 kg of rice produced, and the differences among super hybrid rice varieties were not significant. However, all were significantly lower than the required amount of the common hybrid rice variety Shanyou 63 (20.22 kg) and the conventional rice variety Shengtai 1 (20.09 kg) (Table 11–4). The genotypic differences in phosphorus requirement were significant, with the highest phosphorus requirement at 4.38 kg for Ⅱ Youhang 1, while the phosphorus requirements of all other super hybrid rice varieties (Ⅱ-You 084, D-You 527, Y Liangyou 1, Liangyoupeijiu, Nei-liangyou 6, Zhongzheyou 1, and Zhunliangyou 527) were lower than those of the common hybrid rice variety Shanyou 63 (4.33 kg) and the conventional rice variety Shengtai 1 (4.24 kg). There were significant genotypic differences in the required amount of potassium. Super hybrid rice varieties (Ⅱ-You 084, Ⅱ-Youhang 1, D-You 527, Y Liangyou 1, Liangyoupeijiu, Neiliangyou 6, Zhongzheyou 1 and Zhunliangyou 527) required significantly less potassium than conventional hybrid rice Shanyou 63 (18.26 kg). In addition, except for Ⅱ-Youhang 1 (17.39 kg), D-You 527 (17.08 kg) and Shengtai 1 (17.26 kg), which required similar amounts of potassium, the other six super hybrid rice varieties (Ⅱ-You 084, Y Liangyou 1, Liangyoupeijiu, Neiliangyou 6, Zhongzheyou 1 and Zhunliangyou 527) required significantly less potassium than Shengtai 1. Table 11 - 4 also shows the differences among locations and years in terms of the required amounts of nitrogen, phosphorus, and potassium. Changsha has the highest requirement among the three locations, while there was no significant difference among the years. Another study suggests that with the increase in yield, the amounts of nitrogen, phosphorus, and potassium required for each 1,000 kg of grain produced by Zhunliangyou 527 and Liangyou 293 decrease (Fig. 11 - 2). It can be seen that both high yield and effective use of nutrients can be achieved in super hybrid rice in good coordination.

Table 11 - 4 Amount required of nitrogen, phosphorus and potassium per 1,000 kg of rice produced (Zou Yingbin et al., 2011)

Year/location/variety		Nitrogen (kg)		Phosphorus (kg)		Potassium (kg)	
		Average	Significance	Average	Significance	Average	Significance
Year	2007	16. 95	c	3. 85	b	17. 42	a
	2008	19. 61	b	4. 61	a	15. 88	b
	2009	20. 74	a	3. 59	c	16. 96	a
Location	Changsha	19. 30	a	4. 31	a	17. 55	a
	Guidong	18. 86	b	3. 93	b	15. 76	c
	Nanxian	19. 14	ab	3. 81	c	16. 96	b
Variety	Ⅱ - You 084	19. 27	b	4. 05	bc	16. 42	ode
	Ⅱ - Youhang 1	19. 22	b	4. 38	a	17. 39	b
	D - You 527	19. 24	b	4. 06	bc	17. 08	bcd
	Y Liangyou 1	18. 48	bc	3. 91	cd	16. 35	cde
	Liangyoupeijiu	17. 99	c	3. 78	de	15. 78	e
	Neiliangyou 6	18. 83	b	3. 85	cde	16. 21	de
	Zhongzheyou 1	18. 93	b	3. 89	cde	16. 23	de
	Zhunliangyou 527	18. 77	b	3. 69	e	16. 57	cde
	Shanyou 63	20. 22	a	4. 33	a	18. 26	a
	Shengtai 1	20. 09	a	4. 24	ab	17. 26	bc

Note. The same significance level in one column indicates a significance level below 5%.

Section 2 Characteristics of Photosynthesis of Super Hybrid Rice

Ⅰ. Photosynthetic Physiology of Super Hybrid Rice

1. Photosynthesis efficiency and rice yield potential

The formation of rice yield is actually a process of production and distribution of photosynthetic

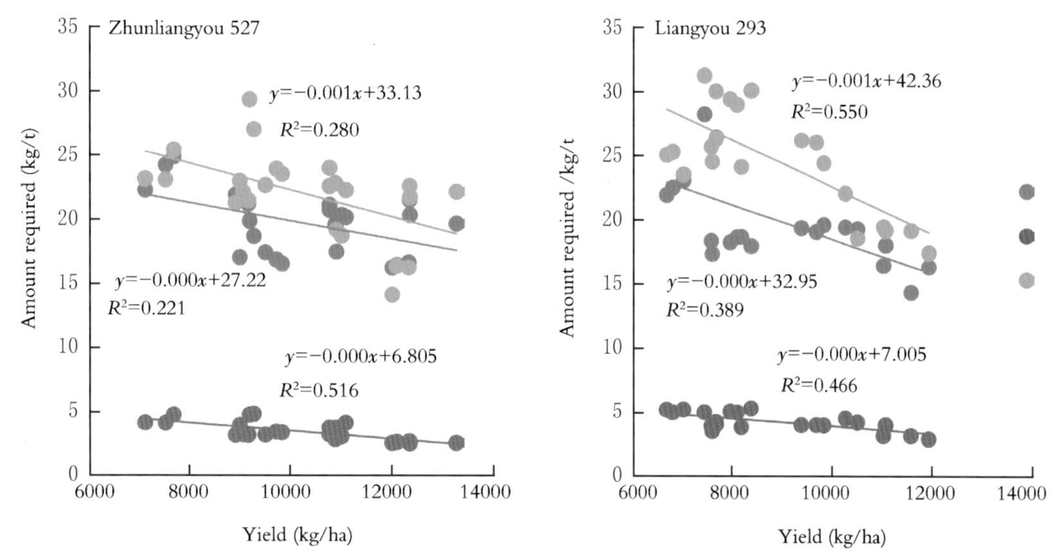

Fig. 11 - 2 Relationship between yield and required amounts of nitrogen, phosphorus and potassium (Ao Hejun et al., 2008)

products. Yield potential (Y) can be calculated with the formula as follows,

$$Y = E \times \varepsilon_i \times \varepsilon_c \times \eta$$

where E refers to the total solar energy radiated to a certain area of land; ε_i refers to the canopy light energy interception efficiency; ε_c refers to the canopy light energy conversion efficiency; and η refers to the harvest index (Monteith, 1977). ε_i can be improved by accelerated canopy development, early ground coverage, increased growth period, improved lodging resistance and more application of nitrogen fertilizer. At present, for superior rice varieties, the ε_i throughout the entire growth period is about 0.9, leaving less room for improvement. Therefore, one effective way to increase yield potential is to increase light energy conversion efficiency (ε_c), which is jointly determined by photosynthesis and respiration in the field, currently only around 1/3 of the theoretical maximum (Zhu et al., 2008b), leaving more room for improvement (Long et al., 2006; Zhu et al., 2008a). The increase in rice yield is highly related to the increase in photosynthesis and biomass (Hubbart et al., 2007). Researchers from Nanjing Agricultural University and other academic institutions investigated the physiological traits of the super high yielding hybrid rice combination Xieyou 9308 before and after heading, and the relationship between flag leaf photosynthesis and panicle substance accumulation, using Xieyou 63 as the control variety. It was found that compared with Xieyou 63, Xieyou 9308 has a stronger substance production capacity before and after heading, especially after heading. In terms of flag leaf photosynthesis, Xieyou 9308 outperforms Xieyou 63 extremely significantly, with better grain filling; the flag leaf photosynthesis of Xieyou 63 deteriorates rapidly 20 days after heading, and the net assimilation products of individual plants fails to meet the needs of grain filling. These results indicate that maintaining high photosynthetic efficiency at the late stage of grain filling is the key to achieving super high yield of rice (Zhai Huqu, 2002).

2. Photosynthetic physiology of super hybrid rice

As a higher plant, the photosynthesis of hybrid rice occurs in the chloroplast, and its photosystem,

Calvin-Benson cycle, photorespiration, and starch and sucrose metabolism are all consistent with what happens in higher plants. However, in super hybrid rice, the photosynthetic organ, photosystem, Calvin-Benson cycle, and photorespiration also show specific characteristics compared with previous rice varieties. The photosynthesis of the flag leaf of the high-yielding hybrid rice Liangyoupeijiu, its male parent 9311, female parent Peiai 64S and the three-line hybrid Shanyou 63 were systematically studied at Hunan Agricultural University. The results showed that the mean values of net photosynthetic rate, initial carboxylation activity of Rubisco, total carboxylation activity of Rubisco and Rubisco activation rate of Liangyoupeijiu were higher than those of Shanyou 63 (Guo Zhaowu, 2008). The number of chloroplasts per chloroplast cell was higher, chlorophyll content was higher and Mg^{2+} content was higher than those of Shanyou 63. Further research shows that the flag leaf of Liangyoupeijiu had stronger resistance to membrane lipid over-oxidation and a longer photosynthetic period; that the photosynthesis I reduction power and photosynthesis Ⅱ oxygen evolution activities of the flag leaf blades and their sheaths were higher than those of Shanyou 63; and that the average cyclic and noncyclic photophosphorylation activities were also higher than those of Shanyou 63 (Guo Zhaowu, 2008). The Institute of Agricultural Biogenetics and Physiology of Jiangsu Academy of Agricultural Sciences found that compared with Shanyou 63, Liangyoupeijiu had higher chlorophyll content, PSII activity and photosynthetic rate; the primary photochemical efficiency (Fv/Fm) was less down-regulated, the photochemical quenching coefficient was higher and photoinhibition was lighter under strong light at noon on sunny days; Liangyoupeijiu had both photooxidation and shade tolerance (Li Xia et al., 2002). Measurements of the activities of C4 cycle-related enzymes (PEPC, NADP-ME, NAD-ME, and PPDK) indicate an over-parent heterosis of Liangyoupeijiu, which may be an important physiological basis for its resistance to photoinhibition (Zhang Yunhua, 2003). Under strong light and high temperature, compared with its parents, Liangyoupeijiu has a higher ability to convert the light energy absorbed during senescence into chemical energy, lower heat dissipation, and is more tolerant to photoinhibition. In addition, Liangyoupeijiu has a more active endogenous reactive oxygen species scavenging enzyme system and has a higher resistance to photooxidation and premature senescence (Wang Rongfu et al., 2004). Moreover, it has a high resistance to low temperature and strong light, and it shows heterosis over both parents and a maternal bias (Zhang Yunhua et al. 2008). In the flag leaves, Liangyoupeijiu has over-parent heterosis in pigment content, net photosynthetic rate, electron transport activity and capacity absorption, transport and transformation (Wang Na et al., 2004).

Li Xiaorui from Chen Guoxiang's team at Nanjing Normal University studied the dynamic changes of chloroplast light energy conversion, chloroplast protein composition and chloroplast ultrastructure during aging, with afocus on the three functional, fully exserted leaves of Liangyoupeijiu. It was found that during aging, the chlorophyll content of the functional leaves increased first and then decreased. The net photosynthetic rate, photophosphorylation activity, Hill activity, Ca^{2+}-ATPase activity and Mg^{2+}-ATPase activity of the functional leaves gradually increased at the early stage, but after reaching the peak on the 14th day, decreased rapidly. In terms of the net photosynthetic rate, flag leaf > penultimate leaf > antepenult leaf. As a result, the exoxygenation activity and ATP content of the three functional leaves reached a high level once they were fully spread and gradually decreased thereafter; among the three functional leaves, the flag leaf had the highest ATP content, followed by the penultimat and antepenult leaves. During functional leaf senescence, SOD and CAT activities showed a decreasing trend in all three functional leaves, and POD increased and then decreased, and all three antioxidant enzymes had the highest activity in the flag leaf. In the early stage of flag leaf senescence, the chloroplasts are generally small and have a narrow oval shape. As senescence progresses, the chloroplasts increase in size, the basal granules become loose, the starch grains increase, and the osmiophilic droplets appear. At the later stage of senescence, the chloroplasts further expand, the biofilm ruptures, the content become less, and even vacuolation occurs; and finally the chloroplast structure disintegrates, the internal structure is destroyed, and the stroma is lost. Similar to what happens to their physiological functions, the cytoarchitectonic changes happen earlier in the penultimate leaf and antepenult leaf than in the flag leaf during senescence.

A comparative study of Liangyoupeijiu, its male parent 9311 and female parent Peiai 64S reveals that the three varieties experience an increase at the early stage and then a decrease at the middle and late stages in the net photosynthetic rate; Liangyoupeijiu has an over-parent heterosis in light energy conversion; the electron transport is most active in PSI and PSII of Liangyoupeijiu, followed by the male parent and then the female parent; and Liangyoupeijiu is more genetically affected by its male parent in this regard. Throughout the growth period of the flag leaf, the ratio of fatty acid content gradually increases (14 : 0, 16 : 0 and 18 : 1), and decreases (16 : 1, 18 : 2 and 18 : 3); Liangyoupeijiu is more genetically similar to its female parent in terms of membrane lipid fatty acid composition (Zhou Quancheng, 2005).

China National Hybrid Rice R&D Center conducted a systematic analysis of the high yield traits of Y Liangyou 900. It turns out that compared with Shanyou 63, the chlorophyll content of all leaf blades in-the canopy of Y Liangyou 900 at each growth stage is higher (Table 11 - 5). Especially, the chlorophyll contents of the three leaves in the canopy at the milk-ripe stage show a significant difference from that of Shanyou 63. The data shows that the chlorophyll contents of the three leaves of Y Liangyou 900 are 8.34%, 13.57% and 19.22% higher than those of Shanyou 63 respectively on average over the two years. The thickness of the three leaves of Y Liangyou 900 is significantly different from that of Shanyou 63 from the tillering stage to the yellow ripening stage, and such a difference is more significant at the milk-ripe stage and the yellow ripening stage.

Table 11 - 5 Chlorophyll content of Y Liangyou 900 and Shanyou 63 at each growth stage

(Chang Shuoqi et al., 2016)

Leaf position	Stage	2013				2014			
		Y Liangyou 900	Shanyou 63	Difference (%)	P	Y Liangyou 900	Shanyou 63	Difference (%)	P
First	Tillering	44.52 ± 1.91	40.42 ± 0.80	10.14	0.08	47.08 ± 0.55	40.0 ± 0.68	17.47	< 0.01
	Young panicle differentiation	45.95 ± 0.38	41.90 ± 2.38	9.67	0.14	47.16 ± 1.38	41.38 ± 1.62	13.97	< 0.05
	Milk ripe	46.48 ± 1.02	43.40 ± 0.74	7.10	< 0.05	47.05 ± 0.52	42.91 ± 0.72	9.65	< 0.01
	Yellow ripening	42.13 ± 1.14	26.08 ± 1.43	61.54	< 0.01	24.88 ± 0.31	24.05 ± 1.41	3.43	0.62

Continued

Leaf position	Stage	2013				2014			
		Y Liangyou 900	Shanyou 63	Difference (%)	*P*	Y Liangyou 900	Shanyou 63	Difference (%)	*P*
Second	Tillering	45.85 ±1.29	43.15 ±2.04	6.26	0.22	47.90 ±0.70	46.30 ±0.52	3.46	0.09
	Milk ripe	44.26 ±0.60	41.74 ±0.73	6.04	< 0.05	50.00 ±0.65	41.29 ±0.13	21.09	< 0.01
	Yellow ripening	43.13 ±1.46	28.83 ±0.89	49.60	< 0.01	26.60 ±1.15	26.67 ±1.65	−0.26	0.97
Third	Young panicle differentiation	44.40 ±2.23	42.55 ±6.79	4.35	0.48	49.30 ±1.75	45.76 ±1.62	7.08	< 0.05
	Milk ripe	42.22 ±1.25	37.36 ±4.05	13.01	< 0.05	50.03 ±1.11	39.89 ±2.15	25.42	< 0.01

From the stage of young panicle differentiation, the netphotosynthetic rate of the leaves of Y Liangyou 900 during the whole growth period is higher than that of Shanyou 63 (Fig. 11 −3); from the heading stage to the yellow ripening stage, Y Liangyou 900 has a significantly higher net photosynthetic rate than Shanyou 63 (Fig. 11 − 3). Besides, Y Liangyou 900 has a higher saturated photosynthetic rate (Asat) than Shanyou 63, and the difference is more significant from the heading stage on (Fig. 11 −3). The photosynthetic rate of rice leaf is related not only to the net photosynthetic rate, but also to the area of the canopy leaf. Compared with Shanyou 63, Y Liangyou 900 has larger leaves at all stages except the tillering stage and the difference is more significant at the mike-ripe stage (Table 11 −6).

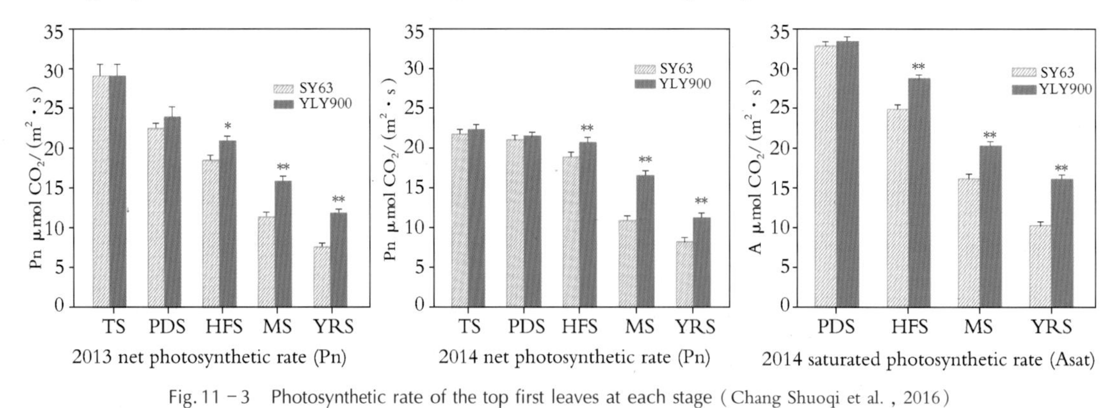

Fig. 11 −3 Photosynthetic rate of the top first leaves at each stage (Chang Shuoqi et al., 2016)

TS: Tillering stage; PDS: Panicle differentiation stage; HFS: Heading stage; MS: Mike-ripe stage; YRS: Yellow ripening stage

Table 11 −6 Area of canopy leaf at each stage (Chang Shuoqi et al., 2016)

Stage	2013				2014			
	Y Liangyou 900	Shanyou 63	Difference (%)	*P*	Y Liangyou 900	Shanyou 63	Difference (%)	*P*
Tillering	127.66 ±13.77	132.36 ±19.44	−3.55	0.89	142.05 ±7.12	156.93 ±5.87	−9.48	0.15
Young panicle differentiation	199.67 ±8.37	171.02 ±10.93	16.75	0.11	407.07 ±6.77	372.22 ±6.78	9.36	< 0.05

Continued

Stage	2013				2014			
	Y Liangyou 900	Shanyou 63	Difference (%)	*P*	Y Liangyou 900	Shanyou 63	Difference (%)	*P*
Milk ripe	222.71 ± 4.17	173.69 ± 7.76	28.22	< 0.05	401.09 ± 15.39	319.85 ± 12.31	25.40	< 0.05
Yellow ripening	170.11 ± 8.09	102.97 ± 14.53	65.20	< 0.05	134.93 ± 2.67	129.50 ± 3.65	4.19	0.41

3. Photorespiration of super hybrid rice

Like all other plants, the Rubisco of hybrid rice species is not only low in catalytic number but also highly nonspecific. It can catalyze not only the carboxylation of RuBP with CO_2 to produce two molecules of 3-phosphoglyceric acid (PGA), but also the reaction of RuBP with O_2 to produce PGA and 2-phosphoglycolic acid. 2-phosphoglycolic acid passes along the photorespiratory pathway through three organelles, cytoplasm, peroxisome and mitochondrion, eventually returning 75% of the carbon contained in 2-phosphoglycolic acid back to the Calvin cycle in the form of PGA, which is then used to regenerate RuBP. Since CO_2 is released in mitochondria during photorespiration, some CO_2 will inevitably be released although some can be re-fixed by Rubisco. The ability to increase the rate of CO_2 released from photorespiration (or even respiration) directly into the chloroplast to be re-fixed theoretically helps improve the efficiency of light energy utilization. Currently, the structural and biochemical characteristics that control the rate of these CO_2 re-fixations are not clear. The photorespiration process is a huge loss to the efficiency of light energy utilization in plants. At a temperature of 25 ℃, C3 plants lose 30% of the CO_2 fixed by Rubisco through the photorespiration pathway (Zhu et al., 2008a). As the temperature increases, the solubility of CO_2 decreases faster than that of O_2, and the specificity (τ) of Rubisco to CO_2 decreases, resulting in a gradual increase in CO_2 loss through the photorespiration pathway (Long, 1991). In the context of global warming, finding new ways to reduce the metabolic flow through the photorespiratory pathway is important for securing food yield.

4. CO_2 diffusion-stomatal conductance and mesophyll conductance

Before CO_2 enters the chloroplast matrix and is fixed by Rubisco, it needs to overcome a series of obstacles, including the leaf boundary layer, stomata, intercellular space, cell walls, cell membranes, cytoplasm, chloroplast membranes and thylakoid membranes. These obstacles and various biochemical factors in the cell determine the final rate of CO_2 fixation. The influence of these obstacles on CO_2 diffusion is normally quantitatively described by stomatal conductance and mesophyll conductance. Stomatal conductance mainly describes the conductance of CO_2 from the leaf boundary layer to the intercellular space, while the mesophyll conductance describes the movement of CO_2 from the intercellular space to be fixed by Rubisco. As increase in mesophyll conductance can increase CO_2 on the leaf blade without increasing stomatal conductance, it is conducive to improving the water use efficiency of the rice plant. Therefore, improving mesophyll conductance is conducive to rice resistance to drought. As cell wall and chloroplast membrane have a huge impact on mesophyll conductance, carbonic anhydrase in the chloroplast stroma significantly affects photosynthesis efficiency and mesophyll conductance; the response of mesophyll conductance to CO_2 is closely related to the permeability of the chloroplast membrane to carbonate ions (Tholen and Zhu, 2011). Given that these structural and biochemical factors that affect mesophyll conductance all have an impact on the diffusion and assimilation of C13 isotope by mesophyll cells (Farquhar et al., 1989), these factors may be related to different assimilation of C13 isotope in different rice varieties. This provides a theoretical basis for using C13 to select drought-resistant rice varieties. In rice, more than 60% of the protoplasts are occupied by chloroplasts, and chloroplasts occupy more than 95% of the cell periphery. These structural features equip rice with high mesophyll conductance, low CO_2 compensation point and low oxygen sensitivity (Sage and Sage, 2009).

5. Respiration and its interaction with photosynthesis

Respiration mainly includes glycolysis, the tricarboxylic acid cycle and respiratory electron transport. Lowering the respiration rate without reducing the photosynthetic rate can help increase crop yields. Field experiments show that there is a significant negative correlation between nighttime temperature and rice yield (Peng et al., 2004). This may be attributed to the increase in the consumption of photosynthetic products by respiration resulted from a high temperature at night.

6. Photosynthetic function of leaf sheath

Hunan Agricultural University investigated the photosynthetic capacity of the flag leaf sheath and the distribution of photosynthetic products in the leaf sheath in high yielding hybrid rice Liangyoupeijiu, its male parent 9311, female parent Peiai 64S and the three-line hybrid rice combination Shanyou 63. It was found that the net photosynthetic rate of the flag leaf sheath of Liangyoupeijiu is higher than that of Shanyou 63, especially during the critical period of grain filling. The chlorophyll content of the leaf sheath and the quantity of photosynthetic products transported to the panicles of the sheath is also higher in Liangyoupeijiu than in Shanyou 63. The photosynthetic products of the leaf sheath contribute 10%−20% of the yield (Guo Zhaowu et al., 2007). The leaf sheath of Liangyoupeijiu exports photosynthetic products faster than that of Shanyou 63, and the quantity of photosynthetic products converted into economic yield is higher for Liangyoupeijiu than for Shanyou 63. The photosynthetic products are mainly transported to the panicles, with only a small quantity stored in the leaf sheath and an even smaller quantity transported to the roots. The leaf sheath contributes 9%−29% of the yield. (Guo Zhaowu, 2008). The peak grain filling period (the third stage of rice flag leaf development) is the key period for the decay and decomposition of mesophyll cells, chloroplasts, and chlorophyll of the flag leaf and its sheath, indicating that the peak grain filling period is when the flag leaf blades and sheaths should be protected in cultivation practice (Guo Zhaowu, 2008). Similar to the flag leaf, the leaf sheath of Liangyoupeijiu has high photosynthetic capacity, strong functional photosynthetic organs, great resistance to membrane lipid over-oxidation, a long photosynthetic function period and strong photosystem assimilation. All these indicate that the high assimilation of the leaf sheath is another important basis for the high yield of Liangyoupeijiu (Guo Zhaowu et al., 2008).

7. Panicle photosynthesis

After measuring the photosynthetic rate of panicles and flag leaves with a self-made panicle photosynthesis measuring chamber, we have found that for most super hybrid rice varieties, such as Xiangliangyou

900, the photosynthetic rate of the panicles is close to that of the flag leaf, and panicle photosynthetic rate differs significantly among varieties, which indicates that improving panicle photosynthesis may be an important factor in increasing rice yield potential.

8. Photosynthetic function period

Existing research shows that early senescence is an important factor limiting the increase of rice yield (Inada et al., 1998; Lee et al., 2001). A study was conducted at Anhui Agricultural University on the photosynthetic characteristics of the high-yielding and high-quality hybrid rice varieties (Fengliangyou 1, Fengliangyouxiang 1 and Fengliangyou 4) with Shanyou 63 as the control in the late growth stages. The results showed that the chlorophyll content of functional leaves was higher, the net photosynthesis of flag leaves was stronger, there was less "photosynthetic siesta", and the superoxide dismutase (SOD) activity was higher for the above hybrids than for the control and decreased slowly. These indicate that the Fengliangyou hybrid series has a strong photosynthetic capacity and the strong photosynthetic function lasts long, providing the basis for high yields (Liu Xiaoqing, 2011). A similar conclusion has also been reached for K You 52 (Wang Hongyan et al., 2010). Research conducted at Nanjing Agricultural University shows that super high yielding hybrid rice Xieyou 9308 maintains a sound photosynthetic function in its flag leaf for about 20 days after heading, and the function then decreases slowly. This is an important basis for maintaining high yields (Zhai Huqu et al., 2002).

China National Hybrid Rice R&D Center found that there is no early senescence of leaf blades in Y Liangyou 900 compared with Shanyou 63. At the tillering stage, the leaves of Y Liangyou 900 had no advantage in photosynthesis, but when the young panicle differentiation stage started and the rice plants shifted from vegetative growth to reproductive growth, the large "sink" of Y Liangyou 900 began to show its strength in photosynthetic response with a lasting high photosynthetic rate, which is significantly different from the situation of the control (Fig. 11-4). After starting reproductive growth, the leaf color of Y Liangyou 900 fades slowly, leaving more green leaves. In essence, there are more functional leaves for photosynthesis and the leaf photosynthetic function period is longer. This phenomenon is more significant at the grain filling stage (Fig. 11-4). In cultivation management, increasing leaf nitrogen content by applying additional panicle fertilizer can both increase the photosynthetic rate of rice after heading and prevent premature leaf senescence. This fertilization technique has become an effective measure to improve rice yield (Ling Qihong, 2000; Yu, Yan et al., 2011). The high photosynthetic rate of leaves and long photosynthetic function period are the basis for the large amount of photosynthetic products required for grain filling, and the higher panicle rate and higher photosynthetic product utilization rate (i. e. less ineffective tillering and less wasted photosynthetic products) of Y Liangyou 900 cannot be ignored. Although the tillering capacity of Shanyou 63 is stronger than that of Y Liangyou 900 at the tillering stage, in the late reproductive growth stage, its effective tillering show no difference from Y Liangyou 900. Maintaining a higher photosynthetic capacity, especially the photosynthetic capacity of the leaves at the base of the canopy, ensures that more photosynthetic products are transported to the roots, maintains higher root vigor, and increases nutrient uptake, which can help prevent plant senescence (Mishra and Salokhe, 2011; Ling Qihong et al., 1982).

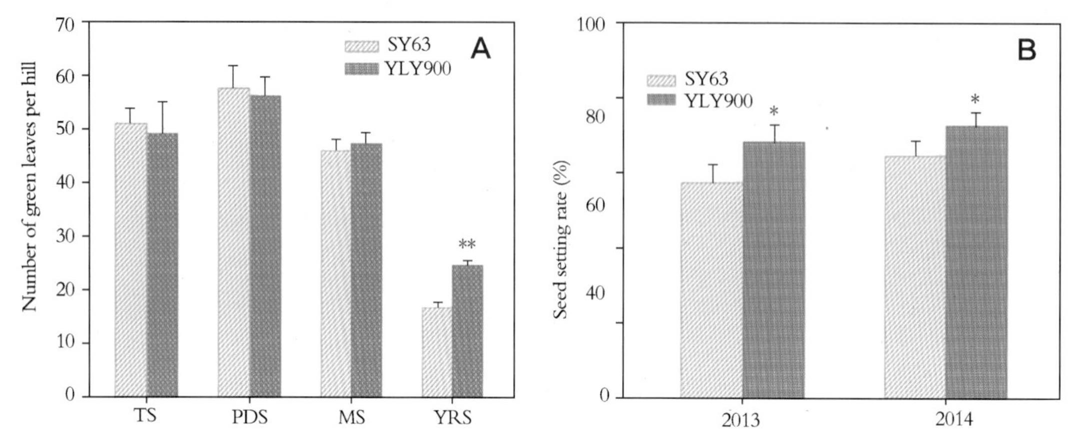

Fig. 11 - 4 Leaf quantity (number of green leaves) and seed setting rate in each stage (Chang Shuoqi et al., 2016)

TS: Tillering stage; PDS: Panicle differentiation stage; MS: Mike-ripe stage; YRS: Yellow ripening stage

9. Canopy photosynthesis

Rice yield is closely related to the photosynthesis of the entire canopy rather than that of individual leaves. Therefore, increasing the total photosynthetic rate of the entire canopy is the key to increasing yield. Different rice varieties have different leaf morphology, leaf area indexes and canopy structures, which leads to great heterogeneity in time and space and in environmental factors such as light intensity, temperature, humidity and CO_2 content in the canopy (Song et al., 2013). There is also great heterogeneity in the photosynthetic traits of the leaves in the canopy (Song et al., 2016a). At present, facilities for measuring canopy photosynthesis have been established (Song et al., 2016b), providing important technical support for studying the canopy photosynthesis of super high yielding hybrid rice and relevant controlling factors. China National Hybrid Rice R&D Center studied the photosynthetic morphology of Shanyou 63 and seven hybrid rice combinations, which showed certain differences in the morphological features of the canopy but all delivered high yields. According to the results, the yield of four combinations is significantly higher than that of control variety, with increased flag leaf length, smaller leaf angle and even light distribution in each part of the canopy (small extinction coefficient). In addition, the light saturation and the photosynthetic rate of each plant at the end of the booting stage, 10 days after heading and 30 days after heading, as well as the photosynthetic rate of the population 10 days after heading increase more significantly in the experimental group than in the control group (Liu Jianfeng et al., 2005). With the same planting density, the average light transmission in the middle of the flag leaf sheath and that in the middle of the penultimate leaf sheath of Liangyoupeijiu are higher than those of Shanyou 63 (Guo Zhaowu, 2008). The length of the internodes from the first node to the fifth of Y Liangyou 900 is shorter than that of Shanyou 63, and the third internode of Shanyou 63 is 18.88%-50.36% longer than that of Y Liangyou 900 (Fig. 11 - 5). The sixth internode (the uppermost internode) of Y Liangyou 900 is 3.90%-19.52% longer than that of Shanyou 63. This character is even more prominent in 2014. The diameter of each stem node of Y Liangyou 900 is also bigger than that of Shanyou 63 (Fig. 11 - 5), and the dry matter of each node weighs higher than that of Shanyou 63. These features help solar radiation

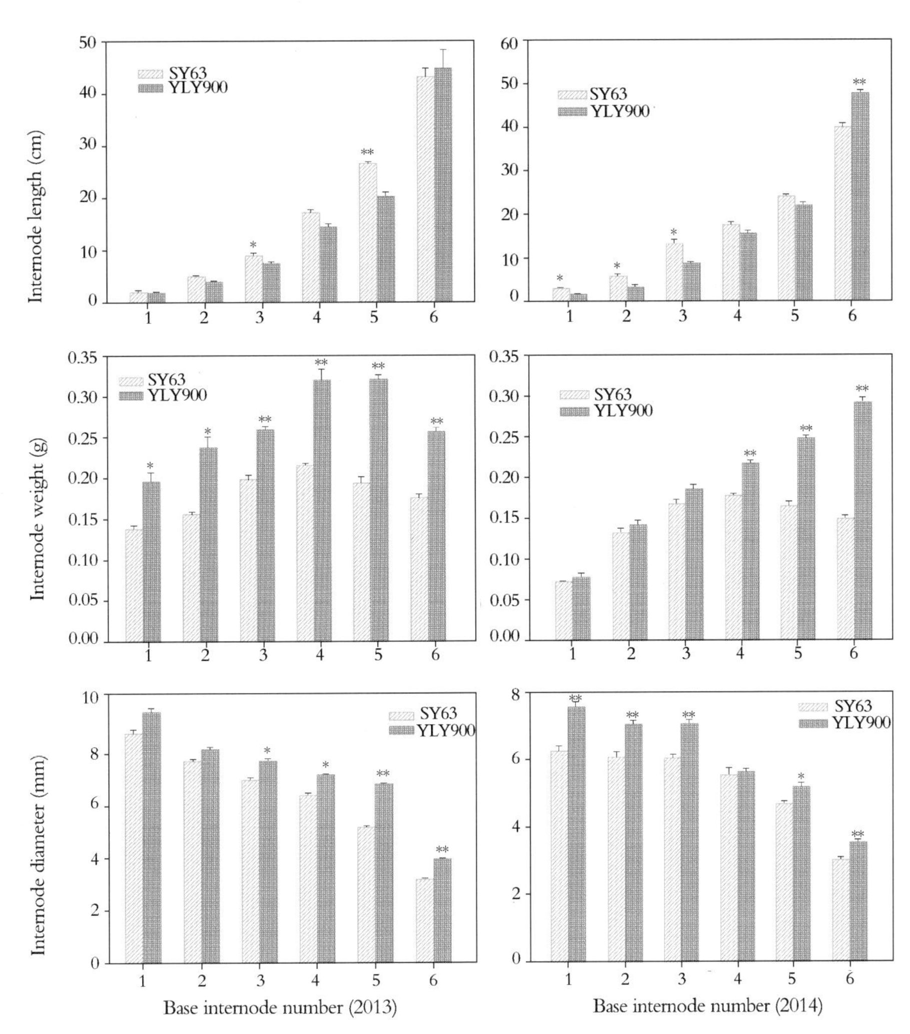

Fig. 11 - 5 Internode length, weight and diameter of Y Liangyou 900 (Chang Shuoqi et al., 2016)

reach the base of the canopy so that the base leaves can be better exposed to light, and that the panicles in the uppermost internode can better intercept photosynthetic radiation by the panicle, thereby increasing their photosynthetic rate (Chang et al., 2016).

Ⅱ. Super Hybrid Rice Photosynthetic Response to the Environment

1. Increasing atmospheric CO_2 concentration

In a study conducted at Yangzhou University, the free air CO_2 enrichment (FACE) experiment was employed to investigate the diurnal changes in photosynthesis during the heading and mid-grain-filling stages of the new hybrid rice combinations Yongyou 2640 and Y Liangyou 2 under two CO_2 levels, a normal ambient CO_2 level and a high CO_2 level (up by 200 μmol/mol). Under the condition of high CO_2 concentration, the net photosynthetic rate of the leaves of both combinations increases significantly at the heading stage (52% on average throughout the day), but the rate is halved at the mid-grain-filling stage and Y Liangyou 2 exhibits a more obvious decrease in its photosynthetic rate. The increase of atmospheric CO_2 concentration greatly reduced leaf stomatal conductance, lowered transpiration, and increased water use efficiency at the heading stage and the mid-grain-filling stage. Research suggests that the ultimate productivity of Y Liangyou 2 benefited less than that of Yongyou 2640 from high CO_2 concentration, which may be explained by the obvious light adaptation at the late growth stage (Jing Liquan et al., 2017).

2. UVB radiation

A systematic study was conducted on the response of Liangyoupeijiu and Jinyou 402 to UVB radiation enhancement at Huazhong Normal University. The results showed that the growth of Liangyoupeijiu was slightly inhibited when treated with five hours of UVB radiation daily for 111 days. The chlorophyll content of the UVB-treated rice plants was lower than plants of the control group at the tillering stage, but higher at the late growth stage. After being treated with UVB radiation, the experimental group had a higher photosynthetic rate which was manifested by 45.2% higher fresh weight and 35.3% higher chlorophyll content than the control group; and the leaves in the experimental group were more resistant to photoinhibition, which could be due to the rapid turnover of the D1 protein in the radiated leaves. During leaf senescence, there was no significant difference in the maximum quantum efficiency in PSII. In conclusion, enhanced UVB radiation promotes the photosynthesis of Liangyoupeijiu. Also, UVB radiation significantly reduces the net photosynthetic rate of Jinyou 402, showing that UVB has different effects on different super hybrid rice combinations (Xu Kai, 2006). A study conducted at Nanjing Normal University on the photosynthesis of Liangyoupeijiu under a low-intensity UVB treatment of 1.6 $kJ/(m^2 \cdot d)$ revealed that UVB could effectively increase the chlorophyll content, net photosynthetic rate, stomatal conductance and oxidation resistance of the variety (Li Wen, 2012).

3. Low-light stress

Using six hybrid rice combinations with different photosensitivity to shade, the Rice Research Institute of Sichuan Agricultural University found that rice yield was significantly reduced after 15 d of shading (80% shading rate) and 15 d of recovery during the field heading stage. During this process, the chlorophyll content of the flag leaf increased, the net photosynthetic rate of leaves decreased, the malondialdehyde content increased, and the leaf area index increased, while SOD activity varied; different treatments responded differently to low light. The maintenance of a high net photosynthetic rate, quantum efficiency, chlorophyll content and lower MDA content and SOD activity in the flag leaf under low light is the physiological basis of low light tolerant rice varieties (Zhu Ping et al., 2008)

4. Heat stress

The high-temperature tolerance of the combinations Peiai 64S/E32 and Liangyoupeijiu were studied at South China Botanical Institute, CAC, using Shanyou 63 as the control. The results showed that high temperature caused a decrease in photosynthetic efficiency and increased photoinhibition; the optimum temperature for photosynthetic carbon assimilation was between 35 ℃ and 40 ℃. At high temperature, the ability of PS Ⅱ linear electron transfer was almost lost, while PS Ⅱ photochemical efficiency decreased less (8.8%–21.0%), indicating that PS Ⅱ photosynthetic linear electron transfer process is more sensitive to high temperature than photochemical energy conversion. Super hybrid rice is more heat tolerant than the control Shanyou 63. The possible mechanisms for its increased heat tolerance are threefold. First, more rapid carotenoid accumulation; second, efficient lutein cycle with increased heat dissipation; and third, higher content of heat-stable protein (Ou Zhiying et al., 2005).

5. Cold stress

The National Hybrid Rice R&D Center found that Y Liangyou 2, Y Liangyou 1 and their parent Yuanhui 2 have strong cold tolerance in the germinating, seedling, heading and flowering stages. The cold tolerance of sterile lines at the germination stage was Y58S > Peiai 64S > Guangzhan 63 -4S, and the reestablishment rate of seedlings treated with low temperature at 5 ℃ for 4 d at the germination stage could be used as an benchmark of cold tolerance in *indica* rice. The cold tolerance of super high-yielding hybrid rice is related to both parents and their hybrid heterosis (Chang Shuoqi et al., 2015).

Ⅲ. Methods to Improve the Utilization Efficiency of Light Energy in Super Hybrid Rice

The effective combination of excellent strains and strong hybrid heterosis is an important way to further improve the light energy utilization efficiency of hybrid rice.

1. Further improving the plant type of super hybrid rice

In the process of hybrid rice breeding, the ideal plant type and the determination of its related plant parameters play an important role in guiding high-yielding rice breeding. For example, compared with Shanyou 63, the super hybrid rice Y Liangyou 900 has a loose and moderate plant type, with a small leaf angle, straight and slightly concave upper three leaves, good population ventilation and light penetration, and high population light energy utilization efficiency (Fig. 11 -6).

Fig. 11 -6 Plant type of Y Liangyou 900 (left) and Shanyou 63 (right)

2. Optimizing the physiological traits for photosynthesis

An important manifestation of strong hybrid heterosis is a high canopy light energy use efficiency in hybrids, where increasing photosynthetic CO_2 assimilation rate is central. The main ways to improve photosynthetic CO_2 fixation are as follows, increasing key enzyme activity in the Calvin-Benson cycle, modifying the photorespiratory pathway, overcoming the leaf diffusion resistance to CO_2 and C4 photosynthetic modification, etc.

(1) Increasing key enzyme activity in the Calvin-Benson cycle

A lot of work has been done in China on the modification of photosynthesis-related enzymes to regulate photosynthesis and yield, especially through a systems biology approach, which identified important genes controlling the Calvin cycle (Zhu et al., 2007). Among these genes, SBPase was overexpressed in the parent of super hybrid rice 9311, which improved the heat resistance of 9311 (Feng et al., 2007), mainly by organizing the Rubisco-activating enzyme from the liquid matrix of chloroplasts to the thylakoid membrane, thus maintaining the activity of Rubisco.

(2) Modifying the photorespiration pathway

In 2007, Christoph Peterhansel, a German scholar, established a photorespiration branch in *Arabidopsis* (Kebeish et al., 2007). This branch used five enzymes associated with glycolate degradation metabolism in *E. coli* to bypass the energy-consuming processes of the photorespiratory pathway in the cytoplasm, peroxisomes and mitochondrion, while releasing CO_2 directly into the chloroplast matrix. Therefore, the transgenic plants showed some improvement in light energy use efficiency compared with the wild type (Kebeish et al., 2007). Veronica et al. established a new branch for photorespiration and proved that it could improve the light energy utilization efficiency and biomass of *Arabidopsis* (Maier et al., 2012). The reduction of photorespiration by genetic modification is based on the assumption that photorespiration has no physiologically important role to play in basic plant metabolism. Recent studies have challenged this conclusion, however. Even in C4 maize plants with low photorespiration, reduced glycolate oxidase activity could lead to difficulties in survival in normal air (Zelitch et al. 2009). Also, when C3 plants were grown at high CO_2 concentrations, the nitrogen content in plant organs was reduced (Bloom et al. 2010). These may indicate that intermediates in the photorespiratory pathway are associated with plant nitrogen metabolism. As for rice, there are no reports on the modification of photorespiratory branching pathways.

(3) Improving mesophyll conductance

In China, in-depth research has been conducted on the mechanisms of mesophyll conductance, and a systems biology model has been established to accurately describe the metabolism and diffusion of CO_2 in three-dimensional leaf cells (Tholen and Zhu, 2011). Using this model, it was found that the cell wall and chloroplast membrane had a strong influence on mesophyll conductance, the carbonic anhydrase activity in the chloroplast matrix affected the photosynthetic efficiency and the mesophyll conductance; and finally the response of mesophyll conductance to CO_2 was related to the permeability of the chloroplast membrane to carbonate ions. Recently, physical models and mechanisms have been proposed to elucidate the variation of leaf conductance at different light intensities, CO_2 and O_2 levels (Tholen et al., 2012), but there has been no report related to mesophyll conductance and its modification in super hybrid rice so far.

(4) Engineering modification of C4

Due to the CO_2 concentration mechanism, C4 photosynthesis has a higher efficiency of light energy utilization than C3 photosynthesis. To ensure an efficient CO_2 concentration mechanism, the C4 photosynthesis pathway utilizes two highly differentiated cells, namely vascular bundle sheath cells and mesophyll cells. For both cell types, photosynthetic organs underwent specific differentiation. In vascular bundle sheath cells, Rubisco content increased, PS Ⅱ content decreased, and only the cyclic electron transport system was present, whereas in mesophyll cells, PEPC content increased, while both PS I and PS Ⅱ were intact (Sage, 2004). In addition to the biochemical differences of the chloroplast, an efficient metabolite transport system has evolved from mesophyll cells and vascular bundle sheath cells, with the Kranz anatomy formed on the wall of the latter (Leegood, 2008). Thus, to transplant the C4 photosynthetic system into C3 plants (e. g. rice), modifications at different levels of biochemistry, leaf structure and cytology are required. The international C4 Rice project funded by the Bill & Melinda Gates Foundation has been in progress for nearly 10 years and has made significant progress, including the systematic analysis of the possible regulatory factors controlling the structure of the Kranz anatomy (Wang et al., 2013) and the development of a C4 system model to guide C4 transformation (Wang et al., 2014). Hunan Agricultural University (HAU) has systematically introduced C4 photosynthetic pathway-related enzymes and transferred bivalent cDNAs of C4 PEPC/PPDK and NADP-MDH/NADP-ME into the super hybrid rice parent Xiangchuai 299 by the agrobacterium-mediated method, and formed tetravalent cDNAs by polymerization of trans-PEPC/PPDK and NADP-MDH/NADP-ME hybrids of transgenic rice. Systematic analysis showed that the exogenous photosynthetic enzyme genes were stably inherited and efficiently expressed in transgenic rice, with higher photosynthetic rates and lower CO_2 and light compensation points than the control during the booting, heading and full heading stages; in both drought and high temperature, the transgenic C4 photosynthetic enzyme rice had higher PS Ⅱ maximum photochemical efficiency (Fv/Fm) and PS Ⅱ potential activity (Fv/F0) (Duan Meijuan, 2010). Jiangsu Academy of Agricultural Sciences also used rice HPTER - 01 with high PEPC expression as the parent and crossed it with a series of sterile and restorer lines to obtain a large quantity of progeny materials and identified several plants with high PEPC activity and photosynthetic efficiency; the photosynthetic rate of the transgenic rice was significantly higher than that of the parent at different times of the day, with lower photoinhibition at noon (Li Xia et al., 2001). These materials are still a long way from successful C4 photosynthesis conversion, but the current C4 conversion is a solid foundation for future C4 conversions.

Section 3 Root Physiology of Super Hybrid Rice

Ⅰ. Characteristics of Super Hybrid Rice Root Growth

1. Morphological and structural characteristics of the root system

(1) Root morphology

Root morphology (number, length, diameter, surface area, volume, and dry weight) is an important aspect reflecting root growth. In general, the root morphology of high-yielding rice varieties is better than that of low-yielding rice varieties. Most studies have also shown that super hybrid rice varieties such as Liangyoupeijiu, Zhunliangyou 527, Y Liangyou 1 and Xieyou 9308 have obvious advantages in root morphology, with the root number, thickness, surface area, volume and dry weight significantly exceeding those of common hybrid rice varieties like Shanyou 63 and 65002 or conventional rice varieties like Tesan'ai 2, Shengtai 1, Huanghuazhan, Yuxiangyouzhan and Yangdao 6. The morphological advantages become more prominent with the growth of the plants and the average increase in root dry weight, in particular, is about 15%. Research also suggested that although the root dry weight of Y Liangyou 087 at the heading and maturity stages was lower than that of Teyou 838, its fine roots (diameter below 0.5mm) had more obvious growth advantages in the late growth stage; and at heading and maturity stages, the length and surface area of the fine roots were significantly higher than those of Teyou 838 (Fig. 11-7). In addition, studies show that the root morphology of super hybrid rice is affected by cultivation, water management, and chemical regulation. No tillage resulted in stunted root growth (decreased root length, root surface area, root dry weight, etc.) during the tillering stage of super hybrid rice, while moist irrigation and foliar spraying of 6-benzyladenine (6-BA) facilitated root growth of super hybrid rice.

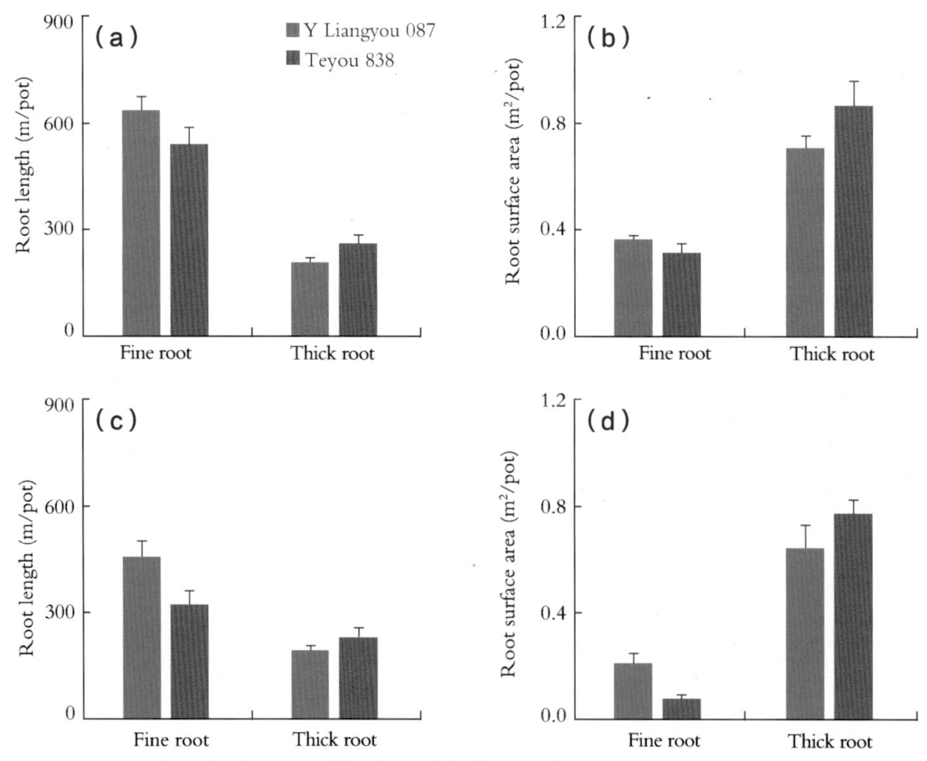

Fig. 11-7 Length and surface area of the fine and thick roots of Y Liangyou 087 and Teyou 838 at heading (a & b) and maturity (c & d) stages (Huang et al., 2015)

(2) Ultrastructure of root tip cells

The root tip (including the root cap and root meristem) is the most active part of the root system physiologically, with functions such as sensing the direction of gravity and responding to and transmitting environmental signals. The endoplasmic reticulum, mitochondria, Golgi apparatus, ribosomes, vacuoles, microbodies and plasma membrane ATPase in root apical cells are important for the performance of root functions. It was shown that the numbers of mitochondria, Golgi apparatus and ribosomes in root tip cells tended to increase significantly with the evolution of rice varieties and the increase of yield. During young panicle differentiation, the root tip cells of super hybrid rice Liangyoupeijiu contained many amyloids and mitochondria, while the above organelles were not obvious in the root tip cells of the conventional rice variety Xudao 2 (Fig. 11 - 8). It has also been proved that cultivating factors such as irrigation and nitrogen fertilization management have an impact on the ultrastructure of the root tip cells of super hybrid rice. Alternate drying and wetting is conducive to increasing the number of mitochondria, Golgi apparatus and amyloid in root tip cells. At the same level of nitrogen application, minor water stress leads to intact root tip cells with clear nuclear membrane boundaries and typical structural features, but severe water stress leads to an increase in osmiophilic granules and amyloid in root tip cells, and the cells are completely distorted and deformed at a later stage, with a significant increase in cell gaps, fracture and degradation of organelles, with only fragments of organelles remaining in the cell matrix. With the same irrigation method, the root cells are more intact and the nuclear membrane is clearer with moderate nitrogen fertilizer treatment, but the degradation of cell walls and nuclear membranes is accelerated in the root system treated with heavy nitrogen fertilizer. There is a coupling effect between irrigation method and the amount of nitrogen fertilizer on the ultrastructure of root tip cells, and the ultrastructure of rice root system treated with mild water stress coupled with moderate amount of nitrogen fertilizer is optimal, showing intact cells, clear nuclear membrane and typical cell structure characteristics.

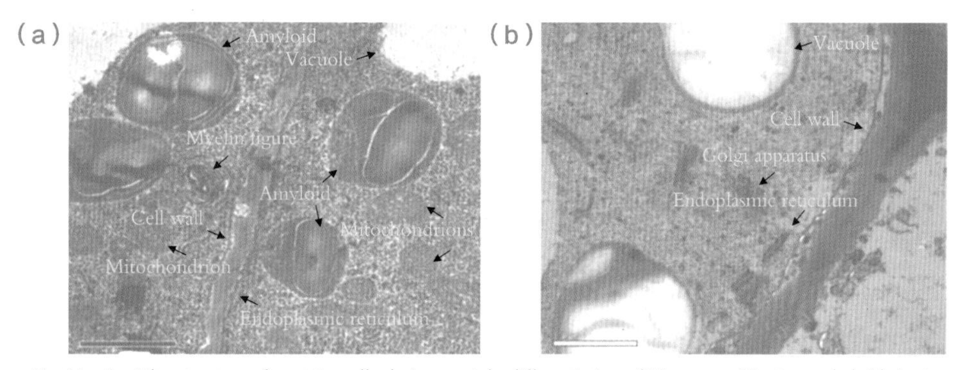

Fig. 11 - 8 Ultrastructure of root tip cells during panicle differentiation of Liangyoupeijiu (super hybrid rice) (a) and Xudao 2 (conventional rice) (b) (Yang et al., 2012)

2. Characteristics of root distribution

Rice roots are shallowly distributed in the soil, mainly in the tillage layer. In general, the fine upper roots are distributed in the upper layer of the soil and their extension direction is lateral or oblique, while the thicker lower roots have different extension directions and distribution areas depending on the location of the internodal units. Studies show that the differences in the lateral distribution of the root system between super hybrid rice (Xieyou 9308), common hybrid rice (Shanyou 63) and conventional rice variety (Tesan'ai 2) were relatively small, and all of them show a tendency of getting away from the plant center as the soil depth increases, but the longitudinal distribution of its root system is more different from that of common hybrid rice and conventional rice, showing that the root system of super hybrid rice gets deeper than that of common hybrid rice and conventional rice (the average depth increases by about 30%), and the proportion of its deep soil root distribution is also larger than that of common hybrid rice and conventional rice (Table 11-7). In addition, it is also pointed out that the distribution of the root system of super hybrid rice in the soil varies from variety to variety. The proportion of deep roots tends to increase with the gradual increase in yield, and the trend is more obvious in large-panicle super hybrid rice. For example, the proportion of root dry weight and root volume of the root system in the soil layer below 10 cm at maturity of the super-high yielding hybrid rice seedling variety Chaoyouqian was about 8% higher than that of the super hybrid rice variety Y Liangyou 1. In addition, the root distribution of super hybrid rice is also influenced by cultivation practices such as tillage, planting methods and water management. No-tillage and direct seedling cultivation can lead to enrichment of super hybrid rice roots to the surface layer, while alternate wet and dry irrigation can increase the proportion of deep soil roots of super hybrid rice.

Table 11-7 Longitudinal distribution of root dry weight of different rice varieties

(Zhu Defeng et al., 2000) (%)

Depth of soil (cm)	Xieyou 9308	Shanyou 63	Tesan'ai 2
0-6	40	46	46
6-12	12	16	14
12-18	10	8	11
18-24	8	7	6
24-36	12	11	12
36-45	18	12	11

Ⅱ. Root Vigor of Super Hybrid Rice

1. Oxidizing force

Root oxidizing force is an important indicator of root vigor, which is usually measured by α-naphthylamine (α-NA) oxidizing force. Studies show that the root α-NA oxidation force of super hybrid rice varieties (Liangyoupeijiu, Ⅱ-You 084 and Yangliangyou 6) was significantly higher than that of semi-dwarf varieties (hybrid rice Shanyou 63, conventional rice Yangdao 6 and Yangdao 2), dwarf varieties (Taichung *indica*, Nanjing 11, Zhengzhu'ai), and early tall varieties (Huangguaxian, Yingtiaoxian, Nanjing 1) in the early and middle stages of growth (initial panicle differentiation and heading stage). However, the decrease in root α-NA oxidazing force was faster in the later stages of growth, which in turn led to significantly lower root α-NA oxidazing force at the grain filling stage than that of the semi-dwarf varieties (Fig. 11-9). However, it was also shown that the decline in root α-NA oxidazing force after heading of the super hybrid rice variety Xieyu 9308 was lower than that of the common hybrid rice variety Shanyou 63. It is also proved that the α-NA oxidizing force of super hybrid rice roots is affected by cultivation factors such as nitrogen fertilization management and planting methods. With field and real-time nitrogen fertilization management, the root α-NA oxidizing force of super hybrid rice is significantly stronger than rice treated with conventional nitrogen fertilization management done by ordinary farmers. it is also stronger under the conditions of ridge cultivation and terrace cultivation than under the condition of flat land.

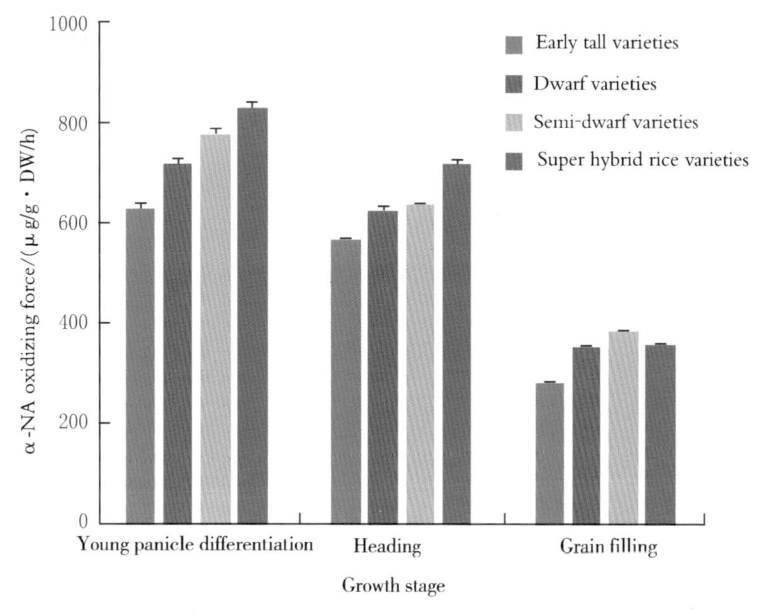

Fig. 11 − 9 Changes in root α-NA oxidizing force during the evolution of mid-season *indica* varieties (Zhang Hao et al., 2011)

2. Root exudation

Root exudation intensity is an important indicator of root vigor. Research shows that the root exudation intensity differs among varieties of super hybrid rice. The root exudation intensity of super hybrid rice variety Liangyoupeijiu was significantly higher than that of common hybrid rice variety Shanyou 63 and conventional rice variety Yangdao 6 in the early stage of grain filling, but the root exudation intensity of Liangyoupeijiu's root system decreased faster in the late stage of grain filling. The root exudation intensity of super hybrid rice variety Xieyou 9308 was similar to that of common hybrid rice variety Xieyou 63 at heading, but after heading, the root exudation intensity of Xieyou 9308 was significantly higher than that of Xieyou 63, and the root exudation intensity of Xieyou 9308 decreased less rapidly than that of Xieyou 63 as growth progressed. In addition, as the selection and breeding of super hybrid rice varieties progressed, the root exudation intensity of super hybrid rice varieties also changed. For example, the intensity of root exudation after full heading of the super hybrid rice pilot variety Chaoyouqian was significantly higher than that of the super hybrid rice variety Y Liangyou 1, with an average increase of more than 65% (Table 11 − 8). It is also proved that the root exudation intensity of super hybrid rice varieties is affected by cultivation factors such as water management and planting methods. Dry farming and direct seeding can reduce the root exudation intensity of super hybrid rice varieties, while wetting irrigation increases the intensity.

Root exudate composition is also an important indicator of root vigor. Sugar is involved in carbon metabolism, and a higher sugar content facilitates root respiratory metabolism, providing energy for root physiological activities and promoting root development. Amino acids are both important substrates and products of nitrogen metabolism, and the amino acid content can reflect the strength of nitrogen metabolism. A study showed that the changes of sugar and amino acid contents in the post-flowering root exudation of the super hybrid rice variety Liangyoupeijiu were basically the same as those of the common hybrid rice variety Shanyou 63 and the conventional rice variety Yangdao 6, all of which showed a single-peak curve. However, the peak of Liangyoupeijiu appeared in the second week after flowering, while the peaks of Shanyou 63 and Yangdao 6 appeared in the first week after flowering. In addition, the study also showed that cultivation factors such as planting methods can also affect the root exudation composition of super hybrid rice varieties. Direct seeding leads to a decrease in soluble sugars and amino acids in the root exudation of super hybrid rice.

Table 11 - 8 Single-stem root exudation intensity of super hybrid rice varieties (mg/h · stem) (Wei Zhongwei et al., 2016)

Year	Variety	Full heading stage	18 days after full heading	35 days after full heading	Average
2014	Chaoyouqian	406	263	195	288
	Y Liangyou 1	267	156	93	172
2015	Chaoyouqian	414	251	189	285
	Y Liangyou 1	273	150	91	171

3. Root enzyme activity

Root enzyme activity isclosely related to root functions. A study showed that the trend of post-flowering root nitrogen metabolism enzymes (glutamine synthetase, glutamate aminotransferase, glutamic acid aminotransferase and glutamate dehydrogenase) activity of super hybrid rice Liangjiupeiyou differed from that of the common hybrid rice variety Shanyou 63 and that of the conventional rice variety Yangdao 6. Generally, the earliest peak of root nitrogen metabolism was observed in Yangdao 6, followed by Liangyoupeijiu (about one week after flowering), and then Shanyou 63 (about two weeks after flowering). In addition, the superoxide dismutase (SOD) activity of the root system of Liangjiupeiyou differed from that of Shanyou 63 and Yangdao 6 in the late stage. Specifically, the SOD activity of the root system of Liangjiupeiyou was between that of Yangdao 6 and that of Shanyou 63 within two weeks after heading, while more than two weeks after heading, the SOD activity of the root system of Liangjiupeiyou was significantly lower than that of Yangdao 6 and that of Shanyou 63, which led to a higher malondialdehyde content in the root system. It is also proved that the root enzyme activity of super hybrid rice is affected by cultivation factors such as planting methods. The activities of antioxidant enzymes (catalase, SOD, and peroxidase) in ridge-planted roots are significantly higher than those of flat land planting (Table 11 - 9).

Table 11 - 9 Effects of planting methods on antioxidant enzyme activities in the roots of super hybrid rice Zhuliangyou 2 (U/mg protein) (Yao, 2015)

Growth stage	Planting method	Catalase	SOD	Peroxidase
Tillering	Flat planting	11. 8	80. 2	86. 9
	Ridge planting	29. 4	92. 4	142. 6
Heading	Flat planting	19. 4	91. 4	96. 7
	Ridge planting	42. 7	110. 6	165. 8
Maturity	Flatplanting	20. 9	86. 7	104. 3
	Ridge planting	40. 8	98. 8	174. 7

4. Root hormone content

The root system is the main organ that synthesizes cytokinin and abscisic acid (ABA) in a rice plant. Research shows that super hybrid rice varieties (Liangyoupeijiu and Ⅱ - You 084) have higher contents of zeatin (Z) and zeatin nucleoside (ZR) in their roots than conventional hybrid rice (Shanyou 63) and conventional rice (Yangdao 6) before heading, but the contents get lower at the middle and late grain filling stages. In contrast, Liangyoupeijiu and Ⅱ - You 084 have lower ABA content than Shanyou 63 before heading, but higher contents at the middle and late grain filling stages. Research also suggests that Xieyou 9308 (super hybrid rice) and Shanyou 63 (conventional hybrid rice) show declining Z+ZR content yet increasing ABA content in their roots after heading; Xieyou 9308 shows a lower rate of decline in its Z+ZR content and increase of ABA content than Shanyou 63 (Fig. 11 - 10). It is also proved that the root hormone content of super hybrid rice is affected by cultivation factors such as water management. Alternate dry and wet irrigation can regulate the root hormone content at the seed setting stage.

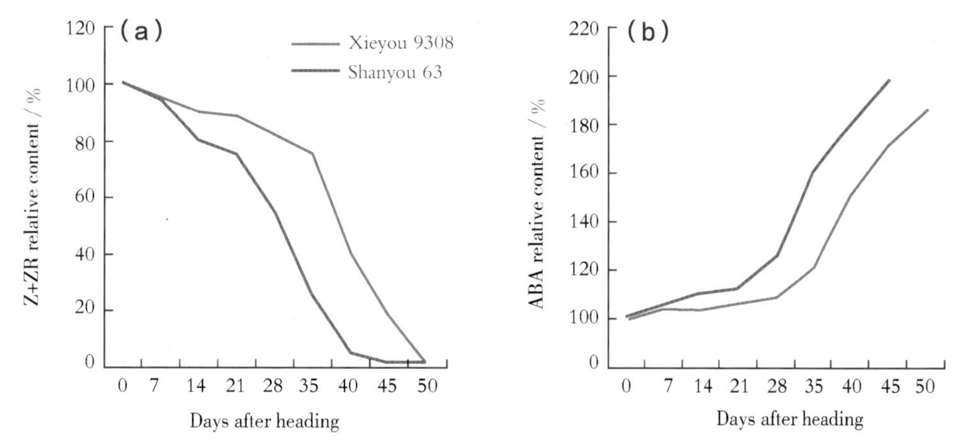

Fig. 11 - 10 Changes of the relative content of Zeatin (Z) + Zeatin Nucleoside (ZR) (a) and Abscisic Acid (ABA) (b) in the roots of Xieyou 9308 and Shanyou 63 after heading (Shu-Qing et al., 2004)

That is to say, the ABA content increases in soils in the dry periods, while Z+ZR can be synthesized when the soils are wet.

Ⅲ. Relationship between the Root System and the Aerial Parts of Super Hybrid Rice

1. Relationship between root structure and aerial parts

(1) Relationship between root morphology and aerial parts

Root morphology not only determines a crop's ability to anchor the plant, but is also closely related to the crop's ability to absorb nutrients and water, which in turn can have an impact on the growth and development of the aerial parts the plant. Most studies show that root morphological indicators such as root number, root length, root diameter, root surface area, root volume and root dry weight of super hybrid rice are generally significantly and positively correlated with yield, but the closeness of such relationship varies from variety to variety. Some studies point out that adventitious root number is most closely related to yield, so it can be used as a selection criterion for high-yielding genetic improvement of super hybrid rice, but some studies also suggest that both adventitious root number and meristematic root surface area play an important role in yield so both should be used as selection criteria for high-yielding rice breeding. In addition, it was also found that the well-developed fine roots of super hybrid rice variety Y Liangyou 087 in the late stage of growth (grain filling stage) were conducive to expanding the contact area and absorption range of the root system and improving the nutrient absorption capacity, which in turn was conducive to maintaining a high chlorophyll content and photosynthetic rate of the leaves, promoting the production of above-ground dry matter and obtaining high yields. A study also showed that the root dry weight of Y Liangyou 087 was relatively small, which could reduce the transport of above-ground dry matter to the root system and thus facilitate the formation of yield. In this regard, it has been shown that the root system is not only an organ for nutrient and water uptake, but also consumes photosynthetic products accumulated above ground to build and maintain growth, and the energy consumed by the root system is twice as much as the dry weight above ground to produce a unit of dry weight. Based on this understanding, the idea of "redundant root growth" has also been proposed, i. e., excessive root growth can cause inefficient consumption and thus have a negative impact on yield.

(2) Relationship between root tip cell ultrastructure and aerial parts

The ultrastructure of root tip cells is closely related to root growth, physiological activity and metabolism, which in turn affects the growth of the aerial parts and yield formation. A study showed that the number of spikelets per panicle was higher in the super hybrid rice variety Liangyoupeijiu (>200) than in the conventional rice variety Xudao 2 (<130), which was associated with more mitochondria and amyloid in its root tip cells. In addition, it was shown that the number of mitochondria and Golgi apparatus in tiller apical cells was significantly or highly significantly positively correlated with seedling dry weight and tiller number; the number of mitochondria, Golgi apparatus and ribosomes in tiller apical cells at seeding stage was highly significantly positively correlated with seeding rate and grain weight of vulnerable grains cells.

2. Relationship between root distribution and aerial parts

The distribution of roots in the soil is related to the use of soil and water resources by the plant, which in turn affects growth and development of the aerial parts and yield formation. Research shows that the contribution of the upper root layer (0 – 5 cm) to yield is 65%, that of the lower root layer (5 – 20 cm) to yield is 35%, and the root system below 20 cm in the soil is not related to yield. It was also suggested that two parameters, root length density and dry weight density, could be used to establish a mathematical model of root system and yield in the upper root layer (0 – 10 cm) at the full heading stage. However, some studies also showed that there was no significant correlation between the quality of the upper roots (0 – 10 cm) and yield in intersubspecific hybrids, while there was a significant positive correlation between the quality of the lower roots (below 10 cm) and yield with a correlation coefficient of more than 0.9. In addition, the root cutting test also showed that cutting the root system at 15 cm below the soil surface during the panicle differentiation stage resulted in shorter panicles and reduced number of grains per panicle, and cutting the root system at 30 cm below the soil surface resulted in reduced total number of grains per panicle and reduced seed setting; cutting the root system at 15 cm and 30 cm below the soil surface during the flowering stage resulted in reduced seed setting, and the effect of root cutting on large-panicle hybrid rice varieties was greater than that on small-panicle hybrid rice varieties. Thus, it is believed that deep root system plays an important role in yield formation of large-panicle hybrid rice.

3. Relationship between root vigor and parts aboveground

Root vigor is related not only to crop nutrient and water uptake capacity, but also to the synthesis of chemical substances such as hormones, amino acids and organic acids, which can play a regulatory role in the growth and development of aerial parts of the plants and are also known as chemical signals of the rice root system. Research suggests that Liangyoupeijiu (super hybrid rice) has a stronger storage capacity than Shanyou 63 (conventional hybrid rice) and Yangdao 6 (conventional rice), which is related to Liangyoupeijiu's strong root α-NA oxidizing force, high exudation intensity, high Z+ZR content and low ABA content in the early and middle stages of growth. However, Liangyoupeijiu has a lower seed setting rate than Shanyou 63 and Yangdao 6, due to the rapid decrease in root α-NA oxidizing force, high exudation intensity and Z+ZR content and the rapid increase in ABA content in the middle and late grain filling stages. However, some research indicates that Xieyou 9308 (super hybrid rice) has a stronger photosynthetic capacity than Shanyou 63 in the late growth stage, which is related to the slow decrease in Z+ZR content and the slow increase in ABA content. It is also indicated that root chemical signals are closely related to rice quality (Fig. 11 – 11). Root Z+ZR, ABA and ACC can affect the processing quality, appearance and cooking quality of rice by regulating endosperm development and the activity of key enzymes for starch synthesis (sucrose synthase, adenosine diphosphate glucose pyrophosphorylase, starch synthase, starch branching enzymes, etc.). Root ACC can affect the processing quality, appearance and cooking quality of rice by regulating the starch structure; root polyamine and amino acid content can affect the nutritional quality of rice by regulating protein synthesis; organic acids secreted by the root can affect the taste and hygiene quality of rice by regulating enzyme activity and heavy metal absorption.

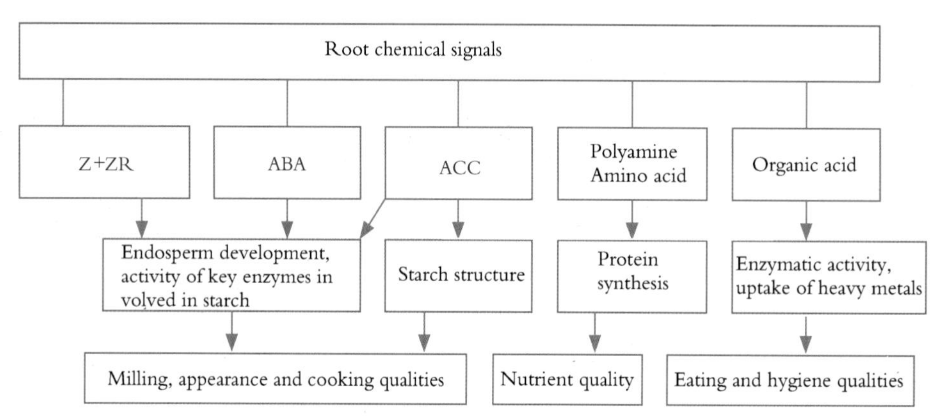

Fig. 11 – 11 Mechanism of root chemical signals on rice quality (Yang Jianchang, 2011)

Section 4 "Source" and "Sink" of Super Hybrid Rice

Ⅰ. Formation and Physiological Traits of the "Source"

1. Morphological traits of leaves

(1) Number of leaves

The total number of leaves of super hybrid rice is related to the variety and the growth period (Table 11 – 10). Generally, there are 11.7 – 12.1 leaves for medium-maturing varieties, 12.7 leaves for late-maturing double-cropping early-season rice, 14.5 leaves for medium-maturing varieties, and 15.1 – 15.3 leaves for late-maturing double-cropping late-season varieties; 15.3 – 15.5 leaves for medium-maturing varieties and 15.7 – 16.2 leaves for late-maturing varieties of single-cropping rice (mid-season rice and single-cropping late-season rice).

Table 11 – 10 Number of leaves on the main stem of super hybrid rice varieties (Zou Yingbin et al., 2011)

Type	Variety	Number of leaves on the main stem
Double-cropping early-season rice	Zhuliangyou 819	11.7
	Luliangyou 996	12.7
	Liangyou 287	12.1
Double-cropping late-season rice	Fengyuanyou 299	15.1
	Ganxin 688	15.3
	Jinyou 299	14.5
	Tianyou-Huazhan	14.3
Mid-season rice	Y You 1	15.7
	Liangyoupeijiu	15.7

Continued

Type	Variety	Number of leaves on the main stem
Mid-season rice	Neiliangyou 6	15. 3
	Zhongzheyou 1	15. 9
Single-cropping late-season rice	Y You 1	16. 0
	Liangyoupeijiu	15. 9
	Neiliangyou 6	15. 5
	Zhongzheyou 1	16. 2

(2) Leaf size

Leaf size is closely related to the capacity of the canopy to intercept light energy. The length and area of the top three leaves of super hybrid rice varieties (Liangyoupeijiu, Zhunliangyou 527, Ⅱ - Youming 86 and Y Liangyou 1) tend to increase and the leaf width tends to decrease as the leaf position gets lower during the heading stage (Table 11 - 11). The lengths of the flag leaf, penultimate leaf and antepenult leaf of the varieties are 33. 5 - 40. 4 cm, 45. 3 - 53. 6 cm and 52. 9 - 58. 6 cm respectively; the widths are 1. 9 - 2. 1 cm, 1. 6 - 1. 8 cm and 1. 4 - 1. 6 cm respectively; and the leaf areas are 50. 5 - 66. 6 cm^2, 56. 5 - 74. 6 cm^2 and 59. 1 - 71. 3 cm^2 respectively. When these were compared with measures of the common hybrid rice variety Shanyou 63, no uniform pattern was found. In addition, it is also shown that the leaf length of super hybrid rice is more influenced by the amount of nitrogen fertilizer and less influenced by planting density. The leaf length increases as nitrogen fertilizer dosage increases, first fast, then slowly. More specifically, when the dosage goes from low nitrogen (90 kg N/ha) to medium nitrogen (135 kg N/ha), the leaf length increases by more than 10%, but when it goes further up to high nitrogen (180 kg N/ha and 225 kg N/ha), the leaf length only increases by about 3%.

Table 11 - 11 Leaf size of super hybrid rice varieties at full heading (Deng Huafeng, 2008)

Trait	Variety	Flag leaf	Penultimate leaf	Antepenult leaf
Leaf length (cm)	Liangyoupeijiu	40. 4	53. 6	58. 6
	Zhunliangyou 527	38. 7	51. 5	56. 8
	Ⅱ - Youming 86	39. 6	52. 7	55. 8
	Y Liangyou 1	33. 5	45. 3	52. 9
	Shanyou 63	38. 6	51. 9	56. 6
Leaf width (cm)	Liangyoupeijiu	2. 0	1. 7	1. 4
	Zhunliangyou 527	2. 0	1. 7	1. 5
	Ⅱ - Youming 86	2. 1	1. 8	1. 6
	Y Liangyou 1	1. 9	1. 6	1. 4
	Shanyou 63	2. 0	1. 7	1. 5

Continued

Trait	Variety	Flag leaf	Penultimate leaf	Antepenult leaf
Leaf area (cm^2)	Liangyoupeijiu	63. 9	69. 4	67. 5
	Zhunliangyou 527	60. 3	70. 6	70. 2
	Ⅱ - Youming 86	66. 6	74. 6	71. 3
	Y Liangyou 1	50. 5	56. 5	59. 1
	Shanyou 63	62. 2	70. 0	68. 5

(3) Blade base angle and pendant angle

Leaf angle is an important indicator of canopy light exposure. Generally, the smaller the leaf angle, the better the canopy is exposed to light. The base angles and pendant angles of the top three leaves of super hybrid rice varieties (Liangyoupeijiu, Zhunliangyou 527, Ⅱ - Youming 86 and Y Liangyou 1) increase as the leaf position goes down at the heading stage (Table 11 - 12). The base angles of the flag leaf, penultimate leaf and antepenult leaf of the varieties are 9. 5° - 12. 1°, 13. 3° - 19. 3° and 18. 0° - 25. 5° respectively; the pendant leaf angles are 0. 5°-2. 5°, 1. 4°-5. 5° and 2. 0°-8. 4° respectively, and they are all smaller than those of Shanyou 63 (common hybrid rice). In addition, research shows that an increase in nitrogen fertilization leads to an increase in the leaf pendant angle.

Table 11 - 12 Blade base and pendant angles of super hybrid rice varieties at full heading
(Deng Huafeng, 2008)

Trait	Variety	Flag leaf	Penultimate leaf	Antepenult leaf
Base leaf angle (°)	Liangyoupeijiu	9. 5	13. 3	18. 7
	Zhunliangyou 527	12. 1	19. 3	25. 5
	Ⅱ - Youming 86	11. 3	18. 4	24. 6
	Y Liangyou 1	11. 4	14. 0	18. 0
	Shanyou 63	15. 6	22. 3	27. 6
Pendant angle (°)	Liangyoupeijiu	1. 1	2. 2	3. 9
	Zhunliangyou 527	1. 8	5. 2	5. 3
	Ⅱ - Youming 86	2. 5	5. 5	8. 4
	Y Liangyou 1	0. 5	1. 4	2. 0
	Shanyou 63	3. 1	6. 5	9. 8

(4) Leaf curl index

The degree of leaf curl is closely related to the maintenance of leaf posture. Moderate leaf curl helps the leaves stay upright. The leaf curl index of the top three leaves of the super hybrid rice varieties (Liangyoupeijiu, Zhunliangyou 527, Ⅱ - Youming 86 and Y Liangyou 1) show a decreasing trend with decreasing leaf position at the heading stage (Table 11 - 13). The indexes of the flag leaf, penultimate leaf, and antepenult leaf of all varieties are 53.2 - 60.0, 51.6 - 58.5 and 51.1 - 58.0 respectively, all higher than those of Shanyou 63.

Table 11 - 13 Leaf curl indexes of super hybrid rice varieties at full heading (Deng Huafeng, 2008)

Variety	Flag leaf	Penultimateleaf	Antepenult leaf
Liangyoupeijiu	57.7	55.1	54.9
Zhunliangyou 527	57.7	53.0	52.7
Ⅱ - Youming 86	53.2	51.6	51.1
Y Liangyou 1	60.0	58.5	58.0
Shanyou 63	52.2	51.4	51.0

(5) Specific leaf weight

Specific leaf weight is usually positively correlated with leaf nitrogen and chlorophyll content and leaf photosynthesis. Therefore, specific leaf weight is often used to measure the photosynthetic performance of crop leaves. A study showed that the specific leaf weight of the top three leaves of super hybrid rice model combination Peiai 64S/E32 and a pilot hybrid Peiai 64S/Changlizhua (at the full heading stage) was significantly higher than that of Shanyou 63 (conventional hybrid rice), with an average increase of about 10% (Table 11 - 14). It is also proved that super hybrid rice varieties Liangyoupeijiu and Y Liangyou 087 have significantly higher specific leaf weight than conventional hybrid rice varieties Shanyou 63 and Teyou 838.

Table 11 - 14 Specific leaf weight of the top three leaves of super hybrid rice combinations at full heading (mg/cm^2) (Deng Qiyun et al., 2006)

Combination/variety	Flag leaf	Penultimate leaf	Antepenult leaf
Peiai 64S/E32	4.51	4.32	4.31
Peiai 64S/Changlizhua	4.82	4.67	4.28
Shanyou 63	4.14	3.97	4.02

2. Leaf growth rate, leaf area index and photosynthetic potential

(1) Leaf growth rate

The growth rate of the leaves can be expressed in terms of leaf age growth rate, which is mainly related to temperature conditions. Generally, the higher the temperature, the faster the growth rate of the leaves. The growth rate of leaves of super hybrid rice varieties is not significantly different from that of common rice varieties, but there are some differences in the growth rate of leaf age of different super hybrid rice varieties even when they are sown on the same date at the same location due to the difference in the total number of main stem leaves and the different responses of different varieties to temperature. In addition, due to year-to-year temperature variations, leaf growth rates can vary from year to year even when the same variety is sown on the same day of the year.

(2) Leaf area index

The leaf area index is an important indicator of the size ("source" intensity) of a crop population, and its dynamic changes are closely related to crop dry matter production. Generally, the leaf area index of high-yielding crop populations is characterized by rapid increase in the early stage of growth, long peak time and slow decline in the later stage of growth. A study showed that the leaf area index of the super hybrid rice variety Xieyou 9308 was smaller than that of the super hybrid rice variety Liangyoupeijiu and the common hybrid rice variety Shanyou 63 in the early stage of growth (from transplanting to 20 d afterwards), but its leaf area index after stem elongation was consistently larger than that of Liangyoupeijiu and Shanyou 63. However, from a different perspective, a high leaf area index is not necessarily beneficial, because when it increases to a certain level, there will be canopy closure, leading to insufficient light for the lower leaves. A study showed that the leaf area index of super hybrid rice varieties (Liangyoupeijiu, Y Liangyou 1 and Liangyou 293) was smaller than that of common hybrid rice varieties (Shanyou 63 and Ⅱ-You 838) at the flowering stage, although it was larger than that of conventional rice varieties (Yangdao 6 and Huanghuazhan) (Table 11-15). Studies also showed that the leaf area index of super hybrid rice varieties is affected by nitrogen fertilizer dosage and planting density and the responses of different types of varieties are not the same. For varieties with strong tillering ability and early and fast growth, the regulatory effect of nitrogen fertilizer dosage is greater than that of planting density; for varieties with weak tillering ability, both nitrogen fertilizer dosage and planting density have a greater regulatory effect. In addition, nitrogen fertilization and planting density have an accumulative effect on the leaf area index of super hybrid rice varieties. The leaf area index often reaches the maximum when there is excessive nitrogen and a high planting density, so reducing nitrogen or planting density can reduce the leaf area index. Therefore, a proper leaf area index can be achieved through appropriate use of water and fertilization.

Table 11-15 Leaf area index of super hybrid rice varieties at the flowering stage (Zhang et al., 2009)

Variety	Liuyang, Hunan		Guidong, Hunan	
	2007	2008	2007	2008
Liangyoupeijiu	6.09	6.99	7.21	5.86
Liangyou 293	5.68	—	7.23	—
Y Liangyou 1	—	5.34	—	5.32
Shanyou 63	6.95	6.99	8.55	6.72
Ⅱ-You 838	7.14	6.80	8.51	6.92
Yangdao 6	5.48	5.70	7.21	5.73
Huanghuazhan	4.99	5.21	7.17	5.92

(3) Photosynthetic potential

Photosynthetic potential refers to the daily accumulation of green leaf area of a crop at a certain stage of growth or in the whole growth period, and is closely related to the accumulation of dry matter. A lot of research shows that the photosynthetic potential of Liangyoupeijiu (super hybrid rice) is significantly higher than that of Shanyou 63 (conventional hybrid rice), especially before heading. In addition, research also shows that the photosynthetic potential of Guiliangyou 2 (super hybrid rice) is higher than that of Yuxiangyouzhan (conventional rice) in both the early and late seasons and at all stages of growth (Fig. 11-12).

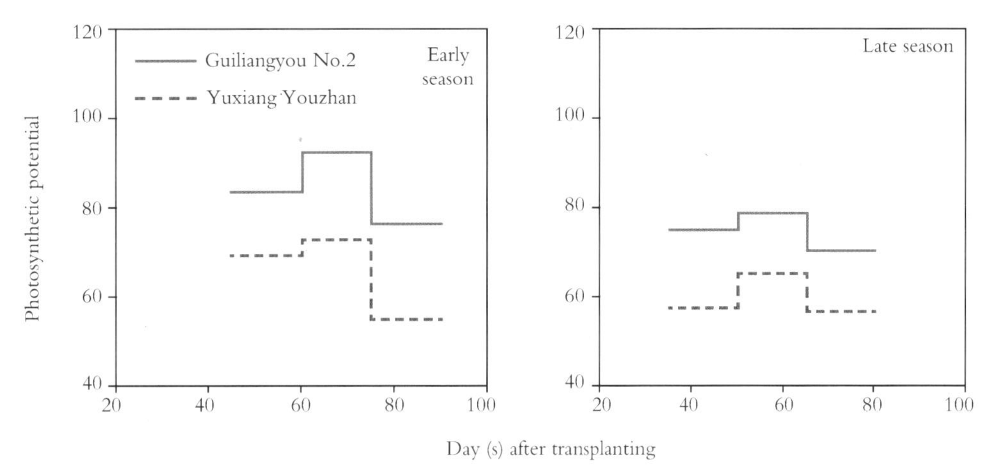

Fig. 11-12 Photosynthetic potential of Guiliangyou 2 and Yuxiangyouzhan (Huang et al., 2015)

3. Physiological traits of the leaves

Leaf physiology covers photosynthetic physiology, senescence physiology and stress resistance physiology. A study showed that compared with the common hybrid rice variety Shanyou 63, the super hybrid rice variety Ⅱ-Youhang 2 showed advantages in physiological activities such as photosynthetic metabolism, stress resistance and response, gene transcription and expression, cell growth, and energy metabolism during the grain filling stage (Table 11-16). The super hybrid rice variety Xieyou 9308 also had a significant advantage over Shanyou 63 in terms of the photosynthetic rate and zeatin (Z) + zeatin nucleoside (ZR) content, while the abscisic acid (ABA) content was lower in Xieyou 9308 after heading. However, some studies showed that although the leaf photosynthetic rate and Z+ZR content of the super hybrid rice variety Liangyoupeijiu were higher and the ABA content was lower than those of Shanyou 63 and conventional rice variety Yangdao 6 in the early and middle stages of growth, the leaf photosynthetic rate and Z+ZR content were lower and the ABA content was higher than those of Shanyou 63 and Yangdao 6 in the middle and late stages of grain filling. In addition, the catalase activity and peroxidase activity of the leaves of Liangyoupeijiu were lower than those of Shanyou 63 in the middle and late stages of grain filling, while the malondialdehyde content was higher than that of Shanyou 63. It can be observed that the physiological features of super hybrid rice differ among varieties. Research also indicates that the leaf physiology of super hybrid rice is affected by cultivation factors such as farming and planting. Compared with plowing, no tillage leads to a decrease in leaf chlorophyll content, net photosynthetic rate and soluble sugar content in the early growth stage of super hybrid rice. Compared with transplanting, direct sowing leads to a decrease in leaf nitrogen content, glutamine synthase activity, soluble protein content and net photosynthetic rate in super hybrid rice.

Table 11 – 16 Shanyou 63 and Ⅱ – Youhang 2 – expressed proteins in leaves during grain filling (Huang Jinwen et al., 2011)

Protein function	Protein name
Photosynthesis	Ribulose diphosphate carboxylase large subunit precursor
	Ribulose diphosphate carboxylase large subunit
	Fe-NADP reductase
Stress resistance	Elongation factor Tu
	Shikimate kinase 2
Protein metabolism	Glycosyltransferase
	Putative helicase SK12W
	Putative Ubiquitin C-terminal Hydrolase 7
	Eukaryotic peptide chain releasing factor subunit 1
Gene transcription regulator cell growth	CCHC zinc finger domain protein
	Retrotransposon
	Microtubule binding protein
	Adenosine kinase protein
Energy metabolism	Fructose 1, 6-bisphosphate aldolase
	ATP synthase CF1β subunit

Ⅱ. Formation and Physiological Characteristics of "Sink"

1. "Sink" size

The size of the rice "sink" is usually indicated by the number of spikelets per unit area. Research suggests that most super hybrid rice varieties (Liangyoupeijiu, Y Liangyou 1, Zhunliangyou 527, Ⅱ – Youhang and Zhongzheyou 1) have an advantage in the number of spikelets per unit area compared with the conventional hybrid rice variety Shanyou 63 (Table 11 – 17). However, research also indicates that the "sink" size should be calculated with the size of a single grain (grain weight) – the number of spikelets per unit area multiplied by grain weight. For example, the number of spikelets per unit area of D-You 527 and Neiliangyou 6 (super hybrid rice) is fewer than that of Shanyou 63, but their grain weights are larger, with around 30g in 1,000-grain weight. If grain weight is taken into account when calculating "sink" size, D-You 527 and Neiliangyou 6 have larger "sink" than Shanyou 63. Therefore, research also indicates that breeding large-grain rice varieties is also an effective way to achieve high yields.

Table 11 - 17 Numbers of spikelets per square meter of super hybrid rice varieties ($\times 10^3$) (Huang et al., 2011)

Variety	2007			2008			2009		
	Changsha	Guidong	Nanxian	Changsha	Guidong	Nanxian	Changsha	Guidong	Nanxian
Liangyoupeijiu	44.6	50.5	36.3	42.0	46.5	42.6	45.5	55.8	37.3
Y Liangyou 1	31.2	46.7	33.2	39.5	46.2	38.0	45.7	47.7	41.9
Zhunliangyou 527	35.4	39.2	27.4	29.1	37.6	33.8	32.7	39.8	38.2
D - You 527	28.5	44.6	28.9	32.7	42.3	36.3	38.3	55.0	43.9
Ⅱ - Youhang 1	36.6	44.4	30.0	36.6	41.1	39.0	46.1	52.8	43.6
Ⅱ - You 084	34.1	47.7	32.0	32.7	39.1	35.7	43.5	48.5	43.3
Neiliangyou 6	31.2	38.5	28.8	32.2	45.2	34.3	37.4	46.4	36.5
Zhongzheyou 1	29.9	40.5	38.6	37.6	46.1	38.4	42.6	45.6	39.6
Shanyou 63	31.1	38.3	28.6	33.2	41.3	36.2	37.6	45.4	47.1

2. Tillering and panicle-bearing traits

The number of panicles is the basis for the number of spikelets per unit area. Research suggests that super hybrid rice (Liangyoupeijiu, Y Liangyou 1, Zhunliangyou 527, D-You 527, Ⅱ - Youhang 1 and Neiliangyou 6) have no significant advantage in the number of panicles per unit area compared with conventional hybrid rice (Shanyou 63) (Table 11 - 18). However, research also shows that the tillering traits of super hybrid rice are different from those of conventional rice. Compared with conventional rice (Shengtai 1, Huanghuazhan and Yuxiangyouzhan), super hybrid rice (Liangyoupeijiu, Zhunliangyou 527 and Y Liangyou 1) have a higher maximum number of tillers but a lower panicle-bearing rate (Fig. 11 - 13).

Table 11 - 18 Numbers of panicles per square meter of super hybrid rice varieties (Huang et al., 2011)

Variety	2007			2008			2009		
	Changsha	Guidong	Nanxian	Changsha	Guidong	Nanxian	Changsha	Guidong	Nanxian
Liangyoupeijiu	241	328	210	226	246	224	238	279	247
Y Liangyou 1	201	294	221	231	246	221	267	282	306
Zhunliangyou 527	206	272	200	205	254	243	244	265	281
D - You 527	189	279	219	208	218	230	244	271	258
Ⅱ - Youhang 1	207	264	192	201	231	228	233	254	233

Continued

Variety	2007			2008			2009		
	Changsha	Guidong	Nanxian	Changsha	Guidong	Nanxian	Changsha	Guidong	Nanxian
Ⅱ - You 084	212	298	204	203	230	226	239	258	243
Neiliangyou 6	220	287	200	201	232	220	241	270	231
Zhongzheyou 1	193	302	231	235	281	243	255	287	293
Shanyou63	209	290	229	206	252	235	256	293	291

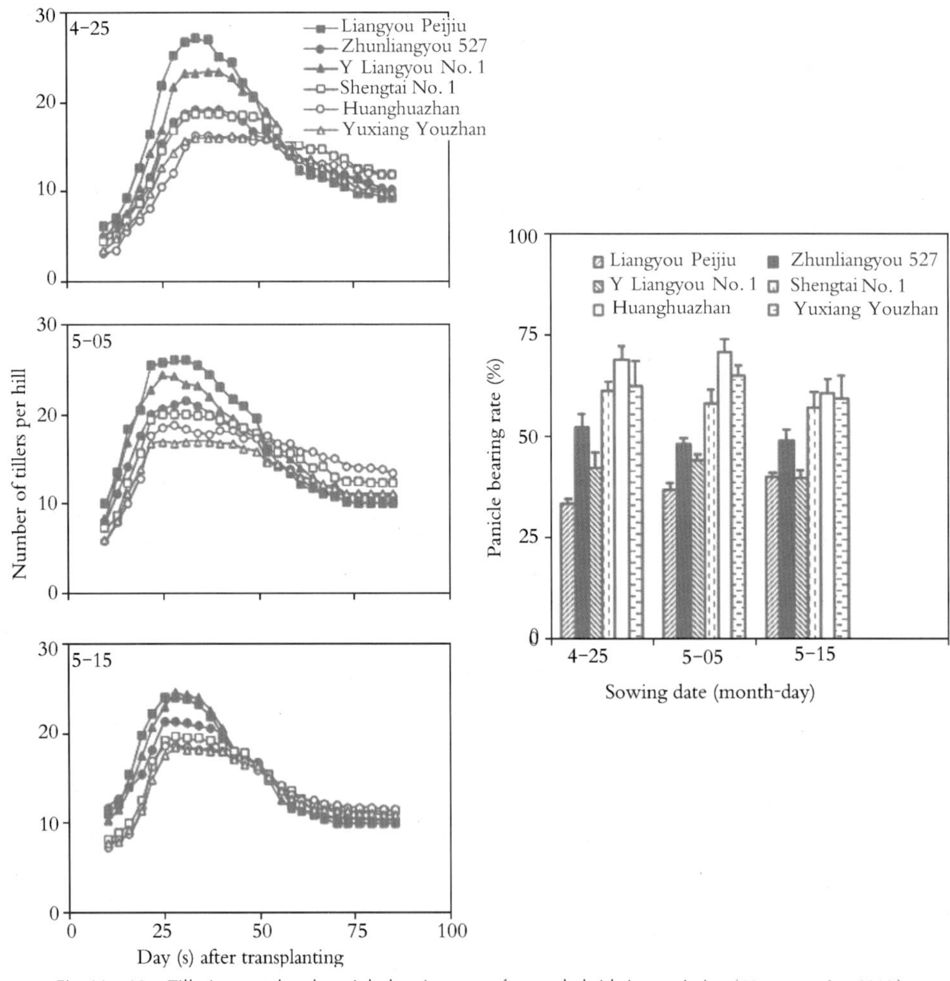

Fig. 11 - 13 Tillering trend and panicle-bearing rate of super hybrid rice varieties (Huang et al., 2012)

3. Panicle traits

(1) Number of grains per panicle

The number of grains perpanicle plays an important role in determining the number of spikelets per unit area. Research shows that super hybrid rice (Liangyoupeijiu, Y Liangyou 1, Zhunliangyou 527, D-You 527, Ⅱ-Youhang 1, Neiliangyou 6 and Zhongzheyou 1) have more grains per panicle than conventional hybrid rice (Shanyou 63) (Table 11-19). The advantage of having large panicles of super hybrid rice is related to the large number of secondary branches (Table 11-20).

Table 11-19 Numbers of grains per panicle of super hybrid rice varieties (Huang et al., 2011)

Variety	2007			2008			2009		
	Changsha	Guidong	Nanxian	Changsha	Guidong	Nanxian	Changsha	Guidong	Nanxian
Liangyoupeijiu	185	154	173	186	189	190	191	200	151
Y Liangyou 1	155	159	150	171	188	172	171	169	137
Zhunliangyou 527	172	144	137	142	148	139	134	150	136
D-You 527	151	160	132	157	194	158	157	203	170
Ⅱ-Youhang 1	177	168	156	182	178	171	198	208	187
Ⅱ-You 084	161	160	157	161	170	158	182	188	178
Neiliangyou 6	142	134	144	160	195	156	155	172	158
Zhongzheyou 1	155	134	167	160	164	158	167	159	135
Shanyou 63	149	132	125	161	164	154	147	155	162

Table 11-20 Panicle traits of super hybrid rice varieties (Huang et al., 2011)

Sowing date	Variety	Number of primary branches per panicle	Number of grains on primary branches	Number of secondary branches per panicle	Number of grains on secondary branches
April 25	Liangyoupeijiu	12.0	5.40	53.4	3.95
	Y Liangyou 1	14.3	5.88	53.5	3.40
	Huanghuazhan	11.9	5.47	34.1	3.43
	Yuxiangyouzhan	11.1	5.70	45.0	3.60
May 5	Liangyoupeijiu	12.4	5.48	50.0	3.68
	Y Liangyou 1	14.7	5.90	45.3	3.38
	Huanghuazhan	11.6	5.63	32.1	3.28
	Yuxiangyouzhan	11.0	5.62	42.6	3.55
May 15	Liangyoupeijiu	12.1	5.38	54.5	3.90
	Y Liangyou 1	13.7	5.92	55.6	3.67
	Huanghuazhan	11.2	6.03	32.9	3.27
	Yuxiangyouzhan	11.3	5.87	45.3	3.62

(2) Panicle branch structure

Rice panicle branch structure is closely related to the storage capacity of the "sink". Research shows that there are significant differences in the structure of the upper primary branches and the lower secondary branches of super hybrid rice. For example, Liangyoupeijiu (super hybrid rice) has one large vascular bundle on the upper primary branch and on the lower secondary branch respectively. However, the duct area and phloem area of the large vascular bundle of the upper primary branch are larger than those of the lower secondary branch. The number of vascular bundles and the area of the phloem of the small vascular bundle on the upper primary branch are smaller than those on the lower secondary branch, and the small vascular bundle duct area on the upper primary branch is larger than that on the lower secondary branch (Table 11 - 21). In addition, the total area of the vascular bundles and the area of the phloem of the upper primary branch are larger than those of the lower secondary branch of Liangyoupeijiu. It can be observed that Liangyoupeijiu has an advantage in the number and area of vascular bundles of the upper primary branch compared with the lower secondary branch. Research also suggests that Liangyoupeijiu has clear vascular bundle ducts, sieve tubes and companion cells of the upper primary branch, and large area of individual duct, sieve tube and companion cell. However, the vascular bundle ducts, sieve tubes and companion cells of the lower secondary branch are unclear, and catheters, screens, and companion cells cover small areas (Fig. 11 - 14). It can be observed that the vascular bundles, ducts and sieve tubes on the upper primary branch of Liangyoupeijiu are better developed and differentiated than those on the lower secondary branch.

Table 11 - 21 Traits of vascular bundles of panicle branches of Liangyoupeijiu (Zou Yingbin et al., 2011)

Branch	Large vascular bundle			Small vascular bundle		
	Number	Duct area (μm^2)	Phloem area (μm^2)	Number	Duct area (μm^2)	Phloem area (μm^2)
Upper primary branch	1	1,120	4,284	3.0	1,064	840
Lower secondary branch	1	840	3,360	3.6	748	1,316

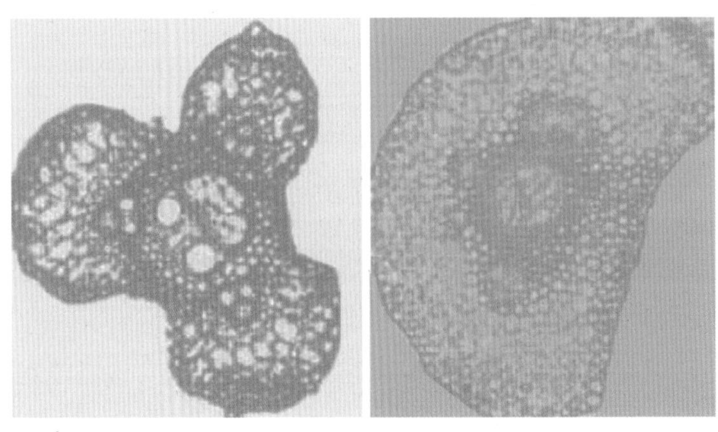

Upper primary branch　　Lower secondary branch

Fig. 11 - 14 Structure of vascular bundles of panicle branches of Liangyoupeijiu (Zou Yingbin and Xia Shengping, 2011)

(3) Physiological traits of the grain

Seed physiological traits are closely related to the ability of seeds to accept substances. Research shows that compared with common rice, super hybrid rice shows great differences in the physiological activities of strong and weak grains. The physiological traits of weak grains are as follows. 1) Low proliferation rate of endosperm cells in the early stage of grain filling, low key enzyme activity in the sucrose-starch metabolic pathway, or low biochemical efficiency of converting sucrose into starch; 2) low total RNA content and mRNA content at the early stage of grain filling (Table 11 - 22); 3) low gene expression of some enzymes (e. g. cell wall invertase, vacuolar invertase, adenosine diphosphate glucose pyrophosphorylase, and starch synthetase) at the early stage of grain filling; 4) high percentage of inhibitory plant hormones (e. g. ABA and ethylene) and promotive plant hormones (e. g. auxin, cytokinin and gibberellin) in the grains; and 5) low concentration of spermidine and spermine and low ratio of spermidine and spermine to putrescine.

Table 11 - 22 Total RNA and mRNA content and total number of grains of super hybrid rice varieties (Zou Yingbin et al., 2011)

Variety	Grain type	Days after heading	Total RNA		mRNA	
			Content ($\times 10^{-2}$ μg/mg)	Total of single grain (mg)	Content ($\times 10^{-2}$ μg/mg)	Total of single grain (mg)
Liangyoupeijiu	Strong	0	48.9	1.36	0.43	0.0121
		5	45.6	2.68	0.40	0.0240
	Weak	0	28.5	0.11	0.35	0.0015
		5	36.6	0.19	0.38	0.0020
Peiliangyou 500	Strong	0	53.2	1.66	0.54	0.0141
		5	45.7	2.51	0.48	0.0272
	Weak	0	30.3	0.12	0.39	0.0015
		5	38.7	0.20	0.41	0.0021

Ⅲ. Relationship between "Source" and "Sink" of Super Hybrid Rice

1. Grain-leaf ratio

The grain-leaf ratio is an indicator often used to measure the relationship between the "source" and "sink" of rice. Research indicates that the grain-leaf ratio of super hybrid rice varieties (Liangyoupeijiu, Liangyou 293 and Y Liangyou 1) is significantly higher than that of conventional hybrid rice varieties (Shanyou 63 and Ⅱ-You 838), but it is not significantly different from that of conventional rice varieties (Yangdao 6 and Huanghuazhan) (Table 11 -23). Thus, compared with common hybrid rice, the improvement of the "sink" is greater than the improvement of the "source" in super hybrid rice. In addition, the study also showed that the grain-leaf ratio of super hybrid rice was affected by cultivation practices such as nitrogen fertilizer management. Increasing nitrogen application results in a decrease in the grain-leaf ratio of super hybrid rice.

Table 11 - 23 Leaf-grain ratio of super hybrid rice varieties at the flowering stage (Zhang et al., 2009)

Variety	Liuyang, Hunan		Guidong, Hunan	
	2007	2008	2007	2008
Liangyoupeijiu	0.82	0.71	0.71	0.85
Liangyou 293	0.84	—	0.71	—
Y Liangyou 1	—	0.83	—	0.96
Shanyou 63	0.54	0.52	0.51	0.63
Ⅱ - You 838	0.50	0.50	0.50	0.61
Yangdao 6	0.65	0.62	0.56	0.65
Huanghuazhan	1.00	0.82	0.67	0.86

2. Grain filling traits

Grain filling traits can reflect the relationship between the "source" and "sink" of rice to a certain extent. Generally, the phenomenon of two-phase filling is obvious in "source"-restricted varieties, but not in "sink"-restricted varieties. A study showed that the two-phase filling phenomenon existed to different degrees in super hybrid rice varieties (e.g. Zhunliangyou 527), common hybrid rice varieties (e.g. Shanyou 63) and conventional rice varieties (e.g. Yuxiangyouzhan and Shengtai 1) (Fig. 11 - 15), and the maximum filling rate of the weaker grains was delayed by more than 10 d compared with that of the stronger grains. This shows that further improvement of "source" is an important aspect to improve rice yield.

Ⅳ. Relationship between "Source-sink" Coordination and Yield of Super Hybrid Rice

The "source" and "sink" of super hybrid rice have been greatly improved compared with common hybrid rice and conventional rice. In terms of the "source", the morphological structure of super hybrid rice leaves has been greatly improved with the top three leaves usually straight, curly and thick, which in turn helps improve canopy solar energy utilization. In addition, super hybrid rice not only has a strong photosynthetic potential, but also has a greater advantage in its leaf physiological activities, such as protein expression, phytohormone content and enzyme activity, and the period of advantageous performance varies from variety to variety, with some varieties such as Liangyoupeijiu having a clear advantage in the early and middle stages of growth, while others such as Xieyou 9308 having a clear advantage in the later stages of growth. In terms of the "sink", most super hybrid rice varieties have the advantage of large panicles, and the formation of large panicles is mainly related to the increase in the number of secondary bran-

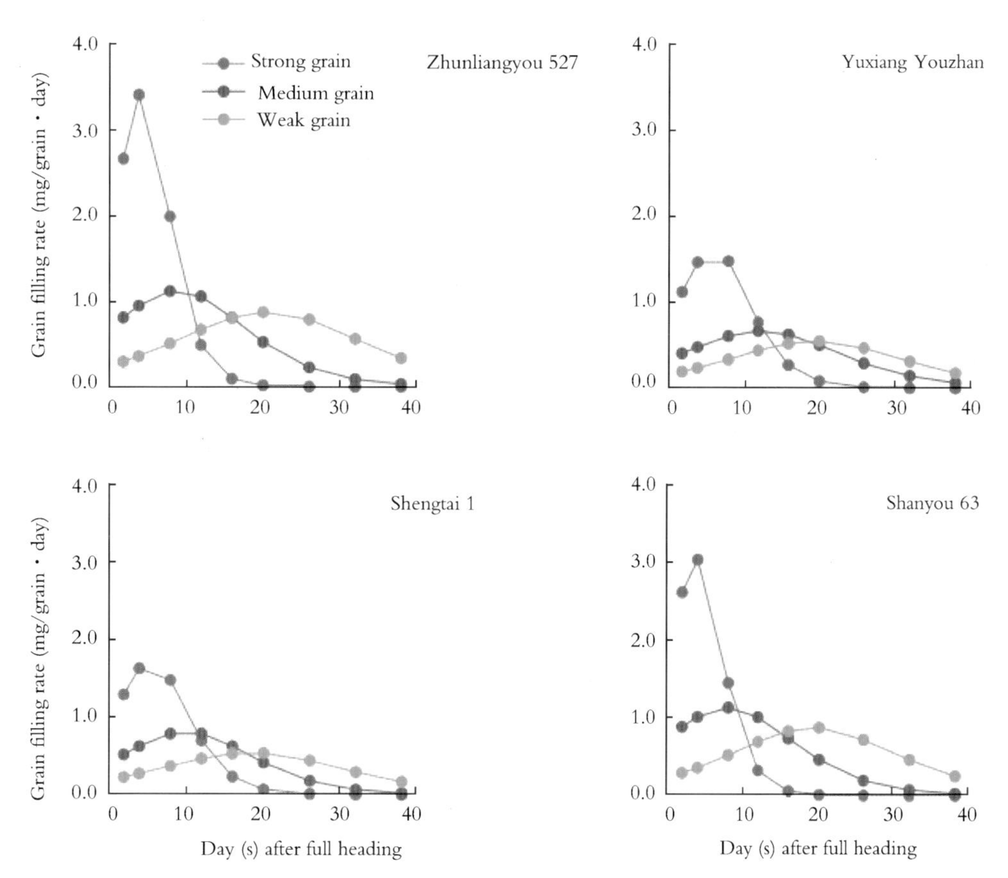

Fig. 11 - 15 Grain filling rate of strong, medium and weak grains of different rice varieties (Huang Min et al., 2009)

ches. However, some super hybrid rice varieties also show the advantage of large grains. Studies showed that the yield potential of super hybrid rice varieties was improved by about 12% compared with that of conventional and hybrid rice varieties due to the improvement of both the "source" and the "sink" (Table 11 - 24).

Table 11 - 24 Yield of super hybrid rice varieties (t/ha) (Zhang et al., 2009)

Year	Variety	Liuyang, Hunan		Guidong, Hunan	
		Moderate nitrogen	High nitrogen	Moderate nitrogen	High nitrogen
2007	Liangyoupeijiu	8.85	9.15	11.22	10.92
	Liangyou 293	8.55	9.01	10.96	10.96
	Shanyou 63	8.29	7.84	9.89	10.17
	Ⅱ - You 838	8.14	7.48	9.89	9.73
	Yangdao 6	8.42	8.34	9.93	10.06
	Huanghuazhan	8.13	8.45	10.16	9.62

Continued

Year	Variety	Liuyang, Hunan		Guidong, Hunan	
		Moderate nitrogen	High nitrogen	Moderate nitrogen	High nitrogen
2008	Liangyoupeijiu	9. 85	10. 15	11. 09	11. 08
	Y Liangyou 1	9. 78	9. 86	11. 48	11. 49
	Shanyou63	8. 17	8. 00	10. 24	9. 88
	Ⅱ - You 838	8. 52	8. 00	10. 45	10. 01
	Yangdao 6	8. 64	8. 82	9. 59	9. 77
	Huanghuazhan	8. 75	8. 75	10. 20	10. 42

Section 5 Dry Matter Accumulation, Transport and Distribution in Super Hybrid Rice

Ⅰ. Basic Concepts and Research Methodology

The accumulation, transport and distribution of dry matter are controlled by the "source", "sink" and "flow" characteristics of rice. "Source" refers to organs or tissues where plants produce or export assimilated products. The "source" of rice plants consists of green stems, sheaths, leaves and roots, among which functional leaves and leaf sheaths are the main part. Research on rice "source" physiology covers the two aspects of "source" morphology and physiological characteristics. "Source" morphology covers plant height, internode configuration, stem and sheath weight, leaf length, width, and thickness, leaf area, leaf area index, leaf posture (leaf angle, leaf curvature, curl), leaf weight, specific leaf weight (dry weight per unit leaf area), leaf stomatal density, root length and weight, root distribution, etc. (Wang Feng et al. 2005, Zhou Wenxin et al. 2004), which can be determined by ruler, vernier caliper, balance, leaf area meter, plant analyzer or other scanning analysis systems. Physiological characteristics include chlorophyll content, individual and population photosynthetic rate, respiration rate, leaf chlorophyll fluorescence characteristics, sucrose phosphate synthase activity, Rubisco enzyme activity, stem sheath material conversion rate, stem sheath material turnover rate, dry matter withdrawal index, sugar synthase activity, R-amylase activity, root vigor, root exudation and exudation composition, and root functional period. These can be analyzed and determined by photosynthesis analysis system, population photosynthesis analysis system, chlorophyll fluorimeter, spectrophotometer, flow analyzer, and enzyme-labeled instrument.

"Sink" refers to the organs or tissues that utilize or store assimilates or other nutrients and the seeds of the panicle are the main "sink" of rice. The "sink" morphological indicators mainly include the total number of spikelets per unit area, the number of effective panicles, the number of grains per panicle, and the seed setting rate, the grain weight, the seed capacity, the grain filling, the endosperm cell matter quality, the number of endosperm cells, and the single cell matter quality (Wang Feng et al., 2005). The total number of spikelets, effective number of panicles, grains per panicle and seed setting rate can be obtained through surveys, while grain weight and grain size (grain length and width) can be measured with scales and seed testing systems, and the number and size of endosperm cells can be observed with microscopy.

The transport and distribution of substances are controlled by the "flow", which includes the structure and performance of all the transporting tissues connecting the "source" and the "sink". Currently, it is mainly analyzed through neck-panicles, neck-panicle internodes, penultimate internodes, number of secondary branches, number of spikelets on secondary branches, and the number of differentiated tube bundles, ducts, sieve tubes, companion cells in the stem base, area of vascular bundles, tube bundle ducts and phloem, single ducts, sieve tubes and companion cells (Wang Feng et al., 2005). These structures and traits can be observed through a microscope. The contents of sucrose, glucose, fructose and zeatin are analyzed and tested by high performance liquid chromatography, and carbon and nitrogen can be measured with an elemental analyzer. The activity of proteins and sucrose synthetase and starch synthase can be measured with an enzyme-labeled instrument. "Flow"-related plant hormones such as ABA can be analyzed with a kit or a high performance liquid chromatography.

Ⅱ. Measures of Substance Accumulation and Distribution Regulation

1. Consistency between photosynthetic product accumulation and substance needs

More than 60% of rice yield comes from photosynthetic products after heading (Yoshida, 1981), but for super high-yielding rice, the proportion can be up to 80% (Zhai Huqu et al., 2002), which is the reason for the great impact of photosynthesis in the later stages on yield. The incongruence between crop material peak demand and photosynthetic product accumulation is the main bottleneck for high yield. Compared with common hybrid rice, super hybrid rice has higher biological yield, larger "sink" and sufficient "source". However, a large "sink" and sufficient "source" do not necessarily lead to super-high yields. The main reason is that there is a mismatch between the material demand of the "sink" and the supply of the "source", i. e., the potential of a sufficient "source" and a large "sink" is not fully exploited. A study showed that Xieyou 9308 did not only have significantly higher photosynthetic carbon assimilation capacity than Xieyou 63, but also had photosynthetic products accumulated that can better meet the material requirements for grain filling, which is an important photosynthetic physiological characteristic of super-high yielding rice and the reason for a super-high yield (Zhai Huqu et al., 2002). According to Feng Jiancheng et al. (2007), a study on the high-yielding hybrid rice Teyouduoxi 1 and Teyou 63 and the control Shanyou 63 showed that the Teyou combinations had a better balance between photosynthesis and material requirements in the late stage, as evidenced by a higher net photosynthetic rate and more dry matter accumulated after heading, and the dry matter accumulated in the stem, leaves and sheath before heading could be more effectively transported to the seeds to better meet the needs of grain filling. The study of Xu Dehai et al. (2010) on large-panicle super hybrid rice Yongyou 6 showed that the photosynthetic advantage of the top three leaves on the canopy after heading was obvious, and the dry matter accumulated in the stem sheaths before heading had a high rate of transport and conversion, and the accumulation of photosynthetic products was in good balance with the material demand, which was an important reason for the high yield of this variety. Therefore, under the premise of large "sink" and sufficient "source", more attentions should be paid to the effective transport of carbohydrates accumulated in the stem sheaths before heading and the balance between photosynthetic product accumulation and grain filling material demand after heading to optimize the "source-sink" relationship.

2. Striving for better grain filling duration and rate

Current super hybrid rice varieties are mostly large-panicle varieties, with a long grain filling period and a moderate grain filling rate, leading to two-phase grain filling (Cao Shuqing, Zhai Huqu, 1999; Zou Yingbin, Xia Shengping, 2011). High-yielding hybrid rice has lower initial grain filling potential, average grain filling rate, and maximum grain filling rate, while the grain filling duration is significantly longer and the final yield is higher than conventional rice (Cheng Wangda et al., 2007; Wang Jianlin et al., 2004). Wang et al. (2012) showed that the near-isogenic line fgl is expected to be a high-yielding rice breeding material because of its longer grain filling period than the recurrent parent Zhefu 802, stable grain filling rate, increased fullness, and high grain weight. It has been shown that the main factor determining the grain filling of rice is the duration of grain filling, followed by the grain filling rate (Wang et al., 2004). Stable temperature and further extension of the grain filling period can increase rice grain filling, improve the seed setting rate, and finally achieve super-high yields (Ao Hejun et al., 2008). Comparing the growth duration of the first phase, second phase, and third phase hybrid rice varieties, and even the fourth phase super hybrid rice Y Liangyou 900, the third-phase super hybrid rice Y Liangyou 2 has a longer grain filling duration than the second phase super hybrid rice Y Liangyou 1 and Liangyou 0293, and Y Liangyou 900 has a longer growth duration than Y Liangyou 2, and the corresponding grain filling period also tends to be gradually extended. It has been proved that the strong and weak grain filling initiation of super hybrid rice Y Liangyou 1 of the second phase is earlier than that of the super hybrid rice Liangyoupeijiu of the first phase and three-line high-yielding hybrid rice Shanyou 63, and the grain-filling termination period is basically the same, i. e. the effective grain filling period of Y Liangyou 1 is longer (Li Cheng, 2013), which is conducive to the coordination between "sink" and "source" to improve grain filling. The photosynthetic product of 70%–80% of the rice yield comes from the photosynthetic product of the late heading stage, and most of the photosynthetic products of the late heading stage is transported to the grains. The higher the rice yield, the greater the quantity of photosynthetic products from the late heading stage, and the correspondingly longer the grain filling duration, to meet the needs of high-yielding rice grain filling, the leaf photosynthetic function period is also correspondingly extended. Based on a daily yield of 100 kg/ha for the whole growth duration of super hybrid rice, the current fourth and fifth phase super hybrid rice growth duration should be more than 150 d, and the grain filling period should be longer. The long photosynthetic period of rice requires the canopy to maintain a relatively long and stable canopy structure, increase basal light transmission, enhance root vigor, strengthen the basal internode resistance, and avoid lodging.

Ⅲ. Impact of External Environment on Substance Accumulation, Transportation and Distribution

Rice growth and development can be severely affected if there is drought stress during the grain filling stage. A study showed that the growth duration of Wuyunjing 3 and Yangdao 4 would be shortened by 2.9 - 5.5 d under conventional nitrogen fertilizer conditions during drought stress at the grain filling stage, and by 5.7 -7.4 d under high nitrogen fertilizer conditions. Compared with normal water management, the grain filling rates under drought stress for conventional and high nitrogen fertilizer treatments ranged from 0.18 -0.29 mg/grain per day to 0.31 -0.37 mg/grain per day, and the rates of non-structural carbohydrate transport stored in the stem sheath increased by 23.8%-27.1% and 19.6%-36.7%, respectively, which is the main reason for the yield reduction with conventional nitrogen fertilizer treatment under drought stress (Yang et al., 2001). It was further shown that drought stress during the grain filling stage would induce early senescence of rice, shorten the grain filling period, increase the transport of non-structural carbohydrates from the nutrient tissues to the grains, and promote the grain filling process (Yang et al., 2006).

Alternate wet and dry water management during the grain filling stage can effectively promote the activity of sucrose synthase (SuSase), adenosine diphosphate glucose pyrophosphorylase (AGPase), starch synthase (StSase) and starch branching enzyme (SBEase) by increasing the content of ABA in the grains, especially the weak grains, thus achieving physiological "sink expansion" (Yang et al., 2009). Bian Jinlong et al. (2017) found that alternate wet and dry irrigation during the grain filling stage of rice significantly enhanced the activities of SuSase, AGPase, StSase and SBEase in the grains of Yangdao 6 and Hangyou 8 during the early and middle grain filling stages, and improved the activities of sucrose-starch metabolism pathway in the grains of Yangdao 6 and Hangyou 8, which ultimately contributed to the increase of the yield of these rice varieties. Under alternate wet and dry irrigation, the enhanced photosynthetic rate, leaf water potential, root oxidizing force, root and leaf cytokinin content, and the activity of key enzymes of the sucrose-starch metabolic pathway in the seeds of Yangdao 6 and Hangyou 8 were physiologically important reasons for the yield increase under this treatment. On the contrary, the decrease in root activity, "source" capacity (leaf photosynthesis) and "sink" strength (cytokinin content and activity of key enzymes of the sucrose-starch metabolic pathway) of Liangyoupeijiu and Zhendao 88 under alternate wet and dry conditions resulted in lower yields.

Ⅳ. Improvement of Substance Accumulation, Transportation and Distribution

1. Increase in stem and sheath dry matter accumulation before heading

Although the accumulation of substances in the stems and sheaths before heading does not contribute as much to rice yield as the photosynthetic products accumulated after heading, the quantity of substances accumulated before heading has a large impact on the yield of super hybrid rice. The photosynthetic products accumulated in rice stems, leaves and sheaths before heading contribute 20%–30% to the grains, while more than 60% of the grain filling substances comes from photosynthetic products after heading (Tong Xiangbing et al.). As the photosynthetic products accumulated before heading make less contribution to yield, they are often ignored. Recent studies have shown, however, that for large-panicle super-high yielding hybrids, especially intersubspecific hybrids, the photosynthetic products accumulated in stems, leaves and sheaths before heading are crucial for a high yield and an important feature of high-yielding varieties (Liu Jianfeng et al., 2005; Yang Jianchang, 2010). If there are more photosynthetic products stored before heading, the sugar-to-spikelet ratio (ratio of non-structural carbohydrate accumulation in the stem and sheath to the number of spikelets at the heading stage) will be higher during heading, which is beneficial to increase the activity of many enzymes in the starch synthesis pathway, and increasing the seed setting rate, grain fullness and yield (Smith et al., 2001; Yang Jianchang, 2010). The peak of sucrose synthase activity in the weak grains of Liangyoupeijiu appears about 10 days later than that of Shanyou 63 (Yang Jianchang, 2010), indicating that super hybrid rice must accumulate more photosynthetic products before heading in order to maintain the physiological activity of weaker grains for good grain filling. Therefore, increasing photosynthetic product accumulation before heading plays an important role in maintaining the activity of weak grains, initiating grain filling, increasing the grain filling rate, increasing grain weight and ultimately achieving high yield.

The accumulation of photosynthetic products in the pre-heading stage not only maintains rice spikelet vigor and improves the seed setting rate, but also facilitates re-transport at the time of grain filling and promotes grain filling. The dry matter weight of rice at the maximum tillering, booting and full heading stages is about 20%, 50% and 70% of that at maturity, respectively (Zou Yingbin, Xia Shengping, 2011). During the heading stage, the dry weight of stems, leaves, and sheaths accounts for 65%–70% of the plant dry weight (Zou Yingbin et al., 2001; Zou Yingbin and Xia Shengping, 2011), and the accumulation of non-structural carbohydrates (NSC) in the stems sheaths during the heading stage shows a significant positive correlation with the grain weight of both strong and weak grains (Dong Minghui et al., 2012). Therefore, high NSC accumulation in the stems and sheaths before heading is conducive not only to improving the seed setting rate, but also to promoting grain filling and increasing grain weight. Katsura et al. (2007) confirmed that Liangyoupeijiu had a higher yield than Nipponbare and Takanari mainly because of a high amount of accumulated dry matter before heading, and of carbohydrates transported to the grains from the stems, leaves and sheaths. Our research indicates that a proper increase in dry matter accumulation before heading can increase the seed setting rate and yield of super hybrid rice.

2. Technical measures for breeding and cultivation

From the perspective of breeding, yield increase should be based on good other yield and quality traits by selecting combinations with the maximal accumulation of non-structural carbohydrates in the stems and sheaths before heading (Wang et al., 2016), a high photosynthetic rate of the canopy leaves at the later stage, and a long photosynthetic function period. In particular, a high accumulation of non-structural carbohydrates before heading is conducive to increasing the sugar-to-flower ratio, maintaining spikelet vigor and initiating grain filling (Yang, 2010). The high photosynthetic rate and long photosynthetic function period of canopy leaves after heading are conducive to maintaining a higher grain filling rate and duration. From the perspective of cultivation technology, under the premise of sound application of basal and tiller fertilizer, additional application of panicle fertilizer during the second to fourth stages of young panicle differentiation can improve the photosynthetic capacity of the late canopy, increase the accumulation of dry matter in the population before heading, improve the content of non-structural carbohydrates in the stems and sheaths, and balance between the demand and supply of photosynthetic products during the period of young panicle differentiation, grain filling and seed setting.

Ⅴ. A Case Study

The average yield of Y Liangyou 900 was 14. 82 t/ha in 2013 in Yanggu'ao, Longhui, Hunan province, and 15. 40 t/ha in 2014 in Xupu, Hunan, both in 6. 67-ha fields. The key technical systems such as "timely and precise sowing for cultivation of strong seedlings", "reasonably dense planting in wide and narrow rows for promoting the formation of large panicles at low positions", "balanced fertilization and strong culm prevention throughout the growth duration to build a high-yielding population", "moist and aerobic irrigation for vigorous root and healthy plants to promote the smooth 'flow' from the 'source'", "early pest and disease prediction and control", etc. are used to balance between the "source" and the "sink" and fully exploit the yield potential of the varieties. The specific measures on nutrient regulation are as follows (Li Jianwu et al., 2014):

1) Apply sufficient base fertilizer to lay a good foundation. Applied 1,050 kg of 45% (15 – 15 – 15) compound fertilizer and 1,500 kg of calcium-magnesium phosphate per hectare based on the use of farm organic fertilizer.

2) Apply tiller fertilizer early to promote big tillers. In the early stage, quick-acting nitrogen fertilizer should mainly be applied to promote the early growth and rapid development of tillers. The purpose is to build the seedling population as early as possible to improve the photosynthetic efficiency of the population and meet the photosynthetic products required for early tillering. Around 5 – 7 days after transplanting, based on manual mid-tillage and weeding, 112. 5 kg of urea and 75. 0 kg of potassium chloride per hectare should be applied; 12 – 15 days after transplanting, balanced fertilizer should be used in different fields, 45. 0 – 75. 0 kg of urea and 112. 5 kg of potassium chloride per hectare, depending on the seedling condition, to ensure the basic uniform growth of the whole plot of 6. 67 ha.

3) Heavy panicle fertilization for large panicles. Y Liangyou 900 is a large-panicle variety, and such an advantage is given full play to so asto achieve super high yield. The most important measure is to apply heavy panicle fertilization to ensure the formation of large panicles. Spikelet-promoting fertilizer and spikelet-protecting fertilizer should be applied at different times in the second and fourth stages of young panicle differentiation on the main stem to increase the photosynthetic rate of the leaves, provide sufficient photosynthetic products for leaf and young panicle differentiation, coordinate between the "source" and the "sink", and fully exploit the potential for super high yield.

References

[1] BLOOM A J, BURGER M, ASENSIO J S R, et al. Carbon dioxide enrichment inhibits nitrate assimilation in wheat and Arabidopsis[J]. Science, 2010, 328: 899 - 903.

[2] SHU Q C, RONG X Z, WEI L, et al. The involvement of cytokinin and abscisic acid levels in roots in the regulation of photosynthesis function in flag leaves during grain filling in super high-yielding rice (*Oryza sativa*)[J]. Journal of Agronomy and Crop Science, 2004, 190: 73 - 80.

[3] CHANG S Q, CHANG T G, SONG Q F, et al. Photosynthetic and agronomic traits of an elite hybrid rice Y Liangyou 900 with a record-high yield[J]. Field Crops Research, 2016, 187: 49 - 57.

[4] FARQUHAR G D, EHLERINGER J R, HUBICK K T. Carbon isotope discrimination and photosynthesis[J]. Annual Review of Plant Physiology and Plant Molecular Biology, 1989, 40: 503 - 537.

[5] FENG L L, WANG K, LI Y, et al. Overexpression of SBPase enhances photosynthesis against high temperature stress in transgenic rice plants[J]. Plant Cell Report, 2007, 26: 1635 - 1646.

[6] HONGTHONG P, HUANG M, XIA B, et al. Yield formation strategies of a loose-panicle super hybrid rice [J]. Research on Crops, 2012, 13 (3): 781 - 789.

[7] HUANG M, CHEN J, CAO F B, et al. Rhizosphere processes associated with the poor nutrient uptake in no-tillage rice (*Oryza sativa* L.) at tillering stage[J]. Soil and Tillage Research, 2016, 163: 10 - 13.

[8] HUANG M, CHEN J N, CAO F B, et al. Root morphology was improved in a late-stage vigor super rice cultivar [J]. Plos One, 2015, 10: e0142977.

[9] HUANG M, SHAN S L, ZHOU X F, et al. Leaf photosynthetic performance related to higher radiation use efficiency and grain yield in hybrid rice[J]. Field Crops Research, 2016, 193: 87 - 93.

[10] HUANG M, XIA B, ZOU Y B, et al. Improvement in super hybrid rice: a comparative study between super hybrid and inbred varieties[J]. Research on Crops, 2012, 13 (1): 1 - 10.

[11] HUANG M, YIN X H, JIANG L G, et al. Raising potential yield of short-duration rice cultivars is possibleby increasing harvest index[J]. Biotechnology, Agronomy, Society and Environment, 2015, 19 (2): 153 - 359.

[12] HUANG M, ZHOU X F, CHEN J N, et al. Factors contributing to the superior post-heading nutrient uptake by no-tillage rice[J]. Field Crops Research, 2016, 185: 40 - 44.

[13] HUANG M, ZOU Y B, FENG Y H, et al. No-tillage and direct seeding for super hybrid rice production in rice-oilseed rape cropping system[J]. European Journal of Agronomy, 2011, 34: 278 - 286.

[14] HUANG M, ZOU Y B, JIANG P, et al. Effect of tillage on soil and crop properties of wet-seeded flooded rice[J]. Field Crops Research, 2012, 129: 28 - 38.

[15] HUANG M, ZOU Y B, JIANG P, et al. Relationship between grain yield and yield components in super hybrid rice[J]. Agricultural Sciences in China, 2011, 10 (10): 1537 - 1544.

[16] HUANG M, ZOU Y B, JIANG P, et al. Yield component differences between direct-seeded and transplanted super hybrid rice[J]. Plant Production Science, 2011, 14 (4): 331 - 338.

[17] HUBBART S, PENG S B, HORTON P, etal. Trends in leaf photosynthesis in historical rice varieties developed in the Philippines since 1966[J]. Journal of Experiment Botany, 2007, 58: 3429 - 3438.

[18] INADA N, SAKAI A, KUROIWA, et al. Three-dimensional analysis of the senescence program in rice (*Oryza sativa* L.) coleoptiles. Investigations of tissues and cells by fluorescence microscopy[J]. Planta, 1998, 205: 153 - 164.

[19] KATSURA K, MAEDA S, HORIE T, et al. Analysis of yield attributes and crop physiological traits of Lian-

gyoupeijiu, a hybrid rice recently bred in China[J]. Field Crops Research, 2007, 103: 170 - 177.

[20] KEBEISH R, NIESSEN M, THIRUVEEDHI K, et al. Chloroplastic photorespiratory bypass increases photosynthesis and biomass production in Arabidopsis thaliana[J]. Nature Biotechnology, 2007, 25: 593 - 599.

[21] LEE R H, WANG C H, HUANG L T, et al. Leaf senescence in rice plants: cloning and characterization of senescence up-regulated genes[J]. Journal of Experimental Botany, 2001, 52: 1117 - 1121.

[22] LONG S P. Modification of the response of photosynthetic productivity to rising temperature by atmospheric CO_2 concentrations: has its importance been underestimated? [J]. Plant, Cell & Environment, 1991, 14: 729 - 739.

[23] LONG S P, ZHU X G, NAIDU S L, et al. Can improvement in photosynthesis increase crop yields? [J]. Plant, Cell & Environment, 2006, 29: 315 - 330.

[24] MAIER A, FAHNENSTICH H, VON CAEMMERER S, et al. Glycolate oxidation in A. thaliana chloroplasts improves biomass production[J]. Frontiers in Plant Science, 2012, 3: 38.

[25] MISHRA A, SALOKHE V M. Rice root growth and physiological responses to SRI water management and implications for crop productivity[J]. Paddy & Water Environment, 2011, 9: 41 - 52.

[26] MONTEITH J L. Climate and the efficiency of crop production in Britain[J]. Philosophical Transations of the Royal Society of London, 1977, 281: 277 - 294.

[27] PENG S B, KHUSH G S, VIRK P, et al. Progress in ideotype breeding to increase rice yield potential[J]. Field Crops Research, 2008, 108: 32 - 38.

[28] PENG S B, HUANG J L, SHEEHY J E, et al. Rice yields decline with higher night temperature from global warming[J]. Proceedings of the National Academy of Sciences U. S. A., 2004, 101: 9971 - 9975.

[29] SAGE T L, SAGE R F. The functional anatomy of rice leaves: implications for refixation of photorespiratory CO_2 and efforts to engineer C4 photosynthesis into rice[J]. Plant Cell Physiology, 2009, 50: 756 - 772.

[30] SONG Q F, CHU C, PARRY MARTIN A J, et al. Genetics based dynamic systems model of canopy photosynthesis: the key to improved light and resource use efficiencies for crops[J]. Food and Energy Security, 2016, 5: 18 - 25.

[31] SONG Q F, XIAO H, XIAO X, et al. A new canopy photosynthesis and transpiration measurement system (CAPTS) for canopy gas exchange research[J]. Agricultural and Forest Meteorology, 2016, 217: 101 - 107.

[32] SONG Q F, ZHANG G, ZHU X G. Optimal crop canopy architecture to maximise canopy photosynthetic CO_2 uptake under elevated CO_2 - a theoretical study using a mechanistic model of canopy photosynthesis[J]. Functional Plant Biology, 2013, 40: 108 - 124.

[33] THOLEN D, ZHU X G. The mechanistic basis of internal conductance: a theoretical analysis ofmesophyll cell photosynthesis and CO_2 diffusion[J]. Plant Physiology, 2011, 156: 90 - 105.

[34] THOLEN D, ETHIER G, GENTY B, et al. Variable mesophyll conductance revisited: theoretical background and experimental implications[J]. Plant, Cell & Environment, 2012, 35: 2087 - 2103.

[35] WANG D R, WOLFRUM E J, VIRK P, et al. Robust phenotyping strategies for evaluation of stem non-structural carbohydrates (NSC) in rice[J]. Journal of Experimental Botany, 2016, 67 (21): 6125 - 6138.

[36] WANG P, KELLY S, FOURACRE J P, et al. Genome-wide transcript analysis of early maize leaf development reveals gene cohorts associated with the differentiation of C4 Kranz anatomy[J]. The Plant Journal, 2013, 75: 656 - 670.

[37] WANG Y, LONG S P, ZHU X G. Elements required for an efficient NADP-malic enzyme type C_4 photo-

synthesis[J]. Plant Physiology, 2014, 164: 2231-2246.
[38] YANG J C, ZHANG H, ZHANG J H. Root morphology and physiology in relation to the yield formation of rice [J]. Journal of Integrative Agriculture, 2012, 11 (6): 920-926.
[39] YANG J C, ZHANG J H. Grain filling of cereals under soil drying[J]. New Phytologist, 2006, 169 (2): 223-236.
[40] YANG J C, ZHANG J H. Grain-filling problem in "super" rice[J]. Journal of Experimental Botany, 2009, 61 (1): 1-5.
[41] YANG J C, ZHANG J H, WANG Z Q, et al. Remobilization of carbon reserves in response to water deficit during grain filling of rice[J]. Field Crops Research, 2001, 71 (1): 47-55.
[42] YAO Y Z. Effects of ridge tillage on photosynthesis and root characters of rice[J]. Chilean Journal of Agricultural Research, 2015, 75 (1): 35-41.
[43] YOSHIDA S. Fundamentals of rice crop science[M]. International Rice Research Institute, 1981.
[44] ZELITCH I, SCHULTES N P, PETERSON, et al. High glycolate oxidase activity is required for survival of maize in normal air[J]. Plant Physiology, 2009, 149: 195-204.
[45] ZHANG H, XUE Y G, WANG Z Q, et al. Morphological and physiological traits of roots and their relationships with shoot growth in "super" rice[J]. Field Crops Research, 2009, 113: 31-40.
[46] ZHANG Y B, TANG Q Y, ZOU Y B, et al. Yield potential and radiation use efficiency of super hybrid rice grown under subtropical conditions[J]. Field Crops Research, 2009, 114: 91-98.
[47] ZHU X G, DE STURLER E, LONG S P. Optimizing the distribution of resources between enzymes of carbon metabolism can dramatically increase photosynthetic rate: a numerical simulation using an evolutionary algorithm [J]. Plant Physiology, 2007, 145: 513-526.
[48] ZHU X G, LONG S P, ORT D R. What is the maximum efficiency with which photosynthesis can convert solar energy into biomass? [J]. Current Opinion in Biotechnology, 2008, 19: 153-159.
[49] SMITH D L, HAMEL C. Crop yield: physiology and processes[M]. Translated by Wang Pu, Wang Zhimin, Zhou Shunli, et al. Beijing: China Agricultural University Press, 2001.
[50] AO HEJUN, WANG SHUHONG, ZOU YINGBIN, et al. Study on the yield stability and dry matter characteristics of super hybrid rice[J]. Scientia Agricultura Sinica, 2008, 41(7): 1927-1936.
[51] AO HEJUN, WANG SHUHONG, ZOU YINGBIN, et al. Characteristics of nutrient uptake and utilization of super hybrid rice under different fertilizer application rates[J]. Scientia Agricultura Sinica, 2008, 41(10): 3123-3132.
[52] BIAN JINLONG, JIANG YULAN, LIU YANYANG, et al. Effects of alternate wet and dry irrigation on grain yield in rice cultivars with different drought resistance and its physiological mechanism[J]. Chinese Journal of Rice Science, 2017, 31(4): 379-390.
[53] CAO SHUQING, ZHAI HUQU, ZHANG HONGSHENG, et al. Leaf "source" capacity and photosynthetic indexes in different types of rice varieties[j]. Chinese Journal of Rice Science, 1999, 13(2): 91-94.
[54] CHANG SHUOQI, DENG QIYUN, LUO YI, et al. Studies on the cold tolerance of super hybrid rice and its parents[J]. Hybrid Rice, 2015, 30(1): 51-57.
[55] CHEN DAGANG, ZHOU XINQIAO, LI LIJUN, et al. Relationship between root morphological characteristics and yield components of major commercial *indica* rice in south China[J]. Acta Agronomica Sinica, 2013, 39(10): 1899-1908.
[56] CHENG WANGDA, YAO HAIGEN, ZHANG HONGMEI. Difference in grain filling properties and leaf photo-

synthetic characteristics at late growth stage between *japonica* hybrid and conventional rice for the late season in southern China[J]. Chinese Journal of Rice Science, 2007, 21(2): 174 - 178.

[57] DENG HUAFENG. Studies on the objective traits of super hybrid rice in the Yangtze river basin[D]. Changsha: Hunan Agricultural University, 2008.

[58] DENG QIYUN, YUAN LONGPING, CAI YIDONG, et al. Photosynthetic advantages of model plant-type in super hybrid rice[J]. Acta Agronomica Sinica, 2006, 32(9): 1287 - 1293.

[59] DONG MINGHUI, CHEN PEIFENG, GU JUNRONG, et al. Effects of wheat straw residue applied to field and nitrogen management on photosynthate transportation of stem and sheath and grain filling characteristics in super hybrid rice[C]. Nanchang: Abstracts of 2012 Annual Conference of Chinese Crop Society, 2012.

[60] DUAN MEIJUAN. Studies on transforming maize C4 photosynthetic enzyme genes of super hybrid rice parental lines and biological characteristics of transgenic rice[D]. Changsha: Hunan Agricultural University, 2010.

[61] FENG JIANCHENG, GUO FUTAI, ZHAO JIANWEN, et al. "Sink", "source" and "flow" characteristics of high yielding *indica* hybrid teyou combinations[J]. Fujian Journal of Agricultural Sciences, 2007, 22(2): 146 - 149.

[62] FU JING, CHEN LU, HUANG ZUANHUA, et al. Relationship of leaf photosynthetic characteristics and root physiological traits with grain yield in super rice[J]. Acta Agronomica Sinica, 2012, 38(7): 1264 - 1276.

[63] FU JING, YANG JIANCHANG. Research advances in physiology of super rice under high-yielding cultivation[J]. Chinese Journal of Rice Science, 2011, 25(4): 343 - 348.

[64] GUO ZHAOWU, XIAO LANGTAO, LUO XIAOHE, et al. Photosynthetic function of the flag leaf sheath of super hybrid rice Liangyoupeijiu[J]. Acta Agronomica Sinica, 2007, 33(9): 1508 - 1515.

[65] GUO ZHAOWU. Photosynthetic function of high-yielding hybrid rice Liangyoupeijiu[D]. Changsha: Hunan Agricultural University, 2008.

[66] HUANG JINWEN, LI ZHONG, CHEN JUN, et al. Analysis of differential expression of leaf proteins in different hybrid rice during grain-filling[J]. Chinese Journal of Eco-Agriculture, 2011, 19(1): 75 - 81.

[67] HUANG MIN, MO RUNXIU, ZOU YINGBIN, et al. Yield components and grain filling characteristics of super hybrid rice[J]. Crop Research, 2008, 22(4): 249 - 253.

[68] LI CHENG. Studies on morphological features and its regularities in super hybrid rice breeding[D]. Changsha: Central South University, 2013.

[69] LI DIQIN, DUAN CHUNQI, QIN JIANQUAN, et al. effects of nitrogen application on root vigor and yield of super hybrid rice in middle and late stages[J]. Crop Research, 2009, 23(2): 71 - 73.

[70] LI JIANWU, ZHANG YUZHU, WU JUN, et al. High-yielding cultivation techniques of super hybrid rice Y Liangyou 900 yielded 15.40 t/ha on a 6.84 ha scale[J]. China Rice, 2014, 20(6): 1 - 4.

[71] LI WEN. Effects of low-intensity UVB radiation on the photosynthetic and physiological characteristics of super high-yielding hybrid rice Liangyoupeijiu[D]. Nanjing: Nanjing Normal University, 2012.

[72] LI XIA, JIAO DEMAO. Photosynthetic and physiological characteristics of super hybrid rice Liangyoupeijiu[J]. Journal of Jiangsu Agricultural Sciences, 2002, 18(1): 9 - 13.

[73] LI XIA, JIAO DEMAO, DAI CHUANCHAO, et al. Photosynthetic characteristics of rice hybrids with transgenic PEPC parent HPTER - 01[J]. Acta Agronomica Sinica, 2001, 27(2): 137 - 143.

[74] LI XIANGLING, FENG YUEHUA. Research progress on the relationship between the aerial part and root traits of rice[J]. Chinese Agricultural Science Bulletin, 2015, 31(6): 1 - 6.

[75] LING QIHONG, GONG JIAN, ZHU QINGSEN. Effect of leaves at different node position on yield formation[J]. Journal of Jiangsu Agricultural College, 1982, 3(2): 9-26.

[76] LING QIHONG. Crop population quality[M]. Shanghai: Shanghai Science and Technology Press, 2000.

[77] LIU FAMOU, ZHU LIANFENG, XU JIAYING, et al. Root growth advantage of hybrid rice and its response and regulation to environmental factors[J]. China Rice, 2011, 17(4): 6-10.

[78] LIU JIANFENG, YUAN LONGPING, DENG QIYUN, et al. A study on characteristics of photosynthesis in super high-yielding hybrid rice[J]. Scientia Agricultura Sinica, 2005, 38(2): 258-264.

[79] LIU TAOJU, QI CHANGHAN, TANG JIANJUN. Studies on relationship between the character parameters of root and yield formation in rice[J]. Scientia Agricultura Sinica, 2002, 35(11): 1416-1419.

[80] LIU XIAOQING. Study on the photosynthetic characteristics of high-yielding and good-quality hybrid rice varieties of the fengliangyou series during the late growth stage and its yield[J]. Journal of Anhui Agricultural Sciences, 2011, 39(29): 17819-17821.

[81] NING SHUJU, DOU HUIJUAN, CHEN XIAOFEI, et al. Dynamics of nitrogen metabolism activity in rice root at the late development stage[J]. Chinese Journal of Eco-Agriculture, 2009, 17(3): 506-511.

[82] OU ZHIYING, LIN GUIZHU, PENG CHANGLIAN. Response of flag leaves of super high-yielding rice Peiai 64S/E32 and Liangyoupeijiu to high temperature[J]. Chinese Journal of Rice Science, 2005, 19(3): 249-254.

[83] TONG XIANGBING, CEN TANGXIAO, WEI ZHANGHUAN, et al. Exploration of super high-yielding cultivation techniques for *indica-japonica* Yongyou 6[J]. Ningbo Agricultural Science and Technology, 2006(2): 29-31.

[84] WANG FENG, ZHANG GUOPING, BAI PU. Achievement and prospects of research on evaluation of the relationship between "source" and "sink" in rice[J]. Chinese Journal of Rice Science, 2005, 19(6): 556-560.

[85] WANG HONGYAN, WU WENGE, LUO ZHIXIANG, et al. Studies on characteristics of main photosynthetic physiology in high-yielding hybrid rice[J]. Journal of Anhui Agricultural Sciences, 2010, 38(15): 7792-7793.

[86] WANG JIANLIN, XU ZHENGJIN, MA DIANRONG. Comparison on grain filling characters between hybrid and conventional rice in northern China[J]. Chinese Journal of Rice Science, 2004, 18(5): 425-430.

[87] WANG NA, CHEN GUOXIANG, LV CHUANGEN. Studies on photosynthetic characteristics of flag leaves in hybrid rice Liangyoupeijiu and its parents[J]. Hybrid Rice, 2004, 19(1): 53-55.

[88] WANG RONGFU, ZHANG YUNHUA, JIAO DEMAO, et al. Characteristics of photoinhibition and early aging in super hybrid rice (*Oryza sativa* L.) Liangyoupeijiu and Its Parents at the Late Development Stage[J]. Acta Agronomica Sinica, 2004, 30(4): 393-397.

[89] WANG XI, TAO LONGXING, YU MEIYU, et al. Physiological model of super hybrid rice Xieyou 9308[J]. Chinese Journal of Rice Science, 2002, 16(1): 38-44.

[90] WEI ZHONGWEI, MA GUOHUI. Studies on the characteristics of the root system of super high-yielding hybrid rice combination Chaoyouqian[J]. Hybrid Rice, 2016, 31(5): 51-55.

[91] XU GUOWEI, SUN HUIZHONG, LU DAKE, et al. Differences in ultrastructure and activity of rice roots under different irrigation and nitrogen supply levels[J]. Journal of Plant Nutrition and Fertilizers, 2017, 23(3): 811-820.

[92] XU DEHAI, WANG XIAOYAN, MA RONGRONG, et al. Analysis on physiological properties of the heavy

panicle *indica-japonica* intersubspecific hybrid rice Yongyou 6[J]. Scientia Agricultura Sinica, 2010, 43(23): 4796–4804.

[93] XU KAI. Studies on the growth and photosynthetic responses of two hybrid rice cultivars to enhanced UVB radiation[D]. Shanghai: CentralChina Normal University, 2006.

[94] XUE YANFENG, LANG YOUZHONG, LV CHUANGEN, et al. Study on leaf and root senescence of Liangyoupeijiu and its male parent Yangdao 6 after heading[J]. Journal of Yangzhou University (Agriculture and Life Sciences Edition), 2008, 29 (3): 7–11.

[95] YANG JIANCHANG. Mechanism and regulation in the filling of inferior spikelets of rice[J]. Acta Agronomica Sinica, 2000, 36(12): 2011–2019.

[96] YANG JIANCHANG. Relationships of rice root morphology and physiology with the formation of grain yield and quality and the nutrient absorption and utilization[J]. Scientia Agricultura Sinica, 2011, 44(1): 36–46.

[97] YANG ZHIJIAN, XU QINGGUO, ZHU CHUNSHENG, et al. Effect of 6-BA on growth in middle and later periods of rice root[J]. Journal of Hunan Agricultural University (Natural Science Edition), 2009, 35(5): 462–465.

[98] YU YAN, PENG XIANLONG, LIU YUANYING, et al. Effects of nitrogen application at the later stage on the absorption capacity of rice root in cold area[J]. Soils, 2011, 43(4): 548–553.

[99] ZHAI HUQU, CAO SHUQING, WAN JIANMIN, et al. Relationship between photosynthetic function and yield of super high-yielding hybrid rice at the grain-filling stage[J]. Science in China (Series C: Life Sciences), 2002, 32(3): 211–217.

[100] ZHANG HUI, HUANG ZUANHUA, WANG JINGCHAO, et al. Changes in morphological and physiological traits of roots and their relationships with grain yield during the evolution of mid-season *indica* rice cultivars in jiangsu province[J]. Acta Agronomica Sinica, 2011, 37(6): 1020–1030.

[101] ZHANG YUNHUA, QIAN LISHENG, WANG RONGFU. Adaptation to chilling temperature and high light of super hybrid rice Liangyoupeijiu and its parents[J]. Acta Laser Biology Sinica, 2008, 17(1): 75–80.

[102] ZHANG YUNHUA. The study of utilization and conversion efficiency for light energy in super high-yielding hybrid Liangyoupeijiu and its parents[D]. Hefei: Anhui Agricultural University, 2003.

[103] ZHENG HUABIN, YAO LIN, LIU JIANXIA, et al. Effect of ridge & terraced cultivation on rice yield and root trait[J]. Acta Agronomica Sinica, 2014, 40(4): 667–677.

[104] ZHENG JINGSHENG, LIN WEN, JIANG ZHAOWEI, et al. Root developmental morphology of super high yielding rice[J]. Fujian Journal of Agricultural Sciences, 1999, 14(3): 1–6.

[105] ZHENG TIANXIANG, TANG XIANGRU, LUO XIWEN, et al. Effect of water-saving irrigation on physiological characteristics of the roots of precision hill-direct-seeding super rice[J]. Journal of Irrigation and Drainage, 2010, 29(2): 85–88.

[106] ZHOU QUANCHENG. Studies on cell biological characteristics of flag leaf light energy conversion during the aging of super high-yield hybrid rice Liangyoupeijiu and its parents[D]. Nanjing: Nanjing Normal University, 2005.

[107] ZHOU WENXIN, LEI CHI, TU NAIMEI. Research trends on source-sink relationship of rice[J]. Journal of Hunan Agricultural University (Natural Science Edition), 2004, 30(4): 389–393.

[108] ZHU DEFENG, LIN XIANQING, CAO WEIXING. Characteristics of root distribution of super high-yielding rice varieties[J]. Journal of Nanjing Agricultural University, 2000, 23(4): 5–8.

[109] ZHU DEFENG, LIN XIANQING, CAO WEIXING. Effects of deep roots on growth and yield in two rice varieties[J]. Scientia Agricultura Sinica, 2001, 34(4): 429-432.

[110] ZHU PING, YANG SHIMIN, MA JUN, et al. Effect of shading on the photosynthetic characteristics and yield at later growth stage of hybrid rice combination[J]. Acta Agronomica Sinica, 2008, 34(11): 2003-2009.

[111] ZOU YINGBIN, HUANG JIANLIANG, TU NAIMEI, et al. Effects of the VSW cultural method on yield formation and physiological characteristics in double-cropping hybrid rice[J]. Acta Agronomica Sinica, 2001, 27(3): 343-350.

[112] ZOU YINGBIN, XIA SHENGPING. Super hybrid rice San-Ding cultivation theory and technology[M]. Changsha: Hunan Science and Technology Press, 2011.

[113] ZOU YINGBIN, WAN KEJIANG. San-Ding cultivation and moderate scale production of rice[M]. Beijing: China Agriculture Press, 2015.

Chapter 12

Super Hybrid Rice Cultivation Techniques

Li Jianwu / Long Jirui

Section 1　Seedling Raising and Transplanting of Super Hybrid Rice

Ⅰ. Overview of Rice Seedling Raising

1. Physiology of raising rice seedlings

Raising rice seedlings refers to the process from seed germination based on water absorption and growth into seedlings suitable for transplanting, which depends heavily on climatic and environmental conditions. It is necessary to create a suitable environment for seed sprouting and seedling growth during the whole process of rice transplanting, which is conducive to cultivating strong seedlings with multi-tillers. Suitable environmental conditions involve water, oxygen and temperature, while ensuring adequate nutrition is also necessary for the raising of strong seedlings.

(1) Water

Water is the primary requirement for rice seed germination and seedling growth. The free water content in dry seed cells is very low, the protoplasm of the cells is in a gel state, the metabolic activity is very weak and the cells are in a dormant state. The physiological effects of rice seeds can only start gradually when they uptake enough water and the seeds germinate well when they take water about 30% of their own weight. At the same time, in the early stage of rice seedling growth, that is, before the period of two adult leaves and one newly emerged leaf, its growth and development is also closely related to moisture. At this time, the seedling has not formed a sound aeration tissue, so the soil should be kept moist for better aeration, and the healthy growth of seedlings.

(2) Oxygen

Oxygen promotes the respiration of rice seeds, releasing enough energy to meet the needs of various physiological processes, ensuring the activity of amylase, promoting the hydrolysis of amylase, and the synthesis of protein. Oxygen deficiency affects protein biosynthesis, cell division and differentiation, and the inability to form new organs; therefore, oxygen is an essential condition for rice seed germination and good seedling development.

(3) Temperature

Rice seed germination and seedling growth are complex processes composed of a series of physiological and biochemical changes carried out with the participation of many enzymes, and the activity of enzymes is closely related to the temperature. Within a certain range, the activity of enzymes is enhanced with the increase of temperature and weakened with the decrease of temperature. The minimum temperature required for rice germination is generally stable, i. e. an average daily temperature above 12 ℃, the optimal temperature is 28 ℃–32 ℃, and the maximal temperature is 36 ℃–38 ℃. A temperature higher than 40 ℃ during germination will do damage to the seed and the germ. The minimal temperature for *indica* seedling growth is 14 ℃, and for healthy growth, the temperature must be higher than 16 ℃.

(4) Nutrition

Before the three-leaf stage, rice plants mainly rely on the consumption of nutrients stored in the endosperm for their own growth and development. After the three-leaf stage, the nutrients in the endosperm are consumed, and seedling growth relies mainly on the photosynthetic products of the leaves. Therefore, the seedlings need to continuously uptake nutrients from the soil to meet the nutrient needs of seedling growth, and only with an adequate supply of soil nutrients can seedlings grow vigorously (Zhou Peijian et al., 2010).

2. Categorization of seedlings raising methods

Methods to cultivate rice seedlings can be categorized in various ways. Based on the water level of the field, there are water seedlings, wetland seedlings and dryland seedlings; based on the seedling size, there are large seedlings, medium-sized seedlings and small seedlings; based on the seedling age, there are aged seedlings, mid-aged seedlings and young seedlings; based on seeding density, there are dense seedlings, sparse seedlings and super-sparse seedlings; based on coverage conditions, there are open-field seedlings, greenhouse seedlings and mulch seedlings; based on soil conditions, there are soil (mud) seedlings and soilless seedlings; based on plate application, there are seedlings with tray and seedlings without tray, and there are seedlings with soft tray (scattered planting or machine transplanting) and seedlings with hard tray; based on whether direct seeding is adopted, there are one-phase seedlings and two-phase seedlings; and based on the transplanting methods, there are hand seedlings, scattered planting and machine transplanted seedlings. In terms of the compound factor of seedling cultivation, there are wet-raised and thinly sown large seedlings (one phase) with multi-tiller wash-planted seedlings, and dry-raised, densely-sown small, moist-foster care two-phase of long-aged seedlings with soil.

Ⅱ. Seedling Raising Techniques of Super Hybrid Rice

Super hybrid rice seedling raising methods to be discussed below mainly include wetland seedlings, dryland seedlings, scattered seedlings, machine-transplanted tray-raised seedlings, direct seeding and their supporting techniques.

1. Wetland seedling cultivation

The wetland method is the most widely applied technique in raising rice seedlings, and also a technique most skillfully mastered by farmers in South China. It is easy and requires no special supplies or treatment. The major characteristics include water-plowed seedling bed, multiple harrows after one plough, ridging-based horizontal sowing, no-water in seedling bed till one-leaf and one newly emerged leaf (dry field management), and shallow irrigation (wet management) from the two-leaf and one newly emerged leaf stage till transplanting. Afterwards, transplanting can be done with soil for small and medium-sized seedlings or large seedlings after washing.

(1) Preparation before sowing

1) Seed drying

Seeds should be sun-dried for one or two days 3 – 4 days before soaking. Sun-dried seeds have enhanced water and air permeability and water absorption capacity so that the seeds can germinate uniformly. In addition, the short-wave light in the sunlight can kill germs on the seed surface, which has a fungicidal and disease prevention effect. It is also suggested that seeds should be stirred frequently during the drying process to ensure that the seeds are dried evenly. Furthermore, it is better to put the seeds on bamboo pans, drying mats or plastic films instead of directly on cement ground or a stone surface to avoid injury.

2) Seed soaking and sterilization

Rice seeds generally germinate well when they absorb water about 30% of their own weight. Therefore, they should be soaked before sowing so that they can absorb sufficient water. After cleaning, seeds are first soaked in clean water for 10 – 12 hours, and then in trichloroisocyanuric acid (TCCA) solution diluted by 300 times for 8 – 10 hours without changing the water, with the seeds stirred every six hours to ensure every seed is completely soaked and disinfected. After that, the seeds need to be rinsed completely to remove the TCCA residues before germination because TCCA inhibits seed germination. Another method is to soak the seeds in 25% Prochloraz emulsifiable concentrates diluted by 2,000 – 3,000 times for 24 hours. During seed soaking, the liquid surface should be 3 – 5 cm above the top of the seeds to keep them securely immersed after they absorb water and swell. After being soaked in Prochloraz, no rinsing is needed and the seeds can be drained off for pre-germination (Li Jianwu et al., 2013).

3) Seed budding and pre-germination

Pre-germination can make seedlings emerge neatly and prevent damage and rotting, especially during early-season rice sowing when the temperature is low (Shi Qinghua et al., 2010). In regular pre-germination, soaked seeds are drained off before being rinsed with 35 ℃–40 ℃ water. After 3 – 5 min of pre-warming, the seeds are put into cloth bags or bamboo pans (or other water- and air-permeable, and moisture-keeping containers), and wrapped up with agro-plastic film or disease-free straw for warmth. Warm water is sprayed on the seeds every 3 – 4 hours to keep the seeds warm at 35 ℃–38 ℃. The seeds need to be stirred to prevent excessive heat. After around 20 hours, seeds budding and pre-germination will start and the temperature should be lowered to 25 ℃–30 ℃. The entire pre-germination process ends when the buds grow to the length of half a grain and the root length to one grain. Note that the seeds should be sprayed with water and/or stirred to prevent the temperature from getting too high or too low during pre-germination. However, in mid-season rice regions, the temperature is slightly higher in early spring or is getting high, so the seeds can also be sown when they are mostly about to bud. The budded seeds are usually put under room temperature for 3 – 6 hours to let them get accustomed to the environment and increase the seedling rate (Wang Chi, 2013).

4) Seedling bed preparation

A seedling bed should be leeward land plots facing the sun, with convenient irrigation and drainage conditions, fertile soil with an appropriate plough layer, and convenience for seedling transportation. Low-lying and cold waterlogged land is to be avoided. Selected land plots should be prepared as paddy field for wetland sowing, and after plowing, 4. 5 t of thoroughly decomposed manure and 450 kg of 45% compound fertilizer (N, P_2O_5 and K_2O each accounting for 15%) should be applied to every hectare of land as base fertilizer. Four to five days later, the land is to be leveled and the seeds can be sown (Li Jianwu et al. , 2013).

(2) Sowing and seedling raising

Good timing for sowing is the key to raising good seedlings, and the determination of the sowing time depends on various factors such as the growth duration of the variety, flexibility of seedling age, temperature, and the harvest time of the prior crop. The sowing time for early-season rice is mainly determined by the average daily temperature during the sowing period (normally stable at above 12 ℃) , and also by the transplanting time and seedling age flexibility. Single-cropping mid- and late-season rice have a more flexible seedling age, which is determined by their heading and flowering periods which are set the avoid the heat from late July to mid-August, i. e. determined by the growth duration of the variety. Normally, initial heading should take place in late August or early September, in order to avoid the high temperature that can cause poor seed setting. The sowing time for double-cropping late-season rice is mainly determined by the safe heading date. At the same time, the flexibility of seedling age of the variety should also be considered to avoid stem elongation in the seedling bed and early heading in the production field, both causing yield reduction, and to prevent late heading in the case of an autumn chill (or "cold dew wind") that may influence the yield (Zhou Peijian et al. , 2010).

Before sowing, the seedling bed should be finely plowed and separated with ridges into plots that are 1. 5 m wide and 15 cm high, with 30 cm ditches in between. The ratio of the area of the seedling bed to that of the production field should be 1 : (10 – 15). Generally, for wetland seedling raising, 12 – 15 kg of seeds are needed for each hectare of production field and 105 – 135 kg of seeds can be sown on each hectare of the seedling bed with a seedling age of 25 – 30 days. In order to increase effective seedlings and reduce pests, seed dressing agents can be applied, including special seed dressing agents and seedling strengthening agents. Instructions must be followed when applying the agents. A fixed quantity of seeds should be sown for each seedling bed, evenly and sparsely distributed without overlap or gap. After sowing, the seeds should be buried shallowly in the soil in order to improve root growth.

Thermal insulation should be used to keep the seedling bed warm in a low-temperature area; usually a plastic-covered greenhouse is used. Temperature inside the greenhouse should be kept at about 30 ℃ from sowing to the one-leaf and one newly emerged leaf stage, about 25 ℃ at the two-leaf stage, and about 20 ℃ after the three-leaf stage, with a minimum of 15 ℃. The plastic cover should not be removed if the temperature inside the greenhouse is lower than 25 ℃ from the sowing to full seedling. However, if the temperature inside the greenhouse is higher than 35 ℃, the two ends of the greenhouse should be opened for ventilation and cooling. At full seedling, hardening should be carried out through film removal, preferably in the morning or evening when there is a small temperature difference between the inside and outside of the greenhouse so that seedlings can adapt to the new environment quickly. If the cover is removed at noon, irrigation must be performed beforehand to prevent excessively rapid transpiration of the seedlings and slow absorption of water by the roots; otherwise, leaf curl may occur due to dehydration physiologically. As the leaf age increases, the interval of hardening can be prolonged. In particular, at the 2.5-leaf stage, the temperature should not exceed 25 ℃; if so, ventilation and cooling are needed to prevent tall seedlings or "burned" seedlings caused by water loss. After the three-leaf stage, gradually enhanced ventilation is needed to keep the temperature basically the same inside and outside the greenhouse. If there is no frost at night, plastic cover is not necessary till transplanting.

Weaning fertilizer should be applied to seedlings at the two adult leaf and one newly emerged leaf stage. Normally, 45 – 75 kg of urea is needed per hectare for topdressing, together with shallow water. Irrigation management starts from the three-leaf stage. From the three-leaf stage to transplanting, measures should be taken to prompt tillering so as to lay a solid foundation for root vigor and resistance to injury after transplanting. First, "relay" fertilization should be performed for seedling beds with less fertile soil, weak seedlings and slow tillering. Normally, 60 – 90 kg of urea per hectare should be used. Before the two-leaf stage, wetting irrigation should be performed to keep the ditch filled with water and the ridge surface wet. After the two-leaf stage, intermittent irrigation is to be performed to promote multiple strong roots as well as early and strong tillers. Three to four days before transplanting, 112.5 kg of urea per hectare should be applied to help the seedlings re-establish themselves and resume growth. The seedling age of wetland seedlings should be within 30 days (4.5 – 6.0 leaves).

Attention should be paid to the prevention of dead seedlings during the seedling raising process. The causes of dead seedlings can be the following. 1) The seeds are not properly disinfected, leading to pests. 2) The seedling bed is not sufficiently leveled so seedling submergence or sunburn happens. 3) The application of fertilizer may be improper. For example, manure may not be completely decomposed, which causes the roots to be "burned". 4) There may be inappropriate management. For example, the plastic cover may be removed too early when the temperature is high in the early stage, and a sudden temperature drop can cause "green withering". For another example, the cover may not be removed in a timely manner after budding and high temperature can result in seedling burning. 5) Compacted soil in the seedling bed causes unhealthy growth and weak seedlings. 6) There can also be pest damage, such as a thrips outbreak when the temperature, humidity and nitrogen levels are high when single-cropping mid-season rice seedlings are raised. Therefore, efforts must be made to prevent thrips, brown planthopper, white-backed planthopper (transmitting black-streaked dwarf disease) and small brown planthopper (transmitting stripe virus disease). Pesticides, including imidacloprid and acetamiprid, should be applied once or twice to prevent pest damage. Spraying pesticides again two days before transplanting can effectively reduce pests in the early stage in the production field.

(3) Transplanting

Before transplanting, the seedling field should be surrounded by a ditch 20 cm deep and 25 cm wide for drainage. Weeds around the production field should be removed to eliminate pathogen and overwintering pests. Before transplanting, the production field should first be finely plowed with a medium-sized tiller to about 25 cm below the surface, and then treated with 450 – 750 kg per hectare of 45% compound fertilizer as base fertilizer (or 7.5 – 15 t of decomposed manure if conditions permit). After that, the field should be flattened so that no mud appears within the 3 cm water layer. Based on the characteristics of super hybrid rice varieties, seedlings should be planted in wide rows with narrower space between plants. Rows should be parallel to the main angle of incidence of the sun or the direction of the prevailing wind. Seedlings can be placed in the east-west direction in the southwest monsoon region to help with ventilation and light transmission in the middle and late stages. As for planting density, in areas with good ecological conditions, seedlings should be arranged generally in 20 cm × 30 cm or 20 cm × 33.3 cm grids, making a total of 150,000 – 165,000 seedlings per hectare, two per hill. According to the planting principle featuring two tillers within one seedling, there should be totally 900,000 – 990,000 basic seedlings in the field. In areas with normal ecological conditions, seedlings should be arranged generally in 16.65 cm × 30 cm or 16.65 cm × 33.3 cm grids, making a total of 180,000 – 190,000 basic seedlings per hectare, two per hill and a total of 1,080,000 – 1,140,000 seedlings planted in the production field.

2. Dryland seedling cultivation

(1) Overview

The technology of dryland seedling raising was introduced from Japan in 1981, and was first applied in Heilongjiang province with remarkable results. It was then widely promoted throughout the country (Gu Zuxin, 1991). The principle for a high yield is that the seedlings should grow early and quickly after planting by raising seedlings with a vigorous root system. The aerial part of cold-tolerant dryland seedlings is short and strong, with more tillers and leaf epidermal hairs, expanded stomatal openings, a greater total number of roots, more white roots, higher root dry weight, root hair and root absorption area, smaller root tips and mesophyll cells, and enhanced root vigor (Zou Yingbin, 2011). This technology is suitable for single- and double-cropping early-season rice production, and can also be used to raise dryland soft-plate seedlings, as well as dryland and wetland two-phase seedlings of mid- and late-season rice.

Dryland raised seedlings are preferred in the cultivation of high-yielding super hybrid rice. The soil on the bed has good permeability, which is conducive to the cultivation of short and strong seedlings with more root hairs and white roots, and the seedlings are transplanted with more tillers and after transplanting to the production field, the seedlings have strong "bursting" power, quick greening and almost no yellowing period. Therefore, this method of raising seedlings is a better approach for the quality of the seedlings and better transplanting, and to produce early tillering and tillering at lower positions. Moreover, this method is easy, saves labor and time (it can be done where there is open space in front of the house), and does not waste water resources (Zhou Lin et al., 2010)

(2) Process

Dryland seedling raising is a method of seedling raising that does not establish a water layer and only keeps the soil moist during the whole process. The process is as follows – fertilizing → sowing → flattening → earthing → chemical removal → mulching. Around 7 – 10 days before sowing, the field should be plowed deep to make the soil fine and the land flat. Before the land is leveled up, 4,500 – 7,500 kg of decomposed manure and 450 kg of 45% compound fertilizer should be applied per hectare. When making the seedling bed, sufficient fine soil should be prepared for coverage of the seeding bed (fine sieved soil mixed with 20% decomposed manure) and the land plot should be covered it with a film for timely sowing. Before sowing, the field should be plowed on a sunny day. After leveling the soil of the seedling bed, work should be done to create ridge-plots 1.3 m wide and 10 cm high with a drainage ditch around the plots. Before sowing, the seedbed needs to be watered to make the soil fully moisturized, then, pre-germinated seeds are to be evenly spread on the ridge-bed surface, slightly pressed with a broom to bury them in the soil, and evenly covered with 1 cm deep of fine soil.

The key concern in dryland seedling management is the temperature and moisture. In terms of temperature management, the seedlings should be kept warm and moisturized to promote evenness. Do not remove the film when the temperature is below 35 ℃. When the temperature is above 35 ℃, the two ends of the film should be opened for ventilation to prevent bud burning, but the openings should be closed at appropriately 3 or 4 pm. Cool-down of the seedlings can last from the full seedling stage to the 1.5-leaf stage. Remove part of the film from 10 am to 3 pm on a sunny day. Keep the temperature in the film at about 25 ℃ and close it after 3 pm. The 1.5 – 2.5-leaf stage is the critical period for hardening the seedlings, and also a period prone to physiological wilt and bacterial wilt. It is necessary to frequently remove the film for ventilation. This can be done from 9 am to 4 pm on a sunny day to dry the seedbed soil. The two ends of the film can be opened to allow ventilation and keep the temperature in the seedling bed at about 25 ℃. In terms of moisture management, warm and moisture preservation is the key from sowing till when the roots are firmly in the soil. After full seedling, water should be strictly controlled to help the roots get deeper down. The film should be removed in the morning and covered at night for hardening. At the two-leaf stage, the film can be temporarily removed. Normally, film removal is performed in the afternoon on sunny days, in the morning on cloudy days or after rainfall. If a cold spell intervenes, the film should not be removed until the cold spell ends. Irrigation is normally performed once at dawn or at night on the 3 cm surface soil when there is no waterdrop on the leaves of seedlings or both in the morning and at night if the seedbed is too dry. It is advisable to wet the 3 cm surface soil. However, it is better to irrigate well the seedbed whose soil is infertile and compacted for higher permeability. Only by strictly controlling the moisture in the seedling bed, can the advantage during the growth stage be enhanced (Zhang Yong, 2011). In terms of fertilizer management, as the seedbed for dryland seedling is covered with richly fertilized soil, it is not necessary to apply fertilizer during the seedling stage. If the seedbed is not fertilized enough, which is indicated by yellowing leaves of the seedlings, fertilizer can be applied with water supplement in the middle and late stages. Normally, 0.5% urea solution is used to prevent burning. After that, clean water should be applied to seedlings. The seedling age of dryland seedlings should be controlled within 30 days (about five leaves).

3. Scattered planting of seedlings

(1) Overview

Rice scattered planting is a simple technique of using plastic plates or scattered planting agents to raise seedlings with soil by scattering them in the field using gravity. Scattered planting is an efficient cultivation technique that saves land, labor and cost. It is widely accepted by farmers because it frees farmers from heavy workload in the field to pull up and transplant the seedlings. In scattered planting, the damage to the root system caused by the pulling is small and the root system carries soil when scattered. Therefore, the resistance to adversity is stronger after scattering, tillering occurs quickly and at lower positions, and the seedlings have the advantages of early development, early maturity and increased yield and efficiency. However, the flexibility of seedling age is limited with soft plates, so it is not an appropriate method for cultivating aged seedlings. Scattered planting is mainly used for early-season rice production. Its field management also applies to middle- and late-season rice production.

(2) Process

1) Material preparation

The quantity of plastic soft plates per hectare depends on the number of soft plate holes and the density of field cultivation. Usually, there are three kinds of soft plates for seedlings based on the number of holes in each plate. 1) 561 holes in 19 rows per plate (10 rows of 30 holes and 9 rows of 29 holes), which is suitable for small seedlings below 3.5-leaf of age; 2) 434 holes in 17 rows per plate (9 rows of 26 holes and 8 rows of 25 holes), which is suitable for small and medium-sized seedlings below 4.5-leaf of age; 3) 353 holes in 15 rows per plate (8 rows of 24 holes and 7 rows of 23 holes), which is suitable for medium-sized late-season rice seedlings with a leaf age of 3.5 – 5.5 leaves (6.5 leaves for dryland aged seedlings with soft plates). The 353-hole soft plate for cultivated seedlings have a length of 60 cm, a width of 33 cm, and a height of 1.8 – 2.0 cm, 50 g per plate. It is a slope cone with an upward opening. Scattered planting for super hybrid rice requires 600 plastic plates with 353 holes or 450 plates with 434 holes per hectare.

In addition, seeds, seedling substrate and agro-plastic film should be prepared. Based on conditions of the production field, about 22.5 kg/ha of super hybrid rice seeds are needed, together with 150 kg of high-quality sieved farm fertilizer or 150 kgof special seedling substrate, 1,200 – 1,500 kg of sieved germ-free dry soil or fertile clay loam, seed disinfectant, isolane or multifunctional seedling-strengthening agents (prepared according to instructions) and agro-plastic film with a width of 2 m and a length as needed.

2) Seed treatment

i. Seed soaking and disinfection

In order to prevent bakanae disease and rice blast, seeds should be soaked in 2,000 – 3,000 times of 25% Prochloraz emulsifiable concentrates for 48 hours or in 300 times diluted TCCA solution for 8 – 10 hours for disinfection.

ii. Pre-germination

Seeds should be disinfected till budding and placed at a temperature of about 32 ℃ for pre-germination. When 90% of the seeds sprout with 2 mm long buds, they can be sown after six hours of dryingat room temperature.

3) Seedling soil

The soil used for seedling raising should be loose loam, free of pathogens, weeds and herbicide residues. It can be pathogen-free dry soil or fertile clay loam soil in rice fields. After drying, the soil should be crushed into small pieces and sieved. If there are 600 soft plates per hectare of the nursery, then 1,500 kg of substrate soil and 300 kg of high-quality manure are required. Both soil and manure need to be free from grass clippings, gravel, hard lumps and debris through a sieve with an aperture of about 4.24 mm, and the soil-fertilizer ratio should be 8 : 2 or 7 : 3.

4) Seedling bed selection and treatment

i. Seedling bed selection

Seedling beds should be in leeward and sun-facing places and have fertile soil, convenient for irrigation and drainage and transportation.

ii. Seedling bed disinfection and acidity adjustment

The soil should undergo fertility treatment before sowing. During soil preparation, 5 – 10 kg of fully decomposed manure per square meter should be applied for soil flattening, and after that, Isolane should be applied for disinfection and acidity adjustment. The purpose of seedling bed disinfection is to eliminate germs in the soil, which is important for the success of seedling cultivation. About 1.0 – 2.0 mL of Isolane in 2 – 3 kg of water should be sprayed per square meter to effectively inhibit the proliferation of *rhizoctonia solani* and enhance the water absorption, fertilizer absorption and disease resistance of the seedlings.

5) Sowing

The sowing period should be determined according to the local temperature and the age of the seedlings. Sowing can be done when the temperature is stably over 12 ℃. First, seedling plates are placed in parallel on the seedling bed, with the long sides of two plates facing each other and the short sides 15 cm from the boundary of the seedling bed. A rope can be used to keep the plates parallel. The plates should be closely together with no gap in between. After that, nutrient substrate (soil, fertilizer, etc.) should be put in the plates and then sowing can start.

Sowing with dry soil. Spread the soil on the soft plates, fill sieved fine soil to 2/3 of the depth of the holes, evenly spread 70 g of pre-germinated seeds into the holes of each plate, cover the seeds with fine soil, irrigate, fill up the holes if the soil sinks after watering, obtain mud by mixing sieved soil, fertilizer, seedling strengthening agent and water, or by digging directly from the bottom of a pond or ditch, scoop mud into the soft plates to make a smooth and even layer, leaving the holes unobstructed, let the thin mud sink for 3 mm, and do the sowing. After that, use a bamboo broom or sorghum broom to press the buds into the mud surface. There is no need to irrigate as long as there is water in the ditch.

Where the temperature is too low, seedling beds should be covered with a film. After sowing, a greenhouse 1 m wide and 40 cm high should be erected, and covered withplastic film. Dig a small ditch 5 cm deep around the seedbed, insert the lower end of the film into the ditch and press down with soil to make it tight.

6) Seedling stage management

i. Water management

As there is enough water before sowing, there is no need for irrigation before seedlings emerge. When the seedling leaves do not have waterdrops in both morning and evening, and the bed surface becomes white, irrigation should be performed in a timely manner. Irrigation should be preferably performed only once before 10 am or after 4 pm. As the temperature and the seedling age increase, the growth conditions of the seedlings should be checked more frequently in case of insufficient water. When the seedlings are in the stage of one adult leaf and one newly emerged leaf, fertilizer and a small quantity of water should be applied. When the seedlings enter the 2.5-leaf stage (10 d before scattering), water shortage may happen due to hot weather and it is necessary to water the seedlings once a day to ensure healthy growth.

ii. Temperature management

No ventilation is needed 3 – 4 d before the seedlings emerge, but high temperature and humidity must be maintained. Once all seedlings emerge (green), the seedbed temperature should be kept at 25 ℃–28 ℃ and the film should be opened a little to let wind in and closed back up two hours before sunset to prevent chill damage. When the seedlings reach the 1.5-leaf stage, ventilation should be enhanced to keep the temperature at 20 ℃– 23 ℃. When the seedlings reach the 2.5-leaf stage, ventilation should be performed both during the day and at night. One week before scattering, if there is no cold spell, there is no need to cover the seedlings with film (Cai Xueju, et al., 2011).

iii. Fertilization in seedling bed

After the 2.5-leaf stage, seedlings must be topdressed in time if defertilization happens (the seedlings turning yellow). About 6 – 10 g of urea per square meter should be applied together with water. The urea can be sprayed after being mixed with water, or applied directly right before water is sprayed.

iv. Seedling lifting

At the 3.0 – 3.5-leaf stage, the seedling age is about 20 days, and the seedling height is about 10 cm. Check the plate soil humidity the day before scattering. If the soil is too wet, the plates should be drained. If the soil is too dry, irrigation should be performed to the ditch to moisturize the soil but not the seedling bed. Lift the seedling plates from the seedling bed to pull off the roots that are inserted into the soil of the seedling bed, and then rolled the seedlings to the production field for transplanting.

7) Seedling scattering

i. Soil preparation

The seedling scattering field should have good conditions for irrigation and drainage to effectively control the water layer and ensure that the seedlings stand and remain moisturized after scattering. It is better to plow the rice field 7 – 10 days before seedling scattering to a depth of 15 – 20 cm, which is conducive to eliminating weeds. Rice fields free of weeds and green manure can be plowed with a rotavator.

ii. Base fertilizer application

Before soil preparation, 15,000 kg of farm manure and 450 – 750 kg of compound fertilizer (or special compound fertilizer for super hybrid rice, applied according to instructions) per hectare should be applied evenly in the field, which should be watered 3 – 5 days in advance to make it free of stubble. The field surface should be leveled till the height difference is no more than 3 cm so that a 3 cm water layer can cover the field surface.

iii. Timing of seedling scattering

Seedling scattering is the most effective after plowing and harrowing and after the mud sinks. Too long after the mud sink, the field surface will be hard and the seedlings cannot go deep into the soil, causing the seedlings to fail and making re-establishment slow. In this case, try scattering several seedlings to check if it works before starting massive scattering. Normally, clay loam soil should be left overnight after being harrowed and seedling scattering can start the next day. For loamy soil, seedling scattering can start 5 – 6 hours after harrowing; while for sandy loam soil, scattering can start 2 – 3 hours after harrowing.

iv. Seedling scattering

The field should be drained off before scattering, leaving only wet surface or footprint water. If the production field is too large, ropes should be pulled to separate ridges before scattering. After scattering, the seedlings within 20 cm from the dividing ropes should be picked up and thrown further into the plots to keep the ridges clear. When scattering the seedlings, hold the seedling plates in one hand, grab a handful of seedlings, move them a bit to separate the roots, and then throw to more than 3 m high. When scattering, seedlings should first be sparsely and then add some more to fill the gaps. Make sure to scatter the seedlings far first and then near, against the wind first and then downwind. There should be two or three rounds of scattering, with 70% of the seedlings scattered in the first round, 30% in the second, and the third round is for some adjustment if needed. Point scattering is another way to scatter seedlings. It requires the division of compartments within the field before scattering or placing the seedlings along aisles before more seedlings are scattered further as people walk along the aisles. This method strengthens the control of distribution and quality, but reduces efficiency.

v. Density

The scattering density depends on the soil fertility of the field. Specifically, the density should be about 195,000 hills per hectare for highly fertile fields, 225,000 hills per hectare for medium-fertility fields, and 270,000 hills per hectare for low-fertility fields.

8) Management for seedling establishment

i. Floating prevention

It is not wise to irrigate the field after seedling scattering. Two to three days after scattering, a thin layer of water should be maintained and light irrigation should be performed if the field is drained off, which is good for early seedling establishment. If it rains during this period, the field should be drained off in time to avoid seedling floating.

ii. Water layer management

Since seedlings are more shallowly planted (1.5 – 2 cm deep only) by scattering than by manual planting, most of the seedlings are small or medium-sized, so deep irrigation should be avoided. From the end of the scattering to when the seedlings stand upright, shallow irrigation of 2 – 3 cm of water is preferred because it is the only way to allow tillering at low nodes and ensure a high effective panicle rate.

iii. Herbicide application

Scattered seedlings are more sensitive to herbicides because they are small and have shallow roots in the early stage. Therefore, herbicide application should be especially careful. Seven days after seedling scattering, one or two new leaves will emerge. If herbicides have not been applied before seedling scattering in the field, weeds will grow and herbicide application should be considered. For example, "Paoyangjing" (18.5% Propolachlor · Benzyl WP) can prevent weeds such as *Echinochloa crusgalli* (L.)

Beauv, *Cyperus rotundus* L., *Monochoria vaginalis* (Burm. F.) Presl ex Kunth, *Rotala indices* (Willd.) Koehne, and *Linderniaprocumbens* (Krock.). Specifically, 7 – 8 days after seedling scattering, 450 – 525 g of herbicides per hectare should be evenly applied, mixed with 112.5 kg of urea or 225 kg of fine fluvo-aquic soil (tidal sand). When the herbicides are applied, keep 3 – 5 cm of water in the field for 5 – 7 days. If there is water shortage during this period, the water should be replenished immediately instead of drainage and shift to normal field management.

4. Machine transplanting

(1) Overview

Machine transplanting is a major change to the traditional manual transplanting which is labor-intensive, costly and less efficient. As the most important part of the mechanization of rice production, it is conducive to freeing up rural labor, promoting the development of the secondary and tertiary industries and accelerating agricultural modernization. It is not only widely accepted by farmers, but also increasingly valued by agricultural technology, machinery and administrative departments at all levels. Compared with direct seeding, machine transplanting uses fewer seeds and generates even, strong, fertilizer-tolerant and lodging-resistant seedlings. In particular, early sowing and planting can be performed to extend the vegetative growth period, promote early tillering, and create large panicles with more grains, so as to increase yield significantly. Machine transplanting gives rice better agronomic traits than seedling scattering. After transplanting, the period of seedling establishment and re-establishment is shortened by 2 – 3 days, with earlier tillering and even distribution in the field, which makes management easier. The seedlings receive equal sunlight so that they grow evenly with well-developed roots, strong capacity for fertilizer absorption, high lodging resistance and less damage from pests. Although machine-transplanted seedlings are small in size, with shorter seedling duration and early transplanting, shorter full growth period and smaller panicles than those sparsely planted in a traditional way, they have great potential for super high yield as long as there are supporting facilities and technology for field management.

High quality of seedlings is the key to the success of machine transplanting. Special soft plates (or hard plates) are used to grow seedlings with soil for machine transplanting. In seedling scattering, balled seedlings are scattered out with soil and with the roots disconnected. Therefore, the seedling scattering soft plates are cone-shaped, also known as "bowl plates". On the contrary, machine transplanted seedlings need connected roots, which can be packed into the transplanting trough in bundles. Therefore, the plates used are flat-bottomed or semi-short square pot holes, without oversized connected roots and excessive quantity of soil (mud). Otherwise, it may affect the quality and efficiency of transplanting. Machine transplanting can also be performed for outdoor seedlings without plates. Before transplanting, the nursery needs to be divided into squares to fit the transplanter (Zhu Jinping et al., 2002). The methods of transplanting and field management are the same as those for plate seedlings.

Machine transplanting can be done in a factory-based manner or outdoors in dry soil or paddy field. Compared with the traditional methods of seedling raising, machine transplanting is characterized by high seeding density, young seedlings, less land required and convenience for management. However, the production field must be fit for a transplanter. Machine transplanting with soft plates is a simple and easy method used in places where there are no conditions for factory-based seedling cultivation.

(2) Process

1) Sowing date

Due to the high seeding density of machine transplanting, the soil is thin in the plates. Therefore, the seedlings should not be too aged, and generally, seedlings with 3.5 – 4.0 leaves are transplanted. If the leaf age is young, the seedlings will have weak roots, which can be easily damaged in machine transplanting and the seedlings tend to re-establish slowly. If the seedlings are aged, there will be more roots and long leaves, which can also be damaged easily in machine transplanting. A recommended seedling age is about 20 days for early-season rice, and about 15 days for mid- and late-season rice. Therefore, the sowing date should be determined based on the actual conditions, especially the harvest time, so as to ensure that the seedlings are planted at the right seedling age.

2) Sowing quantity

Machine transplanting requires slightly more seeds than conventional methods, about 30 kg per hectare.

3) Seed soaking and pre-germination

Disinfection, seed soaking and pre-germination should be carried out in traditional ways. It is advised to dry the seeds before soaking, which helps the seeds absorb water and increase the germinating rate. During pre-germination, ventilation should be enhanced by frequently stirring the seeds and water can be sprayed for cooling when necessary. The purpose is to keep the temperature at around 30 ℃ for pre-germination. When the seed budding rate reaches 90% or the buds become 1 – 2 mm long, sowing can be performed.

4) Seedling bed selection and preparation for sowing

Nurseries that are convenient for irrigation, drainage and transportation are preferred. Soft plate seedling cultivation can be performed on paddy fields or dryland.

i. No-tillage dryland seedling cultivation

No-tillage seedling raise on dryland saves time, labor and effort, so it is suitable for small planting areas where the soil is loose and good for seedling growth. This method is particularly good for raising seedlings in areas where there is no water supply for irrigation in spring. The seedling bed can be vegetable plots, dryland or small available land plots around the house. Soil should be dried and sieved several days in advance together with a mixture of animal manure or compound fertilizer (mixed with approximately 1,500 kg of soil per hectare for production fields). Before sowing, no tillage is needed and seedling plates can be placed adjacent on the plot flatly and straightly. Then, spare dry soil should be evenly spread on the seedling plates, flattened with a wooden board, and then sufficiently irrigated for soil moisture saturation. Where conditions permit, mud from a pond or a clean ditch near the seedling bed can be scooped up and spread on the seedling plates.

ii. Paddy field seedling cultivation

Land plots with soft loam should be selected as the seedling bed and irrigated and harrowed 4 – 5 days before sowing. One day before sowing, 300 kg of multi-nutrient compound fertilizer per hectare should be applied as base fertilizer, and then the land should be harrowed flat (sandy soil field can be fertilized and harrowed at the same time). After the fertilizer settles, the seedling bed should be flattened according to the bed width (1.5 m, slightly larger than the space between two seedling plates), height (0.15 m) and ditch width (0.3 m). The seedling plates should be placed, two, side by side, in each row, on the bed. The weeds, gravel and other debris in the ditch should be removed. The soil in the ditch should be dragged with a hoe back and forth to make mud, which should then be used to cover the seedling plates with the help of a wooden board (or seedling broom) until it is level and even.

5) Sowing

In general, 30 kg of seeds are needed per hectare, so there should be 300 – 375 machine transplanting plates per hectare. Seedlings must be evenly distributed, including in the corners and along the lines. After sowing, the seeds should be pressed slightly down with a broom so that they are covered in muddy water.

6) Seedling management

Temperature, moisture and nutrition are necessary conditions for seedling growth. In particular, temperature is important in spring. Low temperature is unfavorable to the emergence and growth of seedlings. Therefore, it is necessary to control the temperature, strengthen fertilizer and water management and prevent pests and rodents.

i. Temperature adjustment with film

Temperature is generally low in spring. To ensure even emergence and healthy growth, the seedlings must be covered with white agro-plastic film in a bamboo-strip shed with all ends sealed (which can not only keep them warm but also prevent rats). The temperature inside the shed should be kept at 20 ℃–30 ℃. After seedlings emerge, it is good to uncover the film for ventilation and heat dissipation from 10 am to 3 pm on sunny days, but it should not be fully open in order to maintain the temperature inside the shed. Full uncovering for hardening can be performed 3 – 4 days before transplanting. If there is no film, night irrigation and day drainage should be adopted to keep the seedlings warm in low temperature. Use 2% kasugamycin diluted to 500 times to treat rotting caused by low temperature.

ii. Water management

Water management for the seedlings is mainly to keep the soil moisturized in the seedling plates. Early-season rice seedling raising requires dry ditches to keep the soil ventilated, which is conducive to the growth of seedling roots. However, moisture management, especially for dryland seedlings, requires water spray every day at noon to prevent physiological dehydration. Due to the high temperature, mid- and late-season rice seedlings need water up to half of the depth of the ditch. Drainage should be performed on rainy days. Water control should be exercised three days before planting.

iii. Fertilization

Machine-transplanted seedlings have a short seedling age, so base fertilizer is the main fertilizer, and supplementary fertilizer can be applied when necessary. With sufficient base fertilizer, compound fertilizer should be applied once every four days after sowing. Yet fertilization must be performed after the seedling plates are covered with water and get dried, in order to prevent fertilizer damage. If there is no irrigation,

1.0% compound fertilizer can be applied once or twice with a sprayer after the fertilizer fully dissolves in water. Urea is not required during the seedling period so that the leaves would not get too long.

iv. Pest and rodent prevention

The main pests in the seedling stage include rice thrips, planthoppers, stem borers and leaf folder. Pesticides must be used for prevention and control in a timely manner before transplanting. Special attention should also be paid to rodent damage for early-season rice seedlings. Timely administration of high-efficiency and low-toxicity rodenticides is needed as well.

7) Transplanting

When seedlings have three leaves, it is the time for transplanting. Two days before transplanting, enough fertilization should be applied in the production field, which should have been harrowed in advance to avoid deep transplanting which is likely when transplanting and harrowing are done at the same time. When conditions permit, seedlings can be transported to the field together with the seedling plates, or rolled up for transport. Then, the seedlings should be put down immediately in order to avoid deformation and breakage. Early transport is not encouraged in order to prevent water loss and wilting. As for the transplanting density, in areas with good ecological conditions, there should be about 195,000 seedlings per hectare, while in areas of ordinary ecological conditions there should be about 225,000 seedlings, 2 - 3 per hill.

5. Seedling management for direct seeding

(1) Overview

Direct seeding is a rice cultivation technique where seeds are sown in the field directly without a seedling bed or transplanting. There are several advantages. Firstly, compared with seedling and transplanting, direct seeding saves energy and labor and is easy to perform. Secondly, it can lead to high yield. As rice plants directly seeded have more and even lower nodes, they are likely to have a high panicle rate, leading to a high number of panicles for high yields. Thirdly, after direct seeding, plants have a short growth period, due to no injury to plants and no requirement of seedling re-establishment, usually 5 - 7 days shorter than the growth duration of transplanted seedlings. Fourthly, it is suitable for intensive production. Large-scale direct seeding can save a lot of labor and relieve seasonal labor tension. It is of great significance to mechanization, convenience and modernization of rice production. Therefore, direct seeding is worth promoting. However, compared with transplanting, the key to the success of direct seeding lies in uniform and strong seedlings as well as timely weed control. The transplanted seedlings are tall and soaked in water in the field, which is not good for weed growth. Rice directly seeded grows together with weeds and weeds generally have stronger stress resistance than rice. Therefore, weeds are a threat to rice seedlings. In addition, lodging is more likely to happen in direct seeding, because rice seeds directly sown on the surface of the soil with tillering nodes exposed above the ground and shallow roots, coupled with a large number of seedlings, high tillering rate and long base nodes. Therefore, special attention should be paid to technical measures such as "early development of all seedlings, weed control and pest prevention, fertilization to prevent early senescence and robust cultivation to prevent lodging" (Bai Yunhua, 2017).

(2) Process

1) Refined field preparation

Firstly, direct seeding requires high-quality field preparation. Early plowing with a rotavator should be carried out right after the harvest of prior crops. About 450 – 750 kg of 45% (15 – 15 – 15) compound fertilizer should be applied per hectare as base fertilizer for rotary tillage. Secondly, importance should be attached to field preparation. The field must be leveled sufficiently so that a 3 cm water layer can cover the field fully without leaving any bare mud, because the high parts are prone to weeds while the low parts are prone to seedling rot. Thirdly, 30 cm-wide ditches should be dug to 15 – 20 cm deep at an interval of 3 m for fertilizer and pesticide application and other field management practices. Ditches inside the field and surrounding the field should also be dug, with a width of 20 cm and a depth of 20 cm, and all three kinds of ditches should be connected to ensure water drainage.

2) Even sowing

Seeds should be disinfected by soaking in TCCA or Prochloraz. Before sowing, 70% imidacloprid should be applied for seed dressing to prevent damage caused by rice thrips and planthoppers in the early stage of seedling cultivation. The key points in sowing are as follows.

i. Sowing timely

Sowing should be carried out from late April to early May in the south of the middle and lower reaches of the Yangtze River, in order to avoid excessive heat in late July and early August, and start from April 15 – 20 in mountainous areas and areas north of the Yangtze River when the temperature is stably higher than 15 ℃.

ii. Preparing the field

Before sowing, the field should be prepared in such a way that it has moderate hardness. Sowing will be normally carried out one day after plowing, with half-grain seeding in depth.

iii. Even seed distribution

About 22.5 kg/ha of seeds should be sown in direct seeding. In order to achieve uniform sowing, 70% of the seeds can be sown first, followed by the remaining 30% in another round. After sowing, the seeds should be dragged lightly with a wet sack or a large piece of plastic film to cover them with mud.

iv. Gap filling in low density areas

After the emergence of rice seedlings, transplanting can be carried out at the three-leaf stage to move seedlings in high density parts to fill gaps in low density parts. This can be done with soil to increase the survival rate. Field management after sowing mainly centers on water management, fertilization and weed control.

3) Water management

For direct seeding, the field should be moisturized without water logging at the stage of two mature leaves and one newly emerged leaf. It is necessary to fill the ditches with water on sunny days, half-fill them on cloudy days and leave no water in them on rainy days, so as to ensure seedling rooting and establishment. A shallow water layer can be created on the field after this stage. Intermittent irrigation should be carried out after the four-leaf stage until effective tillering. Continuous deep water flooding should be avoided. Instead, alternate drying and wetting irrigation should be carried out, with wetting as the mainstay. When the total number of seedlings reaches 80% of what is expected, the field should be drained off till the initial young panicle differentiation stage. A shallow water layer should be kept during the booting and flowering stages. Alternate drying and wetting irrigation should be carried out during the grain filling stage, and stopped about seven days before harvest.

4) Weed control

Directly sown rice seedlings are small and sensitive to weeds and herbicides, so weed control must adhere to the principle of "agricultural control as the basis and chemical control as the precursor". As for the specific chemical weed control technology, it is important to ensure right timing to kill grass buds, especially barnyardgrass, which must be eliminated before the two leaves and one newly emerged leaf stage, and the weed community dominated by barnyardgrass should be treated with sealed chemical weeding mainly in the weed germination and seedling stage, so as to obtain the best economic benefits with the least input (Liao Kui, 2017). The weeding process is usually "sealing, herbicide and manual weeding". Sealed weeding means that 24 hours after sowing, when there is no water in the field, pre-emergence herbicides such as Sulfotrim and Propaquizafop should be used in direct seeding fields, mixed with water according to instructions and sprayed on the soil surface evenly. The second step is to use herbicides again 15 days after sowing at the 2 - 3-leaf stage, when the field is wet with no ponded water so that weeds are fully exposed. A solution of 900 - 1,200 mL of Daojie (Penoxsulam) and Qianjin (cyhalofop-butyl) mixed with 750 kg of water per hectare is applied on the leaves of the weeds. Generally, 24 hours after spraying, shallow irrigation should be carried out to promote rice growth and improve the efficiency of weeding. The last step is to remove the remaining weeds manually. After the rice enters the "field drying" (sun field) stage, especially before heading, it is necessary to remove all remaining barnyardgrass manually.

Section 2 Field Management of Super Hybrid Rice

Ⅰ. Balanced Fertilization

1. Fertilization requirement characteristics of super hybrid rice

The fertilization of super hybrid rice should be based on the local climate, soil and annual yield level, and the corresponding fertilization amount should be designed to give full play to the potential of the variety and achieve the goal of high-yield, high-efficiency and safe production. In general, about 1.8 kg of pure nitrogen is required for every 100 kg of rice produced. The dosage can thus be calculated according to the yield potential before fertilization. For example, for a variety whose yield potential is 15 t/ha of rice grains, the pure nitrogen needed should be $15,000 \div 100 \times 1.8 = 270$ kg. If the soil can supply 120 kg/ha of nitrogen, and the fertilizer's seasonal utilization rate is 40%, then the pure nitrogen needed for a rice yield of 15,000 kg/ha can be calculated as $(270-120) \div 40\% = 375$ kg. In the same vein, for a variety with a yield of 9.0 t/ha, the quantity of pure nitrogen needed is about 180 kg; for a yield of

10.5 t/ha, it is 210 kg; for a yield of 12 t/ha, it is 240 kg; for a yield of 13.5 t/ha, it is 300 – 330 kg; and for a yield of over 15 t/ha, it is 360 – 375 kg. The ratio of nitrogen, phosphorus, and potassium should be 1 : 0.6 : 1.2 (Ling Qihong, 2007).

2. Fertilization application of super hybrid rice

Super hybrid rice should be provided with sufficient base fertilizer, early tillering fertilizer, appropriate quantity of panicle fertilizer, and supplementary grain fertilizer at the later stages to delay senescence.

(1) Base fertilizer

Base fertilizer, also known as basal fertilizer, provides the basic nutrition for rice growth and must be applied in sufficient quantities to lay the foundation for a high yield. Base fertilizer accounts for about 40% of the total fertilizer application, generally based on organic fertilizer or compound fertilizer, applied once and for all before transplanting and tilling. The base fertilizer plays several functions. Firstly, base fertilizer such as pig and cow manure, and other types of organic fertilizer can increase the content of soil organic matter, improve agglomerate structure, and enhance the ventilation of the soil, which is conducive to developing a huge and deep root system and making the plant leafy and grow healthily. Secondly, it can promote the early development of seedlings. Base fertilizer is mainly organic fertilizer, with the appropriate proportion of nitrogen, phosphorus, potassium fast-acting compound fertilizers, seedlings start to develop new roots with sufficient nutrient supply, which is conducive to early seedlings and quick development. Thirdly, the fertilizing effect of organic fertilizer lasts long and can keep the rice plant growing steadily in the middle and late stages, without losing fertilizer and early senescence.

(2) Tiller fertilizer

Tiller fertilizer accounts for about 30% of the total fertilizer application. Nitrogen plays a leading role in rice tillering, so early application of fast-acting nitrogen tiller fertilizer to make leaf color turn green quickly is the main measure to promote early tillering. Tiller fertilizer should be applied 5 – 7 days after manual transplanting or machine transplanting, seven days after seedling scattering, and 20 – 25 days after direct seeding to promote early, rapid and multiple tillering and reduce ineffective tillers. If tiller fertilizer is applied too early, the seedlings will not grow new roots and the fertilizer will be lost. However, if applied too late, the seedlings will lose tillers on lower nodes, and grow ineffective tillers. It is advisable to have a thin water layer in the field when applying tiller fertilizer and leave the field to naturally drain off, so as to improve fertilizer efficiency. Wetting irrigation and field ventilation should be performed during this period so that the seedlings can develop robust roots, which are conducive to the absorption capacity and early tillering (Li Jianwu et al., 2013).

(3) Panicle fertilizer

Panicle fertilizer accounts for about 30% of the total fertilizer application. Super hybrid rice requires more fertilizer than conventional rice, so in addition to applying sufficient base and tiller fertilizer, the application of panicle fertilizer is also important, and it is the main measure to give full play to the advantage of large panicles for a high yield. Panicle fertilizer is applied when the young panicles on the main stems are differentiated at the second stage, which provides sufficient nutrients to promote branching, prevent spikelets from degenerating, increase the number of spikelets and produce large panicles and big grains.

The application of panicle fertilizer can also help develop roots, healthy leaves, strong stems and lodging resistance (Li Jianwu et al., 2013).

(4) Grain fertilizer

Grain fertilizer is mainly foliar fertilizer, where fertilizer is directly sprayed on the leaves so that all nutrients, micro-fertilizer and plant growth regulators can enter the plant from the leaves and participate in metabolism and organic matter synthesis. Based on pest control, when heading reaches at 80%, 175 g of potassium dihydrogen phosphate and 15 packs of Gulibao (or Daoduoshou, Penshibao, Fengchansu, Zengchansu, Gudazhuang, or Yemianbao, following the instructions) should be used, mixed with 900 kg of water, in the form of foliar spray to promote panicle and grain growth, reduce empty grains, and increase the seed setting rate and grain weight. The supplementary application of foliar fertilizer can effectively enhance the stress resistance and disease resistance of the plants, extend the functional period of leaves, prevent early senescence, increase the oxygen supply of roots to enhance root vigor, speed up grain filling, promote maturity and grain fullness, and thereby improve rice yield and quality (Li Jianwu et al., 2013).

Ⅱ. Scientific Water Management

Water management for super hybrid rice is mainly based on wetting irrigation. After transplanting, deep irrigation should be performed to protect the seedlings. After seedling re-establishment, shallow irrigation should be performed to promote tillering. When the number of seedlings reaches about 2,250,000 per hectare, the field should be drained off and dried until the leaves turn yellow and fall. When it is time to apply panicle fertilizer, but the leaf color has not faded, field drying should continue without further watering and fertilization. After the field is re-watered, repeated wetting irrigation should be performed (1 - 2 cm water first, and then 1 - 2 cm water again a few days after it gets dry). From the booting stage to the heading stage, a shallow water layer should be maintained, but if the temperature is too high or too low, deep irrigation can be carried out. At the gain filling stage, alternate wetting and drying irrigation should be performed, with the former as the mainstay at the flowering stage and the latter to maintain root vigor. This method helps protect the leaves, prolong the life of functional leaves, prevent early senescence, and improve the population photosynthetic capacity and fertilizer utilization, thereby improving the seed setting rate and seed fullness (Li Jianwu et al., 2014).

The specific process of water management is detailed as follows. 1) Thin water transplanting - leave a thin layer of water during transplanting to ensure the quality of transplanted rice seedlings and prevent seedling floating or seedling missing. If a grid rower is used for field plan, the water should be drained off first to make the grid clear. 2) Inch water rejuvenation - 5 - 6 days after transplanting, irrigate with about one inch of water to create a relatively stable environment in terms of temperature and humidity to promote the occurrence of new roots and rapid rejuvenation of live plants. 3) Shallow water and moist tillering - repeatedly perform alternate drying and wetting irrigation, with the latter as the mainstay, and then perform manual weeding and topdressing with 0.5 - 1 cm of thin water; when it gets dry, make the field open and moisturized for 2 - 3 days. No muddy water on sunny days and no water on rainy days can help boost root growth, early tillering and tillering at lower nodes. 4) Light sun-drying for healthy seedlings – when the total number of seedlings reaches about 2,250,000 per hectare, drain the field and sun-dry it until the leaves become yellow. When it is time to apply panicle fertilizer but the leaf color has not faded, field drying should be continued without further watering and fertilization. Generally, the field should be sun-dried until it cracks but does not sink when trod on; white roots appear on the mud, the leaves stand up until the color fades. Fields with a high groundwater level, heavy soil texture and vigorous seedlings should be sun-dried heavily, while where irrigation is inconvenient, soil is sandy and seedling growth is weak, the field should be lightly sun-dried to promote strong stems, increase the panicle formation rate, reduce humidity in the field and reduce pest damage. 5) Embryo development with water – restart irrigation when the plant population in the field enters the early panicle differentiation stage, perform shallow water irrigation, let the field to drain naturally, and irrigate again 1 – 2 days after mud is out. Before and after young panicle differentiation (the 5th – 7th stages of young panicle differentiation) and meiosis, keep a water layer of about 3 cm. 6) Heading with sufficient water – as more water is required during heading and flowering, keep a small quantity of water to create an environment with high relative humidity in the field, which is conducive to successful heading, flowering and pollination. 7) Drying and wetting for strengthening grain-filling – from the end of the flowering stage to the maturity stage, dry and wet the field alternately, with wetting as the mainstay in order to improve root vigor and delay senescence. This can provide enough oxygen for root growth, develop roots to support leaves and increase grain weight. 8) Water drainage at the full ripe stage – 5 – 7 days before harvest, the population enters the full ripe stage and water should be drained off and the field sun dried. Too early water drainage may affect grain filling and the yield (Li Jianwu et al., 2014).

Ⅲ. Integrated Pest Control

Super hybrid rice pest and disease control adheres to the principles of "prevention combined with control". Prevention of pests and diseases is the main focus, with one or two rounds done during the seedling period, mainly to control rice thrips and rice planthoppers and prevent the occurrence of southern black streak dwarf disease. One day before transplanting, the seedling field should be sprayed with long-lasting pesticide, so that the treated seedlings are in the production field, which can effectively reduce the pests and diseases in the early stage in the field. The main pests to be controlled include *Chilo suppressalis*, *Scirpophaga incertulas*, *Cnaphalocrocis medinalis*, rice planthoppers, blasts, sheath blights, and bacterial blights. Pesticides should be used according to instructions, and the control measures depend on the incidence of pests (Li Jianwu et al., 2013).

Section 3 Cultivation Model of Super Hybrid Rice

Ⅰ. Modified Intensive Cultivation Model

1. Overview of the technique

System of rice intensification (SRI) is a new cultivation method proposed by Father Henri de Laulanie of Madagascar in the 1980s. This system advocates "plant-soil-water-nutrient" management, which is characterized by shortening the seedling age, single transplanting of very young plants, reasonably sparse planting, moisture management and intermittent irrigation, diligent mid-tillage, and heavy organic fertilizer. It is the same as the essence of traditional Chinese rice cultivation, which is "fine cultivation, reasonably sparse planting and a good combination of land use and land maintenance". This system helps increase yield and save water, while improving the resistance of rice to biotic and abiotic stress; and it can also save a lot of labor and seeds.

In 1998, Yuan Longping introduced SRI to China (Yuan Longping, 2001), and discussed its application in the cultivation of super high-yielding hybrid rice. After that, HHRRC, CRRI, Nanjing Agricultural University and Sichuan Academy of Agricultural Sciences carried out corresponding experiments based on local conditions and proposed such technologies as dryland seedling raising for transplanting, aerobic irrigation, triangular planting, and combination of inorganic and organic fertilization. They also proposed requirements for suitable varieties and combinations, appropriate planting densities, and integrated pest control measures, which greatly improved and developed the original SRI and formed an improved and intensified rice cultivation technical system suitable for local conditions. A large number of studies have shown that improved SRI is more suitable for China's rice variety portfolio covering multiple croppings, different ecological conditions, and many types and various other characteristics of rice cultivation in China, and have better adaptability, practicality and operability than conventional SRI, bringing significant yield and income increase.

2. Review of yield and efficiency increase

SRIin Madagascar basically doubled rice yield. The improved SRI for super hybrid rice resulted in a 15% yield increase due to a higher base value. At the same time, it has the advantages of saving seeds, seedling fields and costs. Super hybrid rice SRI generally requires 3.0 – 4.5 kg/ha of seeds, which can save 8.3 – 10.5 kg/ha of seeds and more than 80% of seedling fields compared with conventional cultivation techniques, and the total cost can be lowered by about 215 yuan/ha. This system mainly applies compost and animal manure, which helps maintain soil fertility. According to research data, the yield of high-yielding cultivation of production fields is 10.0 – 12.0 t/ha, with 450 – 600 kg/ha of urea and 275 – 450 kg/ha of compound fertilizer, which cost 1,200 yuan/ha. SRI does not establish a water layer in the paddy field, but practices intermittent "light" irrigation throughout the process, so the evaporation from the field surface is only 1/6 – 1/4 that of the conventional irrigation method, which can save about 3,000 t of water per hectare.

3. Technical key points

(1) Heavy organic fertilizer and focus on high quality

Super hybrid rice needs more fertilizer, especially a reasonable mix of nitrogen, phosphorus, and potassium fertilizer. Therefore, heavy organic fertilizer and balance between nitrogen, phosphorus and potassium are important measures to ensure high yields. The amount of organic fertilizer applied should be no less than 30 t/ha, and if compost is insufficient, additional organic fertilizer such as bean cake fertilizer and safflower grass should be applied.

(2) Cultivation of strong seedlings for early transplanting

Improved super hybrid rice SRI requires the cultivation of strong seedlings with a small number of hills per hectare. Dryland or soft plate dryland seedlings are generally used. Dryland seedling cultivation is good for improving root vigor, while soft plates make cultivation easier. Soft plates with large holes are preferred. Seeds should be sown after the soil is mixed with seedling strengthening agent or the seeds should be soaked in uniconazole before sowing, or 15% paclobutrazol in 450 g of water per hectare should be sprayed at the stage of one mature leaf and one newly emerged leaf. The quantity of seeds used should be controlled carefully and sowing should be careful and even. Fertilizer management should be enhanced for the seedling field, and moist irrigation should be the mainstay. At the same time, scattering (transplanting) should be done at the right age, preferably at the leaf age of 2.1 – 3.5 leaves, i.e., 8 – 15 d after sowing.

(3) Reasonably sparse planting for higher quality

Before scattering (transplanting), the field should be leveled and the height difference should not exceed 3 cm. One seedling should be planted per hill with a distance of 20 cm (between plants) ×30 cm (between rows) and 142,500 – 199,500 seedlings per hectare. If scattered, the density should be 15 – 18 hill/m^2. Compared with improved rice cultivation, the density is 15%–25% lower. As the seedlings are small, shallow scattering (transplanting) is mostly adopted.

When scattering (transplanting), the thin, small and weak seedlings should be discarded, and to ensure that the seedlings stand upright, the root should be kept wet and all damages avoided for a shorter recovering period, early, rapid and multiple tillering, and better lower node tillering.

(4) Proper formula and balanced fertilization

1) Quantity

The amount of fertilizer to be applied depends on the rice field fertility and yield indicators, generally 180 – 270 kg/ha of pure nitrogen, with a ratio of nitrogen, phosphorus, potassium of 2 : 1 : 2.

2) Application

i. Base fertilizer

Organic fertilizer is the main fertilizer, supplemented with compound inorganic fertilizer, generally, 15,000 – 18,000 kg/ha of well-decomposed organic fertilizer, 1,500 – 3,000 kg/ha of organic and inorganic compound fertilizer, 600 – 750 kg/ha of calcium superphosphate, 22.5 kg/ha of zinc fertilizer and some other micronutrients.

ii. Tiller fertilizer

It should mainly be nitrogen fertilizer, specifically 75 kg/ha of urea 5 – 7 days after transplanting, and 105 kg/ha of urea and 30 – 45 kg/ha of potassium fertilizer 12 – 15 days after transplanting.

iii. Panicle fertilizer

When the population size reaches the target number of seedlings in the field (about 80% of the intended number of seedlings), the field is to be sun-dried. Then, after another round of irrigation, nitrogen fertilizer should be applied to promote flowering and the amount is 20%-25% of the total nitrogen of the whole growth duration.

iv. Grain fertilizer

At the fourth stage of young panicle differentiation, spikelet-promoting and grain-promoting fertilizer should be applied, with the quantity of 10%-15% of the total amount of nitrogen for the whole growth duration, together with foliar fertilizer, i. e. 6.0-7.5 kg/ha of potassium dihydrogen phosphate sprayed after mixing with 1,500 kg/ha of water.

(5) Inter-tillage ventilation and alternate irrigation

1) Inter-tillage ventilation

SRI requires a small number of planting hills per unit area. Since the paddy field is alternately wet and dry over a long time, weeds grow fast and early and multiple rounds of weeding is required. At the same time, the quality of mid-tillage weeding needs to be improved to make the tillage layer fully loose and well ventilated, so as to improve the environmental conditions for root growth and development. Generally, the first round of inter-tillage weeding should start 10-15 days after transplanting and inter-tillage ventilation should later be done depending on specific conditions.

2) Alternate drying and wetting irrigation

Transplanting should be done when there is only a thin layer of water or when the field is drained. After transplanting, the field should be kept moist and in 3-5 days a 3 cm depth of water layer should be put in place, then with intermittent irrigation, the field will gradually drain off naturally, and replenishment should be done again to 3 cm deep of water after 2-3 d. After the population gets to the expected size, the field is to be sun-dried to control the growth of ineffective tillers and reduce the occurrence of pests. The degree of field sun-drying depends on growth and soil fertility. Normally, the field should be dried until it cracks or white roots can be seen.

After the field is sun-dried, it should be irrigated with 3-5 cm depth of water followed by alternate drying and wetting irrigation till the penultimate leaf stage when panicle fertilizer is to be applied with a shallow layer of water in the field. A thin water layer of about 3 cm should be kept from the booting stage to the heading stage. After that, alternate drying and wetting irrigation should continue so that roots and leaves can be kept to promote grain filling. Alternate drying and wetting irrigation should continue during the seed setting stage, and should not be terminated too early.

(6) Comprehensive disease and pest control for safe production

The key to improved SRI is to control rice blast and prevent stripe blight. Rice blast control should be paid attention to in the three key periods of seedling blast, leaf blast and neck blast. Pest control involves mainly stem borers, leaf folder and planthoppers and should be based on the information provided by the local technical guidance department so as to achieve timely, accurate, and targeted control.

4. Suitable regions

It can be applied nationwide, especially in the middle and lower reaches of the Yangtze River.

Ⅱ. Nitrogen-saving and Lodging-resistant Cultivation

1. Overview

Since 2000, with the widespread use of super hybrid rice in southern China rice areas, the level of nitrogen application on rice has also been on the rise, resulting in a low nitrogen fertilizer utilization efficiency, nitrogen loss, and serious rural non-point source pollution. At the same time, lodging also occurs on a large scale, making it more difficult to produce high yields. Under the guidance of academician Yuan Longping's "three good" technical route and strong support from Yuan, the Cultivation Research Group of HHRRC carried out research on nitrogen-saving, lodging-resistant, high-yielding and high-efficiency cultivation techniques mainly for hybrid rice, and proposed three nitrogen-saving cultivation directions. 1) Select nitrogen-efficient rice varieties through gene mining; 2) Use high efficiency fertilizer such as slow-released or controlled release fertilizers; and 3) Use integrated cultivation management to save nitrogen, that is, through appropriate increase of basic seedlings, scientific management of fertilizer and water, and disease and pest prevention and control. There are two main approaches to lodging-resistance cultivation. 1) Apply chemical inhibitory products, e. g. Lifengling (a plant growth hormone which can shorten the internode between the first and second nodes); 2) Apply chemical promoting products, such as spraying "liquid silica-potassium fertilizer". At present, a hybrid rice cultivation system has been formed, featuring nitrogen-efficient varieties, early sowing for strong seedlings, one-time application of coated slow/controlled-release fertilizer, increasing seedlings and reducing nitrogen, integrated water management and timely pest control. In addition, a lodging-resistant hybrid rice cultivation system has been formed, featuring lodging-resistant variety selection, planting with appropriate spacing, water-saving management in the whole growth duration, slow-released fertilizer and stable growth and new lodging-resistant agent regulation.

2. Impact of new cultivation technology on yield increase and economic benefits

In 2009, the National Agricultural Technology Extension Service Center organized the promotion and application of nitrogen-saving and lodging-resistant cultivation models in the southern China rice growing region covering Hubei, Jiangxi, Zhejiang and Guizhou provinces, with a total rice planting area of 52 million hectares. Compared with conventional technology, the new model reduces the emissions of nitrogen, phosphorus and other nutrients by more than 20%, increases the utilization rate of nitrogen fertilizer by more than 10%, and saves more than 10% of energy due to nitrogen saving. Years of trials and demonstrations in different ecological rice areas in southern China showed that the new model brought a significant yield increase. According to statistics, the yield increase of single-cropping rice ranged from 600.0 kg/ha to 926.6 kg/ha, 2.5%-9.7% annually on average compared with the conventional technology; the average yield of double-cropping rice increased by 500.0 - 595.0 kg/ha, and the average yield increase of late-season rice ranged from 550.0 kg/ha to 670.0 kg/ha, 2.8%-8.5% per year on average compared with the conventional technology.

3. Technical key points

(1) Good varieties that are lodging-resistant and nitrogen-efficient should be selected based on local conditions.

Varieties with high nitrogen efficiency and strong lodging resistance should be selected. Early-season rice varieties such as Zhuliangyou 39 can be selected in the middle and lower reaches of the Yangtze River and Wufengyou 308 can be used as the late-season rice variety.

(2) Timely sowing for strong seedlings

According to experience of high-yielding cultivation of double-cropping rice areas in the middle and lower reaches of the Yangtze River, early-season rice should be sown around March 25 and late-season rice around June 15. Early-season rice adopts dryland seedlings raised in soft plates to ensure that seedlings grow uniformly and strongly, and is covered with argo-plastic film to prevent low temperature damage in early spring; while late-season, including single-cropping rice, is suitable for wetland seedling cultivation for strong multi-tiller seedlings.

(3) Less nitrogen, more seedlings, reasonably dense planting

In order to compensate for the weak tillering in the early stage due to the reduction of total nitrogen and to ensure the number of seedlings required for a high yield, attention should be paid to increasing basic seedlings by about 10% compared with the control when transplanting with the nitrogen-saving cultivation model adopted. The suggested number of seedlings is 300,000 for early-season rice, 255,000 for late-season rice and 195,000 for single-cropping rice per hectare. There should be 2 – 3 seeds per hill for early rice, and two seeds for late-season rice and single-cropping rice.

(4) Combination of quick-acting and slow-acting fertilizer and fertilizer reduce

Slow/controlled-release compound fertilizer (quick-acting and slow-acting) should be applied once, whose total nitrogen content is 15%–20% lower than what is used in conventional cultivation. The nitrogen fertilizer consumption of different types of rice is shown in Table 12 – 1.

The fertilizer is applied in the form of basal fertilizer after plowing and before transplanting. All fertilizer is applied as base fertilizer at one time, and the suggested amount of slow/controlled-release compound fertilizer is 375 kg/ha, with single-element fertilizer as supplement. The proportion of nitrogen, phosphorus and potassium in nitrogen-saving fertilization is as follows. The base fertilizer should provide 70%–75% of the total nitrogen, 100% of phosphate and 50%–60% of potassium; while tiller fertilizer should provide the remaining 25%–30% of nitrogen and 40%–50% of potassium.

Table 12 – 1 Demonstration nitrogen-saving fertilizer consumption of different types of rice (kg/ha)

Rice type	Nitrogen-saving fertilization			Conventional fertilization		
	Nitrogen	P_2O_5	K_2O	Nitrogen	P_2O_5	K_2O
Early-season rice	120.0	60.0	120.0	150.0	60.0	120.0
Late-season rice	144.0	75.0	135.0	180.0	75.0	150.0

(5) Scientific water management for fertilization adjustment

Slow/controlled-release fertilizer releases nutrients slowly. Shallow water and frequent irrigation from transplanting to the first stage of tillering can help achieve the goal of adjusting fertilization and promoting fertilizer efficiency with water management. In the middle stage, irrigation should be intermittent, mainly shallow irrigation; when the number of seedlings reaches 90% of the effective number of spike seedlings, the field can be drained off and sun-dried. In summary, there should be water for heading, alternate drying and wetting irrigation for grain filling, and drain-off seven days before harvesting.

(6) Chemical regulation to combine promotion and inhibition

In order to control the length of internodes at the base of the plant, Lifengling is sprayed before internode elongation. Specifically, for each hectare of land, 600 g of Lifengling and 3 L of liquid silicon and potassium are mixed with 525 - 600 kg of water and sprayed evenly 5 - 7 days before internode elongation to increase the lodging resistance of super hybrid rice in the later stage, without changing of the plant type and panicle structure. In addition, in order to adjust the length of the upper internodes to prevent decrease in biological yield for higher actual yield, it is necessary to apply liquid silicon and potassium fertilizer once or twice. Specifically, spray 3 L per hectare each time when the tip of the flag leaf is emerged and at the full heading stage. It should be 4. 5 L per hectare for late-season rice and single-cropping rice.

(7) Integrated pest and disease control for lower cost and higher efficiency

According to information provided by the local plant disease and pest control authorities, pesticides should be applied in time to prevent diseases and pests.

4. Suitable regions

It is suitable nationwide and especially the middle and lower reaches of the Yangtze River.

Ⅲ. "Three-definition" Cultivation Model

1. Overview

According to the multi-site joint experimental research on the suitable sowing period, transplanting leaf age and density, fertilizer application period and fertilizer application amount of super hybrid rice in recent years, and with reference to the precise and quantitative rice cultivation theory proposed by Ling Qihong, "three-definition" super hybrid rice cultivation method was formed, featuring precise sowing, wide rows and uniform planting, balanced fertilizer application, dry and wet irrigation, integrated pest control and other technical packages. "Three-definition" is a rice cultivation method that sets the target yield, the population indicators and the technical specifications. Rice yield can be decomposed into yield components such as effective panicle number, number of grains per panicle, seed setting rate and grain weight, among which effective panicle number depends on the number of basic seedlings, number of tillers, tiller-to-panicle ratio, etc. Therefore, on the basis of a target yield (local average yield over the previous three years, plus a 15% - 20% yield increase), the population indicators are set to determine the basic number of seedlings and planting density (defined seedlings), as well as the appropriate amount of nitrogen fertilizer (defined nitrogen). The "three-definition" cultivation method for super hybrid rice can also be understood as a method with set yield target based on local conditions, set number of seedlings based on the target yield, and set amount of nitrogen based on the number of seedlings.

2. Impact of "three-definition" cultivation

Since 2007, more than 40 counties (cities) in Hunan including Liling, Youxian, Xiangyin, Xiangtan, Hengnan, Hengyang, Ningxiang, Dingcheng, Nanxian, Yuanjiang, and Datonghu have conducted 667-ha trials and demonstrations of double-cropping super hybrid rice cultivation technology, with average yields of 7,305 - 8,775 kg/ha for early-season rice and 7,455 - 8,970 kg/ha for late-season rice, respectively, which were 11.4% and 13.6% higher than the yield of local non-demonstration areas in the same year.

3. Technical key points

(1) Strong seedlings

1) Cultivation methods

For early-season rice, heat preservation dryseedlings or soft plate seedlings should be used, while for late-season rice and single-cropping late-season rice, wet and sparse seedlings or soft plate seedlings should be used.

2) Sowing dates

The suitable sowing period for early-season rice is March 25 - 30 in Hubei province and northern Hunan province, and March 20 - 25 in Jiangxi province and central and southern Hunan province. The suitable sowing period for late-season rice is June 20 - 25 for mid-maturing varieties, June 15 - 20 for late-maturing varieties, and June 5 - 10 for extra-late maturing varieties. The suitable sowing period for late-season rice is from mid-and late April to mid-May.

3) Sowing quantity

Seeds are sown after sterilization and germination. For dryland raised early-season rice seedlings, the quantity should be 100 - 130 g/m^2; for plastic plate dryland seedlings, 30 - 40 g/plate; and for production fields of early-season hybrid rice, 30.0 - 37.5 kg/ha. For wetland-raised late-season rice seedlings, it is 20 g/m^2; for plastic plate seedlings, 22 - 25 g (353 or 308 holes) per plate; and for production fields, 22.5 kg/ha, preferably with tillers before transplanting. For extra-late maturing varieties seeds should be sown more sparsely.

4) Nursery fertilization

For dryland nursery, 450 kg/ha of 30% compound fertilizer should be applied as base fertilizer during soil preparation. Before sowing, multifunctional seedling strengthening agents mixed with fine soil are to be spread evenly, or the seeds can be put in plastic plates. Weaning fertilizer should be applied to the seedlings at the stage of two mature leaves and one newly emerged leaf. Normally, 60 - 75 kg/ha of urea is used for topdressing. Four days before transplanting, 60 - 75 kg/ha of urea should be used as nitrogen fertilizer.

5) Nursery management

Wetting irrigation is to be performed before the emergence of early-season rice. After emergence, the seedlings should be kept warm to prevent freezing. If continuous low temperature and rain occur, timely ventilation should be ensured to prevent diseases. Before the emergence of late-season rice, wetting irrigation should be performed. If the seeds are not soaked in Uniconazole solution before sowing, or are not treated with coating agents, spray 300 mL/L paclobutrazol solution on the seedling bed at the stage of one mature leaf and one new emerged leaf, with no water layer on the soil surface. After that, irrigation should continue for 12 – 24 hours to control seedling height and promote tillering. During the seeding stage of late-season rice, attention should also be paid to the prevention and control of rice planthopper, rice blasts, rice stem borers and rice thrips.

(2) Even transplanting (placed planting)

The key technology for "defined seedlings" is to determine the appropriate planting density, which is closely related to the rice culm height (the distance from the first elongated internode above ground to the panicle neck internode). The "golden rule" should be followed when determining row spacing and plant spacing: row space (cm) = culm height (cm) × 0.618 ÷ 2, and plant space (cm) = row space (cm) ÷ 1.618 (Table 12 – 2).

Table 12 – 2 Suitable planting density based on culm height

Rice type	Plant height * (cm)	Culm height * * (cm)	Plant spacing (cm)	Row spacing (cm)	Planting density ($\times10^4$ hill/ha)
Double-cropping early-season rice	80	63	12.0	19.5	42.75
	85	67	12.8	20.7	37.80
	90	71	13.6	21.9	33.60
Double-cropping late-season rice	95	75	14.3	23.2	30.15
	100	79	15.1	24.4	27.15
	105	83	15.9	25.6	24.60
Single-cropping rice	110	87	16.6	26.9	22.35
	115	91	17.4	28.1	20.40
	120	95	18.1	29.4	18.75

Notes. * Plant height refers to the distance between the first elongated internode of the culm to the top of the panicle; * * culm height refers to the distance between the first elongated internode of the culm to the neck of the panicle.

1) Early-season rice

After plowing and soil preparation for winter-fallow fields and oil rape fields, dryland seedlings are transplanted in rows, and plastic plate seedlings are planted in separate ridges. The requirements are even planting and sufficient seedlings. Placing rather than scattering should be adopted. The density of transplanting or placing is 30 hill/m^2 with two seedlings per hill for hybrid rice or 5 – 6 seedlings per hill for conventional rice. Generally, the spacing should be 16.7 cm × 20 cm, or 13.3 cm × 23.3 cm. Transplanting should be performed 20 – 25 days after sowing, or at the 3.7 – 4.1-leaf stage of the seedlings.

2) Late-season rice/single-cropping late rice

After early-season rice is harvested, late-season rice can go through no-tillage placed transplanting, or transplanting after tilling. Generally, 3,750 mL/ha of Gramoxone, mixed with 525 kg of water, should be sprayed evenly under waterless conditions to eliminate rice stubble and weeds, then the field is to be soaked for 1 – 2 days to soften the mud before planting or transplanting. For machine harvested rice fields, it is better to return rice straw to the field before transplanting. The same as for early-season rice, double-cropping late-season rice transplanting requires even planting with sufficient seedlings and plastic plate seedlings require placing rather than scattering. The appropriate density is about 25 hills/m^2 with a spacing of 20 cm × 20 cm or 16.7 cm × 23.3 cm, two seedlings per hill for hybrid rice and 3 – 4 seedlings per hill for conventional rice. Transplanting should be performed 25 – 30 days after sowing, or at the 6 – 7-leaf stage of seedlings. The seedling age should not exceed 35 days.

(3) Intermittent aerobic irrigation

Intermittent aerobic irrigation refers to alternate drying and wetting irrigation. Irrigate, leave the field to dry naturally, irrigate again 2 – 3 days later, and then leave it to dry again. This cycle should go on till maturity. During the growth period of super hybrid rice, except for the water-sensitive period and the use of shallow irrigation when fertilizer is applied, the general rule is to have no water layer or just a wet field. In other words, seedlings are transplanted and roots and tillers are promoted in shallow water. When the number of tillers reaches 300/m^2, the field should be dried lightly for several times until the surface of the soil becomes hard (commonly known as "wood skin"). After the budding stage, alternate drying and wetting irrigation is performed, and it should stop 5 – 7 days before maturity. For deep-foot mud fields or fields with high groundwater levels, it is ditches required around the field and across the middle of it for drainage before the land is sun-dried.

(4) Quantitative application of nitrogen fertilizer based on "defined seedlings".

The key technique for "defined nitrogen" is to measure seedlings before application. Take the main rice producing counties in Hunan province as an example. The basic fertility-based yield of super hybrid rice is 3,000 – 4,500 kg/ha for double-cropping rice, and 4,500 – 6,000 kg/ha for single-cropping rice. The nitrogen fertilizer absorption and utilization rate is 40% – 45%. In order to produce 1,000 kg of rice, 16 – 18 kg of nitrogen, 3.0 – 3.5 kg of phosphorus and 16 – 18 kg of potassium are required. The ratio of nitrogen fertilizer as base and tiller fertilizer to nitrogen fertilizer as panicle fertilizer is 7 : 3 for double-cropping rice, and 6 : 4 for single-cropping rice. The threshold determined by the leaf color chart is 3.5 – 4.0. The amount of fertilizer should be is based on the target yield, soil fertilizer supply capacity and fertilizer nutrient utilization rate (Table 12 – 3). The amount of nitrogen fertilizer provided in Table 12 – 3 is based on balanced application. Specifically, base fertilizer should account for 45% – 50% of the total, tiller fertilizer for 20% – 25% and panicle fertilizer for 30%. Phosphorus and potassium fertilizers are used on a compensatory basis, i.e. the amount needed to achieve the target yield is equal to the amount applied.

Table 12 - 3 Recommended time and quantity of fertilizer

Time of application		Fertilizer	Fertilizer consumption by target yield (kg/ha)		
			7,500	8,250	9,000
Base fertilizer	Before transplanting (day 1 - 2)	Urea	135 - 150	150 - 165	165 - 180
		Superphosphate	450 - 600	525 - 675	675 - 750
		Potassium chloride	60 - 75	75 - 90	90 - 105
Tiller fertilizer	Before transplanting (day 7 - 8)	Urea	60 - 90	60 - 90	75 - 105
Panicle fertilizer	Branch spikelet differentiation stage	Urea	60 - 90	75 - 105	90 - 120
	Young paniclewith white hair	Potassium chloride	60 - 75	75 - 90	90 - 105

Note. If compound fertilizer is applied, the nitrogen, phosphorus and potassium contents must be calculated respectively; urea as base fertilizer can be replaced by ammonium bicarbonate.

Due to the differences in soil fertility between fields and the different responses of varieties to fertilizer nutrients, it is also required to use a leaf color chart to determine the color of the next leaf of the newly emerged leaf 1 - 2 days before determining the follow-up application amount of nitrogen fertilizer. That is to say, when the leaf color is dark (a leaf color chart reading above 4.0), less fertilizer should be used (take the lower limit value); when the leaf color is light (a leaf color chart reading below 3.5), more fertilizer should be used (take the upper limit value). Since there is currently no compound fertilizer with slow release of nutrients, compound fertilizer should be used as both base fertilizer and supplementary fertilizer to increase the utilization rate of nutrients.

(5) Integrated prevention and control of diseases, pests and weeds

Long-acting pesticides can be sprayed 3 - 5 days before pulling up the seedlings and the seedlings will be transplanted with pesticide. In the production field, pest prevention and control should be strengthened for stem borer, leaf folders, planthoppers, sheath blights, false smuts, and blasts. In addition, it is important to obtain information on the occurrence and prevention of field diseases and pests, so as to determine the exact time to take prevention and control measures. Chlorpyrifos and buprofezin are the common choices. In production practices, for concurrent pests and diseases integrated control measures should be adopted. Rice false smut control should focus mainly on prevention in the rice panicle "breaking" stage (when the panicle just emerges from the sheath) to the beginning of heading with agri-chemicals. However, the specific timing and selection of pesticide should be determined according to information from the local plant protection authorities.

For weed control, herbicide for transplanting or for scattering rice can be applied, mixed with tiller fertilizer, and sprayed at the tillering stage on a shallow water layer for about five days.

4. Suitable regions

It is suitable for double-cropping early- and late-season rice, and single-cropping late-season rice areas in the middle and lower reaches of the Yangtze River.

Ⅳ. Precise Quantitative Cultivation Model

1. Overview

Precise quantitative rice cultivation is a new cultivating technique based on the theoretical and technical achievements of the rice leaf age model and population quality cultivation, which is adapted to the development trend of modern rice crop. It treats the cultivation process as an engineering project, which better improves the quantification and precision of the cultivation design, the dynamic diagnosis of growth conditions and the implementation of cultivation measures. It can also contribute to high-yielding, high-quality, high-efficiency and environmentally friendly rice production. It has been widely used in the main rice producing areas in China.

2. Impact on yield increase and economic benefits

Since 2005, this technique has been demonstrated and promoted in Jiangxi, Guangxi, Sichuan, Henan, Anhui, Liaoning and Heilongjiang provinces in China. For *indica* and *japonica* rice, or single-and double-cropping rice, the demonstration results all showed a yield increase of more than 10% , and even up to 20%-30% , with the same fertilization level compared with existing cultivating techniques. In addition, seeds, water, fertilizer and labor can all be saved to some extent.

3. Technical key points

(1) Appropriate sowing dates

To improve the photosynthetic productivity of a population from heading to maturity, the seed setting period must be arranged to have the best temperature and light conditions. *Indica* and *japonica* rice require different optimal temperatures and light conditions during the heading and seed setting periods, and such requirements vary with ecological areas. The periods when the optimal temperature and light conditions are in place in different ecological areas are the best heading and seed setting periods, and sowing dates must be set accordingly.

(2) Precise sowing for strong seedlings

The purposes of raising strong seedlings are to make the root system strong, shorten the period of seedling re-establishment, and promote early and low-node tillering. The specific criteria for strong seedlings are to keep four or more green leaves (except three-leaf seedlings) at the time of transplanting, free of pests and diseases, and even seedling age.

For transplanted, machine-transplanted or plastic plate raised seedlings, reducing the number of seedlings in the seedling bed and reasonably determining the number of seedlings are the primary points in raising strong seedlings. According to practice, for seedlings to be transplanted manually at the age of 3 - 4 leaves, generally 600 - 750 kg/ha of seeds should be sown, and the ratio of seedling bed to production field is 1 : (40 - 50); for mid-sized seedlings to be transplanted at the age of five leaves, the sowing quantity is 400 - 650 kg/ha, and the ratio of seedling bed to production field is 1 : (30 - 40); for big seedlings to be transplanted at the age of six leaves or above, the sowing quantity is 300 - 450 kg/ha, and the ratio of seedling bed to production field is 1 : (20 - 30). For seedlings to be machine-transplanted, the quantity of seeds used is controlled at 50 - 60 g/plate when the 1,000-grain weight is less than 25 g and at 70 - 80 g/plate when the 1,000-grain weight is 26 - 28 g.

(3) Improvement of transplanting quality

1) Shallow planting

Shallow planting is an important point to ensure timely tillering of strong seedlings. Whether it is manual or machine transplanting, the planting depth should be controlled at 2 – 3 cm into the mud. In order to ensure shallow planting, it is very important to improve the soil preparation technology, and the soil must be well settled before transplanting.

2) Improvement of seedling scattering quality

In order to improve the efficiency of tillers turning into panicles, first of all, it is necessary to ensure that the base of the seedlings is 1 cm or more intothe soil. This can also prevent lodging in the late stage. Secondly, it is important to plant (scatter) seedlings evenly without gaps to ensure a sufficient basic seedling population.

(4) Wide rows and narrow spacing between plants

When planting seedlings, attention should be given to optimizing spacing, so as to have reasonable row and plant spacing, and to establish a light-efficient population with sufficient and adequate light. Specifications should be determined according to ecological conditions, variety, season and other factors, to ensure no premature row closure or insufficient high-yielding population.

(5) Basic seedlings

According to the formula proposed by Ling Qihong for calculating the number of basic seedlings –

X (suitable number of basic seedlings per unit) = Y (suitable number of panicles per unit area) $/ES$ (number of panicles per plant), and

$$ES = 1 \text{ (main stem)} + (N - n - SN - bn - a)\ Cr$$

where Y is the appropriate number of panicles per unit area for the local variety, ES is the number of panicles per plant, N is the total leaf age of the variety, n is the number of elongated internodes of the variety, SN is the leaf age at transplanting, bn is the leaf age between transplanting and tillering, a is the adjusted value of the seedling leaf age before $N-n$, (which is between 0.5 and 1, mostly 1), C is the theoretical value of effective tillers, and r is the tillering rate.

According to the theory of leaf-tiller concurrent emergence, the effective tillering leaf age and its theoretical value of effective tillering are listed in Table 12 – 4. For example, if the effective tillering leaf age from transplanting to the critical leaf age is five leaves and the theoretical value of effective tillering is eight leaves. If the leaf age is 5.5, the theoretical value of effective tillering should be (8+12)/2=10.

Table 12 – 4 Relationship between effective tillering leaf age of main stem and theoretical value of tillering in the growth stage

Effective tillering leaf age of main stem	1	2	3	4	5	6	7	8	9	10
Theoretical number of primary tillers A	1	2	3	4	5	6	7	8	9	10
Theoretical number of secondary tillers				1	3	6	10	15	21	28
Theoretical value of the third tillers							1	4	10	20

Continued

Effective tillering leaf age of main stem	1	2	3	4	5	6	7	8	9	10
Total number of theoretical tillers *B*	1	2	3	5	8	12	18	27	40	59
C(strain ratio) = *B*/*A*	1	1	1	1. 25	1. 6	2. 0	2. 6	3. 38	4. 44	5. 9

Note. The value of *C* can be included in the formula as the strain parameter for calculation. For example, when (*X*) *C*=3, then (3) *C*=3×1 =3 tillers in theory; when X=5, then (5) *C*=5×1. 6 =8 tillers in theory; when *X*= 7, then (7) *C*=7×2. 6 =18 tillers in theory.

(6) Fertilization in appropriate quantity

1) Total fertilization quantity

The absorption ratio of nitrogen, phosphorus and potassium in high-yielding rice is generally 1 : (0. 45 - 0. 6) : (1 - 1. 2), and this ratio is often regarded as a parameter for fertilization, but it varies with the soil fertility types in practice. In addition, since nitrogen, phosphorus and potassium have the most prominent effect on yield, it is necessary to first determine the reasonable amount of nitrogen application, and then determine the appropriate amount of phosphorus and potassium factors according to the reasonable ratio of the three factors.

The quantity of nitrogen applied can be determined according to the Stanford difference formula.

$$N(kg/ha) = \frac{\text{N quantity required for target yield(kg/ha)-Soil N supply(kg/ha)}}{\text{N Utilization rate in the season(\%)}}$$

The amount of nitrogen required for the target yield is generally obtained from the amount of nitrogen required per 100 kg of rice grains produced. This indicator varies from one rice crop area to another and should be measured according to the actual nitrogen requirement of a specific high-yielding field. The nitrogen supply of soil is based on the rice yield without nitrogen application (ground yield) and its nitrogen requirement to produce 100 kg of rice. There are many factors affecting the seasonal utilization rate of nitrogen fertilizer, generally 30%-45% for rice, and the specific data can be measured through local fertilizer tests.

2) Proportion of nitrogen application

The quantity of nitrogen fertilizer applied can be determined by the number of elongated nodes of the variety. For varieties with five elongated internodes, the nitrogen quantity required for super rice before culm elongation accounts for 30%-35% of the nitrogen fertilizer in the whole growth period, 45%-50% from elongation to heading, and 15%-20% after heading.

3) Precise panicle fertilizer application

Panicle fertilizer should be applied normally at two applications, i. e. , the first application is at the emergence of the last fourth top leaf (spikelet-promoting fertilizer) and the second application is at the emergence of the penultimate leaf (spikelet-protecting fertilizer), and the proportion of these two applications is 60%-70% and 30%-40%.

It is not advisable to keepa deep water layer in the field when applying panicle fertilizer. Instead, the filed should be kept just moist. Two days after fertilizer application, shallow irrigation can be performed to improve the effect of the fertilizer.

(7) Precise irrigation

1) Seedling re-establishment and tillering-culm elongation

For fields with small transplanted seedlings, ventilation to enhance root growth is the main objective. For machine-transplanted rice, there should be no water layer in the field, instead, the field should be drained off naturally and then be irrigated after seedling re-establishment with new leaves. For medium and large transplanted seedlings, shallow irrigation should be performed after transplanting, and then frequent irrigation should follow every time field drains off.

Seedlings scattered with plastic plates have strong roots, so it is not necessary to water them on cloudy days after scattering, but it is necessary to provide a thin layer of water on sunny day. After 2 - 3 days, irrigation should stop and the field is to be drained to promote rooting, and then shallow irrigation shall be performed. After these, repeat the cycle of shallow irrigation and field draining.

2) Timely sun-drying the field

i. Timing for sun-drying

Sun-drying of the field must be carried out when there are two leaves before the occurrence of ineffective tillering. Generally, field drying starts when the number of seedlings reaches about 80% of the panicle number expected. For example, to control ineffective tillering at the $(N-n+1)$-th leaf node, sun-drying should start before the $N-n-1$ leaf age.

ii. Standard of sun-drying

Duration - period of two-leaf growth on average.

Field surface - with cracks but not sinking when trod on.

Plant - dry the field till leaves turn yellow; drying once or twice is enough when base and tiller fertilizers are in place.

3) Culm elongation to heading

Rice will enter the panicle and branch differentiation period around the time of culm elongation. It is the period of the most vigorous above-ground growth, the highest physiological water demand, and peak root system growth and development, lasting till heading. Alternate shallow and wet irrigation should be used to meet the physiological water demand, promote the growth and metabolic vitality of the root system, and increase the synthesis of cytokinin in the root system for better formation of large panicles. Specifically irrigation should be carried out in the following way, keep the field wet without a water layer, irrigate with 2 - 3 cm of water, wait for the field to drain off naturally (3 - 5 days), and then irrigate with a 2 - 3 cm water layer again.

4) Heading to maturity

After heading, rice enters the grain filling stage. Alternate shallow and wet irrigation should be continued. It can improve root vigor and the photosynthetic function of the plant, and also increase the seed setting rate and grain weight.

4. Suitable regions

All rice cultivating areas across China.

Ⅴ. Light and Simplified Machine-transplanting Model

1. Overview

In response to the challenge of declining hybrid rice planting area under the conditions of large-scale mechanized production during the transition period, Hunan Agricultural University and other institutes developed the hybrid rice cultivation technology of "single-hill dense planting plus machine transplanting", which solves the technical problems of traditional machine transplanting of hybrid rice such as large seeding quantity, short seedling age, poor seedling quality and unsuitability for double-cropping rice varieties; and ensures that each seed can grow into a big seedling for dense planting and machine transplanting. This cultivation technology cultivates high panicle rate populations through single-hill, low nitrogen dense planting of large seedlings that can balance between the grain number of grain size. Compared with traditional machine transplanting, the seed consumption is reduced by more than 60%, the seedling age is increased by 10 – 15 days, and the seedling quality and resistance to machine damage are greatly improved. In addition, seedling raising for this method is simple and easy to use, saving 525 – 750 yuan/ha in the costs of substrates. Increasing the planting density and reducing the amount of nitrogen fertilizer give full play to the advantage of hybrid rice in tillering, panicle formation and large panicles.

2. Impact on yield and efficiency increase

In 2015, the "single-hill dense planting plus machine transplanting" model for double-cropping hybrid rice was demonstrated in Liuyang, Hunan, and Zhaoqing, Guangdong. The rate of missing seedlings was 9.8%. The yield of late-season rice increased by 631 kg/ha and 674 kg/ha, up by 10.3% and 14.0%, respectively compared with traditional machine transplanting. In 2016, in Liuyang and Hengnan double-cropping hybrid rice demonstration plots, the missing seedling rate was 7.7% – 9.2%, and the yields of early- and late-season rice were 17.22 t/ha and 16.8 t/ha, which was 9.3% and 15.1% higher than the control group respectively.

3. Technical key points

(1) Seed selection

Based on the selection of commercial hybrid rice seeds, another round of seed selection is needed with a color sorter to remove moldy and discolored seeds, dehulled rice grains and miscellaneous materials, so that only highly vigorous seeds are used. The seeds selected by a color sorter have a 10% higher germinating rate. Normally, 19.5 kg of early-season rice seeds, 12 kg of late-season rice seeds and 7.5 kg of single-cropping rice seeds per hectare are used.

(2) Seed coating

Commercial rice seed coating agents, or self-developed agents, including seed germination initiators, fungicides, pesticides and film-forming agents, should be used to coat the seeds to prevent germs and pests in the early stage, thereby increasing the germinating rate. Within 25 days after sowing the coated seeds, there is no need to take any pest control measure.

(3) Positioned seeding

A machine-printed seeder or manual seeder can be used to sow 16 rows (25 cm row spacing) or 20 rows (30 cm row spacing) horizontally, and 34 rows of coated hybrid rice seeds longitudinally. Early-season rice is seeded with two seeds per hill, and late-season rice and single-cropping rice are seeded with one or two seeds per hill. Positioned seeding can be performed on paper bonded with degradable starch glue, and paper rolls are carried out during sowing, so as to facilitate transportation.

(4) Mud seeding

Fields with convenient irrigation, drainage and transportation conditions, fertile soil and no weed should be selected as nurseries. Fifteen days before sowing, the nursery should be plowed once; 3 -4 days before sowing, it should be flattened, with the application of 900 kg/ha of 45% compound fertilizer. Ridge beds are to be created with a width of 140 cm plus a ditch 50 cm wide. With the ridge bed in the middle, straighten the two ends of the field with a string; place the four plates vertically, with the two plates in the middle aiming at the string, leaving no gap in between; mud dug out from the ditch should be put into the plates with hard blocks, gravel, grass and weeds removed, and the mud thickness should be 2. 0 -2. 5 cm. Smear the mud (it is best to use a mud machine to save labor for high efficiency). For early-season seedlings, the nursery needs to be disinfected with Dexon (or thiophanate methyl) mixed with water.

(5) Field seedling cultivation

Soft and hard plates, special substrates contained in non-woven fabrics, organic fermented bacterial fertilizer, and nutrient liquid fertilizer can be used for simple field seedlings in flat dryland and cement flat fields.

(6) Paper printing sowing

There are two ways for paper printing sowing. First, seeds can be spread on the paper, and covered with commercial substrate or fine dry soil, so that the seeds are not visible. Second, printed seeds can be spread on a seedling plate and rolled slowly; adjustment is needed so that the paper sticks to the mud smoothly, and the seeds get evenly into the plate; one may also give the paper a light press with hands to make it stick better to the mud.

(7) Film covering and uncovering

After the planting paper is placed, an arch is to be made over the nursery of early-season and mid-season rice with bamboo slips to cover the nursery with film. For single-cropping rice and double-cropping late-season rice non-woven fabrics should be fixed tightly on the plates with thick mud at both ends to shelter from wind and rain. After the seeds take root and grow leaves, uncover the film or unwoven fabric in time based on the weather conditions.

(8) Nursery management

After the first leaf grows, the plates should be filled with water to the brims for 20 -24 hours. Then the water should be drained, the paper removed without moving the seeds. When mid- and late-season rice has one mature leaf and one newly emerged leaf, 2. 25 kg/ha of paclobutrazol should be applied with water by spraying to promote root growth. When the seedlings have two mature leaves and one newly emerged leaf, 45 -60 kg/ha of urea should be applied. After seed budding and before roots grow into the soil tightly, there should be no water layer on the surface of the nursery but shallow water in the ditch to prevent high temperature and rain. After the stage of one mature leaf and one newly emerged leaf, the ditch should be kept full of water and the ridge surface should be kept moist without any crack (if it cracks, provide a shallow water layer).

(9) Machine transplanting of seedlings

Around 20 - 25 days after sowing (no more than 30 days), machine transplanting should be carried out when the leaf age is 4.5 - 4.9 leaves. Early-season rice planting requires no less than 360,000 seedlings per hectare, late-season rice no less than 330,000 seedlings per hectare and single-cropping rice no less than 240,000 seedlings per hectare. To take the seedlings, a 30 cm row spacing transplanter should grab the seedlings for 20 times horizontally and 34 times vertically, or with a 25 cm row spacing transplanter, it should be 16 times horizontally and 34 times vertically.

(10) Field management

1) Recommended fertilization

The quantity of nitrogen fertilizer (pure nitrogen) is 120 - 150 kg/ha for early-season or late-season rice, and 150 - 180 kg/ha for single-cropping rice, 50% as base fertilizer, 20% as tiller fertilizer, and the remaining 30% as panicle fertilizer.

2) Field water management

At the tillering stage, shallow irrigation should be performed. For a field with 2.4 - 3 million seedlings per hectare, the field should be sun-dried until it cracks. One week later, the field should be re-irrigated. Shallow irrigation should be carried out from the booting stage to the heading stage. After heading, alternate drying and wetting irrigation is performed. Irrigation stops one week before maturity.

3) Pest control

Pest prevention and control should be performed according to the instructions of the local department of plant protection.

4. Suitable regions

It is suitable for southern China *indica* hybrid rice producing area.

Ⅵ. Cultivation Technology to Tap the Super-high Yield Potential

1. Overview

After the launch of China's super rice breeding program in 1996, the first phase yield target of 10.5 t/ha was achieved in the year 2000, pushing China's hybridrice breeding technology up to a new level; the second phase (12 t/ha), third phase (13.5 t/ha) and fourth phase (15 t/ha) targets were then reached successively in 2004, 2011 and 2015, respectively. In this process, research on super-high yielding cultivation technology systems has also made remarkable progress.

Since the launch of the Super Rice Breeding Program, Yuan Longping's team has attached great importance to the research of super rice varieties' super-high yielding cultivation technology, in order to fully exploit the yield potential of super rice varieties, and carried out a series of in-depth research on super-high yielding theory and technology, which has achieved great results and formed a number of supporting technology systems for tapping the yield potential of each generation of super rice, such as the technical models for the yield targets of 12 t/ha and 13.5 t/ha.

2. Impact on yield and economic benefits

In 2002, super rice Liangyou 0293 was planted in 8.47 ha of land in Longshan county, Hunan province, with an average yield of 12.26 t/ha. In 2003, Liangyou 0293 produced yields of 12.15 t/ha and 12.10 t/ha in Longhui and Xupu counties, Hunan, respectively. In 2004, the Zhunliangyou 527 yielded 12.63 t/ha and 12.14 t/ha, respectively in Guidong and Rucheng counties, Hunan. In 2011, the yield of super hybrid rice Y Liangyou 2 reached 13.90 t/ha on 6.67-ha land plots in Longhui county, Hunan, and in 2012, the yield in Xupu county exceeded 13.76 t/ha on 6.67-ha land plots. In 2013, the average yield of Y Liangyou 900 was 14.82 t/ha on 6.67-ha land plots in Longhui county, Hunan. In 2014, the average yield of Y Liangyou 900 was 15.4 t/ha on 6.67-ha land plots in Hongxing village, Hengbanqiao town, Xupu county, Hunan. In 2015, the average yield of Xiangliangyou 900 reached 16.01 t/ha on 6.67-ha land plots in Gejiu city, Yunnan. In 2016, the average yield of Xiangliangyou 900 reached 16.32 t/ha and 16.23 t/ha respectively in Gejiu city, Yunnan, and Yongnian district, Hebei. In 2017, Xiangliangyou 900 produced a yield of 17.23 t/ha on 6.67-ha land plots in Yongnian district, Hebei, setting a new world record.

The following are the technical points of the super-high yielding cultivation model of super hybrid rice producing a yield of 13.5 t/ha.

3. Technical key points

(1) Fertility

Super-high yielding rice must have four supporting factors, fertile land, excellent variety, suitable cultivation methods and good ecological conditions. A focus solely on the variety and technical improvement, not on soil fertility, will make it impossible for varieties (combinations) to fully unleash their of super-high yield potential.

In super-high yielding cultivation, first of all, base fertilizer should be heavy to improve the soil environment, which is the basis for achieving super-high yield of rice. Generally, base fertilizer is applied twice with tillage. First, 3.0 t of manure fertilizer and 450 – 600 kg of 45% compound fertilizer per hectare are to be mixed and applied with plowing. Then, 750 kg of rapeseed cake fertilizer and 450 kg/ha of 45% compound fertilizer should be applied with second plowing.

(2) Strong seedlings

Strong seedlings are the prerequisite for shaping a population with high light utilization efficiency. In production, strong seedlings with coordinated growth above and under the ground can be generally raised through such measures as dryland raising, sparse sowing, sufficient fertilizer, and spraying of Paclobutrazol. Strong seedlings after transplanting have early and fast growth, laying the foundation for the formation of sufficient and large panicles.

1) Seedbed selection

Leeward and sun-facing fields which are convenient for irrigation and drainage, and has a deep tillage layer and fertile soil should be selected. After plowing, 600 kg of 45% compound fertilizer and 112.5 kg of potassium chloride should be applied as base fertilizer per hectare; and 4 – 5 days later, the seedling ridges can be leveled to make the plot ready for sowing. Generally, 120 – 150 kg/ha of seeds is sown per hectare on the nursery, and 15 kg/ha of seeds are used for the production field.

2) Seedling management

At the stage of two mature leaves and one newly emerged leaf, 60 kg/ha of urea should be applied as weaning fertilizer; and 105 kg/ha of urea is required 3 – 4 days before transplanting. In addition, efforts should be made to the prevention and control of pests and weeds. Generally, pests are controlled once during the seedling stage by spraying high-efficiency farm chemicals one day before transplanting so that the seedlings are transplanted into the field with the chemicals.

(3) Suitable basic seedling population

1) Principle

The basic seedlings are the starting point of a population. To determine a reasonable number of basic seedlings is an extremely important part of the establishment of a population with high light utilization efficiency. Insufficient basic seedlings and panicles will make it difficult to achieve any breakthroughs in yield, while excessive basic seedlings and too large populations may lead to a decline in individual quality and difficulty in achieving a high yield.

2) Planting density

Planting density should be determined based on the target yield and the characteristics of the specific variety. Normally, 142,500 – 165,000 seedlings are planted per hectare, with two seedlings per hill.

3) Spacing

In order to create a population with high photosynthetic efficiency, planting should be carried out in wide and narrow rows, i. e., the wide row space of 33 – 40 cm and the narrow row space of 20 – 23.3 cm should alternate while plant spacing should be consistent at 20 cm. The row direction is set east-west. A special row marker is used to draw or pull the line for transplanting, which is conducive to ventilation and light transmission in the middle and late stages.

4) Seedling quality

Seedlings are transplanted shallowly, with the roots 0.5 – 1 cm into the mud, which can promote early and multiple low-node tillering. Then, 3 – 4 days after transplanting, an examination is to be carried out to check whether there are missing hills or dead seedlings, and any gap thus found must be filled up as soon as possible.

(4) Precise and quantitative fertilization

1) General principles

Organic fertilizer should be combined with inorganic fertilizer to provide sufficient nitrogen with enough phosphorus and potassium. In addition, the total amount of fertilizer applied must be determined according to the fertilizer requirement of the target yield, soil nutrient availability and the fertilizer utilization rate.

2) Total fertilizer quantity

The target yield is 13.5 t/ha and super rice requires 1.7 – 1.9 kg of nitrogen per 100 kg of rice grains, which means a total of 229.5 – 256.5 kg/ha of pure nitrogen is required with a nitrogen, phosphorus and potassium ratio of 1 : 0.6 : 1.2. The soil in high-yielding rice fields provides about 120 kg/ha of pure nitrogen and the fertilizer utilization rate is 40% , so 273. 5 -341. 3 kg/ha of nitrogen is needed as supplement.

3) Methods and proportion of fertilization.

Rice has different nutrient uptake in different growth stages, and it is necessary to determine the reasonable proportion of each nutrient and its application period. The distribution of nitrogen fertilizer in this model is : base and tiller fertilizer : panicle and grain fertilizer=6 : 4; i. e. , 164. 4 -204. 8 kg/ha as base and tiller fertilizer and 109. 4 - 136. 5 kg/ha a panicle and grain fertilizer.

4) Base and tiller fertilizer

Generally, 70% of the nitrogen fertilizer should be applied as base fertilizer and the remaining 30% as tiller fertilizer. Tiller fertilizer should be applied in two rounds, 90 -105 kg/ha of urea and 75 kg/ha of potassium chloride (60% K_2O) in the first round 4 -6 days after transplanting when the seedlings start to turn green; 40 -60 kg/ha of urea and 90 kg/ha of potassium chloride 12 -15 days after transplanting. All phosphorus fertilizer is applied as base fertilizer, while potassium fertilizer is applied as base fertilizer, tiller fertilizer and panicle fertilizer at 30% , 20% and 50% respectively.

5) Panicle and grain fertilizer

Panicle and grain fertilizer is applied in three rounds, first 75 -90 kg of urea, 225 -270 kg of 45% compound fertilizer and 75 -90 kg of potassium chloride per hectare when the main stem enters the second stage of young panicle differentiation, as flowering promotion fertilizer to promote branching and lay the foundation for large panicles; then 45 -60 kg of urea, 180 -195 kg of 45% compound fertilizer and 75 -90 kg of potassium chloride per hectare when the young panicles enter the fourth stage of differentiation as flower preservation fertilizer to provide sufficient nutrients to young panicles to prevent degeneration of the differentiated spikelets and ensure large panicles with more grains; and finally 22. 5 -30. 0 kg/ha of urea as grain fertilizer five days after full heading, depending on the seedling conditions, to reduce empty spikelets and improve seed setting and grain weight. The final round can be combined with foliar spraying of potassium dihydrogen phosphate to strengthen grain filling.

(5) Scientific water management featuring intermittent irrigation

1) Before transplanting

After field preparation, a ditch is to be dug around the field with a depth of 30 cm and a width of 20 cm and another ditch across the middle with the same depth and width. In addition, a ditch is also needed every 300 m^2 or more, with a width of 20 cm and a depth of 20 cm to make it convenient for irrigation and drainage.

2) Tillering stage

After transplanting, the field should be irrigated with shallow water, and from the time of greening to the critical period for effective tillering, intermittent irrigation is to be adopted. Specifically, a 2 -3 cm water layer is formed first, and then the field should be left to drain off naturally in 3 -4 days; this cycle continues till the sun-drying stage.

3) Field sun-drying

The field should be sun-dried when there are sufficient seedlings. At the $N-n-1$ leaf age, the total number of stem tillers in the population reaches about 80% of the estimated 2.25 million seedlings per hectare, i.e., about two million seedlings, drainage should start and the field should be sun-dried. This can be done in several rounds of light sun-drying till the leaf color turns light (the color of the top fourth leaf is lighter than that of the top third leaf). If the leaf age reaches the point for panicle fertilizer application, but the leaf color is still not light, sun-drying should be continued without irrigation and fertilizer.

4) From heading to maturity

The soil should be kept moist and solid to meet the physiological water demand of the plants, enhance root vigor, and improve the population's accumulation capacity for photosynthetic products in the middle and late stages. Wetting irrigation should be the mainstay, with dry intervals, but sufficient water supply should be ensured at the heading stage. At the late stage of full heading, moisture is important to ensure root vigor, prevent early senescence, and increase the seed setting rate and fullness of the grains. In the later period the field should mainly be dry, with clear water and hard soil for ventilation, so that roots can grow to support leaves, help with grain weight, and keep effective tillers alive for a high yield.

(6) Chemical weed and pest control

Prevention and integrated control should be the mainstay, focusing on thrips in the seedling bed, and rice stem borer, yellow stem borer, leaf folder, planthopper, rice blast, sheath blight, etc. in the production field. Firstly, through the selection of disease-resistant varieties, appropriate sparse planting, reasonable fertilization and irrigation and other methods a suitable population structure can be established to improve the resistance. Secondly, the use of biological and chemical pesticides can strengthen the control of streak leaf blight, rice blast, sheath blight, false smut, rice stem borer, yellow stem borer, rice planthopper, leaf folder, etc. For weed control, manual weeding is encouraged, but in order to save labor cost, herbicides such as Miecaowei, Daotianjing and Kecaowei can also be used after transplanting.

4. Suitable regions

It is suitable for southern China *indica* super hybrid rice super high-yielding cultivation areas.

References

[1] ZHOU PEIJIAN, LUO ZANLEI, CHENG FEIHU. 800 questions about Jiangxi planting industry technology[M]. Nanchang: JiangxiScience and Technology Press, 2010: 8-9.

[2] LI JIANWU, DENG QIYUN, ZHANG YUZHU, et al. Cultivation techniques of Y Liangyou 900, a promising hybrid for phase Ⅳ super rice, yielding 14.82 t/ha in high-yielding demonstration[J]. Hybrid Rice, 2013, 28(6):46-48.

[3] SHI QINGHUA, PAN XIAOHUA. Questions and answers on double-cropping rice production technology[M]. Nanchang: Jiangxi Science and Technology Press, 2010: 55-56.

[4] WANG CHI. Practical Technology of hybrid rice seed soaking and pre-germination[J]. Modern Agricultural Science and Technology, 2007 (3): 81-82, 84.

[5] LI JIANWU, DENG QIYUN, WU JUN, et al. Characteristics and high-yielding cultivation techniques of new super hybrid rice combination Y Liangyou 2[J]. Hybrid Rice, 2013, 28(1): 49-51.

[6] GU ZUXIN. Discussion on cold-tolerant seedling cultivation and high-yield cultivation techniques of early rice

in dry land[J]. Jiangxi Agricultural Science and Technology, 1991 (6): 5 - 8.

[7] ZOU YINGBIN. Development of cultivation technology for double cropping rice along the changjiang river valley[J]. Scientia Agricultura Sinica, 2011, 44(2): 254 - 262.

[8] ZHOU LIN, NIU SHEYU, YIN BIWEN, et al. Changes in rice seedling cultivation methods and their advantages and disadvantages[J]. Agricultural Technology Service, 2011, 28(5): 580 - 582.

[9] ZHANG YONG. Techniques for rice yield increase with dryland seedling, thin planting and shallow seedling[J]. Yunnan Agriculture, 201 (12): 39 - 40.

[10] CAI XUEJU, HE HAILIN, ZHOU TINGTING. High-yielding cultivation techniques of seedling throwing[J]. Rural Economy and Science-Technology, 2013, 24 (10): 184 - 185.

[11] BAI YUNHUA. Application of mechanical direct seeding in rice cultivation[J]. South China Agriculture, 2017 (24): 1 - 2.

[12] LIAO KUI. Direct seeding in Pengxi county[J]. Sichuan Agricultural Science and Technology, 2014 (12): 14 - 15.

[13] LING QIHONG. Precise quantitative rice cultivation: theory and application[M]. Beijing: China Agriculture Press, 2007: 92 - 125.

[14] LI JIANWU, DENG QIYUN, ZHANG ZHENHUA, et al. Characteristics of two-line hybrid rice Y Liangyou 488 and its high-yielding cultivation techniques in Hainan[J]. Crop Research, 2014, 28(1): 19 - 21.

[15] LI JIANWU, ZHANG YUZHU, WU JUN, et al. High-yielding cultural techniques of super hybrid rice Y Liangyou 900 yielded 15.40 t/ha on a 6.84 ha ccale[J]. China Rice, 2014, 20(6): 1 - 4.

[16] YUAN LONGPING. The system of rice intensification[J]. Hybrid Rice, 2001, 16(4): 1 - 3.

[17] ZHU DEFENG, LIN XIANQING, TAO LONGXING, et al. The formation and development of system of rice intensification[J]. China Rice, 2003, 9(2): 17 - 18.

[18] MA JUN, LV SHIHUA, LIANG NANSHAN, et al. Research on system of rice intensification in Sichuan [J]. Agriculture and Technology, 2004, 24(3): 89 - 90.

[19] PENG JIMING, LUO RUNLIANG. The international conference on the system of rice intensification held at Sanya, Hainan[J]. Hybrid Rice, 2002, 17(3): 59.

[20] LING QIHONG. Precise quantitative rice cultivation: theory and application[M]. Beijing: China Agriculture Press, 2007.

[21] YE DANJIE, CHEN SHAOTING, HU XUEYING, et al. Key techniques of precise quantitative rice cultivation[J]. Guangdong Agricultural Sciences, 2010, 37(4): 24 - 25.

[22] SONG CHUNFANG, SHU YOULIN, PENG JIMING, et al. High-yielding cultural techniques of super hybrid rice in large-scale demonstration with a yield over 13.5 t/ha in Xupu, Hunan[J]. Hybrid Rice, 2012, 27 (6): 50 - 51.

[23] LI JIANWU, ZHANG YUZHU, WU JUN, et al. Studies on matching technology of "four elite factors" for super hybrid rice with yield 15.0 t/ha[J]. China Rice, 2015, 21(4): 1 - 6.

Chapter 13

Occurrence, Prevention and Control of Main Diseases and Pests of Super Hybrid Rice

Huang Zhinong / Wen Jihui

Section 1 Occurrence, Prevention and Control of Rice Blast

Rice blast, caused by the pathogenic fungus *Magnaporthe oryzae*, is one of the three major rice diseases in China. There are different degrees of rice blast occurrence in the northern China and southern China rice areas. Rice blast can occur in all rice growth stages with the severity varying from year to year, and from region to region. It mostly occurs in mountainous areas with less sunshine and long duration of fog and dew, places along rivers, as well as coastal areas with a mild climate. The loss caused by rice blast is generally 20%–30% and can by up to 50%–70% in serious cases, or even a complete crop loss. It hits more than 3.6 million hectares of rice fields in China, including 350,000 ha in Hunan, with a loss of 200 million kilograms of rice grains.

Ⅰ. Characteristics of Occurrence

The fungus causing rice blast belongs to fungal *Magnaporthaceae*, with conidia and mycelium overwintering on diseased rice grains and grass. Diseased rice straws stacked in the seedling field or production field can also cause the disease. When the temperature and humidity are appropriate, mycelium overwintering in rice grains and straws will produce conidia, which will then spread with wind and rain or airflow, resulting in the first infestation of the following year. After the initial infestation on rice leaves, when conditions are appropriate, a large number of conidia will be produced at the disease spot, and then spread wider to cause re-infestation. Therefore, under favorable conditions, the formation and accumulation of conidia can cause serious re-infestation, resulting in an outbreak of the disease.

Ⅱ. Symptom Identification

Rice blast can affect all above ground parts of a rice plant, such as leaf, node, neck, parts of panicle, etc., classified by different onset period and the affected parts of rice plant. Seedling blast occurs at the 2–3 leaf stage in the seedling field, turning the seedlings yellow and ultimately death and a gray mold layer can be seen on dead seedlings when the humidity is high (Fig. 13 − 1). Leaf blast occurs on leaves of seedlings and adult plants, and the spots produced on the leaves often differ in shape, size, color and gray mold layer due to the influence of climatic conditions and the resistance of specific varieties. According to the four common types of spots, the disease can be classified as chronic, acute, white-spotted and brown-spotted rice blast (Fig. 13 − 2).

Fig. 13 − 1 Seedling blast

Fig. 13 − 2 Leaf blast (brown-spotted)

Fig. 13 − 3 Leaf blast (chronic)

In case of a typical chronic infection, there appears a spindle-shaped spot grayish white in the center and yellow in the periphery. This is the most commonly seen in the field (Fig. 13 − 3). Node blast causes rice nodes to turn dark brown with sunken spots that gradually expand around the nodes, making the panicle break easily. Panicle-neck blast and branch blast cause dark brown spots on the neck of the rice panicle and its branch peduncle (Fig. 13 − 4 and Fig. 13 − 5), which can cause white panicles if the disease occurs early and severe, or increase the number of immature grains if the disease occurs late. Grain blast occurs in the inner and outer glumes, and leads to dark and immature grains at the late stage.

Ⅲ. Environmental Factors

The dominant factors that cause rice blast are climatic factors such as temperature, humidity, rain, fog, dew and light; and the main conditions that can cause the disease to vary between fields are cultivation management practices (fertilizer, water, etc.) and the resistance of the specific variety. The optimal temperature for conidia germination of rice blast is 25 ℃ − 28 ℃ and the best relative humidity is over 90%. Prolonged rain and insufficient sunshine are conducive to an outbreak of the disease.

Fig. 13 - 4 Neck-panicle blast

Fig. 13 - 5 Branch blast

Ⅳ. Prevention and Control Techniques

1. Green prevention and control techniques

Integrated control of rice blast combines green control with chemical control, using a variety of ecological control measures and biological technology to manage the disease in a sustainable manner.

(1) Select resistant varieties and conduct resistance analysis

Planting varieties with high resistance to the disease is the most cost-effective way to control rice blast. The resistance of rice varieties to the blast pathogens differs greatly. This is determined by three categories of factors, genetic factors of a specific variety, factors of the pathogen, and environment factors. A change in any of these can lead to changes in disease resistance. The resistance of a certain rice variety is subject not only to pathogen-related factors such as the physiological races of the fungi, but also to environmental conditions. The resistance of a variety over the whole growth duration and to the same race of pathogen is relatively stable, but the purity and stability of a variety are only relative and more or less subject to variation. Variation in pathogen races is the main cause of rapid loss of resistance in rice varieties over large areas. Therefore, special attention should be paid to chemical control for varieties with poor resistance or susceptibility in production. Variety selection and resistance monitoring should be implemented in accordance with the diversity of the resistance genes. Before promoting a super hybrid rice variety, it is necessary to understand the variety's level of resistance to rice blast based on field resistance monitoring. The focus is to monitor the incidence rate, disease index and loss rate of leaf blast and panicle blast at the end of the tillering and yellow-ripening stages, respectively, according to the main cultivar (combination), and to calculate and evaluate the composite index of resistance.

(2) Eliminate overwintering fungi and treat diseased grains andstraws

At the time of harvest, rice straws in diseased fields should be stacked separately and dealt with properly before sowing in spring. Diseased straws should not be directly used during pre-germination or for bundling up the seedlings, but can be used as compost if thoroughly decomposed.

(3) Strengthen healthy cultivation via proper fertilization and irrigation management.

Sowing the right amount of seeds and raising strong and healthy seedlings is the key to controlling seedling leaf blast. Sound management of fertilization and water is an important measure of ecological control. Nitrogen, phosphorus and potassium fertilizers should be well-balanced. The principles of fertilization include sufficient base fertilizer, early topdressing, appropriate fertilization based on seedling conditions, weather and field conditions in the middle and late growth stages, extra phosphorus and potassium fertilizers, proper use of silicon fertilizer instead of merely focusing on nitrogen, and appropriate application of panicle fertilizer not to aggravate the disease. Scientific and reasonable drainage and irrigation should be conducted so as to use water to adjust fertilizer, and frequent shallow irrigation should be combined with field sun-drying to achieve a balance between promotion and control. Different irrigation methods should be adopted in different growth stages. Timely field sun-drying at the end of the tillering stage can increase plant tolerance and control leaf blast. The disease can be reduced by shallow irrigation at the heading stage, moist irrigation at the milk ripening stage and alternate wet and dry irrigation at the yellow ripening stage.

(4) Apply biological fungicides, especially probiotic fungicides

Actinomycin and its metabolites should be promoted, such as 2%, 4%, 6% kasugamycin wettable powder, ehydroxide, and 2% blasticidin S emulsifiable oil, as well as other biological fungicides and their compounds, such as 13% kasugamycin + tricyclazole wettable powder. Probiotic microbial fungicides such as biodiasmin and photosynthetic bacteria should be used to promote rice growth and development, enhance photosynthesis, inhibit pathogenic microorganisms, and enhance the disease resistance of rice plants.

2. Chemical prevention and control

Chemical prevention and control must be based on scientific, reasonable and correct use of agricultural chemicals. Fungicide spraying methods, equipment and fungicide formulations are becoming more precise, with a lower volume required, higher concentration, better targeted application, and automation.

(1) Seed disinfection

Generally, coated seeds are used, while uncoated seeds need to be sterilized before use. The sterilization should be done with 40% trichloroisocyanuric acid (TCCA) diluted to 300 times, or 20% tricyclazole diluted to 400 times, or 75% tricyclazole diluted to 1,000 times, or 10% 401 and 80% 402 antimicrobial agents diluted to 1,000 - 2,000 times, or 40% isoprothiolane diluted to 1,000 times, or 25% prochloraz diluted to 2,000 times.

(2) Field prevention and control

Fungicides should be timely applied to susceptible varieties and at susceptible growth stages based on the resistance performance of specific varieties and the results of field survey. In areas where rice blast frequently occurs, the main focus should be to control seedling blast and leaf blast at the early stage of the disease, and to treat panicle and neck blast with fungicides. Fungicides are applied at the third or fourth leaf stage of seedlings or 5 - 7 days before transplanting. Control of seedling blast and leaf blast should focus on center of onset, while panicle and neck blast control mainly focuses on the period when panicles emerge from the sheaths (when 10% of the panicles emerge and heading reaches 5%). Common fungicides used include 1,500 g/ha of 20% tricyclozole wettable powder, 600 g/ha of 75% tricyclazole wettable powder, 1,500 mL/ha of 40% isoprothiolane or 25% prochloraz emulsifiable oil, 1,500 mL/ha of 40% carbendazim suspensoid, 1,500 mL/ha of azoxystrobin or pyraclostrobin or 40% fenoxanil, or 1,500 g/ha of 50% FTHALIDE wettable powder. The requirement is to apply the right fungicide at right time with the right amount mixed in the right amount of water.

Section 2 Occurrence, Prevention and Control of Sheath Blight

As one of the three major rice diseases in China, sheath blight is caused by the pathogen fungi *Rhizoctonia solani* Kuhn., which mainly causes reduced seed setting rate, increased sterile and abortive grains, and reduced grain weight, resulting in a yield loss of 10%–20%, or up to 50% in severe cases. It occurs in China to 15 million hectares, with an annual loss of 10 billion kilograms of rice grains.

Ⅰ. Characteristics of Occurrence

With the planting of dwarf rice varieties, the increase of dense planting, the promotion of hybrid rice and the increase of fertilization, sheath blight has become a frequently occurring and serious disease characterized by a wide prevalence, high frequency, severe damage, and high losses. It is especially prominent in high-yielding hybrid rice cultivation areas. The overwintering and spreading of sheath blight mainly depend on the sclerotium, which overwinters in the soil of rice field and on diseased rice straws or weed residues outside rice field. In general, sclerotium falls into the field in hundreds of thousands with extremely strong viability. Most of the sclerotia are in the top soil layer 6–13 cm deep, which is the main source of initial infestation of the following year. The germination rate of overwintering sclerotia in the top soil was measured to be more than 96.0%, and the pathogenicity rate more than 88.0%.

After the rice field is irrigated and harrowed in the following year, overwintering sclerotia float on the water surface, drift with water, and attach to the leaf sheaths at the base of the seedlings. When the temperature and humidity are appropriate, the sclerotia germinate and produce mycelium which grows on the leaf sheath, invades from the leaf sheath gap and through the stomata, or directly penetrates the epidermis. From the peak tillering to the early booting stage, the disease mainly spreads horizontally, which is manifested by the increase in the rate of diseased plants. During the horizontal expansion, lesions will appear on the invaded parts for about five days, and then mycelium will grow on the diseased spots. It has strong pathogenicity and can spread with water flow to neighboring rice plants. From the booting stage to the heading stage, the disease expands vertically from bottom to top, and can rise to the upper part of the stalk and into the panicles. The incidence of leaf sheath blight has also increased sharply with increased damage.

Ⅱ. Symptom Identification

Sheath blight can occur from the seedling stage to the heading stage, more often from the peak and late tillering stage to the heading stage, and most prevalent around the heading stage. The plant is the

Fig. 13 - 6 Early-stage sheath spots at the stem base

most vulnerable at the tillering and booting stages (Fig. 13 - 6). The disease mainly affects the leaf sheaths and leaves (Fig. 13 - 7), and in severe cases, it can extend to the upper part of the stem or the panicles (Fig. 13 - 8). Early after onset, small, dark green, water-stained lesions are first seen on the leaf sheaths near the water surface, gradually expanding into a circle, with brown or dark brown edges, grass yellow to grayish white in the middle, and grayish green to dark green when wet, expanding into large cloud-like lesions. The leaf blade is dirty green when the disease is severe, and soon rots. The stem is first dirty green when the disease occurs and then turns grayish brown. Infection at the booting and heading stages causes dead panicles or increased blighted grains. Under humid weather conditions, white filamentous mycelium appears on the disease spots. Mycelium creeping on the surface of the rice plant tissue and climbing between plants, can be knotted into white loose fluffy mycelial mass, and finally become black-brown sclerotia, flat spherical, 1.5 - 3.5 mm, and mature from the diseased tissue, fall into the field or floating in the water. When the disease is severe in a field, the stems of the affected plants are easily broken and cause lodging and leaf death.

Fig. 13 - 7 Mid-stage sheath spots at the stem base

Fig. 13 - 8 Mid-stage stem and panicle spots

Ⅲ. Environmental Factors

The occurrence and prevalence of sheath blight are influenced by the number of fungal sources, climatic conditions, variety resistance or tolerance, planting density, fertilization level and other environmental factors, while field microclimate and the growth stage of rice are the leading factors affecting the severity of the disease. Onset begins when the temperature is 23 ℃, and the favorable temperature for disease development is 23 ℃–35 ℃. Sclerotium normally germinates and produces mycelium in 1–2 days when the temperature is 27 ℃–30 ℃ with a relative humidity above 95.0%. The favorable temperature for infestation is 28.0 ℃–32.0 ℃ with relative humidity above 96.0%. Infestation is inhibited when the temperature is above 35 ℃ and the relative humidity is below 85.0%. Sheath blight is a high temperature and high humidity disease, but it also favors excessive fertilizer with green and vigorous rice plants. Super hybrid rice has thick stems and lush leaves with more fertilizer used, especially under the environment of high levels of nitrogen, vigorous growth, field depression, and increased humidity and rain. Therefore, sheath blight can be more serious. It is highly conducive to the development and spread of the disease in paddy fields with long-term irrigation, dense planting, excessive nitrogen and less sun-drying.

Ⅳ. Prevention and Control Techniques

1. Green prevention and control techniques

The green prevention and control of sheath blight involves mainly the removal of sclerotia to reduce pathogen at the source, enhanced resistance, application of biological fungicides and formula fertilizer, controlled the application of nitrogen, sound water management, and sun-drying of the field at the right time.

(1) Remove sclerotia to reduce pathogen at the source

Generally, sclerotia scum floating on the water should be scooped out during irrigation and plowing and this should be done in as large an area as possible, annually, and between the early and late seasons. The scum should be buried deep or burned to reduce pathogen at the source and weeds around the fields should also be removed.

(2) Choose tolerant varieties

Tolerance to sheath blight is different among varieties. Generally, broad-leaved varieties are more vulnerable than narrow-leaved varieties. Rice germplasm resources with good resistance to sheath blight are relatively few and no germplasm with immunity or high resistance has been found so far. The preventive measures to guard against or delay the invasion of the pathogen include applying quick-acting silicon in production, increasing the content of SiO_2 in the rice field, and strengthening the silicified cells on the surface of the rice plant.

(3) Strengthen health cultivation through proper fertilization and water management

Reasonably dense planting is required. The wide- and narrow-row planting pattern is to be adopted, with the row spacing widened according to local conditions to improve the ventilation and light conditions of the rice population and reduce the humidity in the field. In the application of fertilizer, sufficient base fertilizer, early topdressing, a good combination of nitrogen, phosphorus and potassium rather than excessive nitrogen, and silicon application are the key points. Formula fertilization should be promoted, so that rice plants will not have drooping leaves in the early stage, not growing excessively in the middle stage, and not green-greedy in the later stage. As for water management, the principle is shallow irrigation in the early stage (tillering stage), sun-drying the field in the middle stage and wetting irrigation in the late stage. The field should be sun-dried when the number of seedlings is sufficient, to promote root development. Fertile soil should be heavily sun-dried and less fertile fields less sun-dried. Shallow water should be kept for panicle initiation and wetting is to be done for large panicles. Irrigation should stop timely to prevent lodging. In particular, it is vital to sun-dry the fields at the right time to promote healthy root growth, control diseases, and improve tolerance to the disease.

2. Fungicides for prevention and control

Fungicides should be timely applied to prevent and control the disease based on specific disease conditions in the field and relevant control criteria. As for the selection of fungicides, biological fungicide such as validamycin, is preferred. The time to do the control in the field is from the peak tillering stage to the heading stage, with the best effect at the end of tillering and the early booting stages. The criterion is generally to launch control measures when the diseased plants are 15%–20% of the total at the end of tillering or 20%–25% at the early booting stage.

(1) Biological fungicides

Validamycin is effective, harmless, and residue-free due with systematic, protective and therapeutic effects. Therefore, water-soluble validamycin and its compound are the most commonly used in rice production. For each hectare, options include 1,500 g of 5% validamycin water-soluble powder, or 750 g of 15% validamycin water-soluble powder, or 450 g of 20% validamycin water-soluble powder, or 1,500 ml of 10% validamycin solution, or 3,000 ml of 5% validamycin solution, or 1.05 kg of 20% Wenquning wettable powder, or 4.5 L of 2.5% Wenquning water agent, or 3 L of 12.5% Wenmeiqing or Kewenmei water agent, or 1.5 L of 20% Wenzhenqing suspoemulsion.

(2) Chemical fungicides

For each hectare, the option can be 300 mL of 30% difenoconazole + propiconazol EC, or 600 mL of 10% hexaconazole EC, or 450 mL of 25% propiconazole EC (dilibut), or 750 g of 12.5% diniconazole powder (or thifluzamide, or azoxystrobin), or 1.5 kg of 50% carbendazim wettable powder, or 1.2 kg of 30% dimethachlon wettable powder. These should be used in rotation, all to be mixed with 750 – 900 kg of water and applied at the middle and lower parts of the rice plants. When temperature and humidity are both high, the fungicide should be applied for 2 – 3 times with an interval of 7 – 10 days. It can be combined with control measures for stem borer, leaf folder, etc.

Section 3 Occurrence, Prevention and Control of False Smut and Rice Kernel Smut

False smut is caused by the pathogenic fungus [*Ustilaginoidea virens* (*Patou.*) bref = *U. Virens* (Cooke) Tak.], and is one of the most serious diseases for hybrid rice. Since the 1980s, with the promotion of hybrid rice, the area of rice fields that suffer from rice false smut in China has increased significantly, and it is becoming a common disease of mid-season and late-season hybrid rice in the south of the country. In 1982, the disease occurred on 6.7 million hectares of rice fields in Hunan province and in Jiangxi province, the diseased area accounted for 30% of the total rice planting area. In 1983, the disease plagued 40% of the rice fields in Guizhou province. In the early 21st century, hybrid rice was cultivated as single-cropping or mid-season rice in most areas of Hunan. In 2004, false smut went rampant, infecting 40,000 ha out of a total rice planting area of 1.25 million hectares in Chengde, 31.9% of the total, leading to a loss of 126.26 million kilograms of rice grains. Rice kernel smut is a major disease that commonly occurs to the female parents (sterile lines) of hybrid rice in seed production. For example, for Peiai 64S, the incidence could be over 40%.

Ⅰ. Characteristics of Occurrence

The fungus of false smut overwinters from sclerotium that fall into the soil and from chlamydospores that attach to the seed surface. In July and August of the following year, sclerotium germinates and produces ascospores. Chlamydospore can also overwinter on affected rice grains and glumes, and can germinate and produce conidiophore at any time, which is the main source of initial infestation. The spores are dispersed on the leaves and panicles of rice plants by airflow and the period of infestation is from the booting stage to the flowering stage. They mainly invade the floral apparatus and young panicles during the panicle emerging stage, but the disease can also occur from heading to maturity. The chlamydospores produced on the diseased parts are spread by wind and rain to reinfest. Fungus of kernel smut (*Tilletia horrida*) also overwinters in soil and on the surface of diseased grains and the spores are dispersed by wind, which is the source of initial infestation. False smut not only reduces rice seed setting and grain weight, thus reducing the yield, but also severely affects rice quality and is toxic to human, animals and poultry because it contains the pathogenic pigment C9H6O7, which can cause chronic poisoning and is teratogenic.

Ⅱ. Symptom Identification

False smut is a disease that affects rice panicles and is prevalent from the flowering stage to the milk-ripening stage. The fungus grows inside rice glumes. Initially, the infested grain glumes open slightly to reveal small yellowish green tuberous protrusions and the sporodochidia gather on the grain glumes, then gradually expand to wrap up the glumes and form a mass of spore ball (Fig. 13 - 9). The mass of spore ball is several times larger than a grain, nearly spherical with a smooth surface, yellowish green or dark green, and covered by a film. With the growth of such mass balls, the film ruptures, and the cracked surface produces chlamydospores that disperse dark green powder. There are usually one or several mass balls in one panicle, but in severe cases, there can be dozens (Fig. 13 - 10 and Fig. 13 - 11). False smut not only destroys the infested grains, but also con-

Fig. 13 - 9 False smut (less severe)

Fig. 13 − 10 False smut (severe)

Fig. 13 − 11 False smut (extremely severe)

Fig. 13 − 12 Rice kernel smut (affected grains)

sumes the nutrition of the diseased panicle, resulting in blight. As the diseased grains increase, the rate of blighted grain rises. Generally, it occurs more in hybrid rice than in conventional rice, and more in two-line hybrid rice than in three-line hybrid rice. The difference between false smut and rice kernel smut is that the former deforms the whole grain through the wrapping and expansion of the pathogen, while the latter keeps the grain shape, with only a few black tongue-like protrusions at the joint of the glume (Fig. 13 − 12) with black powder (teleutosorus) inside. Hybrid rice used for large-scale production generally has less severe kernel smut, while the sterile line is more severely affected than the restorer line and the maintainer line.

Ⅲ. Environmental Factors

Climatic conditions are an important factor affecting the occurrence of the disease, especially rainfall and temperature are the most closely related to the occurrence. From booting to heading, the weather is favorable for the disease because the temperature and humidity are both high. The disease develops well at 24 ℃− 32 ℃, and the most favorable conditions are when the temperature is 26 ℃− 28 ℃. Long periods of low temperature, less sunlight and more rainfall tend to weaken the disease resistance of rice plants, especially during the heading and flowering stages, and the ascospores and conidia of false smut invade the floral apparatus with wind and rain. Therefore, rain is the main climatic factor affecting the development and infestation of false smut. In addition, if nitrogen is excessively used as panicle fertilizer, rice plants will grow vigorously after heading, resulting in a "green-greedy" and delayed maturity, or too deep irrigation water and poor drainage will exacerbate the condition. The environmental factors conducive to rice kernel smut are similar to those for false smut.

Ⅳ. Prevention and Control Techniques

1. Green prevention and control techniques

(1) Disease-resistant varieties

There are certain differences in resistance between rice varieties. In general, varieties with a long booting period may suffer more of the disease. Varieties with loose-panicle and early-maturity tend to suffer less, while large-panicle and dense-panicle late-maturing varieties may suffer more. At present, most hybrid rice varieties are more susceptible to the disease, and those less susceptible should be selected based on local conditions.

(2) Improved field management

Diseased grains found in the field at an early stage should be promptly removed and burned, and-heavily diseased fields should be tilled deep after harvest. Attention should be paid to keeping the fields clean, and removing diseased residues and pathogens from the fields before sowing to reduce fungi at the source.

(3) Strengthened fertilization and water management

Fertilizer should be applied reasonably with more organic fertilizer and focus should be on the uptake and utilization of nitrogen, phosphorus, potassium and mineral elements, Excessive and late application of nitrogen fertilizer should be avoided. In terms of water management, the field should be sun-dried timely and then irrigated with alternate wetting and drying irrigation.

(4) Biological fungicides

In high yielding and high quality cultivation, the following fungicides can be used. 750 g/ha of 15% validamycin soluble powder, or 2.25 L/ha of 12.5% of Wenmeiqing, or 3.75 L/ha of 5% validamycin water agent, or 4.5 L/ha of 2.5% of Wenquning water agent, or 750 g/ha of 20% of Wenmeixing wettable powder.

2. Chemical prevention and control

Field surveys should be combined with meteorological forecasts for accurate measurement and reporting. Studies have shown that the invasion and prevalence of false smut in late-season rice in Hunan and neighboring provinces generally occurin the first half of September, and the magnitude of rain and temperature coefficient in this period has a great impact on the number of diseased grains (false smut "balls") in the first half of October. If there is more rain in mid-September and the temperature is 25 ℃-30 ℃, the disease is most likely to occur, so fungicides should be prepared for prevention and control.

The first application should be at the late booting stage (young panicle differentiation phase Ⅶ), i. e. 5-7 days before panicle emergence, and the second application can be at the peak time of the panicle emergence and the heading stage (about 50% panicle emergence and heading). Common fungicides options include 300 mL/ha of 30% benzalkonazole EC (Aimiao), or 300 mL/ha of 43% tebuconazole EC, or 600 mL/ha of 10% hexaconazole EC, or 900 mL/ha of 23% kresoxim-methyl epoxiconazole, or 300 mL/ha of 25% folicur (tebuconazole) emulsion, or 1,200 mL/ha of 20% triadimefon (triazolone) emulsion, or 1,500 g/ha of 50% copper (succinate + glutarate + adipate) (DT) wettable powder, or 1,500 g/ha of 50% carbendazim wettable powder. Most of the above can also be used for the control of kernel smut, but mainly for hybrid rice seed production. Therefore, when selecting fungicides the application period and method, full consideration should be given to the complexity of the outcrossing performance of the parents to ensure the effectiveness of disease prevention and seed production.

Section 4 Occurrence, Prevention and Control of Bacterial Blight and Bacterial Leaf Streak

Ⅰ. Bacterial Blight

Bacterial blight is caused by the pathogenic bacteria *Xanthomonas campestris PV. Oryzae* (Ishiya-ma) Dye, which used to be one of the three major rice diseases in China. In the 1950s, the disease only occurred to rice fields to the south of the Yangtze River and in the eastern and southern coastal regions of China. In the 1960s, the disease spread with transported seeds to more areas and was a disease under quarantine. In the 1980s, the country could be divided into three zones according to the occurrence and prevalence of the disease. One is the year-round occurrence zone, such as the area south of the Leizhou Peninsula, which has a warm climate and bacterial blight occurrence all year round. The second is the perennial epidemic area, such as the southern double-cropping rice area, including Guangdong, Guangxi, Jiangxi, Hunan and some other provinces. The disease mainly hits late-season rice and has a great impact on yield. In recent years, the damage has been less severe, but can still be serious in some years. The third is endemic areas, such as single-cropping rice areas to the north of the Huaihe River, where the disease occurs mainly in the rainy season from July to August. With the promotion of super rice over the past years, the occurrence of the disease in South China, East China and some areas in Central China has been on the rise (Fig. 13 − 13).

Fig. 13 − 13 Bacterial blight damage

1. Characteristics of occurrence

The pathogenic bacteria mainly overwinter on rice seeds and straws, which are the main sources of initial infestation. The bacteria can enter the glume tissue or embryo and settle on the endosperm surface to overwinter. The source of bacteria in previously infected areas is mainly diseased straws and straw residues, and the bacteria can survive for more than six months on diseased straws. The bacteria in new disease areas are rice seeds and the bacteria on diseased grains can survive for more than eight months. Transmission of bacteria-bearing seeds at the time of sowing in the following year becomes the source of initial infestation and long-distance seed transportation is the main reason for the expansion of the disease to new areas. However, the quantity of infected seeds is not significantly correlated to the prevalence of the disease. The pathogenic bacteria that exist in straws and husks spread to seedlings with water. The secretions of rice roots can attract surrounding pathogenic bacteria to accumulate at the rhizome, invade through the roots, stem bases, and leaf wounds or stomas on the leaves to the vasculature tissue before multiplying in the vessel, causing typical symptoms. Infestation of highly susceptible varieties can cause acute symptoms when environmental conditions are favorable. The pathogens that invade the vascular bundles proliferate and expand to other sites, forming a systemic infection. Generally, the disease sets on in the stem elongation stage, late in the booting stage or the panicle emergence stage, first in the lower leaves and then the upper leaves. The first plant to suffer the disease is called the central disease plant, which shows leaf blight, acute and chlorotic symptoms in rice leaves.

2. Symptom identification

Rice can be infested throughout the whole growth durationwith various symptoms, mainly on the leaves, due to differences in environmental conditions and variety resistance and the parts of infestation. The diseased leaves often have yellow bead-like bacterial pus overflow. There are five main types of symptoms, i. e. , leaf blight, acute, midvein, withering, and yellowing.

(1) Leaf blight

Leaf blight is a common symptom on the leaves. It sets on from the leaf tip, or the edge of the leaf, where dark green water stain appears at first, then short stripes, and finally long and corrugate stripes. The stripes change from yellowish brown to off-white or yellowish white. When the field is highly humid, there will be yellow bead-like bacterial pus on the diseased parts (Fig. 13 - 14).

Fig. 13 - 14 Leaf blight

(2) Acute symptoms

Leaf blade produces dark green lesions, looking like hot water burns, and the leaf becomes withered due to water loss with bacterial pus in the diseased parts. The disease mainly occurs to susceptible varieties and in high temperature, high humidity conditions, as well as nitrogen excessively applied in the field, which indicates that there will be an outbreak of the disease.

(3) Midvein symptoms

The pathogenic bacteria invade from the wound in the midvein of the leaf and gradually spread along the midvein up and down as long yellowish lesions. Yellow pus overflows when the diseased leaf is folded longitudinally and squeezed. Such symptoms can be seen at the heading stage of susceptible hybrid rice (Fig. 13 - 15).

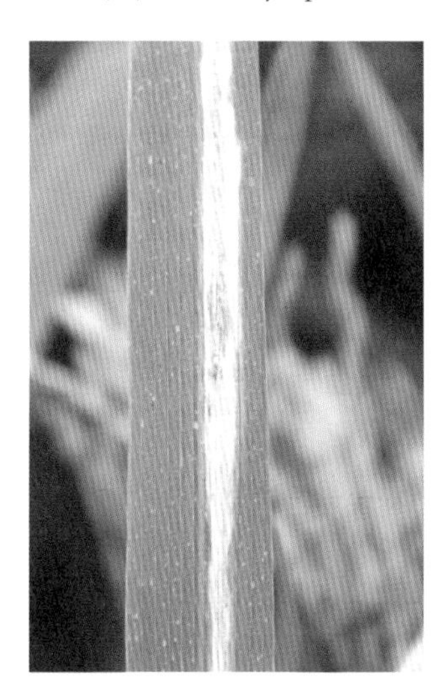

Fig. 13 - 15 Midvein symptoms and bacterial pus

(4) Withering

The newly emerged leaves or one to two leaves below the newly emerged leaves of the diseased plant first show water loss, greening and

Fig. 13 – 16 Withering symptoms

curling, then withering until death. A large amount of bacterial pus overflows on the diseased part, which is different from the withered center caused by rice stem borer. Such symptoms are mostly seen in hybrid rice and susceptible varieties (combinations), often 20 – 30 days after transplanting. The symptoms are mostly likely seen at the end of the peak tillering stage, and a large number of leaves will curl and die at the panicle stage (Fig. 13 – 16).

(5) Yellowing generally occurs on new leaves of adult plants, producing yellowish or yellowish green streaks, causing poor growth of the plant, with few pus, easily confused with physiological yellow leaves. This symptom is rare and has only been found in Guangdong province.

3. Environmental factors

The occurrence and prevalence of the disease are closely related to climatic factors, fertilization, water management, variety resistance, etc., especially water management. The optimal temperature for the development of the disease is 26.0 ℃–30.0 ℃ and the optimal relative humidity is above 90%. In comparison, the disease is inhibited when the temperature is above 33.0 ℃ or below 17.0 ℃. When conditions are favorable, the pathogenic bacteria multiply in the vascular bundle of the diseased plant, and pus overflows from the leaf surface or water stoma, which can spread with running water, wind, rain, or dew drops for re-infestation. Irrigation water, storms and floods are the main vectors for the spread of the disease in the field. In addition, wounds and mechanical damage caused by the friction between rice leaves are conducive to the invasion of the disease, and long-time flooding, string or diffusion irrigation are conducive to the spread of the disease. As for fertilization, nitrogen fertilizer has the greatest influence on the disease. Late or excessive application of nitrogen can aggravate the disease. In one growing season of rice, re-infestation may occur under favorable conditions, causing an outbreak. The disease is prevalent in the Yangtze River Basin from June to July for early-season rice, from July to August for mid-season rice, and from mid-August to mid-September for late-season rice.

4. Prevention and control techniques

(1) Green prevention and control techniques

The basis of green prevention and control of bacterial blight is the selection of disease-free rice seeds and the planting of disease-resistant varieties. Seedling prevention is the key, while fertilization and water management should be strengthened and supplemented by the use of agricultural chemicals.

1) Selection of resistant variety or disease-free seeds

The resistance to the disease varies greatly among rice varieties, and resistant varieties generally have a relatively stable performance of resistance. Planting of disease-resistant varieties is an economical and effective control to the disease and two to three main resistant varieties favorable for local planting can be used in disease areas. In addition, it is the best to produce hybrid rice seeds in disease-free areas. In production, disease-free seeds should be introduced to eliminate the source of disease, to comply with the quarantine system, not to transfer seeds from diseased areas, and to strictly prevent the introduction of disease.

2) Properly dealing with diseased plants and weeds

This requires protection of seedling fields from flooding and raising disease-free and healthy seedlings. The straw residues should be disposed of in a timely manner. Diseased straws should not be used to bind seedlings, cover the seedling bed, or block the field water entrance. Seedling nursery should be set at sites facing the sun and at a high level and where it is convenient for irrigation and drainage, so as to prevent flooding in the field.

3) Fertilization management and fitness cultivation

For seedlings to grow healthily and stably, a good drainage and irrigation system should be put in place with drainage and irrigation separate to prevent string or diffuse irrigation and avoid flooding. Fertilizer should be used reasonably based on the change of leaf color, preferably formulas featuring a good balance of nitrogen, phosphorus, potassium and micro elements.

(2) Chemical prevention and control

The key to chemical control is early detection, prevention and control. The focus is to spray bactericide in the diseased areaat the seedling stage for isolation of the center of the disease and protection of the rest of the production field. In chemical prevention and control for production fields, the principle of treating a patch upon finding a point of infection, and treating the whole field upon finding a patch of infection. It is necessary to apply bactericide immediately, mainly in the primary disease period of the tillering stage and booting stage, to control effectively the spread of the disease, especially when the spots of the acute type appears in the field and the climate is favorable for the spread of the disease.

1) Seeds should be soaked for disinfectionin 40% trichloroisocyanuric acid diluted to 300 times, or 20% bismerthiazol wettable powder diluted to 500 times, or 70% antibiotic 402 or 10% phenazine-N-oxide diluted to 2,000 times.

2) Bismerthiazol, or phenazine-N-oxide, or streptomycin sulfate should be sprayed once before irrigating and pulling out the late-season rice seedlings.

3) For production fields, options include 1.5 kg/ha of 20% bismerthlazol wettable powder, or 3 kg/ha of 25% Yekuling (Chongqing 7802) wettable powder, or 25% propuazole wettable powder, or 3.75 kg/ha of 10% Yekujing wettable powder, or 1.5 kg/ha of 70% Yekujing (shakujing) suspensoid, or 1.2 kg/ha of 90% NF-133 soluble powder, or 1.8 kg/ha of 77% kocide wettable powder, or 375 g/ha of 24% streptomycini sulfas wettable powder, or 1.5 L/ha of 20% thiazolium (thiosen copper) suspensoid. The fungicides should be applied with 750-900 kg of water per hectare on the leaves every 7-10 days, 2-3 times in total.

Ⅱ. Bacterial Leaf Streak

Rice bacterial leaf streak, [*Xanthomonas oryzae pv. oryzicola* (Fang et al.) Swing et al.)], caused by *Xanthomonas*, is still one of the quarantined diseases in China. Since the 1980s, due to the popularization and application of hybrid rice and the south-to-north transportation of rice seeds in China, the disease

Fig. 13 – 17 Bacterial leaf streak in field

occurred first in the south China rice area, and then spread rapidly to the central, east and southwest rice areas of China, now seen in more than 10 provinces and municipalities in the country. In recent years, with the promotion of super hybrid rice, the disease areas and damage have both been gradually increasing (Fig. 13 – 17).

1. Occurrence characteristics

The pathogenic bacteria also overwinter in rice seeds and straws, and become the source of initial infestation in the following year, and the transport of diseased seeds is also a major reason for long-distance transmission. The pathogen can be spread by wind and rain, mainly irrigation water and rain. It generally invades through stomata or wounds, and after invasion, multiplies in stomata and expands into the intercellular space of the parenchyma, forming strips due to the blockage of leaf veins (Fig. 13 – 18). Under high humidity conditions, pus overflows from the spots and then reinfests with wind, rain, water and agricultural practices. The characteristics of this disease are basically the same as those of bacterial blight.

2. Symptom identification

Rice can be affected throughout the entire growth period, mainly on the leaves, specifically the young leaves. The pathogen mostly invades through stomata, first forming small watery lesions on diseased leaves, then expanding into short, thin, dark green to yellowish brown strips between leaf veins, transparent to light, and the strips can be connected (Fig. 13 – 19), merging into dead patches (Fig. 13 – 20). The surface of the lesions often secretes bead-like yellow pus, which dries up into yellow gelatinous pellets and adheres to the surface of the diseased spot.

Fig. 13 – 18 Early-stage diseased leaves

Fig. 13 – 19 Mid-and Late-stage diseased leaves

Fig. 13 – 20 Late-stage diseased leaves

3. Environmental factors

With the bacteria in existence, the occurrence and prevalence of this disease is mainly affected by climatic conditions, variety resistance, cultivation practices and other factors.

At a temperature of 25 ℃–30 ℃ and a relative humidity over 85%, heavy rain and high wind are conducive to the spread of the disease, especially typhoon and floods, which can cause a large number of wounds on rice leaves, easily leading to an outbreak. Long-time deep irrigation, as well as excessive and late nitrogen application may exacerbate the disease. In general, mid- and late-season rice suffers more severely than early-season rice. The disease is the most prevalent in the middle and lower reaches of the Yangtze River from June to September. Conventional rice generally has mild symptoms, while hybrid rice is prone to the disease, specifically for super-high yielding hybrid rice. Special attention is required as for observation for the prevention and control of this disease in rice production.

4. Prevention and control techniques

(1) Green prevention and control techniques

1) Plant quarantine

Effective quarantine must be ensured at the seed origin and seeds without quarantine should not to be transported.

2) Resistant varieties

Suitable varieties should be selected according to local conditions, with attention paid to the resistance of the selected varieties and their reactions to the disease.

3) Disease-free seedlings

Dryland-raised seedlings, wetland-raised seedlings and greenhouse seedlings should be used. Seedlings should also be strictly protected from deep water.

4) Field management

Diseased straws should be carefully dealt with. Fertilization should be balanced to avoid excessive use of nitrogen.

(2) Chemical prevention and control

The chemical control method of bacterial leaf streak is basically the same as that of bacterial blight.

Section 5 Occurrence, Prevention and Control of Rice Stem Borers [*Chilo suppressalis* (Walker) and *Scirpophaga incertulas*]

Rice striped borer (*Chilo suppressalis*) and rice yellow borer (*Scirpophaga incertulas*) are two important pests for hybrid rice, and both of them belong to the family *Pyralidae* of *Lepidoptera*. Hybrid rice plant has a sturdy stalk with a large medulla cavity, green dark leaves, rich nutrients, high starch content and more soluble sugars, which provides favorable conditions for rice stem borers, especially striped borer, to cause severe damage.

Ⅰ. Damage Characteristics

1. Rice striped stem borer (***Chilo suppressalis***)

Rice striped stem borer exists in all rice regions in China. There can be 3 – 4 generations a year in the Yangtze River basin (26° – 32°N latitude). As an omnivorous pest, except rice, it also feeds on sugarcane, corn, sorghum, wheat, *Zizania aquatica* L. and *Leersia hexandra* Sw. The larvae of striped stem borer overwinter in rice stubble, straws and *Zizania aquatica* L., and have strong tolerance to cold. Fourth instar larvae can overwinter successfully. When the temperature rises to 11 ℃ the following year, the adult larvae begin to pupate, and they become imagoes at 15 ℃ – 16 ℃. The imagoes are phototactic and prefer laying eggs on tall, verdant rice plants. Since the 1980s, with the development of hybrid rice, striped stem borer has become a major rice pest, first damaging early-season rice and now plaguing mid- and late-season rice too. The larvae infestation causes dead sheaths and deadhearts at the tillering stage (Fig. 13 – 21), dead panicles at the booting stage, and whiteheads at the heading stage (Fig. 13 – 22). Damaged rice plants and half-dead panicles at the grain filling and maturity stages reduce yield by 5% – 10%, or more than 30% in severe cases. The imagoes are phototactic and have fresh-green taxis. Each female lays two to three egg masses, each with 50 – 80 eggs. After hatching, the borers cluster inside leaf sheaths and feed on leaf sheath tissues, causing the sheaths to wither. Second instar larvae then disperse to inside the rice plant, causing deadhearts or whiteheads. The larvae move among the plants frequently and infested plants in clusters. Sixth and seventh instar larvae pupate in lower stems or inside sheaths, usually about 3 cm from the water surface.

Fig. 13 – 21 Dead sheath caused by rice striped stem borer

Fig. 13 – 22 Dead panicles caused by rice striped stem borer

2. Rice yellow stem borer (***Scirpophaga incertulas***)

Rice yellow stem borer only occurs locally inthe Yangtze River basin and areas to its south in China. The population of yellow stem borer has decreased and there can be 3 – 4 generations each year in the central China rice area and the Yangtze River basin. It is a monophagous pest that only feeds on rice. Adult larvae overwinter in rice stubble, and begin to pupate when the temperature rises to about 16 ℃ the following year. Larvae infestation causes deadhearts at the seedling and tillering stages, dead panicles at the booting stage, a large number of whiteheads at the panicle emergence and heading stages (Fig. 13 – 23), and damaged rice plants at the milk-ripe and maturity stages. An outbreak undermines rice production and may lead to crop loss. The imagoes are strongly phototactic and the moths will mate on the night of emergence and start laying eggs the next day, with the most eggs laid on the second and third days. Each female moth lays 1 – 5 egg masses, with two being the most common. Each mass contains 50 – 100 eggs. After hatching, some of them crawl down along the leaf, while others crawl to the tip of the leaf to spit out silk and droop, or drift away with the wind. After about 30 minutes, they each choose a suitable part of the rice plant to enter the stalk and the larvae can move between plants to cause damage. From tillering to late booting and panicle emergence is the period most suitable for the invasion of yellow stem borer, known as the "critical growth period" in stem borer prevention and control. Stem borers that hatch from the same egg mass in the field often spread and infest nearby rice plants, causing dozens or even more than a hundred plants to have deadheart or deadwhite mass. After the larvae mature, they move into the stem of healthy plants and bite an emergence hole in the stem wall, leaving only a membrane. Then they pupate, and the imagoes will break the membrane to emerge.

Fig. 13 – 23 Dead panicles caused by yellow stem borer

In the past 30 years, hybrid rice has been planted in large areas in Hunan, Jiangxi and other provinces of China. Based on local temperature and light conditions, the principle of early sowing, early planting, and safe heading before September 15 has been followed in cultivation practice to avoid the damage of "cold dew wind". The peak hatching period of the fourth generation of rice yellow stem borer is after September 15, therefore, when a large number of stem borer eggs hatch, late-season hybrid rice plants are mostly already full-headed or mature. Thus, the critical growth period does not meet the peak hatching period. In addition, because the stalk tissues of hybrid rice are hard, wrapped in multiple layers of leaf sheaths, it is difficult for the stem borer to invade and survive. These are the main reason for the decline of the yellow stem borer populations over the past years.

Ⅱ. Morphological Identification

1. Striped stem borer (*Chilo suppressalis*)

An adult striped stem borer has a body length of 10 – 15 mm and is grayish yellowish brown. The forewings are nearly rectangular with seven small black spots on the outer edge. Male moths are smaller than female ones and have darker body and wings. The egg mass is composed of multiple or dozens of oval and flat eggs arranged into scales and covered with glue. The larva is light brown and the body length of a high instar is 20 – 30 mm, with five brown vertical lines on the back (Fig. 13 – 24). The pupa is

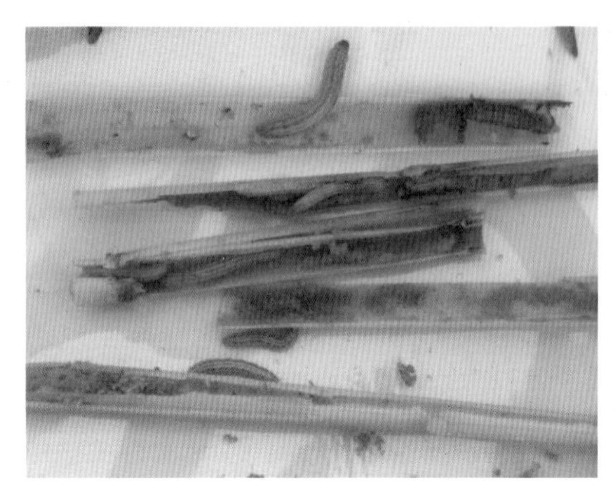
Fig. 13 – 24 High-instar larvae and pupae of striped stem borer

yellowish brown with five dark brown longitudinal lines visible on the back at the early stage, and the ends of the hind feet are as long as the wing buds.

2. Yellow stem borer (*Scirpophaga incertulas*)

An adult has a body length of 8 – 13 mm with triangular forewings. The female is larger, light yellow, with a small black spot in the center of the forewing, and a bunch of yellowish brown hairs at the end of the abdomen. The male is smaller, light grayish brown, and in addition to black spots, there is also a dark brown twill between the wing tip and the center of the wing. The egg mass is oval, with yellowish brown downy on the surface, like half a moldy soybean, with dozens to hundreds of eggs arranged in layers. The larva is grayish black and a newly hatched larva is called ant borer. Afterwards, the larvae become milky white or light yellow in different instar, with a transparent vertical line at the center of the back. The pupa is long cylindrical and brown with a length of 12 – 13 mm. The hindfeet of a female pupa extend to the sixth abdominal segment, and those of a male pupa extend beyond the eighth abdominal segment.

Ⅲ. Environmental Factors

Climatic factors, mainly temperature, have a direct impact on the occurrence period and the amount of rice stem borer. If the required temperature is not reached, overwintered larvae will not be able to pupate and emerge properly. When temperåture is highin spring, the overwintering generation of stem borer can occur earlier and vice versa. Humidity and rainwater also have a certain influence on the occurrence of rice stem borer. If the borer encounters heavy rainfall during the pupation period in the field, floodwater in the field can drown a large number of stem borers to reduce the damage. Stem borer damage has been increasing as the rice planting structure changes, such as from single or double cropping to multi-cropping staggered sowing and mixed planting, which provides rich food and favorable living conditions for stem borer. In terms of field management, due to excessive use of nitrogen fertilizer for super hybrid rice, the plants grow vigorously with dark green leaves, which are attractive to striped stem borer and can lead to more severe damage. The long growth duration of some varieties and the variety portfolio also affect seedling conditions and pest situation, which in turn affects the stem borer populations.

Ⅳ. Prevention and Control Techniques

1. Green prevention and control techniques

Green prevention and control requires a combination of prevention, control, avoidance and treatment. Agricultural and ecological control remains as the basis and specific methods include adjusting the variety portfolio, reducing bridge fields, carrying out deep-water pupal eradication, reducing overwintering pests, protecting natural enemies, using insects to control pests, and promoting physical and chemical trapping techniques such as striped borer sex attractants and insecticidal lamps.

(1) Reduce pests at the source through deep-water pupal eradication

Winter fallow fields should be plowed and flooded before winter, or flooded for 3 - 5 days from the end of March to early April in the following spring. No-tillage fields in lake areas can be flooded for a longer time after spring and green manure reserved fields should be flooded for 2 - 3 days in early April, because deep water can drown most mature larvae and pupae. Seedling-scattering fields can be irrigated and plowed earlier, plus spring plowing to bury rice straws into the soil, which can help eliminate overwintering pests on rice stubble, thereby effectively reducing the pest population.

(2) Adjust the rice variety portfolio to avoid stem borer damage through cultivation

Light and simple rice cultivation techniques are recommended so as to avoid mixed planting of single- and double-cropping rice. Instead, double-cropping or single-cropping rice can be planted separately in large areas to effectively control the pest at the source and reduce "bridge fields" for stem borer reproduction. Proper adjustment should be made to balance between early-, mid- and late-maturing varieties of early-season rice, plant more early- and mid-maturing varieties while appropriately reducing the area of late-maturing varieties, and transplant seedlings in time to separate the damage period of first-generation stem borer and the critical growth period of rice. During the growth period, efforts should be made to reduce the numbers of first-generation stem borers and the annual occurrence base to achieve avoidance through cultivation.

(3) Choose resistant varieties

Resistance of rice varieties to stem borer also affects the damage severity. Varieties (combinations) with good resistance or tolerance to stem borers generally have thicker stem walls, smaller medullary cavities, smaller distance between vascular bundles and leaf sheath air cavity, and increased silicified cells in stem and leaf sheaths. Super hybrid rice is vulnerable in the early stage, but in the later stage, it becomes more difficult for the ant borer to invade as it will have thicker stems and stem walls.

(4) Protect natural enemies and use insects for pest control

There are a variety of hunting spiders in the field preying on the newly hatched larvae of the rice borer. In particular, *Lycosidae*, a single *Pardosa pseudoannulata* (spider) can eat all ant borers hatched from a single borer egg mass in a day. There are many natural predators in late-season rice fields, and in addition to spiders, there are other natural predators. Parasitic natural enemies include *Trichogramma japonicun* Ashmead, *Trichogramma confusum* Viggiani, *Apanteles ruficrus* (Haliday), *Itoplectis naranyae* (Ashmead), *Brachymeria lasus* Walker, and *Pseudoperichaeta insidiosa* Robineau-Desvoidy. Where conditions permit, artificial release of *Trichogramma japonicun* Ashmead can also be carried out.

(5) Use insect sex pheromone (sex attractant) to carry out field trapping

At present, the main types used commonly in production are water basin trap, cage trap and cylindrical trap. The water basin trap is a plastic basin with an upper mouth diameter of 20 - 30 cm and a depth of 8.0 - 12.0 cm. The mouth of the basin traverses a thin iron wire, and a *Chilo suppressalis* sex attractant

Fig. 13 – 25 *Chilo suppressalis* sex attractant

core is hung in the middle. The basin is about 80% full of water mixed with washing powder. The distance between the core and the water surface is about 1. 0 cm, and then the basin is placed on a tripod supporter of wooden sticks or bamboo poles (Fig. 13 – 25). When the adult striped stem borer gradually emerges, the male moth is lured by the female pheromone released by the lure core in the trap and plunges into the water basin to drown. The effective period of the lure core in a basin is generally about 30 days. It is necessary to keep the basin filled with water. Generally, 45 – 60 basins are placed per hectare because basins placed in large areas can have a better result. In recent years, white cylindrical traps and cage traps have been widely used, which are simple and convenient to use.

(6) Use moth trap as insecticidal lamp

Phototaxis of the rice stem borer can be used to trap and kill adults, especially in the peak moth period which is very effective. Each lamp can accommodate 2. 5 – 3. 5 ha of rice field and generally can trap and kill more than 200 adults of *Chilo suppressalis* one night. With the lamps in place, there are 70% fewer borer eggs in rice fields with alleviated damage to the plants. In recent years, Henan Jiaduo PS – 15 Ⅱ ordinary and frequency-vibration solar insecticidal lamp and Hunan Shenbu solar suck-in pest separation insecticidal lamps have been widely used in production (Fig. 13 – 26).

Frequency-vibration solar insecticidal lamp

Solar insecticidal lamp

Fan suck-in pest separation insecticidal lamp [solar (left), electric (right)]

Fig. 13 – 26 Insecticidal lamp

2. Chemical prevention and control

Before carrying out chemical control, it is necessary to do a field survey to check the progress of egg masshatching to determine the appropriate period of chemical control, and to check the seedling situation, egg mass density, and deadheart rate to determine the target fields. Chemical control adopts the strategy of "treating the first generation hard, treating the second generation selectively, treating the third generation cleverly, and focusing mainly on the peak period". Accordingly, stem borer treatment should mainly be carried out at the tillering stage and the panicle stage with equal emphasis. Prevention and control measures should start when there are 3.0%–5.0% of deadhearts at the tillering stage, or 2.0%–3.0% at the heading stage so as to keep the borer damage rate within the economically permissible range. In the past, mainly insecticide monosultap, disosultap and triazophos, were used in production and generally for each hectare, 3.75 mL of 18% disosultap solution, or 750 g of 90% monosultap powder, or 2.25 mL of 20% triazophos emulsion, or 5% disosultap granules mixed with fine soil should be used. However, rice stem borer resistance to insecticides has clearly developed due to the long-term use of these pesticides in large quantities. In recent years, application of high-efficiency, low-toxicity pesticides have been promoted. For each hectare, 150 mL of 20% chlorantraniliprole (Kang Kuan), or 150 g of 20% flubendiamide (Longge), or 2.25 mL of 2% avermectin, or 450 g of 5.7% emamectin benzoate, or bacillus thuringiensis of Miao Nong series can be used. These above-mentioned options can be used alternately, and sprayed with 900 kg of water per hectare. Other possible options include flubendiamide, cyantraniliprole, methoxyfenozide and abamectin+chlorantraniliprole.

Section 6 Occurrence, Prevention and Control of Rice Leaf Folder

Rice leaf folder (*Cnaphalocrocis medinalis*) belongs to *Crambidae* of *Lepidoptera*, and is a migratory pest. It mainly feeds on rice, and also on wheat, millet, corn, sugarcane and *Leersia parviflora*. It occurs in the northern and southern rice areas of China, especially the southern area. The larva feeds on the leaves of rice and spins leaves to form bracts, which affects rice growth, reduces grain weight, increases the rate of empty and immature grains, and can reduce yield by 10%–20%, or up to 50% in severe cases (Fig. 13–27).

Fig. 13–27 Rice leaf folder damage

Ⅰ. Damage Characteristics

Five to six generations occur annually to a wide range of rice areas in southern China, including Yunnan, Guizhou, Hunan, Jiangxi, northern Guangdong and Guangxi, Sichuan, and southern Zhejiang. The first generation is mild, but the second generation can cause heavy damage during the heading stage of early-season rice in mid- to late June, the third generation causes damage to single-cropping rice from late July to early August, and the fourth generation causes damage to double-cropping late-season rice in early to mid-September. The imagoes prefer to hide during the day and come out at night in clusters and in shade. They are phototactic and fresh-green taxis. They can mate at many times, and each female can produce 50 –80 eggs on both sides of upper and middle leaves. The newly hatched larvae first crawl into newly emerged leaves and leaf tips to feed without forming bracts. Generally, the second instar larva starts to spin and curl the leaves at 3 cm from the leaf tip, forming a small bract. After the third instar, the leaves are rolled to a tube shape. Generally, one borer forms one bract on one leaf. The larva hides in the bract to feed itself (Fig. 13 – 28). As it eats more, the bract becomes larger. The fourth and fifth instar is the binge eating period. They damage 5 – 8 leaves throughout the larval period. Generally, the egg stage lasts 3 –6 days, and the larval stage 20 – 25 days. In the last instar, larvae usually pupate on withered yellow leaves at the base of a thicket, lasting 5 –7 days. The imago stage lasts 5 – 15 days, and they usually choose green fields to lay their eggs.

Fig. 13 – 28　Single-leaf damage

Ⅱ. Morphological Identification

The imago of rice leaf folder has a body length of about 8 mm and a wingspan of about 18 mm, with the color of yellowish brown. The forewing is subtriangular, with two dark brown horizontal lines from the anterior to the posterior margin and a short line between the two lines. The hindwing has two horizontal lines, the inner one of which does not reach the posterior margin. Both the forewing and the hindwing have dark brown broad margins on their outer edges. Males are smaller, more brightly colored, with a tuft of dark brown hairs in the center of the forewing's front edge. The frontwing and hindwing diagonally spread on the sides of the back when an adult moth stays still, and the male moth has its tails raised (Fig. 13 – 29). Eggs are oval, about 1 mm long, slightly elevated in the center, with white reticulation on the surface, white and translucent when newly laid, gradually becoming yellowish. There are five-instar and six-instar larvae, mostly five-instar, and they are 14 – 19 mm in length at the last instar, green or yellowish green (Fig. 13 – 30). The black spots on the dorsal plate of the anterior thorax are bracketed and the dorsal plate of the middle and posterior thorax have two horizontal rows of black circles, two in the posterior row, and the three-order missing rings of the gastropod toe hooks. The pupa is 9 – 11 mm long, slightly spindle-shaped, pointed at the end, with 8 – 10 hip spines.

Fig. 13 – 29 Rice leaf folder imago

Fig. 13 – 30 Rice leaf folder larva

Ⅲ. Environmental Factors

The occurrence of rice leaf folder is closely related to climatic conditions. The favorable temperature for the growth, development and reproduction of the pest is 22 ℃– 28 ℃, with the relative humidity above 80%. Temperatures above 30 ℃ or relative humidity below 70% are not conducive to the pest's activity, spawning and survival. Planting of various rice varieties with different growth durations and a mix of early-, mid- and late-season rice provide rich food for each generation of the pest. Improper fertilizer and water management also lead to unfavorable delayed rice maturity, contributing to more severe occurrence of leaf folder.

Ⅳ. Prevention and Control Technology

1. Green prevention and control techniques

Green prevention and control must adhere to coordinated control technology that combines health cultivation with high yielding cultivation and biological control with chemical control, while optimizing super hybrid rice cultivation techniques.

(1) Promoting health cultivation featuring nitrogen-saving and pest-control

The status of nitrogen nutrient in rice is closely related to the occurrence of rice stem borer, leaf folder and sheath blight. Excessive nitrogen fertilizer causes rice to grow luxuriantly, with more green leaves, closed field shade, poor ventilation and light penetration, which is conducive to the occurrence and reproduction of major diseases and pests. The nitrogen-saving and pest-control cultivation technology for high-yielding rice focuses on scientific fertilization, combination of organic and chemical fertilizers, and mastering the appropriate dosage of common nitrogen and chemical fertilizers (urea, ammonium acid carbonate, etc.). The economical standard of nitrogen application is 120 – 150 kg/ha for conventional rice, 165 – 195 kg/ha for ordinary hybrid rice, and 210 – 240 kg/ha for super hybrid rice (Fig. 13 – 31). The key is to promote and apply slow-release fertilizer, controlled-release fertilizer, microbial fertilizer, and compound nitrogen, phosphorus and potassium fertilizer. Nitrogen from slow-release and controlled-release fertilizer is slowly and controllably released in the rice field. The release rate and the amount required for rice growth can maintain a dynamic balance, and a single application can meet the fertilizer demand of the whole growth period of rice, which not only fully ensures high and stable yield of rice, but also reduces the loss of nitrogen fertilizer, and pest and disease infestation.

Fig. 13 - 31 Nitrogen-saving and damage-controlling experiment for super hybrid rice

(2) Optimizing high-yielding cultivation techniques for superhybrid rice

Efforts should be made to promote light and simplified cultivation, "three definition" cultivation, nitrogen-saving and lodging-resistant cultivation, precise quantitative cultivation and super high-yielding SRI cultivation. As super hybrid rice has tall and sturdy stalks, wide, thick and hard leaves, and compact main veins, it is more difficult for rice leaf folder larvae to roll the leaves, reducing their survival rate. Thus, super hybrid rice combinations are less likely to suffer from this.

(3) Protecting and leveraging natural enemies

There are more than 60 kinds of natural enemies of rice leaf folder, and most of them are parasitic or predatory in all stages, including *Trichogramma japonicun* Ashmead in the egg stage, *Apanteles cypris* Nixon in the larval stage, and other predatory natural enemies such as frogs and spiders. Protecting and leveraging these natural enemies can improve the effectiveness of the ecological control of the pest. Application should be scientifically and rationally coordinated with the time, types and methods of chemical control. If a pesticide needs to be applied at the conventional time, but will cause big damage to the natural enemy, the pesticide should be applied earlier or later. If the amount of the pest has reached the level required for the initiation of control measures, but the parasitism rate of natural enemies is high, pesticide should not be used. Pesticides should not harm the natural enemies or should only have minimal impact, and should be applied with advanced equipment.

(4) Manual release of *Trichogramma japonicun* Ashmead

Trichogramma japonicun Ashmead is an egg parasitic wasp. Before releasing, the time and number of the wasp release are to be determined based on understanding of the occurrence period and amount of rice leaf folder in the field, mainly using the egg-card release method. Rainproof and windproof disposable plastic cups or paper cups can be used as the releaser. The egg card is pasted into the cup with sellotape, so that the imagoes can emerge inside the releaser and fly out to look for the eggs of rice leaf folder and other target pests in the rice field and reproduce. Normally, the release should be carried out on sunny days, with 150 – 180 releasers per hectare (Fig. 13 – 32). A long bamboo pole with egg card cups hung upside down should be installed 40 – 60 cm from the rice plants, depending on the rice growth period. The wasps are released during the oviposition period of rice leaf folder every 2 – 3 days, with 150,000 – 300,000 wasps per hectare. The wasps can be released 2 –3 times for the leaf folder generation which causes severe damage. The rate of parasitic eggs of field pests should be surveyed. Normally, the eggs parasitized by *Trichogramma japonicun* are black.

Fig. 13 – 32 Manual release of *Trichogramma japonicun* Ashmead

(5) Using insect sex pheromone

Sex attractant (sex pheromone) can be used to lure and kill the adult leaf folder and interfere with the mating of male and female moths. Attention should be paid to keeping the water basin trap always about 20 cm higher than the rice plants and, the trap core 0.5 – 1.0 cm from the surface of the water in the basin. The water should contain 0.3% washing powder to enhance stickiness. The water and washing powder should be replaced every seven days, and the trap core every 20 – 30 days to achieve effective prevention and control. For simplicity, white tube traps should be promoted.

2. Pesticides for prevention and control

(1) Biological insecticides and their modifiers

Avermectin is an antibiotic insecticide. Its 1.8%, 2% Avermectin emulsion and 0.05%, 0.12% wettable powder have good effect on the low instar larvae of rice leaf folder. *Bacillus thuringiensis* is also a microbial insecticide that has been used for a long time. In recent years, Hunan Miaonong Technology Company has improved the plant protection insecticides such as Miaonong series of *Bacillus thuringiensis*, which is effective in the rice tillering and booting periods. *Bacillus cereus* and Entobacterin are also effective. *Empedobacter brevis* and *Beauveria bassiana* can be promoted as well.

(2) Chemical pesticides and compounding pesticides

For chemical control, the strategy of "treating the second instar generally and the third and fourth instar heavily" should be adopted. The most suitable period for chemical control is the peak time of the second and third instar stages when a large number of leaf tips are rolled. Pesticide is to be applied when the control threshold is reached, i.e., 50 – 60 larvae per 100 seedlings at the tillering stage or 30 – 40 larvae per 100 seedlings at the heading stage. Common pesticides used for each hectare include 1.5 L of 48%

chlorpyrifos EC, or 2.25 L of 20% triazophos EC, or 450 g of 5.7% emamectin benzoate. There are four preferred options, 150 mL/ha of 20% chlorantraniliprole (Kangkuan), 150 g/ha of 20% flubendiamide (Longge), 150 g/ha of 40% chlorantraniliprole + thiamethoxam (Fuge), or 450 mL/ha of 10% flubendiamide+avermectin (Daoteng).

Section 7 Occurrence, Prevention and Control of Rice Planthoppers (Brown Planthopper and Whitebacked Planthopper)

Rice planthopper belongs to *Delphacidae* of *Homoptera*. In the south of China, brown planthoppers (BPH) and whitebacked planthoppers (WBPH) are the main causes of damage to rice. Specifically, BPH, which has the largest area of occurrence and serious damage, occurs almost occurs every year in area to the south of the Yangtze River. In general, early-season rice is mainly plagued by WBPH, mid-season rice by both WBHP and BPH with BPH rising quickly in the later stage, and late-season rice by BPH. When the pest population grows rapidly, an outbreak can develop into a disaster.

Ⅰ. Damage Characteristics

1. Brow Planthopper

BPH is a migratory and temperature-loving pest, which likes to swarm and cause damage to rice in patches. Six to seven generations occur every year in Guizhou, Jiangsu, Zhejiang, Hunan, and Jiangxi. There are two types of BPH adults, long-winged and short-winged. Long-winged adults are migratory and phototactic, while short-winged ones are resident and have strong reproductive ability, able to reproduce within 20 days or so in the rice-growing season under favorable climatic conditions. Each female lays 200 – 500 eggs. The field proliferation rate is 10 – 30 times per generation, but for late-season rice from mid and late September to early October, it can usually be 40 – 50 times that in early and mid-August. Adults and nymphs of BPH cluster at the base of rice plants (Fig. 13 – 33), sucking out sap with their stinging mouthparts and secreting toxic substances from their salivary glands, poisoning and withering the rice plants. When the damage is light, the lower leaves turn yellow, affecting grain weight. When the damage is severe, the affected tissues necrotize, the leaves turn yellow, and the plant withers and even die and lodge in blocks. This is commonly known as "hopperburn", which can cause serious yield reduction or even complete crop loss (Fig. 13 – 34).

2. Whitebacked Planthopper

WBPH is also a migratory pest, with adults moving in earlier than BPH and advancing from south to north. Six generations per year can occur in areas to the south of the Yangtze River. WBPH males are long-winged and have strong flying ability, while females have lower reproductive ability than BPH, laying an average of about 85 eggs per female. The distribution in the field is relatively even and hardly causes rice plants to die in patches. The habits of the WBPH adults and larvae are basically the same as those of

Fig. 13 - 33 BPH cluster damage

Fig. 13 - 34 Field damage by BPH

BPH and the two often occur in overlap, yet WBHP tends to cause damage and take habitat at positions slightly higher than those of BPH. However, what is important is that WBPH transmits the southern rice black-streaked dwarf virus (SRBSDV). The main symptoms of SRBSDV are dwarf plant, stiff, dark green and crinkling leaves, barbed roots and high-node branches at the base, waxy white, short tumor-like protrusions on the stalk (which turns brown later), and no heading or small neck-panicles or poor seed setting (Fig. 13 - 35). In recent years, the disease has spread rapidly in southern China, and the damage is getting more serious.

Fig. 13 - 35 Black-streaked dwarf virus disease

Ⅱ. Morphological Identification

1. BPH

Long-winged adults are 4 - 5 mm long, yellowish brown or blackish brown, with the tip of the forewing exceeding the end of the abdomen. There are three obvious raised vertical lines on both pronotum and scutellum. The short-winged type is similar to the long-winged type, but with short wings which do not reach the end of the abdomen. The body shape is stubby and the female is obese in the abdomen, 3.5 -4 mm long. The eggs are loofah-shaped at the early stages and banana-shaped in the middle and late stages with 5 - 20 eggs arranged in rows, a single row at the front and double rows at the back, showing the egg-laying mark in the egg cap. The nymphs are pale yellow when hatched, and then brownish, nearly oval. The third instar nymphs are yellowish brown to dark brown (Fig. 13 -36), with obvious wing bracts. The fifth instar insects have a white-shaped pattern on the abdomen of the third to the fourth abdominal segments. The two hind legs of the nymph are horizontal on the surface of the water.

2. WBPH

Long-winged adults have a body length of 3. 8 –4. 6 mm. The body is yellowish with brown spots, yellowish white on the dorsal plate of the prothorax, yellowish in the middle of the small scutellum, black on both sides of the males, dark brown on both sides of the females. Short-winged females are fat, grayish yellow or yellowish, about 3. 5 mm in length, with short wings only to the middle of the abdomen in length. The eggs are crescent-shaped, creamy white when laid, turning yellowish later, with two red eyes, 3 – 10 in a single row, no egg-laying mark visible. The nymphs are olive-shaped, creamy white with gray spots when hatched, but grayish white or grayish brown after the second instar, grayish black and creamy white mosaic at the fifth instar, with irregular dark brown spots on the back of the thorax. The hind legs spread wide when on the water (Fig. 13 – 37).

Fig. 13 – 37 WBPH adult and nymph

Fig. 13 – 36 BPH nymph

Ⅲ. Environmental Factors

BPH adults and nymphs prefer a shady and humid environment, generally inhabiting the moist base of a rice bush. Generation overlapis obvious. They grow and develop well under a temperature of 20 ℃–30 ℃, relative humidity of 80% or above, with the most favorable temperature at 26 ℃–28 ℃. If it is rainy, not very hot in summer, and not cool in late autumn, BPH infestation is more likely to occur. An increase in the number of short-winged BPH in the field and an exponential increase in reproduction will be a harbinger of a serious infestation. The appropriate temperature for the development of WBPH is 22 ℃–28 ℃ with a relative humidity of 80%–90%. The most favorable temperature for egg-laying is 28 ℃, and the highest survival rate of nymphs is at 25 ℃–30 ℃. The rice booting to flowering period is suitable for the reproduction of BPH due to the increased water-soluble protein content in the rice plants, while WBPH has the reproductive peak at the tillering and stem elongation stages. Excessive nitrogen, shade and high humidity in the field are all conducive to the occurrence of BPH and WBPH.

Ⅳ. Prevention and Control Technology

1. Green prevention and control techniques

Promoting resistant varieties, raising ducks and frogs, using nitrogen saving cultivation for pest control, light trapping, biodiversity and other green prevention and control practices should beadopted to reduce the pest population density. There are many kinds of predatory and parasitic natural enemies of rice planthoppers. In addition to spiders, the predatory natural enemies include *Cyrtorhinus lividipennis* Reuter, rove beetles, ground beetles and tiger beetles, and the parasitic natural enemies mainly include *Paracentrobia andoi* and *Anagrus nilaparvatae* at the egg stage, and *Haplogonatopus japonicus* Esaki et Hashimoto and nematodes at the nymph and adult stages. It is necessary to protect and leverage these natural enemies in order to prevent and control the pests.

(1) Promoting pest-resistant (tolerant) varieties

The resistance or tolerance of the rice variety is closely related to the amount of BPH occurrence and the severity of infestation, andit plays a decisive role in the reproductive growth of the BPH population. In the absence of farming chemical application, the varieties with less resistance to BPH tend to suffer from more short-winged BPH, leading to a high density of the pest and heavy damage, while the more resistant varieties generally suffer less (Fig. 13 - 38). At present, few super hybrid rice varieties planted in large areas have a high resistance to rice planthoppers, but most of the hybrid rice varieties generally have a moderate resistance, or are between moderately resistant and moderately susceptible. In particular, super hybrid rice has strong stalks, thicker stem walls, and a high level of silicification of the epidermis, all of which lead to tolerance to rice planthoppers. In production, resistance analysis is conducted based on the population density of super hybrid rice varieties and susceptible control varieties that are widely planted. Statistical analysis can be carried out using the formula $FC = Nc - NtNc$ to measure the field control effect against BPH (where, FC is the effectiveness of field control of BPH, Nc is the pest population density on the susceptible control variety, and Nt is the pest population density of the promoted variety). The field BPH control effectiveness on the susceptible control variety TN1 in the field is set as 0 and a complete control as 1. In general, when the field control effectiveness of variety resistance plus natural enemies is 0.85 - 0.99, the variety should be determined as having certain resistance to rice planthoppers under natural conditions.

Fig. 13 - 38 Differences in resistance of resistant and susceptible rice varieties to BPH

(2) Protecting and leveraging predatory natural enemies

Rice field spiders and *Cyrtorhinus lividipennis* are important predatory natural enemies of rice planthoppers. There are more than 10 dominant types of rice field spiders in Hunan (Fig. 13 -39), including *Coleosoma octomaculatum*, *Ummeliata insecticeps*, *Pardosa pseudoannulata*, *Pirata piratoides*, *Pirata subpiraticus*, *Hylyphantes graminicola*, *Tetragnatha maxillosa*, *Tetragnatha shikokiana* and *Clubiona japonicola*. When early-season rice reaches maturity, ridge weeds and other non-paddy habitats have become green corridors for natural

Fig. 13 - 39 Predatory spiders in rice fields

enemies such as spiders, so care should be taken to protect these sites during production. In addition to protecting the green corridors, artificially assisted migration techniques for the safe transfer of predatory natural enemies such as spiders from early-season rice fields to late-season rice fields should be adopted. In the double-cropping rice area, watermelons, vegetables and other cash crops can be planted with late-season rice, leaving bridge fields for spiders and other natural enemies. Relocating paddy straws from early-season rice fields to late-season rice fields after harvesting and planting beans on field ridges are conducive to the inhabitation and migration of spiders and other predatory natural enemies.

(3) Raising ducks and frogs for ecological rice farming

Raising ducks and frogs in rice fields is a green prevention and control technology based on the principle of symbiosis and mutual benefit between animals and plants, making full use of spatial and temporal ecological niches and the biological characteristics of ducks and frogs to prevent and control pests. As ducks have strong foraging power and flocking and water-loving characteristics, they are suitable for rice field. Medium-sized and small ducks with strong vitality, high growth rate, fast growth and high egg-laying rate can be selected (Fig. 13 - 40). Rice seedlings can be planted in a wide- and narrow-row pattern for walking and foraging convenience of ducks in the field. Generally, about 20 days after rice transplanting, duck flocks with 150 - 225 ducklings or 120 - 150 adults per hectare can be put in the field with a 5 - 7 cm layer of water. In addition, frogs are also important natural enemies and should be protected and leveraged. Adult frogs, such as *Fejervarya limnocharis*, *Rana nigromaculata*, *Hylachinensis* and *Pelophylax nigromaculatus*, live mainly in rice paddies or along ditch ponds, and have a high reproductive capacity with a lot of eggs (Fig. 13 - 41). The hind limbs of frogs are strong for jumping. Frogs can prey on not only rice planthoppers at the base of the rice plants, but also the adult rice borers on the leaves. In high-quality

Fig. 13 - 40 Ducks in rice fields for pest control

Fig. 13 - 41 Frogs in rice fields for pest control

rice production bases, such as Chunhua town in Changsha and Beisheng town in Liuyang, there are demonstration plot of where frogs are raised in paddy fields. Frog-raising paddy fields need to be ditched, with the ditches connected to keep some water. The effect of pest control using frogs is better for late-season rice when the frogs are stocked during mid- to late August when the number of pests such as rice planthoppers rises.

2. Chemical prevention and control

Chemical control should be carried out in a timely manner according to the rice varieties and the occurrence of rice planthoppers. For BPH and WBPH, the peak periods of the second and the third instar nymphs are the appropriate time for chemical prevention and control with pesticides of long-lasting, high-efficiency and low-toxicity. The prevention and control criteria vary with rice type and time, but are generally based on the number of rice planthoppers per 100 rice plants, i. e. , 1,000 – 1,500 planthoppers at the heading stage for conventional rice, 1,500 – 2,000 for hybrid rice, and 2,500 – 3,000 for super hybrid rice. The major pesticides used on each hectare of land include 750 g of 25% buprofezin WP, or 300 g of 10% imidacloprid WP, or 300 g of 25% pymetrozine WP, or 75 g of 25% thiamethoxam (Actara), or 450 mL of 10% nitenpyram water agent, or 150 g of 40% chlorantraniliprole + thiamethoxam (VIRTAKO), as well as triflumezopyrim and dinotefuran. If BPH breaks out, 2. 25 L of quick-acting pesticides such as DDVP, MTMC or isoprocarb EC can be applied. The above pesticides should be used in rotation and liquids should be spread evenly to the base of the stem of rice roots. In addition, because the toxic WBPH is a virulent agent for the transmission of SRBSDV disease, pest control and disease prevention are the objectives, and the strategy of controlling WBPH and preventing SRBSDV should be adopted, focusing on the key control period before the seventh leaf stage of rice. As WBPH is more sensitive to imidacloprid, it is recommended to use imidacloprid or pymetrozine to prevent SRBSDV during the seedling and the early tillering stages in the field.

Section 8 Occurrence, Prevention and Control of *Lissorhoptrus oryzophilus* Kuschel and *Echinocnemus squameus* Billberg

Lissorhoptrus oryzophilus Keschel and *Echinocnemus squameus* Billberg cause damage to rice production in China at present.

I. Damage Characteristics

1. *Lissorhoptrus oryzophilus* Kuschel

Lissorhoptrus oryzophilus Kuschel is a major international quarantine pest, an invasive alien pest in China, and a new pest of rice. It has occurred in more than 10 provinces and municipalities across China. Adults of the pest mainly eat rice leaves while larvae eat the roots. The damage caused by the larvae can reduce rice yield by 20% , or up to 50% in serious cases (Fig. 13 – 42). Adults have many ways to spread. They have strong reproductive ability and the female can reproduce in parthenogenesis. It has a wide adaptability with rapid development and spread. In addition to rice, it also feeds on sweet potatoes, corn, sugarcane and *Triticeae* crops. Two generations occur annually in southern China. The adults overwinter in the grass, tree leaf litter and topsoil layer at the edge of the field, and start to move into the seedling field or early-planted production field of early-season rice in mid- and late April of the following year, and gradually spread to large areas. Adults gnaw on the upper epidermis and mesophyll of young rice leaves, leaving white strips of feeding spots (Fig. 13 –43). Adults lay eggs in the roots of rice plants or in the leaf sheaths below the water surface. Each female can lay 50 – 100 eggs. The larvae worm invades the rice roots before the third instar, resulting in broken and destroyed roots. After that, they chew on the rice roots, causing severe damage.

Fig. 13 – 42 *Lissorhoptrus oryzophilus* Kuschel larvae damage in the field

Fig. 13 – 43 Imago feeding streaks

2. *Echinocnemus squameus* Billberg

Echinocnemussquameus Billberg is a local pest and occurs in different rice areas in China with local outbreaks in some areas. It mainly affects rice, and also corn, *Triticeae* crops and weeds. The adults feed on the newly emerged leaves of seedlings near the water surface, causing a horizontal row of round holes to appear on the newly emerged leaves, and making the leaves break easily. The larvae feed on young and tender fibrous roots, which can cause yellowing and poor growth of rice plants, or failure to produce panicles, or blighted grains. One to two generations of the pest occur throughout the year, with the greatest damage in the seedling reestablishment and tillering stages of early-season rice. The pest damages early-season rice seedlings and production fields from April to May in the southern parts of China. It lays eggs on rice stalks near the surface of the water with 3 –20 eggs at each spot. After hatching, the larvae feed on the young fibrous roots of rice and will pupate in a soil chamber near the rice roots after they mature.

Ⅱ. Morphological Identification

1. *Lissorhoptrus oryzophilus* Kuschel

The adult body has a length of 2. 5 –3. 5 mm with a grayish brown to grayish black color, and the body surface is densely covered with grayish brown scales. From the end of the dorsal plate of the prothorax to the base, there is a wide mouth vase-like dark spot composed of black scales and along the base of the elytra, down to 3/4 the length of which there is a dark spot. The antennae are reddish brown, borne in front of the middle of the rostrum, with three rod-shaped stripe segments, densely hairy only at the end, smooth at the base, leg segments rod-shaped, not dentate, tibial segments slender and curved with one row of long swimming hairs on each side of the tibial segment. The eggs are white, about 0.8 mm long, in a long cylindrical shape. The larvae have four instars, with brown heads and white bodies without legs, and the mature larvae are 8 – 10 mm long with one pair of cone-like protrusions on the dorsal surface of the second to the seventh abdominal segments, six pairs in total, each pair of protrusions with a central hook-like valve extending forward. Mature larvae pupate on rice roots as earthen cocoons, which are mud-yellow, filled with air and connected to the aeration tissue of rice roots for gas exchange, with white pupal bodies (Fig. 13 – 44).

Fig. 13 – 44 Various forms of *Lissorhoptrus oryzophilus* Kuschel

2. *Echinocnemus squameus* Billberg

The adult body has a length of 5 – 6 mm, dark brown to black, with dense grayish yellow scale hairs. Each elytrum has 10 fine longitudinal grooves with a pair of oblong spots composed of white scales at 1/3 of the posterior part. The dorsal plate of prothorax has many small incised spots with yellow hairs forming longitudinal strips on both sides. The antennae are dark brown, borne at the proximal part of rostrum, finely velutinous, and the tibial segments of each foot have one row of setae in rostrum. The eggs are white or gray, shiny, oval-shaped. Senior larvae are creamy white, footless, about 9 mm long, slightly curved to the ventral surface without protrusion on the back. The matured larvae pupate in a soil chamber, and are about 5 mm long, white at first, then grayish, with one pair of fleshy spines at the end.

Ⅲ. Environmental Factors

1. *Lissorhoptrus oryzophilus* Kuschel

Overwintering adults emerge from the soil in early April to early May each year, first feeding on new leaves of weeds at the overwintering site. The appropriate temperature for activity is 20 ℃−24 ℃. It moves into the early-season seedling or production fields to reproduce and infest when the temperature reaches 20 ℃ or higher. During this period, it is mainly affected by climatic factors. In addition to temperature, wind and rain have a greater impact on the number and distance of migration of overwintering adults. If it is windy with heavy rainfall, the migration of overwintering adults is blocked, and those who have not migrated will continue to hibernate, and move when the weather gets better. Rice seedlings are conducive to the spawning and reproduction of the pest, especially scattered seedlings of early-season rice because of the exposed roots. They can lay a large number of eggs and cause great damage. Long-term deep irrigation in the field is conducive to the occurrence of the pest. Sun-drying the field timely has a certain controlling effect on the development of the population. From late August to early October each year, the second generation of adults in late-season rice field emerges and then migrates to different places nearby for overwintering. The migration distance and the number of migrated pests are also affected by temperature, humidity, wind and rain.

2. *Echinocnemus squameus* Billberg

Adults overwinter in the weeds or the deciduous layers of the field, or larvae live in the soil under rice roots. Adults migrate and damage early-season rice seedlings from April to May every year when the temperature is appropriate, andthey lay eggs and reproduce from May to June. Generally, occurrence is more likely in hills and mountains than in plains or lake areas, especially in high fields with good aeration and low water content. Dry seedling fields suffer more than water seedling fields, and damage is more serious in sandy soil than in clay soil.

Ⅳ. Prevention and Control Techniques

1. Green prevention and control techniques

(1) Agricultural prevention and control

After rice is harvested, some of the adult pests that have not yet migrated overwinter in the rice roots or soil layer of the rice field. Timely tilling or plowing of the rice field can greatly reduce the survival rate of overwintering adults in the field. Practice of fertilizer accumulation combined with field management to remove weeds can help eliminate overwintering pests, and remove sheath blight sclerotia and overwintering pests floating on the water surface before spring production of the following year.

(2) Quarantine

The transfer of seedlings, rice seeds, rice grains and other agricultural products from infected areas should be prohibited and host plants should not be used as filling materials. Quarantine inspection should be strictly implemented for rice seedlings, rice seeds and rice grains shipped from infected areas for adult *Lissorhoptrus oryzophilus* Kuschel. Scientific fertilization should be carried out to maintain the root vitality of super hybrid rice.

(3) Physical prevention and control

Insecticidal lamps can be used for light trapping to reduce the pests at the source, set up nets to stop adult *Lissorhoptrus oryzophilus* Kuschel from moving into rice fields or the covers over seedling beds and reduce the number of eggs dropping on rice plants.

(4) Biological prevention and control

Predatory natural enemies such as birds, frogs, fish, netting and hunting spiders and ground beetles, should be protected and leveraged. If conditions permit, ducks can be raised in rice fields.

2. Chemical prevention and control

Chemical control should mainly focus on the overwintering generation. The control strategy is divided into three stages based on the seedling and production stages and overwintering sites - overwintering adults should be treated strictly, the first-generation larvae generally, and the first-generation adults concurrently. Trials of insecticide showed that 20% triazophos EC mixed with appropriate amounts of thiamethoxam, or imidacloprid or avermectin can be over 90% effective. For single chemical application, 300 g/ha of 10% imidacloprid WP, or 1.5 L/ha of 40% chlorpyrifos EC, or 2.25 L/ha of 20% triazophos EC can be 75%, 91% and 85% effective respectively; while 75 g/ha of 25% thiamethoxam water dispersible granules can be 92% effective. All are effective in preventing and controlling both *Lissorhoptrus oryzophilus* Kuschel and *Echinocnemus squameus* Billberg.

References

[1] YUAN LONGPING. Key techniques for 12 t/ha super hybrid rice[M]. Beijing: China Three Gorges Publishing House, 2006: 80-103.

[2] HUANG ZHINONG. Comprehensive management of hybrid rice pests and diseases[M]. Changsha: Hunan Science and Technology Press, 2011: 33-64.

[3] XIA SHENGGUANG, TANG QIYI. Primary color ecological atlas of rice disease, pest and weed prevention and control[M]. Beijing: China Agriculture Press, 2006: 1-65.

[4] FU QIANG, HUANG SHIWEN. Primary color atlas of diagnosis, prevention and control of rice diseases and pests[M]. Beijing: Golden Shield Press, 2005: 73-96.

[5] LEI HUIZHI, LI HONGKE, LI XUANKENG. The occurrence, prevention and control of hybrid rice pests and diseases[M]. Shanghai: Shanghai Science and Technology Press, 1986: 2-34.

[6] CAI ZHUNAN, WU WEIWEN, GAO JUNCHUAN. Rice pest prevention and control[M]. Beijing: Jindun Press, 2005: 73-105.

[7] XIAO QIMING, LIU ERMING, GAO BIDA. Agricultural phytopathology[M]. Beijing: China Education and Culture Press, 2007: 2-30.

[8] JIN CHENZHONG. Integrated prevention and control of rice diseases, pests and weeds: theory and Practice[M]. Chengdu: Southwest Jiaotong University Press, 2016: 65-98.

Part 4

Seeds

Chapter 14 Foundation Seed Production of Super Hybrid Rice Sterile Lines

Liu Aimin / Li Xiaohua

Seed production of super hybrid rice involves the seed production of CMS lines and their maintainer lines in the case of three-line hybrids, PTGMS lines of two-line hybrids, and restorer lines. The process includes three closely related links, i. e. , foundation seed production, parental seed production and hybrid seed production. Foundation seed production is related to the trait stability of hybrid rice and the performance of heterosis. In the process of foundation seed production, it is necessary not only to ensure the stability and consistency of the typical traits of the parents and the seed purity, but also to maintain the stable relationships between the sterile and maintainer lines, and the sterile and restorer lines.

Section 1 Foundation Seed Production of Three-line Sterile Lines

Ⅰ. Performance and Inheritance of Cytoplasmic-genic Male Sterility

The genetic tool used for hybrid heterosis in the three-line system is the rice cytoplasm-nucleus interactive male sterility gene(s), which consists of a set of sterile genes in the nucleus and the cytoplasm, which interacts to express male sterility. A set of three-line hybrid rice consists of a male sterile line (A), a male sterile maintainer line (B), and a male sterile restorer line (R). The offspring produced by the cross between a CMS line and its B line keeps male sterilility; while the hybrid offspring produced by a CMS line and its R line have normal fertility and could possess heterosis and be used for hybrid rice production. The male organs (anther and pollen) of a male sterile line develop abnormally, which is completely abortive without pollination ability and selfed seeds, while the female organs develop normally and can accept pollen to get fertilized and set seeds. The B and R lines are normal rice varieties with fertile male and female organs, with which they can not only perform selfing but also provide pollen for the male sterile line for fertilization.

1. Performance of male sterility in rice sterile lines

Rice pollen can be divided into four types according to its morphology, size and staining reaction to iodine-potassium iodide solution, typical abortive pollen, spherical abortive pollen, semi-stained pollen, and black-stained pollen, with the following characteristics respectively.

Typical abortive pollen – The pollen grains are irregular in shape and can be prismatic, triangular, semicircular, circular, etc. The pollen grains are small, hollow and do not stain with iodine-potassium iodide solution.

Spherical abortive pollen – The pollen grains are round, large, hollow, without contents and do not stain with iodine-potassium iodide solution.

Semi-stained pollen – The pollen grains are round or irregularly rounded, large, and can be stained blue-black to iodine-potassium iodide solution. The stain is light or partially (less than 2/3), or in the case of a full stain, the pollen grains are not round in shape (pear-shaped, oval or other irregularly rounded shaped).

Black-stained pollen – The pollen grains are round and large, and can be stained blue-black by iodine-potassium iodide solution.

Typical abortive pollen and spherical abortive pollen are abortive and have no fertilization ability. Black-stained pollen is normal fertile pollen with the ability to fertilize and set seeds. Semi-stained pollen could be abortive or fertile. Semi-stained pollen and black-stained pollen are collectively referred to as stained pollen. When there is stained pollen in the sterile line, it indicates that the pollen abortion of the sterile line is incomplete and the sterility is unstable.

These four kinds of pollen are basically present in the anthers of normal fertile rice, but in indifferent proportions, which determines the male fertility of rice. In normal circumstances, when the proportion of black-stained pollen grains is above 30%, the male plants are normally fertile, and the self-fertility rate is about 50%. When the stained pollen grains account for 0.5%–30% of the total, it is semi-sterile, and the self-seeding rate is less than 50%. When the stained pollen rate is 0%–0.5%, the male is sterile and the self-seeding rate is 0%–0.1%, showing male sterility.

The male sterility of sterile lines is classified into four types based on the amount of typical abortion, spherical abortion and semi-stained pollen, pollen-free, typical abortion, spherical abortion and stained abortion.

Pollen-free abortion – The sterile lines do not contain any pollen grains or only a few very small pollen grains in the anthers which are thin and milky white, water-stained, completely indehiscent.

Typical abortion – The pollen in the anthers of the sterile line is mostly typically abortive, with few spherical abortion pollen, thin anthers, white or yellowish, completely indehiscent, such as the WA CMS lines.

Spherical abortion – The sterile lines have mostly spherically abortive pollen grains with some typically abortive pollen grains in the anthers which are small, thin, white or yellowish, and generally indehiscent.

Stained abortion – The pollen grains in the anthers of the sterile lines are composed of semi-stained pollen, typical abortion pollen and spherical abortion pollen grains with different proportions. The anthers are plump, light yellowish or yellow, generally dehiscent and disperse pollen, and the self-fertility rate is less than 0.5%.

The above four types of male sterility are all presented in *indica* or *japonica* CMS lines in China. Regardless of the type of sterile lines, the criteria for sterile lines with practical value in production are 100% sterile plants, 99.5% or higher pollen abortion rate, 0.1% or less of self-fertility rate, least affected by environmental conditions (mainly temperature), which means stable sterility.

2. Genetic characterization of rice CMS

The inheritance of the CMS lines is relatively simple, i.e., the expression of sterility is controlled by the interaction of nuclear and cytoplasmic sterility genes, male sterility is expressed when both the nucleus and cytoplasm have sterility genes, male sterility is maintained when the nucleus has sterility genes and the cytoplasm has no sterility genes, and fertility is restored when neither the nucleus nor the cytoplasm has sterility genes. However, different genetic backgrounds of CMS lines results in differences in the expression of male sterility in the sterile lines. There are two main phenomena. Firstly, the degree of pollen abortion in some sterile lines is affected by the ambient temperature, i.e., pollen development during the pollen formation and maturity periods is affected by the environmental high temperature, which causes some anthers to become yellow and plump, with a certain percentage (up to 70%) of normal stained pollen grains that and can be dispersed with self-fertilization. Different CMS lines have different sensitivity to high temperature, generally an average daily temperature of 30 ℃ or more and a highest daily temperature of 36 ℃ or more. Secondly, some sterile maintainer lines contain minor restorer genes, resulting in sterile lines with incomplete pollen abortion, which have no practical value in production.

3. Major CMS lines used in super hybrid rice production

Rice CMS lines can be classified into different types by the origin of cytoplasmic sterility genes. There are some differences in the sterility performance and the relationship between the maintainer line and restorer line for different CMS lines.

(1) *Indica* sterile lines

There are six types of *indica* CMS lines currently used in hybrid rice production in China.

1) WA-type CMS lines

This type has the largest planting area in China and account for about 95% of the sterile lines in use, including Longtepu A, Fengyuan A, Tianfeng A, Wufeng A, Zhongzhe A, Quanfeng A, Gufeng A, and Chuanxiang 29A.

2) Wild rice CMS lines

This type of CMS lines are developed by crossing common wild rice and cultivated varieties. They can be divided into two subtypes by the relationships between maintaining and restoring abilities, i.e., one subtype has a similar restoration-maintenance relationship as the WA CMS lines, such as Xieqingzao A. The other subtype is the Honglian type which has a different restoration-maintenance relationship from WA CMS, such as Yuetai A and Luohong A.

3) G-type sterile lines

The cytoplasm of this type was from the late *indica* cultivar Gambiaka kokum of West Africa, with the representatives of Gang 64A and Gang 46A.

4) D-Type sterile lines

The cytoplasm of this type comes from the rice cultivar Dissi D52/37, with representatives of D Shan A, Yixiang 1A, and Hongai A.

5) Indonesian paddy rice CMS lines

The cytoplasm of this type comes from the *indica* Indonesian paddy rice, with the representatives of Ⅱ-32A, You 1A, and T98A.

6) Other CMS lines

Yue 4A, as a representative of this type, is in production.

Except for the HL-CMS lines which has the cytoplasm from a common wild rice (red-awn wild rice) with a unique restoration-maintenance relationship and spherical abortion pollen, all of the other five types of CMS have the same restoration-maintenance relationship and typical abortion pollen, with the maintainer lines genetically derived from dwarf early-season *indica* varieties in the Yangtze River Basin of China, while the restorer lines are mostly from mid-season *indica* varieties in low-latitude tropical and subtropical Southeast Asia.

(2) *Japonica* sterile lines

There are two main types of *japonica* CMS lines used in production in China. One is the Dian-CMS lines developed in Yunnan, represented by Fengjin A, Tudao 4A, Yanjing 902A, Tai 2A (Tai 96-27A) and Dianyu 1A. The other is the BT-CMS lines with the cytoplasm (Chinsuran Boro Ⅱ) introduced from Japan. The representatives include Liming A, Zhongzuo 59A, Liuqianxin A, 80-4A, and Jingyin 66A. These two types of CMS lines have the same restoration-maintenance relationship.

The pollen of Dian-type CMS lines are mainly semi-stained, with a small amount of spherical abortion pollen and even few typical abortion pollen, belonging to the type of stained abortion with yellowish, plump anthers indehiscent and unable to dispersing pollen.

The pollen of BT-CMS lines features mainly spherical abortion, with few semi-stained pollen and typical abortion pollen, belonging to the type of spherical abortion with thin, tiny, yellowish anthers that do not dehiscent to disperse pollen.

Ⅱ. Mixture and Degeneration of CMS Lines and Maintainer Lines

1. Specific characteristics

Parental foundation seed production of three-line hybrid rice includes foundation seed production for the A line (CMS line), B line and R line. The parents involved in the foundation seed production are independent but also interconnected and mutually constrained.

During the production of the foundation seeds, it is necessary to ensure the stability and consistency of each parent's traits between generations and within the population, and to maintain their typical characteristics. Yet it is also essential to consider the stability of the interrelationship between the parents, i. e., the ability of the B line to maintain male sterility of the CMS line, the ability of the R line to restore fertility of the CMS line, and the stability of hybrid heterosis. Meanwhile, the foundation seed production of CMS lines requires the pollination and seed setting of the B line, which is an outcrossing process.

Therefore, the procedure of foundation seed production of three-line parents is quite complicated and technically demanding.

Years of research and practice in the production of three-line parental foundation seeds show that only by maintaining the typical traits of each parent and the stable consistency of all traits between generations and within the population can the interrelationship between the parents and the heterosis be stabilized.

2. Performance of mixture and degradation of three-line parents and F_1 hybrids

The mixture and degradation of CMS lines exhibit segregation and variation in sterility and morphological traits such as leaf, panicle and grain shapes, growth duration, etc. , resulting in the reduced sterility and sterile plant rate, stained pollen and self-fertilization, reduced restorability and combining ability, degradation of outcrossing and flowering habits, and reduced stigma exsertion. The mixture and degradation of a B line or a R line are characterized by segregation and variation in morphological traits such as plant type, leaf and grain shapes, growth duration, weakened maintenance ability or restorability, reduced combining ability, insufficient pollen load, poor pollen dispersal, weakened growth vigor and reduced resistance. In particular, the traits of R lines derived from an *indica-japonica* cross appear to segregate leaning towards *indica* or *japonica* after being in use for many years. The production of hybrid rice seeds using parents of mixture and degradation, or in segregation not only affects the seed production yield of the parents and the hybrids, but also more seriously affects the heterotic performance of hybrids.

The off-types in the F_1 hybrids mainly include plants of A, B and R lines, as well as semi-sterile, ever- green and other variants. The off-types in a CMS population mainly are the plants of the B line, followed by plants of early- or late-maturity, different plant heights, different plant and leaf types, semi-sterility, and other variants. The off-types in a B-line population are CMS plants and other variants produced by mechanical or biological mixtures, while the off-types in an R-line population is mainly those plants produced by mechanical or biological mixtures.

In addition to the above-mentioned off-types in a CMS population, the male sterility of a CMS line also could be changed to self-fertility in a small quantity, which is caused by nuclear-cytoplasmic interaction and also influenced by climatic conditions or intrinsic physiological function. Generally, the next generation of self-fertilized seeds is still sterile with a self-fertilization rate of only a few thousandths, which does not have significant impact on production. The small number of cases of self-fertilization are randomly distributed in a population and cannot be eliminated completely. More serious self-fertility in a CMS population is mainly due to the migration or accumulation of restorer genes, that is, the production of the so-called homo-cytoplasmic R line or partially restored R line, which is essentially the products of biological mixing and must be eliminated fully using the procedures of mixture prevention and purity maintenance. As for the composition of sterile pollen in a CMS line, it is a quantitative trait with a relatively stable distribution and a range of variation. Any internal or external conditions favoring pollen development will generally increase the rate of spherical abortion and stained abortion, but it will not affect the CMS's productive use, nor does it imply a change in fertility. The pollen composition and selfing rate are the main criteria for the evaluation of a CMS line.

3. Causes of parental and F_1 generation mixture and degradation

The causes of parental and F_1 generation mixture and degradation are mainly mechanical and biological contamination, followed by trait variation.

(1) Mechanical mixture

During the CMS and hybrid seed productions, the two parents are planted in the same field, which is easy to cause mechanical mixture during the operation. The off-types in a CMS line or hybrid population are mostly results of mechanical mixture, accounting for 70%–90%, mainly plants of the B line.

(2) Biological contamination

Another major source of mixture in a CMS line or hybrid population is off-types from biological contamination, which is mainly caused by pollination from other rice varieties in the same field due to mechanical mixture, or pollination from other varieties in other fields due to poor isolation during seed production. The sterile plants in a hybrid population are also caused by pollination of other B lines during CMS seed production, and "ever-green", semi-sterile, and other off-type plants in a hybrid population are mainly caused by pollination of other varieties, rather than its R line.

(3) Trait variation

A restorer line or maintainer line is a self-crossed pure variety with relatively stable traits. However, variations are always possible, but with a small probability. The variation of sterile and maintainer lines is mainly expressed in two aspects, stained pollen or even some self-crossed plants appear in the sterile line, and changes in the traits of the line, such as maturity, plant height, number of leaves, reduced panicle neck wrapping or even no wrapping. Variation of these traits is often correlated with the sterility variation of the CMS line. The variations in an R line are usually manifested in reduced restorability and combining ability, reduced seed setting in the F_1 hybrids, followed by the variation of plant traits such as reduced number of leaves on the main stem, reduced growth duration, change of plant type, leaf and grain types, and reduced resistance to diseases or other stress.

Ⅲ. Production of Foundation Seeds of CMS Lines and Maintainer Lines

Seed production of a CMS line relies on the B line which maintains male sterility, so the foundation seed production of a CMS line and its B line are carriedout simultaneously. Genetically, a CMS line and its B line are only different in the cytoplasm and the reproduction of a CMS line is in fact a process of continuous nuclear substitution of the CMS line by its B line. Therefore, the B line plays a decisive role in the long-term stability of the CMS line. Variation in a CMS line is constantly "corrected" through nuclear substitution by its B line.

1. Methods of foundation seed production

Research and practice over the past 40 years have formed a set of techniques for foundation seed production, which can be described as "single plant selection and paired backcrosses→ inspection and selection of paired crosses → family comparison → foundation seeds". The purification methods include step-by-step purification, set-based purification and simplified purification, all of which have been proved effective. It has been proved that the systems of "three-line seven-nursery", "modified purification", and "modified mixture selection" are more effective and less costly than the methods of "two-crosses and four-nursery" and "set-based purification". Therefore, considering the quality and economic benefits of foundation seed production, it is advisable to use the systems of "three-line seven-nursery", "modified purification", and "modified mixture selection" or "simplified purification" with the following theoretical bases.

1) The nuclear genes of a CMS line and its B line are basically homozygous and relatively stable. The B line is a self-pollinated pure variety. Selfing keeps the genes homozygous with stable heritability and very low probability of mutation. Meanwhile, the maintaining ability of the B line is mainly controlled by one or two pairs of major genes in the nucleus, which are relatively stable and not easily changed by external influences. The selfing rate of the CMS is a few per million.

2) The main reasons for the mixture and degradation of the three-lines and the F_1 hybrid are biological contamination and mechanical mixture, rather than heterosis degradation due to the variation of the CMS line. In addition, only the F_1 hybrids are used so there is no transmission or accumulation of variation.

3) In a three-line set, most traits are controlled by the nuclear genes of the B and R lines. As long as the typical traits of the B and R lines are maintained, the degradation of the three-line set can be prevented, and the fertility of the CMS line and the heterosis of the hybrid can be stabilized.

4) There is a small quantity of stained pollen or some selfed seeds in a CMS line, which may be caused by genetic or environmental factors. However, the plants from self-pollinated seeds are sterile in the next generation, and a small number of selfed seeds of a CMS line have no significant impact on the hybrid yield.

Generally, the method of "simplified purification" can be employed for those CMS lines with complete pollen abortion and stable sterility, while the methods of "set-based" or "step-by-steppurification" should be adopted for CMS lines with incomplete pollen abortion and unstable sterility.

"Simplified purification" is characterized by neither paired backcrossing nor testcrossing by families and heterosis inspection, but by purification based on the typical traits and fertility of the three-lines. The main methods include "three-line seven-nursery", "modified mixture selection", or "modified purification", and "three-line nine-nursery". The "three-line seven-nursery" and the "three-line nine-nursery" methods are the most widely adopted and more effective in foundation seed production.

(1) "Three-line seven-nursery" method

This method includes single plant selection, line comparison, and mixed line multiplication. There are three nurseries forthe CMS line, i. e. the plant-to-row nursery, row-to-family nursery and foundation seed nursery; and two nurseries each for the B line and the R line, namely the plant-to-row and row-to-family nurseries (Fig. 14 - 1).

(2) "Three-line nine-nursery" method

This method is a modified version of mixed selection with plant-to-row, row-to-family, and foundation seeds nurseries for all three lines, making a total of nine nurseries. Single plant selection, row comparison, family inspection and mixed multiplication are carried out in this method.

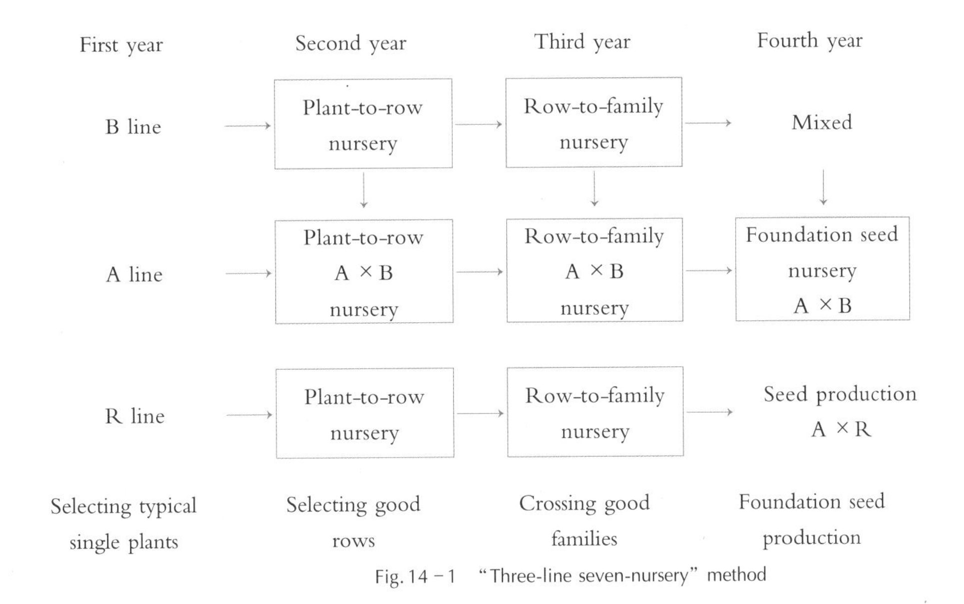

Fig. 14 - 1 "Three-line seven-nursery" method

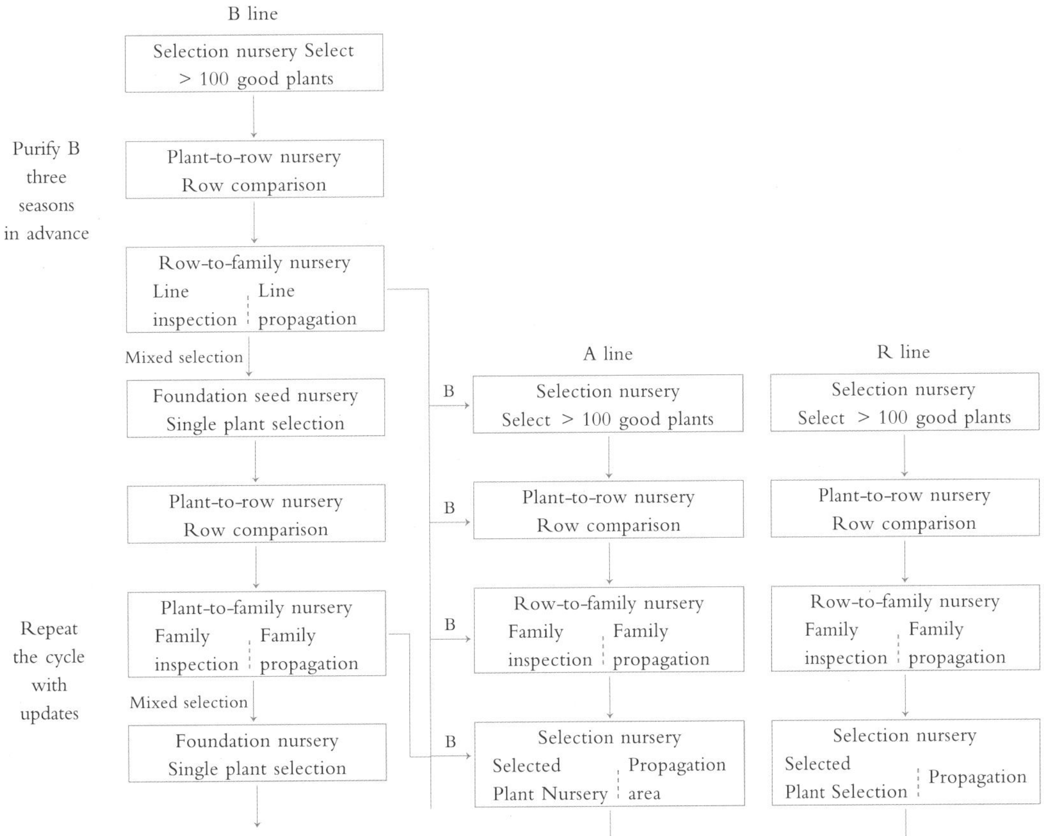

B line

Selection nursery Select > 100 good plants

Purify B three seasons in advance

Plant-to-row nursery Row comparison

Row-to-family nursery Line inspection | Line propagation

Mixed selection

Foundation seed nursery Single plant selection

Plant-to-row nursery Row comparison

Repeat the cycle with updates

Plant-to-family nursery Family inspection | Family propagation

Mixed selection

Foundation nursery Single plant selection

A line

B

Selection nursery Select > 100 good plants

B

Plant-to-row nursery Row comparison

B

Row-to-family nursery Family inspection | Family propagation

B

Selection nursery Selected Plant Nursery | Propagation area

R line

Selection nursery Select > 100 good plants

Plant-to-row nursery Row comparison

Row-to-family nursery Family inspection | Family propagation

Selection nursery Selected Plant Selection | Propagation

Fig. 14 - 2 Three-line nine-nursery method

The detailed steps for the "three-line nine-nursery" method are as follows.

1) Starting B-line purification three seasons in advance.

First, select single B-line plants and put them in the three nurseries for purification following regular procedures. In other words, in the first season, about 100 single plants with typical traits are selected from the foundation seed nursery. In the second season, about 30% of the plants are selected through line comparison. In the third season, about 50% of the plants are selected through family comparison and mixed for foundation seed production.

2) Purification of the CMS line

Select 100 sterile plants in the foundation nursery and plant them in the plant-to-row nursery in the next season, then the rows will be compared and backcrossed using the best B-line plants from the previous season as the male parents. Specifically, about 30% of selected CMS plants will be planted in the row-to-family nursery and the families will be compared and the best B line families of the previous season will serve as the male parents for backcrossing. Then about 50% of the plants will be planted in the foundation seed nursery in the next season and the foundation seeds of the B line from the previous season will be used as the male parent for backcrossing.

3) Purification of the R line

Select about 100 single plants in the foundation seed nursery, then plant them plant-to-row and row-to-family for comparison using regular purification procedures, and then mix the selected families for multiplication.

The "three-line nine-nursery" method has the following characteristics.

1) B line is purified three seasons in advance because the B line determines the fertility and typicality of the CMS line, i. e. a CMS line can only purified after the B line is purified.

2) Three nurseries are set for each of the three lines (A, B and R). By the genetic law, there is no segregation in the F_1 population, but the F_2 population would have segregation if the parent is outcrossed. Therefore, only through two generations of comparison and characterization in the plant-to-row and row-to-family nurseries can the biologically contaminated plants be identified and eliminated, so as to select the good plants of the three lines accurately.

3) There is no requirement for paired backcrossing and testcrossing for R lines, and isolation for rows and families. The core of purification is the selection of typical traits of the lines and the fertility of the CMS line, which provides a large population for selection.

4) Foundation seed production is performed every three years.

5) The plants of a B line in the CMS nurseries will be harvested in advance, and not used as the seeds for the next generation so as to ensure CMS seed purity.

(3) Modified purification method

This method involves only four nurseries – the plant-to-family and foundation seed production for the CMS line and the R line (Fig. 14 – 3).

The modified purification method is simpler than the three-line seven-nursery and three-line nine-nursery methods because there is no plant-to-row or row-to-family nurseries of the B line. The B line is purified by single-plant mixed selection and propagated as the backcross parent of the CMS line in the same AxB nursery. It also has no plant-to-row nursery for the CMS line and the R line, and selected single plants go directly into the family nursery. This is the easiest way to purify three-lines. The key to the implementation of this method is very accurate and strict single plant selection and comparative identification of the families, especially in the selection of B-line plants because there is only one chance for the selection and it must be accurate.

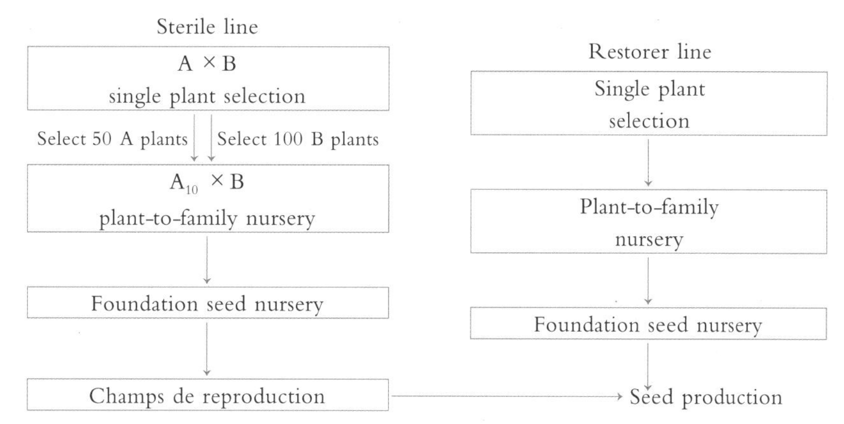

Fig. 14 – 3 Modified purification

(4) "Family cycle" method

The "family cycle" method is based on the "three-line seven-nursery" borrows parts of the purification and multiplication procedures for conventional rice varieties. This method combines the plant-to-row and row-to-family selection into one step. In the selected rows, 10 single plants are selected as a planting unit. If the reproduction coefficient of rice is regarded as 100, then the plot is 10 times larger than the plant-to-row nursery and 10 times smaller than the row-to-family nursery, thus known as the "small family". The seeds reproduced are subject to massive propagation as foundation seeds (Fig. 14 – 4)

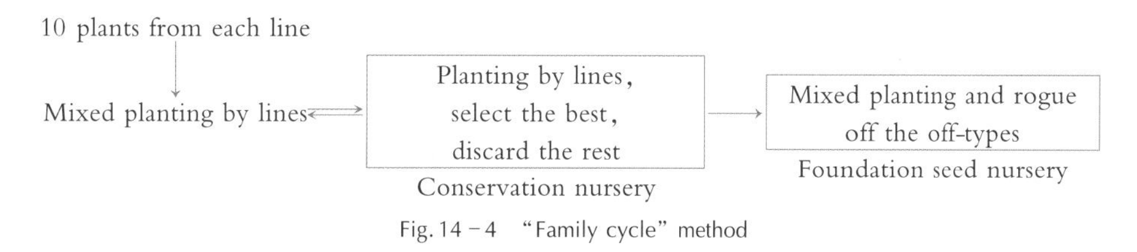

Fig. 14 – 4 "Family cycle" method

This method is characterized by stable basic materials, continual and trackable genealogy, and reliability for line characterization and selection as it has a larger population than the plant-to-row nursery. It also requires one year less than the "three-line seven-nursery" method and delivers seeds with high and stable quality. The families in the conservation nursery can still be obtained from the selected rows in the plant-to-row nursery. Any homo-cytoplasmic restorer line can be ruled out by one-time paired testcrossing between the B line and the CMS line. Other technical requirements are the same as those in the "three line seven nursery" method.

(5) "Simplified purification" method

"Simplified purification" can be performed for new hybrid parents just provided by the breeder seeds or when the parents have not started foundation seed production. The specific practice is to select a parental production field with uniform fertility and uniform growth of rice pants in the current year according to the needs of the next year's production quantity, as a pre-foundation seed production field. All suspicious off-types in the field are rogued off in several rounds from panicle emergence to harvesting. Gibberellic acid is not applied (especially for the CMS line with good outcrossing) so as to ensure the trait performance of the CMS line and the detection of any off-type plants. The seeds harvested in this way are used as pre-foundation seeds.

2. Key technical procedures of foundation seed production

(1) Single plant selection and paired backcrossing

This is the first step in the production of foundation seeds, which must be strict and accurate so as to ensure good quality and sufficient quantity.

1) Selection range

The selection shall cover a wide range, not confined to the plant-to-row nursery. The foundation seed nursery or the first-generation of the multiplication field should be covered too.

2) Selection criteria

The CMS line traits, such as the traits of plant, leave, panicle and grain types, anther, pollen, flowering time, glume-opening angle, panicle exsertion, stigma exsertion and maturity, all have to be typical for the line. The same is true for the selection of B line plants, which should also be highly consistent in their traits, with a large pollen load, fewer abortive pollen grains, and high resistance. High importance must be given to typicality, especially in the selection of B and R line plants.

3) Selection methods

Field selection is the main focus, supplemented by indoor seed test. Preliminary selection is made at the initial heading stage, and then single plants are selected through indoor seed test. All selected seeds are numbered for safe storage.

4) Quantity of selected plants

The selected population should be large enough to prevent loss of minor genes. If the "simplified purification" method is used, 100 – 150 plants should be selected for a CMS line, and 100 plants for its B line and R line, respectively. For the "set-based" method, 60 – 80 sterile plants of a CMS line and 20 plants each of its B line or R line should be selected, respectively.

5) Bagging of AxB pairs for paired backcrossing

Selected B line plants are to be moved to the CMS nursery and paired-crossed with the selected CMS plants in bags (Fig. 14 – 5). No gibberellic acid should be applied. Flag leaves can beclipped and panicles of the CMS plants can be manually unwrapped to increase seed setting.

Fig. 14 - 5 CMS AxB bagged pair-crossing

(2) Strict isolation

The production of foundation seeds requires multiple generations of backcrossing and comparison, so the isolation of the three nurseries must be strict, which is the key to ensuring the quality of the purification. Methods of isolation include as follows.

1) Flowering time isolation

Arrange the flowering time of the three nurseries in the period when there is no other rice flowering around and the actual difference of initial flowering from other rice should be more than 25 days.

2) Natural barrier isolation

3) Distance isolation

The three nurseries of the CMS line should be more than 500 m and the three nurseries for the B or R lines should be more than 20 m apart.

4) Cloth isolation

Cloth isolation is used for the plant-to-row nursery.

5) Isolation cover

This is used for single plant testcrossing. It is a wooden frame covered by white cloth 1.2 m in height, 1 m in length and 0.7 m in width. The wood frame is set up before initial heading, and is covered before flowering and uncovered after flowering every day.

(3) Sterility inspection

As an important means for the purification of a sterile line, sterility inspection is required in single plant selection and the plant-to-row, row-to-family and foundation seed production nurseries. It can be done in the following three ways.

1) Pollen inspection by microscopy

Pollen inspection using microscopy is a major method for sterility inspection. When the main panicle is heading, check the pollen sterility under a microscopy for three spikelets taking from the upper, middle and lower parts of the main panicle and record the numbers of typical, spherical, semi-stained and black-stained abortion pollen grains. Every selected single plant must be examined. Specifically, microscopy inspection should be done for 20 plants per line in the plant-to-row nursery, 30 plants per family in the row-to-family nursery, and more than 450 plants per hectare in the foundation seed production nursery so as to eliminate plants with atypical pollen traits.

2) Selfed seeding inspection on isolated pots

This produces accurate and reliable results and should be carried out together with pollen inspection by microscopy.

3) Selfed seeding on baggings

(4) Setup and management of the three nurseries

1) Two comparisons of rows and families

By the genetic laws, the traits in biologically mixed off-types will have segregated in the F_2 population, which means the segregation will not appear in the plant-to-row nursery as the off-type is an F_1 hybrid. However, the segregation will appear in the row-to-family nursery, which is the F_2 and the off-types will be eliminated. Two comparisons can thus ensure the quality of the selected plants.

2) Foundation seed nurseries for B and R lines

Three nurseries are usually set up for a CMS line, and also for B and R lines, which not only is synchronized with the CMS line, but also increases the reproduction index without delaying the cycle of foundation seed production. Foundation seed production nurseries of B and R lines are set up differently from those of the CMS line in separate fields and this can help improve the seed quality of the B and R lines. The setup of a separate nursery for the foundation seeds of the B line also facilitates the early harvesting of the parents in the foundation seed nursery of the CMS line and more thorough roguing.

3) Separate plant-to-row and row-to-family nurseries for the B line

There are many benefits of having separate plant-to-row and row-to-family nurseries for a B line. It allows the CMS line to use only one backcross male parent in the plant-to-row and row-to-family nurseries and eliminates the need for isolation. Since it is not necessary to select the same quantity of CMS and B line plants, the selected CMS population can be increased. A separate B line nursery does not require leaf clipping and gibberellic acid application, retaining its original state and facilitating the selection. It is possible to have the rows and families of the CMS line backcrossed with only one B line or mixed B line plants, so that the male parents of backcrosses are consistent to facilitate the correct selection of the CMS line. Therefore, it is advisable to set up separate plant-to-row and row-to-family nurseries for the B and CMS lines.

4) Management of the selection, plant-to-row and row-to-family nurseries

The field with the previous cropping of grains or rapeseeds should not be used as the nurseries, but field of previous cropping with green manure is preferred to ensure consistent ground fertility to facilitate comparative characterization, and to ensure the precision and consistency of field management and avoid human error.

(5) Seasonal schedule

The planting season of each nursery for the foundation seed production of the three lines should be as consistent as possible. The three nurseries of a CMS line and its B line can be propagated in spring, summer or autumn. Spring propagation leads to a high yield, but there could be cross pollination with other early-season rice. Autumn propagation often leads to concurrent flowering with the plants from seeds dropped off from early-season rice and late-season *japonica*, which would result in cross-pollination and unstable yield. Summer propagation allows for free choice of the best flowering period, which can avoid cross-pollination and ensure safe pollination. Therefore, summer propagation is better for the plant-to-row and row-to-family nurseries with the flowering period set around mid-July to mid-August.

(6) Mixed line and single family line

There are two methods of propagation in the foundation seed nursery, i. e. , mixed line and single family line. Mixed line is better than single family line because mixed line can maintain the genetic pool for the three-line parents and complementary traits, making the population more stable and more adaptive, also facilitating the arrangement of propagation.

(7) Selection criteria and methods

Selection, comparison and inspection of single plants, rows and families must be based on the original characteristics and typicality of each parent. However, because the genetic characteristics of each trait are different in terms of trait performance and variation, traits should be selected by different methods.

1) Mode selection

Mode selection can be adopted for traits of the initial heading stage, the number of leaves on the main stem, and the number of panicles per plant. For example, if there are 100 plant rows, 60 of them start heading on August 20, the selection for initial heading should be set up for August 19 – 21. If there are 12 leaves on the main stem, 11 – 13 (preferably 12) leaves should be selected on the main stem. The number of panicles per plant varies greatly, so the range of selection can be wide.

2) Mean selection

Mean selection can be adopted for traits such as plant height, panicle length, and grain weight (per plant or per 1,000 grains). Plots whose parameters are close to the mean value with a plus or minus one-fifth of the mean value should be selected.

3) Optimal selection

Optimal selection can be adopted for fertility, seed setting rate, resistance and yield per unitarea. Plants with the highest typical abortion rate should be selected. Plants with the highest seed setting rate, resistance and yield per unit area should be selected following the high to low order.

4) Overall selection

Overall selection is based on observation records, seed lab test and yield per unit area, focusing on typicality, fertility and uniformity. For a sterile line, sterility should be above 99.9%. For B and R lines, attention should be paid to typicality, resistance and maintaining ability (100%) or restorability (> 85%). Typicality should be examined strictly for single plants in the selection process.

(8) Requirement for field cultivation and management

In terms of field cultivation and management of foundation seed production, there are some special requirements in addition to the general ones for hybrid rice seed production.

1) Field selection for the nurseries

All nurseries for the three lines, as well as the heterosis identification nursery, must be selected from fields with very uniform ground fertility, drought and flood protection, easy drainage and irrigation, boosting seedling growth, and also conducive to strict isolation.

2) Layout

For the plant-to-row and row-to-family nurseries of a sterile line, the row ratio of male to female-parents is 2 : (6 –8) with a space of 17 cm×(17 –23) cm. One single plant is planted in each hill. The plot has a fixed length but a flexible width to accommodate all the CMS plants. There are aisles (65 cm) between rows and families. At the initial heading stage, a small part at the front of each plot should be kept without gibberellic acid application or leaf clipping for the observation of normal heading performance.

For the plant-to-row and plant-to-line nurseries of B and R lines, seedlings are planted with one single plant per hill and a control in every 5 or 10 entries. Each plot may have two rows or five rows with a space of 13 cm×17 cm. For the row-to-family characterization nursery, the plots are in a random layout with three replications. A standard variety is used as the control. All plots have uniform areas and the same size.

3) Strict maintenance for purity

Mechanical mixture is strictly prevented in all operation processes including sowing, planting, harvesting, drying and storage. The whole row or family must be discarded if biologically mixed plants are found. The foundation seed nursery should be thoroughly rogued at the initial heading stage.

4) Traits for observation

a. Growth duration – dates for sowing, transplanting, panicle emergence, initial heading, full heading and maturity.

b. Typical traits – plant type, leaf shape, panicle shape, grain type, awn and its length if any, color of basal leaf sheath and lemma apiculus.

c. Plant uniformity, morphological and trait uniformity, and growth duration uniformity are all classified into three grades as good, intermediate and poor.

d. Heading and flowering status – anther size of the B and R lines (large, intermediate, or small); pollen dispersal (good, intermediate, or poor), CMS flowering time, stigma exsertion and panicle enclosure (serious, intermediate, or light).

e. Sterility – pollen composition under microscopy (proportion of plants with stained pollen in the population and selfing rate).

f. Number of leaves on the main stem – Select more than 10 plants, observe every three days to count the number of leaves and check the tillering status.

g. Resistance – resistance to rice blast and bacterial blight mainly.

h. Seed lab test – number of effective panicles per plant, plant height, panicle length, total number of grainsper panicle, filled grains per panicle, seed setting rate, grain weight (per plant and per 1,000 grains).

(9) Comparison test of foundation seeds (mixed-line plants)

Comparative tests are performed on the foundation seeds and their hybrids to identify stability and consistency and to improve the quality of the foundation seed. Make an overall evaluation based on the results regarding yield, purity in the field and seed lab tests.

Ⅳ. CMS Lines Reproduction

The multiplication of a CMS line is an outcrossing operation, requiring the B line to serve as the male parent and the A line as the female parent. The parental lines are planted alternately according to a certain row ratio, heading and flowering synchronized so that the A line plants can receive the pollen from the B line plants.

The principles and techniques for the multiplication of a CMS line are the same as those for hybrid seed production, but there are three special points.

1) The difference in the growth duration of the A and B lines are very small, so the difference in sowing dates for both parents is very small. It is easy to solve the problem of synchronization for the parents.

2) The tillering ability and growth potential of a CMS line is much stronger than its B line. Because the A line is sown earlier than its B line, in the process of multiplication, special attention should be paid to the cultivation of the B line so that it can have sufficient pollen to meet the needs of the A line for outcrossing.

3) The multiplication of a CMS line is to provide seeds for hybrid rice seed production, which requires the seed purity of the CMS line to be higher than 99.5%. Therefore, it is more important to eliminate the off-types. However, the morphological similarity of a CMS line and its B line makes roguing very difficult.

1. Isolation of reproduction base

Reproduction of a CMS line requires more stringent isolation than hybrid seed production. Natural isolation is strongly preferred, especially since with the expansion of some companies, large reproduction areas are more common and natural isolation is more important. In terms of distance, the isolation requirement is more than 200 m. As for time isolation, it requires a difference of 25 days or more between the flowering periods of the reproduction field and other rice fields. If isolation is done by planting the B line around, the B line must be planted within 200 m around the reproduction area. Isolation by barriers requires the barriers to be above 2.5 m in height; but if other rice production fields are in the upwind direction of the reproduction field, barrier isolation should not be used.

2. Season and sowing date for CMS reproduction

CMS line reproduction can be carried out in spring, summer, autumn, or winter in Hainan. In the rice areas of the Yangtze River Basin, the flowering period of spring reproduction is from late June to early July, that of summer reproduction is from late July to mid-August, that of autumn reproduction is from late August to early September, and that of winter reproduction in Hainan is from mid-March to early April. CMS line reproduction is mainly in spring and summer, followed by autumn and winter in Hainan. The early-season *indica* CMS line is strongly sensitive to temperature with a short vegetative growth period, so it is more appropriate for spring reproduction and for a mid-season *indica* CMS line, it is more appropriate to carry out summer reproduction.

Spring reproduction is appropriate for an early-season *indica* CMS line because the season provides appropriate temperature and light conditions for the full expression of the growth duration, as well as the plant, leaf, panicle and grain types. The weather facilitates the plants to form a high yielding seedling structure for reproduction. The heading and flowering stages are from late June to early July, when there are the most appropriate temperature and humidity for flowering and pollination, which is beneficial to increasing CMS seed setting from outcrossing and the yield. Also, the fields for spring reproduction is usually a previous winter cropping or fallow without ratooned rice or grains dropped from the previous rice cropping, which is conducive to the prevention of off-types and the maintenance of good purity.

Years of practice shows that the climate from heading to the maturity stage has a great impact on seed germination and growth vigor, therefore, a climate with favorable conditions for increasing seed germination should be considered in the selection of the reproduction season.

The sowing period for CMS reproduction must be in line with the requirements of safe heading and flowering to ensure the safety of flowering and pollination.

The dates from sowing to initial heading of a B line is 2 – 3 days shorter than its corresponding CMS line, but the flowering period of a B line is 2 – 3 days shorter than its CMS line. To synchronize the full flowering of the B and A lines, the initial heading date of the B line should be 2 – 3 days later than that of its CMS line, so the sowing of the B line should be 4 – 6 days later than that of the A line in the same season.

The B line used in CMS reproduction can be sown once (one-stage male parent) or twice (two-stage male parent). If one-stage planting of the male parent is adopted, the interval between the sowing of the male and female parents should be 5 – 6 days with a difference of 1.0 – 1.2 leaves in leaf age. If two-stage planting of the male parent is adopted, the first sowing of the B line should be 2 – 4 days later than that of the A line and the second sowing 6 – 8 days later than that of the A line. One-stage male parent reproduction can lead to uniform and vigorous growth of the B line, more seedlings, panicles and spikelets in the population with a greater pollen load for high reproduction yield. Two-stage planting of the male parent could result in less sufficient growth with fewer and smaller panicles for the second batch of male plants, which could reduce the pollen load in the reproduction field due to a lower pollen density than that for one-stage planting, resulting in a low yield. However, two-stage male parent planting can prolong the flowering duration and secure synchronization with CMS line flowering, specifically, for those CMS lines with long flowering periods, it is advisable to use the two-stage planting pattern.

3. Cultivation management

(1) Raising strong seedlings with multiple tillers and ensuring the number of basic seedlings

Strong seedlings with multiple tillers are the basis of a good seedling and panicle structure for a high yield in CMS line reproduction. The parents have a short vegetative growth period in the production field, and that of the B line is shorter than that of the CMS line. To establish a high-yielding population structure for reproduction, it is necessary to raise good seedlings. The quantity of A line and B line seeds must be sufficient, usually 30.0 kg and 15.0 kg per hectare for the A and B lines, respectively. The ratio of seedling field to production field is 1:10. The seedling field is leveled with sufficient base fertilizer. Sowing should be spare and even, with the seeds covered with mud. The seedling nursery is managed with moisture during the germinating stage. At the three-leaf stage, weaning fertilizer is applied and seedlings in the dense areas are moved to the sparse areas. Tillering starts with balanced growth at the four-leaf stage. Transplanting is carried out at the 5.0 – 5.5-leaf stage. The seedlings of the male parent can be raised in dryland or with soft plates and transplanted. The male to female row ratio in the CMS reproduction field is usually 1 : (6 – 8) or 2 : (8 – 10), which is smaller than that in hybrid seed production.

The parents are transplanted at the same time with the B line transplanted first followed by the A line. For the male parent to grow better, it is advisable to adopt narrow double-row planting for the male plants with a plant distance of 13.3 – 16.7 cm. The row distance between the A and B parents is 23.4 – 27.7 cm, which can also serve as working aisle in the field. The row distance between two rows of the male parent is 13.3 – 16.7 cm with 60,000 – 75,000 seedlings per hectare and 3 – 4 grains per hill. The density of CMS line transplanting should be (10 – 13.3) cm×13.3 cm, with 375,000 – 430,000 hills per hectare, 2 – 3 grains per hill. The basic seedling population per hectare is 22.500 hills for the B and 1,500,000 hills for the A parents.

(2) Directional cultivation of the parents

Fertilizer and water management for sterile line reproduction is in line with hybrid seed production in principle. Base fertilizer is the main fertilizer applied. Fertilizer is generally no longer applied to an early-season *indica* CMS line after transplanting, but some fertilizer may be applied after the fifth or sixth stage of panicle differentiation or after field sun-drying depending on the growth status of the CMS line. In terms of water management, shallow irrigation promotes tillering in the early stage, the field should be sun-dried in the middle stage, a water layer should be kept at the booting and heading stages, and the field should be kept moist after pollination.

During CMS reproduction, emphasis should always be placed on the cultivation of the male parent. Fertilizer should be applied for the B line after transplanting. In order to make the fertilizer work fast on the male parent, two approaches can be used. One is to make ball fertilizer and apply it deep in the parent rows 3 – 4 days after transplanting and the other is to put the fertilizer inside the ridges when making the ridges for male parent rows.

Through the directional cultivation of the parents, the maximum number of seedlings per hectare of the A line is about 4.5 million with more than three million effective panicles, and the maximum number of seedlings per hectare of the B line is 1.2 – 1.5 million with more than 0.9 million effective panicles.

4. Flowering prediction and adjustment

The two parents have minor differences in their growth duration and panicle differentiation. In order to synchronize the peak flowering periods of the two parents, the initial heading of the B line should be 2 – 3 days later than that of the A line. The criteria for the prediction of the flowering period are as follows. In the early stage of panicle differentiation, the B line should be one stage (2 – 4 days) later than the A line, that is, when the A line is at the second stage of panicle differentiation, the B line should be in the first stage, or when the A line is at the third stage, the B line should be at the second stage. This status should remain till the middle and late stages of panicle differentiation, i.e., the B line should be one stage behind the A line. If the progress of parental panicle differentiation does not match the standard of flowering synchronization, the flowering dates should be adjusted in time with methods basically the same as those used in hybrid seed production.

5. Spraying gibberellic acid

Different sterile lines have different sensitivities to gibberellic acid and the dosage and spraying methods should also be different. First of all, the dosage and the spraying time based on the initial heading rate are determined according to the sensitivity of the sterile lines to gibberellic acid and the temperature of the spraying time. Secondly, the spraying time should be adjusted according to the status of parental flowering synchronization. If it is an ideal flowering date, i. e. , the initial heading of the B line is 2 – 3 days later than that of the A line, then gibberellic acid can be sprayed on both parents at the same time. If both parents have the same initial heading date, or the B line starts initial heading earlier than the A line, then gibberellic acid is applied 1 – 2 days in advance. If the initial heading of the B line is 4 – 6 days later than that of the A line, then gibberellic acid is sprayed on the female parent first. If the temperature during the spraying period is high (daily maximum temperature > 33 ℃), the dosage should be reduced, or if the temperature is low (daily maximum temperature < 30 ℃), the dosage should be increased. In short, the goal of applying gibberellic acid is to have moderate plant height (about 100 cm), exposure of panicles and grains of the A line, and a plant height of the B line about 10 cm higher than that of the A line.

6. Preventing off-types and maintaining high purity

During the reproduction process, off-types need to be removed in time in the stages of seedling, tillering and heading. The identification and roguing methods are the same as those used in hybrid seed production. The key stage for roguing is the 1 – 2 days before and after gibberellic acid spraying. After repeated roguing, off-types should account for less than 0. 02% at the initial heading stage. All off-types should be removed 3 – 4 days after pollination. Field inspection should be carried out again 3 – 5 days before harvesting to make sure that the panicles of off-types account for less than 0. 01%. Every plot and row should be checked before harvesting.

To prevent mechanical mixing of A line and B line seeds in the reproduction field, B line plants should be cut out immediately after pollination and moved out of the reproduction field, which is beneficial for roguing and the field and purity inspection.

Section 2 Foundation Seed Production and Multiplication of Rice PTGMS Lines

Since the photosensitive genic male sterile (PTGMS) plant was found in the late-season *japonica* variety Nongken 58 by Shi Mingsong in Hubei province in 1973, Chinese rice breeders have found and developed a large number of PTGMS lines from different sources and with various methods. Since 1995, two-line hybrid rice has been studied and cultivated on a large scale. The two-line varieties used in production in China are basically high-temperature sterile and low-temperature fertile (TGMS) or photothermo-sensitive nuclear sterile PTGMS lines.

The main PTGMS lines used for large scale productionin China are mainly Peiai 64S, Zhu 1S, Lu 18S, P88S, Zhun S, 1892S, C815S, Guangzhan 63 -2S, Anxiang S, Guangzhan 63 -4S, Y58S, Shen 08S, Xiangling 628S, Longke 638S, Jing 155S, etc.

I. Characterization of Fertility Transition and Its Genetic Variation in Rice Two-line Sterile Lines

1. Performance of fertility transition and genetic variation

The fertility of PTGMS lines often exhibits the following features. 1) When the environment temperature during the fertility temperature-sensitive period is 0.5 ℃ higher than the sterility temperature (the critical temperature for fertility transition), the plants are completely sterile (sterility period); 2) When the environment temperature during the fertility temperature-sensitive period is 0.5 ℃ lower than the sterility temperature, the plants are fertile, with the stained pollen rate varying among individual plants (fertility period); 3) When the environment temperature during the fertility temperature-sensitive period is between ±0.5 ℃ of the sterility temperature, the stained pollen rate of individual plants varies greatly, ranging from 0 to 80% (fertility fluctuation period). Using the sterile lines Y58S, P88S, Guangzhan 63S, 1892S and 095S, Li Xiaohua et al. submerged plants at the fifth stage of panicle differentiation (0.5 remaining leaves) in cold water at 22.5 ℃ for six days. Then, on the 17th day after cold water treatment, anthers were taken from flower that blossomed on that day for pollen inspection under microscopy. The stained pollen rates were analyzed for individual plants of different sterile lines. The results are shown in Table 14 - 1.

Table 14 - 1 Stained pollen rate of individual plants of different sterile lines in the fertility fluctuation period

Line	Stained pollen rate (%)									
	0	0.1 - 9.9	10 - 19.9	20 - 29.9	30 - 39.9	40 - 49.9	50 - 59.9	60 - 69.9	70 - 79.9	≥80
Y58S	36.2	20.2	14.3	8.6	5.7	6.7	2.9	2.9	1.9	1.0
P88S	21.5	13.6	11.4	10.3	10.1	8.4	6.1	5.7	5.7	6.7
Guangzhan 63S	9.6	77.8	6.1	3.5	2.0	1.0	0	0	0	0
1892S	25.0	28.7	11.4	8.7	6.3	5.7	3.0	3.3	3.0	4.8
095S	59.2	35.7	1	2.0	1.0	1.0	0	0	0	0

The results reveal that the differences in pollen fertility among single plants of sterile lines during the period of fertility fluctuation were quite large. The performance also differs among sterile lines, suggesting that the stability of population sterility differs between sterile lines depending on the genetic background.

Studies of rice PTGMS genes show that, on the one hand, rice PTGMS is regulated by one to two pairs of major sterility genes, which is a qualitative trait, and the characteristic of fertility transition is genetically stable, manifesting sterility under high temperature and fertility under low temperature; on the other hand, the critical sterility transition temperature is modified by a large number of minor genes that are difficult to be homozygous and sensitive to light and temperature, showing quantitative characters, which is continuously recombined in continuous selfing to produce different types and numbers of individuals with synergistic minor genes, so that the critical fertility transition temperature is different among sterile lines and between generations. This is the genetic basis of the "genetic drift" of PTGMS critical sterility temperature. He Yuqing et al. suggested that PTGMS is controlled by major genes, while its fertility stability and transition are also affected by a set of photo-thermal sensitivity QTLs. Xue Guangxing et al. deduced that the male fertility of the PTGMS lines is the result of the combined effect and expression of the GMS gene and the genetic factors that control the photoperiod effect (PE) and temperature effect (TE), and concluded that the inheritance of sterility in all male sterile individuals is based on nuclear sterility genes, and the magnitude of sterility variation is determined by the cumulative effect of PE and TE genetic factors. Therefore, the sterility of the PTGMS lines currently used in China shows a typical genetic model of quality-quantity characters, which contributes to the stable inheritance of PTGMS fertility traits, but the critical temperature for fertility transition is subject to change.

Table 14 - 1 shows that a large difference in the rate of stained pollen among different individual plants of rice PTGMS lines, and thus there are individuals with small differences in the critical temperature for fertility transition. It is relatively easy to meet the temperature and photoperiod conditions required for individual plants with high stained pollen rate and high critical temperature for fertility restoration during the reproduction process, resulting in a higher seed setting rate. After years of selfing, the number of plants with a high seed setting rate increases from generation to generation, which leads to an increase in the critical temperature for fertility transition of the entire sterile line population, and thus produces the "genetic drift" of the critical temperature for sterility. The critical temperature for fertility transition of Peiai 64S was 23.3 ℃ when it went through technical appraisal in Hunan province in 1991. With conventional rice seed production and reproduction procedures and methods, its critical temperature for fertility transition increased from generation to generation, to 24.2 ℃ by 1993 and even partially to 26 ℃ in 1994. Hengnong S - 1 passed the technical appraisal in 1989, and after many generations of conventional reproduction, about 15% of fertile plants appeared in 1993 when seed production suffered from low temperature, and the critical sterility temperature was identified to be 26 ℃, and the self-fertility rate reached 70%. Guangzhan 63S, which passed the technical appraisal in 2021, has been applied in large-scale production. After nearly 10 years of reproduction, there were seven families with the same typical agronomic traits but different critical fertility transition temperatures among breeders and in production. These families are going further apart in terms of their fertility transition temperatures. These examples show that it is very difficult to eliminate the "genetic drift" of PTGMS lines. In other words, the critical temperature for fertility transition of PTGMS lines can only be controlled within a desired effective temperature range during reproduction with technical means, and the problem of "genetic draft" cannot be genetically eradicated. Such a "genetic drift" can increase the risk of seed production and even lead to the loss of practical value for application in production of the PTGMS.

2. Major off-types and their performance in a population of rice PTGMS line

In addition to the off-types usually produced by mechanical mixture and genetic variation during the multiplication of PTGMS lines, there are two special off-types among PTGMS plants, i. e. , high-thermo-sensitive plants and isomorphic fertile plants. High-thermosensitive plants are those with a high critical temperature for fertility transition caused by the "genetic drift", while isomorphic fertile plants are those that have undergone biological hybridization due to natural outcrossing and multiple generations of back-crossing during the reproduction of the sterile line (Fig. 14 - 6).

Fig. 14 - 6 Iso-morphic fertile off-types in the PTGMS line

(1) High-thermosensitive plants

The morphological traits of high-thermosensitive plants are completely identical to those of plants with normal sterility, and their fertility performance in reproduction depends on the temperature during the sensitive period for fertility transition. When the ambient temperature is at or slightly above the critical sterility temperature of the sterile line, the high-thermosensitive plants are fertile or semi-sterile, with anther dehiscence, pollen dispersal and self-fertilization. When the ambient temperature is higher or lower than the critical sterility temperature, the fertility performance of high-thermosensitive plants has no difference from that of normal sterile plants. Therefore, it is important to observe and analyze the temperature of the fertility temperature-sensitive period during propagation to determine the emergence of the possible high-thermosensitive plants and remove them in time, or deliberately set a critical high temperature of fertility transition to identify and remove these plants from the population.

(2) Isomorphic fertile plants

The plant, leaf, panicle and grain shapes of iso-morphic fertile plants are basically the same as those of sterile plants, but they show normal fertility and selfing regardless of the ambient temperature during the temperature-sensitive period for fertility transition. It is easy to identify and remove those isomorphic fertile plants in seed production, but difficult to do so in the reproduction of the sterile line, especially when the sterile line has good reproducibility with a high seed setting rate. Moreover, it is difficult to completely eradicate isomorphic fertile plants by roguing. Only by producing the foundation seeds following the seed production procedure and strengthening the isolation in reproduction can they be completely eradicated.

Ⅱ. Foundation Seed Production of PTGMS Lines

1. Rationale

Because the critical sterility temperature of PTGMS lines is subject to the "genetic drift", a certain amount of low temperature pressure must be given to screen single plants and control the reproduction generations in order to ensure the uniformity and purity of the sterile line seeds produced and to maintain the stability of the critical sterility temperature during reproduction. Breeders and seed producers have proposed feasible methods and technologies for the foundation seed production of PTGMS lines. Yuan Longping (1994) first proposed the procedures for the production of core seeds and foundation seeds of PTGMS lines, i. e., single plant selection-low temperature or long daylength low temperature treatment-seeds from ratooned plants (breeder seeds)-core seeds (plant-to-row core seeds)-foundation seeds (row-to-family foundation seeds)-improved seeds (seeds for hybrid seed production). This procedure can effectively control the "genetic drift" of the critical sterility temperature. Since there are many minor genes that control the critical temperature of fertility transition in PTGMS lines, and the "drift" of the critical sterility temperature cannot be fundamentally solved yet (Liao Fuming et al., 1994), so the improved seeds for large-scale seed production must be obtained through the process of "breeder seed-core seeds-foundation seeds".

Core individual plants can be selected by low temperature treatment in phytotron or cold water pool (Fig. 14 – 7), or from individual plants with low critical sterility temperature using natural ecological conditions under high selection pressure from off-site by shuttle breeding, or from a double haploid population. Through these methods and technologies, a PTGMS population with stable and relatively low sterility transition temperature can be selected, which can basically solve the problem of the "genetic drift" of the critical temperature for sterility transition.

Fig. 14 – 7 Cold water pool for selecting two-line sterile lines

In 2010, He Qiang etal. treated three sterile lines, Annong 810S, P88S and C815S using cold water 23.4 ℃, 22.1 ℃ and 22.2 ℃, respectively, at the young panicle differentiation stage. Plants with zero stained pollen were selected as the breeder seeds for propagation. In 2011, the breeder seeds were planted in plant-to-row nurseries and treated with 23. 3 ℃, 22. 3 ℃, and 22. 3 ℃ cold water, respectively, during the stages of pistil formation and young panicle differentiation. Statistics of the individual plants with different stained pollen rates after two years (two generations) of the three PTGMS lines are given in Table 14 – 2.

Table 14 – 2 Effect of cold water treatment for core plants of three PTGMS lines

Line	Year	Water temperature (℃)	Stained pollen rate (%)										
			0	0. 1 – 10. 0	10. 1 – 20. 0	20. 1 – 30. 0	30. 1 – 40. 0	40. 1 – 50. 0	50. 1 – 60. 0	60. 1 – 70. 0	70. 1 – 80. 0	80. 1 – 90. 0	> 90. 0
Annong 810S	2010	23. 4	41. 28	24. 2	23. 5	3. 02	2. 01	3. 02	2. 01	0. 67	0. 00	0. 34	0. 00
	2011	23. 3	71. 67	7. 00	2. 67	3. 33	0. 67	4. 67	4. 33	2. 33	3. 00	0. 33	0. 00
P88S	2010	22. 1	0. 71	0. 71	0. 35	0. 00	1. 41	6. 36	12. 0	18. 37	24. 03	34. 63	1. 41
	2011	22. 3	66. 00	5. 33	3. 67	2. 67	7. 00	5. 00	5. 67	3. 00	1. 33	0. 00	1. 41
C815S	2010	22. 2	0. 33	6. 62	2. 65	3. 97	4. 97	5. 63	6. 62	13. 58	22. 19	26. 82	6. 62
	2011	22. 3	68. 67	4. 00	5. 33	2. 00	7. 67	2. 33	4. 67	3. 33	2. 00	0. 00	0. 00

The results in Table 14 – 2 show that the percentage of plants with zero stained pollen in the progeny of core plants screened by one round of cold waterlow-temperature treatment increased significantly. Specifically, the rates are up by 30. 4% , 65. 3% and 68. 3% , respectively for Annong 810S, P88S and C815S. Most of the plants have a zero stained pollen rate. The number of semi-sterile plants with a stained pollen rate of 0. 1%–50% decreases by 37. 3% , 15. 9% , and 3. 3% respectively. The number of plants with a stained pollen rate above 50% decreases by 4. 6% , 79. 0% , and 65. 8% respectively. It shows that after one round of low-temperature selection, the proportion of plants with zero stained pollen rates and low critical sterility transition temperature increases significantly, however, there is still a certain proportion of sterile plants with a relatively high critical temperature, so it is necessary to conduct multiple rounds of low-temperature selection.

2. Procedures of foundation seed production

There are four steps for the production of PTGMS core seeds and foundation seeds.

Step 1. Selection of standard individual plants. Select individual plants with typical traits of the sterile line at the fourth stage of panicle differentiation in the breeder seed or core seed nurseries.

Step 2. Low temperature treatment for individual plants selected from the breeder seeds for reproduction. The standard plants are moved to a phytotron or cold water pool during the critical sterility temperature-sensitive stage, i. e. , the fifth to sixth stage of young panicle differentiation, which is from when the plant has 0. 5 remaining leaves to when the flag leaf collar exserts by about 2 cm. Perform pollen inspection twice under microscopy during the flowering period, and select plants with zero or low stained pollen rate as individual plants of breeder seeds.

Step 3. Propagation of breeder seeds from the selected individual plants. Cut the individual plants of the breeder seeds to ratoon or cultivate the plants with late-tillering panicles and propagate the breeder seeds through low temperature treatment as the core seeds.

Step 4. Propagation of plant-to-row core seeds harvested from the breeder seeds to produce the foundation seeds planting in row-to-family pattern of the core seeds. The core seeds should be sown in plant-to-row nurseries and the propagation can be done in cold water treatment, or in winter at Hainan, or in a location of low-latitude and high-altitude. In the plant-to-row nurseries, select plant rows with typical traits and consistency and collected the seeds in a mixed manner to serve as foundation seeds for the row-to-family nurseries. In order to ensure the stability of the critical temperature for sterility in seed production, it is advisable to only use foundation seeds harvested from the plant-to-family seeds for seed production.

3. Core seed production

(1) Low temperature treatment and individual plant selection

The key to the selection of core individual plants is low temperature treatment during the fertility temperature-sensitive period.

First, a source of low temperature should be selected. There are currently three main sources of low temperature. The first is a phytotron (or artificial climate chamber), where the temperature is lowered by cooling the air for standard sterile plants. The second is a cold water pool, where water is cooled for standard sterile plants. The third is natural low temperature. Due to the high thermal conductivity of the air and rapid temperature changes, air conditioner may lead to uneven temperature distribution in a room. In comparison, with low thermal conductivity and slow temperature changes, water is more likely to produce even temperature for the plants. Natural low temperature is less controllable. Therefore, cold water treatment is the best among the three, followed by natural cold temperature.

Second, the low temperature should be set at an appropriate level. Phytotron treatment is to place the plants in a low temperature environment that imitates natural low temperature, using a variable temperature treatment of 18 ℃–27 ℃ within 24 h. Two different levels of daily average temperature are set for such treatment according to the different sterility transition temperatures of the selected sterile lines, 23.5 ℃ for those with a sterility transition temperature of 23.5 ℃–24 ℃ and 23.0 ℃ for those with a sterility transition temperature of 23.0 ℃–23.5 ℃. The plants treated in a cold water pool are only submerged to the base of the young panicles, but a constant low temperature of the water must be maintained. Most parts of the plant leaves are still in a natural high temperature state, so the water temperature should be set 0.5 ℃ or 1.0 ℃ below the critical sterility transition temperature of the sterile line. For those sterile lines with a critical sterility transition temperature below 23.5 ℃, the water temperature should be set at 22.5 ℃, and for those sterile lines with a critical sterility transition temperature above 23.5 ℃, the water temperature should be set at 23.0 ℃.

Third, the duration of low temperature treatment should be determined. The fertility temperature-sensitive period of a single panicle is in the fifth or sixth stage of young panicle differentiation and it lasts about 5–6 days. The low temperature treatment can last 6–10 days. The exact duration of cold water treatment should be determined by the treatment temperature and the critical sterility transition temperature of the sterile line.

Fourth, criteria should be set for individual plant selection. Pollen examination under microscopy should be carried out on the 17th – 20th days after the cold water treatment, and the examination should be done twice in total. The criteria of selection is to be determined by the stained pollen rates of the individual plants. Those plants with zero stained pollen are preferred, followed by those with the lowest stained pollen rate and 10%–20% of the plants can be selected.

(2) Core seed production

There are two ways to reproduce the core seeds from those selected individual plants treated with low temperature, one is through ratooned plants, and the other is through plants with late-tillering panicles.

Ratooned propagation is to cut off the selected core individual plants (leaving rice stubble about 15 – 20 cm tall) for seedling regeneration. The ratooned seedlings are subject to low temperature treatment (21 ℃–22 ℃ in a phytotron or a cold water pool during the fourth stage of young panicle differentiation, so as to make it fertile and reproduce the core seeds.

As for late-tillering panicle propagation, special techniques are used to cultivate standard individual plants with a heading and flowering period of more than 15 days. After finishing the first low temperature treatment (for pollen examination and individual plant selection), those selected plants are to be cold-water treated again with a temperature of 21 ℃– 22 ℃ to make the late-tillering panicles set core seeds.

4. Methods of foundation seed production

In order to prevent the "genetic drift" of the critical sterility transition temperature of PTGMS lines in reproduction, Yuan Longping proposed the key procedures of PTGMS seed production, i. e., individual plant selection-low-temperature or long daylength and low temperature treatment-ratooned seeds (breeder seeds)-core seeds-foundation seeds-improved production seeds. This procedure can ensure that the critical temperature for sterility of a PTGMS line is always at the same level, and can effectively control the "genetic drift" of the critical sterility transition temperature within a foundation seed production cycle. It has been successful with sterile lines such as Peiai 64S. Based on this, other breeders have proposed more practical methods for foundation seed production.

(1) Three-layer panicle method

Liu Aimin et al. proposed to produce foundation seeds by cultivating a sterile line population with late-tillering and a long heading period through two-stage low temperature treatment duringthe sterility temperature-sensitive period. This is to produce sterile line populations with a heading and flowering period of more than 15 days with special cultivation technology. The details are as follows. Firstly, check the morphological and agronomic traits of the sterile line during main panicle heading and select typical individual plants. Secondly, using cold water to treat the plants and induce fertility fluctuation in the second layer of late-tillering panicles, so as to select core sterile plants. Thirdly, carry out cold water treatment again for the third layer of late-tillering panicles to restore fertility, and harvest the selfed seeds as the core seeds. The method is characterized by a simple production procedure, reduced costs, and a higher yield of core seeds. The key to this method is to cultivate a population of single plants in the selection nursery with a long heading and flowering period, which is rather difficult in practice. Otherwise, the core seeds cannot be produced in large quantities unable to streamline the procedure and reduce costs. This method has been patented in China (No. ZL200810101298. X), and has contributed to the agricultural technical specifications of Hunan province.

(2) Plant-to-row characterization and selection

Chen Liyun proposed and practiced a method of plant-to-row foundation seed production, which reduces the generations required for reproduction, lowers the chance of mixture, improves the accuracy of eliminating plants not meeting the requirements, reduces production costs, and ensures seed purity. The specific procedure is described as follows. Select 50 individual plants with typical traits of the sterile line from the breeder seed production field and plant them in about 50 rows. During the fertility temperature-sensitive period, randomly select six plants from each row and place them in a cold water pool (with temperature set 0. 5 ℃ below the critical temperature for sterility) for five days, then check pollen sterility with microscopy. Eliminate rows not meeting the requirements, and carry out cold water treatment for the remaining rows under strict isolation, and harvest the remaining rows in bulk as the foundation seeds or as improved seeds for hybrid seed production. The number of rows and the area covered are determined by the number of sterile line plants required for seed production.

(3) Natural light and temperature conditions for breeder seed production

Deng Huafeng et al. proposed procedures and methods to produce foundation seeds using natural light and temperature conditions, such as the natural low temperature in spring in southern Hainan to select single plants. This method has two main features. First, it effectively controls the proportion of high thermosensitive sterile plants, so that they are stable and no longer "drift". Second, the use of natural light and temperature reduces production costs and lowers technical requirements. However, there is a high risk because of the uncontrollable natural conditions. For example, in February 2009, in Sanya, Hainan, continuous high temperature led to the failure of most PTGMS seed reproduction. The natural fluctuations of temperature cannot be accurately controlled. If the daily average temperature is above 23. 5 ℃, the critical temperature for fertility cannot meet the requirement of production. If the temperature is below 19 ℃, which is close to the lower limit of physiological sterility, the efficiency of reproduction or selection of the core individual plants may be undermined. Therefore, this method still needs to be further improved.

(4) Anther culture

The use of anther culture technology for the production of PTGMS lines is considered as an effective way to solve the problem of unstable fertility. Rapid purification can be achieved by obtaining genotypically pure PTGMS lines through doubling of anther-induced healing tissue. This technique has the advantages of being fast and efficient, requiring less time. However, in actual operation, it is difficult or impossible to effectively control the somatic asexual variation generated during anther culture, and at the same time, the target traits such as sterility transition temperature, typical traits of sterile lines and heterotic level of PTGMS lines cannot be selected in a directional manner, i. e. with a certain degree of randomness. This technology needs further practical testing.

Ⅲ. PTGMS Line Multiplication

The necessary condition for PTGMS fertility transition is to provide a low temperature of 20 ℃–22 ℃ for young panicles during the fertility temperature-sensitive period of the sterile line. Three methods have been developed for this purpose, including cold water irrigation, southern winter multiplication, and at a low-latitude and high-altitude location. They are different in terms of promoting stable high yield, but the same in maintaining purity.

1. Stable and high yield propagation techniques

(1) Cold water irrigation

Cold water irrigation multiplication is a method to reproduce seeds of the sterile line by submerging the young panicles in cold water in the fertility sensitive period of the sterile line (Fig. 14 – 8 and Fig. 14 – 9). The key technical points include as follows.

Fig. 14 – 8 Cold water irrigation for PTGMS line multiplication

Fig. 14 – 9 Cold water irrigation base in Lingchuan, Guangxi

1) The multiplication base has sufficient supply of cold water at 16 ℃–18 ℃.

2) Build irrigation ditches and drainage ditches. The size of irrigation and drainage ditches is determined by the multiplication area and the amount of water required for cold water irrigation. The ditches can be constructed at a flow rate of about 0.5 m^3/s per hectare for the field to ensure smooth irrigation and drainage. Set up multiple irrigation and drainage outlets for each field to ensure consistent water temperature between fields. Build a high ridge of 20 – 25 cm to ensure 20 cm depth of irrigation.

3) Sowing at the right time. Sowing dates are from late March to early April in double-croppping rice areas, and mid-to late April in single-cropping rice areas.

4) Cultivate of a more uniform population with sufficient seedlings and panicles, usually 4.5 million seedlings and 3.0 – 3.75 millioneffective panicles per hectare.

5) Carry out cold water irrigation in a timely manner with good water temperature control. Cold water irrigation starts at the stage of 0.9 – 1.1 remaining leaves of the tillers (the fourth stage of young panicle differentiation) in more than 50% of the seedlings and less than 0.5 leaf remaining in the main panicle. When more than 90% of the seedlings are at the early seventh stage of panicle differentiation (flag leaf collar 2 cm inside the sheath), terminate cold water irrigation. The water temperature is 18 ℃–

20 ℃ at the water entrance and 22 ℃–23 ℃ at the exit during the irrigation. The water depth should be 3–5 cm to submerge young panicles and increase as the panicles grow.

6) Ensure good observation and records. The temperature of the cold water should be carefully monitored and the inbound and outbound flow adjusted to ensure an appropriate water temperature.

Cold water multiplication is less risky and only requires a sufficient amount of low temperature water at 20 ℃–22 ℃ during the cold water irrigation period. The reproduction yield is generally 3.0–3.75 t/ha.

(2) Southern winter multiplication

Southern winter multiplication is a method of seed reproduction using the natural low temperature in winter in the southern region of Hainan to restore the fertility of PTGMS lines (Fig. 14–10 and Fig. 14–11). Key technical points for a stable and high yield include as follows.

Fig. 14–10 Southern winter multiplication base for PTGMS lines

Fig. 14–11 Y58S southern winter multiplication

1) In the rice growing areas of Sanya, Ledong and Lingshui of Hainan, choose a field base with sufficient water supply for irrigation, contiguous plots, good isolation and no quarantined rice disease and pest.

2) The fertility temperature-sensitive period of the sterile lines (the fourth to sixth stages of panicle differentiation) is planned to be from mid-January to early February, i.e., around the Lesser Cold and Greater Cold periods in the 24 Solar Terms, and the heading and flowering period is around mid-to-late February.

3) Determine the sowing date by the days from sowing to initial heading of the sterile line. For example, Shen 08S and Longke 638S need 100 days from sowing to initial heading and can be sown in late October of the previous year. Y58S and P88S need about 90 days and can be sown around November 10 of the previous year. Zhu 1S and Xiangling 628S need 70 days and can be sown around November 30 of the previous year. The same sterile line can be sown twice with an interval of 8–10 days.

4) Raise seedlings in a irrigated nursery to construct a vigorous population with sufficient seedlings and panicles and a long duration of flowering.

5) Monitor the temperature during the fertility temperature-sensitive period and check pollen fertility at panicle differentiation and flowering related to daily average temperature, so as to determine the date

and criteria for roguing isomorphic fertile plants and high-temperature sensitive plants.

There is high risk for PTGMS multiplication using this method, because unusual high temperature or low temperature could happen during the period from mid-January to early February. The risk factor is about 0.3. The yield in a season with normal temperature is 3.0 t/ha.

(3) Low-latitude and high-altitude propagation

Low-latitude and high-altitude multiplication is a method using the natural low temperature in a low-latitude and high-altitude area to restore the fertility of PTGMS lines (Fig. 14-12 and Fig. 14-13). Key technical points for stable and high yield include:

1) Baoshan city, Yunnan province (99.0°-99.2°E, 24.8°-25.2°N) is at an altitude of 1,500-1,650 m. It has the rice cultivation areas characterized by sufficient irrigation water, concentrated and contiguous plots, favorable isolation conditions and no quarantined rice disease or pest.

Fig. 14-12 Two-line sterile line multiplication base in Shidian, Yunnan

Fig. 14-13 Meng S multiplication and seed setting in Shidian, Yunnan

2) The fertility temperature-sensitive period of a sterile line (the fourth to sixth stages of panicle differentiation) should be planned to be from mid-July to early August with heading and flowering from late July to mid-August.

3) The sowing dates of a sterile line are planned by the days from sowing to initial heading and thefertility temperature-sensitive period. For example, Shen 08S and Longke 638S require 120 days from sowing to initial heading in Baoshan, Yunnan (90 days in summer sowing in Hunan), so the seeds should be sown around April 5. Y58S, 1892S and Guangzhan 63S need 100 days from sowing to initial heading in Basoshan (75 days in summer in Hunan), and they should be sown around April 20. Zhu 1S, Lu 18S and Xiangling 628S have 80 days of sowing to initial heading (60 days in summer in Hunan), so they should be sown around May 20. The same sterile line should be sown twice with an interval of 6-8 days.

4) Adopt water-raised seedlings of a vigorous growth with a longer heading and flowering period.

5) Monitor the temperature of the sterile line during the fertility temperature-sensitive period and examine pollen fertility at the flowering stage, so as to determine the date and standard for roguing isomorphic fertile plants and high-temperature sensitive plants.

Because of the instability of natural low temperature in high altitude areas, there is a certain risk of multiplication for a PTGMS line by this method. The risk is mainly from the unusually high temperature and low temperature from mid-July to early August. The risk factor is 0. 1 – 0. 2. The yield is as high as around 6. 0 t/ha regularly or 9. 0 t/ha in high-yielding fields. The seeds produced can have a germinating rate of about 90% .

2. Purity and quality maintenance techniques

The following techniques for purity and quality maintenance apply to the multiplication method mentioned above.

(1) The multiplication base requires strict natural isolation for more than 30 days from other pollen sources. Isolation by space requires the multiplication fields to be 100 m upwind or 200 m downwind away from other rice pollen sources.

(2) The seeds for multiplication are the plant-to-row core seeds or row-to-family foundation seeds with a purity higher than 99. 9% .

(3) In preparing the field, perform submergence of the field twice and plow twice in advance to eliminate previous seedlings or dropped grains of the previous rice cropping.

(4) Remove all abnormal plants or off-types at the seedling stage and in the production field (at the three-or four-leaf stage, tillering, heading and maturity, respectively) for four times.

(5) Remove high-temperature sensitive plants and isomorphic fertile plants at the stages of heading, flowering, seed setting and maturity.

(6) Harvest and dry the field in time to ensure good seed germinating rate. Clean all equipment in advance to prevent mechanical mixture.

References

[1] HE QIANG, PANG ZHENYU, SUN PINGYONG, et al. Effects of low temperature treatment in screening core plants of PTGMS lines in rice[J]. Hybrid Rice, 2016 (1): 18 – 20.

[2] HE QIANG, DENG HUAFENG, PANG ZHENYU, et al. Research progress on purifying and multiplying technologies of photo- and thermo-sensitive genic male sterile lines in rice[J]. Hybrid Rice, 2010(6): 1 – 3, 7.

[3] LI XIAOHUA, LIU AIMIN, ZHANG HAIQING, et al. A preliminary study on population fertility performance of PTGMS lines in rice[J]. Hybrid Rice, 2016, 31(3): 23 – 26.

[4] LIU AIMIN, XIAO CENGLIN. Super hybrid rice seed production technology[M]. Beijing: China Agriculture Press, 2011.

[5] TU ZHIYE, FU CHENJIAN, ZHANG ZHANG, et al. High-yielding and quality seed multiplying techniques for rice PTGMS lines in high-altitude area of Yunnan[J]. Hybrid Rice, 2016, 31(1): 23 – 25.

[6] LIU AIMIN, LI XIAOHUA, XIAO CENGLIN, et al. Method of producing core seeds of thermo-sensitive male sterile line of rice by "three-layer panicle". China, ZL200810101298. X, 2012 – 01 – 11.

Chapter 15
Super Hybrid Rice High-yielding Seed Production Techniques

Liu Aimin / Zhang Haiqing / Zhang Qing

Super hybrid rice seed production techniques are basically the same as ordinary hybrid rice seed production techniques, including techniques of ecological condition (base and season) selection, flowering synchronization, construction of male and female parent populations, outcrossing trait improvement (with gibberellic acid application as the key), supplementary pollination, maturity and harvesting, mixture prevention and purity maintenance, etc. However, the techniques of super hybrid rice seed production are more difficult because the genetic background of super hybrid rice parents is more complicated, which combines *indica*, *japonica* and *javanica* rice parents and multiple types of sterility genes. The parents are more sensitive to temperature and light during the reproductive period. The parental plants are tall with larger panicles, high grain density, strong tillering, low panicle forming rate, more dispersed or more concentrated flowering period, poor population stability and narrower adaptability to the ecological climate for seed production. Meanwhile, most super hybrid rice is two-line hybrids involving PTGMS lines whose sterility expression is critically sensitive to daylength and temperature. The fertility of some three-line super hybrid rice is also greatly affected by the environmental temperature with incomplete pollen abortion.

Based on the research and practice of two-line hybrid rice seed production in the past ten years, the national technical standard "Two-line Hybrid Rice Seed Production Technical Specification" (GB/T 29371.4 - 2012) was compiled mainly by Hunan Hybrid Rice Research Center, Yuan Longping Agricultural High-Tech Co. Ltd. at al. This technical specification plays a guiding role for the seed production of two-line super hybrid rice.

Section 1 Selection of Climatic and Ecological Conditions for Seed Production

Ⅰ. Climatic Conditions

Rice is a typical self-pollinated crop. After a long history of natural selection, its floral apparatus, flowering, pollination and seed setting have been adapted to the climatic (mainly temperature and humidity) conditions of the rice growing regions and seasons. However, hybrid rice seed production is a process of outcrossing and seed setting. A female parent needs to accept pollen from a male parent to produce hybrid seeds, which is a complete change of the reproductive mode of self-pollination of rice. At the same time, the male sterility of the female parent brings about obstacles in the morphology, physiology, biochemistry and metabolism of the sterile lines. For example, there can be panicle enclosure at heading, dispersed flowering time, stigma exsertion, glume splitting, germination in the panicles, and powdery endosperm. Therefore, hybrid rice seed production demands better ecological and climatic conditions. To be specific, there should be favorable temperature, humidity and light during the periods of fertility sensitivity (the fourth to sixth stages of young panicle differentiation), flowering and pollination, maturity and harvest. Therefore, although hybrid rice seed production is possible in areas where rice can be grown, but it is not always possible to obtain high yield and high quality seeds.

1. Climatic conditions required for fertility-sensitive safety period

(1) Three-line hybrid seed production

The major CMS lines used in three-line super hybrid rice are Tianfeng A, Yuetai A, Wufeng A, Zhongzhe A, Zhongjiu A, Ⅱ-32A, T98A, etc. The sterility of some of the above CMS lines is sensitive to high temperature. In the process of pollen differentiation and development, one or two anthers in a small number of glumes will have a high proportion of stained pollen under a high temperature, capable of dehiscence with pollen dispersal and self-fertilization, resulting in different degrees of reduction in seed purity. However, there are few reports on the effect of high temperature on the fertility of such sterile lines. After years of practice, many seed companies have managed to find the most appropriate locations and seasons for the seed production of each sterile line, with the daily average temperature being 26 ℃-28 ℃ and the daily maximum temperature below 32 ℃ during pollen development. If the daily average temperature is higher than 28 ℃, there would be a high quantity of plump anthers with a high rate of stained pollen and a high selfed seeding rate as the temperature rises, resulting in reduced purity of seeds.

(2) Two-line hybrid seed production

The sterile lines used in two-line super hybrid rice are basically PTGMS lines and temperature is the main factor controlling the fertility expression during the fertility-sensitive period. The critical sterility temperature among current sterile lines varies. For example, the critical sterility temperature of Zhu 1S and Longke 638S is around 23.0 ℃, that of Peiai 64S, Y58S and Shen 08S is 23.5 ℃, and that of P88S, 1892S and Guangzhan 63-2S is around 24.0 ℃. Therefore, the seed production using sterile lines with different critical sterility temperature is based on different temperature conditions during the fertility temperature-sensitive period. By collecting historical meteorological data of seed production bases, analyzing the average daily temperature of the safety period of possible fertility sensitivity, and finding the number of years in which the average daily temperature for three consecutive days is lower than the fertility transition temperature of a sterile line used for seed production, a fertility safety factor for seed production can be calculated by the following formula.

$$\text{Fertility safety factor} = 1 - \frac{\text{Number of years when the average daily temperature for three consecutive days is lower than the critical sterility temperature}}{\text{Total number of years}}$$

When the fertility safety factor is 1, it is safe for seed production, when the fertility safety factor is higher than 0.95, the female parent has a low risk and could be selected for seed production, when the fertility safety factor isbelow 0.95, the female parent is more risky and cannot be used for seed production under local conditions.

When the environment temperature in the fertility temperature-sensitive period is around the critical sterility temperature, the sterile line shows a period of fertility fluctuation, and the fertility performance of the sterile line population is complicated with some single plants showing a high stained pollen rate and fertility, while some others are sterile with a low stained pollen rate. Therefore, in order to ensure complete pollen abortion in the female population for seed production, it is advisable to increase the critical sterility temperature of the sterile line by 0.5 ℃ as the fertility safety temperature when selecting and analyzing the fertility safety factor.

When the ambient temperature in the fertility temperature-sensitive period is higher than the fertility safety temperature, the pollen abortion degree of the sterile lines increases from typical abortion sterility to pollen-free sterility with the gradual increase of the temperature, which is safer in terms of sterility. However, many years of seed production practice and research have shown that the higher the temperature of the fertility temperature-sensitive period and the more complete the pollen abortion, the poorer the outcrossing and seed setting in most sterile lines, manifesting as reduced stigma vigor, increased split glumes or deformed glumes. Therefore, seed production of two-line hybrid rice should not be conducted in locations and seasons where the environment temperature is too high during the fertility temperature-sensitive period, but where the environment temperature is 2 ℃ higher than the critical sterility temperature.

2. Climatic conditions for safe heading, flowering and pollination

During heading, flowering and pollination (from panicle emergence to flowering termination), gibberellic acid (GA3) application and supplementary pollination are carried out. The climatic conditions of this period affect the performance of both male and female parents in heading, flowering, pollination, fertilization and thus seed yield. Seed production for both two-line and three-line hybrids requires the following four essential climatic conditions.

1) Sufficient sunlight, especially during the peak flowering period of the male and female parents, and no more than two days of continuous rainfall.

2) An appropriate daily average temperature of 26 ℃–28 ℃ with a temperature difference more than 10 ℃ between day and night, no more than three consecutive days with a daily average temperature above 30 ℃ or below 24 ℃, and no more than three consecutive days with maximum daily temperature above 35 ℃ or minimum daily temperature below 22 ℃.

3) Relative humidity of 75%–85% without three consecutive days when it is above 95% or below 70%.

4) Natural wind below Level 3 during pollination time of the day.

3. Climatic conditions for safe maturing and harvesting

Hybrid rice seed production requires application of GA3 and supplementary pollination. Due to the scattered flowering without an obvious peak period in the female in a day, seed setting on the female can result from timely or untimely pollination, leading to poor closure of the inner and outer glumes of the seeds and different degrees of split glume. If humidity and temperature are both high with rainfall during the seed maturity period, the seeds would not dry in time, or naturally in the event of consecutive cloudy days. Such seeds would be prone to germination on panicles or powdery endosperm and mold, with reduced seed vigor or even vitality. Therefore, dry sunny days without rain, or low-humidity days are required during the maturity and harvesting periods, which is 15 – 25 d after the end of pollination at the seed production base.

4. Coordination of the "three safety periods"

In the selection of hybrid rice seed production locations and seasons, it is necessary to comprehensively consider and effectively coordinate between the three climate safety periods for temperature-sensitive fertility, flowering and pollination, and maturity and harvesting. The first of these is the priority because it determines the purity of the hybrid seeds. Then, the safety of flowering and pollination determines the yield of hybrid seeds, while the safety of maturity and harvesting determines the seed germinating rate and vitality. Therefore, the climate safety of the three periods plays an important role in hybrid rice seed production. It is necessary to analyze the climate safety of these "three safety periods" overall in selecting seed production locations and seasons. A seed production base should have a fertility safety factor over 0.95, with the probability of abnormal climate below 0.2 during the flowering and pollination period and rain probability below 0.3 during the maturity and harvesting period.

Ⅱ. Ecological Conditions for Rice Cultivation

In addition to appropriate climatic conditions, a seed production location should also have sound ecological conditions in the following aspects.

1) The plots should be contiguous for convenient isolation, and large and square with appropriate access to roads to facilitate mechanized seed production.

2) The soil should have a sound structure, uniformly fertile without cold water, or deep mud for the convenience of mechanized seed production.

3) There should be a sound irrigation and drainage system.

4) There should be no quarantined pest or disease (e.g. bacterial leaf streak and *Lissorhoptrus oryzophilus* Kuschel).

5) There should be no devastating disasters such as strong storms, mountain flood, hailstorms, or persistent drought.

Ⅲ. Main Seed Production Locations and Seasons in China

Research and practice of hybrid rice seed production has been going on for more than 40 years and there are currently seven major ecological seed production areas in China, southern Hainan, early-season and late-season in southern China, Xuefeng Mountain Range, Luoxiao Mountain Range, Wuyi Mountain Range, and summer seed production in Mianyang, Sichun, and Yancheng, Jiangsu.

1. Southern Hainan

The region mainly consists of six counties in southern Hainan, namely Lingshui, Sanya, Ledong, Dongfang, Changjiang and Lingao, covering an annual seed production area of about 6,777 ha (Fig. 15 - 1 and Fig. 15 - 2). It is suitable for both three-line and two-line hybrid seed production with many types of hybrid combinations. Before 2010, it was basically a seed production base to make up for the shortage of hybrid rice seeds in the mainland market, but after 2010, it became one of the four major seed production regions for two-line super hybrid rice in China.

Fig. 15 - 1 Seed production base in Baowang village, Jiusuo town, Ledong county, Hainan

Fig. 15 - 2 Seed production base in Ledong hilly area, Hainan

It is the best seed production base in terms of the coordination between the "three safety periods" resulting in high yield, high quality in appearance and vitality of hybrid seeds.

For seed production of three-line super hybrid rice using sterile lines that are susceptible to high temperature, the fertility safety period is scheduled for early to mid-March when there tend to be appropriate temperatures, no hot weather, and complete pollen abortion of the female. The flowering and pollination period is arranged in late March when there is a low probability of low temperature or rain, and the maturity and harvesting period is set in mid-to late April, right at the end of the dry season, but before the wet season.

For seed production of three-line hybrid rice using sterile lines not affected by high temperature, the "three safety periods" can enjoy higher flexibility. Flowering and pollination can be set for late March to April, while maturity and harvesting can happen from late April to May.

The first priority in two-line super hybrid rice seed production is the safety of the fertility temperature-sensitive period. Generally, this period in Sanya, Lingshui and Ledong is after April 10, and the corresponding flowering and pollination period is from April 25 to May 20 when brief showers are possible but would not have a great impact on pollination. Maturity and seed drying are from late May to mid-June when the dry season ends and the wet season starts, with a high probability of brief showers.

The "three safety periods" of seed production in Dongfang and Changjiang are 5 - 7 days later than those in Ledong.

Currently, the highest risk for two-line seed production in southern Hainan is showers during the maturity and harvesting period, which occur so frequently and irregularly, resulting in poor seed quality due to germination on panicles, powdery endosperm and mildew. Therefore, for better seed production of two-line hybrids in southern Hainan drying equipment is required to dry seeds on a large scale.

The hybrids currently produced in the seed production area of Lingao are mainly photosensitive late-season rice from southern China, including the Boyou and Teyou series. The flowering and pollination period is from late April to early May, and the maturity and harvesting period is from late May to early June.

2. Seed production area for early-and late-season rice of southern China

At present, the bases are mainly located in Bobai, Nanning, Wuming, Tianyang, Beiliu and Yulin of Guangxi; and Lianjiang, Suixi, Huazhou and Gaozhou of Guangdong, with a total area of about 6,667 ha.

The early season is good for only seed production of three-line hybrids. Normally, seeds produced in the early season can be used for late-season rice production. The heading and flowering period is in late May, and the maturity and harvesting period is in late June.

Seeds of both three-line and two-line hybrids can be produced in the late season with the heading and flowering period from mid-to late September and the maturity and harvesting period in late October.

3. Seed production areas in Xuefeng, Luoxiao and Wuyi mountain ranges

These regions include Shaoyang, Huaihua, Yongzhou and Chenzhou of southern Hunan; Guilin and Hezhou in northwestern Guangxi; and Jianning, Shaowu and Shaxian in northwestern Fujian. The seed production bases are mainly located in hilly and mountainous areas at low and middle altitudes and production mainly occurs in spring, summer and autumn (Fig. 15 - 3 and Fig. 15 - 4).

Fig. 15 - 3 National seed production base in Suining, Hunan

Fig. 15 - 4 Large-scale seed production base in Zixing, Hunan

Spring production bases are areas below 350 m in elevation, including Chenzhou, Shaoyang, Huaihua, Yongzhou and Zhuzhou of Hunan; Jianning of Fujian; Yichun, Suichuan, Le'an and Nanfeng of Jiangxi; and Pingle, Xing'an and Quanzhou of Guilin, Guangxi, covering about 20,000 ha annually. The most suitable hybrids for seed production here are early- or late-season three-line hybrid rice for the middle and lower reaches of the Yangtze River. Seed sowing lasts from mid-March to early April, and the heading, flowering and pollination period is from mid-June to early July.

The summer seed production bases are in Chenzhou, Shaoyang and Huaihua of Hunan; and Jianning and its surrounding areas of Fujian, with an elevation of 350 – 600 m, about 20,000 ha in area for seed production annually. They can serve seed production either for three-line or for two-line hybrids. Areas for two-line hybrid seed production should be below 500 m in elevation. The safe heading, flowering and pollination period is in early and mid-August, while the maturity and harvesting period is in early and mid-September.

The autumn seed production bases are mainly in the low-altitude rice areas of Yichun in Jiangxi, Yongzhou in Hunan, Guilin in Guangxi, and Jianning and its surrounding areas in Fujian, covering an annual area of about 3,333 ha. It is mainly suitable for late-season three-line hybrid rice, and early-season two-line super hybrid rice. Sowing is normally in June with the flowering and pollination period from late August to early September, and the maturity and harvesting period from late September to early October.

This area is the earliest and largest seed production area in China, with high risks regarding the "three safety periods". Therefore, in selecting location and season, local climate data should be collected and analyzed carefully to determine the risk coefficient and put the "three safety periods" at times when the risk is the lowest. In addition, mechanized seed dryers should be used to reduce the impact of germination on panicles, powdery endosperm and mildew caused by rainy weather during the harvesting period.

4. Summer seed production area in Mianyang

This area mainly includes seed production bases in Mianyang and its surrounding areas in Sichuan, as well as some areas in Chongqing. They are typical single-cropping rice cultivation areas only suitable for summer seed production. The bases are mainly located in Mianyang, Deyang, Nanchong and Suining of Sichuan; and Bishan, Jiangjin, Fuling, and Zhongxian of Chongqing. This area, about 26,667 ha annually, is mainly suitable for mid-season three-line super hybrid rice in the Yangtze River Basin. Sowing is normally in April, transplanting is from late May to mid-June, and the flowering and pollination period is from mid-July to early August.

5. Seed production area in Yancheng

This area mainly includes Dafeng, Jianhu, Funing and Xiangshui of Yancheng, Jiangsu (Fig. 15 – 5). Theregion, featuring a maritime climate with moderate temperature and high humidity, is a suitable base for summer seed production and the total area has reached more than 13,333 ha for three-line and two-line hybrids. The fertility safety period is from July 25 to August 15, the safe initial heading period is from August 15 to August 20, the flowering and pollination period is from August 15 to late August, and the maturity and harvesting period is late September.

Fig. 15 – 5 Yancheng seed production base in Jiangsu

There are certain risks for both the fertility safety period and the flowering and pollination period for two-line hybrid seed production in this area with a fertility safety factor of about 0. 95. Sterile lines with a critical sterility temperature of 23.5 ℃ or below have a higher safety factor, while those with a critical sterility temperature higher than 24 ℃ are not suitable for this area. The risks of the flowering and pollination period mainly include high and low temperatures plus rainfall. Abnormal high temperature causes a significant reduction in the seed output of the female parent, while low temperature and rainfall affects the female parent's seed setting rate and induces rice kernel smut. Therefore, for seed production in Yancheng, firstly, suitable hybrids are those two-line combinations using a female with a low critical sterility temperature, high tolerance to high temperature and high resistance to kernel smut; and secondly, sowing and transplanting should be done in time to ensure safe heading, flowering and pollination; and thirdly, sound management is required to prevent and control rice kernel smut.

In addition, it is a rice and wheat double-cropping area. Wheat is harvested in early June and the rice field can only be prepared after mid-June. Therefore, the transplanting period for seed production is around June 20. The aged seedlings and delayed transplanting of parents may affect the safety of heading and flowering.

Section 2 Flowering Synchronization of Male and Female Parents

Flowering synchronization refers to the concurrent heading and flowering of the male and female parents in hybrid rice seed production, which is the prerequisite for seed yield. Rice has a relatively short flowering period, only about 10days. The synchronization of male and female flowering can be roughly divided into five categories depending on its extent. 1) Ideal synchronization means that the heading and flowering stages of both parents are perfectly synchronized during the whole flowering period; 2) good synchronization means that the peak flowering period is 70% synchronized; 3) basic synchronization is only means that the peak flowering period is 50% synchronized; 4) poor synchronization means the flowering period is basically not synchronized, but initial heading and the end of heading are synchronized; and 5) unsynchronization means that there are eight or more days between the flowering periods of the male and female parents, in which case seed production is very likely to fail.

Flowering synchronization of male and female parents involves three aspects.

1) Determination of the difference of parental seeding period (seeding split) by their sensitivity to nutrition, temperature and daylength, basic vegetative growth duration, and heading and flowering characteristics.

2) Prediction and regulation of the heading and flowering period. In the process of parental growth and development, evaluation of the degree of flowering synchronization should be done in a timely manner and the flowering period should be regulated if there is a deviation in the flowering period of the parents.

3) Standardized cultivation and management. Standards should be formulated and strictly implemented for cultivation and management from parental sowing to heading to ensure normal and controllable growth and development of the parents, and avoid the deviation of the growth duration caused by im-

proper cultivation and management that may affect the flowering period.

Ⅰ. Determination of Seeding Split and Seeding Period of the Parents

1. Determination of the sowing number of the male parent

(1) Sowing number of the male parent

In terms of the sowing number of the male parent, there are the following three options.

1) All male parent seeds are sown at one time.

2) The male seeds are sown in two batches, equally divided, with a split of 7 – 10 days, or 1.3 – 1.5 leaves. The two batches of male seedlings can be transplanted alternately in the same row or in alternate rows. If the female parent is found to develop faster than scheduled, more seedlings of the first batch of the male parent should be transplanted or only this batch will be transplanted. If the female parent is developing slower than scheduled, more or all of the second-batch seedlings of the male parent should be used.

3) Seeds of the male parent are sown in three batches with a split of 5 – 7 days or 0.5 – 0.7 leaves, each batch containing 1/3 of the seeds, or the first and third batches containing 1/4 while the second batch having 1/2 of the seeds. The three batches should be transplanted alternately, with those from each batch taking up 1/3 of a row, or in such a way that second-batch seedlings form one row and first-and third-batch seedlings together form another row. The growth of the female parent should be checked when transplanting the male plants so that adjustment can be made by selecting male seedlings from different batches to ensure synchronization.

There are big differences in heading and pollination, and pollen load in the field for different batch systems of the male parent. The flowering period is the longest for three-batch parent seeding, followed by two-batch and then one-batch seeding. One-batch male has 10% more spikelets per unit area than two-batch male which has 5% more spikelets than three-batch male. Although the one-batch system has a short flowering period, it has a large pollen load in the field and a high pollen density in space and time per unit, which increases the pollination probability for the female parent. The three-batch system, in contrast, although providing a long flowering period to ensure flowering synchronization, has a reduced amount of pollen in the field with a low pollen density, which could reduce the pollination probability for the female parent.

(2) Determination of the sowing batch of the male parent

The following factors should be considered in determining the number batches of male seeding in seed production.

If the characteristics of the parents used in the seed production is well known and the flowering period of the parents is certain, two-batch male can be used. If the male parent has a strong tillering capacity and long flowering period (3 – 4 days longer than the female), one-batch male can be used.

If it is a new hybrid in seed production without much knowledge about the parents' characteristics, or if the flowering period of the parents is no certain, then the three-batch system is suggested.

If the male parent has a weak tillering capacity and short flowering period, the three-batch system is more appropriate.

If the male parent or female parent is sensitive to temperature and nutrition conditions, the three-batch system is also more suitable.

(3) Determination of the sowing time of the male parent

There can be two scenarios when determining the sowing time of the male parent. When the male's duration from sowing to initial heading is longer than that of the female, the male is sown before the female and the sowing time of the male is determined by the "three safety periods"; if the male's duration from sowing to initial heading is shorter than that of the female, the female is sown before the male, while the sowing time of the male is determined by the sowing time of the female and the difference between the sowing time of the two.

2. Determination of parental seeding split

Because of the difference in the duration from sowing to initial heading of the parents, they cannot be sown at the same time in seed production and the interval between the sowing dates of the two parents is the seeding split, which is determined by the characteristics of the two parents (sensitivity to daylength and temperature and vegetative growth) and the predicted data of initial heading of the two parents for ideal synchronization. A staged seeding test is a must in order to understand the parents' growing patterns, heading and flowering.

The seeding split can be determined based on the differences in leaf number, growth duration and effective accumulated temperature.

(1) Leaf number method

The method of projecting the seeding split between the parents based on the leaf emergence rate of the main stem at different growth and development periods is called the leaf number (or age) method. It is worth pointing out that the difference in the total number of leaves on the main stem of the parents is not the seeding difference used for seed production. The parental sowing leaf difference has two meanings. One is the age of the main stem leaves of the first parent when the later parent is sown, and the second is in the symbiotic stage of the later parent and the first parent after sowing. The two parents have different leaf emergence rates due to different growth periods, so the difference in the total number of main stem leaves of the two parents cannot be used as the sowing leaf difference of the two parents.

For example, Fengyuanyou 299 is used in summer seed production at the Suining production base in Hunan. The total number of leaves on the main stem of the female parent is 12 and that of the male parent is 16. However, the seeding split measured by leaf number is not four leaves, but 6.5 – 7.0 leaves, because the duration of the female to develop 12 leaves is the duration of the male to develop 9.0 – 9.5 leaves during symbiosis. In addition, due to the difference in the period required from flag leaf exsertion to initial heading of the male and female (i.e., the "rupturing" period), the seeding split needs to be regulated for flowering synchronization.

There are differences in the total leaf number on the main stem and their leaf emergence rate among hybrid rice parents. However, the leaf number on the main stem of the same parent is relatively stable under normal climatic conditions and cultivation management in the same location and season. The leaf number on the main stem of a parent varies with the length of the growth period. Early-season *indica* sterile lines such as Xieqingzao A, T98A, Fengyuan A, Zhu 1S, Lu 18S and Zhun S have a short growth period with 11 - 13 leaves on their main stems. Mid-season *indica* sterile lines such as Ⅱ-32A, Peiai 64S, P88S, Y58S, Shen 08S and Longke 638S have a long growth period with 15 - 17 leaves on their main stems.

In the same sowing season and under the same cultivation conditions, the leaf number on the main stem of the same variety is mostly the same, but in years with different climatic conditions and cultivation techniques, there may be a difference of 1 - 2 leaves. For seed production of the same hybrid in the same location, the same season and different years, it is accurate to use leaf difference to arrange the parental seeding split. However, there are differences in leaf number between different regions and seasons for the same parent, especially for those parents with strong daylength and temperature sensitivity. For example, Liangyoupeijiu is used in summer seed production in Yancheng, Jiangsu, and the seeding split is 32 d with a leaf difference of 7.0 - 7.5 leaves; however, in Suining, Hunan, it is 18 d with a leaf difference of 3.8 - 4.0 leaves, while in Mayang, Hunan, it is 22 d with a leaf difference of 5.8 leaves. Therefore, leaf number difference should be used by the season and local conditions.

(2) Seeding split by growth duration difference (time difference)

The time difference method set the seeding split based on the difference in the duration from sowing to initial heading of the parents. The number of days from sowing to initial heading is relatively stable for the parents in the same season, with the same cultivation management and in the same location with similar ecological conditions. The seeding split by time difference is based on the above principle.

For example, seed production of Fengyuanyou 299 is conducted at Suining, Hunan. The male parent, Xianghui 299, is sown on April 10 and the date of initial heading is July 20, which is 100 days from sowing. In comparison, the female parent, Fengyuan A, needs 66 days from sowing to heading and is sown in mid-May with initial heading also on July 20. The seeding split is 100 - 66 = 34 (days). The criterion for ideal synchronization of this combination is that the female flowers 2 - 3 d earlier than the male, so the seeding split is 31 - 32 d for summer seed production in Suining.

The seeding split by time difference is only suitable for those locations with minor variation of temperature between years and seasons and it is commonly used in summer and autumn seed production of the same hybrid in different years. In seasons and areas with significant temperature variation, such as in the middle and lower reaches of the Yangtze River in spring and summer seed production, the duration from seeding to heading is often affected by the spring temperature, which may lead to poor flowering synchronization or non-synchronizion if the time difference method is used.

(3) Effective accumulated temperature (EAT) difference method

The biological lower and upper limit temperatures of *indica* rice are respectively 12 ℃ and 27 ℃. Based on this, the sum of all daily temperature values between 12 ℃ and 27 ℃ from sowing to initial heading is the EAT. The EAT from sowing to initial heading is relatively stable for a temperature-sensitive rice variety sown in the same area but at different sowing dates and it can be used for determining the seeding split of the parents. This is the EAT difference method. For example, for a hybrid in Hunan summer seed production, the EAT difference of the parents is 300 ℃. The daily EAT is recorded for the male from the second day after sowing to initial heading and the female parent is sown on the day when the EAT of the male reaches 300 ℃. Although the EAT method can prevent errors caused by temperature variation between years, it cannot avoid errors caused by cultivation management and different field conditions.

In determining the parental seeding split, the above three methods should be combined with the parental characteristics and the climatic conditions of the seed production season. The leaf difference is used as the basis, the EAT as reference, and the time difference is only used in the seed production season when the temperature is more stable. In spring and summer seed production, the leaf difference method is mostly used, and the EAT and time differences are for reference due to the unstable temperature. In autumn, the temperature is stable and the time difference method is used mostly, and the leaf difference and EAT are the reference.

(4) Adjustment of seeding split

In addition to the above three methods, time and the method of sowing and transplanting, the quality of the seedlings, the temperature change of the year and the water and fertilizer conditions in seed production also need to be considered for minor adjustment of seeding split.

1) Adjustment by weather conditions after sowing of the first parent. If the sowing time of the year is earlier and the temperature is lower than usual, the duration from sowing to initial heading will be prolonged with more leaves on the main stem. Therefore, the seeding split should be extended and vice versa.

2) Adjustment by parental sowing and transplanting methods. The seeding split between the parents is usually determined based on wetland seedlings transplanted at mid-size. If the method of seedling raising and transplanting is changed for one parent, such as direct seeding for the female parent, the seeding split should be prolonged for 2 – 3 d. If machine transplanting is adopted for seedlings with soft plates or scattered seedlings for the female, then the seeding split of the parents should be shortened by 3 – 5 d.

3) Adjustment by the coincidence of the leaf number difference and time difference. If there is a good coincidence between the two, sowing should be done with the originally designed seeding split. If the leaf number difference comes before the time difference, the seeding split should be shortened according to the time difference. If the leaf number difference is behind the time difference, the seeding split should be based on the reduction of the leaf number difference and the increase of the time difference.

4) Adjustment by the quality of the first-sown parent. If the seedlings of the male parent are of high quality and grow well, the female parent should be planted 1 – 3 d ahead of the schedule, and vice versa.

5) Adjustment by the weather after scheduled sowing ofthe late-sown parent. Attention should be paid to the weather forecast. If a low temperature and rainfall are expected when or after the scheduled sowing of the late-sown parent, the sowing should be adjusted 1 – 3 d earlier than scheduled. If the weather is normal, no adjustment is needed.

6) Adjustment by quality and quantity of the female's seeds. If the female's seeds are of high quality and high quantity per unit area, the sowing can be postponed for 1 – 2 d. If the female's seeds are of low

quality and low quantity per unit area, the sowing should take place 2 – 3 d earlier.

7) Adjustment by the sowing batch of the male parent. The heading and flowering period of the one-batch male parent population is shorter than that of a two-batch population. Therefore, the seeding split of the female parent should be shortened by 0.5 leaves or 2 – 3 d of time difference than for the two-batch male parent system. For the three-batch male system, the sowing date of the one-batch male parent should be 2 – 3 d earlier than that of the first-sown male in the two-batch male system.

Ⅱ. Determination of the Parental Seedling Age

On the basis of the seeding split and flowering period of the parents, an appropriate seedling age should be determined according to parental characteristics. A suitable seedling age involves not only the number of days required from sowing to transplanting, but also the leaf number of the seedlings at the time of transplanting. The season, methods of seedling raising and transplanting, and the parental growth type are the main factors for determining the appropriate seedling age. For wetland raised seedlings of early-season rice and early- or mid-maturing late-season rice in spring or summer seed production, the seedlings should be transplanted within 45% of the total number of leaves on the main stem. For mid-season rice in summer seed production, the transplanting leaf age should be controlled within 40% of the total number of leaves on the main stem. Aged seedlings have fewer tillers after transplanting in a seed production field, and premature heading is likely to happen. Transplanting at an appropriate seedling age can ensure sufficient vegetative growth in the field to promote the growth and development of the root system and tillers, laying a solid foundation for young panicle development, heading and seed setting.

The seedling age has a great influence on the parents' sowing-to-heading duration, the longer the seedling age, the longer the duration, and vice versa. Therefore, when arranging the seeding split of the parents, the seedling age of the parents should also be determined according to cultivation technology factors. The seeding split should be adjusted if aged seedlings are used because the previous crop is harvested late, or timely transplanting cannot be ensured due to limited labor. For example, in summer seed production in Hunan, the duration from seeding to heading of the male parent of Xiangliangyou 68 is 48 – 50 d if transplanted at a leaf age of 3.0, or 50 – 52 d with a leaf age of 4.0, or 52 – 54 d with a leaf age of 5.0. For seed production of Liangyoupeijiu in Nanning, Guangxi, and the duration from seeding to heading should be increased by 0.8 – 1 d when the seedling age of the male parent 9311 is increased by 1 d (Table 15 – 1).

Table 15 – 1 Duration from seeding to heading and duration from transplanting to heading of 9311 at different seedling ages (Nanning, 2006)

date (m/d)	Transplanting date (m/d)	Seedling age (d)	Heading date (m/d)	Seeding to heading duration (d)	Transplanting to heading duration (d)
6/3	7/7	34	9/2	91	57
6/3	7/8	35	9/3	92	57

Continued

date (m/d)	Transplanting date (m/d)	Seedling age (d)	Heading date (m/d)	Seeding to heading duration (d)	Transplanting to heading duration (d)
6/3	7/12	39	9/7	96	57
6/10	6/30	20	8/28	79	59
6/20	7/8	18	9/6	78	60

Ⅲ. Flowering Date Prediction

Flowering date prediction determines the flowering synchronization degree of parents byobserving and analyzing the parents' morphological performance, leaf age and leaf emergence rate, and the progress of young spikelet differentiation, and estimate the number of days till parental initial heading. In addition to the genetic characteristics of the parents, the growth duration of a parent is also influenced by climate, soil, seedling quality, transplanting leaf age, fertilizer and water management, which may lead to an earlier or later flowering date of the parents, resulting in a deviation in flowering synchronization. In particular, for seed production of a new hybrid or at a new production base, in which case there is less knowledge of the appropriate seeding split and cultivation techniques, flowering non-synchronization is more likely to happen. Therefore, prediction of flowering date is an important step for hybrid rice seed production with the purpose of accurately determining the initial heading dates of the parents as early as possible. Once the parents are found to have deviated in their flowering time, corresponding measures should be taken early to regulate the growth and development process of the parents so as to ensure flowering synchronization.

There are many ways to predict flowering date and the corresponding adjustment methods can be used at different growth and development stages. Commonly used methods include panicle stripping, leaf age remainder, corresponding leaf number, EAT and sowing-to-heading duration. The leaf age remainder method and EAT method can be used throughout the entire growth and development stages. Panicle stripping applies only during young panicle differentiation. The most common used methods are panicle stripping and the leaf age remainder method (Table 15 - 2).

1. Young panicle stripping

Observation of the young panicle development of the parents is performed based on the morphology of the eight development stages in order to predict the flowering synchronization of the parents. The specific practice is to strip and check 10 - 20 young panicles from the main stem of both parents. The population-wide young panicle development stage is determined by the status of 50% - 60% of the checked plants. Panicle stripping and checking is to be done during young panicle differentiation, specifically every 1 - 2 days at the early stage of young panicle differentiation and every 3 - 5 days at the middle and late stages.

Table 15 - 2 and Table 15 - 3 outline the panicle differentiation stages and their corresponding morphological characteristics and leaf age remainder. If there are four more leaves on the main stem of the

Table 15 – 2 Simplified method and traditional eight-stage classification of young panicle differentiation

Simplified Method						Eight-Stage Classification		
Panicle Differentiation period		Panicle morphological characters	Relationship with the 4th leaf from the top	Emerging leaves	Remaining leaves	Panicle Differentiation period	Emerging leaves	Remaining leaves
Stage 1	Bract differentiation stage	Spike-stalk differentiation Spike-stalk merogenesis	Late stage of the fourth leaf from the top	0.5	3.5 – 3.0	Stage 1	0.5	3.5 – 3.0
Stage 2	Branch differentiation stage	Primary branch differentiation secondary branchdifferentiation	Emergence of theantepenult leaf	1.0	3.0 – 2.0	Stage 2	0.5	3.0 – 2.5
						Stage 3	1.0	2.5 – 1.5
Stage 3	Spikelet differentiation stage	Spikelet differentiation pistil and stamen formation	Early emergence stage of the flag leaf and the penultimate leaf	1.2	2.0 – 0.8	Stage 4	0.7	1.5 – 0.8
Stage 4	Sexual cell differentiation and formation period	Pollen mother cell formation Mother cell meiosis	Middle and late stage of flag leaf emergence	0.8	0.8 – 0	Stage 5	0.5	0.8 – 0.3
						Stage 6	0.3	0.3 – 0
Stage 5	Pollen grain filling and maturity period	Pollen grain filling Pollen grain maturity	Flag leaf sheath elongation and expansion (booting)	1.0+2 d		Stage 7	5 – 6 d	
						Stage 8	2 d	

Table 15 – 3 Young panicle differentiation stages of some sterile lines and restorer lines

Line		Panicle differentiation duration (d)								Seeding to heading duration (d)	Number of leaves on the main stem
		Stage 1	Stage 2	Stage 3	Stage 4	Stage 5	Stage 6	Stage 7	Stage 8		
		First bract primordium differentiation	First branch primordium differentiation	Second bract and spikelet primordium differentiation	Pistil and stamen primordium formation	Pollen mother cell formation	Pollen mother cell meiosis	Pollen content enrichment	Pollen maturity		
Jin 23A	Days of differentiation	2	2	4	5	3	2	8		51 – 60 (Changsha)	10 – 12
Xinxiang A	Days from initial heading	26 – 25	24 – 23	22 – 19	18 – 14	13 – 11	10 – 9	/			
T98A	Days of differentiation	2	2	4	5	3	2	8 – 9		55 – 70	11 – 13
Zhun S	Days from initial heading	27 – 26	25 – 24	23 – 20	19 – 15	14 – 12	11 – 9	/			
Zhenshan 97A	Days of differentiation	2	3	5	5	3	2	9		60 – 75	12 – 14
Fengyuan A	Days from initial heading	28 – 27	26 – 24	24 – 20	19 – 15	14 – 12	11 – 10	/			
Xianghui 299	Days of differentiation	2	3	5	6	3	2	7	2	95 – 120	15 – 17
H – 32A	Days from initial heading	28 – 27	26 – 24	23 – 20	19 – 14	13 – 11	10 – 9	9 – 3	2		
Milyang 46	Days of differentiation	2	3	5	7	3	2	7	2	90 – 110	16 – 18
IR 26	Days from initial heading	30 – 29	28 – 26	25 – 22	21 – 15	14 – 12	11 – 10	9 – 3	2		
Shuhui 527	Daysof differentiation	2	3	5	7	3	2	8	2	85 – 110	17 – 19
9311	Days from initial heading	31 – 30	29 – 27	26 – 22	21 – 15	14 – 12	11 – 10	9 – 2	2		

male parent than on that of the female parent, the young panicle differentiation duration of the male is longer than that of the female. To achieve ideal flowering synchronization, the male parent should be 1 – 2 stages ahead of the female parent before the third stage of young panicle differentiation. When the young panicle differentiation is in the fourth-sixth stages, the male parent should be 0. 5 – 1 stages ahead of the female parent. When young panicle differentiation is at the seventh and eighth stages, the young panicles of both parents should be very close or at the same development stage. For the male parent with 2. 0 – 3. 0 leaves on the main stem more than that of the female, the young panicle differentiation period of the male parent is slightly longer than that of the female. For an ideal flowering synchronization, the young panicle differentiation of both parents can have a same development progress or the male is slightly later than the female parent.

For hybrid parents with the same leaf number on the main stem, the rate of young panicle differentiation and the rate of heading and flowering of the male parent arefaster than those of the female, so the young panicle differentiation on the female should be 1 – 1. 5 stages ahead of that on the male parent.

2. Leaf age remainder

Leaf number remainder refers to the number of leaves that have not been exserted onthe main stem (i. e. the total number of leaves on the main stem minus the number of leaves that have exserted). At a late stage of young panicle differentiation, leaf emergence is significantly slower than during the vegetative growth period but with a stable rate, so the residual leaf number can be used to predict initial heading. Zhou Chengjie's chart of rice leaf age and young panicle development (Table 15 – 4) shows the temporal relationship between the exsertion of the last few leaves and young panicle development and initial heading, indicating the relationship between the male and female parents with different numbers of leaves on the main stem and young panicle differentiation.

Anatomical observation shows that there are four young leaves and leaf primordia in a newly emerged leaf before young panicle differentiation. The leaf primordia terminate and the bract primordia start differentiation as young panicle differentiation begins. Since a newly emerged leaf normally has four young leaves and leaf primordia, the differentiation of the fourth leaf primordium is replaced by that of bract primordium when the fourth leaf from the top is exserted. This means that panicle differentiation always starts when the fourth leaf from the top is exserted, and hence is closely related to the top 3. 5 leaves. From the late stage of the fourth leaf from the top, panicle differentiation progresses every one stage with one new leaf emergence. The stage of the fourth and fifth leaf from the top is the most important period to predict and adjust the flowering date. Anatomical observation with a stereo microscope can be used for an accurate evaluation one month before panicle differentiation.

3. Leaf emergence rate

When young panicle differentiation starts (the reproductive growth period), the leaf emergence rate is significantly lower than during the vegetative growth period, which can be used to predict the flowering period. Under normal weather conditions, it takes 2 – 3 d more for each leaf to emerge in the young panicle differentiation period than in the vegetative growth period. For late-maturing parents (e. g. Minghui 63) with a long growth period, the leaf emergence rate during the vegetative growth period is 4 – 6 d/leaf,

Table 15 - 4 Rice leaf number and young panicle development (Zhou Chengjie, 1989)

Number of leaves on the main stem								Young panicle development stage	Days of differentiation (d)	Leaf number remainder	Days from heading *
11	12	13	14	15	16	17	18				
8.2	8.5 - 9.0	9.5 - 10.1	10.5 - 11.2	11.5 - 12.0	12.5 - 13.0	13.5 - 14.0	14.5 - 15.0	First bract differentiation (invisible at Stage 1)	2 - 3	3.5 - 3.1	24 - 32
8.3 - 8.9	9.1 - 9.7	10.2 - 10.9	11.3 - 12.0	12.1 - 12.7	13.1 - 13.7	14.1 - 14.6	15.1 - 15.6	Primary branch differentiation (Stage - 2 bract hair emergence)	3 - 4	3 - 2.6	22 - 29
9.0 - 9.6	9.9 - 10.4	11.0 - 11.5	12.2 - 12.7	12.8 - 13.4	13.8 - 14.4	14.7 - 15.3	15.8 - 16.3	Secondary branch differentiation (Stage - 3 bract hair multiplication)	5 - 6	2.5 - 2.1	19 - 25
9.7 - 10.0	10.5 - 10.9	11.6 - 12.0	12.8 - 13.1	13.6 - 13.9	14.6 - 14.9	15.5 - 15.9	16.5 - 16.9	Pistil and stamen formation (Stage - 4 grain emergence)	2 - 3	1.5 - 0.9	14 - 19
10.2 - 10.5	11.0 - 11.4	12.1 - 12.5	13.2 - 13.6	14.0 - 14.3	15.0 - 15.3	16.0 - 16.3	17.0 - 17.3	Mother cell formation (Stage - 5 glume split)	2 - 3	0.7 - 0.5	12 - 16
10.6 - 11	11.5 - 12	12.6 - 13.0	13.6 - 14	14.4 - 15	15.4 - 16	16.4 - 17	17.4 - 18	Meiosis (Stage - 6 phyllula flattening)	3 - 4		7 - 9
								Pollen filling (with green panicles)	4 - 5		7 - 9
								Pollen maturity (panicle emergence)	2 - 3		3 - 4

Note. * The varieties with ashort growth period have fewer main stem leaves and shorter young panicle differentiation duration; otherwise, longer.

but 7 − 9 d/leaf during young panicle differentiation. The parents of early- and mid-maturing rice have a leaf emergence rate of 3 − 5 d/leaf in the vegetative growth period and 5 − 7 d/leaf during young panicle differentiation. Therefore, when a rice plant transitions from vegetative growth to reproductive growth, there will be a turning point in its leaf emergence rate, which marks the beginning of young panicle differentiation.

4. Seeding to heading duration estimation

The method of predicting initial heading based on the duration from seeding to heading is based on the relative stability of the sowing-to-heading duration of a parent between years in the same location and same season under the same cultivation and management conditions. For the same combination planted in the same location and same season, treated with the same cultivation techniques, the late-sown parent is planted based on the prediction of the seeding-to-heading duration after the early-sown parent is sown and is adjusted according to climate conditions. Combining young panicle stripping, leaf emergence rate and the number of elongated nodes, the dates of young panicle differentiation and initial heading of the parents can be primarily predicted.

5. Corresponding leaf age prediction

Based on records of the parent leaf age of the same combination in the same location and the same season in previous years, we can make a chart and compare the parental development of the current seed production with historical data to predict flowering synchronization.

Ⅳ. Regulation of the Flowering Period

Regulation of the flowering period is a unique technique in hybrid rice seed production. Based on the parents' different growth characteristics and sensitivity to water and fertilizer, various cultivation and management measures are adopted for parents with deviations in flowering synchronization, promoting or delaying their growth and development and prolonging or shortening their heading and flowering duration to achieve synchronization.

Flowering period regulation should be performed if the parents are found to be more than three days apart from ideal flowering synchronization.

The role of flowering period regulation is manifested in two ways. First, it can promote plant growth and development to advance heading or shorten the duration from sowing to heading; second, it can delay plant growth and development to delay heading or extend the flowering duration from sowing to heading. Retarding regulation is for fast-growing parents and facilitating regulation is for slow-growing parents. Flowering regulation should be done early rather than late, focusing on control, and more on the male than on the female.

In the practice of seed production, one or more of the following regulating methods can be adopted to regulate the flowering period of parents according to the extent of the parents' unpredictable flowering period, the growth and development characteristics such as tillering into panicles, fertilizer tolerance, lodging resistance, soil fertility, and the growth and development status of parents, respectively.

1. Agronomic regulation

(1) Regulation by planting density or basic seedling population

The growth and development of a parent varies with the cultivation density or the size of the seedling population. Dense planting and multiple seedlings per hill increase the number of basic seedlings per unit area, advance panicle heading, promote uniform heading and concentrated flowering; while sparse planting and one seedling per hill reduce the number of basic seedlings per unit area, delay initial heading, scatter the heading of the population and prolong the flowering period. This method is effective in regulating parents with strong tillering ability and a long growth duration. When using the density adjustment method, it is advisable to adopt dense planting with multiple seedlings per hill and low fertilizer to shorten the period from sowing to heading and the flowering period, and thin planting with a single seedling per hill and high fertilizer to delay heading and prolong the flowering period.

(2) Regulation by seedling age

Seedling age has a great impact on the initial heading of the parents, which is related to the parental growth duration and seedling quality. The restorer line IR26 is seven days earlier in initial heading when the seedling age is 25 d than when the seedling age is 40 days, while it starts heading six days earlier when the seedling age is 30 d than when the seedling age is 40 days. When the seedling age is over 40 days, the heading is not uniform. When seedlings of Zhenshan 97A are 13 d in seedling age, initial heading happens about four days earlier than when the seedling age is 28 d, but when the seedling age is 18 d, initial heading happens only one day earlier than when the seedling age is 28 d. When the seedling age is over 35 d, premature heading appears and heading is not uniform. Regulation by seedling age has a good regulation effect on seedlings of medium or poor quality, but is less effective on high-quality seedlings. The sowing-to-heading duration of late-season rice seed production in autumn in southern China increases with the extension of the seedling age. When the seedling age is 20 – 45 d, the sowing-to-heading duration is extended by 0. 8 d when the seedling age extends for 1 d.

(3) Regulation byinter-tillage

Inter-tillage combined with a certain amount of nitrogen fertilizer can significantly delay initial heading and extend the flowering period. This method is effective for parents with a low number of seedlings, below the expected number per unit area, and weak growth potential. It is not advisable to fertilize parents with high growth potential by only inter-tillage, but inter-tillage can be combined with leaf cutting for better results. Therefore, the use of this method depends on the seedlings.

(4) Regulation by fertilizer and water management

For parents with faster development and less vigorous growth, 75 – 150 kg/ha of urea can be applied. Depending on the condition of the seedlings, urea may be applied only to the female, plus inter-tillage, to delay the growth and development of the female and postpone flowering by about 3 d. For slow-growing parents, potassium dihydrogen phosphate can be sprayed with water once a day for 2 – 3 d, to move the flowering period by 2 – 3 d. At the late stage of young panicle development, if the flowering period is found unsynchronized and regulation can be done through field water control because some restorer lines are sensitive to water yet sterile lines are less responsive to water. If the male flowers early and the female flowers late, the field can be drained to slow down male and boost female development. If the female flowers early and the male flowers late, deep-water irrigation can be done to boost the male and contain the female development. Effective water control can help postpone or advance the flowering period by 3 - 4 d.

2. Regulation by chemical

(1) Regulation by GA3

Two to five days before heading, foliar spraying of about 7.5 g/ha of GA3 with 450 kg of water and 1.5 - 2.25 kg of potassium dihydrogen phosphate can advance heading by 2 - 3 d.

The use of GA3 to regulate the flowering period should be done late rather than early and the dosage should be low rather than high, and it should be done only at the end of the seventh stage and beginning of the eighth stage of panicle differentiation. Too early or too much GA3 spraying will only elongate the middle and lower internodes, leaves and leaf sheaths, resulting in tall plants and heading failure.

Given the characteristics of sterile lines with high stigma exsertion and strong vitality, the effect is obvious when using growth regulators or hormones to enhance stigma exsertion and vitality, and extend stigma life, which can somewhat reduce the flowering split when the female flowers early and the male late. During the flowering period of the female, this can be done by spraying 15 - 30 g/ha of GA3 in the afternoon every day with 600 - 750 kg of water for three or four days, deep water shall be kept in the field, which can prolong the viability of the stigma by 2 - 3 d.

(2) Regulation by paclobutrazol

When the parents are predicted to be more than 5 d apart for initial heading, paclobutrazol can be applied on the fast-growing parent to delay heading. If paclobutrazol is used on the female, it is better to use it early than late and before the fourth stage of panicle differentiation. If it is used at the middle and late stages of panicle differentiation, panicle enclosure will increase. Before the fourth stage of panicle differentiation, 1.5 - 2.25 kg/ha of paclobutrazol with 450 kg of water can be applied, followed by fertilizer depending on the seedling growth to promote later tiller growth and prolong the flowering period of the population. For the parent which has received paclobutrazol, GA3 spraying should be advanced by 2 d with 30 - 45 g/ha. For the parent with premature growth and development, paclobutrazol can be sprayed at a dosage of 450 - 600 g/ha with water.

3. Remedies for severe failure of flowering synchronization

(1) Pulling out bracts and young panicles

If there is a difference of 7 - 10 days or more in heading synchronization, the bracts at the seventh panicle differentiation stage or young panicles at panicle emergence of the fast-growing parent can be pulled out to regulate the heading period. The bracts or young panicles that are pulled out are usually those that emerge more than 5 d earlier than the late-heading parent, mainly from the main stem and the first tiller panicles. If bract and panicle pulling is adopted, fertilizer must be heavily applied in the early stage of young panicle differentiation to promote late tillers.

(2) Regulation by mechanical damage

Measures such as cutting the leaves, lifting the plants and injuring the roots can be used to delay the growth and development of the fast-growing parent through heavy injury, thus delaying the flowering period by about 5 d. However, due to damage to the plant, it tends to cause abnormal heading and flowering. Therefore, it is only used when the flowering periods are severely unsynchronized (by more than 7 d), and must be combined with fertilization to restore the growth of the plants.

(3) Regulation by ratooning

A difference of more than 10 days in initial heading between the parents is considered complete non-synchronization. When the late-growing parent enters panicle differentiation, the plants of the fast-growing parent can be cut before panicle differentiation, and the stubble left is based on the degree of difference between the parents and the regeneration ability of the ratooned parent. After cutting, appropriate amount of fertilizer should be applied to generate more ratooned seedlings, and predict flowering synchronization with the late-growing parent. Ratooning regulation is the only remedy for complete non-synchronization in seed production, and it is usually used for small-scale seed production of new hybrids.

Section 3 Cultivation Techniques for Male and Female Parent Populations

Seeds harvested in hybrid rice seed production are the seeds produced by the female parent population, which relies on the pollen of the male parent to set seeds. Therefore, in terms of the relationshipbetween parental groups in hybrid rice seed production, first of all, the dominant position of the female parent must be established, but a certain number of parents must also be ensured to provide sufficient pollen for the seed setting required. In terms of the proportion of parental planting, the land occupation ratio of the male parent should be reduced as much as possible, while that of the female parent should be expanded. Only by establishing a coordinated parental group structure can high seed yield be obtained. The key indicators of the parental group structure are parental row ratio, parental planting density and spikelet ratio.

Ⅰ. Planting Design in the Field

1. Row ratio of male and female parents

The male and female parents in seed production are planted in alternate rows at a certain ratio. The ratio of the row number of the male parent to that of the female parent is the basis of the male and female parent population per unit area. There are three factors that determine this ratio.

The first isthe traits of the male parent. If the male parent has a long growth period, e. g. large seeding split between the parents, strong tillering capacity, high panicle formation rate, a large amount of pollen and a long flowering and pollination period, it should have a large parental row ratio; otherwise, a small row ratio.

The second is the pollination method and planting pattern of the male parent. For example, for supplementary pollination, the male parent can only be planted in single or double rows because of the small force acting on the male parent with a close distance of pollen dispersal. If the male parent is planted in a large double-row pattern (30 cm between two male rows), the row ratio of the male to the female is

2 : (12 - 16). If the male parent is planted in a small double-row pattern (20 cm between two male rows) or in a false double-row pattern (10 - 13 cm between two male rows, cross-planted), the row ratio is 2 : (10 - 14). If the male parent is planted in the single-row pattern, the row ratio is 1 : (8 - 12).

If agricultural drones are used to assist pollination, the male parent can be planted in 6 - 10 rows with the compartment 180 cm wide because the strong wind generated by the drone rotor spreads the pollen far. For male parents with a long growth duration and strong tillering capacity, six rows can be planted. The male parent can be planted with a seedling transplanter with a row spacing of 30 cm, or a seedling transplanter with a row spacing of 20 cm, in two rows with an empty row in between. For male parents with a short growth duration and weak tillering capacity, 8 - 10 rows can be planted, with male compartment of 180 - 200 cm, equal row space. The female parents can be planted in 30 - 40 rows, with a compartment of 700 - 800 cm. The exact number of rows should be determined by the female seedling transplanter. For example, if the transplanter has 25 cm row spacing with eight rows transplanted at the same time, then the female should be planted in 32 rows with a compartment 700 - 800 cm wide. If the transplanter has a row spacing of 18 cm, transplanting 10 rows at a time, then the female should be planted in 40 rows with a compartment 720 cm wide. Therefore, when agricultural drones are used to assist pollination, the male-to-female row ratio can be (6 - 10) : (30 - 40), which is conducive to mechanized planting and harvesting (Fig. 15 - 6).

Fig. 15 - 6 Largemale-to-female row ratio planting

The third is the outcrossing performance of the female parent. If the female has a high stigma exsertion rate, strong stigma viability and high compatibility with the male parent pollen, more rows of the female can be planted, otherwise, fewer.

2. Row direction

Two principles should be followed when determining the row direction of the male and female parents. First, planting rows should be oriented to facilitate light between the rows so that the plants are easily exposed to light and grow well. Second, the row direction should be set in such a way that ensures the wind direction during the flowering and pollination period is favorable for the transmission of the male pollen to the female parent compartment, and natural wind has a considerable influence on the transmission of male parent pollen. Natural wind should be used to improve pollen utilization. Therefore, the best row direction of the parents is parallel to the light direction, perpendicular to monsoon wind at the seed production base during the flowering and pollination period or with an angle of 45° or more. However, the wind direction varies in different regions, terrain and seasons. For example, in Hunan and other areas of central China, there are mostly south wind in summer and north wind in autumn, and the rows are oriented in an east-west direction, which is conducive to light transmission and pollination by wind. For mountainous areas, the row direction is set according to the wind direction in the valley. In coastal areas, the row direction is based on the direction of the sea breeze. A row direction plan should give priority to the natural wind direction at the time of pollination.

3. Methods of male parent planting

For human-assisted pollination, there are four methods for male parent planting, i. e., single row, false double row (narrow double row), small double row and large double row planting (Fig. 15 - 7 to Fig. 15 - 10). Single-row planting means that only one row of the male parent is planted in each compartment with a row ratio of 1 : *n* (*n* refers to the row number of the female) and a row spacing of 25 - 30 cm between the male and female rows. The male parent occupies a width of 50 - 60 cm with a spacing of 20 cm between plants and the female parent is planted in 14 cm×16 cm grids. False double row, small double row and large double row all have a row ratio of 2 : *n*.

Fig. 15 - 7 Single rows of male parent

Fig. 15 - 8 Narrow double rows of male parente

Fig. 15 - 9 Small double rows of male parent

Fig. 15 - 10 Large double rows of male parents

When pollination is assisted by agricultural drones, the male parent is planted in 6 – 10 rows. A parent with a long growth duration is planted in six rows; with either 30 cm equal row spacing or 20 cm wide-and-narrow rows, with a plant spacing of 25 cm. A parent with a short growth duration is planted in 8 – 10 rows, with 20 cm or 18 cm equal or wide-and-narrow row spacing (Fig. 15 – 11 and Fig. 15 – 12).

Fig. 15 – 11 Machine transplanted male parents

Fig. 15 – 12 Wide-and-narrow row machine transplanted male parents

The planting of male and female parent populations is shown in Table 15 – 5.

Ⅱ. Techniques for Targeted Cultivation of Parental Populations

1. Sowing and seedling raising techniques for the male parent

The entire growth and development process of the parents is divided into two stages during seed production due to the difference of parental growth duration. The first stage is independent growth and development for each parent and the second stage is when the parents grow and develop together. In seed production, in response to the growth and development duration of the male parent or the female parent, i. e. the seeding split between male and female parents, seedlings of the male parent are raised by the wetland method or the two-stage seedling raising method.

(1) Wetland seedlings

Wetland seedling raising can be used for hybrid parents with a short growth duration and the seeding split is within 20 d (less than 5.0 leaves inleaf age difference), because of a short growth period after transplanting into the seed production field for the male parent. About 7.5 – 15.0 kg/ha of male parent seeds are to be used. For early-maturing combinations whose male parent has a short growth duration of a male parent or who is sown basically at the same time with the female parent, or whose growth duration is shorter than the female parent, the seed quantity of the male should be no less than 15.0 kg/ha. For a male parent with a long growth duration, 7.5 kg/ha of seeds can be used. For a male parent with weak tillering capacity and low panicle formation rate, more seeds are needed. The quantity of seeds sown depends on the transplanting leaf age of the male parent. The higher the leaf age is (five leaves or more), the fewer seeds sown (150 kg/ha). The lower the leaf age, (less than 4.5 leaves), the more seeds sown

Table 15 – 5 Population structure of male and female parents forseed production

Pollination	Male parent planting method	Row ratio	Planting density (cm)		Compartment width (cm)		Number of seedlings (10,000/ha)		Hybrids suitable for seed production
			Male parent	Female parent	Male	Female parent	Male	Female	
Supplementary pollination	Single row	1 : 8	14×(25−25) *	14×18	50	126	4. 058	32. 469	Early-maturing hybrids (with a short growth duration of parent), mid-and late-maturing hybrids
		1 : 10	20×(30−30)	14×18	60	162	2. 252	32. 177	
	Small double-row	2 : 10	16×(25−14−25) **	14×18	64	162	5. 531	31. 607	Early-maturing hybrids (with a short growth duration of parent), mid-and late-maturing hybrids
		2 : 12	25×(27−14−27) **	14×18	68	198	3. 008	32. 225	
	Large double-row	2 : 10	16×(14−30−14) **	14×18	58	162	5. 682	32. 469	Early-maturing hybrids (with a short growth duration of parent), late-maturing hybrids
		2 : 12	25×(15−33−15) **	14×18	63	198	3. 065	32. 842	
Agricultural drone pollination	6 – 10 rows	6 : 40	25×30 or 25×(20−40) ***	14×18	210 or 200	702	2. 632 2. 661	31. 330 31. 677	Late-maturing hybrids
		6 : 32	25×30 or 25×(20−40)	12×25	210 or 200	775	2. 444 2. 462	27. 157 27. 352	Late-maturing hybrids
		8 : 40	18×20	14×18	200	702	4. 928	31. 677	Mid-and late-maturing hybrids
		8 : 32	18×20	12×25	200	775	4. 559	27. 352	Mid-andlate-maturing hybrids
		10 : 40	16×18	14×18	222	702	6. 764	30. 923	Early-maturing hybrids (with a short growth duration of parent)

Notes. Male parent planting density: * indicates plant spacing × (male and female parent spacing−male and female parentspacing); ** indicates plant spacing × (male and female parent spacing−two rows of male parent spacing−male and female parent spacing); *** indicates plant spacing × (male parent narrow row spacing−wide row spacing). Just as large row ratio planting, the spacing between male and female parents is 30cm for agricultural drone pollination.

(180 - 225 kg/ha). In terms of management of the seedling field, water and fertilizer management and disease and pest control are the same as those for ordinary wetland rice seedling raising.

(2) Two-stage seedlings

For the combination with a long growth duration of a male parent and a seeding split of over 20 d (>5.0 leaves in leaf age difference), two-stage seedling raising is adopted for the male to facilitate fertilizer and water management, weeding and pest control because of long independent growth for the male parent.

The first stage is to raise seedlings in dryland or plastic plates. The seedbed is in a dryland or dried rice field leeward and sun-facing. The seedbed base is flattened with compartments 1.5 m wide. The surface is compacted and covered with a 3 cm layer of mud or disinfected fine soil. Pre-germinated seeds are evenly and densely sown on the seedbed and covered with fine soil. In early spring, a plastic film or shading net or non-woven fabrics is used to cover the seedling bed to protect it from rain. Water should be sprayed regularly to keep the bed wet. Seedlings are moved to a temporary paddy field at 2.5 - 3.0 leaf stage. The area for temporary planting, which is a fertile field with sufficient base fertilizer, is prepared according to the number of male parent seedlings and planting density. The planting density can be 10 cm× 10 cm or 10 cm×(13 - 14) cm with two or three seedlings per hill. The seedlings in the temporary field should be no more than 7 - 8 leaves in leaf age (i.e. about 50% of the total number of leaves) and will be transplanted to the production field with mud to reduce injury and the re-establishment duration. After planting in the temporary field, shallow irrigation or wet management should be carried out, the field is to be topdressed with fertilizer early to promote tillering on the basis of sufficient base fertilizer, and 105 - 120 kg of urea and 75 kg of potash per hectare should be applied about 7 d after planting.

2. Sowing and seedling raising techniques for the female parent

For combinations with a small seeding split between the two parents, the seedlings of the female parent are mostly wetland-raised, while for combinations with a large seeding split, the seedlings of the female parent are wetland-raised or soft plate-raised.

(1) Wetland seedlings

Strong female parent seedlings with multiple tillers are the basis for a high yield in seed production. Strong seedlings are characterized by tillering with three leaves and one newly emerged leaf, two tillers at the five-leaf stage, flat stem bases, green leaves and strong white roots. The key techniques for raising wetland seedlings are as follows.

1) Choose a paddy field at a ratio of 1/10 to the production field as the seedling field, with uniform fertility, convenience for irrigation and drainage, sound soil texture and sufficient sunlight.

2) Level up the seedling fields, apply base fertilizer, and make compartments and drainage ditches.

3) Soak the seeds with TCCA to disinfect and perform pre-germination with less soaking and sufficient temperature and moisture for ventilation. Dress the seeds with agents before sowing and then sow them with an appropriate seeding split. When sowing, spread the seeds evenly by compartments.

4) From sowing to the 2.5-leaf stage, keep the seedbed surface moist but without a water layer. Perform shallow irrigation from the 2.5-leaf stage to transplanting.

5) Perform shallow irrigation and fertilization at the 2.5-leaf stage and 5 - 7 d before transplanting.

6) Pesticides are used to control rice thrips, planthoppers, *Chlorops oryzae*, blast and other diseases or pests.

(2) Soft-plate seedlings

It is used to raise seedlings of female parents with a short growth duration, to be transplanted at a low leaf age. Soft plates and mud are prepared as required for seedlings with soft plates in a production field. Pre-germination of female parent seeds is done the same way as raising wetland seedlings. After seed budding, the seeds are to be evenly sown in the holes of the plastic plates with 2 – 3 budded seeds per hole. The seedlings are raised as wetland seedlings or dryland seedlings. Seedlings are scattered in the production field at 3.0 – 3.5 leaves with a shallow layer of water or no water. Keep shallow water after scattering to benefit the seedlings.

If female seedlings are scattered and male seedlings are wetland raised, the seeding-to-heading duration of the female parent is extended by 2 – 3 d, and the seedling split between the parents is reduced accordingly.

(3) Direct seeding of the female parent

Direct seeding, i. e., direct sowing of pre-germinated seeds (manually or with machine) of the female parent into the female compartments of the seed production field, needs no seedling raising or transplanting (Fig. 15 – 13 and Fig. 15 – 14), and it can be done in the following three steps.

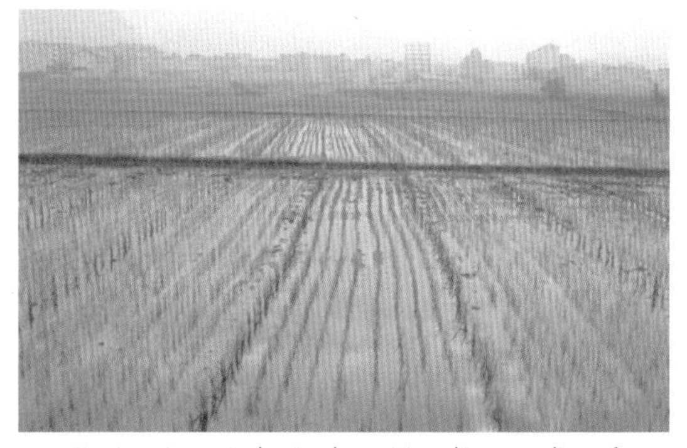

Fig. 15 – 13 Mechanized precision direct seeding of the female parent in seed production

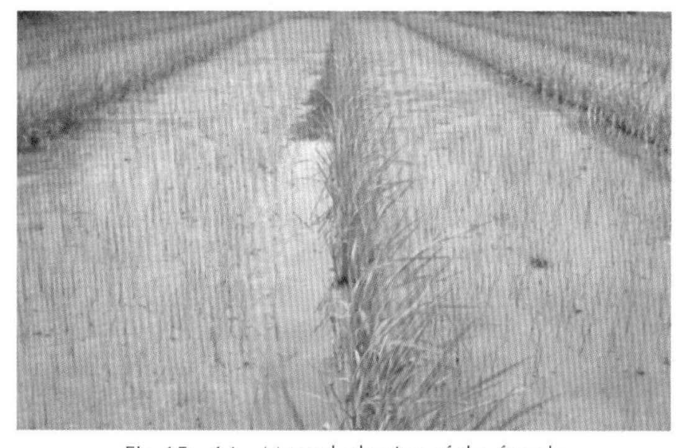

Fig. 15 – 14 Manual planting of the female parent in seed production

1) Irrigate the field in advance and plow (ordinary tillage and rotary tillage) and irrigate repeatedly to let the dropped seeds in previous rice crop fully germinate.

2) Apply herbicide in time to eliminate different types of weeds.

3) Adjust the seeding split. Direct seeding may reduce the sowing-to-heading duration of the female parent by 2 – 4 days. Therefore, if the male is transplanted, but the female is directly seeded, the seeding split between the male and female should be shortened by around 3 d.

3. Cultivation techniques for male and female populations

(1) Determination of the basic population of the male and female parents

Hybrid rice seed production is a process of outcrossing and seed setting in which the female parent receives pollen from the male parent. The outcrossing rate of the female parent depends on the coordination and synchronization between the two parents in heading and flowering. Due to the differences in the flowering characteristics of the parents, the requirements for the parent populations are different. The male parent is required to have a long heading and flowering period with sufficient pollen per unit time and space, while the female parent is required to have sufficient panicles and spikelets per unit area and a short heading and flowering period to ensure flowering synchronization. Therefore, the cultivation techniques cannot be the same for the two parents. In the late 1980s, Xu Shijue proposed the technical principle of targeted cultivation as the male population mainly comes from the tillers, while the female population mainly comes from the basic seedlings, which plays an effective role in increasing the yield of hybrid rice seed production.

With the increase in the types of super hybrid rice parents, the cultivation techniques for the parents need to be diversified. For example, large-panicle super rice parents tend to have weak tillering capacity with fewer effective panicles per plant, a structure of more compact panicles, high grain density and longer flowering period of a single panicle, the number of plants per hill should be increased when using large-panicle parents, regardless of the length of the growth duration. If the parent has weak tillering ability or low panicle formation rate, four seedlings per hill should be planted. For early-maturing combinations, or combinations with "reversed" the seeding split (the seeding-to-heading duration of the male parent is shorter than that of the female parent), there should be more male parent plants per hill with the spacing reduced to 14 - 17 cm. The female parent is required to be planted uniformly and densely, like with a transplanting spacing of 14 cm×17 cm and 2 - 3 plants, or 6 - 9 basic seedlings, per hill, so a total of 1.5 million basic seedlings per hectare is required for the female parent.

(2) Targeted cultivation for the female parent

The goal of cultivating a female parentin seed production is to have a robust population with appropriate panicle size, many and uniform panicles, short canopy leaves and no early senescence at the late stage. Years of high-yielding seed production proves it important to promote early growth and rapid development in the early stage, ensure normal growth in the middle stage and control the vigorous growth in the late stage as the general direction of female parent cultivation.

Firstly, there should be sufficient basic seedlings of the female parent. On the one hand, dense planting is required. From Table 15 - 5, it can be seen that the female parent can generally be planted 310,000 hills per hectare. On the other hand, 2 - 3 seedlings are to be planted per hill to ensure that more than 1.5 million basic seedlings are transplanted per hectare based on strong seedlings with multi-tillers.

Secondly, the requirement for fertilizer application is "heavy base fertilizer, light topdressing, supplementary fertilizer if necessary, appropriate nitrogen and high phosphorus and potassium". The key is to apply more base fertilizer, less or no topdressing, i. e., fertilizer is applied once only. Especially for early-maturing combinations, due to a short growth duration of the parents and a short effective tillering period, 80% of nitrogen and phosphorus, and 100% of potassium can be applied at one time as base fertilizer before transplanting, leaving only about 20% of nitrogen and phosphorus for follow-up application one week after transplanting for good, fertile fields with good water retention performance. If the field is poor in water and fertilizer retention and the parent has a long growth duration, 60%-70% of nitrogen and potassium and 100% of phosphorus are used as base fertilizer, leaving 30%-40% of nitrogen and potassium for follow-up application after transplanting and seedling re-establishment. In the fifth to sixth stages of young panicle differentiation, appropriate supplementary application of nitrogen and potassium or foliar fertilizer containing multi-nutrients may be done depending on the leaf color of the plants.

Thirdly, in terms of water management, shallow and wet irrigation is required to promote tillering in the early stage (from transplanting to the tillering peak), the field needs to be sun-dried in the mid-stage to promote deep root growth and control the number of seedlings and leaf length, and deep water is needed in the late stage for booting. The key is to sun-dry the field at the mid-stage. In the early stage, efforts are needed to promote early tillering and fast development, and when the number of seedlings reaches the target, the field should be sun-dried. Field sun-drying has four objectives. One is to shorten the canopy leaf length, specifically to shorten the flag leaf to 20 - 25 cm. The second is to promote the expansion of roots and deep rooting, which is conductive to the absorption and utilization of the fertilizer. The third is to strengthen the culm to prevent lodging. Due to the tall plants after GA3 application, the plants are prone lodging. Field sun-drying enhance the lodging resistance of plants by shortening and thickening the internodes at plant base. The four is to reduce ineffective tillering, promote uniform heading, improve field ventilation and light penetration in the field, and reduce pest and disease infestation. The suitable period for field-drying is determined based on the targeted number of female seedlings. Generally, it starts before panicle differentiation and ends at the third to fourth stages of panicle differentiation, which is about 7 - 10 days. The standard of drying the field is that the field surface is open on the edge of the field, the mud in the field is hard and does not sink to the feet, the white root runs on the surface, and the leaves are upright. The degree and time of field drying is determined by the growth and development of the female parent, as well as the irrigation. Cold water logged fields with deep mud need to be sun-dried heavily. Fields with late tillering and insufficient seedlings should be sun-dried later. Fields with insufficient water sources are to be lightly dried or not dried, or it would affect the growth and development of the female parent, resulting in a failure of flowering synchronization.

(3) Targeted cultivation for the male parent

The goal of targeted cultivation for a male parent is to have a robust population with large and many panicles, long flowering duration, and high pollen density per unit of time and space.

Firstly, a male parent should be planted sparsely. Different planting densities are used for parents with different growth durations and tillering traits (Table 15 - 5). The spacing of the male plants is 14 -16 cm if the growth duration is short, with 50,000 hills/ha, and is 20 - 25 cm if the growth duration is long, with 25,000 hills/ha. Generally, there are 2 - 3 seedlings per hill. The number of basic seedlings planted per unit area varies greatly among different male parents due to their different growth durations, and tiller-to-panicle formation, as well as differences in planting density.

Secondly, compared with the female, one or two more fertilizer applications to the male are the key technical measures to develop a robust population. The first fertilizer application takes place 5 - 7 d after transplanting of the female, while the second fertilizer application for male with a long growth duration is usually when the field is waterless but wet, from the tillering peak to the early stage of panicle differentiation. The amount of fertilizer for each application depends on the growth duration of the male and its characteristics of tillering-to-panicle formation. For parents with a long growth duration, fewer plants should be transplanted per hill while there are more panicles per plant. In this case, the amount of fertilizer should be higher, and otherwise lower. Normally, 30 – 50 kg of urea and 30 – 50 kg of 45% compound fertilizer are applied per hectare each time. There are two ways of focused fertilization for the male. One is to spread the fertilizer between the male rows through inter-tillage when the female parent is in shallow water or just transplanted, especially for the parent with a short growth duration. The second is to apply ball fertilizer, which is a mixture of urea, compound fertilizer and fine soil, deep in the middle of two-per-hill or four-per-hill male parents.

4. Machine transplanting techniques

(1) Raising seedlings for machine transplanting

Cultivation of even seedlings with multiple roots is the key to the machine transplanted seedling technique.

At present, there are two main methods of raising seedlings for transplanting, i. e. , factory-based seedlings and field seedlings (Fig. 15 – 15 and Fig. 15 – 16). There are three main types of bed soil used in the seedling plates, i. e. , special substrate, sieved dry fine soil and filtered mud, and two types of seedling plates, i. e. , hard plastic plate and soft plate.

Fig. 15 – 15 Machine-transplanted water-land seedling

Fig. 15 – 16 Machine-transplanted field seedling

Factory-based seedling cultivation is highly automated, with temperature and humidity controlled. The seedlings are neat and uniform, and the planting outcome is good. However, the cost is high and the scope of application is limited. Field seedling cultivation is much affected by the environment and climate, but the cost is low with fewer restriction, and convenient water and fertilizer management. Therefore, to realize machine transplanting on large areas of land, raising seedlings for machine transplanting in the field is an effective method.

Special substrate or sieved dry andfine soil can be used as soil in the seedling plate. Good water management is required to prevent uneven growth caused by insufficient water. Mud-raised seedlings make water management easier at the budding and seedling stages. In order to prevent the seeds from being washed away and disrupted by heavy rainfall after sowing, which may result in uneven seedling growth and a high empty hole rate when planting, large field seedlings should be covered with non-woven fabrics or shade net after sowing.

Based on the practical experience of machine transplanting over the years, it is appropriate to adopt the field seedling raising method of "mud + substrate +non-woven fabric (or film)" (Fig. 15 - 17) with the following key technical elements.

Making ridges Setting plates Filtering and loading mud

Even seeding Covering the seeds with substratet Non-woven fabric covering

Fig. 15 - 17 "Mud + substrate + non-woven fabric" for seedlings

1) Choose a good seedling field, and plow and level the seedling field to high standards.

2) Prepare in advance seedling compartments 140 - 150 cm wide with a spacing of 120 - 140 cm between compartments for mudding. Fill the compartment to ensure a flat surface with a height difference no more than 2 cm.

3) When the surface is hard, pull strings to arrange plates in an orderly way, with two plates horizontally placed in each ridge. The size of the plate should be based on the rice transplanter to be used.

4) Pound the mud by hand, filter it and fill the plates. After the mud settles, make sure that it occupies 2/3 of the plate holes. Apply 300 - 450 kg/ha of compound fertilizer as base fertilizer for seedlings at the place where the mud is taken before pounding.

5) After the mud has settled, sow the seeds manually or mechanically. Before sowing, determine thequantity of seeds per plate and correspondingly the sowing quantity for each compartment to ensure consistency. With 2 - 3 grains per hill at a 1,000-grain weight of 25 g, the seed quantity is 40 - 45 g per plate for 15.5 cm×56 cm×2 cm plates, 60 - 65 g per plate for 23 cm×56 cm×2 cm plates, and 75 - 80 g per plate for 28 cm×56 cm×2 cm plates, Perform seed dressing with agents before sowing.

6) After sowing, cover the plates with seedling-specialized substrate and then cover the seedling compartments with non-woven fabric to prevent washing from heavy rain. A film is used to cover seedlings to preserve the heat in the middle and lower reaches of the Yangtze River from late March to early April.

7) Keep the seedling plates moist after sowing and ensure normal seedling growth to promote roo-

ting. Water control is strengthened throughout the seedling period. When it is sunny and hot, keep water in the ditch. When the surface substrate in the seedling plates is dry, water it in time and drain the field in rainy days.

8) Apply agents to control bacterial wilt and basal rot when the seedlings are at the stage of one mature leaf andone newly emerged leaf, and also at the stage of two mature leaves and one newly emerged leaf to control planthoppers and rice thrips.

(2) Machine transplanting techniques for the female parent

1) Transplanter selection

There are four transplanter types for the female parent (Fig. 15 – 18).

Single-wheel ride-on rice transplanter (row spacing 18 cm, 10 rows)

High-speed rice transplanter (row spacing 25 cm, 7 rows)

Walk-behind rice transplanter (row spacing 20 cm, 8 rows)

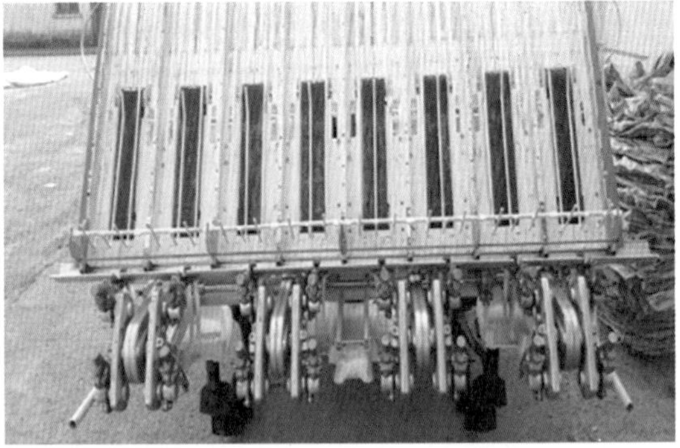

High-speed rice transplanter (row spacing of 20 cm, 8 rows)

Fig. 15 – 18 Transplanters for female parent in seed production

Type Ⅰ is a single-wheel ride-on and walk-behind rice transplanter, with a row spacing of 18 cm, plant spacing of 12 cm, 14 cm, 16 cm and 18 cm (adjustable), and 10 rows.

Type Ⅱ is a four-wheel high-speed andwalk-behind rice transplanter, with a row spacing of 20 cm, plant spacing of 12 cm, 14 cm, 16 cm and 18 cm (adjustable), and eight rows.

Type Ⅲ is a four-wheel high-speed rice transplanter, with a row spacing of 25 cm, a plant spacing ranging from 12 cmto 24 cm (adjustable with multiple gears), and seven or eight rows.

Type Ⅳ is a four-wheel high-speed rice transplanter, with a row spacing of 30 cm, a plant spacing ranging from 12 cm to 24 cm (adjustable with multiple gears), and six rows.

By the technical standards of targeted cultivation of female parent, as well as test results and the practice of female parent seedling cultivation, it is advisable to choose a 10 – row or 8 – row transplanter with a row spacing of 18 cm or 20 cm and a plant spacing of 14 cm, or one with a row spacing of 25 cm and a plant spacing of 12 cm. For some female parents with long growth duration, strong tillering capacity and high panicle formation rate, a rice transplanter with a row spacing of 30 cm and a plant spacing of 12 cm can also be used.

2) Characteristics of machine-transplanted female parent seedlings

Compared with manual transplanting of wetland seedlings, female parent seedlings for machine transplanting have the following characteristics.

a. The seeding-to-heading duration is extended by 1 – 4 days and the number of leaves on the main stem increased by 0. 1 – 1. 1 leaves. The extended days are different among sterile lines.

b. The numbers of basic seedlingsper unit area, maximum seedlings, effective panicles and spikelets are higher than those of manual transplanting, varying with sterile lines. The total number of spikelets per unit area increases by 1. 3%– 48% with an average of 15. 1% , indicating that female parent seedlings for machine transplanting are conducive to high yield in seed production (Table 15 – 6).

Table 15 – 6 Panicle formation difference between machine transplanted seedlings and manually cultivated seedlings of eight female sterile lines

Parent	Number of basic seedlings (m^2)		Number of highest seedlings (m^2)		Number of effective panicles (m^2)		Number of spikelets (m^2)	
	Machine-transplanted seedlings	Increase over CK (%)	Machine-transplanted seedlings	Increase over CK (%)	Machine-transplanted seedlings	Increase over CK (%)	Machine-transplanted seedlings	Increase over CK (%)
Y58S	102	16. 6	1,166	1. 3	414. 8	0. 4	6. 29	9. 6
P88S	132. 6	30. 6	1,125. 4	-1. 7	384. 2	3. 5	5. 55	4. 1
Guangzhan 63S	142. 9	36. 1	839. 8	0. 4	394. 4	23. 8	5. 14	29
Xiangling 628S	108. 8	7. 2	642. 6	6. 1	428. 4	2. 8	4. 04	19. 2
Zhun S	102	0. 5	642. 6	6. 1	394. 4	1. 5	2. 97	1. 3
Shen 95A	125. 8	56. 3	928. 2	34	394. 2	28. 0	4. 72	48. 0
Fengyuan A	129. 2	19. 1	839. 8	26. 3	411. 4	0. 5	3. 97	2. 1
T98A	125. 8	33. 1	703. 8	26. 5	425. 0	12. 4	4. 85	7. 3
Mean	121. 1	24. 9	861	12. 4	405. 9	9. 1	4. 7	15. 1

c. There is no significant difference in the heading status and heading duration between the methods.

d. The seeding-to-heading duration of machine-transplanted female seedlings increases significantly with the increase of seedling age. The seedling age increases by one day within the range of 15 – 27 d, the seeding-to-heading duration increases by 0. 42 – 0. 75 d, varying with different sterile lines.

e. The seedling quantity has a minor effect on the characteristics of machine transplanted female seedlings, such as the seeding-to-heading duration, number of seedlings, tiller-to-panicle formation and heading when the quantity seeds used is within the range of 80%−150% of the number used in manual transplanting.

3) Supporting techniques for machine transplanted female seedlings

The technical key points are as follows based on the characteristics of machine transplanted female seedlings.

a. Adjust the seeding split of the parents, e. g. , the seeding split is shortened by 3 −5 days when the female is machine transplanted and the male is manually transplanted compared with when both parents are transplanted manually.

b. The seed quantity can be 100%−120% that of manual transplanting and seeds with a germinating rate over 85% should be used.

c. For machine transplanted seedlings, the suitable leaf age is 2. 5 −3. 5 leaves, 15 −18 d for spring sowing and 12 −16 d for summer sowing.

d. By adjusting the quantity of seedlings for the rice transplanter, 2. 5 −3 seedlings should be planted per hill on average.

e. For machine transplanting, the field should be leveled and the field surface kept at about 5 cm in height. A ditch is to be dug to dry the field and perform shallow irrigation before machine transplanting.

f. After machine transplanting the field should be kept moist. If there is drought and dehydration in high fields, shallow water should be provided.

g. Just as in the case of manual transplanting, fertilization and pest prevention and control should be strengthened by weeding and applying chemical agents.

(3) Machine transplanting techniques for the male parent

Supplementary pollination and single-row or double-row planting are adopted for male parent seedlings instead of machine transplanting. If agricultural drones are used to assist in pollination, 6 −10 rows of male parent will be planted per ridge. Under this circumstance, machine transplanted seedlings can be used (Fig. 15 −19).

Fig. 15 −19 Male parent machine transplanted seedlings (row spacing 30 cm, plant spacing 25 cm, 6 rows)

1) Transplanter selection

There are two types of seedling cultivation for machine transplanting of a male parent, namely same-row planting, and wide-and-narrow row planting, which requires different transplanters.

For male parents with a long growthduration, a six-row transplanter with a row spacing of 30 cm and a plant spacing of 20 – 24 cm is used for same-row planting. For a male parent with a short growth duration, a 10 – row transplanter with a row spacing of 18 cm or 20 cm and a plant spacing of 16 – 20 cm is used.

Wide-and-narrow row planting can be used for a male parent with a long growth duration, using an eight-row transplanter with a row spacing of 20 cm and a plant spacing of 20 – 24 cm; while for a male with a short growth duration, a 10 – row transplanter with a row spacing of 18 cm and a plant spacing of 16 – 20 cm is used. Wide-and-narrow row planting means that a row is left blank after every two rows.

2) Characteristics of machine transplanted seedlings

According to the male parent cultivation objective in seed production, the male parent is sown in two batches with a seeding split of 8 – 10 d manually. The two batches of male parents can also be transplanted separately to extend the heading and flowering period. The male parent can also be sown in two-batches if machine transplanting is used, but in this case, the transplanting must be done once and for all. Considering the effects of the seedling age on the seeding-to-heading duration, the seeding split between the two batches of the male parent should be 5 – 6 d. The seedlings of machine-transplanted parent have a heading period 1 – 3 d shorter and 10% more panicles per unit area than manually transplanted seedlings. Those transplanted in a wide-and-narrow row pattern have a slightly longer heading period and 5% more panicles per unit area than those with single-row transplanting. Therefore, machine transplanting should be combined with wide-and-narrow row transplanting.

3) Cultivation techniques for machine transplanted seedlings

Different from thosefor machine transplanted female, key technical points for machine transplanted male seedlings are as follows.

a. The seeding split between the first and second batches is advisable to be 5 – 6 d.

b. The two batches of the male are machine-planted at the same time. When the first batch reaches a leaf age of 3.5 – 4.0 leaves and the second 2.7 – 3.3 leaves, they are planted alternately within 1 – 2 d.

c. The seedling quantity is adjusted for the transplanter to meet the standard of 2 – 3 seedlings per hill.

d. If machine transplanting is adopted for both male and female parents, the seeding split should be 2 d shorter than manual transplanting.

Section 4 Techniques to Improve Parental Outcrossing and Pollination

Ⅰ. Techniques to Improve Outcrossing Characteristics of the Male and Female Parents

Rice is a typical self-pollinated crop with a suitable floral structure and flowering habit for self-fertilization. However, hybrid rice seed production produces seeds from male sterile lines and normal fertile lines through complete outcrossing pollination. Rice male sterile lines have a series of physiological and biochemical responses to male sterility, which not only results in the loss of the ability to self-pollinate and set seeds, but also show specificity in certain aspects, such as enclosed panicle neck nodes in the leaf sheaths (Fig. 15 - 20), long and scattered flowering period, high stigma exsertion rate, poor closure or even splitting of the inner and outer glumes after flowering, etc. Both CMS and PTGMS lines have the phenomenon of panicle enclosure and in general, the panicle enclosure rate can reach 100% and the enclosed grain rate is about 30%. Such abnormal characteristics seriously affect the pollination and seed setting on the female parent during seed production, resulting in panicle enclosure and scattered flowering, but they also help the pollination of the female parent by delivering high stigma exsertion rate and stigma vigor. In particular, panicle enclosure is the key factor that hinders the outcrossing of the female parent. In the early days of hybrid rice seed production, manual leaf clipping and panicle exposure were done to improve outcrossing characteristics, but it was not very effective and the seed yield hovered long at about 750 kg/ha. Later, with the use of GA3 and other supporting techniques, the outcrossing of the female parent is greatly improved with an average yield of more than 2,250 kg/ha. GA3 plays a key role in improving outcrossing of the female parent.

Fig. 15 - 20 Panicle enclosure of the sterile line during heading

1. Techniques for improving parental outcrossing posture

The posture of the parental outcrossing refers to the spatial composition of the plant, leaves, panicles and spikelets when the male parent pollen is dispersed andthe female spikelets receive pollen, which includes three structures of plant, panicle layer and spikelets, and is a very important outcrossing characteristic. Improving the posture of parental outcrossing is a very important technical aspect of hybrid rice seed production.

In large-scale seed production, the differences in parental characteristics and their sensitivity to GA3, GA3 application techniques and seed cultivation management in the production field cause the diversity of parental outcrossing posture, resulting in considerable disparity in seed production yield in different fields of the same combination at the same location and at the same time. Therefore, it is very important to understand and establish the best pollination posture to ensure high and stable seed yield.

(1) Main technical criteria of outcrossing posture

The main technical criteria to evaluate parental outcrossing posture include the height of the female panicle layer, the height difference of the male and female panicle layers, the length of panicle enclosure, the rate of panicle exposure, the rate of fully exserted panicles, and the distance from the panicle tip to the leaf tip. The best outcrossing posture is when the female height is 90 – 100 cm, the male is 10 cm higher, the panicle enclosure length is 0 – 2 cm, the panicle exposure rate is ≥95%, the fully exserted panicle rate is ≥80%, and the distance from the panicle tip to the flag leaf tip is >5 cm but < 3/4 panicle length.

By examining morphological traits such as panicle height, penultimate internode length, antepenult internode length, length of panicle neck internode, leaf sheath length, leaf width, panicle length, distance from neck to grain (the distance from the panicle neck to the first spikelet at the panicle base), flag leaf extension angle, number of enclosed spikelets, and total number of spikelets in the panicle, the following criteria can be calculated and analyzed as follows, the average height of the panicle layers, enclosed neck length, spikelet exposure rate (or enclosed spikelet rate), fully exposed spikelet rate and the distance from the panicle tip to the leaf tip.

1) Panicle layer height is the mean of the effective panicle height of each parent. The height of panicles varies greatly among plants after GA3 is applied and it cannot be used to describe the panicle layer height of the parents. An effective panicle of the female parent is a panicle that has three or more grains.

2) Length of panicle enclosure = length of leaf sheath – length of panicle internode.

The length of panicle enclosure can be either a positive or a negative value. A negative value indicates that the panicle is fully exserted without enclosure. The greater the absolute value of a negative value, the longer the panicle is exserted from the leaf sheath. A positive value indicates that the panicle is enclosed without full exsertion. The greater the value is, the more serious the panicle enclosure is.

3) Panicle exposure rate refers to the ratio of panicle exsertion from leaf sheath, calculated with the following formula.

$$\text{Panicle exposure rate} = \frac{[(\text{panicle length} - \text{distance of neck-to-the 1st spikelet at panice base}) - (\text{length of neck enclosed} - \text{distance of neck-to-the 1st spikelet at panicle base})]/2}{\text{length} - \text{distance of nect-to-the 1st spikelet at panicle base}} \times 100\%$$

(when neck exclosure ≤ distance of neck-to-the 1st spikelet at panice base, panicle exposure rate = 100%)

4) Spikelet enclosure rate refers to the ratio of the number of spikelets enclosed in the flag leaf sheath to the total number of spikelets of the panicle at full heading, calculated with the following formula.

$$\text{Spikelet enclosure rate} = \frac{\text{number of spikelets enclosed}}{\text{total number of spikelets}} \times 100\%$$

5) Fully exserted panicle rate refers to the ratio of the number of panicles with 100% panicle exposure to the total number of panicles surveyed in the female parent population, calculated with the following formula.

$$\text{Fully exserted panicle} = \frac{\text{number of panicles with 100\% of panicle exposure}}{\text{total number of panicle surveyed}} \times 100\%$$

6) Distance from the panicle tip to the leaf tip refers to the vertical distance from the tip of the panicle to the tip of the flag leaf and is calculated as following.

$$R = L - N - H \times \cos \alpha$$

Where, R is the distance from the panicle tip to the leaf tip, L is the panicle length, N is the length of panicle enclosure, H is the flag leaf length and α is the flag leaf extension angle.

The R value can be positive or negative. When R is positive, it means that the panicle layer is higher than the leaf layer. The larger the R value is, the higher the panicle layer is than the leaf layer. When R is negative, the panicle layer is lower than the leaf layer, with panicles fully covered by the leaves. When R equals 0, the panicle layer and the leaf layer are at the same height.

(2) Outcrossing diversity

The construction of good parental outcrossing posture is mainly dependent on the spraying of GA3 and targeted cultivation of population panicles.

Due to the different responses of different sterile lines to GA3, the difference of sensitive nodes, the difference of cultivation and management between fields, the influence of weather conditions, and the difference in GA3 application methods, there are many variations for the outcrossing postures in terms of sterile lines, seasons and locations. The diversity of outcrossing postures of sterile lines can be analyzed from three aspects, i. e., plant structure, panicle layer structure, and panicle-spikelet structure.

1) Plant structure

Plant structure refers to the composition of a plant's panicle height and internode length at each node. Panicle height is composed of panicle length, panicle neck internode length, and the lengths of the penultimate, antepenult, and fourth-from-top internodes. Since the panicle neck node and the penultimate internode (internode of flag leaf) basically elongate simultaneously after exogenous GA3 application, even the penultimate internode elongate later than the panicle neck internode and the two internodes are in the middle and upper parts of the plant. These two internodes are called upper internodes, and the internodes below them are called lower internodes.

The plant structure is mainly determined by the ratio of the length of the upper internodes to that of the lower internode as well as the plant height. There are three types of plant structures grouped by the ratio of the upper internode length to the lower internode length.

First, the upper internode elongation type, characterized by long upper internodes and short lower internodes, a high length ratio of upper internodes to lower internodes (generally >3), moderate plant height and almost no panicle enclosure, which is beneficial to female outcrossing.

Second, the low internode elongation type, which has long lower and upper internodes, a small ratio of upper to low internode, fairly high plant height, thin stem, high probability of lodging, and generally low outcrossing.

Third, the non-elongation type, which is characterized by short upper and lower internodes, small elongation, and medium ratio of upper to lower internode lengths. This plant structure is often caused by late or insufficient use of GA3, which results in failure to significantly change the plant structure and avoid serious panicle enclosure. This type thus has the worst outcrossing performance.

Many years of practice of using *indica* sterile lines in seed production with a panicle enclosure rate of about 30% shows that the female parent with a good pollination posture requires more than 40% of panicle neck internode elongation. The elongation is controlled at 100% and 60% for the penultimate and the antepenult internodes, respectively, and the fourth internode from the top should not elongate. The panicle height increase is controlled at about 80%.

Plant structure plays three roles on the outcrossing and seed setting of the female parent. One involves the panicle layer height and the difference between the panicle layers of the parents, which directly affects pollen dispersal. The male panicle layer is required to be appropriately higher than the female parent panicle layer, so as to facilitate pollen dispersal from the male parent to the female compartment. It is not desirable for the female parent's panicle layer to be higher than the male parent's. The second is that it can determine the superiority of the spikelet structure of the panicle layer. Third, the internal nutritional status affects the high outcrossing and yield performance. If the plant is too high, it would have poor resistance to lodging and consume more nutrients, affecting the grain weight, and also stigma exsertion and vigor.

2) Panicle layer structure

The panicle layer structure refers to the composition and density of the panicle layer. In general, the panicle layer is mainly composed of panicles, flag leaves, and the upper and middle parts of the penultimate leaves, and also the flag leaf sheaths because of the panicle enclosure in the female parent.

The structure of the panicle layer depends mainly on the proportion and relative position of the panicle to the leaf. There are three types of panicle layer structure, the panicle type, the panicle-leaf type and the leaf type, classified according to the different proportions and position of the leaves.

The panicle type has a long panicle-neck internode without panicle enclosure. The panicle layer has basically no leaf sheath and penultimate leaf. The flag leaf is flat with a large extension angle. There are basically no leaves in the upper and middle parts of the panicle layer. The distance from the panicle tip to the leaf tip is 10 cm. The panicles are loose, with fewer obstacles to pollen dispersal and high dispersal efficiency; ample space allows for good ventilation, so the temperature rises and dew evaporates quickly in the panicle layer, which is conducive to early flowering and outcrossing of the sterile line. It helps prevent rice kernel smut and seed germination on panicles at the maturity stage (Fig. 15-21). Cultivation of a female population with neat and uniform panicles, a short heading period and short canopy leaves is the basis for the formation of a panicle-type structure.

Fig. 15-21 Panicle type of panicle layer structure

In the panicle-leaf type structure, the panicle neck nodes generally exsert out of the flag leaf sheaths, and there is basically no or only minor panicle enclosure. Panicles intertwine with flag leaves and part of the penultimate leaves. There are two cases. One is that the panicle is not evenly exserted and the heading period is long, resulting in rice panicle layering (commonly known as "three layers") after GA3 spraying. The panicle layer is thicker and the panicles, leaves and stems are interlaced in the panicle layer

Fig. 15 – 22 "Three layers" of the panicle-leaf type

(Fig. 15 – 22). Such a structure can stand a thick pollen layer and contribute to efficient pollen utilization. Second, the panicles and leaves are at a same layer due to the long lengths of the flag leaves and penultimate leaves (canopy leaves) and the small extension angle of the flag leaves. There is obstruction and adsorption of the pollen by the leaves. Leaf clipping is normally adopted to improve the panicle layer structure

The leaf-type structure is characterized by a low panicle layer, severe panicle enclosure, leaf layer above the panicle layer, less than 5 cm from the panicle tip to the leaf tip, panicles hidden in the leaves and a compact panicle shape. This structure is not conducive to pollen dispersal, resulting in delayed flowering, low stigma exsertion, tight panicle layer, poor ventilation and light transmission, and susceptibility to diseases and pests, especially rice kernel smut. The main causes of this structure are late spraying or low dosage of GA3, or insensitivity of the sterile line to GA3 combined with long, stiff, non-pendulous flag leaves.

3) Panicle-spikelet structure

Panicle-spikelet structure refers to the composition of spikelets in the female parent. Generally, the spikelets of a sterile line are categorized into those that can accept pollen and those that cannot accept pollen at all. There are two types of spikelets that cannot accept pollen at all. One is when the spikelets are enclosed in the sheaths, and the other is that the spikelets cannot bloom normally (closed spikelets).

Closed spikelets are common in rice, more or less in different varieties, but more severe in rice sterile lines, generally with an occurrence rate of 2%–20%, mainly subject to the weather at the time of flowering, and the physiological and biochemical characteristics of the sterile lines themselves.

Ⅱ. Gibberellic Acid Application Techniques

GA3 is a hormone that regulates plant growth and development with the effect of promoting cell elongation. GA3 can promote the elongation of panicle neck internodes, reduce the panicle enclosure of sterile lines, promote the spikelet exposure, and build a good outcrossing posture. However, GA3 can also promote the elongation of all young internodes. If sprayed improperly, it can easily cause the plant to be too tall or have a poor effect, so the spraying of GA3 during seed production is time-sensitive and technically demanding.

The effect of GA3 application in seed production is reflected in the following three aspects. 1) It promotes the elongation of the panicle neck internode, opens up enclosed panicles of sterile lines, increases the angle (extension angle) of the stem with the upper layer of leaves (mainly flag leaves) so that the panicle layer is higher than the leaf layer, exposes the spikelets, forms a panicle layer structure, and improves outcrossing of the female parent. 2) It can also increase the rate of stigma exsertion of the female, enhance the vitality of the stigma and prolong the life of the stigma. 3) It can advance the flowering time in the day and improve the flowering rate before noon.

The sensitivity of different sterile lines to GA3 is different and can be classified into three types, i. e., insensitive, moderately sensitive, and sensitive, based on the differences in the appropriate amount and period of GA3 spraying. The insensitive type of sterile lines requires GA3 at a dosage of 750 g/ha or more, and the spraying period is from 0 to 5% heading; while the sensitive type requires GA3 at a dosage of 225 g/ha or less and the spraying period is about 30% heading. The moderately sensitive type is between the two.

There are three main technical indicators regarding the application of GA3 in seed production, initial spraying (panicle emergence), dosage, and frequency.

1. Determination of GA3 application

The timing of the first GA3 spraying is called the initial spraying period and the panicle heading rate of the parent in the field is called the panicle emergence index, which is obtained by observing the population in the field. The exact date is determined by the following factors.

(1) Sensitivity of the sterile line to GA3

For sterile lines that are sensitive to GA3, such as Y58S, Longke 638S, Zhu 1S, Xiangling 628S, Zhun S, T98A and Tianfeng A, the initial spraying should be postponed when the panicle emergence index is at 20%–40%. For the sterile lines that are insensitive to GA3, such as Peiai 64S and P88S, the initial spraying is advanced to 0 – 5% panicle emergence. For the sterile lines with moderate sensitivity to GA3, such as Shen 08S, C815S, Jing 4155S and Fengyuan A, their panicle emergence index is 10%–20%.

(2) Flowering synchronization between male and female parents

GA3 is sprayed at the most suitable time for both parents if they have a good flowering synchronization. If there is a deviation, for instance, late female flowering and early male flowering, the initial GA3 spraying can be advanced by 1 – 2 d for the female, but with a lower dosage. In the case of early female and late male flowering, if the female is insensitive to GA3, the initial spraying of GA3 is advanced to 10% panicle emergence; while if the female is sensitive to GA3, the initial spraying can be postponed to 50% panicle emergence or delayed for 2 – 3 d. If the initial GA3 spraying is delayed, the frequency of spraying should be reduced to only once or twice with increased dosage.

(3) Heading evenness of the female parent

For fields with uniform heading of the female parent population, the panicle emergence rate for the initial spraying of GA3 can be reduced by 5%–10%, and the number of sprays and total dosage can be reduced accordingly. For fields with uneven heading of the female population, such as the fields with slow and late tillering at the early stage, or pre-maturity heading due to aged seedlings, the spraying of GA3 should be postponed subject to the heading status of the majority of the female population or done separately in more rounds.

2. GA3 dosage

(1) Determining the basic dosage according to the sensitivity of the sterile line to GA3

For the sterile lines that are sensitive to GA3, such as T98A, Zhu 1S and Annong 810S, the dosage of GA3 should be 150 – 225 g/ha, because the plants are prone to lodging if the dosage is high. For sterile lines with average sensitivity to GA3, such as Y58S, P88S, C815S, Ⅱ-32A, and Fengyuan A, the dosage of GA3 should be 300 – 600 g/ha; while for sterile lines insensitive to GA3, such as Peiai 64S, the dosage should be 750 g/ha or higher.

(2) Adjusting the dosage according to the temperature at the heading period

The temperature during the heading has a great influence on the GA3 effect, so the dosage of GA3 varies greatly with different locations and seasons of seed production due to the temperature changes. Years of practice shows that the dosage of GA3 sprayed at a low-temperature weather (average daily temperature of 24 ℃– 26 ℃) increases about double than that at a high-temperature (average daily temperature of 28 ℃–30 ℃). Therefore, the basic dosage should be determined by the sensitivity of the sterile line to GA3 under appropriate weather conditions (average daily temperature 26 ℃–28 ℃), and is adjusted according to the actual temperature at the heading stage.

(3) Adjusting the dosage according to other factors

In the case of good flowering synchronization, the spikelet composition is reasonable with uniform heading and normal growth, the basic dosage of GA3 can be used for seed production. However, the dosage of GA3 often needs to be adjusted due to other factors.

First, the initial spraying time of GA3 is subject to the level of flowering synchronization.

When spraying GA3 early on sterile lines, the elongation value of each internode increases because the plants are young and tender. Although the panicles are not completely released from enclosure, plant height increases and the spraying dosage should be properly reduced to avoid excessive plant height and lodging. On the contrary, GA3 spraying should be delayed when some stem internodes tend to age, and the dosage should be increased appropriately in order to reduce the panicle enclosure.

Second, adjust the GA3 dosage according to the growth of the female population. If the number of seedlings per unit area is too large and the upper leaves are long, the dosage of GA3 should be increased. On the contrary, if the structure off female population is reasonable with appropriate length of leaves, and dark green leaves, the dosage should be reduced.

Third, the dosage should be adjusted based on the weather conditions. If GA3 is sprayed on rainy and low-temperature days, not only will it be washed away by rain, but also the stomata and hydathodes of the leaves will not be well open and GA3 would be poorly absorbed. Therefore, GA3 should be sprayed at a dosage 50%–100% higher than usual during rain intermission or on days with drizzle. If GA3 is sprayed in hot, dry and high-temperature weather, it is easy to evaporate, so the dosage also needs to increase.

Fourth, the dosage should be adjusted according to the planting method ofthe female parent. If the female parent is planted by direct seeding or scattering, the population usually has uniform growth but with shallow roots, and GA3 may induce lodging. Therefore, the dosage should be reduced appropriately.

(4) The number of GA3 spraying and the dosage ratio

GA3 is normally sprayed two or three times during seed production. The exact number of spraying is determined by the following two factors.

One is the heading evenness of the population. For seed production fields with high heading evenness, there can be two or even only one spraying, while for those with less heading evenness there should be three or even four.

The other is the spraying time. If spraying occurs early, the number of spraying should be increased; on the contrary, if the spraying is late, then the number of spraying should be reduced. Only one spraying is needed if the panicle emergence index is large (over 50%).

The spraying of GA3 can help solve the problem of panicle enclosure in different times with different dosages according to the degree of difference in the growth and development progress of the female population, and the general principle is "light at the early, heavy in the middle and low at the late stages". If GA3 is sprayed twice, the dosage ratio is 2 : 8 or 3 : 7; if it is sprayed for three times, the ratio is 2 : 6 : 2 or 2 : 5 : 3; if four times, 1 : 4 : 3 : 2 or 1 : 3 : 4 : 2.

When spraying GA3, the interval between two spraying varies, 24 hours under normal circumstances, 12 hours when the difference of panicle layer is small (once in the morning and once in the afternoon)

(5) Time of application

Daily spraying can be arranged at 07:30 – 09:30 in the morning or when dew is almost dry at 4:00–6:00 pm. It should not be sprayed at noon when there are high temperature and strong sunlight.

(6) Tools and water volume

There is no strict requirement for the amount of water used, regardless of how much GA3 is sprayed each time. The point is that GA3 is uniformly sprayed on the leaves of the parent plants at an appropriate dosage. It is better to have less water than more. Based on the different spraying tools, the amount of GA3 per unit area varies greatly.

1) Knapsack sprayer

Knapsack sprayer is a traditional tool used for GA3 spraying, about 300 kg/ha of water should be used with a nozzle that produces fine droplets (Fig. 15 – 23).

In addition, there are powered plant protection misters and ground self-propelled plant protection machines that can be used for GA3 spraying (Fig. 15 – 24 and Fig. 15 – 25).

Fig. 15 – 23 Spraying of GA3 with Knapsack sprayer

Fig. 15 – 24 GA3 spraying with Knapsack mist sprayer

Fig. 15 – 25 GA3 spraying with self-propelled plant protection machine

2) Agricultural plant protection drone

The result of GA3 application with agricultural drones for riceseed production is shown in Table 15 –

7, which indicates a better effect than with backpack sprayers, with a short panicle layer, reduced panicle enclosure and increased rate of fully exposed panicles. Agricultural plant protection drone, therefore, is a good choice for spraying GA3 (Fig. 15 - 26, Fig. 15 - 27 and Fig. 15 - 28).

Table 15 - 7 Analysis of the effect of spraying different doses of GA3 by agricultural plant protection drones (Y Liangyou 1 seed production in Sanya, Hainan, 2016)

Tool	Dosage (g/ha)	Panicle layer height (cm)	Panicle exposure rate (%)	Full panicle exposure rate (%)
Agricultural drone	576	99.77 ± 0.40	1.27 ± 0.23	84.64 ± 0.40
	480	97.14 ± 0.17	1.08 ± 0.08	85.48 ± 1.11
	384	95.50 ± 0.08	1.74 ± 0.35	76.26 ± 1.33
Knapsack sprayer	480	108.78 ± 0.14	1.91 ± 0.05	66.57 ± 2.39

Fig. 15 - 26 Application of GA3 with agricultural plant protection drone

When a drone is used for GA3 spraying, the amount of liquid sprayed per hectare is about 15 L, with a small amount of water and high spraying concentration. If the dosage of GA3 on the female population reaches 900 g/ha (equal to22.5 L of e-mulsion) or more in two spraying, 11.25 L/ha each time, only 3.75 L of water is needed, which is equivalent to the spraying of GA3 original solution and the effect of spraying has a certain impact on the spraying. The spraying time should be longer with reduced amount each time.

Fig. 15 - 27 Spraying of GA3 with agricultural drone on male parent

Fig. 15 - 28 Effect of GA3 spraying with agricultural plant-protectingdrone

The dosage and the spraying time when using a droneare basically the same as using a backpack sprayer, but for parents insensitive to GA3, the amount can be reduced by 20% and the spraying should be done once a day on two consecutive days.

(7) GA3 application on the male parent

There are differences in GA3 sensitivity between the male and the female parents and among different male parents. In order to prepare the male parent well for pollination, the panicle layer of the male parent should be 10 – 15 cm higher than that of the female parent, so that pollen can be dispersed evenly over a certain distance to improve pollen utilization. Therefore, it is necessary to determine if the spraying is only needed for the male parent according to its sensitivity to GA3. The amount and time of spraying on the male parent also depend on its sensitivity to GA3.

Ⅲ. Supplementary Pollination Techniques

The floral, flowering and pollination characteristics of rice are adapted to self-pollination rather than to cross-pollination. Hybrid rice seed production is a process of complete cross-pollination. Seed setting on the female parent depends on the pollen of the male parent scattered on the stigma of the female parent. There are two conditions for this process. First, the density of male parent pollen in unit time and space, the higher the pollen density, the greater the probability of reaching the stigma of the female parent. Second, rice pollen grains are small and light but still a certain level of wind is required to help with complete dispersal and spread of pollen from dehisced anthers. However, the force of natural wind is uncertain when the male parent blooms and disperses pollen. Supplementary pollination is therefore needed during the peak flowering and pollination period of the male parent, so that the pollen from the male parent can be concentrated, spread farther and evenly scattered on the stigma of the female parent. There are two ways to do this – manually or with an agricultural drone.

1. Supplementary pollination methods

(1) Human-assisted pollination

Human-assisted pollination is a common practice used in China and Southeast Asia for seed production. Ropes, bamboo poles or other tools are used to vibrate the parent plants, so that the male's pollen is dispersed by the elasticity of the vibration to the female. However, because the elasticity acting on the male is small and pollen does not spread far, this method suits only cases where the parents are planted at a small row ratio, specifically one or two rows of the male and 8 – 12 rows of the female. Pollination can be done manually in four ways, rope, single-pole vibration, single-pole push and double-pole push, depending on the tools used.

1) Rope pollination

Two persons hold the two ends of a long rope (0.3 – 0.5 cm in diameter) parallel to the male parent row, pull the rope fast along the compartment perpendicular to the row (at a speed of 1m/s or higher) to vibrate the panicle layer so that the pollen of the male parent is spread to the female parent compartment (Fig. 15 – 29). It is fast and efficient, and can catch the pollen at the peak of parental pollen dispersal. However, it has two shortcomings. One is that the vibration force on the male is too small to fully disperse male pollen or to spread the pollen grains far. Second, it spreads male pollen mainly in one direction so the pollen load is high along the direction of the rope and can easily cause uneven distribution of pollen in the field resulting in a low pollen utilization rate. Therefore, a smooth rope with an appropriate

Fig. 15 – 29 Rope pollination

length should be used, preferably 20 – 30 m. In addition, the rope holders should walk fast enough for more effective supplementary pollination. This method applies to male parents planted in single, false-double or small-double rows.

2) Single-pole vibration pollination

One person, holding a 3 – 4 m long bamboo pole or wooden pole, walks between male parent rows, or between a male row and a female row, or in the female parent compartment, puts the pole at the height of the base of the panicle layer, and vibrates the panicle layer in a fan shape, causing the male pollen to disperse into the female compartment (Fig. 15 – 30). This method is slower and less efficient than rope pollination. However, it has stronger vibration force on the male parent, so that the pollen can fully disperse from the male and spread far. The disadvantage is that the pollen only spreads in one direction and the distribution is uneven. This method applies to male parents planted in single, false-double and small-double rows.

3) Single-pole push pollination

This pollination method requires a special parental transplanting pattern in seed production. A pollination working aisle is set up about 30 cm wide in the direction perpendicular to the rows, divided by the length of the pollination pole (5 – 6 m). A person holds the pole in the middle, walks in the working aisle, places the pole in the upper middle part of the male plants, and pushes the male row at the time of flowering, so that the male's pollen drifts into the female compartment (Fig. 15 – 31). The advantage of this method lies in the effectiveness of driving the pollen without driving the female and that it is fast. The disadvantage is that the pollen spreads in one direction, so the pole needs to be pushed back and forth to spread the pollen evenly. It is suitable for single-row, false double-row and small double-row parent planting patterns.

Fig. 15 – 30 Single-pole vibration pollination

Fig. 15 – 31 Single-pole push pollination

4) Double-pole push pollination

Fig. 15 – 32 Double-pole push pollination

A person holds a short pole (1.8 – 2.0 m) in each hand, walks between two male rows with the poles placed at the upper middle part of the two male rows, and vibrates the male plants 2 – 3 times so that the pollen can be fully spread to the female compartments on both sides. The most important is to push gently, vibrate forcefully and move back slowly (Fig. 15 – 32). It can spread the pollen sufficiently and far, but it is slow and less efficient, making it difficult to ensure timely pollination during the peak flowering period of the male parent. This method only applies to male parents planted in large double or small double rows.

(2) Agricultural drone-assisted pollination

By using agricultural plant protection helicopter drones to assist seed pollination, the wind generated by the drone rotor (single-rotor or multi-rotor) can drive the male's pollen farther (Fig. 15 – 33 and Fig. 15 – 34), thus changing the planting pattern of the parents, and the row ratio of the parents can be increased to (6 – 8) : (30 – 40), which is convenient for mechanized planting and harvesting, facilitating the mechanization of hybrid rice seed production.

Fig. 15 – 33 Single-rotor drone-assisted pollination

Fig. 15 – 34 Four-rotor drone-assisted pollination

Liu Aimin et al. used a single-rotor drone in an experiment designed with three row ratios (6:40, 6:50 and 6:60) to test the effect of drone-assisted pollination. The results (Table 15 – 8) show that the seed setting and yield of the female parent with drone-assisted pollination and a large row ratio were normal with no significant difference from those with manual pollination and a small row ratio. There was also no significant difference between the three row ratios. This indicates that drones can be used in seed production for pollination assistance and can significantly increase the parental row ratio.

The results (Table 15 – 9) of the survey on the seed setting rate of females in different parts of the female compartment showed that there were some differences, but without significance or a clear pattern. This means that the seed setting rate of the female is not likely to be correlated with the distance from the male.

Table 15 - 8 Pollination assisted by electric single-rotor agricultural drone

Pollination method	Male and female parent row ratio	Ledong, Hainan (2015)			Wugang, Hunan (2015)		
		Combination	Seed setting rate (%)	Yield (kg/ha)	Combination	Seed setting rate (%)	Yield (kg/ha)
Agricultural drone	6 : 40	Longke 638S/R534	50.0	3,448.5	Guangzhan 63S/R1813	32.93	3,044.25
	6 : 50		46.5	3,427.0		34.36	2,901.75
	6 : 60		45.5	3,258.0		32.52	2,927.55
Human-assisted pollination	2 : 12		42.9	3,549.0			
Agricultural drone	6 : 40	33S/ Huanghuazhan	53.3	4,318.5	Shen 08S/ R1813	36.61	3,044.25
	6 : 50		49.5	4,084.5		28.91	2,901.75
	6 : 60		51.5	4,150.5		32.05	2,927.55
Human-assisted pollination	2 : 12		53.4	3,945.0			

Table 15 - 9 Seed setting rate of female parent via agricultural drone-assisted pollination (Longke 638S/R1813, Wugang, Hunan)

Year	Row ratio	Seed setting rate at different observation points in the female parent compartment (%)								
		1	2	3	4	5	6	7	8	9
2014	6 : 40	47.6	52.1	51.3	53.9	45.7	50.9	42.1	48.1	46.0
	6 : 60	44.4	48.8	47.3	45.2	45.2	43.2	44.9	39.4	36.7
2015	6 : 40	27.3	34.6	31.3	35.6	35.7	35.8	35.9		
	6 : 60	38.6	37.6	27.3	35.0	30.3	28.0	33.0		

Table 15 - 10 shows results of the multi-location demonstration of the mechanized seed production technology with drone-assisted pollination. The seed setting rate and yield with drone-assisted pollination can fully reach or even exceed the levels with manual pollination.

The drone-assisted pollination is to use the wind generated by the rotor blades in flight to drive up the male's pollen and spread it to the female parent compartments. The force of the wind and the width of the wind field generated by a drone are directly correlated to the height and speed of the drone. When the drone flies high, the width of the wind field increases but the wind speed decreases, while when it flies low, the width of the wind field is smaller but the wind speed increases. When it flies fast, the wind force acting on the male decreases, while when it flies slow, the wind force acting on the male parent increases. The wind force acting on the panicle layer of the male parent is better to be about 3 m/s and the flight parameters of a drone can be set accordingly.

According to the wind speed measurement based on different flight parameters of a drone and the pollination test results, the main technical points of drone-assisted pollination are as follows.

Table 15 - 10 Yield of mechanized seed production with agricultural drone-assisted pollination

Year	Location	Combination	Female parent seed setting rate (%)		Predicted yield (kg/ha)	
			Agricultural drone-assisted pollination	Human-assisted pollination	Agricultural drone-assisted pollination	Human-assisted pollination
2015	Wugang	H638S/R1813	41. 5	36. 6	3,630	3,210
2016	Ledong *	Y58S/R900	39. 3	43. 8	3,228	3,720
2016	Wugang	03S/R1813	43. 0		3,176	
		Shen 08S/R1813	46. 1		3,117	
2016	Suining * *	Y58S/R302	50. 0	42. 8	4,029	3,383

Notes. * In 2016, the male parent planting direction was in line with the monsoon direction, resulting in a low seed setting rate in the female parent compartments in Ledong, Hainan. * * Four-rotor agricultural drones were used for pollination.

1) Single-rotor or multi-rotor plant protection drones can be used to assist in seed pollination; a dedicated pollination drone is not necessary.

2) When a drone flies over the male parent compartment, 1. 5 - 2. 0 m above the panicle layer of the male parent, considering the drift effect of natural wind on the wind field, the flight path should be adjusted according to the draft of the wind field, so that the vortex of the wind field falls completely on the male parent compartment.

3) The flying speed should be 4 - 4. 5 m/s.

4) To ensure stable speed, height and distance trajectory, it is advisable to use autonomous or intelligent drones.

5) When flying the drones manually, each drone can cover 15 - 15. 5 ha of seed production field per day.

2. Time and frequency of supplementary pollination

The flowering period of rice is short, normally about 10 d for each population, 1. 5 - 2 h/d in clear weather, and suitable temperature and humidity conditions before noon. The female parent receives pollen from the male parent, but the flowering habits are quite different between them. Therefore, the period, time and frequency of supplementary pollination must be carefully planned based on the flowering period of the female parent. The pollination period includes the flowering date of the female parent and the three days that follow.

The time for pollination depends on the flowering time of the male parent. Upon arrival of the male's flowering peak when the pollen density in the field is the highest, pollination will be carried out. The time of the day for pollination can be determined separately for two stages. Firstly, before the female parent enters the peak flowering stage (4 - 5 d after initial flowering), the female has a small number of flowers, so the time of the first pollination every day is mainly based on the time of female flowering. Secondly, after the female parent has entered the peak flowering period, the numbers of flowers and flowered spikelets increase per day and there are also more spikelets with exserted stigma, so the time of the first pollination per day should be based on the flowering time of the male parent. After the first pollination, the second and third pollinations are carried out when the male parent is at the second flowering peak. The time for each pollination is controlled within 30 minutes, with a general interval of about 10 minutes. During the flowering period of the male parent, visible pollen fog should be formed every day when the supplementary pollination is carried out and the pollen density in the field is high, so that the female parent can get more pollen.

References

[1] LIU AIMIN, XIAO CENGLIN. Super hybrid rice seed production technique[M]. Beijing: China Agriculture Press, 2011.

[2] XIONG CHAO, TANG RONG, LIU AIMIN, et al. Effect of machine transplanting row ratio on yield and its relative traits in seed production of Ke S/Huazhan[J]. Crop Research, 2015, 29(4): 362-365, 373.

[3] LIU AIMIN, SHE XUEQING, YI TUHUA, et al. Studies on characteristics of mechanized seedling transplanting of the seed parent in hybrid rice seed production[J]. Hybrid Rice, 2015, 30(1): 19-24.

[4] LIU AIMIN, ZHANG HAIQING, LIAO CUIMENG, et al. Effects of supplementary pollination by single-rotor agricultural unmanned aerial vehicle in hybrid rice seed production[J]. Hybrid Rice, 2016, 31(6): 19-23.

[5] TANG RONG, ZHANG HAIQING, LIU AIMIN, et al. Study of the techniques of gibberellin spraying by agricultural unmanned aerial vehicle in hybrid rice seed production[J]. Crop Research, 2017, 31(4): 360-368.

[6] YANG YONGBIAO, LIU AIMIN, ZHANG HAIQING, et al. Effect of machine transplanting density on the growth and development characteristics of the female parent population in hybrid rice seed production[J]. Crop Research, 2017, 31(4): 342-348, 372.

[7] CHEN YONG, ZHANG HAIQING, LIU AIMIN, et al. Effects of mechine transplanting and fertilization mode on the population growth and development of male parent in hybrid rice seed production[J]. Crop Research, 2017, 31(4): 355-359, 376.

[8] WANG MING, LIU YE, ZHANG HAIQING, et al. Effect of high temperature on outcrossing characteristics at the fertile sensitive stage of photo thermo sensitive genic male sterile (PTGMS) rice lines[J]. Journal of Hunan Agricultural University (Natural Sciences), 2017, 43(4): 347-352.

[9] LIU FUREN, LIU AIMIN, HE CHANGQING, et al. Demonstration of key techniques in hybrid rice seed production with whole process mechanization[J]. Hybrid Rice, 2017, 32(1): 34-36.

[10] LIU AIMIN, ZHANG HAIQING, LUO XIWEN, et al. A fully mechanized method of hybrid rice seed production. China, XL201210438297.0, 2012-11-06.

[11] LIU AIMIN, XIAO CENGLIN, SHE XUEQING, et al. A hybrid rice seed production method. China, XL201210417925.7, 2012-10-26.

Chapter 16

Quality Control of Super Hybrid Rice Seeds

Liu Aimin / Xiao Cenglin / He Jiwai

The quality requirements for super hybrid rice seeds are the same as those for ordinary hybrid rice seeds. But in addition to purity, cleanliness, germination rate, moisture content, and no quarantined diseases and pests, it is also essential to check the percentages of glume-split, panicle budded, diseased, dehulled, powdery and mildewed, colored and other abnormal seeds. In addition to high combining ability and heterosis, super hybrid rice seeds have some specific characteristics because of the complex and diverse genetic background. Moreover, a higher quality is required for hybrid rice seeds with the development of mechanized and large-scale production of rice. Therefore, it is necessary to take appropriate technical measures in the process of seed production, processing, inspection, storage and use so as to maintain high vitality and commercial quality of the seeds.

Section 1 Characteristics of Super Hybrid Rice Seeds

Hybrid rice seeds are products of outcrossing, which changes the inherent self-pollination of rice. This, coupled with male sterility of the female parent, results in many physiological and biochemical abnormalities of the female parent in flowering and grain-filling, such as low gibberellin content, panicle enclosure, scattered flowering and poorly closed glumes, which in turn cause deterioration in the seeds such as glume splitting, budding on panicle, powdering and mildew, resulting in large differences in the vitality of seed in a same batch. The deterioration of hybrid rice seeds is caused by glume-splitting, budding on panicle, powdering, mildew, and dead seeds.

Ⅰ. Glume-splitting Seeds

1. Glume-splitting of hybrid rice seeds

Glume-splitting is a specific and universal characteristic of outcrossed products of rice male sterile lines. The degree and rate of glume-splitting vary greatly among sterile lines and also with different climatic and ecological conditions for seed production of the same sterile line. A survey shows that the glume-splitting seed rate in hybrid rice is 1%–70%, with an average of 26.1%, while for conventional rice seeds, generally there is no glume-splitting or a very low rate of glume-splitting seeds, at about 2% on average. The rate of glume-splitting seeds varies with sterile lines. For example, Longke 638S, 33S, Xinxiang A and Sanxiang A have a glume-splitting seed rate of over 50% , and Ⅱ - 32A, T98A and Y58S have a rate of 20% - 30% , while Peiai 64S and Jing 4155S have a rate below 5% . Glume-splitting seeds have normal embryos with germination ability, but a small endosperm with insufficient content. The grain weight is low, the seeds are less tolerant to storage and moisture and show abnormal budding in germination, a low effective seedling rate and poor quality of seedlings (Fig. 16 - 1 and Fig. 16 - 2).

Fig. 16 - 1 Glume-splitting seeds

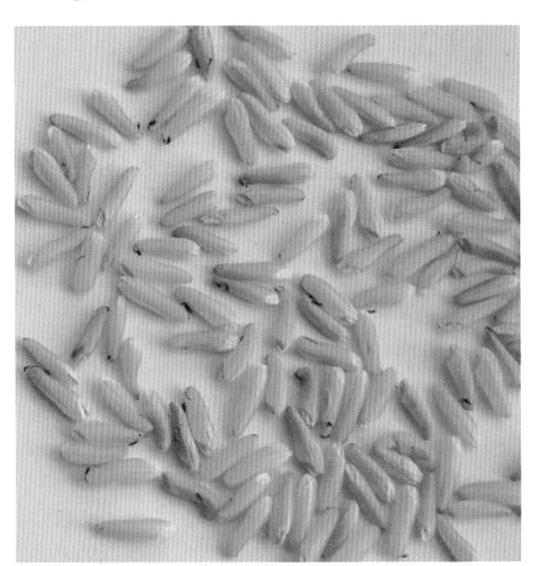

Fig. 16 - 2 Grains of glume-splitting seeds

2. Causes

The female parents used for hybrid rice seed production generally have a long period of glume opening after flowering. During the blooming process, the inner and outer glumes gradually age and even shrink due to external conditions such as water loss, reducing the closing ability of the glumes after fertilization, or making some glumes unable to close, forming a split. In the most serious case, both the inner and outer glumes of Longke 638S would remain open after flowering when it rains, so the spikelets can be fertilized but cannot set seeds. This phenomenon can be observed from the morphological traits of the floral organs of the sterile line. Due to the poor development of the lodicules and the small spike-stalk vascular bundles of the sterile line, the lodicules absorb and lose water slowly, and remain swollen after flowering. The small spike-stalk and the cell structure at the lemma base grow and the small spike-stalk loses the elasticity to restore the lemma, causing the inner and outer glumes to fail to close normally and leave a split. Due to the poor closure of the inner and outer glumes, the seeds are deformed and not fully filled. The shape and fullness of the seeds vary with the degree of glume split.

The glume-splitting rate is related not only to the traits of the sterile line but also to the climatic conditions, GA3 application, and the tools and methods of supplementary pollination during flowering and pollination. Practice shows that if the pollination period meets with rainy days with low temperature or extremely hot (maximum daily temperature>35 ℃) and dry days, the recovery elasticity of the lodicule would becomes poor after flowering, and the glume closure would be obstructed. Too early application of GA3 makes the glumes delicate and white after exsertion, and weakly closed after flowering. If a rough rope is used across the female panicle layer for supplementary pollination and damages the spikelets to a certain extent. All these will increase the glume-splitting rate. Seed production practice shows that hybrids with the same female parent produced in southern Hainan in spring have a low rate of glume-splitting seeds because of moderate temperature and relative humidity during flowering and pollination. In contract, when they are produced elsewhere, there may be a high rate of glume-splitting seeds because of the high temperature, dry weather and hot wind in summer or rainy weather and low temperature in autumn.

3. Classification

According to the opening of the inner and outer glumes and the size and shape of rice grains, glume-splitting seeds can be divided into four categories.

1) Severely glume-splitting seeds are seeds with more than 2/3 of the inner and outer glumes open. The rice grains are small, in a triangular cone shape, and easily sorted out by a general winnowing machine.

2) Moderately glume-splitting seeds have more than half of the inner and outer glumes open, and the rice grains are half the size of the normal grains and can be sorted out by a gravity sorter.

3) Mildly glume-splitting seeds have only a small proportion of the inner and outer glumes open. The grains are 2/3 the normal size and can sprout, but with abnormal seedlings, mostly can be sorted out by a gravity sorter.

4) Slightly glume-splitting seeds have the inner and outer glumes well closed at the top, but with a split in the middle, and the rice grains are visibly full.

Ⅱ. Panicle Budding

1. Performance of panicle budding

Panicle budding means that the seed germinate on the panicles at maturity or during drying (Fig. 16-3). Pre-harvest sprouting means that the seed embryo is swollen with the sprout hole split, but the radicle and germ have not emerged from the seed coat. The radicle and germ are visible from the outside of the seed with varying lengths in the case of panicle budding, which is likely to occur in hybrid rice seed production when it is rainy or the humidity is high in the field during the maturity and harvesting period. The paicle budding rate could be more than 30% in a serious case. Panicle budding not only reduces the seed yield, but also seriously affects the germination rate. Research shows that the embryos of panicle-budding seeds expand and the endosperm content begins to decompose. If the seeds are harvested and stored for a sales period (about six months) under ordinary conditions, the budded seeds would lose the ability to germinate. It not only affects the germination capacity of the seed itself, but also causes the whole batch of seeds containing the budded seeds to turn sour and odorous with slippery husk, and fail

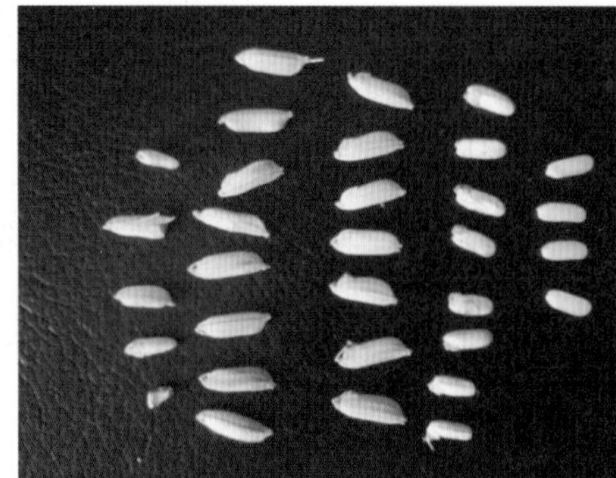

Fig. 16 – 3 Panicle budding

during seed soaking and pre-germination, resulting in the loss of an entire batch of seeds in terms of value. Panicle budding has become a key obstacle to maintaining and improving the quality of hybrid rice seeds.

2. Causes

(1) Characteristics of the sterile line

Panicle budding and its extent are related to the sterile lines used in seed production. Under the same weather conditions at maturity and in the drying period, some sterile lines are not prone to panicle budding, such as Peiai 64S, while others, such as Jing 4155S, are very prone to it, budding when there is a slight rainfall or excessive humidity in the field or when the seeds are drenched during drying. Most of the sterile lines will develop panicle buds after two or three days of continuous rainfall at maturity or after wet seeds have been piled up for a long time after threshing and drenched. Panicle budding is related to the dormancy of the sterile line seeds, and dormant seeds are not prone to panicle budding. However, most of the sterile lines currently used in production do not have dormancy.

(2) Weather conditions during maturity and harvest

Most of the sterile line parents are susceptible to panicle budding when they encounter high temperature, rainfall or high humidity during the maturity and harvest periods. Generally, both high temperature and high humidity are required for panicle budding. Specifically, late May and early June is the period of maturity and harvest for two-line hybrid rice seed production in Hainan, but it is also the end of a dry season and the start of a wet season with high temperature and high humidity, coupled with passing shower or local artificial rainfall. Thus, panicle budding is more likely to occur in the field or during seed drying. However, the maturity and harvest of three-line hybrid seed production in Hainan are mostly from late April to early May, when the dry season is coming to an end with low rainfall, so panicle budding is not very likely, which is conducive to hybrid rice seed drying and harvesting. In the hilly and mountainous areas of the middle reaches of the Yangtze River for spring and summer seed production, such as those in Hunan and Fujian, the maturity and harvesting period is August and early to mid-September when there are high temperature and high humidity with a high probability of rainfall, so panicle budding is more likely to occur. In comparison, in autumn seed production in this region, the maturity and harvesting period is early October when the temperature is low and rainfall is less, so panicle budding is less likely to occur. In the late summer seed production area of Yancheng, Jiangsu, heading and pollination occur in mid-to-late August, and the maturity and harvesting period is from late September to early October when the temperature is low and the probability of rainfall is low, so there is little risk of panicle budding in seed production.

(3) Seed maturity

Seeds harvested from a female parent can germinate 10 days after flowering and pollination, and the flowering and pollination period of hybrid parents is also around 10 days, which means when pollination ends, the seeds that are fertilized early have the germination capacity already, but the late fertilized seed ovary just starts to swell, so there is a big difference in the time of pollination, fertilization and maturity of the seeds in the field and the early mature seeds are prone to panicle budding. Therefore, early harvesting is required in hybrid rice seed production, appropriately at 80% maturity.

The grain filling speed differs with sterile lines. Some fill fast, reaching 80% maturity about 15 days after fertilization, such as Guangzhan 63S and Longxiang 634A, while others fill slowly, reaching 80% maturity 25 days after fertilization, such as Fengyuan A. Therefore, tests are required to determine the best harvesting time for each sterile line to prevent panicle budding and maintain a high seed vitality.

(4) Gibberellin application

Gibberellin was originally found as a natural hormone from the metabolites of bakanae pathogen, and now is produced in factories. The most significant effects of gibberellin on the plants are activation of DNA, promotion of mRNA and protein synthesis, induction of a-amylase, protease and ribonuclease production, release and synthesis of these enzymes, enhancement of organic matter movement and metabolism, promotion of cell elongation, resulting in elongated internodes and elevated plants. In hybrid rice seed production, it is an essential agent to relieve the panicle enclosure of sterile lines and improve the outcrossing. But, gibberellin also has the effects of breaking seed dormancy and promoting seed germination, which is the main cause of panicle budding.

Ⅲ. Seed Powdering

Seed powdering means that the endosperm (rice grain) of hybrid rice seeds turns milky white and limey to varying degrees (Fig. 16 – 4), and can be classified as 1/3 powdering, 1/2 powdering, 2/3 powdering and complete powdering. There is no research report on the cause of hybrid rice seed powdering. It may be caused by starch hydrolysis due to high moisture content and high starch hydrolase activity during seed maturity and harvest, which is a seed deterioration phenomenon.

Preliminary studies show the following observations.

1) Powdering is common in hybrid rice seeds, but its extent and rate varywith sterile lines (Fig. 16 – 5).

2) Endosperm powdering has a great impact on the seed germination rate (Table 16 – 1).

According to Table 16 – 1, the highest germination rate was observed in non-powdering seeds, and the germination rate decreased as the degree of seed powdering increased. The germination rate of 1/3 powdering seeds of four female lines is 2.4 – 7 percentage points lower than that of non-powdering seeds,

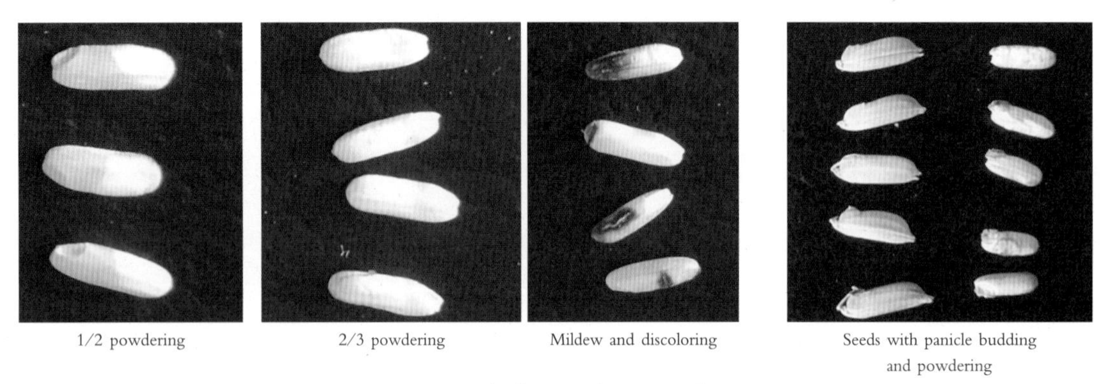

Fig. 16 – 4 Hybrid rice seed grain powdering

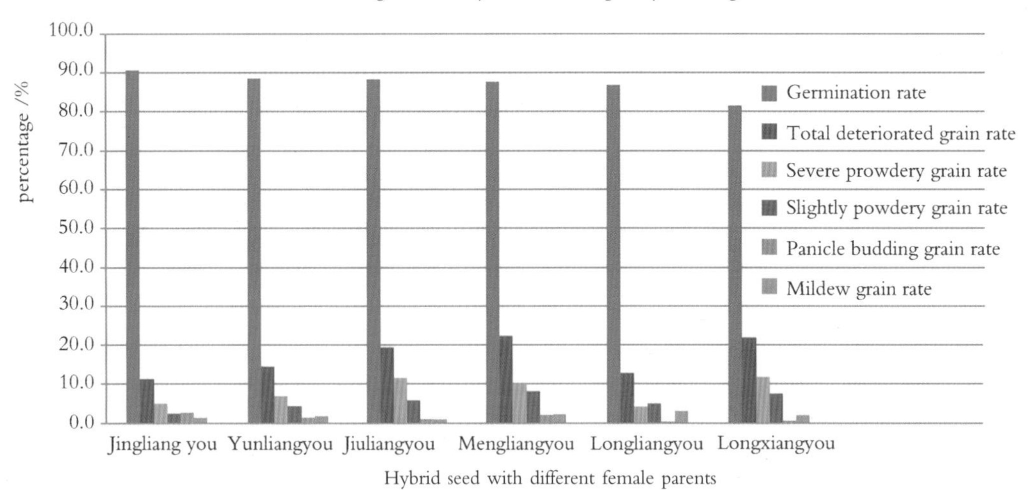

Fig. 16 – 5 Grain powdering and germination rates of hybrid rice seed from six different sterile lines

Table 16 – 1 Seed germination rate with different extents of seed powdering

Line	Non-powdering seed germination rate (%)	1/3 powdering seeds		1/2 powdering seeds		2/3 powdering seeds	
		Germination rate (%)	Reduction (percentage points)	Germination rate (%)	Reduction (percentage points)	Germination rate (%)	Reduction (percentage points)
Mengliangyou	92. 7	89. 4	3. 3	82. 8	9. 9	76. 4	16. 3
Jingliangyou	93. 1	88. 5	4. 7	84. 3	8. 9	67. 1	26. 1
Longxiangyou	87. 2	84. 8	2. 4	77. 8	9. 4	43. 3	43. 9
Longliangyou	92. 8	85. 7	7. 0	80. 0	12. 8	54. 2	38. 6
Average	91. 4	87. 1	4. 4	81. 2	10. 3	60. 2	31. 2

with an average of 4. 4 percentage points; that of 1/2 powdering seeds is 8. 9 – 14. 1 percentage points lower than that of non-powdering seeds, with an average of 10. 3 percentage points; and that of 2/3 powdering seeds is 16. 3 –43. 9 percentage points lower than that of non-powdering seeds, with an average of 31. 2 percentage points. This indicates that seed powdering is an important factor that affects the germination rate of hybrid rice seeds.

3) The seeds produced with the same sterile line as a parent differ in the rate of seed powdering between production locations or seasons (Fig. 16-6), indicating that differences in ecological and climatic conditions can result in differences in seed powdering and germination rates.

4) Panicle budding seeds are those with powdery endosperm.

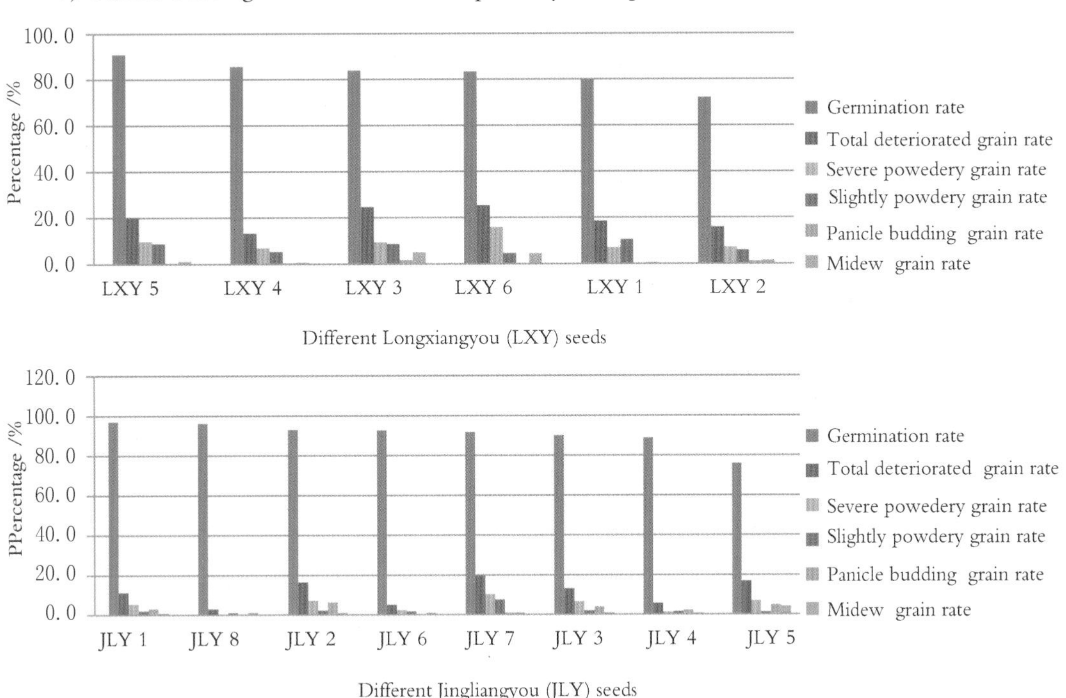

Fig. 16-6 Grain powdering and germination rates of Longxiangyou and Jingliangyou hybrid seeds from different places

Ⅳ. Pathogen-carrying Seeds

From flowering to seed maturity in seed production, pathogens invade the seeds through the floral organs, and reproduce and adhere inside and outside the glumes. Hybrid seeds are prone to infestation of pathogens because a long period of spikelet opening in sterile lines provides a high probability of invasion by micro organism spores (*Trichoconis padwickii*, *Alternaria* and *Fusarium*, etc.), while exserted stigma, elongated filaments and remaining anthers lay the ground for putrefactive and parasitic bacteria. These microorganisms have no impact on seed yield, but the toxins they secrete reduce seed germination and seedling success, and affect seedling quality. In particular, hybrid rice seeds are highly susceptible to kernel smut, precisely because of their scattered flowering, big glume opening angle, long glume opening period, poor glume closure and high exserted stigma. The occurrence rate of kernel smut is also related to the resistance of specific varieties, cultivation and climatic conditions. Peiai 64S and Y58S are more prone to the disease and can have a diseased grain rate of over 50%. The locations of seed production have a large base of pathogens and rapid changes in physiological races, increasing the probability of the disease. Large plant populations in seed production fields, poor ventilation and light penetration, improper use of GA3, insufficient panicle layer exposure, and high temperature and high humidity during heading and flowering all contribute to the prevalence of the disease. The bakanae pathogens carried by hybrid rice seeds are also hazardous in that they leave the seedlings prone to bakanae disease in early spring when temperature is low. Panicle enclosure of the sterile lines contributes to the transmission of the pathogens from the leaf sheaths to the spikelets, thus increasing the quantity of seeds with the pathogens.

In summary, hybrid rice seeds are sensitive to heat and humidity during storage due to glume split, panicle budding, powdering and pathogens. If the temperature and humidity are not properly controlled during storage, it will be easy for the seeds to lose vitality. Hybrid rice seeds absorb water faster than conventional rice seeds, and can reach saturation in a short period of time. Therefore, if seed soaking lasts too long, substances will leak from the cell, and harmful substances will easily penetrate into the cells to cause seed immersion injury. Drying the seeds before soaking has the functions of breaking dormancy and sterilization, but it can aggravate the damage to the cell membrane system, resulting in the negative effect of reduced germination rates. In pre-germination, due to respiration and microbial activity, the seeds consume a lot of oxygen, leading to hypoxia and failure to bud and germinate. In short, in seed soaking and pre-germination, less soaking and appropriate heat and moisture should be ensured.

Section 2 Purity Maintenance in Hybrid Rice Seed Production

Hybrid rice seed purity is the primary criterion in seed quality evaluation. In addition to the genetic factors of the parents, the main factors affecting seed purity include biological and mechanical mixture throughout the seed production process. Through a long period of research and practice, China has formed a comprehensive and systematic technical system to control the risk of hybrid rice seed purity, which can ensure that the seed purity meets the standard specified in GB4404. 1—2008.

Ⅰ. Parent Seeds with High Seed Purity and Genetic Purity

The purity of parental seeds is the basis for producing high-purity hybrid rice seeds. The purity of parental seeds involves both seed purity and genetic purity. Cases of substandard seed purity or seed production failure due to substandard genetic purity of the parental seeds occur often in hybrid rice seed production.

The national seed quality standard GB4404. 1—2008 in China requires a 99. 5% or higher purity for sterile lines and restorer lines, and a 99. 9% or higher purity for foundation seeds in hybrid rice seed production. However, as large-scale and mechanized seed production spreads and the cost of roguing off-types rises, the requirements for seed purity have gone up to 99. 8% for sterile line and restorer line parents and 99. 95% for foundation seeds.

Although three-line CMS lines have stable sterility, there still are variations in sterility after multiple generations of reproduction. The sterility of two-line PTGMS lines is affected by temperature and daylength and there is a genetic drift in the critical sterility temperature after multiple generations of propagation. The sterile lines Y58S, Zhu 1S, Lu 18S, P88S, C815S, Xiangling 628S, Guangzhan 63S, T98A, Ⅱ-32A and Tianfeng A have the characteristics of good outcrossing performance and high seed setting rate, but they are prone to biological mixture. In particular, the genetic backgrounds of super hybrid rice parents are complicated and diverse, including *indica*, *japonica* and even *javanica* elements, so the progenies are prone to genetic variation. Therefore, attention should be paid to the genetic purity of the parents during super hybrid rice seed production.

Ⅱ. Strict Isolation in Seed Production

Rice pollen can survive for 5 - 10 minutes after leaving the mother plant under natural conditions and can travel a distance of more than 100 m with the wind. Therefore, a hybrid rice seed production base should be strictly isolated to prevent cross-pollination from pollen not intended for seed production.

1. Isolation methods

The following measures can be adopted according to the specific conditions of the seed production areas and fields.

(1) Natural barrier isolation

Mountains, buildings, rivers (50 - 100 m wide), and other crops can be used for isolation in seed production.

(2) Distance isolation

A 50 - 100 m isolation belt can be set around the seed production area. The exact distance is subject to the natural wind direction during the flowering and pollination period. Specifically, it should be more than 100 m downwind and more than 50 m upwind. Other crops or restorer lines can be planted as an isolation belt.

(3) Flowering isolation

When planting a non-restorer variety in the isolation area, it is required that the initial heading stage of the variety is more than 20 days apart from the initial heading of the female parent in the seed production area. The rice varieties planted in the isolation area should be those planted locally before with known maturity. Corresponding planting and cultivation management plans should be formulated and implemented.

2. Inspection

Two rounds of isolation inspection should be carried out to ensure complete isolation for the seed production area.

The first is at the time of transplanting the female parent to check if the varieties, sowing date and transplanting are consistent with the original isolation plan.

The second is at the time when the female parent is in the fourth to sixth stages of young panicle differentiation, which is carried out by stripping the young panicles of the variety used for isolation to predicate if its growth is in sync with that of the female parent.

Any issue identified must be properly handled immediately.

Ⅲ. Roguing Off-types

1. Paddy-dropped rice growth

In rice fields or seed production fields, there are always some seeds dropped in the field during harvesting, especially in the case of mechanized harvesting. Some of these seeds can germinate and grow into seedlings in the next or future planting seasons, usually referred to "paddy-dropped rice". This has become an increasingly serious problem in mechanized harvesting and farming, bringing a huge problem to hybrid rice seed production. Therefore, plants from paddy-dropped rice should be removed in time in hybrid rice seed production.

At present, there is no effective method to deal with paddy-dropped rice plants. Two practices are mainly adopted. One is to plow the field deep to bury the dropped rice grains into the soil. The other is to perform irrigation, drain-off, and rotary tillage repeatedly when preparing the seed production field, letting the dropped seeds germinate and forming seedlings first, then mechanically remove them.

2. Timely roguing

Roguing off-types is a practice that must be repeated through the entire process of seed production, with special focus on four periods.

(1) During the 3.5 – 4.5 leaf stage of seedlings, mainly remove plants with different color and shape.

(2) During tillering, remove plants with different color and shape, paddy-dropped rice and regenerated seedlings.

(3) Heading and flowering period from panicle rupturing to full heading is the critical time for off-type removal. Remove the plants with different shape and color (e.g. colors of leaf sheath, apiculus, stigma and leaf), individual plants more than three days from normal plants during the heading stage, and fertile and semi-sterile plants of in the female population during the flowering stage. The rate of off-types should be controlled within 0.2% or that of abnormal panicles within 0.01%.

(4) During the seed maturity stage, one to five days before harvesting, remove plants that set normal seeds, or have different shape and color (different apiculus color in the female parent population). If mechanized harvesting is used, the male parent plants should be removed first, including the male parent panicles dropped off in the field.

3. Mechanical mixture prevention

Mechanical mixing is a purity problem that occurs more often during the selection and processing of rice seed production, mainly in sowing, transplanting, harvesting, drying, selection and processing. All equipment and packages involved, such as sacks, fiber bags, rice threshers, harvesters, drying mats, dryers, sorting machines and warehouses, should be cleaned in advance so that no rice grains or other impurities are left.

Ⅳ. Monitoring the Purity of Two-line Hybrid Rice Seed Production

The technical operations for purity control are the same for both two-line and three-line hybrid rice seeds in terms of parental purity, isolation and roguing of off-types. However, since the expression of female sterility in two-line hybrid rice is influenced by temperature and daylength, the purity risk is greater for two-line hybrid rice seeds than for three-line hybrid rice seeds, so special technical measures are required for the former.

1. Determination of female parent fertility

Although an appropriate seed production base is selected and a temperature sensitive period for sterility safety is arranged in accordance with the technical requirements of safe two-line hybrid rice seed production, it is still necessary to observe and evaluate the fertility changes of the parents in time during the actual operation of seed production due to following reasons. First, on a seed production site, the temperature in the same time period of the year varies greatly, and abnormal low temperature may still occur during the fertility sensitive period of the parent, resulting in different degrees of fertility fluctuations in the sterile line. Second, in large seed production areas, especially in hilly and mountainous areas, microclimate differs among fields in different areas, especially in fields with cold water and shady places, the temperature is even lower and the female is more prone to sterility fluctuations at low temperature. Third, there are plants with a high critical sterility-sensitive temperature in the female population and these plants are prone to sterility fluctuations and self-pollination when exposed to cold weather. Therefore, in two-line hybrid rice seed production, it is necessary to observe the sterility of the female parent so as to promptly sort out the seeds grown in fields with sterility fluctuations or fertility transition and store them as suspicious seeds.

The main methods for confirming the safety of female parent sterility are temperature analysis, anther pollen observation, and planting in isolation for selfing.

(1) Temperature analysis

During the fertility sensitive period, set up temperature observation stations, record the temperature (daily minimum, daily maximum and daily average temperatures) and other weather parameters at the production base daily. If a low temperature occurs, analyze the data in combination with temperature data from the local weather station, compare the data with the fertility threshold temperature of the female, determine if the low temperature has an impact on the fertility fluctuations of the female, and speculate the date of fertility fluctuations of the female according to the time of low temperature.

For different areas and types of seed production fields, temperature observation should be done at the young panicle position of the female plant or with irrigation water. If the air or water temperature is found lower than the temperature of the sterility threshold of the female parent, pollen must be inspected with microscope at a fixed point during the flowering period of the female parent and samples taken for isolated cultivation.

(2) Observation of anther and pollen

Throughout the flowering period, observation should be done for the shape, color and anther dehiscence of the female parent in different areas and different types of fields. Pollen microscope examination should be carried out for the fields on a daily basis.

(3) Selfing examination in isolated cultivation

About 2 – 3 days before panicle emergence of the female parent, take 5 – 10 sample plants from three or five sites in different areas and types of seed production fields and planted them in a safe and isolated area with mud, with labels showing the sampling site. GA3 should be applied to the isolated plants as do in their original fields and the environmental conditions should be the same as those in the production field. The plants should be protected from livestock or poultry damage. Check if there is any selfed seeding in the isolated plants 15 – 20 days after flowering.

If the sterility of the female parent during the fertility sensitive period is projected to fluctuate through temperature observation and analysis, the number of sample plants should be increased by 3 – 5 times for each site. This is particularly true for fields where a small number of female plants are found to have a high critical sterility temperature. If necessary, more than 100 plants can be selected at each site to ensure the typicality and accuracy of the test results.

In short, it is necessary to fully understand the female parent sterility in seed production. In determining the safety of female sterility, it is necessary to use the above three methods together as cross-reference to obtain a more accurate result.

2. Water temperature control

Shaded, cold-water soaked or irrigated fields cannot be used for producing two-line hybrid rice seeds. All of the fields in seed production bases must be protected from low temperature and cold-water irrigation during the female parent sterility sensitive period. When the female parent is at the fourth to sixth stage of young panicle differentiation, measure the temperature of the irrigation water to ensure it is above 25 ℃.

Due to the slow heat conduct rate of water, when a cold airflow comes, the air temperature decreases quicker than the water temperature. Water can remain "warm" for some time. If low air temperature (less than 0.5 ℃ from the critical fertility temperature of the PTGMS line) is projected for a short spell during the sterility sensitive period, an option is to irrigate the field deep (submerge the young panicles of the female plants) with water 0.5 ℃ higher than the critical fertility temperature of the female parent before the onset of the low temperature, so as to keep the temperature above the critical fertility temperature at the part of the young panicles that is temperature sensitive. The water can be drained after the low temperature spell passes. Practice has shown that this method can reduce the degree of fertility fluctuations during the fertility sensitive period when it encounters low temperature, and can even avoid fertility fluctuations and ensure seed purity. To ensure effective use of this method, consideration should be given to having sufficient water to irrigate the seed production field during the fertility-sensitive safe period of the parent when selecting a two-line hybrid rice seed production base.

3. Seed purity evaluation

Seed purity can be evaluated from the results of selfed seed setting of the female parent planted in isolation, and the seed setting and the impurity rate of the female parent in the seed production field with the following formula.

$$X(\%) = 100 - \left(a + \frac{n}{m}\right) \times 100$$

where X is seed purity (%), a is the off-type rate (%) of non-sterility fluctuation of the line (including the off-types from male and female parents in seed production, cross-pollinated plants in the female parent and those caused from loose isolation and mechanical mixture), which can be calculated from the results of field inspection, n is the selfed seeding rate (%) of isolated female plants, and m is the seed setting rate of the female parent (%).

Seeds with a purity below 98% should be subject to evaluation and only when they are evaluated as qualified, can the seeds be further processed, packaged and offered for sale.

Section 3 Hybrid Rice Seed Vitality Maintenance

Seed vitality is a comprehensive indicator of seed germination and seedling success under field conditions, mainly reflected in seed germination rate, germination potential, seedling root length, fresh weight and germination capacity under stress. It is an important indicator of hybrid rice seed quality and sowing quality, which is related to the seedling emergence rate and seedling success rate after sowing. As hybrid rice seed production is a process of outcrossed seeding, with scattered flowering, long time of glume opening, glume splitting, panicle budding and seed powdering, the vitality of hybrid rice seeds is generally lower than that of self-pollinated conventional rice seeds. In GB4404. 1—2008, the germination rate of hybrid rice seeds is set at over 80%, while that of conventional rice seeds is over 85%. The factors affecting the vitality of hybrid rice seeds are related not only to the genetic traits of the parents, but also to the climate, nutrition, harvesting period, drying method and storage conditions during seed production. Therefore, there are factors that can help maintain seed vitality in all aspects.

Ⅰ. Control of Panicle Budding

1. Regulating field temperature and humidity during seed maturity

The vast majority of sterile lines used in hybrid rice production does not have dormancy and are prone to panicle budding after GA3 application. Seeds on female parent panicles mature gradually after pollination. When the field temperature and humidity meet the requirements of seed germination, the seeds fertilized early may sprout and germinate on the panicles. Therefore, controlling field temperature and humidity during seed maturity is the key to controlling panicle budding and the following are the main measures.

1) Select a seed production base and season with moderate temperature, no rain, and low humidity during the seed maturity and harvesting period and a low probability of rainfall.

2) Cultivate a robust seedling panicle structure of the female parent to prevent seedlings from growing too vigorously, forming too large a population and too long and too large canopy leaves, which will result in shade for the panicles, poor ventilation and light penetration, and high humidity at the panicle layer.

3) Cut or trample the male parent immediately after pollination to reduce the shade caused by the male parent to the female and reduce humidity in the field.

4) Keep the soil moist during grain filling, drain and dry the field during the maturity period to re-

duce humidity.

5) Spray agent 3 – 4 days before harvesting, dry the canopy leaves and stems so that they are quickly dehydrated and the seeds have low moisture to prevent panicle budding.

2. Construction of optimal outcrossing status of parents and control of GA3 dosage

Applying GA3 to build a good outcrossing posture is key to ensuring high yield of hybrid rice seed production, but research shows that applying GA3 tends to aggravate panicle budding. Therefore, in order to reduce panicle budding, GA3 should be applied according to the sensitivity of the sterile lines, the parental panicle budding characteristics, with appropriate amount, time and method. If the seed maturity period meets high temperature and high humidity suitable for seed germination, for the effect of GA3 spraying, the female panicle layer should be fully exposed to form a panicle-type structure in order to reduce the humidity of the panicle layer. Drone-spraying of GA3 is a method to apply an ultra-low volume which enables the GA3 solution to be evenly distributed on the plant foliage and absorbed sufficiently by the plant at a lowered dosage. If common spraying equipment is used for GA3 application, the amount increases because the particles of the solution are large and not uniform. On the basis of uniform spraying, the addition of micro-elements or synergists will help improve the effect of GA3 and reduce the dosage.

3. Application of panicle budding inhibitor

In order to control panicle budding in hybrid rice seed production, trials of panicle budding inhibitors have been carried out in China and some progress has been made. In 2001, Zhou Xinguo applied panicle budding inhibitors in different types of fields within three days after final flowering in Jinyou 207 spring seed production, and the results showed that the panicle budding rate was 0.92%–1.52% for the treated plants and 7.88%–11.05% for the control, indicating a significant reduction. The germination potential and germination rate of inhibitor-treated seeds were no different from normal seeds tested in 2002, indicating no adverse effect of the panicle budding inhibitor for germination potential and germination rate. The experiment was repeated in 2002, when it rained continuously with high temperature and high humidity, and the results were consistent across different plots, with an average panicle budding rate of 1.23% for those treated with panicle budding inhibitor, down by 7.94 percentage points compared with the 9.17% of the control.

As research on the physiological mechanism of panicle budding deepens, the technique of using exogenous abscisic acid (ABA) to control panicle budding has increasingly been recognized andvalued. It has been shown that the use of ABA at the seed filling and maturity stage of hybrid rice seed production can bring about a significant reduction in panicle budding at seed maturity. Paclobutrazol, a triazoles compound, is an efficient and low-toxic plant growth retardant and broad-spectrum fungicide effective in cultivating strong seedlings, regulating the flowering period, inhibiting panicle budding and increasing seed yield. In the early yellow ripening stage, 1.5 – 2.2 kg/ha of 15% paclobutrazol with 1,500 kg of water can be sprayed to effectively prevent seed panicle budding. With the increase of dosage, the panicle budding rate decreases. The effect of paclobutrazol at 750 g/ha is the most significant.

Ⅱ. Optimal Seed Harvest Period

The outcrossing rate of the female parent in hybrid rice seed production is about 40% , which is about half that of ordinary cultivated rice. After pollination, the plant has sufficient nutrient supply during the grain filling stage and the seed fills and matures quickly. Cao Wenliang et al. studied the seed germination of Zhu 1S and Lu 18S series hybrid seeds harvested on different days after pollination and found that the seeds harvested 13 - 18 days after pollination were completely mature with fully filled grains and normal germination potential and germination rate. In seeds harvested from the 19th day on after pollination, the starch in the endosperm of some seeds was gradually hydrolyzed with reduced endosperm transparency, more seed powdering, and reduced seed vitality (Table 16 - 2), indicating there is an optimal harvesting period in hybrid rice seed production. Table 16 - 2 shows that the optimal harvesting period for Zhu 1S and Lu 18S series of hybrid seeds are 13 - 15 days after pollination.

Table 16 - 2 Seed characteristics of Luliangyou 996 at different harvest dates after pollination (2009, Suining, Hunan)

Days after pollination	Germination rate (%)	Germinating vigor (%)	Germinating index	1,000 - grain weight (g)	Glume splitting rate (%)	Starch content (mg/g)	Protein content (mg/g)	Cyan grains (%)	Transparent grains (%)	Yellow grains (%)
8	32.50g	18.50e	10.63g	25.73	37.33	55.38	9.78	93	7	0
9	49.00f	35.50d	17.65f	26.53	41.33	56.06	7.51	89	11	0
10	71.50e	41.50d	22.77e	27.13	43.00	61.03	8.76	86	14	0
11	73.00e	60.00c	29.23d	27.70	54.67	63.26	8.03	53	45	2
12	79.50cde	69.50bc	32.98cd	27.76	59.67	65.26	7.47	41	47	12
13	91.50a	83.50a	39.05ab	28.74	60.67	61.94	8.32	39	46	15
14	88.50bcd	80.50ab	38.35ab	28.73	64.67	83.45	8.71	12	71	17
15	90.00abc	81.50ab	41.41a	30.81	66.67	91.19	8.03	8	82	10
16	89.00bcd	81.00ab	41.15a	28.65	67.33	56.70	8.68	9	81	10
17	90.50ab	84.00a	42.85a	28.82	67.67	67.45	8.56	3	78	19
18	91.50a	82.00ab	42.48a	28.41	70.00	67.27	8.32	2	86	12
20	78.50cde	71.00abc	36.11b	28.39	70.33	57.02	7.95	0	79	21
22	78.00de	68.50bc	32.62cd	28.44	74.67	48.18	9.20	0	70	30

Practice shows that the speed of grain filling and maturing of different sterile lines vary greatly when they are used as the female parent in different seed production bases and in different seasons, which means that different sterile lines have different optimal harvest dates in different production bases and different seasons. Therefore, a test of seed vitality on different harvest dates 10 - 30 days after pollination should be carried out to find out the best harvest date for maintaining seed vitality for each sterile line produced in different seed production bases and seasons.

Ⅲ. Safe and Fast Drying

Fresh-harvested hybrid rice seeds have a high content of moisture at about 30% and a mixture of impurities such as straw, grass clippings, empty spikelets and diseased grains with a volume rate of impurities about 0.2. If the seeds are not quickly dried, it is easy to cause seed deterioration such as powdering, mildew and heat, resulting in vitality loss. Therefore, once the hybrid rice seeds are harvested, they should be immediately transported to a drying yard, quickly spreading out to dry naturally under the sun or by a dryer, lowering the moisture content to 11%–12%.

1. Natural drying

Natural drying is a traditional method of drying seeds and it has been adopted in hybrid rice seed production since the 1970s. Seeds are spread out on cement ground or on a mat on a hot and sunny day. The seeds can be completely dried in 2 – 3 days, while the seeds produced in southern Hainan can be dried to about 12% moisture within a day. Natural drying is quite risky because of its dependence on the climate, and the viability of the seeds is not guaranteed in the case of rainfall, especially in the rainy season. Decades of experience shows that natural dying can produce qualified seeds with a germination rate of 92% in the best year, but the figure can be down to only 70% in some production bases if the weather is unfavorable. Natural drying is no longer suitable for large-scale seed production today. Instead, mechanical drying can be adopted.

2. Mechanical drying

Fig. 16 – 7 Cross-flow circulation vertical dryer (with fuel oil hot airstove)

Mechanical drying is a method first adopted in 2010 and widely promoted since 2015. There is currently no specialized equipment for mechanical seed drying, instead, low-temperature grain dryers and improved tobacco flue-curing houses are used.

At present, there are three types of low-temperature grain dryers, i. e., cross-flow circulating vertical dryer, mixed-flow circulating vertical dryer and static horizontal dryer (Fig. 16 – 7, Fig. 16 – 8 and Fig. 16 – 9). Liu Aimin et al. studied the characteristics of hybrid rice seeds dried by the three kinds of dryers and found advantages and disadvantages for each of them. The vitality of hybrid rice seeds dried by the three kinds of dryers did not differ from those naturally dried when the dryers were set at a constant temperature of 40 ℃–45 ℃. The germination rate could reach more than 85%, but the drying and dehydration rates differed greatly (Table 16 – 4).

Fig. 16 – 8 Static horizontal dryer (with fuel oil hot air stove)

Fig. 16 – 9 Mixed flow circulation vertical dryer (with hot air stove)

Table 16 – 3 Seed vitality after drying in the three kinds of dryers

Dryer	Combination	Treatment	Germination rate (%)	Germinating vigor (%)	Germinating index	Vigor index
Static horizontal dryer	Jingliangyou 534	Mechanical drying	87b	86a	49. 7a	20. 6a
		sun-drying	94a	88a	46. 6ab	21. 8a
	Longxiangyou-Huazhan	Mechanical drying	89ab	80ab	41. 2b	9. 3b
		sun-drying	91ab	80ab	42. 1b	11. 1b
	Mengliangyou-Huanglizhan	Mechanical drying	83b	80ab	36. 0b	7. 9bc
		sun-drying	87b	69b	30. 0c	5. 4c
Cross-flow circulation vertical dryer	Guangliangyou 1128	Mechanical drying	88ab	59c	31. 1c	6. 5c
		sun-drying	87ab	57c	29. 2c	6. 6c
	Jingliangyou 534	Mechanical drying	92a	91a	46. 1a	14. 7b
		sun-drying	94a	88a	46. 6a	21. 8a
	Mengliangyou-Huanglizhan	Mechanical drying	84b	71b	37. 4b	8. 3c
		sun-drying	87ab	68b	29. 7c	5. 4c
Mixed-flow circulation vertical dryer	Guangliangyou 1128	Mechanical drying	88a	59b	29. 2b	6. 0b
		sun-drying	88a	57b	29. 1b	6. 6b
	Keliangyou 889	Mechanical drying	86a	72a	36. 0a	7. 0b
		sun-drying	85a	76a	39. 3a	10. 3a

Table 16 – 4 Dehydration rate with the three kinds of dryers

Dryer	Combination	Quantity of dried seeds (kg)	Moisture content before drying (%)	Moisture content after drying (%)	Time (h)	Dehydrationrate (%/h)
Static horizontal dryer	Jingliangyou 534	3,500	27.0	11.0	24	0.67
	Longxiangyou-Huazhan	3,100	32.6	11.2	28	0.76
	Mengliangyou-Huanglizhan	2,800	24.6	11.3	15	0.90
Cross-flow circulation vertical dryer	Jingliangyou 534	6,600	28.1	12.4	68	0.23
	Guangliangyou 1128	6,000	30.0	12.5	57	0.30
	Mengliangyou-Huanglizhan	4,500	33.0	12.4	51	0.40
Mixed-flow circulation vertical dryer	Guangliangyou 1128	8,500	30.5	12.4	34	0.51
	Keliangyou 889	6,000	30.0	12.2	29	0.61

Table 16 – 4 shows that the dehydration rate produced by the static horizontal dryer is the highest at 0.67%–0.9% per hour, while that with the cross-flow circulation vertical dryer is the lowest at 0.23%–0.4% per hour, and that with the mixed-flow circulation vertical dryer is in between at 0.51%–0.61% per hour. Both cross-flow and mixed-flow circulation vertical dryers have the problems of more dust and breakage, while the static horizontal dryer has the problems of more work and inconvenient discharging. Although these three types of low-temperature grain dryers can all be used for drying hybrid rice seeds, they cannot achieve the purpose of safe, fast, green and efficient drying. The technique principle and method of improved tobacco flue-curing houses are the same with the static horizontal dryer and it has been promoted for application in seed production areas.

Based on years of practice, the following are the technical key points for mechanical drying of hybrid rice seeds.

1) Pre-cleaning of seeds. Because the fresh-harvested hybrid rice seeds by a harvester contain a lot of impurities that may cause the problems of blockage and slow initial circulation speed for the circulating dryer after the seeds enter the dryer bin and increase the cost because impurities are also dried. So a dryer plant should be equipped with pre-cleaning machines to get rid of most of the impurities before starting seed drying.

2) Quick harvest, transport and input. After harvesting, the seeds should be transported to the drying site as fast as possible. Pre-screening should start immediately to feed the seed into the dryer at the earliest time possible. Circulation should go on while feeding and then switch to the drying mode once the feeding is completed. The time from the end of threshing to the start of pre-cleaning should be no more than three hour in a high-temperature season, which is the key to seed vitality in the case of mechanical seed drying.

3) Control of the drying temperature. The temperature of the air in a static horizontal dryer should

be 38 ℃-40 ℃. Temperature-changing drying can be adopted in a cross-flow or mixed-flow circulation vertical dryer. The drying temperatures should be 50 ℃, 45 ℃ and 50 ℃ respectively when the seed moisture is >20%, 15%-20%, and <15%.

4) Timely inspection of the changes of seed temperature and moisture content. During the drying process, check the temperature and moisture content of the seedpile in different parts of the drying chamber once every 2-3 hours to ensure that the temperature remains at about 30 ℃ when the moisture content is over 20%, and at 35 ℃-40 ℃ when the moisture content decreases.

5) Two-stage drying. If seeds need to be harvested fast in the field in case of imminent rain but there is a shortage of equipment, two-stage drying can be adopted. In the first stage, the harvested seeds are dried to 16 ± 0.5% of moisture content and then discharged from the dryer and stored temporarily for about five days, during which time the seeds still in the field can be harvested in time and dried. Then, the seeds stored temporarily after the first drying can be divided into groups for further drying till the moisture content reaches 11%-12%.

Ⅳ. Seed Selection and Processing

The dried hybrid rice seeds still contain empty spikelets, stalks, grass clippings, and diseased, glume-split, panicle budded, mildew, discolored, powdery and dehulled seeds, as well as weeds, grass seeds and sand. Therefore, the seeds must be carefully selected and processed with a variety of equipment and procedures. The equipment or machinery used for hybrid rice seed selection and processing mainly includes the wind sifter, gravity selector, nest-eye sorter, optical sifter, etc.

1. Wind sifter

A wind sifter is a cleaning machine that integrates wind power and screens of various impurities. It can remove empty spikelets, straws, grass clippings, weed seeds, sand, gravel, severely glume-split grains, and dehulled, panicle budded and diseased grains. The best cleaning effect can be obtained by adjusting the air volume and replacing screens when appropriate.

2. Gravity selector

There are two types of gravity selectors, i. e., wind negative pressure gravity selector and vibrating screen plate gravity selector. A gravity selector can sort out moderately glume-split grains, grains with kernel smut, and powdery, panicle budded and dehulled grains from the seeds. The best selection effect is obtained by adjusting the wind pressure or vibration frequency and the inclination angle of the screen plate.

3. Nest-eye sorter

It is a sorting machine used for seeds containing grains or seeds of other varieties with different sizes and lengths. By changing the size of the nests, the seeds of different grain types can be sorted out.

4. Optical sorter

Seeds with panicle budding, powdering mildew and discolor are not significantly different from normal seeds and it is difficult to sort them out by wind sifter, gravity selector or nest-eye sorter. The optical sorter uses special light-wave light source that can penetrate the glumes and rice grains to sort out deteriorated seeds. It can sort out most panicle budded, powdery, moldy and discolored grains from the seeds, thus greatly improving the seed germination rate from about 70% to more than 80% generally, and can also increase the germination rate from about 60% to more than 80% through multiple sorting.

Hybrid rice seeds from drying to finished commercial products need to go through a variety of selection and processing procedures. Seed companies assemble these selection processing equipment and coating, sub-packaging and sample mixing devices into a seed production and processing line. Seeds going through a complete set of processing procedures along the line not only have a cleanliness of almost 100% with uniform moisture content, but also contain very few seeds with no or low vitality, improving seed germination rate, seed vitality and the sowing quality, reducing the quantity of seeds required in the field, improving the quality of seedlings, giving full play to the high-yielding heterosis of hybrid rice, and meeting the demand for precise and accurate production of rice mechanization.

With the advancement of science and technology in China, the sorting and processing technology of hybrid rice seeds will be further improved. In recent years, color sorting and optical sorting for hybrid rice seeds, as well as electro magnetic treatment have been experimented and applied by some seed companies. The development and application of these technologies will effectively solve the difficulties in hybrid rice seed production and storage, which not only can fully ensure and increase the earnings of seed farmers and seed companies, but also will further improve the quality of hybrid rice production.

Section 4 Rice Kernel Smut Prevention and Control

Fig. 16 – 10 Rice kernel smut

Rice kernel smut is a common and specific disease in hybrid rice seed production (Fig. 16 – 10), but rare in rice production. Both rice kernel smut and false smut are fungal diseases. The spores of the disease enter the ovary through the rice stigma or residual anthers, and multiply by absorbing nutrients in the ovary, eventually filling the entire ovary. Then the ovary wall splits under suitable conditions. The spores are dispersed as black powder or green powder, which often causes severe loss in seed production and seed quality and seriously affects the appearance of seeds and the seed yield, causing difficulties in seed selection and processing. Therefore, the control of rice kernel smut and false smut has become an essential technical aspect of hybrid rice seed production. Combination of agronomic measures and chemical control are adopted for this purpose.

I. Agronomic Measures

1. Selection of seed production base and season

A moderate temperature (28 ℃−30 ℃) and high humidity are favorable conditions to rice kernel smut infestation and incidence. The practice of hybrid rice seed production shows that kernel smut is more severe in autumn than in spring and summer seed production, in mountainous areas than in plain areas, and in mountain shaded fields than in sunny fields; but the incidence varies greatly between years, bases and fields. If the pollination and ripening stage meets with sunny weather, kernel smut would be minor or would not occur at all. On the contrary, if pollination and ripening happen in a period with a lot of rain and little sunshine (cloudy and wet weather), the disease would be more serious, and agri-chemical control would not be very effective. The grain-filling and maturity stage of seed production in autumn in the Yangtze River basin is from late August to mid-September when the weather is cold and rainy, prone to kernel smut. The incidence of kernel smut is lower in spring and summer seed production. Therefore, in addition to sufficient sunlight for the seed production bases and fields, the season should be selected according to the duration of the growth period so that the flowering to maturity period can be away from smut-prone climatic conditions.

2. Cultivation of a robust female parent population

It is found that where kernel smut is more severe, there is usually excessive use of nitrogen fertilizer, resulting in excessive vegetative growth, long leaves, dark leaf color, over-populated seedlings, and poor ventilation and light penetration in the female population in hybrid rice seed production. The female parent population with a moderate seedling structure, robust plant growth, moderate or short leaf length and light leaf color, is less prone to the disease. From this, it can be seen that targeted cultivation of the female parent, which is characterized by basic fertilizer as the mainstay, early topdressing, well-controlled nitrogen fertilizer application in the middle and late stages, increased application of phosphorus and potassium, and other medium and trace element fertilizers, helps cultivate a robust female parent population, which is effective in preventing kernel smut.

The invasion pathway of kernel smut fungus is the stigma and residual anthers after flowering of the sterile lines. When the parents synchronize well at flowering, the pollen density at the female panicle layer is high during pollination with sufficient pollen from the male parent and a high probability of the female's stigma receiving the male's pollen in time, resulting in a high outcrossing rate, which can inhibit the invasion of kernel smut fungus spores. Therefore, reasonable arrangement of parental planting row ratio, cultivation of strong male parent, ensuring parental flowering synchronization or regulating parental flowering synchronization well, prompting the female stigma to receive sufficient amount of the male's pollen in time, and improving the outcrossing rate and seed setting rate, are also effective measures to control kernel smut.

3. Application of gibberellin in proper time and amount

Spraying GA3 is to grow the panicle layer higher than the leaf layer. A panicle layer featuring a loose structure, good ventilation and light penetration with a low humidity is not conductive to the development of rice kernel smut. On the contrary, If GA3 spraying is too late or insufficient, the panicle layer has a low degree of spikelet exposure, poorly spread flag leaves, and a leaf type or spikelet-leaf type structure, kernel smut is more likely to occur. Therefore, GA3 should be sprayed in time. Too early spraying makes the plant too tall and prone to lodging, prone to severe kernel smut. However, spraying it too late is also conducive to the occurrence of kernel smut because the spikelets would be unable to fully expose due to aged upper internodes. In addition, for fields with long parent canopy leaves, the spike layer is still not above the leaf layer after GA3 spraying, leaf clipping should be carried out to raise the spike layer above the leaf layer, which is conducive to pollination, can improve the female outcrossing rate, and reduces kernel smut incidence.

Ⅱ. Chemical Prevention and Control

1. Seed disinfection

Rice kernel smut can spread through seeds, so disinfection of parental seeds is one of the effectiveways to control the disease. Commonly used chemicals for seed disinfection are 500 – 1,000 times of 20% triadimefon EC, 500 times of 50% carbendazim wettable powder and 500 times of 20% trichloroisocyanuric acid wettable powder. Soak the seeds in clean water for 6 – 10 hours, wash and drain the seeds, then soak the seeds in the agent for 8 – 12 hours. After disinfection, the seeds should be washed with clean water several times to clean the agent.

2. Chemical prevention and control

Commonly used agents for the control of rice kernel smut include triadimefon, Mieheiling, Miehei No. 1, Keheijing, Miebingwei and Aimiao. Recent studies have shown that 240 – 360 mL of 25% Kairun (pyraclostrobine) per hectare, or 450 mL of 40% Xiangle (coumoxystrobin+tebuconazole) per hectare, or 450 mL of 30% difenoconazole per hectare, can effectively control smut. Since the smut pathogens invade through the stigma of the spikelets during the flowering period, Therefore, the control of rice kernel smut should be carried out before and during flowering and 1 – 2 days after pollination. Spray the highly effective agents for 2 – 3 times. In case of rainy days, add two more rounds of spraying. The application time should be 16:00—18:00.

Section 5 Hybrid Rice Seed Storage and Treatment

Ⅰ. Hybrid Rice Seed Storage

Seed storage refers to the preservation of seeds after they are processed and packaged and before use. During this process, seed vitality must be maintained to make them good to use when they are to be used.

Hybrid rice seeds are less storable than conventional rice seeds, posing higher requirements for storage conditions. Suitable environmental conditions such as temperature and humidity must be ensured to maintain seed vitality. There are two ways to store hybrid rice seeds, i. e., room-temperature storage and low-temperature and low-humidity storage. Due to changes in the seed market and uncontrollable factors in seed production, it has been common for seeds to be stored for 3 – 5 years in low temperature and low humidity in the seed industry. It is also a strategic need of meeting market demand in the seed industry and plays a very important role in regulating demand and supply in the market.

1. Conditions and treatments of warehouses

(1) Storage conditions

Warehouses for storing hybrid rice seeds must be airtight, moisture-proof, leak-proof, rodent-proof and insect-proof. Cold storage warehouses must be in constant low temperature and low humidity. There are two types of medium- and short-term cold storage warehouses for hybrid rice seeds in China, i. e. low-temperature caverns, which are the low-temperature, low-humidity warehouses originally used for the storage of ammunition and weapons for national defense purposes; and self-built low-temperature and low-humidity warehouses. Years of practice have shown that the low-temperature and low-humidity cold storage for medium- and short-term storage of hybrid rice seeds requires the warehouse temperature to be controlled at 8 ℃–10 ℃ with a relative humidity of 50% throughout the year, equipped with dehumidification equipment.

(2) Warehouse treatments

The following steps should be taken before seeds are put into a warehouse.

1) Remove foreign objects, garbage, etc. in the warehouse to keep it clean and tidy. Clean the tools stored in the warehouse. Remove eggs and nests on the walls, door frames, doors and windows, and in the corners to prevent the propagation of pests.

2) Check warehouse cooling, proof against moisture and rodent and sealing the door and windows to ensure that the seeds are not damaged by insects, rats, birds and moisture during storage.

3) Fumigation for disinfection. Select appropriate agents to sterilize the warehouse. For an empty warehouse, a 0.2% solution made with 2 g of 80% dichlorvos EC and 1 kg of water, or 3 g/m^3 of 56% aluminum phosphide can be evenly applied on the upper, middle and lower levels of the warehouse for fumigation. Close the doors and windows during disinfection. Ventilate for more than 24 hours after disinfection and clean the chemical residues.

2. Seed storage standards

In order to ensure the safe storage and maintain the vitality of seeds, quality inspection should be carried out before storage. The main factors affecting the safe storage of hybrid rice seeds are seed moisture-content and seed cleanliness, of which, seed moisture content is a key factor. With high moisture content, the seed's internal physiological activity is strong, consuming seed nutrients and accelerating the propagation of micro organisms and insects in the warehouse. Therefore, the moisture content of hybrid rice seeds must be strictly controlled below 12% before storage. In addition, seed cleanliness should be above 98.5%. If this standard is not met, there will be abnormal plants in the seed pile with different physical and chemical characteristics, affecting storage safety.

3. Seed storage in different batches

The seeds in storage should be stored and stacked in batches, and each batch should be packed uniformly in sacks or fiber bags, with a uniform weight standard. Seeds in the warehouse should be stored by varieties, grades, moisture content and age, with labels indicating the variety, place of origin, year of production, harvest time, storage time and other related information.

The seeds are stacked neatly with the ends of the bags facing each other and the bag openings facing outwards. The stacked seeds should be 0.5 m from the wall, and 0.6 m from adjacent piles, placed in parallel to windows and doors for ventilation and heat dissipation. A certain distance should be kept between piles to facilitate heat and moisture dissipation during storage, as well as regular inspection.

4. Management during storage

Seeds in storage, due to the low moisture content and the controlled warehouse temperature and humidity, have a low respiration and consume very few internal substances. The seeds are in a dormant state but can maintain vitality. The change of seed moisture content mainly depends on the relative humidity in the air in the warehouse. To ensure storage safety, the relative humidity in the warehouse must be controlled below 65%. Warehouse temperature is also an important factor affecting seed storage. Increased temperature enhances seed respiration, and increases pest and disease infestation.

Seed moisture content and air moisture content are in a dynamic equilibrium which is related to the temperature. If both of the seed moisture content and the temperature in the seed pile is low, but the temperature and humidity outside are high, the moisture in the wet and warm air is absorbed by the seeds, which raises the temperature and moisture of the seeds. On the contrary, when the external temperature and humidity are lower than that in the seed pile, the moisture of the seeds will be dissipated into the air, which helps keep the seeds dry. The temperature and humidity in a warehouse are affected by changes in the temperature and humidity in the atmosphere. In late autumn and winter in southern China, the temperature inside the warehouse is higher than that outside and hot air radiates outwards, which helps lower seed temperature and moisture content. In spring and summer, the opposite happens. Therefore, doors and windows of a warehouse can be open in dry and cold weather in late autumn and winter to lower the warehouse temperature and release seed moisture. On the contrary, in spring and summer, doors and windows should be kept close to prevent hot and humid air from flowing into the warehouse. Damp seeds are prone to clumping, mold and even sprouting and germinating, seriously lowering seed quality. In the case of high humidity in the warehouse, moisture removal and cooling equipment can be installed to control the warehouse temperature at below 15 ℃, which is conducive to seed preservation.

Warehouse pests are also a major threat to the quality of seeds in storage. Therefore, regular inspection should be carried out to spot pests during seed storage. Screening is a common way of inspection. Specifically, pests are to be detected with special tools, and samples are taken to find out the type and prevalence. If pests are found, chemical agents such as aluminum phosphide and other fumigation insecticides need to be applied, or tablets packed in bags can be placed in a ventilated place or stuffed into a slit. When insecticides are used, the warehouse must be kept shut for 7 – 10 days.

In short, prevention is the focus and comprehensive control should be practiced when necessary in seed storage. The warehouse manager must regularly check the seeds and keep the warehouse clean. Once pests and mildew are found, control measures should be taken immediately. Otherwise, the seed quality will be undermined, thereby affecting seed germination. It has been proved that hybrid rice seeds are qualified for further use with vigor for one year when they have a germination rate of over 85% , a moisture content of less than 12% and qualified purity, disinfected before storage, and regularly inspected during storage in the warehouse where the temperature is controlled at 15 ℃ and the humidity below 60% .

Ⅱ. Hybrid Rice Seed Vigor Improvement Technology

There are many researches on seed treatment across the world, and achievements have been made in terms of relevant principle and techniques. Seed treatment techniques mainly include specific gravity grading, coating, seed chemical dressing, corona field and dielectric sorting etc. Some of them have been applied to hybrid rice seeds.

1. Specific gravity grading

Specific gravity grading is a traditional method of seed processing, such as specific gravity selection and water selection. Today, seed grading by specific gravity is also a processing method to improve seed vitality. Liu Jiexiang et al. processed Y58S seeds with a high germination rate (88%) and Fengyuan A seeds with a low germination rate (67. 5%) by the specific gravity method at four levels, and compared them with the unprocessed control seeds for germination rate and vitality. The test results showed that there was an obvious grading effect regardless of the original germination rate of the seeds. The germination rate of Y58S with the highest specific weight (grade Ⅳ) and the lowest (grade Ⅰ) had a 5% difference in germination rate and a 23. 4% difference in vigor index, while, for Fengyuan A seeds, the difference was 28. 5% for the germination rate and 34. 1% for the vigor index, indicating a better grading effect on the seeds with a low germination rate. Therefore, specific gravity grading can realize the grading and sorting of seeds with different vigor in the same seed batch, and sort out the seeds with high germination rate and vitality (Table 16 - 5).

2. Coating

Seed coating technology originated in some Western developed countries, and is an important technology to improve the technological content of seeds and achieve seed quality standardization. Seed coating is a mixture of pesticides, micro-nutrients and plant growth regulators at a certain ratio. Seed coating technology is to cover the seed surface with a layer of seed coating agent containing different active ingredients by means of mechanical or manual processing. The seed coating agent only absorbs water and swell but does not dissolve, so it does not affect the normal germination of the seeds but enables the slow release of agents and fertilizers, so as to improve the vitality of seeds, prevent diseases and pests, and ensure the safe and normal growth of seedlings. Therefore, seed coating technology is a multi-disciplinary technology that integrates good varieties, plant protection and soil fertilization, and is a high-tech achievement that has been promoted in the agricultural sector in China in recent years. At the same time, as seed coating uses less agent and has low toxicity and high efficiency, it is an effective way to save resources and protect the environment.

Since the 1990s, hybrid rice seed coating has been required in small packages in China. However, the coating technology has not been further promoted due to three reasons. First, the requirements for seed purity, germination rate and fullness are high for coating. However, glume-split seeds in hybrid rice

Table 16 - 5 Seed germination rate and vigor of sterile lines after specific gravity grading

Variety	Weighting	Germinating vigor (%)	Germinating index	Germination rate (%)	Young seedling length (cm)	Vigor index
Y58S	Ⅰ	88.50	21.60	88.7	6.27	135.43
	Ⅱ	89.0	23.05	93.5	6.46	148.88
	Ⅲ	98.5	23.07	93.7	6.60	152.26
	Ⅳ	96.2	24.35	97.7	6.49	158.03
	CK	88.0	23.12	93.7	6.39	147.74
FengyuanA	Ⅰ	45.75	13.28	57.5	6.54	86.85
	Ⅱ	69.0	16.90	69.7	6.95	117.46
	Ⅲ	77.25	19.97	80.7	7.17	143.46
	Ⅳ	81.0	22.64	86.0	6.95	157.23
	CK	67.5	17.99	75.3	6.73	120.98

seeds cause non-plumpness and low germination rate and a high loss in seed selection before coating. Second, the seed moisture content increases after coating and they must be dried to safe storage moisture content before small-packing. Third, the coated seeds must be sold out in the seeding season of the year. Once overstocked, they can no longer be kept in storage, further processed or used for other purposes.

There are three types of seed coating machines.

The first is the churning type. Loading and unloading boxes are fixed on the same shaft and measured separately, while the bucket is turned over to lower the agents and seeds. In the coating chamber, the rotating churn advances the seeds and agents into a double-channel packer for packaging.

The second is the spray drum type, which coats the seeds by spraying agent solution compressed in the tank to the rotating circulation drum.

The third is the dispersion disk type which is a combination and improvement of the above two types. The agent is not directly compressed and sprayed onto the drum, but gently sprayed onto the seeds by the dispersion disk, and the churn can be used to roll, and the liquid does not atomize into droplets and splash into the air.

At present, some large Chinese seed companies are equipped with complete sets of seed processing equipment. After selection, the seeds are coated and dried to a moisture content level safe for storage.

However, to further improve the coating rate of hybrid rice seeds, attention must be paid to the following aspects. First, with genetic improvement and seed production technology, glume-split seeds need to be reduced and seed cleanliness, fullness and germination rate improve. Second, the coating process and the drying process must be matched so that the coated seeds are dried sufficiently for safe storage. Third, coating the seeds according to market demands, so that the coated seeds can be sold and used in the seeding season of the year, and the packaging technology needs to be improved so that the coated seeds can be stored across the year.

3. Seed dressing

Seed dressing can be traced back to the Western Han Dynasty (206 B. C. —A. D. 24) in Chinese history, when silkworm or sheep manure were used to wrap or soak seeds. However, chemical treatment of seeds was originated from the copper sulfate powder technology and new techniques invented by Mr. Damell-Smith in 1901. Seed dressing technology, coming from early disease prevention and treatment, now can improve seed quality, increase seed vigor, and promote seed germination and early seedling growth. After 2006, with the development and promotion of the direct seeding technology in China, seed dressing technology has been developing rapidly. Companies represented by Hunan Haili promoted 35% carbosulfan dry powder (mixed) with bird-repellent and rodent-repellent features as the selling point, and it was used in seed dressing. In 2011, Hunan Qiandu International Plant Protection Model Communication Center began to promote Huanglong Xiufeng (Imidacloprid + Tebuconazole) from Yancheng, Jiangsu province, in Hunan. This product can prevent the damage of birds, rodents and rice thrips.

With more convenience than the coating technology, seed dressing has been widely promoted in rice seedling cultivation and direct seeding in recent years. High-quality and high-efficiency seed dressing agents continue to emerge. It is more appropriate to make seed dressing agents with micro-fertilizers, pesticides and growth regulators, which are characterized by low toxicity, high efficiency, long efficacy and convenience for use. Seed dressing agents deliver effects in a comprehensive way, resulting in uniform and dwarfed seedlings with well-developed roots, quick tillering, strong stem bases and few or no pest damage. This significantly improves the quality of seedlings and lays the foundation for high yield and high quality rice.

The procedures of seed dressing with agents are to wash the seeds and soak them in a seed disinfectant for 8 - 12 h, specifically, wash the seeds and then soak them in clear water for water absorption; control the temperature and moisture for pre-germination (in a greenhouse in early spring with a constant temperature of 28 ℃-30 ℃, and in summer, soaking, washing and drying in alternation for pre-germination); when a certain seed germination rate (commonly known as "rupturing") is reached, spread the seeds, lower the moisture of the seed husks, and dress the seeds with agents such as 35% carbosulfan powder, with an agent to seed ratio of 1 : (80 - 110), i. e., 9 - 12 g of agent mixed with 1 kg of seeds. Sow the seeds 30 min after soaking.

References

[1] MA HAO, SUN QINGQUAN. Seed processing and storage[M]. Beijing: China Agriculture Press, 2007.
[2] DENG RONGSHENG, LIANG WEI, ZHUO JING, et al. Rice seed storage: features and methods[J]. Nong-

min Zhifu Zhiyou, 2016 (22): 73.

[3] ZHU XIANHAO, QIAN XIANGYANG, ZHANG GUANGYIN, et al. Relationship between rice seed plumpness and Yield[J]. Seeds, 1986(3): 12 – 17.

[4] SU LINGFANG, LIU JINMING. Effect of different grain proportions on the quality of rice seedlings[J]. Jiangsu Agricultural Sciences, 1987(2): 4 – 6.

[5] LIU JIEXIANG, ZHANG HAIQING, LIU AIMIN, et al. Effect of seed density classification of sterile lines on seed vigor and population characteristics[D]. Changsha: Hunan Agricultural University, 2014.

[6] XIAO CENGLIN. Seed pre-sowing processing technology: theory and practice[J]. Seeds, 1991(6): 41 – 43.

[7] CHEN HUIZHE, ZHU DEFENG, LIN XIANQING, et al. Effect of seed plumpness on germination rate, seedling rate and growth of hybrid rice[J]. Fujian Journal of Agricultural Sciences, 2004(2): 65 – 67.

[8] LI YANLI, JIA YUMIN, MENG LINGJUN, et al. Effect of seed selection with different saline specific gravity on yield and quality of rice[J]. Jilin Agricultural Sciences, 2011, 36 (1): 8 – 10.

[9] ZHANG FAN, ZHANG HAIQING. Current situations and prospects of research on seed coating technology[J]. Crop Research, 2007 (S1): 531 – 535.

[10] LIU AIMIN, ZHANG HAIQING, ZHANG QING, et al. A two-stage drying method for hybrid rice seeds. China, 201610443431. 4, 2016 – 11 – 09.

Part 5

Achievements

Chapter 17
Overview of Super Hybrid Rice Promotion and Application

Hu Zhongxiao / Xu Qiusheng / Xin Yeyun

Section 1 Typical Cases of High Yield Demonstration Super Hybrid Rice Cultivation

In order to enhance China's grain science and technology capacity and increase China's rice yields per unit area significantly, China launched the Super Rice Research Project in 1996, organized and implemented by the Ministry of Agriculture (MOA) with the breeding goal of a 15% yield increase by 2000 and a 30% yield increase by 2005, so that China's rice yield can achieve the third leap after the success of dwarf rice breeding and hybrid rice research. The Super Rice Project was planned to be carried out in three phases. Taking mid-season rice in the Yangtze River Basin as an example, the goal was to develop rice varieties in large-scale demonstration with an average yield of 10.5 t/ha, 12.0 t/ha and 13.5 t/ha in the three phases, respectively. After successively achieving the goal of the three phases, the MOA launched the fourth phase of the super rice breeding program with a goal of 15.0 t/ha in 2013.

The development of hybrid rice has gone through a process of continuous innovation, progress and improvement from three-line to two-line and then to super hybrid rice. The average yield of three-line hybrid rice is 20% higher than that of conventional rice with a yield increase of about 1,500 kg/ha. Two-line hybrid rice was successfully put into production in 1995 with a yield increase of 10% over that of three-line hybrids. The research on super hybrid rice started in 1997, and so far, major breakthroughs have been made by combining the ideal plant type with the utilization of *indica-japonica* intersubspecific heterosis, and taking into account both quality and resistance. The goals established by the MOA were 10.5 t/ha by 2000, 12.0 t/ha by 2004, 13.5 t/ha by 2012 and 15.0 t/ha by 2014. After that, average yields of 16.0 t/ha and 17.0 h/ha were achieved in 6.67-ha demonstration programs in 2015 and 2017 successively, breaking the record of yield per unit area.

I. First-phase Demonstration – 10.5 t/ha

In 1999, a new two-line hybrid rice, Liangyouppeijiu (Peiai 64S/9311), was planted in 14 provinces in China, including Hunan, Jiangsu and Henan, with a total area of 66,700 ha and a yield range of 9.75 –

10.5 t/ha, about 1.5 t/ha higher than the main control hybrids. In Hunan and Jiangsu, there were 14 6.67-ha demonstration plots (>6.67 ha demonstration) and one 66.7-ha demonstration plot (>66.7 ha demonstration) yielding more than 10.5 t/ha. The highest yield on 6.67-ha demo plots was 11.67 t/ha (Chenzhou, Hunan, 9.43 ha), and the highest yield of 66.7-ha demo plots was 10.97 t/ha (Jianhu, Jiangsu, 110.4 ha), which was 2.25 t/ha more than the control yield of Shanyou 63. Among them, the areas and yields of the hybrid were 7.1 ha with a yield of 10.53 t/ha in Fenghuang county, 6.83 ha with a yield of 10.56 t/ha in Longshan county, and 7.41 ha with a yield of 10.86 t/ha in Suining county of Hunan; 39 ha with a yield of 10.99 t/ha in Yandu county and 8 ha with a yield of 10.7 t/ha in Gaoyou city of Jiangsu, respectively.

In 2000, Liangyoupeijiu was promoted to a total of 233,300 ha in the 16 provinces (autonomous regions and municipalities) of Jiangsu, Anhui, Zhejiang, Fujian, Guangdong, Guangxi, Yunnan, Guizhou, Henan, Hubei, Hunan, Jiangxi, Sichuan, Chongqing, Shaanxi and Hainan. Among them, the planting area in Hubei, Jiangxi, Anhui, southern Henan and northern Jiangsu exceeded or was close to 33,300 ha. In Hunan province alone, 17 6.67-ha demo and four 66.7-ha demo yielded 10.5 t/ha or more, which was an increase of 2.25 t/ha over the control Shanyou 63, including 6.77 ha with 11.23 t/ha in Fenghuang county, 71.67 ha with 10.55 t/ha in Longshan county, and 67.93 ha with 10.62 t/ha in Suining county, respectively. The 6.67-ha demo in these three counties yielded more than 10.5 t/ha for two consecutive years. In addition, the demo in Xixian county of Henan province had a yield of 10.63 t/ha in a plot of 8.67 ha. Longshan county is located in the northwest of Hunan province, connected to Youyang county and Xiushan county of Chongqing Municipality in the west, bordered by Laifeng county and Xuanen county of Hubei province in the north, adjacent to Sangzhi county and Yongshun county in the east, and adjacent to Baojing county across the Youshui River in the south. It is located at 109°13′–109°46′E, 28°46′–29°38′N with an average annual temperature of 15.8 ℃ and an altitude of about 460 m. In October 2000, a nationally renowned rice expert group organized jointly by the Bioengineering Center of the Ministry of Science and Technology (MOST), the 863 Plan Biological Field Expert Committee of MOST, and the Science and Education Department of MOA conducted an appraisal of Liangyoupejiu planted in a 71.67 ha land plot with an average yield of 10.55 t/ha in the Guandu village, Huatang township, Longshan county, Hunan. In 1999 and 2000, the average yield of 6.67-ha demo plots in the same ecological zone exceeded 10.5 t/ha, marking the achievement of the Phase-I goal of China's Super Rice Project. Liangyoupeijiu, a pioneer super hybrid rice variety, is a two-line intersubspecific combination selected and bred jointly by the Jiangsu Academy of Agricultural Sciences and Hunan Hybrid Rice Research Center (HHRRC). In 1999, it was approved by Jiangsu provincial authorities (No. 313), and then by the national authorities in 2001 (No. 2001001), by Hubei (No. 006—2001), Guangxi (No. 2001117), Fujian (No. 2001007), Shaanxi (No. 429), and Hunan (No. 300), respectively.

The average yield of Xieyou 9308 was 11.84 t/ha with the highest yield of 12.28 t/ha in 2000 and 11.95 t/ha with the highest yield of 12.4 t/ha in 2001 (a record high in Zhejiang province), in 6.67-ha demo fields in Xinchang county, Zhejiang province, respectively, and passed appraisals organized by the

Department of Science and Education of MOA. In 2000, Ⅱ Youming 86 yielded an average of 12. 42 t/ha in a 6. 67ha demo field in Youxi county, Fujian province. Around 2000, China successfully developed a group of new super rice hybrids, represented by Liangyoupeijiu, Xieyou 9308, Ⅱ Youming 86 and Fengliangyou 4.

Ⅱ. Second-phase Demonstration - 12. 0 t/ha

In 2002, a pilot super rice hybrid, Liangyou 0293, was demonstrated in an 8. 1 ha plot in Longshan county, Hunan province, with an average yield of 12. 26 t/ha, which was the first time that the average yield of 6. 67 ha demo in the Yangtze River Basin exceeded 12 t/ha. This hybrid was planted in four 6. 67-ha demo fields in Hunan with a yield of over 12. 0 t/ha in 2003 and seven 6. 67-ha demo (two in Hainan, four in Hunan, and one in Anhui) with an output of 12 t/ha in 2004, respectively, of which Hunan 6. 67-ha demo in the three counties of Zhongfang, Longhui and Rucheng exceeded 12 t/ha for two consecutive years.

In 2003, the hybrid Zhunliangyou 527 was planted in Guidong county, Hunan, in a 6. 67-ha demo field with an average yield ofover 12 t/ha. Guidong county is located at the southeastern border of Hunan province and at the southern end of the Luoxiao Mountains and the northern foot of the Nanling Mountains (113. 37° - 114. 14°E, 25. 44° - 26. 13°N). It has a humid monsoon climate of the mid-subtropical zone. The annual average sunshine duration is 1,440. 4 h and the annual average temperature is 15. 8 ℃. The average summer temperature is 23. 6 ℃ with an extreme maximum temperature of 36. 7 ℃. The annual precipitation is 1,742. 4 mm with a maximum of 2,444. 2 mm and a minimum of 1,572. 5 mm, the annual evaporation is 1,205. 1 mm, and the frost-free period is 249 d. In 2004, the hybrid Zhunliangyou 527 was planted in Hunan, Guangxi, Jiangxi, Hubei Guizhou and other provinces (autonomous regions) for high-yielding cultivation as 6. 67-ha demo. It had an average yield of 12. 14 t/ha in Rucheng county, Hunan, and 12. 19 t/ha in Zunyi city, as approved by experts from the Department of Agriculture of Guizhou province. Zhunliangyou 527 is a two-line hybrid developed by HHRRC and approved in Hunan in 2003 (XS006—2003), at the national level in 2005 (for the middle and lower reaches of the Yangtze River and the Wuling Mountains, GS2005026), and in Fujian (MS2006024) and at the national level (for the South China rice area, GS2006004) in 2006.

According to the goal of the second phase of China's super rice program set by the MOA, i. e., to achieve a yield of 12 t/ha in two 6. 67-ha demo plots in the same ecological zone for two consecutive years by 2005, China has achieved the goal one year ahead of schedule. Super hybrid rice varieties of the second phase include Zhunliangyou 527, Y Liangyou 1, Ⅱ Youhang 1 and Shenliangyou 5814.

Ⅲ. Third-phase Demonstration - 13. 5 t/ha

In 2011, HHRRC demonstrated the hybrid Y Liangyou 2 in a 7. 2-ha field in Yanggu'ao town, Longhui county, Hunan. The site is in the central part of Hunan, slightly southwest of the northern bank of the upper reaches of the Zishui River. It has a mild subtropical monsoon humid climate with four distinct seasons, concentrated rainfall, wet springs and dry autumns, and large differences between the north and the south. The demonstration base has an altitude of about 500 m, and the annual average daily temperature is 11 ℃-17 ℃, and the annual average frost-free period is 281.2 d, and the annual average precipitation is 1,427.5 mm. On September 18, the MOA organized an expert group to conduct on-site yield tests and approved the 6.67-ha demo with an average yield of 13.90 t/ha. Y Liangyou 2 is a two-line hybrid rice bred by HHRRC and approved in Hunan in 2011 (XS2011020), in Honghe, Yunnan [DS (Honghe) 2012017] in 2012, at the state level (GS2013027) in 2013, and in Anhui (WS2014016) in 2014.

In 2012, HHRRC conducted a cultivation trial for third-phase super rice with a target yield of 13.5 t/ha in five counties of Hunan, i.e. Xupu, Longhui, Rucheng, Longshan and Hengyang. Xupu county (110°15′-111°01E, 27°19′-28°17′N) is located in western Hunan, northeast of Huaihua city, and in the middle reaches of the Yuanshui River with a humid subtropical monsoon climate. The demonstration base is about 500 m above sea level with an average annual temperature of 16.9 ℃, an average annual precipitation of 1,539.1 mm, and an average annual frost-free period of 286 d. All of the seven 6.67-ha demo fields delivered high yields despite many unfavorable factors such as a rice blast outbreak. On September 20, the Hunan provincial Department of Agriculture organized experts from Wuhan University, Hunan Academy of Agricultural Sciences and Hunan Agricultural University to conduct an on-site yield inspection of the 6.91-ha Y Liangyou 8188 high-yielding demo plot in Xinglong village, Hengbanqiao town, Xupu county, which had an average yield of 13.77 t/ha in the 6.67-ha demo field, achieving the goal of 13.5 t/ha in the same ecological zone for two consecutive years. Y Liangyou 8188 is a new two-line rice hybrid bred by Hunan Aopulong Science Technology Co., Ltd. It approved in Honghe, Yunnan, in 2012 [DT (Honghe) 2012021], in Hunan in 2014 (XS2014005), and at the state level in 2015 (GS2015017).

On November 27, 2012, the highest yield of Yongyou 12 in Bailiangqiao village, Dongqiao, Yinzhou, and Ningbo, Zhejiang reached 15.21 t/ha in a 6.67-ha demo field with an average yield of 14.45 t/ha.

Ⅳ. Fourth-phase Demonstration - 15.0 t/ha

In April 2013, the MOA launched the fourth phase of China's super rice program with a yield target of 15.0 t/ha. Led by academician Yuan Longping, a collaborative team was formed for the project, with members from many research institutes, universities and some seed companies in China to implement the "four good" research strategy (good seed, good method, good field and good conditions), insisting on good coordination and combination of breeding, cultivation, soil fertilization and plant protection, this is a continuation from previous research of super high-yielding hybrid rice. The project team formulated a scientific and reasonable technical plan and strengthened the technical management and implementation in the field. After overcoming the unfavorable climatic factors such as low temperature, rain and low sunshine during the peak growth season, 6.75 ha of Y Liangyou 900 planted in Hongxing village, Hengbanqiao township, Xupu county, Hunan, was certified to have an average yield of 15.4 t/ha by experts organized by the MOA on October 10, 2014, achieving the goal of 15.0 t/ha for the fourth phase of the super rice project.

On September 17, 2015, experts from the Hunan Provincial Department of Science and Technology tested the yield of a 6. 8-ha demonstration plot of Xiangliangyou 900 in Xinwafang village, Datun town, Gejiu city, Yunnan province, and the average yield was 16. 01 t/ha, setting a new world record for large-area rice yield and breaking the 15. 9 t/ha limit for rice yield in tropical regions recognized by the rice industry. On October 12, experts organized by the MOA measured the yield of a 7. 2-ha Xiangliang-you 900 plot in Leifeng village, Yanggu'ao township, Longhui county, Hunan, with an average yield of 15. 06 t/ha, thus achieving in the Phase -Ⅳ goal of 15 t/ha in the same ecological zone for two consecutive years. On May 9, Xiangliangyou 900 was harvested and tested in Sanya, Hainan, with an average yield of 14. 12 t/ha and the highest field yield reached 15. 15 t/ha in 6. 67-ha demo, creating a record of yield per unit and in large scale production in tropical rice areas. In addition, three Xiangliangyou 900 6. 67-ha demo sites, respectively in Suizhou city, Hubei province, Yongnian district, Hebei province, and Guangshan county, Henan province all broke the high-yield record with an average yield of over 15 t/ha.

In 2016, the average yield of the single-cropping rice Xiangliangyou 900 on 6. 67-ha demo plots in Junan county, Shandong province, was 15. 21 t/ha, setting a new record of yield per unit area in the province. In Xingning city, Guangdong province, the average yield of double-cropping early-season rice Xiangliangyou 900 in 6. 67-ha demo plots was 12. 48 t/ha, and the average yield of double-cropping late-season rice Xiangliangyou 900 in 6. 67-ha demo plots was 10. 59 t/ha, making a total of 23. 07 t/ha from the two seasons, setting the record of double-cropping rice yield in the world. In Qichun county, Hubei province, Xiangliangyou 900 was tested in the cropping pattern of single season + ratooning rice cultivation in 6. 67-ha demo plots and the average yield of ratooning rice manually harvested in the first season was 7. 65 t/ha, while that of mechanically harvested was 5. 92 t/ha, setting a high-yield record of ratooning rice in the middle and lower reaches of the Yangtze River. The average yield of manually harvested single-cropping and ratooning rice was 18. 80 t/ha, while that of mechanically harvested ratooning was 17. 08 t/ha, setting a new record for the single-cropping + ratooning rice cropping pattern in the middle and lower reaches of the Yangtze River. In Guanyang county, Guangxi, the average yield of ratooning rice Xiangliangyou 900 was 7. 46 t/ha, setting the record for ratoning rice yield in South China. The average yield of single-cropping + ratooning rice was 21. 72 t/ha, also a new record for this cropping pattern in the area.

Ⅴ. Super Hybrid Rice Demonstration - 16. 0 t/ha

In 2016, the average yield of Xiangliangyou 900 in 6. 67-ha demo plots in Yongnian district, Hebei province, was 16. 23 t/ha, creating a new record in the northern rice area and a world record of high yield in high-latitude areas. In Gejiu city, Yunnan province, the average yield of single-cropping Xiangliangyou 900 was 16. 32 t/ha, breaking the record of 16. 01 t/ha set in 2015 and was the highest yield per unit area in large-scale planting in the world.

In 2017, HHRRC planted Xiangliangyou 900 in a 6. 93 ha demonstration plot in Yongnian district,

Hebei province. The site is at 114°20′E, 36°33′N, in the southern part of the province and at the eastern foot of the Taihang Mountains, with a warm semi-humid continental monsoon climate. The location has a mild climate with sufficient rainfall and sunshine, and the altitude is 41 m. The average annual precipitation is 549.4 mm, the average annual accumulated temperature is 4,371.4 ℃, and the average annual frost-free period is 205 d. On November 15, experts from the Hebei Provincial Department of Science and Technology inspected the demonstration plots and certified an average yield of 17.24 t/ha. In the same year, the average yield in 6.67-ha demo plots of Yongyou 12 in Quantang village, Shimen town, Jiangshan city, Zhejiang province, was 15.16 t/ha, reaching beyond the 15 t/ha mark and setting a record for rice yield in Zhejiang province.

Ⅵ. Experience in Super Hybrid Rice Research and Promotion in China

1. High priority of the party and government

In 1996, the MOA launched China Super Rice Project, while the MOST and other departments also set up special projects to support the research and application of super hybrid rice in China. On October 3, 2003, General Secretary Hu Jintao went to the National Hybrid Rice Engineering and Technology Research Center (NHRETRC) to inspect the progress of the super hybrid rice research project and gave full recognition. On August 13, 2005, Premier Wen Jiabao also visited the NHRETRC to inspect the progress of super hybrid rice research.

2. Technology innovation

Japan started its super high-yielding rice research project in 1981, and the International Rice Research Institute (IRRI) started the new plant type breeding project in 1989 with the aim of greatly improving yield potential. Some new lines were developed with high yield potential, but have not been widely applied.

Chinese scientists did not follow in their footsteps. Yuan Longping creatively proposed the technical route of combining ideal plant type with heterosis utilization for China's efforts. Specifically, this involves *indica-japonica* crossing to widen the genetic distance, reforming the plant type based on biological heterosis, and improving the quality and resistance in addition to pursuing higher yields. For this purpose, variety improvement is integrated with other supporting technologies to achieve the "four good".

3. Joint efforts to tackle key technical difficulties

China's super hybrid rice research features multi-disciplinary collaboration. When the China Super Rice Project was launched in 1996, the MOA established the Super Rice Breeding Expert Group, and in 2015 it established the national Super Rice Research and Extension Expert Group in response to the needs of super rice research and promotion. A multi-disciplinary research team was established to work with local agricultural technology promotion departments for super hybrid rice demonstration and promotion, with a clear division of tasks, stable support, moderate competition and a rolling support mechanism. A collaborative network based on ecological zones, research focus, and demonstration and promotion has been established. According to the characteristics of the growth and yield formation of super hybrid rice combinations in China's main rice planting areas, various research institutes, universities, agricultural promotion departments and agricultural enterprises have carried out extensive joint research initiatives, supporting high-yielding and high-quality seed production with comprehensive efforts.

4. Oriented to production and market demands

Meeting production and market demands is the fundamental goal of China's super hybrid rice breeding efforts, and is also the fundamental difference between Chinese super rice breeding and Japanese super-high yield rice breeding, and the new plant type breeding of IRRI. The research and application of super rice in China not only clarifies the yield indicators for different types of super rice in different rice cropping areas, but also has high requirements for rice quality and resistance, which makes the research of super hybrid rice in China difficult, but the results easy to be accepted by production. For example, super rice Liangyoupeijiu, co-bred by Jiangsu Academy of Agricultural Sciences and HHRRC, has good resistances and high yield, and six of its quality parameters have reached the standards for first-grade quality rice. With its promotion and application in production, super hybrid rice has become a new growth driver in China's rice seed industry.

Section 2 Super Hybrid Rice Breeding and Promotion

Take mid-season rice as an example. In 2000, 2004, 2012 and 2014, the goals for the Phase Ⅰ (10.5 t/ha), Phase Ⅱ (12.0 t/ha), Phase Ⅲ (13.5 t/ha) and Phase Ⅳ (15.0 t/ha) of China's super hybrid rice project set by the MOA were achieved successively. In 2017, a new yield record was set for large-scale rice cultivation at 17.24 t/ha. Super hybrid rice breeding has played a key role in the progress towards higher yields in China's super hybrid rice research work. Only varieties with super high yield potential can achieve super high yield in high-yielding demonstration plots. The promotion and application of new super hybrid rice varieties have also made great contributions to China's food security and the increase in farmers' harvest and income. Statistics from the annual *Super Rice Certification Announcement* issued by the MOA and the *Statistical Table for the Promotion of Major Crop Cultivars in China* published by the National Agricultural Technology Promotion Service Center in terms of super hybrid rice breeding and promotion are presented in Table 17 - 1.

Table 17 - 1 Number of super hybrid rice varieties recognized in China by year

Year	Super hybrid rice	Three-line super hybrid rice	Two-line super hybrid rice	*Indica* super hybrid rice	*Indica-japonica* super hybrid rice
2005	22	20	2	19	3
2006	12	7	5	11	1
2007	5	3	2	5	0
2008	0	0	0	0	0
2009	7	4	3	7	0

Continued

Year	Super hybrid rice	Three-line super hybrid rice	Two-line super hybrid rice	*Indica* super hybrid rice	*Indica-japonica* super hybrid rice
2010	6	4	2	6	0
2011	6	3	3	5	1
2012	9	6	3	9	0
2013	5	4	1	4	1
2014	12	5	7	12	0
2015	7	5	2	4	3
2016	8	6	2	8	0
2017	8	4	4	7	1
Total	107	71	36	97	10

Ⅰ. Breeding of Super Hybrid Rice Varieties

Before 2005, the varieties used in high-yielding demonstration projects were all selected from existing varieties with high yield potential, without standardized technical criteria and official recognition. The MOA promulgated and implemented the *Measures for the Confirmation of Super Rice Cultivars* (*Trial*) in 2005, and later revised it in 2008, laying out clear technical criteria for the test, approval, naming and cancellation of super hybrid rice varieties. The official recognition of super rice varieties greatly promoted the development of super rice breeding in China. Starting from 2005 (except for 2008), the MOA granted official recognition to super rice varieties every year. By the end of 2017, a total of 166 super rice varieties had been officially approved, including 165 super hybrid rice varieties (Tianyou-Huazhan was recognized twice in 2012 and 2013, respectively). Among all the recognized varieties, 108 recognitions were granted to 107 super hybrid rice varieties (Table 17 - 1), accounting for 64.85% of all super rice varieties (165). At the same time, starting from 2009, a succession of super hybrid rice varieties were removed from the list of super rice varieties, and by 2017, a total of 14 had been removed.

The year of 2005 had the largest number (22) of super rice varieties recognized, then the number gradually went down between 2006 and 2008, hitting zero in 2008. From 2009 on, super rice recognition has been advanced steadily and the number of super hybrid rice varieties recognized each year has been stable at 5 to 12.

Among the 107 super hybrid rice varieties recognized by the MOA from 2005 to 2017, 71 were three-line super hybrid rice, accounting for 66.36%, and 36 were two-line super hybrid rice, accounting for 33.64%. There are 97 *indica* super hybrid rice varieties, accounting for 90.65% and 10 *indica-japonica* super hybrid rice, taking up only 9.35%, which were Liaoyou 5218, Liaoyou 1052, Ⅲ You 98, Yongyou 6, Yongyou 12, Yongyou 15, Yongyou 538, Chunyou 84, Zheyou 18 and Yongyou 2640.

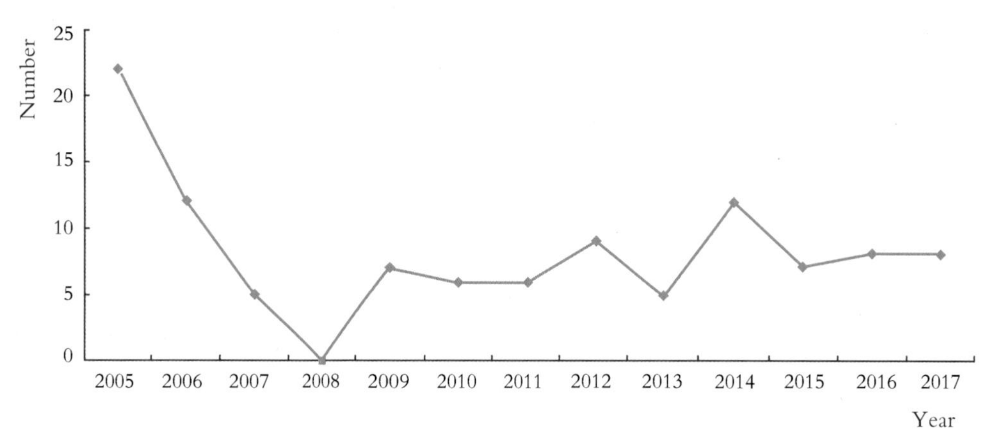

Fig. 17 − 1 Changes in the number of super hybrid rice varieties recognized in China by year

Among these, the Liaoyou series were bred by the Liaoning Academy of Agricultural Sciences and the Yongyou series were developed by the Ningbo Academy of Agricultural Sciences and Ningbo Seed Co., Ltd.

Ⅱ. Promotion of Super Hybrid Rice Varieties

1. Changes in planting area of super hybrid rice

Some super rice recognized had been widely promoted before the official recognition in 2005 (Table 17 − 2). Specifically, Ⅱ You 162 was widely promoted as early as 1997, with a planting area of 16,700 ha, and Xieyou 9308 was promoted to 8,000 ha in 1998. From 1997 to 2015, the accumulative planting area of super hybrid rice reached 46,293,000 ha, including 3,756,700 ha added to the cohort in 2015.

The accumulative planting area of super hybrid rice in China had reached 46.384 million hectares as of 2015 (Fig. 17 − 2). The annual planting area of super hybrid rice increased year by year from 1997 to 2015. Specifically, a significant increase happened from 1999 to 2000 due to the large-scale planting of Liangyoupeijiu on 324,700 ha and Ⅱ You 7 on 90,700 ha of land in their first year of promotion, and also with You 162 planted on 104,700 ha of land. From 2001 to 2002, with the rapid expansion of Liangyoupeijiu, the annual planting area of super hybrid rice had a rapid increase. From 2003 to 2005, as the planting area of Liangyoupeijiu stabilized with a slight decline, there was no increase in the annual planting area of super hybrid rice. However, due to the promotion of D You 527, Ⅱ Youming 86, Tianyou 998 and Yangliangyou 6, the planting area of super rice increased again due to the promotion of these new hybrids. In particular, in 2005, the promotion of two-line hybrids Liangyoupeijiu, Yangliangyou 6, Zhunliangyou 527, Xinliangyou 6, Zhuliangyou 819, and Liangyou 287 in 2005 injected new impetus into the second round of expansion of super hybrid rice planting.

In 2006, with the official recognition system for super rice launched, the annual planting area had a rapid increase of 35.25% over 2005. From 2006 to 2012, there was another significant increase in the annual planting area of super hybrid rice, leading to a peak at 4,282,700 ha in 2012, despite the small increase and slight decline in 2008 and 2011. From 2012 to 2015, there was a decline in the annual planting area of hybrid rice and super hybrid rice across China, with the most drastic decrease occurring in 2015.

2. Changes in the planting area of three-line and two-line super hybrid rice

All super hybrid rice varieties were three-line varieties before 1999 (Table 17 - 2 and Fig. 17 - 3). The planting area of two-line varieties exceeded that of three-line hybrids, accounting for 55.28% of the total due to the planting of Liangyoupeijiu on a large scale, with a total area of 324,700 ha in the first year of promotion in 2000. From 2000 to 2002, the annual planting area of two-line super hybrid rice was higher than that of three-line super hybrids, but it dropped below that of three-line hybrids from 2003 to 2015. However, the proportion of the annual planting area of two-line super hybrids was on the rise on the whole, reaching 47.88% in 2015.

From 2000 to 2005, two-line hybrids took a larger share in the total annual planting area of super hybrid rice with the promotion of Liangyoupeijiu. From 2006 to 2009, with the large-scale promotion of Yangliangyou 6 and Xinliangyou 6, the top three super hybrid rice varieties in terms of annual planting area were all two-line hybrids, i. e. Liangyoupeijiu, Yangliangyou 6, and Xinliangyou 6. From 2010 to 2011, with the large-scale promotion of Y Liangyou 1, the top three of super hybrid rice in the annual planting area were still all two-line hybrids, i. e. Y Liangyou 1, Xinliangyou 6 and Yangliangyou 6. In 2012, three-line super hybrid rice Wufengyou 308 joined the top three with a large total planting area. From 2012 to 2015, in the top three super hybrids, i. e. Y Liangyou 1, Xinliangyou 6 and Shenliangyou 5814, two were two-line hybrids.

Table 17 - 2 Planting area of super hybrid rice in China

Year	Super hybrid rice area (10,000 ha)	Three-line		Two-line		*Indica*		*Indica-japonica*	
		Area (10,000 ha)	Proportion (%)	Area (10,000 ha)	Proportion (%)	Area (10,000 ha)	Proportion (%)	Area (10,000 ha)	Proportion (%)
1997	1.67	1.67	100.00	0.00	0.00	1.67	100.00	0.00	0.00
1998	2.87	2.87	100.00	0.00	0.00	2.87	100.00	0.00	0.00
1999	2.40	2.40	100.00	0.00	0.00	2.40	100.00	0.00	0.00
2000	58.73	26.27	44.72	32.47	55.28	58.73	100.00	0.00	0.00
2001	105.20	46.93	44.61	58.27	55.39	105.20	100.00	0.00	0.00
2002	153.73	71.20	46.31	82.53	53.69	151.00	98.22	2.73	1.78
2003	174.00	100.93	58.01	73.07	41.99	170.93	98.24	3.07	1.76
2004	198.20	131.07	66.13	67.13	33.87	195.13	98.45	3.07	1.55
2005	207.27	128.53	62.01	78.73	37.99	202.60	97.75	4.67	2.25
2006	280.33	156.67	55.89	123.67	44.11	276.53	98.64	3.80	1.36
2007	316.40	181.47	57.35	134.93	42.65	311.53	98.46	4.87	1.54
2008	332.53	192.27	57.82	140.27	42.18	326.80	98.28	5.73	1.72
2009	371.80	221.00	59.44	150.80	40.56	367.47	98.83	4.33	1.17
2010	410.53	256.53	62.49	154.00	37.51	406.33	98.98	4.20	1.02

Continued

Year	Super hybrid rice area (10,000 ha)	Three-line		Two-line		*Indica*		*Indica-japonica*	
		Area (10,000 ha)	Proportion (%)	Area (10,000 ha)	Proportion (%)	Area (10,000 ha)	Proportion (%)	Area (10,000 ha)	Proportion (%)
2011	378.73	212.93	56.22	165.80	43.78	373.40	98.59	5.33	1.41
2012	428.27	247.87	57.88	180.40	42.12	421.13	98.33	7.13	1.67
2013	424.33	236.07	55.63	188.27	44.37	413.33	97.41	11.00	2.59
2014	415.80	224.47	53.98	191.33	46.02	401.27	96.50	14.53	3.50
2015	375.67	195.80	52.12	179.87	47.88	359.93	95.81	15.73	4.19
Total	4,638.47	2,636.93		2,001.53		4,548.27		90.20	

Note. Data in the table are from incomplete statistics, and the actual figure may be larger.

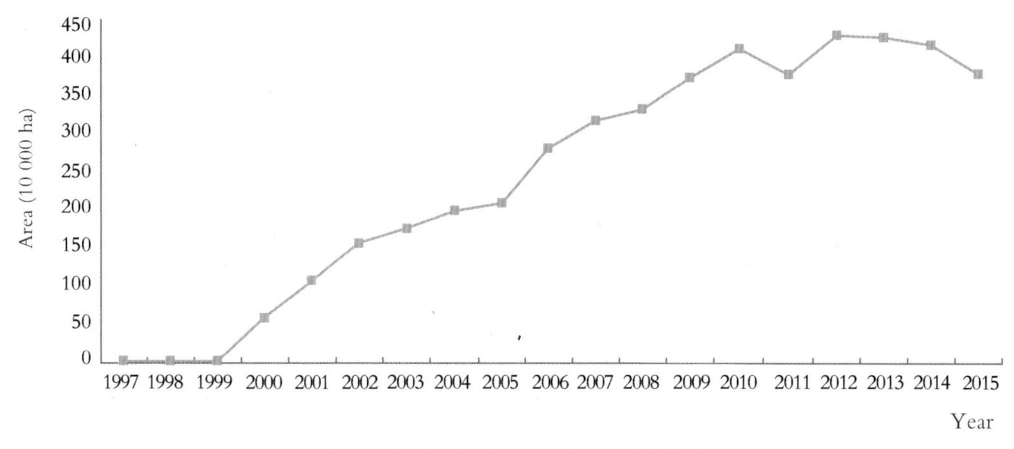

Fig. 17 - 2 Changes in the planting area of super hybrid rice in China

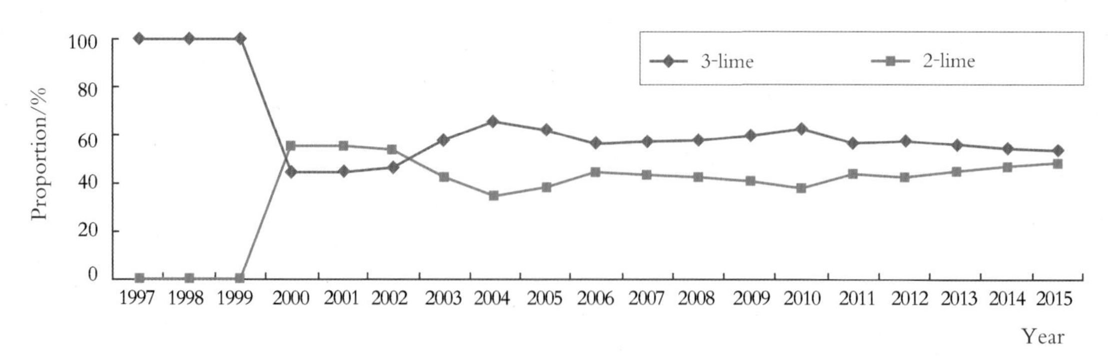

Fig. 17 - 3 Changes in the proportion of annual planting area of three-line and two-line super hybrid rice

3. Changes in the planting area of *indica* and *indica-japonica* super hybrid rice

The annual planting area of *indica-japonica* super hybrid rice accounts for only a small proportion in the total since its first promotion in 2002 (Table 17 - 2 and Fig. 17 - 4). Though the planting area increased to more 100,000 ha in 2013 with the large-scale promotion of Yongyou 12 and Yongyou 15, reaching a maximum of 157,300 ha in 2015, the proportion in the portfolio was still less than 5%.

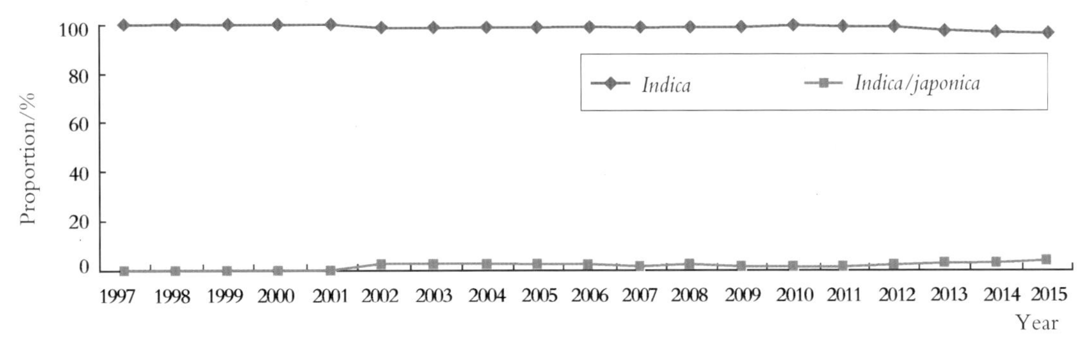

Fig. 17 - 4 Changes in the proportion of annual planting area of indica and indica-japonica super hybrid rice

4. Planting area of main super hybrid rice varieties

A total of 97 super hybrid rice varieties were promoted on a large scale (varieties with an annual planting area of over 6,700 ha) from 1997 to 2015, of which Liangyoupeijiu had the largest accumulative planting area, 6.0313 million hectares, followed by Yangliangyou 6 (2,668,700 ha), Y Liangyou 1 (2,153,300 ha) and Xinliangyou 6 (2.06 million hectares). The top four were all two-line super hybrid rice. The three-line super hybrid rice varieties with the largest accumulative planting area were Tianyou 998 (1,746,700 ha), D You 527, H You 084, Zhongzheyou 1, Q You 6, Ⅱ Youming 86, Wuyou 308 and Tianyou-Huazhan, all reaching over one million hectares. Ten super hybrid rice varieties were not widely promoted, accounting for 9.35% of the total of the 107 varieties.

Table 17 - 3 Planting area of super hybrid rice (1997 - 2015)

Ranking	Combination	Area (10,000 ha)	Ranking	Combination	Area (10,000 ha)
1	Liangyoupeijiu	603.13	11	Wuyou 308	132.93
2	Yangliangyou 6	266.87	12	Shenliangyou 5814	115.20
3	Y Liangyou 1	215.33	13	Tianyou-Huazhan	108.47
4	Xinliangyou 6	206.00	14	Ⅱ You 7	99.00
5	Tianyou 998	174.67	15	Fengliangyouxiang 1	97.00
6	D You 527	164.00	16	Fengyuanyou 299	90.60
7	Ⅱ You 084	144.80	17	Ⅱ You162	88.33
8	Zhongzheyou 1	140.80	18	Fengliangyou 4	82.67
9	Q You 6	140.20	19	Ⅱ Youhang 1	69.07
10	Ⅱ Youming 86	137.07	20	Ganxin 688	67.07

Continued 1

Ranking	Combination	Area (10,000ha)	Ranking	Combination	Area (10,000ha)
21	Ⅱ You 7954	65. 00	48	Yongyou 12	20. 33
22	Liangyou 287	62. 73	49	Tianyou 3301	19. 47
23	Ganxin 203	61. 20	50	Zhunliangyou 608	19. 27
24	Wufengyou T025	60. 60	51	F You 498	18. 80
25	Zhunliangyou 527	55. 27	52	Peiliangyou 3076	17. 40
26	Jinyou 458	50. 53	53	Lingliangyou 268	17. 33
27	Luoyou 8	49. 20	54	Yongyou 15	16. 80
28	Guodao 1	47. 93	55	D You 202	16. 47
29	Zhuliangyou 819	47. 80	56	Wufengyou 615	14. 87
30	Jinyou 527	43. 13	57	C Liangyou-Huazhan	14. 73
31	Yiyou 673	40. 13	58	Y Liangyou 2	14. 53
32	Ⅱ You 602	41. 07	59	Nei 5 You 8015	13. 87
33	Rongyou 225	38. 93	60	Ⅱ You 98	13. 40
34	Guangliangyouxiang 66	37. 80	61	Xieyou 527	13. 07
35	Shenyou 9516	36. 67	62	Huiliangyou 6	13. 00
36	Dexiang 4103	34. 47	63	Guodao 3	11. 80
37	Peizataifeng	33. 07	64	Guiliangyou 2	11. 40
38	Teyouhang 1	31. 53	65	Huiliangyou 996	10. 80
39	Tianyou 122	30. 20	66	Wuyou 662	10. 60
40	Xieyou 9308	29. 20	67	Teyou 582	10. 13
41	Xinliangyou 6380	29. 13	68	Jinyou 299	8. 13
42	Y Liangyou 5867	29. 00	69	Zhong 9 You 8012	8. 00
43	Neiliangyou 6	28. 60	70	H Liangyou 991	7. 93
44	Yongyou 6	28. 00	71	Xinfengyou 22	7. 73
45	Yixiangyou 2115	24. 87	72	Tianyou 3618	7. 60
46	Ⅱ Youhang 2	21. 33	73	Chunguang 1	6. 67
47	H You 518	21. 07	74	Yixiang 4245	6. 60

Continued 2

Ranking	Combination	Area (10,000ha)	Ranking	Combination	Area (10,000ha)
75	Y Liangyou 087	5.73	87	Liaoyou 5218	2.53
76	03 You 66	5.53	88	Liangyou 038	2.40
77	Yongyou 538	5.47	89	Zhunliangyou 1141	2.20
78	Q You 8	4.93	90	Shengtaiyou 722	1.93
79	Zhuliangyou 819	4.60	91	Wufengyou286	1.93
80	Deyou 4727	3.80	92	Longliangyou-Huazhan	1.60
81	Yifeng 8	3.60	93	Wuyouhang 1573	1.27
82	Liangyou 616	3.53	94	Y Liangyou 900	1.20
83	Guangliangyou 272	3.20	95	Jifengyou 1002	1.00
84	Shenliangyou 870	2.73	96	Liaoyou 1052	1.00
85	Chunyou 84	2.67	97	Fengtianyou 553	0.67
86	Rongyou 225	2.53			

Note. Data in the table are from incomplete statistics, and the actual figure may be larger.

5. Annual planting area of super hybrid rice in different provinces (autonomous regions and municipalities)

Table 17 - 4 Annual planting area of super hybrid rice in different provinces (autonomous regions and municipalities) (10,000ha)

Year	Jiangsu	Zhejiang	Fujian	Anhui	Jiangxi	Henan	Hubei	Hunan	Guangdong	Guangxi	Chongqing	Sichuan	Guizhou	Yunnan	Shanxi
1997	0.00	0.00	0.00	0.00	0.00	0.00	0.00	0.00	0.00	0.00	0.00	1.67	0.00	0.00	0.00
1998	0.00	0.80	0.00	0.00	0.00	0.00	0.00	0.00	0.00	0.00	0.00	2.07	0.00	0.00	0.00
1999	0.00	1.47	0.00	0.00	0.00	0.00	0.00	0.00	0.00	0.00	0.93	0.00	0.00	0.00	0.00
2000	1.47	5.47	0.67	3.13	4.00	4.40	19.13	3.47	0.13	0.00	2.00	14.87	0.00	0.00	0.00
2001	8.27	9.40	1.80	0.87	18.00	8.87	18.27	9.67	0.00	0.00	7.33	20.67	2.07	0.00	0.00
2002	12.67	11.40	10.40	18.60	16.47	6.33	21.33	13.20	0.73	1.27	9.67	27.33	1.47	1.33	1.47
2003	12.13	4.93	10.60	22.60	14.07	4.13	24.00	13.40	0.00	2.80	14.33	36.67	3.33	0.67	2.33
2004	15.87	13.60	10.20	10.20	17.47	5.00	25.13	15.00	1.00	2.67	12.07	39.33	4.87	0.00	2.33
2005	17.40	15.67	10.73	32.93	19.67	2.67	38.07	12.80	7.20	1.87	10.20	34.07	4.07	0.00	0.00
2006	18.20	22.47	10.60	42.53	26.67	5.40	50.93	18.80	12.67	10.33	9.13	40.87	10.67	0.00	0.00

Continued

Year	Jiang su	Zhe jiang	Fu jian	An hui	Jiang xi	He nan	Hu bei	Hu nan	Guang dong	Guang xi	Chong qing	Si chuan	Gui zhou	Yun nan	Shan xi
2007	16. 13	23. 40	13. 13	45. 27	49. 33	18. 20	55. 13	24. 60	18. 80	21. 00	6. 07	17. 93	4. 67	0. 87	0. 47
2008	14. 27	23. 20	10. 67	40. 93	47. 67	16. 40	58. 67	27. 87	23. 53	31. 00	5. 40	15. 80	5. 40	1. 33	0. 53
2009	14. 73	19. 07	16. 20	54. 53	72. 87	14. 27	61. 13	28. 80	21. 40	37. 07	5. 33	17. 80	6. 73	1. 33	0. 33
2010	12. 13	16. 07	16. 80	55. 93	79. 27	18. 67	60. 40	42. 27	18. 13	52. 07	6. 67	24. 13	7. 00	1. 53	0. 27
2011	10. 27	14. 93	16. 53	54. 27	72. 80	21. 00	63. 27	47. 40	20. 33	32. 27	6. 07	15. 93	2. 87	1. 33	0. 00
2012	7. 93	6. 20	15. 67	53. 73	66. 87	19. 27	60. 93	60. 67	30. 80	72. 53	6. 67	19. 53	3. 73	2. 40	1. 00
2013	7. 80	17. 67	19. 47	48. 00	70. 80	17. 87	62. 67	56. 53	30. 67	56. 93	8. 07	19. 27	3. 33	2. 40	1. 20
2014	7. 07	14. 80	17. 07	36. 60	63. 13	21. 07	63. 73	65. 73	28. 27	50. 60	9. 40	27. 47	2. 87	2. 20	0. 13
2015	5. 47	13. 80	19. 60	35. 93	47. 67	35. 67	45. 67	66. 53	26. 47	27. 00	9. 27	36. 53	4. 00	2. 53	0. 33
Total	181. 80	234. 33	200. 13	579. 60	686. 73	219. 20	728. 47	506. 73	240. 13	399. 40	128. 60	411. 93	67. 07	17. 93	10. 40

Note: Data in the table are from incomplete statistics, and the actual figure may be larger.

Super hybrid rice in China is mainly distributed in 15 provinces (autonomous regions and municipalities), including Jiangsu, Zhejiang, Fujian, Anhui, Jiangxi, Henan, Hubei, Hunan, Guangdong, Guangxi, Chongqing, Sichuan, Guizhou, Yunnan and Shaanxi, with the accumulative planting area reaching over 100,000 ha (Table 17 - 4 and Fig. 17 - 5). Hubei has the largest planting area of 7,284,700 ha, followed by Jiangxi (6,867,300 ha), Anhui (5,796,000 ha), and Hunan (5,067,300 ha). Guangxi and Sichuan also have large planting areas of about four million hectares. Guangdong, Zhejiang, Henan and Fujian each have more than two million hectares. Jiangsu, Chongqing and three other provinces (municipalities) each have more than one million hectares. Guizhou was in the range of 0. 5 - 1. 0 million hectares, and Yunnan and Shaanxi have a planting area of 100,000.- 200,000 ha.

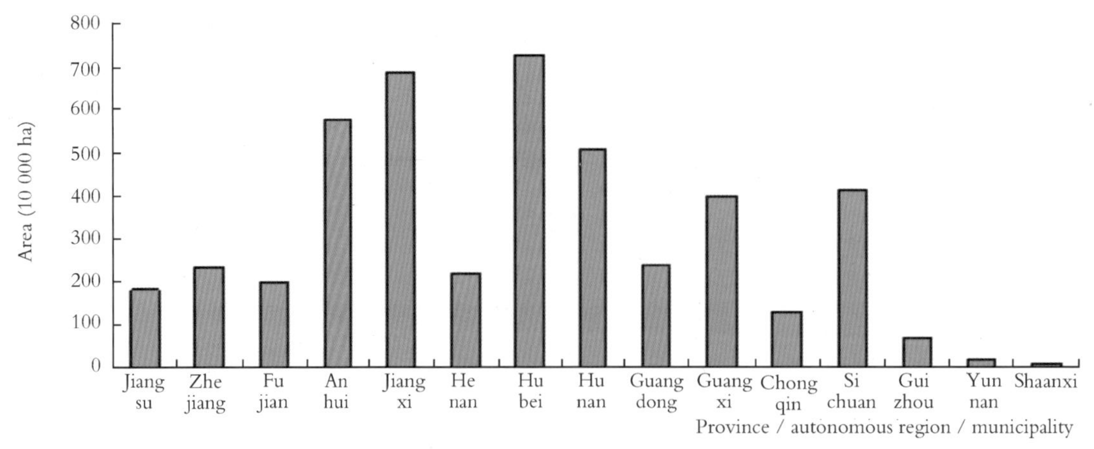

Fig. 17 - 5 Cumulative planting area of super hybrid rice by region (1997—2015)

Hunan was ranked fourth after Hubei, Jiangxi and Anhui in the accumulative planting area of super hybrid rice (Fig. 17 - 5), but had the largest planting area in 2015 (Fig. 17 - 6) with a total of 665,300 ha, which was 45.68%, 39.56% and 85.17% higher than that of Hubei, Jiangxi and Anhui respectively.

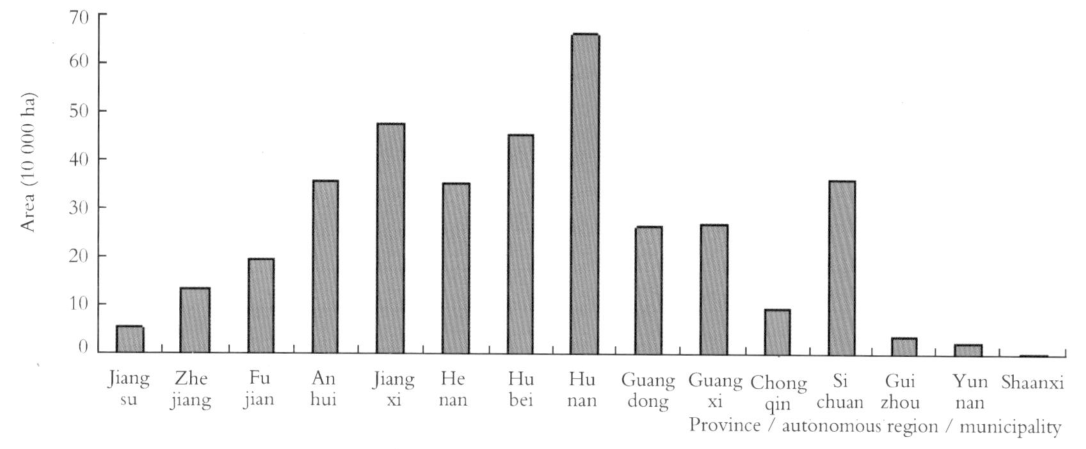

Fig. 17 - 6 Planting area of super hybrid rice by region, 2015

Among the six provinces having the largest share of super hybrid rice planting area, only Hunan saw a constant increase in the planting area from 2000 to 2015 (Fig. 17 - 7). The stability and good momentum of super hybrid rice development in Hunan province is inseparable from the promotion and implementation of the super hybrid rice projects launched by the province since 2007.

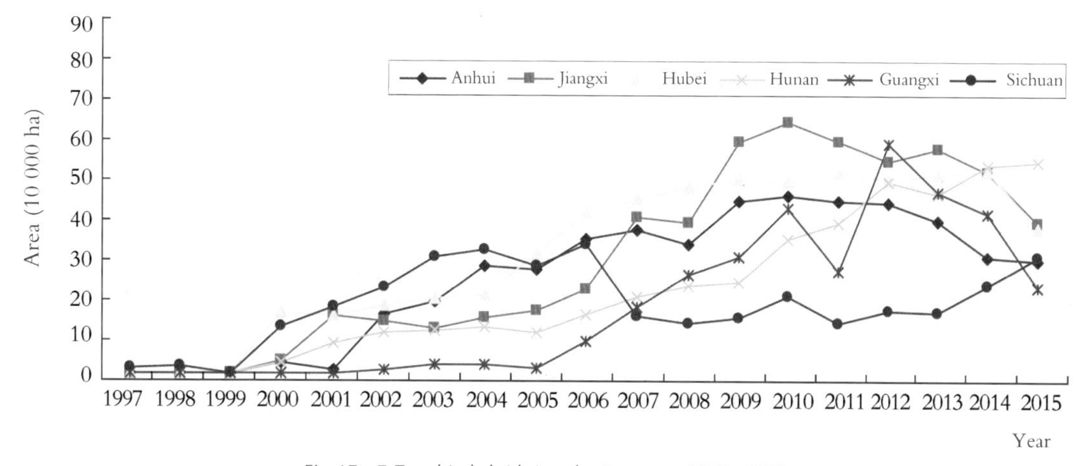

Fig. 17 - 7 Trend in hybrid rice planting area, 1997—2015

References

[1] QUAN YONGMING. An overview of demonstration and extension of pioneer super hybrid rice Liangyoupeijiu[J]. Hybrid Rice, 2005, 20(3): 1 - 5.

[2] HU ZHONGXIAO, HE JUN, LI XIMEI, et al. Construction of chinese hybrid rice variety and germplasm resource database[J]. Hybrid Rice, 2013, 28(6): 1-6.

[3] HU ZHONGXIAO, TIAN YAN, YANG HEHUA, et al. Construction and application of Chinese super rice varieties database[J]. Hybrid Rice, 2017, 32(3): 5-9.

[4] HU ZHONGXIAO, TIAN YAN, XU QIUSHENG. Review of extension and analysis on current status of hybrid rice in China[J]. Hybrid Rice, 2016, 31(2): 1-8.

Chapter 18
Elite Parental Lines of Super Hybrid Rice

Wang Weiping

Super hybrid rice parents include sterile lines and restorer lines, and most of the sterile lines are bred by transformation and have relatively simple pedigrees, while the restorer lines have a wide variety of sources and types, and are more complex. More than 50 elite parents of super hybrid rice have so far been applied in production according to the combination and application area.

Section 1 Elite Male Sterile Lines of Super Hybrid Rice

The male sterile linesof super hybrid rice are divided into cytoplasmic-genetic male sterile (CMS) lines and photo- and thermo-sensitive genic male sterile (PTGMS) lines. There are more than 20 elite CMS lines, in the types of Indonesia Paddy, wild abortive (WA), D, K, dwarf abortive (DA), and Honglian (HL), and more than 10 elite PTGMS lines, which are mainly derived from Nongken 58S, Annong S-1 and others.

I. CMS Lines

1. II-32A

II-32B was a progeny derived from the cross of Zhenshan 97A/IR665, and then it was used as the recurrent parent to cross and backcross with Zhending 28A to develop II-32A after multi-generations of backcrossing and selection at the HHRRC. II-32A has the cytoplasm of Indonesia Paddy rice and belongs to the type of sporophytic CMS line. It has a plant height of 100 cm, compact plant type, strong stems and high tillering capacity, good cold tolerance at the seedling stage, usually 8-10 effective panicles per plant with an average panicle length of about 22 cm, and 150-160 spikelets per panicle. Pollen abortion mainly occurs before the mononuclear stage with the rates of typical abortion, spherical abortion and stained abortion being 89.1%, 9.5% and 1.4%, respectively plus 0.043% of bagged selfed seeding, showing stable sterility. The CMS line is moderately resistant to sheath blight and bacteria blight, but moderately susceptible to rice blast. II-32A is one of the three most important CMS lines in China, characterized by remarkable growth increment, high combining ability, good flowering habit, high stigma exsertion rate, and high outcrossing seed setting rate. More than 210 hybrids using Ⅱ - 32A have been approved and used in rice production, including the eight super hybrid varieties of Ⅱ Youming 86, Ⅱ Youhang 1, Ⅱ Youhang 2, Ⅱ You 162, Ⅱ You 7, Ⅱ You 602, Ⅱ You 084 and Ⅱ You 7954. The accumulative total planting area is more than 6,656,700 ha.

2. Tianfeng A

An F_4 plant was selected from the cross of Bo B/G9248 in the Rice Research Institute of Guangdong Academy of Agricultural Sciences, and then used as the recurrent parent to backcross with Guang 23A to develop Tianfeng A, which passed the technical evaluation of Guangdong province in January 2003. Tianfeng A has a plant height of 78. 0 cm with compact plant type, strong stems, slightly broad flag leaves, intermediate tillering ability, 66 - 85 days from seeding to heading, and 13. 4 - 14. 3 leaves on the main stem. It has a high rate of stigma exsertion, which is 82. 3% under normal weather conditions, and a dual stigma exsertion rate of 62. 8%. Its stigmas are large and vigorous with good outcrossing performance. Tests show a frequency of 97% in terms of resistance to representative isolates of Guangdong rice blast; 100% resistance to the blast dominant groups ZB and ZC; and 66. 6%, 100% and 100% resistance to the secondary groups ZA, ZF and ZG, respectively. The parameters of rice quality are measured as 82. 8% brown rice, 74. 48% milled rice, 45. 8% head rice, 6. 6 mm in grain length, 3. 0 in length to width ratio, 30% in chalky grain rate with a chalkiness degree of 8. 8%, 44 mm in gel consistency, 6. 0 in alkali spreading value, 24. 7% in amylose content, and 13. 5% in protein content. Tianfeng A has a strong combining ability. A total of 97 hybrids derived from the CMS line have been approved for production, including the six super rice hybrids of Tianyou 998, Tianyou 122, Ganxin 688, Tianyou 3301, Tianyou-Huazhan and Tianyou 3618, with a total planting area of more than 4. 0747 million hectares.

3. Wufeng A

An individual F_4 plant derived from the cross of You IB/G9248 was selected and used as the recurrent parent to backcross with the WA CMS line of Guang 23A, and developed the Wufeng A in 2003 at the Rice Research Institute of Guangdong Academy of Agricultural Sciences. It passed the technical validation of Guangdong province. It has a plant height of 80 - 90 cm, needs 61 - 78 d from seeding to heading, and has 12. 2 leaves on the main stem on average. It is compact in plant architecture, loose at the early growing stages, but compact at the late stage. It has vigorous growth with moderate tillering capacity, strong stems, and high lodging tolerance. The flag leaves are straight and upright, dark green in color, and have slightly above-average respiration and moderate length. Its apiculus, sheath and stigma are all purplish red. A panicle has 150 - 160 spikelets with a length of 19 - 21 cm. The 1,000-grain weight is 22 - 24 g. The sterile plant has 100% sterility with a 99. 96% pollen sterility, and typical pollen abortion accounts for 95. 26%. The rice quality parameters are 6. 0 mm in grain length, 2. 7 in length-to-width ratio, 1% in chalky grain rate, 0. 4% in chalkiness degree, 4. 0 in the alkali spreading value, 75 mm in gel consistency and 10. 9% in amylose content. The CMS line has a fairly concentrated flowering time with good glume closure after flowering. The stigma exsertion rate is as high as 83. 1% with a dual exsertion rate of 62. 3%. It is sensitive to GA3 and 205 - 225 g/ha is generally used for seed production. Wufeng A has a high combining ability with more than 70 hybrids approved for production, including seven super hybrid varieties, namely Wuyou 308, Wufengyou T025, Wuyou 615, Wuyou 662, Wufengyou 286, Wuyouhang 1573 and Wuyou 116, with a total planting area of 2. 222 million hectares.

4. D62A

A progeny of the cross D297B/Hongtu 31 was selected and backcrossed with D(Shan) A to develop D62A at the Rice Research Institute of Sichuan Agricultural University. D62A is an early-season *indica* CMS line which has yellow awn-less glume and purple apiculus, with moderate length of the sterile lemma, intermediate grain length and small grain width forming an oval grain shape, and moderate grain weight. The grain pericarp is white, of moderate length and low width. The plant has a large number of grains per panicle, with moderate shattering. The sterility can be easily restored. The stigma exsertion rate is 75% with a high outcrossing rate. Comparing to Zhengshan 97A, it has more leaves on the main stem, better panicle exsertion, and a higher number of secondary branches. A total of 32 hybrids using D62A have been approved, including two super hybrid varieties, D You 527 and D You 202, with a total of planting area of more than 1,804,700 ha.

5. Fengyuan A

An F_3 progeny from the cross Jin 23B/V20B was selected and backcrossed for five generations with V20A to develop Fengyuang A at HHRRC. It is an early-maturing mid-season *indica* CMS line which passed the technical validation of Hunan province in 1998. The plant has a height of 70 cm, needs 76 - 81 d from seeding to heading, and has 12 - 13 leaves on the main stem, with strong tillering, moderate plant architecture, uniform heading, a high seedling-to-panicle ratio, 26. 5 g in 1,000-grain weight, 100% sterility with 99. 99% pollen sterility, mostly typical abortion. The stigma exsertion and dual exsertion rates are 57. 4% and 28. 5%, respectively. The rice quality indicators are 81. 9% brown rice, 75. 4% milled rice, 39. 5% head rice, 33% in chalky grain rate, 4. 6% in chalkiness degree, 3. 1 in length to width ratio, 6. 5 in alkali spreading value, 30 mm in gel consistency, 23. 2% in amylose content and 9. 1% in protein content. The CMS line has a high combining ability and 29 hybrids derived from it have been approved for production, including super hybrid rice varieties Fengyuanyou 299, with a total planting area of more than 906,000 ha.

6. Jin 23A

An F_5 progeny was selected from the cross of Feigai B/Ruanmi M (a Yunnan local soft rice variety) and crossed with high quality rice variety Huangjin 3 (Jin 3), and then a progeny from the three-way cross was backcrossed with V20A to develop the CMS line Jin 23A at the Changde Agriculture Science Academy, Hunan. Jin 23A is a mid-maturing early-season *indica* CMS line with WA cytoplasm. It is highly thermo-sensitive, with a period of 52 - 74 d from seeding to heading and 10. 5 - 11. 5 leaves on the main stem. The plant is 57 cm in height, with a compact plant type and thin stems. Its leaves are virid, with purple leaf sheaths, auricle, pulvinus and apiculus. It has a high tillering capacity, a high tiller-to-panicle ratio, and 7 - 9 effective panicles per plant. The panicle enclosure length is 43. 4%, with 18% enclosed grains. The panicle is 17. 5 cm long with about 85 grains per panicle and 25. 5 g in 1,000-grain weight. The grain is 9. 9 mm in length, 2. 75 mm in width and 3. 6 mm in length to width ratio. The pollen of the CMS line is completely abortive with typical, spherical and stained abortion respectively accounting for 76. 14% , 23. 65% and 0. 16% . The pollen abortion rate is 99. 95% , with fertile spikelets accounting for 0. 05% and the selfed seeding rate is 0. 0165% . Jin 23A has an early and concentrated flowering time with a high rate of stigma exsertion, which could be higher than 90% , and a dual exsertion rate above 70% . It has a high combining ability and 162 hybrids derived from it have been approved for production, including the four super rice hybrids of Jinyou 299, Jinyou 458, Jinyou 527 and Jinyou 785, with a total planting area of more than 1,018,000 ha.

7. Rongfeng A

An F_4 progeny was selected from the cross of You IB/Bo B and crossed with You IA and backcrossed for 13 generations to develop Rongfeng A at the Rice Research Institute of Guangdong Academy of Agricultural Sciences. It passed the technical evaluation of Guangdong province in October 2005. The plant is 63 – 70 cm in height, needs 55 – 73 d from seeding to heading, and has 13 – 14 leaves on the main stem. It has a compact plant architecture, thick stems, dark green leaves, and purple leaf margins and sheaths. The panicles are about 20 cm in length with 140 – 160 grains per panicle. The enclosed grain rate is 21% , and the 1,000-grain weight is 24. 4 g. There are a few short awns on the panicle tips. The forenoon flowering rate is 85. 84% . The stigma is black and big with a stigma exsertion rate of 83. 93% and a dual exsertion rate of about 40% . The outcrossing seeding rate is over 60% . The plant is sensitive to GA3. The seed yield is generally 3. 0 t/ha and could exceed 4. 5 t/ha. The sterile plant rate is 100% . The proportions of typical or spherical, stained, and dark-stained abortive pollen are 99. 32% , 0. 63% and 0. 05% respectively. The sterility rate is 99. 95% and the bagged selfed seeding rate is 0% . The parameters of rice quality are 77. 9% brown rice, 69. 1% milled rice, 66. 2% head rice, 6. 6 mm in grain length with a 3. 3 length-to-width ratio, 8. 8% in chalky grain rate, 1. 4% in chalkiness degree, grade 2 grain translucency, 2. 3 in alkali spreading value, 26. 8% in amylose content, 36 mm in gel consistency and 10. 8% in protein content. Tests conducted at the Plant Protection Research Institute of Guangdong Academy of Agricultural Sciences show a frequency of 82. 3% in terms of resistance to the representative isolates of Guangdong rice blast, 100% and 76. 5% resistance to dominant group C and secondary group B, respectively. Overall, the CMS line is blast resistant. It has a high combining ability and 27 hybrids derived from it have been approved for production, including two super hybrid varieties, Ganxin 203 and Rongyou 225, with a total planting area of more than 1,001,300 ha.

8. Zhong 9A

A progeny from the cross of You IB/Feigai B was selected and backcrossed with You IA at CNRRI to develop Zhong 9A. It is an early-season *indica* CMS line with Indonesia Paddy cytoplasm. It passed technical evaluation in September 1997. The plant is 65 – 82. 4 cm in height with moderate tillering capacity, moderate plant architecture, well-shaped leaves, and narrow, long, and straight flag leaves. The plant has a total growth period of 84 – 95 d and 12 – 13 leaves on the main stem. The panicles are 19. 4 cm in length with 105. 1 grains per panicle. Its flowering habit is good with an early and concentrated flowering time. The stigmas are colorless with a stigma exsertion rate of 82. 3% and a dual exsertion rate of 55. 0% , and the outcrossing seeding rate could exceed 80% . The grain length is 6. 7 mm with a length-to-width ratio of 3. 1 and the 1,000-grain weight is 24. 4 g. The rice quality is generally good with

80.4% brown rice, 71.1% milled rice, 31.3% head rice, 23.7% in amylose content, 6.0 in alkali spreading value, 32 mm in gel consistency, 8% in chalky grain rate, 0.6% in chalkiness degree, and grade 3.0 translucency. It shows moderate resistance to bacterial blight (grade 3.5 on average and 5.0 at maximum). The CMS line features typical pollen abortion with 100% sterile plants, 99.93% pollen sterility, and 0.01% bagged selfed seeding. It has a good combining ability and 122 hybrids derived from it have been approved for production, including three super hybrid varieties, Guodao 1, Zhong 9, and You 8012. The total planting area is over 559,300 ha.

9. Xieqingzao A

A progeny plant from Junxie/Wenxuanqing//Qiutangzao 5 was crossed and backcrossed with a sterile plant from Dwarf A/Zhujun//Xiezhen 1 to develop this CMS line at the Guangde Institute of Agricultural Sciences, Anhui. It has the Dwarf abortive cytoplasm, needs 60.4 – 70.2 d from seeding to heading, and has 12 – 13 leaves on the main stem. The CMS line has a high combining ability and 88 hybrids derived from it have passed variety validations, including two super rice hybrids, Xieyou 9308 and Xieyou 527, with a total planting area of more than 422,700 ha.

10. Longtefu A

A progeny plant from the cross of Nongwan/Tetepu was crossed and backcrossed with the WA-type CMS line V41A (an *indica* sporophyte CMS line) at the Zhangzhou Institute of Agricultural Sciences, Fujian. The plant is 85 – 90 cm in height with a compact plant architecture, strong stems, and high tolerances to fertilizer and lodging. Its duration from seeding to heading is 79 – 91 d, and it has 13 – 14 leaves on the main stem with an average of 13.6 leaves. The leaves are dark green, thick and upright. The flag leaf is 25 – 30 cm in length. Purplish red are the leaf sheaths, leaf margins and auricle. It has a moderate tillering capacity with 8 – 10 effective panicles per plant, 21.8 cm in average panicle length, with 125.6 – 148.5 grains per panicle. The plant has oval-shaped grains, well-developed purple stigma and purplish apiculus, and well-closed glumes. Its 1,000-grain weight is about 28.0 g and the panicle enclosure rate is 100% with a low degree of enclosure. Pollen abortion mainly occurs before the mononuclear stage. The anthers are mostly water-stained milky-white, and a few of them are water- or oil-stained yellowish. The pollen abortion rates are 78.7%, 16.4% and 4.9% for typical abortion, spherical abortion, and stained abortion, respectively, with 0.902% of bagged selfed seeding. The changes in temperature and daylength may cause fertility fluctuation. The total, single and dual stigma exsertion rates are 61.02%, 34.62% and 26.4%, respectively. The color is good at maturity and in the late stage of growth, with green culms. The plant shows fairly good resistance to sheath blight, but is moderately susceptible to blast and kernel smut. The CMS line has a high combining ability and 118 hybrids derived from it have been approved for production, including two super hybrid varieties, Teyouhang 1 and Teyou 582, with a total planting area of more than 416,700 ha.

11. Shen 95A

A progeny plant from the cross of Boro II/Zhenshan 97B//(Cypress/V20B//Bengali wild rice///Fengyuan B) was crossed and backcrossed with Jin 23A at Tsinghua Shenzhen Longgang Research Institute of China National Hybrid Rice Research and Development Center to develop Shen 95A. Cypress is an environment-insensitive variety from the U. S. , and Boro II is a high quality variety from Bangladesh. The plant architecture of Shenzhen 95A is moderate. It has a high tillering capacity and a growth period of 97 – 118 d. The number of effective panicles is 2. 55 – 3. 60 million per hectare, with 85 – 150 grains per panicle, and 35 – 38 g in 1,000-grain weight. It is resistant to panicle blast (grade 3 – 7). It meets the ministerial rice quality standards of grade 2 – 4. The plant flowers early with a concentrated flowering time. The panicle enclosure rate is 30%–45% and the outcrossing seeding rate is 30%–65%. It has a stable fertility. More than 30 hybrids derived from it have been approved for production, including two super hybrid rice varieties, Shenyou 9516 and Shenyou 1029, with a total planting area of more than 366,700 ha.

12. Dexiang 074A

An F_3 progeny from the cross of Luxiang 90B/Yixiang 1B, aromatic, was selected and named 6474 – 2, crossed and backcrossed with K17A for 10 generations at Rice and Sorghum Research Institute, Sichuan Academy of Agricultural Sciences, to develop Dexiang 074A, a fine aromatic CMS line. It passed the technical evaluation of Sichuan province in August 2007. The plant has a height of about 85 cm with a moderate plant architecture and remarkable growth increment. The stem is strong and the leaves are green with short, wide and erect flag leaves. In the case spring sowing, the plant has about 15 leaves on the main stem. The period from seeding to heading is about five days shorter than that of Ⅱ – 32A. Its stigma and ligule are white, with green leaf sheath and yellow glume tip and internodes. The stigma exsertion rate is 70%– 80% with a dual exsertion rate of 40%– 50%. The stigma is thick with well-developed pinnate branches. The flowering habits are good. The sterility is complete with typical abortive pollen. Both the sterile plant rate and pollen abortion rate are 100% with a bagged selfed seeding rate of zero. The plants and seeds are aromatic. The grain size is moderate without awn, and the 1,000-grain weigh is 28 g. Five rice hybrids were approved for production, including two super hybrids, Dexiang 4103 and Deyou 4727, with a total planting area of more than 382,700 ha.

13. Jifeng A

A progeny from the cross of Rongfeng B/BL 122, which carries the broad-spectrum blast resistance gene *Pi-1*, was selected and backcrossed using the MAS technique, and selected with the target genes with desirable agronomic traits at the Rice Research Institute of Guangdong Academy of Agricultural Sciences, then backcrossed with Rongfeng A to develop Jifeng A. It passed the technical evaluation of Guangdong province in June 2011. The plant height is about 66 cm, with a duration of 57 – 75 d from seeding to heading, and 11 – 13 leaves on the main stem. The plant type is moderate with thick stems, dark green leaves, and purple leaf margins and leaf sheaths. The panicle length is about 20 cm with a few short awns at the top. The number of grains per panicle is 102 – 125, and the panicle enclosure rate is 27. 1%. The 1,000-grain weight is 24 g and the sterile plant rate is 100%. The CMS sterility is 100% with 99. 65% typical or spherical pollen abortion and 0. 35% stained pollen abortion. The flowering period is concentrated with a forenoon flowering rate of 78. 11%. Its purple stigma is big and thick with a total exsertion rate of 59. 75% and a dual exsertion rate of 31. 62%. The outcrossing seeding rate exceeds 60%. The seed yield could generally reach 3 t/ha or up to 4. 5 t/ha. In terms of rice quality, it is 73. 2% brown rice, 68.8% milled rice, 60.8% head rice, 7.2 mm in grain length with a length to width ratio of 3.3, 24.0% in chalky grain rate, 6.4% in chalkiness degree, grade 2 translucency, 3.3 in alkali spreading value, 24.7% in amylose content, 44 mm in gel consistency, and 10.3% in protein content. Artificial inoculation conducted at the Plant Protection Research Institute of Guangdong Academy of Agricultural Sciences showed a frequency of 100% resistance to 30 representative isolates of Guangdong rice blast and 100% resistance to both dominant group C and secondary group B. Natural inducement tests in areas with a high incidence of rice blast show a high resistance with a "0" scale of invasion of leaf blast and panicle blast. The CMS line has a high combining ability and a total of 25 hybrids have been approved for production, including two super hybrids, Jiyou 225 and Jifengyou 1002, with a total planting area of more than 35,300 ha.

14. Neixiang 2A

A progeny of 1521B, which was derived from the cross of IR8S///Digu B/Dali B//Digu B, was crossed and backcrossed with Yixiang 1B, and then a selected elite plant was crossed and backcrossed with 88A to develop Neixiang 2A after multiple generations of selection and backcrossing at the Neijiang Hybrid Rice Science and Technology Development Center. It passed the technical evaluation of Sichuan province in August 2003. Neixiang 2A is a mid- to late-maturing CMS line with the same duration from seeding to heading as that of Ⅱ-32A. It has compact plant architecture, a high tillering capacity, strong stems and high tolerance to fertilizer and lodging. The leaf sheath, auricle, leaf margin and apiculus are all purple. The leaves are wide and upright. A panicle has about 170 grains and the 1,000-grain weight is 30.0 g. Pollen abortion is complete with 100% sterile plants and 100% pollen sterility, and 0.02% bagged selfed seeding. The sterility is dominantly typical pollen abortion with some stained pollen. The CMS line flowers early with a wide glume opening angle for a fairly long time, and has a high stigma exsertion rate and good outcrossing performance. The seed yield is generally 2.25 t/ha, and could exceed 3.0 t/ha. The grains meet 12 grade 2 premium rice parameters as tested at the Rice Quality Testing Center of the MOA. The resistance to rice blast is equivalent to that of Zhenshan 97A. A total of 12 hybrids derived from this CMS line have been approved for production, including super hybrid variety Nei 2 You 6 (Guodao 6), with a total planting area of more than 286,000 ha.

15. Neixiang 5A

A maintainer line N7B, which was developed at the Neijiang Hybrid Rice Science and Technology Development Center, with good outcrossing performance, resistance to diseases and big grains, was crossed with Yixiang 1B, and a progeny with elite agronomic and aromatic traits was selected and crossed with 88A. After multiple backcrossing, Neixiang 5A, a CMS line with high quality and high combining ability was developed and passed the technical evaluation of Sichuan province in August 2005. The plant has a height of 62 cm, needs 90-94 d from seeding to heading, and has 13.0 leaves on the main stem. It has a compact plant architecture, high tillering capacity, and about 12 effective panicles per plant. The leaves are dark green and the leaf blade is narrow and upright and slightly revolute. The leaf sheath, auricle, leaf margin, and apiculus are all purple. The grains are long without awn with a length to width ratio of 3.5. The number of grains per panicle is 135 and the 1,000-grain weight is about 30.0 g. The CMS line has 100% sterile plant rate and pollen abortion rate with 98.5% typical pollen abortion, 1.5% of spherical pollen abortion, and 0% of bagged selfed seeding. The total stigma exsertion rate is 70.43% including 27.82% of dual exsertion. The stigma has strong viability and long opening time. The CMS line is sensitive to GA3 with generally 225 g/ha applied in seed production, and the seed yield is generally 3 t/ha and could be as high as 4 t/ha. The parameters of rice quality are 79.2% brown rice, 72.4% milled rice, 67.4% head rice, 5% in chalky grain rate, 0.4% in chalkiness degree, grade 1 translucency, 7.0 in alkali spreading value, 70 mm in gel consistency, 13.5% in amylose content and 9.3% in protein content. Its resistance to leaf blast is graded 3 – 6, and the resistance to panicle blast is 3 – 5, classed as resistant or moderately susceptible to rice blast. The CMS line has been used for 22 hybrids, including the super hybrid variety Wuyou 8015, which has a total planting area of more than 138,700 ha.

16. Yixiang 1A

A fragrant and glutinous progeny plant from the cross of D44B (female) and N542, which was a mutant induced by ^{60}Co radiation from a Yunnan local aromatic variety, was selected and crossed with D44A. After multiple generations of backcrossing, Yixiang 1A was developed at the Yibin Academy of Agricultural Sciences and passed the technical evaluation in Sichuan province in July 2000. The CMS line is aromatic with D-type cytoplasm. The plant height is 80 – 90 cm. It takes 85 – 95 d from seeding to heading and there are 14.0 – 15.0 leaves on the main stem. The plant architecture is straight with long flag leaves, narrow and erect, dark green leaves, and colorless leaf sheath, stigma and apiculus. It has a high tillering capacity with about 12 effective tillers per plant, 24.3 – 28.5 cm in panicle length. The length to width ratio of the grains is 3.2 and there are 120 – 130 grains per panicle. The 1,000-grain weight is 30.5 g. The pollen abortion rate is 100% with 96.0% typical pollen abortion, and 4.0% spherical pollen abortion. The pollen abortion is complete with stable sterility. The stigma is colorless, of a moderate size, and remarkably viable. The glume opening lasts fairly long. The stigma exsertion rate is 62.1% and the dual exsertion rate is 29.2%. Yixiang 1A is sensitive to GA3 and a dosage of 225 g/ha is preferred in seed production with a general seed yield of 2.95 t/ha and a maximum of 4.50 t/ha. Resistance to leaf blast is graded 7 and resistance to panicle blast is graded 5. The rice quality indicators are 78.9% brown rice, 72.6% milled rice, 64.3% head rice, 7.3 mm in grain length with a length to width ratio of 3.2, 1% in chalky grain rate, 0.1% in chalkiness degree, grade 1 translucency, 7.0 in alkali spreading value, 82 mm in gel consistency, 14.7% in amylose content and 11.6% in protein content. The CMS line has a high combining ability with 74 hybrids approved for production, including three super hybrids, Yiyou 673, Yixiangyou 2115 and Yixiang 4245, with a total planting area of more than 716,000 ha.

17. Q2A

An F_2 plant from the cross of Jin 23B/Zhongjiu B was crossed with another F_2 plant from the cross of 58B/Ⅱ-32B, and a F_2 progeny was selected and crossed and backcrossed with Zhenshan 97A to develop Q2A through joint efforts of Chongqing Academy of Agricultural Sciences and Chongqing Seed Company. Q2A is an *indica* WA CMS line and passed the technical evaluation of Chongqing in July 2003. It has a growth duration of about 120 d with 14.5 leaves on the main stem. The plant is about 90 cm in height and has a loose plant architecture, upright leaves, strong stem, and dark green leaf color. The leaf sheath, auricle, pulvinus, stigma and apiculus are all purple. The plant has an intermediate tillering capacity and a high tiller-to-panicle ratio with about 10 effective panicles per plant. The panicle is about 25 cm in length, large and contains many grains with a moderate density, about 240 grains per panicle. The panicle enclosure rate is 31.12% and the rate of enclosed grains is 15.75%. The grain is 7.3 mm in length and 2.4 mm in width, with a length to width ratio of 3.0. Spherical pollen abortion is the dominant type of abortion, and the rates of spherical pollen abortion, typical pollen abortion, and stained pollen abortion are 58.10%, 41.72% and 0.18%, respectively, 100% abortive. Flowering is concentrated with 73.2% forenoon flowering. Without spraying GA3, its total stigma exsertion rate is 96.6%, with a dual exsertion rate of 78.4% and a single stigma exsertion rate of 18.2%. It is superior to Zhenshan 97A in terms of comprehensive resistance to rice blast. The rice quality is measured as 79.8% brown rice, 71.4% milled rice, 69.4% head rice, 6.3 mm in length with a length to width ratio of 2.9, 4.0% in chalky grain rate with a 0.2% chalkiness degree, grade 1 translucency, 7.0 in alkali spreading value, 77 mm in gel consistency, 14.7% in amylose content and 12.3% in protein content. Three hybrids derived from this CMS line, Q2 You 3, Q You 6 and Q You 5, have been approved for production. The super hybrid rice Q You 6 has a total planting area of more than 1.402 million hectares.

18. K22A

A progeny plant from the cross of Ⅱ-32B/02428//Xieqingzao/Keqing B was selected and crossed with Keqing A, and the male parent was anther-cultured to speed up stabilization and was used for backcrossing to develop the new CMS line K22A at the Rice and Sorghum Institute of Sichuan Academy of Agriculture Sciences. K22A has a moderate growth period lasting 70 d from seeding to heading and has 14.0 leaves on the main stem. It has a plant height of about 68 cm, erect leaves with moderate width, a compact plant type, and a high tillering capacity. The leaf margin, leaf sheath, ligule, apiculus and stigma are all purple. The grain is yellow with a length to width ratio of about 2.85. The pollen abortion is dominantly typical abortion with a sterile plant rate of 100%, a sterility degree of over 99.98%, and bagged selfed seeding less than 0.01%. Without GA3 spraying, the total stigma exsertion rate is about 48% with a high outcrossing rate. It is suitable for the southern *indica* rice growing region in China. Four hybrids derived from this CMS line have been approved for production, including two super hybrid rice varieties, Yifeng 8 and Qiannanyou 2058, with a total planting area of more than 36,000 ha.

19. Luohong 3A

An early-maturing plant, T08B, which was obtained through radiation-induced mutation of Yuetai B, was crossed and backcrossed with Yuetai A to develop Luohong 3A at the College of Life Sciences of Wuhan University. It is a CMS line with HL-type cytoplasm and passed the technical evaluation of Hubei province in 2006. The CMS plant is 86 cm in height, has a compact plant type, a high tillering capacity, dark green, narrow, long and straight leaves, and medium-sized flag leaves. The panicle is large with slender grains with a few spikelets degenerated on the panicle top. The leaf sheath, lemma tip, and stigma are colorless, and the anthers are thin and pale yellow. The panicle is 23.8 cm in length with about 189 spikelets per panicle. The 1,000-grain weight is 24.5 g. If it is seeded in early May in Hubei, the duration from seeding to heading is about 72 d with an average of 14. 6 leaves on the main stem. The CMS line has a good flowering habit with a total stigma exsertion rate of 89. 6% , including 61. 6% of dual exsertion. The sterility is stable with 100% sterile plant rate in a 1,000-plant population and the pollen sterility rate is 99. 96% and the bagged selfed seeding is 99. 97% . Its pollen abortion is of the gametophyte type, dominated by spherical pollen abortion with a few spikelets of stained pollen abortion. The grain quality meets the national standard for premium rice grade 2. One super rice hybrid derived from this CMS line, Luoyou 8 (Honglianyou 8), has been in rice production with a total planting area of more than 492,000 ha.

20. Yongjing 2A

An F_5 plant from Ningbo 2 was selected and crossed and backcrossed with Ning 67A to develop Yongjing 2A through joint efforts of the Ningbo Agricultural Science Research Institute and Ningbo Seed Company. It passed the technical evaluation of Zhejiang province in September 2000. It is a mid-maturing late-season *japonica* CMS line with Dian-1 cytoplasm. The CMS line has uniform agronomic traits with a semi-dwarf plant architecture. The sterile pollen rate is 99. 99% , dominated by stained pollen abortion with a bagged selfed seeding rate of 99. 98% . The stigma exsertion rate is 30. 94% . Flowering is concentrated with a single panicle flowering duration of 6 d, a single plant flowering duration of 11 d, and a field outcrossing seed setting rate of 59. 97% . The grain quality meets the ministerial standards for premium rice grade 1 in terms of brown rice, milled rice, chalky grain rate, chalkiness degree, translucency, gelatinization temperature and amylose content. It is resistant to rice blast and moderately resistant to bacterial blight and bacterial leaf streak. Nine *indica-japonica* hybrids derived from this CMS line have been approved for production, including two super hybrid rice varieties, Yongyou 6 and Yongyou 12, with a total planting area of more than 483,300 ha.

21. Shengtai A

Hunan Dongting Hi-Tech Seed Co. , Ltd. crossed and backcrossed Shen 21232—Q3B, which was a progeny from the cross of Yue 4B/Wulingxiangsi//Yue 4B, with Yue 4A to develop Shengtai A, an early-season *indica* CMS line which passed technical evaluation in Hunan in 2010. It needs 67 – 68 d from seeding to heading. The plant is 69. 0 cm in height, with a loose plant architecture, remarkable tillering capacity, and straight flag leaves. The leaf sheath, lemma and stigma are colorless. There are 14. 9 – 15. 3 leaves on the main stem and 10 – 11 effective panicles per plant with an average panicle length of 21. 6 cm. The number of spikelets per panicle is 113. 6 and the 1,000-grain weight is 25. 4 g. The CMS line has 100% sterile plants and 99. 99% sterility, dominated by typical pollen abortion. The bagged selfed seeding rate is 0. 006% . The total stigma exsertion rate is 70. 0% with 28. 9% of dual exsertion. The rate of enclosed grains per panicle is 40. 5% and the outcrossing seed setting rate is 50. 4% . The resistances to leaf blast, panicle blast and bacterial blight are graded 6, graded 7 and graded 3, respectively. The rice quality measures are 82. 0% brown rice, 68. 1% head rice, 3. 4 in length to width ratio, 4% in chalky grain rate, 0. 3% in chalkiness degree, 72 mm in gel consistency, and 16. 0% in amylose content. Four rice hybrids derived from this CMS line have been approved for production, including the super hybrid rice Shengtaiyou 722, with a total planting area of more than 19,300 ha.

Ⅱ. PTGMS Lines

1. Zhun S

HHRRC crossed N8S, which was derived from Annong S-1, with Xiang 2B, Huaizao 4 and Zaoyou 1 through two rounds of random multi-crossing and then went through pedigree selection to develop the S line and passed the technical evaluation of Hunan in March 2003. The male sterile plant has a height of 65-70 cm, loose plant architecture, light green leaves, and colorless leaf sheath, lemma tip and stigma. The duration from seeding to heading is 65-80 d. It has 11-13 leaves on the main stem, and a moderate tillering capacity. Each panicle is about 23 cm in length with 120 spikelets. The 1,000-grain weight is 28 g. The critical sterility inducing temperature is 23.5 ℃-24 ℃. Both the sterile plant rate and the pollen abortion rate are 100%, with typical pollen abortion. It flowers early with more than 75% of stigma exsertion and 50% of outcrossing seeding. Its grade of resistance to bacterial blight is 3. Three mid- and late-season *indica* super hybrids derived from this male sterile line, Zhunliangyou 527, Zhunliangyou 1141 and Zhunliangyou 608, have been used in rice production with a total planting area of more than 767,300 ha.

2. HD9802S

A plant of Huda 51 from the cross of 92010/Zaoyou 4 was crossed with Hongfuzao and the progenies went through low-temperature selection in phytotron at Hubei University to develop this early-season *indica* PTGMS line, with a critical sterility inducing temperature of 23 ℃-24 ℃, and a cold-tolerance period of five days. The sterility is controlled by a pair of recessive nuclear genes which are allelic to the TGMS genes of Xiang 125S and Zhu 1S. HD9802S has a moderate plant architecture, short and straight flag leaves in dark green, a duration of 60-78 d from seeding to heading, 12-13 leaves on the main stem, a plant height of 60.6 cm, and a panicle length of 21.3 cm. It usually has 8.5 effective panicles, each with 140-150 spikelets, and the 1,000-grain weight is 24.2 g. The male sterile line is sensitive to GA3 and 180-250 g/ha of GA3 is appropriate in seed production. Its stigma exsertion rate is 74.7% with a dual exsertion rate of 25.3%. The outcrossing seed setting rate is more than 45%. The grain is slender with a length of 6.8 mm and a length to width ratio of 3.2. The apiculus is colorless. The rice quality indicators are 76.8% brown rice, 56.2% head rice, 0% in chalky grain rate, 0% in chalkiness degree, 11.8% in amylose content, and 65 mm in gel consistency. Three early- and late-season super rice hybrids, Liangyou 287, Liangyou 6 and H Liangyou 991, have been approved at provincial level for production, with a total planting area of more than 706,700 ha.

3. Xiangling 628S

A dwarf mutant, SV14S, was obtained through somatic cell culture from a young panicle of the PTGMS line Zhu 1S, and used as the female parent to cross with ZR02, which is a variety with high quality and high resistance to blast. The progenies went through pressure selection and was developed as Xiangling 628S at Yahua Seeds Science Academy of Hunan and passed the technical evaluation of Hunan in 2008. It is a mid- to late-maturing early-season rice 58-84 d in whole growth duration. The plant has 12 leaves on the main stem and is 63-65 cm in plant height with a compact plant type, upright leaves and high tillering capacity. A single plant usually has 12-13 tillers, and 9-10 effective panicles with about 136 spikelets per panicle. The panicle is large and straight with a grain enclosure rate of 16. 6% . Its forenoon flowering rate is more than 75% , and the stigma exsertion rate is 82. 6% , including a dual exsertion of 32. 6% , and an outcrossing seeding rate more than 45% . The parameters of rice quality are 3. 0 in length to width ratio, 25 g in 1,000-grain weight, 81. 3% brown rice, 73. 6% milled rice, 68. 6% head rice, 4% in chalky grain rate, 1% in chalkiness degree, grade 1 translucency, 5. 9 in alkali spreading value, 62 mm in gel consistency and 12. 8% in amylose content. The S line has a high combining ability and 26 hybrids derived from it have been approved for production, including the super early-season rice hybrid Lingliangyou 268, with a total planting area of more than 173,300 ha.

4. Zhu 1S

Zhu 1Swas selected from the F_2 population of the cross of Kangluozao//4342/02428, which is a distant cross of different ecotypes, at the Zhuzhou Institute of Agricultural Science of Hunan province. It is a TGMS line with a critical fertility-inducing temperature below 23 ℃ and it passed the technical evaluation of Hunan province in 1998. The plant is 75 - 80 cm in height, with a moderate plant architecture. In the early growth stage, the leaves are slightly flat. The last three leaves are upright with a thick flag leaf 25 cm in length, 1. 65 cm in width, and 30 degrees in leaf angle. The leaf color is light green while the leaf sheath and apiculus are colorless. The stem is thick with four elongated internodes above ground and a short basal node. Generally, there are 11 effective tillers per plant and 7. 5 panicles per plant, 100 - 130 spikelets per panicle. The glume is light green with less glume pubescence. The grains are plump and weighted 28. 5 g per 1,000. It is resistant to rice blast and bacterial blight. The rate of enclosed panicle during the sterility period is low with a spikelets exsertion rate of 75%-80% . The parameters of rice quality are measured as 80. 4% brown rice, 72. 3% milled rice, 44. 43% head rice, 7. 1 mm in grain length with a length to width ratio of 3. 2, 40% in chalky grain rate with a chalkiness degree of 5. 5% , Grade 2 translucency, 5. 7 in alkali spreading value, 26. 3% in amylose content, 43 mm in gel consistency and 11. 2% in protein content. The rice quality meets the ministerial standards for premium rice Grade 2 or higher. The male sterile line has broad compatibility and high combining ability. More than 50 hybrids derived from the sterile line have passed the technical evaluation above the provincial level, including one super early-season hybrid, Zhuliangyou 819, with a total planting area of more than 478,000 ha.

5. Peiai 64S

Peiai 64S was developed from a cross with Nongken 58S as the female parent and Peiai 64 as the male parent, the latter being a *javanica* rice line from the cross of Peidi/Aihuangmi//Ce 64, through backcrossing at the HHRRC. It passed the technical evaluation of Hunan province in 1991. The sterile line plant is 65 - 70 cm in height, with a high tillering capacity, moderate plant architecture, moderate thickness of stems, colorless sheath and auricle, light purple stigma and apiculus, and dark green leaves. It has 13 - 15 leaves on the main stem. The flag leaf is 30 - 35 cm in length and 1. 6 - 2. 0 cm in width, with an angle of 15° - 30° between the flag leaf and the panicle. The panicle is below the canopy. The critical fertility-inducing temperature is 23. 5 ℃ with a daylength of 13 h, or over 24 ℃ when the daylength is 12 h in Hainan. It is moderately resistant to rice blast and bacterial blight, and is susceptible to rice kernel smut and sheath blight. The grain is long with a length to width ratio of 3. 1 and weighs 21 g per 1,000 grains. Peiai 64S has a high broad-spectrum compatibility. More than 50 hybrids derived from it have passed the evaluations above the provincial level, including three super rice hybrids, Liangyoupeijiu, Peizataifeng and Peiliangyou 3076, with a total planting area of more than 6.536 million hectares.

6. Y58S

Y58S was developed from the multi-crosses of Annong S-1, which was used as the TGMS donor parent, and Changfei 22B, Lemont and Ce64S, and then went through pedigree selection and testcrosses at HHRRC. It passed technical evaluation of Hunan province in 2005. It is an *indica* PTGMS line with 76-97 d from seeding to heading. The plant is 65-85 cm in height with a moderate plant architecture, light green leaf color, green leaf sheath, yellow lemma tip, and white stigma. The leaf is long, straight, narrow, concave and thick. The plant has 12-15 leaves on the main stem, usually 9-12 effective panicles per plant, 26 cm in length and about 150 spikelets per panicle. The 1,000-grain weight is 25 g. Both sterile plant rate and sterile degree are 100% with typical abortive pollen. The male sterile line has a critical fertility-inducing temperature lower than 23 ℃. The total stigma exsertion rate is 88.9%, including a dual exsertion rate of 59.6%. The outcrossing seeding rate is 53.9%. The resistance to both leaf blast and panicle blast are graded 3, and that to bacterial blight is graded 5. The parameters of rice quality are 79.3% brown rice 70.9% milled rice, 66.8% head rice, 6.2 mm in grain length with a length to width ratio of 2.9, 5% in chalky grain rate with a chalkiness degree of 0.8%, grade 2 translucency, 7.0 in alkali spreading value, 66 mm in gel consistency, 13.7% in amylose content and 11% in protein content. The yield of parent seed production is usually 4.5 t/ha. The sterile line has a high general combining ability with more than 100 Y Liangyou series hybrids approved for production, including seven super hybrid rice varieties recognized by the MOA, namely Y Liangyou 1, Shenliangyou 5814, Y Liangyou 087, Y Liangyou 2, Y Liangyou 5867, Y Liangyou 900, and Y Liangyou 1173, with a total planting area of more than 3.81 million hectares.

7. Shen 08S

Shen 08S was developed through a cross of Y58S/Zaoyou 143 at the Tsinghua Shenzhen Longgang Institute of China National Hybrid Rice Research and Development Center. It is a PTGMS line developed through multi-generations of pedigree selection with a critical fertility-inducing temperature of 23.5 ℃. It passed the technical evaluation of both Guangdong and Anhui provinces in 2009 and 2012, respectively. The plant is about 70 cm in height, with strong stems, straight and inward curling leaves, and a good plant architecture. There are 14 leaves on the main stem. The duration from seeding to heading in summer is 85-94 d. The leaf blade is light green, and the leaf sheath and tip of the lemma are colorless. There are short awns on the grains. The total grain number per panicle is about 180 and the 1,000-grain weight is about 24 g. Shen 08S has a high stigma exsertion rate of about 85% with a dual exsertion rate of about 60%, and an outcrossing seeding rate over 50%. Sterility is dominated by pollen-free abortion. It is sensitive to GA3 and the dosage is usually 180-225 g/ha in seed production. The male sterile line has a high combining ability and 25 hybrids derived from it have passed validations at provincial and national levels, including two super rice hybrids, Shenliangyou 870 and Shenlianyou 8386, with a total planting area of more than 27,300 ha.

8. Guangzhan 63S

Guangzhan 63S was developed from the cross of N422S/Guangzhan 63 after 11 generations of selection at the North China Japonica Hybrid Rice Engineering Technology Center. It passed the technical evaluation in Anhui in August 2001. N422S is a widely compatible variety and was used as the female parent, while Guangzhan 63 is a high-quality *indica* variety from Guangdong province. Guangzhan 63S is a PTGMS line with a critical fertility-inducing temperature of 23.5 ℃, under a daylength over 14 h. Its selfed seeding rate is below 0.05%. The plant is about 80 cm in height, and the duration from seeding to heading is 69 – 78 d with 12.8 – 14.1 leaves on the main stem. The panicle is 22.5 cm in length and there are 7 – 9 panicles per plant and 140.6 – 165.2 grains per panicle. The 1,000-grain weight is 25.0 g. The forenoon flowering rate is about 65%. The stigma is colorless with an exsertion rate of 74.2%, including a dual exsertion of 45%. The outcrossing seed setting rate could reach 48% and the sterility type is pollen-free sterility. The parameters of rice quality are 79.6% brown rice, 72.6% milled rice, 64.5% head rice, 6.5 mm in grain length with a length to width ratio of 2.9, 4% in chalky grain rate with a 0.2% chalkiness degree, Grade 1 translucency, 7 in alkali spreading value, 78 mm in gel consistency, and 12.7% in amylose content. It is highly resistant to bacterial blight (lesion area 3%–5%) and resistant to rice blast. It has a high combining ability and 19 rice hybrids derived from it passed validations above the provincial level, including one super rice hybrid, Fengliangyouxiang 1, with a total planting area of more than 970,000 ha.

9. Guangzhan 63 – 4S

Guangzhan 63 – 4S was re-selected from Guangzhan 63S through pedigree selection at the North China Japonica Hybrid Rice Engineering Technology Center. It passed the technical evaluation in Jiangsu province in 2003 (introduced by the Institute of Agricultural Sciences of Lixiahe). The duration from seeding to heading is 75 d. The plant has a moderate architecture with a plant height of 85 cm and a high tillering capacity. It has narrow and deep-colored leaves. There are five elongated internodes above the ground and 14 – 15 leaves on the main stem. The sterility of the line is stable. When the daylength is 14.5 h, the critical fertility-inducing temperature is below 23.5 ℃ and below 24 ℃ when the daylength is less than 12.5 h, as tested at the Nanjing Institute of Meteorology and Huazhong Agricultural University. During the sterility period, the sterile plant rate is 100% and the selfed seeding rate is 0%. The sterility is pollen-free sterility. It has good fertility restoration in the fertility period. Compared with Guangzhan 63S, its sterility, quality and resistance are similar, but the growth duration is 4 – 5 d longer, the plant is 3 – 5 cm taller, and there are 5 – 10 more spikelets per panicle. It has 160 spikelets per panicle and the 1,000-grain weight is 35 g. The parameters of rice quality are measured as 64.5% head rice, 2.9 in grain length to width ratio, 4% in chalky grain rate with a 0.2% chalkiness degree, grade 1 translucency, 12.7% in amylose content, 7 in alkali spreading value, and 78 mm in gel consistency. Four *indica* super hybrid rice varieties derived from the male sterile line, Yangliangyou 6, Guangliangyouxiang 66, Guangliangyou 272 and Liangyou 616, have been approved for production with a total planting area of more than 3.114 million hectares.

10. 1892S

The PTGMS line 1892S is a variant of Peiai 64S after eight generations of selection at the Rice Research Institute of Anhui Academy of Agricultural Sciences. It is an *indica* PTGMS line with a lower fertility-inducing temperature and high combining ability. It passed the technical evaluation of Anhui province in August 2004. It is a mid-season rice suitable for the Yangtze River Basin. The plant is 62. 7 cm in height and the duration from seeding to heading is 70 – 87 d with 15. 0 – 16. 3 leaves on the main stem. The panicle is 15. 8 cm in length and there are 8. 6 panicles per plant and 136. 3 spikelets per panicle. The 1,000-grain weight is 22 g. Its flowering time is concentrated. The stigma is purple with an exsertion rate of 87% and a dual exsertion rate of 46% . The outcrossing seed setting rate is 62% . The panicle enclosure rate is 50% . It is sensitive to GA3 and generally 375 – 450 g/ha is applied in seed production. It is resistant to bacterial blight (Grade 3, moderately resistant) and to rice blast (Grade 2, resistant) as tested by artificial inoculation at the Plant Protection Institute of Anhui Academy of Agricultural Sciences. The parameters of rice quality are measured as 79. 5% brown rice, 74. 1% milled rice, 72. 6% head rice, 3. 1 in length to width ratio, 19% in chalky grain rate with a 1. 6% chalkiness degree, grade 3 translucency, 4. 2 in alkali spreading value, 15. 1% in amylose content, 96 mm in gel consistency and 9. 8% in protein content. It is up to grade 3 standards in the national standards for premium rice. The male sterile line has a high combining ability with 24 rice hybrids derived from it passing validations at the provincial or national level, including two super rice hybrids, Huiliangyou 6 and Huiliangyou 996, with a total planting area of more than 238,000 ha.

11. C815S

C815S was developed by using 5SH038 as the female parent, which is an F_6 plant from the cross of Anxiang S/Xiandang//02428, Peiai 64S as the male parent in Hunan Agricultural University, and then carrying out selection under the pressure of short daylength and low temperature in Hainan, long daylength and low temperature in summer in Changsha, short daylength and low temperature in autumn in Changsha, and in temperature-controlled water pool over 10 generations in five years. C815S is an *indica* PTGMS line with a critical fertility-inducing temperature of 23 ℃. It passed the technical evaluation in Hunan province in 2004. The plant is 71 – 75 cm in height with a compact plant type. The leaf color is dark green, while the leaf sheath, apiculus and stigma are purple. The leaves are long, straight, narrow, concave and thick. The duration from seeding to heading is 65 – 95 d. There are 13 – 16 leaves on the main stem. It has a moderate tillering capacity, 11 – 12 effective panicles per plant, 24 cm in length, and about 165 spikelets per panicle. The 1,000-grain weigh is 24 g. The sterile plant rate and sterility degree are 100% and 99. 99% , respectively. The sterility is mainly typical pollen abortion. The sigma exsertion rate is 90. 5% , with a dual exsertion rate of 62. 0% . The outcrossing rate is 55% – 60% . It is a widely compatible parent, and the seed setting rates of the hybrids are more than 81. 0% when tested with IR36, Nanjing 11, Akihikari and Balilla testers. Its resistance to both leaf blast and panicle blast are graded 7 and that to bacterial blight is graded 3. The rice quality parameters are measured as 78. 5% brown rice, 72. 5% milled rice, 71. 5% head rice, 2. 7 in grain length to width ratio, 6% in chalky grain rate, and 0. 4% in a chalkiness rate. The yield of parental seed production is generally 4. 5 t/ha. It has a high combining ability with 33 derived hybrids passing validations above the provincial level, including the super hybrid rice C Liangyou-Huazhan with a total planting area of more than 147. 300 ha.

12. Longke 638S

Longke 638S was developed by pedigree breeding from the cross of Xiangling 628S, an early-season *indica* PTGMS line, and C815S, a mid-season *indica* PTGMS line, at Yahua Seeds Science Academy of Hunan. It is a mid-season PTGMS line with a critical fertility-inducing temperature of 23. 5 ℃. The duration from seeding to heading is 103 – 109 d for spring sowing or 80 – 90 d for autumn sowing in Hunan. The number of leaves on the main stem is 15. 3. The plant is 91. 8 cm in height, and the plant architecture is compact, having strong stems, dark green and straight leaves, and green leaf sheaths. The plant tillers early with a high tillering capacity. The flag leaf is upright, long and wide. The grain is long with a length to width ratio of 2. 9. The glume is green and the apiculus and stigma are colorless. Some of its grains have awns. There are 12 – 14 effective tillers and 8 – 9 panicles per plant with the tiller-to-panicle rate over 65% . The panicle is 22. 9 cm in length with about 200 spikelets per panicle. The 1,000-grain weight is about 25 g. The sterility is complete with typical abortive pollen and the sterile plant rate and the sterility degree are both 100% . The panicle enclosure rate is 13. 8% . It is sensitive to GA3. The stigma is large and vigorous, with a stigma exsertion rate of 78. 3% , including a dual exsertion rate of 51. 5% without GA3 spraying. The forenoon flowering rate is over 86% and the outcrossing seed setting rate is over 40% . The resistance to blast at the seedling stage is graded 3, that to panicle blast is graded 5, and that to bacterial blight is graded 5. The parameters of rice quality are measured as 78. 9% brown rice, 68. 6% milled rice, 65. 2% head rice, 6. 3 mm in grain length with a length to width ratio of 2. 7, 20% in chalky grain rate with a chalkiness degree of 2. 4% , grade 3 translucency, 3. 0 in alkali spreading value, 88 mm in gel consistency, and 12. 4% in amylose content. The male sterile line has a high combining ability with 33 hybrids derived from it passing the validations above provincial level, including the super hybrid rice Longliangyou-Huazhan which has a total planting area of more than 16,000 ha.

13. Xin'an S

Xin'an S, a mid-season *indica* PTGMS line, was developed through seven generations over five years from the cross of Guangzhan 63 – 4S and M95, is a *javanica* rice with light brown markers, at the Quanyin Agricultural High-tech Research Institute, Anhui province, Xin'an S passed the technical evaluation of Anhui province in 2004, and passed the validation of Anhui province in 2005. The plant is about 80 cm in height. The leaves are straight and concave in dark green with 14. 1 – 14. 5 leaves on the main stem, and the exposed culm is reddish. Sown in Hefei, Anhui province, in May or June, the duration from seeding to heading is 73 – 89 d, with a whole growth period of about 120 d. The male sterility is stable with 100% of sterile plant, a sterility degree of 99. 99% , and a 0% bagged selfed seeding rate. With a daylength of 14. 5 h daylength and a temperature of 23. 5 ℃, 99. 57% its pollen are abortive. The number of spikelets per panicle is 165 – 185. The stigma is colorless with a stigma exsertion rate of 79. 5% , including 42. 5% of dual exsertion. It is sensitive to GA3 with a high outcrossing rate. The 1,000-grain weight is about 26 g and the glume is brown. The parameters of rice quality are measured as 78. 2% brown rice, 71. 0% milled rice, 65. 2% head rice, 12% in chalky grain rate with 1. 0% chalkiness degree, grade 1 translucency, 7.0 in alkali spreading value, 14.1% in protein content, 68 mm in gel consistency, 10.1% in amylose content, and 6.2 mm in grain length with a length to width ratio of 2.7. Its resistance to bacterial blight is graded 5 (moderately susceptible) and that to rice blast is graded 1 (resistant). The sterile line is suitable for developing mid-season *indica* hybrid rice, and 12 hybrids derived from it have passed validations, including the super hybrid rice Xinliangyou 6 (Wandao 147) with a total planting area of more than 2.06 million hectares.

14. 03S

03S is a PTGMS line selected after six generations from the cross of Guangzhan 63 - 4S/Duoxi 1 developed in the Quanyin Agricultural High-tech Research Institute of Anhui province. Two super hybrid rice varieties derived from the sterile line, Xinliangyou 6380 and Liangyou 038, have passed variety validations, with a total planting area of more than 315,300 ha.

Section 2 Major Restorer Lines of Super Hybrid Rice

The restorer lines of super hybrid rice are classified as CMS restorer lines and PTGMS restorer lines. The CMS restorer lines are mainly derived from Minghui 63, Milyang 46, Teqing and Gui 99, while the PTGMS restorer lines are widely sourced, mainly from conventional varieties or lines and from CMS restorer lines. At present, there are more than 20 elite restorer lines of super hybrid rice according to the combination selection and application area.

Ⅰ. CMS Restorer Lines

1. Shuhui 527

Shuhui 527 was selected through pedigree breeding from the cross of Hui 1318/88 - R3360 at the Rice Research Institute of Sichuan Agricultural University. Hui 1318 was a progeny from the cross of Gui 630/Gu 154//IR1544 - 28 - 2 - 3, and 88 - R3360 was derived from the cross of Fu 36 - 2/IR24 with good grain quality. Shuhui 527 has green leaves with green sheaths, erect flag leaves, moderate glume opening and time, early and long flowering, plump anthers, strong stems, a moderate number of leaves on the main stem, moderate panicle size, long and big grains, a large number of grains per panicle, a high seed setting rate and strong restorability. More than 40 hybrids derived from it have been approved for preproduction, including five super rice hybrids, You 527, Xieyou 527, Zhunliangyou 527, Yifeng 8 and Jinyou 527, with a total planting area of more than 2.7907 million hectares.

2. Huazhan

Huazhan was developed at the CNRRI using SC02 - S6, which was introduced from Malaysia, as the basic material through system selection and repeated testcrosses. It has strong restorability and has been a popular elite restorer line in China with 57 CMS and PTGMS hybrids approved for production, including three super rice hybrids, Tianyou-Huazhan, C Liangyou-Huazhan and Longliangyou-Huazhan, with a total planting area of more than 1.248 million hectares.

3. Xianghui 299

HHRRC crossed the restorer line R402, which is early-season rice with high combining ability, resistance to rice blast and high grain quality, as the female parent, with the male parent Xianghui 20, which has affinity with *japonica* rice. Then, Xianghui 299 was developed through multi-generations of pressure selection and testcrossing. The restorer line has a good plant type with high restorability, a wide restoring spectrum, high combining ability, sufficient pollen load, and high seed yield in seed production. Four rice hybrids have been approved for production, including two late-season super rice hybrids, Fengyuanyou 299 and Jinyou 299, with a total planting area of over 987,300 ha.

4. Zhonghui 8006

An F_4 progeny from the cross of Duoxi 1/Minghui 63 as the female parent was crossed with the male parent IRBB60 to develop Zhonghui 8006 at CNRRI through multi-generations of selection and MAS. Four rice hybrids derived from this R line have been approved for production, including three super rice hybrids, Guodao 1, Guodao 3 and Guodao 6, with a total planting area of more than 883,300 ha.

5. R225

R225 was bred at the Rice Research Institute of Jiangxi Academy of Agricultural Sciences fromthe mutant line R998. Three late-season hybrids have been approved for production, including two super rice hybrids, Rongyou 225 and Jiyou 225, with a total planting area of over 414,700 ha.

6. R7116

R7116 was developed at the Shenzhen Zhaonong Agricultural Technology Co. from the cross of R468 and an F_2 progeny of Lunhui 422/Shuhui527, after six generations of selection, of which, R468 was an F_8 progeny from the cross of Minghui 63/Tetepu. Nine hybrids using R7116 as the male parents have been approved for production, including two super rice hybrids, Shenyou 9516 and Wuyou 116 with a total planting area of more than 366,700 ha.

7. F5032

F5032 is a *japonica* restorer line of hybrid rice developed at the Crop Research Institute of Ningbo Academy of Agricultural Sciences. It has been used in 15 rice hybrids for production, such as Yongyou 12, Yongyou 13 and Yongyou 15. The two super rice hybrids, Yongyou 12 and Yongyou 15, have a total of planting area of more than 371,300 ha.

8. Hang 1

A mutant from Minghui 86 through satellite-carried dried seeds was selected and went through multi-generations of shuttle breeding in different ecological zonesin Fuzhou, Sanya, Shanghang and Nanjing, to develop this R line at the Rice Research Institute of Fujian Academy of Agricultural Sciences. Hang 1 has a greater plant height, longer panicles, higher grain quality and higher resistance to rice blast, comparing with the original Minghui 86. Four hybrids derived from this R line, such as I Youhang 1, Guyouhang 1, Ⅱ Youhang 1 and Teyouhang 1, have been approved for production. The two super rice hybrids Ⅱ Youhang 1 and Teyouhang 1 have a total planting area of more than 1.006 million hectares.

9. Minghui 86

P18, which was from the cross of IR54/Minghui 63//IR60/Gui 630, as the female parent, was crossed with Minghui 75, an *indica-japonica* restorer line from the cross of *japonica* 187/IR30//Minghui 63, to develop Minghui 86 at the Sanming Institute of Agricultural Sciences of Fujian after eight generations and six years of plant selection and testcrosses. It is resistant to rice blast and moderately resistant to bacterial blight. It has a high combining ability and strong restorability. A total of 15 rice hybrids derived from this R line have been approved for production, including the super hybrid rice Ⅱ Youming 86 with a total planting area of more than 1.3707 million hectares.

10. Hang 2

Hang 2 was derived from dried seeds of the restorer line Minghui 86, which was mutagenized by high-altitude radiation ona satellite, through shuttle and targeted breeding in different ecological zones and various locations in Fujian and Sanya in Hainan. Its overall traits are better than those of Minghui 86. Hybrid rice varieties, such as Liangyouhang 2, Teyouhang 2 and Ⅱ Youhang 2, have been approved for production. The super rice hybrid Ⅱ Youhang 2 has a total planting area of more than 213,300 ha.

11. Luhui 17

Luhui 17 was developed from the cross ofthe widely compatible *japonica* line 02428 and Gui 630 at the Rice and Sorghum Institute of Sichuan Academy of Agricultural Sciences. It has been used in five hybrids, including Xieyou 117, Chuannong 2, B You 817, K You 17 and Ⅱ You 7, in production. The super rice hybrid Ⅱ You 7 has a total of planting area of more than 990,000 ha.

12. Luhui 602

A progeny plant from the cross of 02428, a widely compatible *japonica* line, and Gui 630, an R line with high restorability and combining ability, was selected and crossed with IR244, which is tolerant to stem borer, the progeny then went through six generations over 10 years of selection and shuttle breeding, and five generations of anther culture to develop Luhui 602 at the Rice and Sorghum Institute of Sichuan Academy of Agricultural Sciences in 1996. This R line needs 105 d from seeding to heading. The plant is 115 cm in height, with a regular plant architecture, high tillering ability and strong stems. There are 15 – 16 leaves on the main stem, with upright and wide leaves in dark green, and wide flag leaves. The anther is large and plump with a huge load of pollen. The panicle is spindle-shaped and long, with an average length of 25 cm, and about 150 grains on each. The seed setting rate is 90%, and the 1,000-grain weight is 35.5 g. The grain is light yellow, long and oval, with a length to width ratio of 2.79. It is moderately resistant to rice blast, suffers less from sheath blight, and has high resistance to lodging. The R line is closer to the *indica* type from *indica/japonica* cross, with high combining ability. The rice hybrids derived from it, such as K You 8602 and Ⅱ You 602, have been approved for production. The super rice hybrid Ⅱ You 602 has a total planting area of more than 410,700 ha.

13. Zhenhui 084

An F_5 progeny, 91 – 2156, derived from the cross of Minghui63/Teqing, was used as the female parent to cross with R16 and developed the R line Zhenhui 084, which is an *indica* restorer line with high quality and high resistance to diseases, at the Jiangsu Qiuling Area of Zhenjiang Research Institute of Agricultural Sciences in 1997. The plant is about 115 cm in height, with a whole growth period of 141 d. It has a compact plant architecture, upright leaves with light leaf color, strong stems, high tillering ability, high tiller-to-panicle ratio, small panicles, high lodging resistance, good appearance at maturity in the late growth period, and early maturity. It is resistant to panicle blight and bacterial blight, and moderately resistant to sheath blight. The parameters of rice quality are measured as 58.2% head rice, 3.1 in length to width ratio, 8% in chalky grain rate with a 0.6% chalkiness degree, 98 mm in gel consistency, and 13.3% in amylose content. The hybrids derived from this R line are Zhenshanyou 184, Shanyou 084, Xieyou 084, Fengyou 084, Tianfengyou 084 and Ⅱ You 084, which all have been approved for production. The super rice hybrid Ⅱ You 084 has a total planting area of more than 1.448 million hectares.

14. Zhehui 7954

Zhehui 7954 was developed by pedigree breeding from the cross of R9516, an R line derived from Peiai 64S/Teqing, as the female parent and M105, derived from the cross of Milyang 46/Lunhui 422, at the Zhejiang Academy of Agricultural Sciences. It has dark green leaves, moderate length and thickness of the stems, long and upright flag leaves, many leaves on the main stem, moderate panicle length with compact seeding density, high numbers of secondary branches and grains per panicle, high seed setting rate, intermediate grain length, big grain width, and large grain weight. It is resistant to bacterial blight. The hybrid rice with Zhehui 7954 as the male parents and have been approved for production are Gangyou 7954, Qianyou 1, Xieyou 7954 and Ⅱ You 7954. The super hybrid rice Ⅱ You 7954 has a total planting area of over 650,000 ha.

15. Shuhui 162

Milyang 46, as the female parent, was crossed with an F_8 progeny of the cross of 707/Minghui 63, then the F_1 was anther-cultured and the progenies were selected for good maturity color, high resistance to lodging without premature senescence, to develop the R line Shuhui 162 at Sichuan Agricultural University. Shuhui 162 has a kinship with Milyang 46 and African rice. The rice hybrids of D You 162, Ⅱ You 162 and Chiyou S162 have been approved for production. Ⅱ You 162 was recognized as super hybrid rice with a total of planting area of more than 883,300 ha.

16. Guihui 582

The restorer line Guihui 582 was developed through pedigree breeding from the cross of Gui 99, as the female parent and a progeny of the cross of Calotoc/02428 at the Rice Research Institute of Guangxi Academy of Agriculture Sciences. Two super rice hybrids derived from this R line, Teyou 582 and Guiliangyou 2, have been used in production with a total planting area of more than 215,300 ha.

17. Fuhui 673

A mutant derived from the dried seeds of the restorer line Minghui 86, which was mutagenized by high-altitude radiation on a satellite, as the female parent, was crossed with Tainong 67, then an F_2 plant was selected as the female parent to crossed with N175, and then the progenies went through five generations of selection to develop Fuhui 673 at the Rice Research Institute of Fujian Academy of Agricultural Sciences. The R line has been used in 10 rice hybrids approved for production, including the super rice hybrid Yiyou 673 with a total planting area of more than 401,300 ha.

18. Chenghui 727

This R line was developed from the cross of Chenghui 177/Shuhui 527 at the Crop Research Institute of Sichuan Academy of Agricultural Sciences. Twenty-two rice hybrids using this R line as the male parents have been approved for production, including the super hybrid rice Deyou 4727 with a total planting area of more than 38,000 ha.

19. Guanghui 998

Guanghui 998 was developed from an F_7 progeny from the cross of R1333/R1361 at the Rice Research Institute of Guangdong Academy of Agricultural Sciences. The female parent R1333 was bred by using an F_1 progeny from the cross of Guanghui 3550/518 and Zhenguiai, then crossing an F_4 progeny of Guanghui 3550/518//Zhenguiai with Minghui 63. The male parent R1361 was from the cross of 836 - 1 and BG35, which was introduced from Sri Lanka. Guanghui 998 has narrow and straight upper functional leaves in the shape of an inward-curling tube, dark green leaves, moderate leaf length and width, thick mesophyll and strong leaf veins. It has a high combining ability with the CMS lines of Zhenshan 97A, Ⅱ-32A, You 1A, Yuefeng A, Huanong A, Zhongjiu A, etc., with high seed setting rates. Sixteen rice hybrids using this R line have been approved for production, including the super rice hybrid Tianyou 998, with a total planting area of more than 1.7467 million hectares.

20. Guanghui 308

The Rice Research Institute of Guangdong Academy of Agricultural Sciences developed Guanghui 308 bycrossing a progeny of Zhaoliuzhan/Sanhezhan, which has high grain quality and resistance to diseases, as the female parent, with Guanghui 122 and going through eight generations of selection on quality, resistance and testcrossing results over four years. Five rice hybrids derived from this R line have been approved for production, including the late-season super rice hybrid Wuyou 308, with a total planting area of over 1.3293 million hectares.

21. Guanghui 122

The Rice Research Institute of Guangdong Academy of Agricultural Sciences developed Guanghui 122 by crossing a late-maturing F_5 progeny with excellent overall traits from the cross of Minghui 63/Guanghui 3550, as the female parent, with 836 - 1, which is highly resistant to rice blast, and going through multi-generations of selection and testcrossing. Seven rice hybrids derived from this R line have been approved for production, including the super hybrid rice Tianyou 122 with a total planting area of more than 302,000 ha.

Ⅱ. PTGMS Restorer Lines

1. Yangdao 6 (9311)

An F_1 progeny from the cross of Yangdao 4/3021 (mid-season *indica*) was treated with ^{60}Co-γ irradiation for mutagenesis, and then went through the conventional breeding process to develop the R line Yangdao 6 (also called "9311") at the Institute of Agricultural Sciences of Lixiahe Region, Jiangsu province. It passed the technical evaluations of Jiangsu, Anhui and Hubei provinces in April 1997, November 2000 and March 2001, respectively. It is a late-maturing mid-season variety with a growth duration of 145 d. The plant is short and strong at the seedling stage, with remarkable growth increment and moderate tillering ability. The plant has a height of 115 cm, with strong stems, five elongating internodes above the ground, and 17 – 18 leaves on the main stem. The panicle is 24 cm in length, forming a uniform panicle layer, good maturing appearance, and high resistance to lodging. Generally, the number of panicles is about 2.25 million per hectare, with more than 165 spikelets per panicle. The seed setting rate is over 90% and the 1,000-grain weight is about 31 g. The parameters of rice quality are measured as 80.9% brown rice, 74.7% milled rice, 3.0 in length to width ratio, 5% in chalky grain rate, grade 2 translucency, 17.6% in amylose content, 7.0 in alkali spreading value, 97 mm in gel consistency and 11.3% in protein content. The grain quality meets the Grade 1 requirements in the ministerial standards for premium rice. When cooked, the rice is loose and soft, and will not harden after cooling down, delivering good taste quality. It is resistant to bacterial blight (grade 3), highly resistant to rice blast (grade R), and highly tolerant to heat and cold at the seedling stage (grade 3). Yangdao 6 is a famous restorer line. Three mid-season *indica* super rice hybrids Liangyoupeijiu, Y Liangyou 1 and Yangliangyou 6 have been planted in a total area of more than 10.8533 million hectares.

2. Anxuan 6

Anxuan 6 was selected from a natural mutant of mid-season *indica* 9311 in 1998 at Anhui Agricultural University. The plant has a whole growth period of 142 d, which is 3 – 4 d longer than that of Shanyou 63. The plant is 110 – 115 cm in height with a compact plant architecture, strong stems, and straight green leaves. The average number of spikelets per panicle is about 150. The seed setting rate is over 80%. The grain is slender without awn, and weights 29 g per 1,000 grains. Among the 12 parameters of grain quality, it has 10 meeting the grade 2 requirements of the ministerial standards for premium rice and the brown rice and milled rice are close to the grade 2 requirements. It is highly resistant to bacterial blight but moderately susceptible to rice blast. Six rice hybrids using this R line have been approved for production, including the super hybrid rice Xinliangyou 6, with a total planting area of more than 2.06 million hectares.

3. Bing 4114

Bing 4114 was developed from an F_{10} progeny from the cross of Yangdao 6/Shuhui 527 at the Tsinghua Shenzhen Longgang Institute of China National Hybrid Rice Research and Development Center. Four rice hybrids using this R line, namely Shenliangyou 814, Huiliangyou 114, Liangyou 1 and Shenliangyou 5814, have been approved for production. The super hybrid rice Shenliangyou 5814 has a total planting area of over 1.152 million hectares.

4. Hua 819

Hua 819 was developed from an F_5 progeny of the cross ZR02/Zhong 94 – 4 at Yahua Seeds Science Academy. Two early-season super rice hybrids, Zhuliangyou 819 and Luliangyou 819, have been planted in a total area of more than 524,000 ha.

5. Yuanhui 2

Yuanhui 2 was developed from the cross of R163/Shuhui 527 at the HHRRC. The R line has large panicles with a large number of grains, thick stems and high restoring ability. Two rice hybrids, Y Liangyou 2 and Xiangliangyou 2, have been approved for production. The super rice hybrid Y Liangyou 2 has been planted in a total area of more than 145,300 ha.

6. R900

R900 was developed at Hunan Yuanchuang Super Rice Technology Co. , Ltd. It has moderate plant height, compact plant architecture, strong stems, high density of grain setting, and large panicles with a large number of grains. It has a high combining ability. Three rice hybrids, namely Y Liangyou 900, Xiangliangyou 900 and Yuanliangyou 1000, have been approved for production. The super hybrid rice Y Liangyou 900, a pilot hybrid of the Phase Ⅳ project of the Super Rice Project, has been planted in a total area of over 12,000 ha.

References

[1] WAN JIANMIN. Genetic breeding and pedigree of rice in China (1986 - 2005) [M]. Beijing: China Agriculture Press, 2010:485 - 495.

[2] YUAN LONGPING. Hybrid rice studies[M]. Beijing: China Agriculture Press, 2002:119 - 124.

[3] DENG QIYUN. Breeding of the wide-compatibility PTGMS line Y58S[J]. Hybrid Rice, 2005, 20(2):15 - 18.

[4] DENG YINGDE, TANG CHUANDAO, LI JIARONG, et. al. Breeding of *indica* three-line sterile line Fengyuan A[J]. Hybrid Rice, 1999, 14(2):6 - 7.

[5] MA RONGRONG, WANG XIAOYAN, LU YONGFA, et al. Breeding and application of late *japonica* sterile line Yongjing 2A and related *indica-japonica* late rice combinations[J]. Hybrid Rice, 2010, 25(S1):185 - 189.

[6] LIU DINGYOU, PENG TAO, XIANG ZUFEN, et al. Dexiangyou 146: a high-yielding mid-season *indica* hybrid combination[J]. Hybrid Rice, 2017, 32(3):87 - 89.

[7] YANG LIANSONG, BAI YISONG. Study of the Breeding and application of *indica* PTGMS line 1892S[J]. Journal of Anhui Agricultural Sciences, 2012, 40(26):12808 - 12810.

[8] TANG WENBANG,CHEN LIYUN, XIAO YINGHUI, et al. Breeding and utilization of the dual-purpose GMS line C815S[J]. Journal of Hunan Agricultural University (Natural Sciences), 2007, 33:26 - 31.

[9] ZHOU YONG, JU CHAOMING, XU GUOCHENG, et al. Breeding and application of high-quality early *indica* PTGMS line HD9802S[J]. Hybrid Rice, 2008, 23 (2):7 - 10.

[10] FU CHENJIAN, QIN PENG, HU XIAOCHUN, et al. Breeding of PTGMS line Xiangling 628S[J]. Journal of Agricultural Science and Technology, 2010, 2(6):90 - 97.

[11] XIA SHENGPING, LI YILIANG, JIA XIANYONG, et al. Breeding of high-quality *indica* sterile line Jin 23A[J]. Hybrid Rice, 1992, 7(5):29 - 31.

[12] JIANG KAIFENG, ZHENG JIAKUI, YANG QIANHUA, et al. Breeding and application of high quality, high combining ability sterile line Dexiang C74A. Bulletin of Agricultural Science and Technology, 2008 (10):115 - 116.

[13] LU XIANJUN, REN GUANGJUN, LI QINGMAO, et al. Breeding and utilization of the high-quality blast-resistant restorer line Chenghui 177[J]. Hybrid Rice, 2007, 22(2):18 - 21.

[14] LI SHUGUANG, LIANG SHIHU, LI CHUANGUO, et al. Characteristics, properties, and technology for high-yielding and high-quality reproduction of the high-quality *indica* sterile line Wufeng A[J]. Guangdong Agricultural Sciences, 2009(8):29 - 30.

[15] YANG ZHENYU, ZHANG GUOLIANG, ZHANG CONGHE, et al. Breeding of high-quality mid-season *indica* PTGMS line Guangzhan 63S[J]. Hybrid Rice, 2002, 17(4):4-6.

[16] ZHANG CONGHE, CHEN JINJIE, JIANG JIAYUE, et al. Breeding of the PTGMS line with light brown mark on the shell, Xin'an S[J]. Hybrid Rice, 2007, 22(4):4-6.

[17] CHEN ZHIYUAN, LI CHUANGUO, SUN YING, et al. Characteristics, properties, and utilization of *indica* sterile line Tianfeng A[J]. Guangdong Agricultural Sciences, 2006(9):54-55.

[18] CHENCHENZHOU, TIAN JIWEI, MENG XIANGLUN, et al. Breeding and application of wide compatibility high quality super hybrid rice Shenyou 9516[J]. Hybrid Rice, 2016, 31(3):15-17.

[19] XU TONGJI, YANG DONG, HUANG DABIAO, et al. Characteristics and properties of longtepu a and the relevant purification and high-yielding reproduction technology[J]. Hybrid Rice, 2010, 25(6):21-23.

[20] LIU WUGE, WANG FENG, LIU ZHENRONG, et al. Breeding and application of *indica* early-maturing blast-resistant three-line sterile line Jifeng A[J]. Hybrid Rice, 2014, 29(6):16-18.

[21] LIU ZHENRONG, LIU WUGE, WANG FENG, et al. Breeding and utilization of *indica* early-maturing blast-resistant three-line sterile line Rongfeng A[J]. Hybrid Rice, 2006, 21(6):17-18.

[22] CHEN YONG, XIAO PEICUN, XIE CONGJIAN, et al. Breeding and application of new cytoplasm high-quality *indica* fragrant sterile line Neixiang 5A[J]. Hybrid Rice, 2008, 23(1):13-15.

[23] ZHANG ZHIXING, ZHANG WEICHUN, BAO YANHONG, et al. Characteristics, properties, and high-yielding seed production technology of high-quality sterile line Xieqingzao A[J]. Modern Agricultural Science and Technology, 2011 (4):80-82.

[24] JIANG QINGSHAN, LIN GANG, ZHAO DEMING, et al. Breeding and utilization of high-quality fragrant rice sterile line Yixiang 1A[J]. Hybrid Rice, 2008, 23(2):11-14.

[25] LI XIANYONG, WANG CHUTAO, LI SHUNWU, et al. Breeding of high-quality *indica* sterile line Q2A[J]. Hybrid Rice, 2004, 19(5):6-8.

Chapter 19

Super Hybrid Rice Combinations

Yang Yishan

In order to further clarify the concept and indicators of super rice varieties, standardize the procedures and methods for certifying super rice varieties, and promote the selection and application of super rice varieties, China launched the Super Rice Demonstration and Promotion Project in 2005, released the first list of 28 varieties that met the criteria for super rice, and formulated *Super Rice Variety Confirmation Measures* (*Trial*) (General Office of the MOA [2005] No. 39). Then, in July 2008, *Super Rice Variety Confirmation Measures* (General Office of the MOA [2008] No. 38) was released after revision to the trial version, and the main indicators for different rice regions and types of super rice varieties are shown in Table 19 – 1.

As of 2017, a total of 107 super hybrid rice demonstration and promotion projects have been confirmed and named bythe MOA, of which 14 have been removed from the list of super rice varieties due to variety degradation or failure to meet the requirement of annual production area. There are a total of 93 super rice hybrids, including 52 three-line *indica* hybrids, 34 two-line *indica* hybrids, and 7 three-line *indica-japonica* hybrids currently used in rice production. This chapter mainly introduces the basic information and characteristics of these 93 super hybrid rice varieties, with data mainly obtained from the National Rice Data Center (www.ricedata.cn/variety/superice.htm).

Section 1 Three-line *Indica* Hybrid Rice

Yixiang 4245

Institution (s) – Yibin Academy of Agricultural Sciences

Parental lines – Yixiang 1A/Yihui 4245

Approvals – State in 2012 (GS2012008), Sichuan in 2009 (CS2009004); super rice recognition by the MOA in 2017

Characters – It is a three-line mid-season *indica* hybrid rice with a compact plant type, fairly high tillering capacity, straight and stiff flag leaves, and pale green leaf color. Tested in late-maturing mid-season *indica* rice trials in the upper reaches of the Yangtze River in 2009 and 2010, it yielded 8.77 t/ha, higher than the yield of the control variety Ⅱ You 838 by 3.8%. It is 159.2 d in whole growth duration, 0.5 d longer than the control variety Ⅱ You 838. It had 2.28 million effective panicles per hectare, is 117.2 cm in plant height and 26.2 cm in panicle length with 175.5 spikelets per panicle, 79.5% in seed setting rate and 28.4 g in 1,000-grain weight; susceptible to rice blast and highly susceptible to brown planthopper, and of Grade 2 grain quality as per the national standard *High Quality Paddy*.

Suitable areas – Single-cropping mid-season rice in low-to-medium altitude *indica* rice areas in Yunnan and Guizhou, excluding the Wuling Mountains, hilly areas in Pingba of Sichuan and southern Shaanxi without severe incidence of rice blast. By 2015, the accumulated planting area exceeded 66,000 ha.

Table 19 – 1 Major indicators of super hybrid rice varieties

Region	Early-maturing early-season rice in the Yangtze River Basin	Mid-and late-maturing early-season rice in the Yangtze River Basin	Mid-maturing late-season rice in the Yangtze River Basin, photosensitive late-seasonrice in south China	Early-or late-season rice in south China, late-maturing late-season rice in the Yangtze River Basin, early-maturing rice in northeast China	Single-cropping rice in the Yangtze River Basin, mid-maturing *japonica* rice in northeast China	Late-maturing single-cropping rice in the upper reaches of the Yangtze River, late-maturing *japonica* rice in northeast China
Growth period (d)	≤105	≤115	≤125	≤132	≤158	≤170
Yield (t/ha)	≥8.25	≥9.00	≥9.90	≥10.80	≥11.70	≥12.75
Grain quality	*Japonica* rice in North China meets minimum Grade 2 standards, late-season *indica* rice in south China meets minimum Grade 3 standards, and early-season *indica* rice and single-cropping rice meet minimum Grade 4 standards as stipulated by the MOA.					
Resistance	Resistance to one or two major local pests and/or diseases					
Planting area	More than 3,333.33ha within two years after approval					

Jifengyou 1002

Institution (s) – Rice Research Institute of Guangdong Academy of Agricultural Sciences and Guangdong Golden RiceSeeds Co., Ltd.

Parental lines – JifengA/Guanghui 1002

Approvals – Guangdong in 2013 (YS2013040), super rice recognition by the MOA in 2017

Characters – It is a three-line *indica* hybrid rice with weak photosensitity, a moderately compact plant type, medium-to-high tillering capacity, high resistance to lodging, weak-to-medium tolerance to cold during flowering, high yield potential, low grain quality falling short of the national standard *High Quality Paddy*, high resistance to rice blast, and susceptibility to bacterial blight. Tested in weak-photosensitive late-season rice trials in Guangdong in 2011 and 2012, it yielded 7.42 t/ha and 7.59 t/ha, higher than the yield of the control variety by 14.3% and 8.16%, respectively. It is 120 – 122 d in whole growth duration, and 99.5 – 102.0 cm in plant height. It has 2.595 – 2.730 million effective panicles per hectare, 20.1 – 21.3 cm in panicle length, and 131 – 142 spikelets per panicle. The seed setting rate is 85.5% – 85.6% and 1,000-grain weight 25.2 – 26.5 g.

Suitable areas – Late-season rice in the plain areas of south-central and southwestern Guangdong. By 2015, the accumulated planting area exceeded 10,000 ha.

Wuyou 116

Institution (s) - Guangdong Modern Agriculture Group Co., Ltd. and Rice Research Institute of Guangdong Academy of Agricultural Sciences

Parental lines - Wufeng A/R7116

Approvals - Guangdong in 2015 (YS2015045), super rice recognition by the MOA in 2017

Characters - It is athermo-sensitive three-line *indica* hybrid rice with a moderately compact plant type, intermediate tillering capacity, medium-to-strong resistance to lodging, intermediate tolerance to cold during booting and flowering, and high yield potential. Tested in the late-season rice trials of Guangdong in 2013 and 2014, it yielded 7.13 t/ha and 8.18 t/ha, higher than the yields of the control variety by 11.36% and 8.87%, respectively. It is 114 d in whole growth period, 107.8 - 114.0 cm in plant height, 2.490 - 2.775 million in effective panicles per hectare, 22.5 - 22.8 cm in panicle length with 149 - 152 spikelets per panicle, 77.7% - 86.8% in seed setting rate and 25.9 - 26.5 g in 1,000-grain weight. It is resistant to rice blast but highly susceptible to bacterial blight, with Grade 3 grain quality as per relevant national and provincial standards.

Suitable areas - Late-season rice in northern Guangdong, early- and late-season rice in north-central Guangdong

Deyou 4727

Institution (s) - Rice and Sorghum Research Institute of Sichuan Academy of Agricultural Sciences and Crops Research Institute of Sichuan Academy of Agricultural Sciences

Parental lines - Dexiang 074A/Chenghui 727

Approvals - State in 2014 (GS2014019), Sichuan in 2014 (CS2014004), Yunnan in 2013 (DS2013007), super rice recognition by the MOA in 2016

Characters - It is a late-maturing mid-season three-line *indica* hybrid rice with a moderate plant type, green leaf sheaths, colorless stigma, and good maturing color. Tested in the late-maturing mid-season *indica* rice trials of the upper reaches of the Yangtze River in 2011 and 2012, it yielded 9.19 t/ha, higher than the yield of the control variety Ⅱ You 838 by 5.6%. It is 158.4 d in whole growth duration, 1.4 d longer than the control variety Ⅱ You 838. It is 113.7 cm in plant height, 24.5 cm in panicle length, with 2.235 million effective panicles per hectare and 160.0 spikelets per panicle, 82.2% in seed setting rate and 32.0 g in 1,000-grain weight. It is susceptible to rice blast and brown planthopper, moderately tolerant to heat during the heading stage, and has Grade 2 grain quality as per relevant national and provincial standards.

Suitable areas - Single-cropping mid-season rice in low-to-medium altitude *indica* rice areas in Yunnan and Guizhou, excluding the Wuling Mountains; *indica* rice in areas below 800 m of altitude; and hilly rice areas in Pingba of Sichuan and southern Shaanxi. By 2015, the accumulated planting area exceeded 38,000 ha.

Fengtianyou 553

Institution (s) - Rice Research Institute of Guangxi Academy of Agricultural Sciences

Parental lines - Fengtian 1A/Guihui 553

Approvals – Guangdong in 2016 (YS2016052), Guangxi in 2013 (GS2013027), super rice recognition by the MOA in 2016

Characters – It is a three-line *indica* hybrid rice with weak photosensitivity, a moderately compact plant type, intermediate tillering capacity, large panicles, long grains, yellow apiculus, short awns, high tolerance to lodging, intermediate tolerance to cold, and high yield potential. Tested in the late-season photosensitive rice trials of southern Guangxi in 2011 and 2012, it yielded 7.37 t/ha, higher than the yield of the control variety Boyou 253 by 5.23%. It is 120 d in whole growth duration, similar to the control variety Boyou 253. It has 2.79 million effective panicles per hectare, and is 109.1 cm in plant height, 23.0 cm in panicle length with 135.8 spikelets per panicle, 86.1% in seed setting rate, 23.3 g in 1,000-grain weight, susceptible to rice blast and highly susceptible to bacterial blight. Tested in the late-season rice trials of Guangdong in 2014 and 2015, it yielded 7.22 t/ha and 6.91 t/ha, higher than the yields of the control variety by 5.44% and 7.87%, respectively. The whole growth duration is 115 d, similar to the control variety, and has Grade 3 grain quality as per relevant national and provincial standards.

Suitable areas – Late-season rice in southern Guangxi and Guangdong, excluding the northern part. By2015, the accumulated planting area exceeded 6,700 ha.

Wuyou 662 (Wufengyou 662)

Institution (s) – Jiangxi Huinong Seeds Co. Ltd. and Rice Research Institute of Guangdong Academy ofAgricultural Sciences.

Parental lines – Wufeng A/R662

Approvals – Jiangxi in 2012 (GS2012010), super rice recognition by the MOA in 2016

Characters – It is a three-line late-season *indica* hybrid rice with a moderate plant type, dark green leaves, wide and stiff flag leaves, vigorous growth potential, high tillering capacity, purple apiculus, large numbers of panicles per hectare and spikelets per panicle, high seed setting rate, and good maturing color. Tested in the late-season rice trials of Jiangxi in 2010 and 2011, it had a yield of 7.43 t/ha, which is 5.18% higher than that of the control variety Yueyou 9113 and the whole growth duration is 119.2 d, which is 0.2 d shorter than that of Yueyou 9113. It is 96.1 cm in plant height, with 3.12 million effective panicles per hectare, 127.2 spikelets per panicle and 93.1 filled grains per panicle, 73.2% in seed setting rate, 27.2 g in 1,000-grain weight, highly susceptible to rice blast.

Suitable areas – Late-season rice in Jiangxi where there is no severe occurrence of rice blast. By2015, the accumulated planting area exceeded 106,000 ha.

Jiyou 225

Institution (s) – Rice Research Institute of Jiangxi Academy of Agricultural Sciences, Jiangxi Super Rice Research and Development Center, and Rice Research Institute of Guangdong Academy of Agricultural Sciences

Parental lines – Jifeng A/R225

Approvals – Jiangxi in 2014(GS2014014), super rice recognition by the MOA in 2016

Characters – It is a three-line late-season *indica* hybrid with a moderate plant type, short and straight flag leaves, intermediate tillering capacity, purple apiculus, large numbers of panicles and spikelets per panicle, high seed setting rate, and good maturing color. It is highly susceptible to rice blast and has high grain quality. Tested in the late-season rice trials of Jiangxi in 2012 and 2013, it yielded 8. 13 t/ha, higher than the yield of the control variety Yueyou 9113 by 4. 00% . It is 116. 8 d in whole growth duration, which is 0. 3 d shorter than that of Yueyou 9113. It is 96. 5 cm in plant height, with 2. 88 million effective panicles per hectare, 144. 6 spikelets per panicle including 116. 2 filled grains per panicle. It is 80. 4% in seed setting rate and 24. 8 g in 1,000-grain weight, with Grade 2 grain quality as per the national standard *High Quality Paddy*.

Suitable areas – Late-season rice in Jiangxi where there is no severe occurrence of rice blast. By 2015, the accumulated planting area exceeded 25,300 ha.

Wufengyou 286

Institution (s) – Jiangxi Modern Seed Industry Co. , Ltd. and China National Rice Research Institute

Parental lines – Wufeng A/Zhonghui 286

Approvals – State in 2015 (GS2015002),Jiangxi in 2014 (GS2014005), super rice recognition by the MOA in 2016

Characters – It is a three-line early-season *indica* hybrid rice with a moderate plant type, straight and stiff leaves, strong and thick stems, vigorous grow potential, relatively high tillering capacity, purple apiculus, large numbers of panicles and spikelets per panicle, high seed setting rate, and good maturing color. It is highly susceptible to rice blast and brown planthopper, and susceptible to bacterial blight and white-backed planthopper. Tested in the late-maturing early-season *indica* rice trials of the middle and lower reaches of the Yangtze River in 2012 and 2013, it yielded 8. 06 t/ha, higher than the yield of the control variety Luliangyou 996 by 7. 0% . It is 113. 0 d in whole growth duration, which is 0. 3 d longer than that of the control variety, 84. 1 cm in plant height and 18. 9 cm in panicle length, with 3. 015 million effective panicles per hectare and 144. 3 spikelets per panicle. It is 82. 9% in seed setting rate and 24. 5 g in 1,000-grain weight.

Suitable areas – Early-season rice in the double-cropping rice areas in Jiangxi, Hunan, northern Guangxi, northern Fujian, and south-central Zhejiang where there is no frequent occurrence of rice blast. By 2015 the accumulated planting area exceeded 19,300 ha.

Wuyou 1573 (Wuyouhang 1573)

Institution (s) – Jiangxi Super Rice Research & Development Center, Jiangxi Huifengyuan Seed Co. , Ltd. , and Rice Research Institute of Guangdong Academy of Agricultural Sciences

Parental lines – Wufeng A/Yuehui 1573

Approvals – Jiangxi in 2014 (GS2014020), super rice recognition by the MOA in 2016

Characters – It is a three-line late-season *indica* hybrid rice with a moderate plant type, straight and stiff leaves, good-looking growth, high tillering capacity, purple apiculus, large numbers of panicles and spikelets per panicle, a high seed setting rate, low grain weight, and good maturing color. Tested in the late-season rice trials of Jiangxi in 2012 and 2013, it yielded 8. 42 t/ha, higher than the yield of the control variety Tianyou 998 by 3. 70% . It is 123. 1 d in whole growth duration, which is 0. 8 d shorter than that of the control variety, 98. 9 cm in plant height, with 3. 18 million effective panicles per hectare and 146. 9 spikelets per panicle including 121. 9 filled grains per panicle. It is 83. 0% in seed setting rate and 23. 0 g in 1,000-grain weight, highly susceptible to rice blast, and with Grade 2 grain quality as per the national standard *High Quality Paddy*.

Suitable areas – Late-season rice in Jiangxi where there is no severe occurrence of rice blast. By 2015, the accumulated planting area exceeded 12,700 ha.

Yixiangyou 2115

Institution (s) – College of Agronomy of Sichuan Agricultural University, Yibin Academy of Agricultural Sciences, and Sichuan Lvdan Seed Co., Ltd.

Parental lines – Yixiang 1A/Yahui 2115

Approvals – State in 2012 (GS2012003), Sichuan in 2011 (CS2011001), super rice recognition by the MOA in 2015

Characters – It is a three-line mid-season *indica* hybrid rice with a moderate plant type, straight, stiff flag leaf and light green leaves, green leaf sheaths, light green auricles, and high tillering capacity. Tested in the late-maturing mid-season *indica* rice trials of the upper reaches of the Yangtze River in 2010 and 2011, it yielded 9. 06 t/ha, higher than the yield of the control variety Ⅱ You 838 by 5. 6%. It is 156. 7 d in whole growth duration, which is 1. 5 d shorter than the control variety. It has 2. 25 million effective panicles per hectare, is 117. 4 cm in plant height, and 26. 8 cm in panicle length with 156. 5 spikelets per panicle. It is 82. 2% in seed setting rate and 32. 9 g in 1,000-grain weight, with intermediate susceptibility to rice blast, high susceptibility to brown planthopper, and Grade 2 grain quality as per the national standard *High Quality Paddy*.

Suitable areas – Mid-season rice in low-to-medium altitude *indica* rice areas in Yunnan, Guizhou, and Chongqing; hilly rice areas in Pingba of Sichuan; and single-cropping rice areas in southern Shaanxi, excluding the Wuling Mountains. By 2015, the accumulated planting area exceeded 248,700 ha.

Shenyou 1029

Institution (s) – Jiangxi Modern Seed Industry Co., Ltd.

Parental lines – Shen 95A/R1029

Approvals – State in 2013 (GS2013031), super rice recognition by the MOA in 2015

Characters – It is a three-line late-season *indica* hybrid rice. Tested in the early-maturing late-season *indica* rice trials of the middle and lower reaches of the Yangtze River in 2010 and 2011, it yielded 7. 53 t/ha, higher than the yield of the control variety Jinyou 207 by 3. 5%. It is 118. 4 d in whole growth duration, which is 2. 5 d longer than that of the control variety. It is 103. 9 cm in plant height and 22. 1 cm in panicle length, with 3. 015 million effective panicles per hectare and 149. 3 spikelets per panicle. It is 78. 0% in seed setting rate and 24. 1 g in 1,000-grain weight, highly susceptible to rice blast, bacterial blight and brown planthopper, with low tolerance to cold during heading, and Grade 3 grain quality as per the national standard *High Quality Paddy*.

Suitable areas – Late-season rice in double-cropping rice areas in Jiangxi, Hubei, Zhejiang, and Anhui, excluding areas with frequent occurrence of rice blast

F You 498

Institution (s) – Rice Research Institute of Sichuan Agricultural University, and Chuanjiang Rice Research Institute of Jiangyou, Sichuan

Parental lines – Jiangyu F32A/Shuhui 498

Approvals – State in 2011 (GS2011006), Hunan in 2009 (XS2009019), super rice recognition by the MOA in 2014

Characters – It is a three-line mid-season *indica* hybrid rice with a moderate plant type, thick and strong stems, high tillering capacity, vigorous growth potential, light green and long leaves, purple leaf sheaths and apiculus, large panicles, good maturing colored, high tolerance to cold, poor tolerance to heat, vulnerability to sheath blight, and susceptibility to rice blast and brown planthopper. Tested in the late-maturing mid-season rice trials of the upper reaches of the Yangtze River in 2008 and 2009, it yielded 9. 32 t/ha, higher than the yield of the control variety Ⅱ You 838 by 5. 9%. It is 155. 2 d in whole growth duration, which is 2. 7 d shorter than that of the control variety, 111. 9 cm in plant height with 2. 25 million effective panicles per hectare, 25. 6 cm in panicle length with 189. 0 spikelets per panicle, 81. 2% in seed setting rate, and 28. 9 g in 1,000-grain weight, and Grade 3 grain quality as per the national standard *High Quality Paddy*.

Suitable areas – Mid-season rice in low-to-medium altitude *indica* rice areas in Yunnan, Guizhou, and Chongqing, excluding the Wuling Mountains; hilly rice areas in Pingba of Sichuan, southern Shaanxi where there is no severe occurrence of rice blast; and hilly areas below 600 m in altitude in Hunan where there is no rice blast. By 2015, the accumulated planting area exceeded 188,000 ha.

Rongyou 225

Institution (s) – Rice Research Institute of Jiangxi Academy of Agricultural Sciences, and Rice Research Institute of Guangdong Academy of Agricultural Sciences

Parental lines – Rongfeng A/R225

Approvals – State in 2012 (GS2012029), Jiangxi in 2009 (GS2009017), super rice recognition by the MOA in 2014

Characters – It is a three-line late-season *indica* hybrid rice with a moderate plant type, dark green leaves, average tillering capacity, purple apiculus, large numbers of panicles and spikelets per panicle, relatively high seed setting rate and good maturing color. It is highly susceptible to rice blast, rice black-streaked dwarf virus and brown planthopper, moderately susceptible to bacterial blight, with low tolerance to cold during heading. Tested in the early-maturing late-season *indica* trials of the middle and lower reaches of the Yangtze River in 2009 and 2010, it yielded 7. 75 t/ha, higher than the yield of the control variety Jinyou 207 by 10. 1%, with 116. 5 d in whole growth duration, which is 3. 6 d longer than that of Jinyou 207. It has 2. 895 million effective panicles per hectare, is 101. 4 cm in plant height and 21. 8 cm in panicle length, and has 157. 7 spikelets per panicle. It is 74. 9% in seed setting rate and 25. 7 g in 1,000-grain weight. Tested in the late-season rice trials of Jiangxi in 2007 and 2008, it yielded 7. 02 t/ha, higher than the yield of the control variety Jinyou 207 by 6. 56%. It is 114. 1 d in whole growth duration, which is 3. 4 d longer than that of Jinyou 207. The grain quality is rated as Grade 2 as per the national standard *High Quality Paddy*.

Suitable areas – Late-season rice in double-cropping rice areas in Jiangxi and Hunan where there is no severe occurrence of rice blast and rice black-streaked dwarf virus. By 2015, the accumulated planting area exceeded 389,300 ha.

Nei 5 You (Guodao 7)

Institution (s) – China National Rice Research Institute and Zhejiang Nongke Seed Co., Ltd.

Parental lines – Neixiang 5A/Zhonghui 8015

Approvals – State in 2010 (GS2010020), super rice recognition by the MOA in 2014

Characters – It is a three-line mid-season *indica* hybrid rice with a moderate plant type, thick and strong stems, good maturing color, colorless apiculus, no awn, two-phase grain filling, high yield potential, high susceptibility to rice blast, bacterial blight and brown planthopper, and high grain quality. Tested in the late-maturing mid-season rice trials of the middle and lower reaches of the Yangtze River in 2007 and 2008, it yielded 8.86 t/ha, higher than the yield of the control variety Ⅱ You 838 by 3.3%. It is 133.1 d in whole growth duration, which is 1.6 d shorter than that of the control variety. It has 2.415 million effective panicles per hectare, is 122.2 cm in plant height and 26.8 cm in panicle length, and has 157.0 spikelets per panicle. It is 80.8% in seed setting rate and 32.0 g in 1,000-grain weight, and has Grade 3 grain quality as per the national standard *High Quality Paddy*.

Suitable areas – Single-cropping mid-season rice along the Yangtze River in Jiangxi, Hunan, Hubei, Anhui, Zhejiang and Jiangsu, excluding the Wuling Mountains; and northern Fujian and southern Henan where there is no severe occurrence of rice blast and bacterial blight. By 2015, the accumulated planting area exceeded 138,700 ha.

Shengtaiyou 722

Institution (s) – Hunan Dongting Hi-Tech Seed Co., Ltd. and Yueyang Research Institute of Agricultural Sciences.

Parental lines – Shengtai A/Yuehui 9722.

Approvals – Hunan in 2012 (XS2012016), super rice recognition by the MOA in 2014.

Characters – It is a three-line mid-maturing late-season *indica* hybrid rice with a moderate plant type, vigorous growth potential, elastic stems, high tillering capacity, upright flag leaves, greenish leaves, colorless leaf sheaths, auricles and pulvinus, good maturing color, high yield potential, high grain quality, susceptibility to rice blast and moderate tolerance to cold. Tested in the mid-maturing late-season rice trials of Hunan in 2010 and 2011, it had a yield of 7.52 t/ha, which was 9.31% higher than that of the contro l variety, and the whole growth duration was 112.6 d. It is 94.8 cm in plant height, with 3.30 million effective panicles per hectare and 119.7 spikelets per panicle, and 75.3% in seed setting rate and 26.1 g in 1,000-grain weight, with Grade 3 grain quality as per the national standard *High Quality Paddy*.

Suitable areas – Late-season rice in double-cropping rice areas of Hunan where there is no severe occurrence of rice blast. By 2015, the accumulated planting area exceeded 19,300 ha.

Wufengyou 615

Institution (s) – Rice Research Institute of Guangdong Academy of Agricultural Sciences

Parental lines – Wufeng A/Guanghui 615

Approvals – Guangdong in 2012 (YS2012011), super rice recognition by the MOA in 2014

Characters – It is a thermo-sensitive three-line *indica* hybrid rice with a moderately compact plant type, intermediate tillering capacity, large panicles with a large number of grains, intermediate-to-strong tolerance to lodging, intermediate-to-strong tolerance to cold, good maturing color, high yield potential, low grain quality falling short of the standards for high quality grains, intermediate resistance to rice blast, and susceptibility to bacterial blight. Tested in the early-season rice trials of Guangdong in 2010 and 2011, it yielded 6.71 t/ha and 8.15 t/ha, higher than the yields of the control variety Yuexiangzhan by 14.79% and 18.78%, respectively. It is 129 d in whole growth duration, similar to that of the control variety. It is 98.6 – 102.1 cm in plant height with 2.655 – 2.715 million effective panicles per hectare, 21.4 – 21.7 cm in panicle length with 157 – 168 spikelets per panicle, 80.3% – 85.0% in seed setting rate and 22.2 – 22.9 g in 1,000-grain weight.

Suitable areas – Early- or late-season rice in Guangdong, excluding the northern part. By 2015, the accumulated planting area exceeded 148,700 ha.

Tianyou 3618

Institution (s) – The Rice Research Institute of Guangdong Academy of Agricultural Sciences

Parental lines – Tianfeng A/Guanghui 3618

Approvals – Guangdong in 2009 (YS2009004), super rice recognition by the MOA in 2013

Characters – It is a thermo-sensitive three-line *indica* hybrid rice with a moderately compact plant type, intermediate tillering capacity, tolerance to lodging, large panicles with a large number of spikelets per panicle, good maturing color, intermediate-to-strong tolerance to cold during booting and flowering, high yield potential, low grain quality as early-season rice, falling short of the standards of high quality rice, resistance to rice blast, and intermediate susceptibility to bacterial blight. Tested in the early-season rice trials of Guangdong in 2007 and 2008, it yielded 6.93 t/ha and 7.08 t/ha, higher than the yields of the control variety Yuexiangzhan by 13.18% and 16.08%, respectively. It is 126 – 127 d in whole growth duration, the same as the control variety, 96.6 – 98.2 cm in plant height, and 19.6 cm in panicle length with 143 spikelets per panicle, 76.1% – 79.3% in seed setting rate and 23.8 – 24.9 g in 1,000-grain weight.

Suitable areas – Early- or late-season rice in Guangdong, excluding the northern part. By 2015, the accumulated planting area exceeded 76,000 ha.

Zhong 9 You 8012

Institution (s) – China National Rice Research Institute

Parental lines – Zhong 9A/Zhonghui 8012

Approvals – State in 2009 (GS2009019), super rice recognition by the MOA in 2013

Characters – It is a three-line mid-season *indica* hybrid rice with a moderate plant type, thick and strong stems, wide and long flag leaves, light green leaves, good maturing color, colorless apiculus, no awn, intermediate growth duration, high yield potential, high susceptibility to rice blast, susceptibility to bacterial blight and brown planthopper, and average grain quality. Tested in the late-maturing mid-season *indica* rice trials of the middle and lower reaches of the Yangtze River in 2006 and 2007, it yielded 8.51 t/ha, higher than the yield of the control variety Ⅱ You 838 by 3.02%. It is 133.1 d in whole growth duration, which is 0.1 d shorter than that of Ⅱ You 838. It has 2.34 million effective panicles per hectare, is 125.7 cm in plant height and 26.0 cm in panicle length with 184.5 spikelets per panicle, 79.9% of seed setting rate and 26.6 g of 1,000-grain weight.

Suitable areas – Mid-season rice along the Yangtze River in Jiangxi, Hunan, Hubei, Anhui, Zhejiang, and Jiangsu, excluding the Wuling Mountains; northern Fujian and southern Henan, where there is no severe occurrence of rice blast and bacterial blight. By 2015, the accumulated planting area exceeded 80,000 ha.

H You 518

Institution (s) – Hunan Agricultural University and HengyangResearch Institute of Agricultural Sciences

Parental lines – H28A/51084

Approvals – State in 2011 (GS2011020), Hunan in 2010 (XS2010032), super rice recognition by the MOA in 2013

Characters – It is a three-line late-season *indica* hybrid rice with a relatively loose plant type, intermediate-to-long flag leaves, colorless sheaths and apiculus, awns on some grains, good maturing color, moderate tolerance to cold, high susceptibility to rice blast and brown planthopper, susceptibility to bacterial blight, and good grain quality. Tested in the early-maturing late-season *indica* rice trials of the middle and lower reaches of the Yangtze River in 2009 and 2010, it yielded 7.49 t/ha, higher than the yield of the control variety Jinyou 207 by 6.8%. It is 112.9 d in whole growth duration, which is 0.5 d shorter than that of Jinyou 207. It is 96.2 cm in plant height with 3.615 million effective panicles per hectare, 22.3 cm in panicle length with 113.6 spikelets per panicle, 80.7% in seed setting rate, 25.8 g in 1,000-grain weight, and Grade 3 in grain quality as per the national standard *High Quality Paddy*.

Suitable areas – Late-season rice in double-cropping rice areas in Jiangxi, Hunan, Hubei, Zhejiang, and south of the Yangtze River in Anhui where there is no severe occurrence of rice blast or bacterial blight. By 2015, the accumulated planting area exceeded 210,700 ha.

Tianyou-Huazhan

Institution (s) – China National Rice Research Institute, Institute of Genetics and Developmental Biology of Chinese Academy of Sciences, and Rice Research Institute of Guangdong Academy of Agricultural Sciences.

Parental lines – Tianfeng A/Huazhan.

Approvals – State in 2012 (GS2012001, as early *indica* rice in south China), Guizhou in 2012 (QS2012009), Guangdong in 2011 (YS2011036), State in 2011 (GS2011008, as late-maturing mid-season *indica* rice in the upper, middle and lower Yangtze River), Hubei in 2011 (ES2011006), State in 2008 (GS2008020, as late-maturing late-season *indica* rice in the middle and lower Yangtze River), super rice recognition by the MOA as late-season rice in the middle and lower Yangtze River in 2012, and as mid-season rice in Hubei in 2013.

Characters – It is a three-line *indica* hybrid rice with relatively low plant height, a moderately compact plant type, straight and stiff leaves, soft stems, average tolerance to lodging, high tillering capacity, intermediate-to-large panicle size, large number of spikelets per panicle, good maturing color, intermediate growth duration, high yield potential, good grain quality, intermediate tolerance to cold, low tolerance to heat, various resistance or susceptibility to rice blast depending on specific rice areas, susceptibility to bacterial blight, normal or high susceptibility to brown planthopper, and moderate resistance to white-backed planthopper. Tested in the late-maturing late-season rice trials of the middle and lower reaches of the Yangtze River in 2006 and 2007, it yielded 7.86 t/ha, higher than the yield of the control variety Shanyou 46 by 10.32%. It is 119.2 d in whole growth duration, which is 0.3 d shorter than that of the control variety. It has 2.835 million effective panicles per hectare, is 101.3 cm in plant height and 21.1 cm in panicle length, and has 155.1 spikelets per panicle. It is 76.8% in seed setting rate and 24.9 g in 1,000-grain weight with Grade 1 grain quality as per the national standard *High Quality Paddy*. Tested in the late-maturing mid-season rice trials of the upper reaches of the Yangtze River in 2008 and 2009, it yielded 8.95 t/ha, higher than the yield of the control variety Ⅱ You 838 by 2.8%. The whole growth duration is 4.9 d shorter than that of control variety. The grain quality is rated as Grade 2 as per the national standard *High Quality Paddy*. Tested in the late-maturing mid-season rice trials of the middle and lower reaches of the Yangtze River in 2009 and 2010, it yielded 8.86 t/ha, higher than the yield of the control variety Ⅱ You 838 by 7.4% and the whole growth duration is 2.9 d shorter than that of the control variety. Tested in the early-season *indica* rice trials of south China in 2009 and 2010, it yielded 7.54 t/ha, higher than the yield of the control variety Tianyou 998 by 6.9% with the whole growth duration 0.1 d shorter than that of control variety. The grain quality is rated as Grade 3 as per the national standard *High Quality Paddy*.

Suitable areas – Late-season rice in double-cropping rice areas in north-central Guangxi, north-central Fujian, south-central Jiangxi, south-central Hunan, and southern Zhejiang where there is no severe occurrence of bacterial blight; single-cropping mid-season rice along the Yangtze River in Jiangxi, Hunan, Hubei, Anhui, Zhejiang and Jiangsu, excluding the Wuling Mountains; northern Fujian and southern Henan where there is no severe occurrence of bacterial blight; low-to-medium altitude rice areas in Yunnan, Guizhou and Chongqing, excluding the Wuling Mountains; hilly rice areas in Pingba of Sichuan, southwest Shaanxi with medium soil fertility; early-season rice in double-cropping rice areas in southern Guangxi and Hainan where there is no severe occurrence of bacterial blight; and early- or late-season rice in Guangdong. By 2015, the accumulated planting area exceeded 1,084,700 ha.

Jinyou 785

Institution (s) – Guizhou Rice Research Institute

Parental lines – Jin 23A/Qianhui 785

Approvals – Guizhou in 2010 (QS2010002), super rice recognition by the MOA in 2012

Characters – It is a three-line mid-season *indica* hybrid rice with a moderate plant type, relatively thick and strong stems, average tillering capacity, large panicle size, long grains, purple apiculus, no awn, relatively high tolerance to cold, and susceptibility to rice blast. Tested in the late-maturing rice trials of Guizhou in 2008 and 2009, it yielded 9.62 t/ha, higher than the yield of the control variety of Ⅱ You 838 by 9.27%. It is 157.1 d in whole growth duration, the same as that of Ⅱ You 838. It is 112.1 cm in plant height with 2.325 million effective panicles per hectare including 147.4 filled grains per panicle, 80% in seed setting rate, and 29.2 g in 1,000-grain weight.

Suitable areas – Late-maturing mid-season rice in Guizhou

Dexiang 4103

Institution (s) – Rice and Sorghum Research Institute of Sichuan Academy of Agricultural Sciences

Parental lines – Dexiang 074A/Luhui H103

Approvals – State in 2012 (GS2012024), Honghe of Yunnan in 2012 (DTS [Honghe] 2012016), Chongqing in 2011 (YY2011001), Pu'er and Wenshan of Yunnan (DTS [Pu'er and Wenshan] 2011003), Sichuan in 2008 (CS2008001), super rice recognition by the MOA in 2012

Characters – It is a three-line *indica* hybrid rice with a moderate plant type, upright flag leaves, intermediate to high tillering capacity, good maturing color, large panicles with a large number of grains, intermediate numbers of spikelets per panicle, high tolerance to cold, and average tolerance to heat. Tested in the late-maturing mid-season *indica* rice trials of the middle and lower reaches of the Yangtze River in 2009 and 2010, it yielded 8.60 t/ha, higher than the yield of the control variety Ⅱ You 838 by 5.2%. It is 134.4 d in whole growth duration, which is 0.8 d longer than that of the control variety Ⅱ You 838. It has 2.325 million effective panicles per hectare, is 125.0 cm in plant height and 25.9 cm in panicle length, and has 162.1 spikelets per panicle. It is 79.9% in seed setting rate and 31.1 g in 1,000-grain weight, highly susceptible to rice blast and bacterial blight, and susceptible to brown planthopper.

Suitable areas – Rice areas along the Yangtze River in Jiangxi, Hunan, Hubei, Anhui, Zhejiang, and Jiangsu, excluding the Wuling Mountains; rice areas in northern Fujian and southern Henan where there is no severe occurrence of rice blast and bacterial blight; Pingba and hilly areas in Sichuan; areas below 800 m in altitude in Chongqing; *indica* rice areas in Pu'er excluding Mojiang, Jingdong and Lancang; areas below 1,300 m in altitude in Wenshan excluding Yanshan, Xichou and Guangnan; and hybrid rice areas below 1,400 m in altitude in Honghe. By 2015, the accumulated planting area exceeded 344,700 ha.

Yiyou 673 (**Yixiangyou 673**)

Institution (s) – Rice Research Institute of Fujian Academy of Agricultural Sciences

Parental lines – Yixiang 1A/Fuhui 673

Approvals – Yunnan in 2010 (DS2010005), state in 2009 (GS2009018), Guangdong in 2009 (YS2009041), Fujian in 2006 (MS2006021), super rice recognition by the MOA in 2012

Characters – It is a three-line *indica* hybrid rice with a moderate plant type, relatively tall plants, high tillering capacity, vigorous growth potential, average tolerance to lodging, intermediate tolerance to cold during booting and flowering, intermediate growth duration, good maturing color, large grains, good grain quality, and high yield potential. Tested in the late-maturing mid-season *indica* rice trials of the middle and lower reaches of the Yangtze River in 2006 and 2007, it yielded 8.51 t/ha, higher than the yield of the control variety Ⅱ You 838 by 3.02%. It is 133.8 d in whole growth duration, which is 0.5 d longer than that of the control variety Ⅱ You 838. It has 2.49 million effective panicles per hectare, is 132.4 cm in plant height and 28.1 cm in panicle length, and has 152.6 spikelets per panicle. It is 75.8% in seed setting rate and 30.9 g in 1,000-grain weight; highly susceptible to rice blast and bacterial blight; and susceptible to rice brown planthopper. Tested in the late-season rice trials of Guangdong in 2007 and 2008, it yielded 6.90 t/ha and 6.85 t/ha, higher than the yields of the control variety Jingxian 89 by 3.07% and 7.63%, respectively. It is 110 – 113 d in whole growth duration, which is 1 – 3 d shorter than that of Jingxian 89; highly resistant to rice blast, and moderately susceptible to bacterial blight.

Suitable areas – Single-cropping mid-season rice in areas along the Yangtze River in Jiangxi, Hunan, Hubei, Anhui, Zhejiang, and Jiangsu, excluding the Wuling Mountains; northern Fujian and southern Henan where there is no severe occurrence of rice blast and bacterial blight; *indica* rice areas below 1,300 m in altitude in Yunnan; early- or late-season rice in Guangdong, excluding the northern part; and late-season rice in Fujian where there is no severe occurrence of rice blast. By 2015, the accumulated planting area exceeded 401,300 ha.

Shenyou 9516

Institution (s) – Graduate School at Shenzhen, Tsinghua University

Parental lines – Shen 95A/R7116

Approvals – Shaoguan of Guangdong in 2012 (SS201207), Guangdong in 2010 (YS2010042), super rice recognition by the MOA in 2012

Characters – It is a thermo-sensitive three-line hybrid rice with relatively tall plants, a moderately compact plant type, intermediate to high tillering capacity, large panicles with large numbers of grains, high seed setting rate, high resistance to lodging, intermediate tolerance to cold, high yield potential, high grain quality, resistance to rice blast, and moderate susceptibility to bacterial blight. Tested in the late-season rice trials of Guangdong in 2008 and 2009, it yielded 7.78 t/ha and 7.21 t/ha, higher than the yields of the control variety Jingxian 89 by 22.22% and 18.24%, respectively. It is 112 – 116 d in whole growth duration, which is the same as that of the control variety. It is 112.0 – 113.2 cm in plant height, has 2.49 – 2.61 million effective panicles per hectare, 23.0 – 23.3 cm of panicle length with 137 – 149 spikelets per panicle, 84.1% – 85.0% in seed setting rate and 27.1 – 27.3 g in 1,000-grain weight, with Grade 3 grain quality as per the national standard *High Quality Paddy*.

Suitable areas – Early- or late-season rice in Guangdong, excluding the northern part; and late-season rice in Shaoguan of Guangdong. By 2015, the accumulated planting area exceeded 366,700 ha.

Teyou 582

Institution (s) – The Rice Research Institute of Guangxi Academy of Agricultural Sciences

Parental lines – LongtepuA/Gui 582

Approvals – Guangxi in 2009 (GSD 2009010), super rice recognition by the MOA in 2011

Characters – It is a thermo-sensitive three-line hybrid rice with a compact plant type, dark green leaves, colorless leaf sheaths, stigmas and apiculus, and upright and stiff flag leaves. Tested in the late-maturing early-season rice trials of northern Guangxi in 2007 and 2008, it yielded 7.97 t/ha, higher than the yield of the control variety Teyou 63 by 7.66%. It is 124 d in whole growth duration, which is 2 – 3 d shorter than that of Teyou 63. It has 2.475 million effective panicles per hectare, is 108.0 cm in plant height and 23.2 cm in panicle length with 167.4 spikelets per panicle, 82.6% in seed setting rate and 24.9 g in 1,000-grain weight, with Grade 5 resistance to rice seedling blast, Grade 9 to rice panicle blast, 42.8% of rice blast loss index, and Grade 6.5 of overall resistance to blast, Grade 7 and Grade 9 resistance to bacterial blight Ⅳ and Ⅴ, respectively.

Suitable areas – Early-season rice in southern Guangxi or early-season rice in central Guangxi. By 2015, the accumulated planting area exceeded 101,300 ha.

Wuyou 308

Institution (s) – Rice Research Institute of Guangdong Academy of Agricultural Sciences

Parental lines – Wufeng A/Guanghui 308

Approvals – State in 2008 (GS2008014), Guangdong in 2006 (YS2006059), Meizhou of Guangdong in 2004 (MS2004005), super rice recognition by the MOA in 2010

Characters – It is a thermo-sensitive three-line *indica* hybrid rice with a moderate plant type, intermediate to high tillering capacity, thick and strong stems, high tolerance to lodging, a large number of effective panicles, short flag leaves, large panicles with a large number of grains, intermediate tolerance to cold at the late growth stage, slight panicle enclosure in a low temperature, moderate growth duration, high yield potential, and good grain quality. Tested in the early-maturing late-season *indica* rice trials of the middle and lower reaches of the Yangtze River in 2006 and 2007, it yielded 7.57 t/ha, higher than the yield of the control variety Jinyou 207 by 6.68%. It is 112.2 d in whole growth duration, which is 1.7 d longer than that of Jinyou 207. It has 2.91 million effective panicles per hectare, is 99.6 cm in plant height and 21.7 cm in panicle length with 157.3 spikelets per panicle. It is 73.3% in seed setting rate and 23.6 g in 1,000-grain weight; highly susceptible to rice blast, susceptible to bacterial blight, and moderately susceptible to brown planthopper, with Grade 1 grain quality as per the national standard *High Quality Paddy*. Tested in early-season rice trials of Guangdong in 2005 and 2006, it yielded 7.37 t/ha and 6.58 t/ha, higher than the yields of the control variety Zhong 9 You 207 by 17.30% and 13.84%, respectively. It is 125 – 127 d in whole growth duration, the same as that of Zhong 9 You 207, and highly resistant to rice blast and bacterial blight.

Suitable areas – Late-season rice in double-cropping rice areas in Jiangxi, Hunan, Zhejiang, Hubei, and south of the Yangtze River in Anhui where there is no severe occurrence of rice blast and bacterial blight; and early- or late-season rice in Guangdong. By 2015, the accumulated planting area exceeded 1,329,300 ha.

Wufengyou T025

Institution (s) – Agronomy College of Jiangxi Agricultural University

Parental lines – Wufeng A/Changhui T025

Approvals – State in 2010 (GS2010024), Jiangxi in 2008 (GS2008013), super rice recognition by the MOA in 2010

Characters – It is a three-line late-season *indica* hybrid rice with a moderate plant type, straight and stiff leaves, high tillering capacity, relatively high number of effective panicles, vigorous growth potential, large numbers of panicles and spikelets per panicle, small grain weight, moderate growth duration, good maturing color, intermediate yield potential, high susceptibility to rice blast and brown planthopper, susceptibility to bacterial blight, and high grain quality. Tested in the early-maturing late-season *indica* rice trials of the middle and lower reaches of the Yangtze River in 2007 and 2008, it yielded 7.52 t/ha, higher than the yield of the control variety Jinyou 207 by 2.0%. It is 112.3 d in whole growth duration, which is 1.4 d longer than that of Jinyou 207. It has 2.82 million effective panicles per hectare, is 103.3 cm in plant height and 22.8 cm in panicle length with 174.6 spikelets per panicle. It is 77.7% in seed setting rate and 22.8 g in 1,000-grain weight, with Grade 3 grain quality as per the national standard *High Quality Paddy*. Tested in the late-season rice trials of Jiangxi in 2006 and 2007, it yielded 6.90 t/ha, higher than the yield of the control variety Jinyou 207 by 8.77%. It is 114.7 d in whole growth duration, which is 3.5 d longer than that of Jinyou 207, with Grade 1 grain quality as per the national standard *High Quality Paddy*.

Suitable areas – Late-season rice in double-cropping rice areas in Jiangxi, Hunan, Zhejiang, Hubei, and south of the Yangtze River in Anhui where there is no severe occurrence of rice blast and bacterial blight. By 2015, the accumulated planting area exceeded 606,000 ha.

Tianyou 3301

Institution (s) – Research Institute of Biotechnology of Fujian Academy of Agricultural Sciences and Rice Research Institute of Guangdong Academy of Agricultural Sciences

Parental lines – Tianfeng A/Minhui 3301

Approvals – Hainan in 2011 (QS2011015), State in 2010 (GS2010016), Fujian in 2008 (MS2008023), super rice recognition by the MOA in 2010

Characters – It is a thermo-sensitive three-line *indica* hybrid rice with a moderate plant type, vigorous growth potential, moderate growth duration, good maturing color, average tolerance to cold, intermediate susceptibility to rice blast, susceptibility to bacterial blight and brown planthopper, high yield potential, and average grain quality. Tested in the late-maturing mid-season *indica* rice trials of the middle and lower reaches of the Yangtze River in 2007 and 2008, it yielded 8.97 t/ha, higher than the yield of the control variety Ⅱ You 838 by 6.19%. It is 133.3 d in whole growth duration, which is 1.7 d shorter than that of Ⅱ You 838. It has 2.475 million effective panicles per hectare, is 118.9 cm in plant height and 24.3 cm in panicle length with 165.2 spikelets per panicle. It is 81.3% in seed setting rate and 29.7 g in 1,000-grain weight.

Suitable areas – Single-cropping mid-season rice in areas along the Yangtze River in Jiangxi, Hunan, Hubei, Anhui, Zhejiang, and Jiangsu, excluding the Wuling Mountains; northern Fujian and southern Henan where there is no severe occurrence of bacterial blight; late-season rice in Fujian where there is no severe occurrence of rice blast; and early-season rice in Hainan. By 2015, the accumulated planting area exceeded 194,700 ha.

Luoyou 8 (Honglianyou 8)

Institution (s) – Wuhan University

Parental lines – Luohong 3A/R8108

Approvals – State in 2007 (GS2007023), Hubei in 2006 (ES2006005), super rice recognition by the MOA in 2009

Characters – Itis a three-line mid-season *indica* hybrid rice with a moderate plant type, vigorous growth potential, partially exposed internodes, strong stems, dark green leaves, colorless leaf sheaths, narrow and upright flag leaves, late maturity, two-phase grain filling, panicle enclosure and pocking mark in low temperature, average maturing color, high yield potential, average yield stability, high grain quality, high susceptibility to rice blast, susceptibility to bacterial blight, and vulnerability to rice false smut. Tested in the late-maturing mid-season *indica* rice trials of the middle and lower reaches of the Yangtze River in 2004 and 2005, it yielded 8.53 t/ha, higher than the yield of the control variety Shanyou 63 by 3.48%. It is 138.8 d in whole growth duration, which is 4.2 d longer than that of Shanyou 63. It has 2.58 million effective panicles per hectare, is 122.1 cm in plant height and 23.1 cm in panicle length with 174.7 spikelets per panicle. It is 74% in seed setting rate and 26.9 g in 1,000-grain weight, with Grade 3 grain quality as per the national standard *High Quality Paddy*.

Suitable areas – Single-cropping mid-season rice in areasalong the Yangtze River in Jiangxi, Hunan, Hubei, Anhui, Zhejiang, and Jiangsu, excluding the Wuling Mountains; and northern Fujian and southern Henan where there is no severe occurrence of rice blast and bacterial blight. By 2015, the accumulated planting area exceeded 492,000 ha.

Ganxin 203 (**Rongyou 3**)

Institution (s) – Rice Research Institute of Guangdong Academy of Agricultural Sciences, Jiangxi Modern Seed Industry Co., Ltd., and Agronomy College of Jiangxi Agricultural University

Parental lines – Rongfeng A/R3

Approvals – Shaoguan of Guangdong in 2010 (SS201001), State in 2009 (GS2009009), Jiangxi in 2006 (GS2006062), super rice recognition by the MOA in 2009

Characters – It is a three-line early-season *indica* hybrid rice with a moderate plant type, light green straight leaves, short, wide and stiff flag leaves, high tillering capacity, a large number of effective panicles, high seed setting rate, high grain weight, good maturing color, moderate growth duration, relatively high yield potential, intermediate tolerance to cold, susceptibility to rice blast, intermediate susceptibility to bacterial blight, high susceptibility to brown planthopper and whitebacked planthopper, and average grain quality. Tested in the late-maturing early-season *indica* rice trials of the middle and lower reaches of the Yangtze River in 2007 and 2008, it yielded 7.81 t/ha, higher than the yield of the control variety Jinyou 402 by 4.66%. It is 114.4 d in whole growth duration, which is 1.7 d longer than that of Jinyou 402. It has 3.27 million effective panicles per hectare, is 95.5 cm in plant height and 18.4 cm in panicle length with 103.5 spikelets per panicle. It is 86.3% in seed setting rate and 28.3 g in 1,000-grain weight.

Suitable areas – Early-season rice in double-cropping rice areas in plain regions of Jiangxi, Hunan, northern Fujian, south-central Zhejiang, and Shaoguan of Guangdong where there is no severe occurrence of rice blast. By 2015, the accumulated planting area exceeded 612,000 ha.

Guodao 6 (Nei 2 You 6)

Institution (s) – China National Rice Research Institute

Parental lines – Neixiang 2A/Zhonghui 8006

Approvals – State in 2007 (GS2007011, upper Yangtze rice area), Chongqing in 2007 (YS2007007), state in 2006 (GS2006034, rice areas of the middle and lower reaches of Yangtze River), super rice recognition by the MOA in 2007

Characters – It is a three-line mid-season *indica* hybrid rice with a moderate plant type, thick and strong stems, straight and stiff leaves, high tolerance to lodging, moderate growth duration, high yield potential, high grain quality, and high susceptibility to rice blast and bacterial blight. Tested in the late-maturing mid-season *indica* rice trials of the middle and lower reaches of the Yangtze River in 2004 and 2005, it yielded 8. 68 t/ha, higher than the yield of the control variety Shanyou 63 by 5. 38% . It is 137. 8 d in whole growth duration, which is 3. 2 d longer than that of Shanyou 63. It has 2. 475 million effective panicles per hectare, is 114. 2 cm in plant height and 26. 1 cm in panicle length with 159. 7 spikelets per panicle. It is 73. 3% in seed setting rate and 31. 5 g in 1,000-grain weight, with Grade 3 grain quality as per the national standard *High Quality Paddy*. Tested in the late-maturing mid-season *indica* rice trials of the upper reaches of the Yangtze River in 2005 and 2006, it yielded 8. 84 t/ha, lower than the yield of the control variety Ⅱ You 838 by 0. 10% . It is 154. 4 d in whole growth duration, which is 0. 2 d longer than that of Ⅱ You 838.

Suitable areas – Single-cropping mid-season rice in Fujian, Jiangxi, Hunan, Hubei, Anhui, Zhejiang, Jiangsu, and southern Henan; low-to-intermediate altitude *indica* rice areas in Yunnan, Guizhou, and Chongqing, excluding the Wuling Mountains; and hilly rice areas in Pingba of Sichuan and southern Shaanxi where there is no severe occurrence of rice blast. By 2015, the accumulated planting area exceeded 286,000 ha.

Ganxin 688 (Changyou 11)

Institution (s) – School of Agricultural Sciences, Jiangxi Agricultural University

Parental lines – Tianfeng A/Changhui 121

Approvals – Hunan Introduction in 2010 (XYZ201026), Jiangxi in 2006 (GS2006032), super rice recognition by the MOA in 2007

Characters – It is a three-line late-season *indica* hybrid rice with a compact plant type, dark green leaves, wide and stiff flag leaves, vigorous growth potential, thick and strong stems, high tillering capacity, relatively high number of effective panicles, large number of spikelets per panicle, high seed setting rate and good maturing color. Tested in the late-season rice trials of Jiangxi late rice in 2004 and 2005, it yielded 7. 90 t/ha and 7. 03 t/ha, higher than the yields of the control variety Shanyou 46 by 1. 57% and 4. 98% , respectively. It is 123. 7 d in whole growth duration, which is 1. 4 d longer than that of Shanyou 46. It is 101. 6 cm in plant height, has 2. 95 million effective panicles per hectare and 146. 6 spikelets per panicle including 112. 2 filled grains, and is 76. 5% in seed setting rate and 24. 9 g in 1,000-grain weight, with Grade 5 resistance to both rice seedling blast and leaf blast, plus Grade 3 resistance to panicle blast.

Suitable areas – Late-season rice areas in Jiangxi where there is no severe occurrence of rice blast. By 2015, the accumulated planting area exceeded 670,700 ha.

Ⅱ Youhang 2

Institution (s) – Rice Research Institute of Fujian Academy of Agricultural Sciences

Parental lines – Ⅱ – 32A/Hang 2

Approvals – Guizhou Introduction in 2008 (QY2008012), state in 2007 (GS2007020), Fujian in 2006 (MS2006017), Anhui in 2006 (WPS060104970), super rice recognition by the MOA in 2007

Characters – It is a three-line mid-season *indica* hybrid rice with a moderate plant type, thick and strong stems, vigorous growth potential, intermediate tillering capacity, large numbers of panicles and grains, good maturing color, moderate growth duration, high yield potential, average grain quality, and susceptibility to rice blast and bacterial blight. Tested in the late-maturing mid-season *indica* rice trials of the middle and lower reaches of the Yangtze River in 2005 and 2006, it yielded 8.55 t/ha, higher than the yield of the control variety Ⅱ You 838 by 7.01%. It is 134.5 d in whole growth duration, which is 0.8 d longer than that of Ⅱ You 838. It has 2.43 million effective panicles per hectare, is 129.9 cm in plant height and 25.8 cm in panicle length with 159.7 grains per panicle. It is 79.0% in seed setting rate and 28.5 g in 1,000-grain weight.

Suitable areas – Single-cropping mid-season rice in areas along the Yangtze River in Jiangxi, Hunan, Hubei, Anhui, Zhejiang and Jiangsu, excluding the Wuling Mountains; rice areas in Fujian and southern Henan where there is no severe occurrence of rice blast and bacterial blight. By 2015, the accumulated planting area exceeded 213,300 ha.

Tianyou 122 (Tianfengyou 122)

Institution (s) – Rice Research Institute of Guangdong Academy of Agricultural Sciences

Parental lines – Tianfeng A/Guanghui 122

Approvals – State in 2009 (GS2009029), Guangdong in 2005 (YS2005022), super rice recognition by the MOA in 2007

Characters – It is a three-line late-season *indica* hybrid rice with a moderately compact plant type, high tillering capacity, thin stems with low tolerance to lodging, short and straight flag leaves, light green leaves, good maturing color, purple apiculus, short awns at the panicle top, moderate growth duration, intermediate yield potential, intermediate susceptibility or high resistance to rice blast, susceptibility to intermediate resistance to bacterial blight, high susceptibility to brown planthopper, and good grain quality. Tested in the late-maturing *indica* late-season rice trials of the middle and lower reaches of the Yangtze River in 2006 and 2007, it yielded 7.29 t/ha, higher than the yield of the control variety Shanyou 46 by 2.35%. It is 116.6 d in whole growth duration, which is 2.9 d shorter than that of Shanyou 46. It has 2.82 million effective panicles per hectare, is 101.8 cm in plant height and 21.5 cm in panicle length with 141.9 spikelets per panicle. It is 77.6% in seed setting rate and 25.4 g in 1,000-grain weight, with Grade 1 grain quality as per the national standard *High Quality Paddy*. Tested in the early-season rice trials of Guangdong in 2003 and 2004, it yielded 7.24 t/ha and 7.88 t/ha, higher than the yields of the control variety by 12.53% and 7.79%, respectively. It is 124 – 125 d in whole growth duration, which is

3 – 5 d longer than that of the control variety.

Suitable areas – Late-season rice in double-cropping rice areas in north-central Guangxi, north-central Fujian, south-central Jiangxi, south-central Hunan, and southern Zhejiang where there is no severe occurrence of bacterial blight; and early- or late-season rice in Guangdong. By 2015, the accumulated planting area exceeded 302,000 ha.

Jinyou 527

Institution (s) – Rice Research Institute of Sichuan Agricultural University

Parental lines – Jin 23A/Shuhui 527

Approvals – State in 2004 (GS2004012), Shaanxi Introduction in 2003 (SY2003003), Sichuan in 2002 (CS2002002), super rice recognition by the MOA in 2006

Characters – It is a three-line mid-season *indica* hybrid rice with a moderate plant type, vigorous growth potential, dark green and straight leaves, intermediate tillering capacity, high tiller-to-panicle formation rate, large numbers of panicles and grains, low tolerance to cold, good maturing color, moderate growth duration, high yield potential, high susceptibility to rice blast, intermediate susceptibility to bacterial blight, susceptibility to brown planthopper, and good grain quality. Tested in the late-maturing mid-season *indica* rice trials of the upper reaches of the Yangtze River in 2002 and 2003, it yielded 9.14 t/ha, higher than the yield of the control variety Shanyou 63 by 8.78%. It is 151.2 d in whole growth duration, which is 1.4 d shorter than that of Shanyou 63. It is 115.5 cm in plant height, has 2.475 million effective panicles per hectare, and is 25.7 cm in panicle length with 161.7 grains per panicle. It is 80.9% in seed setting rate and 29.5 g in 1,000-grain weight, with Grade 3 grain quality as per the national standard *High Quality Paddy*.

Suitable areas – Single-cropping mid-season rice in Yunnan and Guizhou; low-to-intermediate altitude rice areas in Chongqing, excluding the Wuling Mountains; Pingba of Sichuan and southern Shaanxi where there is no severe occurrence of rice blast. By 2015, the accumulated planting area exceeded 431,300 ha.

D You 202 (Taiyou 1)

Institution (s) – Rice Research Institute of Sichuan Agricultural University and Sichuan Nongda Hi-Tech Agriculture Co., Ltd.

Parental lines – D62A/Shuhui 202

Approvals – State in 2007 (GS2007007), Hubei in 2007 (ES2007010), Anhui in 2006 (WPS06010503), Fujian Sanming in 2006 (MS2006G02 [Sanming]), Zhejiang in 2005 (ZS2005001), Guangxi in 2005 (GS2005010), Sichuan in 2004 (CS2004010), super rice recognition by the MOA in 2006

Characters – It is a three-line mid-season *indica* hybrid rice with a moderate plant type, thick and strong stems, high tillering capacity, vigorous growth potential, low tolerance to lodging, straight and stiff leaves, panicles below the canopy, uniform panicle layer, intermediate panicle size, even and scattered seeding, good maturing color, moderate growth duration, high yield potential, good grain quality, high susceptibility or resistance to rice blast, high susceptibility or susceptibility to bacterial blight, and susceptibility to brown planthopper. Tested in the late-maturing mid-season *indica* rice trials of the upper reaches of the Yangtze River in 2005 and 2006, it yielded 8.76 t/ha, higher than the yield of the control variety Ⅱ You 838 by 1.89%. It is 155.0 d in whole growth duration, which is 1.4 d longer than that of the control variety Ⅱ You 838. It has 2.43 million effective panicles per hectare, is 115.1 cm in plant height and 25.8 cm in panicle length with 158.4 spikelets per panicle. It is 80.3% in seed setting rate and 29.6 g in 1,000-grain weight, with Grade 3 grain quality as per the national standard *High Quality Paddy*.

Suitable areas - Single-cropping mid-season *indica* rice in low-to-medium altitude areas in Yunnan and Guizhou, excluding the Wuling Mountains; hilly areas in Pingba of Sichuan, southern Shaanxi, Hubei (excluding southwestern part), Anhui and Zhejiang where there is no severe occurrence of rice blast; late-season rice areas in Sanming of Fujian where there is no severe occurrence of rice blast; early-season rice areas in southern Guangxi or mid-season rice in cold hilly areas of southern Guangxi. By 2015, the accumulated planting area exceeded 164,700 ha.

Q You 6 (**Qingyou 6**)

Institution (s) - Chongqing Seeds Company

Parental lines - Q2A/R1005

Approvals - State in 2006 (GS2006028), Hubei in 2006 (ES2006008), Hunan in 2006 (XS2006032), Chongqing in 2005 (YS2005001), Guizhou in 2005 (QS2005014), super rice recognition by the MOA in 2006

Characters - It is a three-line mid-season *indica* hybrid rice with a moderately compact plant type, thick and strong stems, exposed internodes, upright flag leaves, purple auricles, sheaths and apiculus, high tillering capacity, large panicles with scattered grains, moderate growth duration, good maturing color, high yield potential, good grain quality, high tolerance to heat, intermediate tolerance to cold, and high susceptibility to rice blast and bacterial blight. Tested in the late-maturing mid-season rice trials of the upper reaches of the Yangtze River in 2004 and 2005, it yielded 8.98 t/ha, higher than the yields of the control variety Shanyou 63 by 5.43%. It is 153.7 d in whole growth duration, which is 0.8 d longer than that of Shanyou 63. It has 2.40 million effective panicles per hectare, is 112.6 cm in plant height 112.6 cm and 25.1 cm in panicle length with 176.6 spikelets per panicle. It is 77.2% in seed setting rate and 29.0 g in 1,000-grain weight, with Grade 3 grain quality as per the national standard *High Quality Paddy*.

Suitable areas - Single-cropping mid-season rice in low-to-medium altitude rice areas in Yunnan, Guizhou, Hubei, Hunan, and Chongqing, excluding the Wuling Mountains; mid-season rice in Pingba of Sichuan and southern Shaanxi where there is no severe occurrence of rice blast. By 2015, the accumulated planting area exceeded 1,402,000 ha.

Guodao 1 (**Zhong 9 You 6**, **Zhongyou 6**)

Institution (s) - China National Rice Research Institute

Parental lines - Zhong 9A/R8006

Approvals - Shaanxi Introduction in 2007 (SY2007001), Guangdong in 2006 (YS2006050), Jiangxi in 2004 (GS2004009), state in 2004 (GS2004032), super rice recognition by the MOA in 2005

Characters - It is a thermo-sensitive three-line hybrid rice with a moderate plant type, thick and strong stems, vigorous growth potential, relatively flat flag leaves, intermediate tillering capacity and tolerance to lodging, moderate to high tolerance tocold in the late growth stage, moderate growth duration, intermediate yield potential, good grain quality, high susceptibility or intermediate resistance to rice blast, susceptibility to intermediate susceptibility to bacterial blight, and high susceptibility to brown planthopper. Tested in the late-season mid- to late-maturing *indica* rice trials of the middle and lower reaches of the Yangtze River in 2002 and 2003, it yielded 6.87 t/ha, higher than the yields of the control variety Shanyou 46 by 1.43%. It is 120.6 d in whole growth duration, which is 2.6 d longer than that of Shanyou 46. It is 107.8 cm in plant height and 25.6 cm in panicle length with 142.0 spikelets per panicle. It has 2.67 million effective panicles per hectare and is 73.5% in seed setting rate and 27.9 g in 1,000-grain weight, with Grade 3 grain quality as per the national standard *High Quality Paddy*.

Suitable areas - Late-season rice in double-cropping rice areas in north-central Guangxi, north-central Fujian, south-central Jiangxi, south-central Hunan, and southern Zhejiang where there is no severe occurrence of rice blast and bacterial blight; late-season rice in Guangdong; and early-season rice in Guangdong excluding the northern part. By 2015, the accumulated planting area exceeded 479,300 ha.

Zhongzheyou 1

Institution (s) - China National Rice Research Institute and Zhejiang Wuwangnong Seeds Shareholding Co., Ltd.

Parental lines - ZhongzheA/Hanghui 570

Approvals - Hainan in 2012 (QS2012004), Guizhou in 2011 (QSD 2011005), Hunan in 2008 (XSD 2008026), Zhejiang in 2004 (ZSD 2004009), super rice recognition by the MOA in 2005

Characters - It is a three-line late-maturing mid-season *indica* hybrid rice with a moderate plant type, high tillering capacity, vigorous growth potential, thick and strong stems, moderate length and width of leaves, involute leaf edges, upright leaves, large number of effective panicles, high seed setting rate, good maturing color, colorless apiculus, high yield potential, good grain quality, average tolerance to cold and heat, high susceptibility to rice blast, vulnerability to sheath blight, intermediate susceptibility to bacterial blight, and susceptibility to brown planthopper. Tested in the single-cropping rice trials of Zhejiang in 2002 and 2003, it yielded 8.03 t/ha and 7.31 t/ha, higher than the yields of the control variety Shanyou 63 by 10.7% and 1.9%, respectively. It is 136.8 d in whole growth duration, which is 5.5 d longer than that of Shanyou 63. It is 115 - 120 cm in plant height and 25 - 28 cm in panicle length with 180 - 300 spikelets per panicle and 17 leaves on the main stem. It has 2.25 - 2.55 million effective panicles per hectare, and is 70% in tiller-to-panicle formation rate, 85% - 90% in seed setting rate and 27 g in 1,000-grain weight.

Suitable areas - Single-cropping mid-season rice in Zhejiang, Hunan, and Guizhou where there is no severe occurrence of rice blast; and late-season rice in Hainan. By 2015, the accumulated planting area exceeded 1,408,000 ha.

Fengyuanyou 299

Institution (s) - Hunan Hybrid Rice Research Center

Parental lines－Fengyuan A/Xianghui 299

Approvals－Hunan in 2004 (XS2004011), super rice recognition by the MOA in 2005

Characters－It is a three-line late-season *indica* hybrid rice with a moderate compact plant type, stiff stems, light green leaves, purple leaf sheaths, good maturing color, high yield potential, good grain quality, intermediate tolerance to cold, susceptibility to rice blast, and intermediate resistance to bacterial blight. The hybrid was tested in the late-season rice trials of Hunan in 2002 and 2003, and it had yields of 7.04 t/ha and 7.11 t/ha, higher than the yields of the control variety by 7.55% and 2.66%, respectively. It is 114 d in whole growth duration, which is 4 d longer than that of the control variety Jinyou 207. It is 97 cm in plant height, has 2.85 million effective panicles per hectare, and is 22 cm in panicle length with 135 grains per panicle, 80% in seed setting rate and 29.5 g in 1,000-grain weight.

Suitable areas－Late-season rice in Hunan where there is no severe occurrence of rice blast. By 2015, the accumulated planting area exceeded 906,000 ha.

Jinyou 299

Institution (s)－Hunan Hybrid Rice Research Center.

Parental lines－Jin 23A/Xianghui 299.

Approvals－Shaanxi in 2009 (SS2009005), Jiangxi in 2005 (GS2005091), Guangxi in 2005 (GS2005002), super rice recognition by the MOA in 2005.

Characters－It is a thermo-sensitive three-line hybrid rice with a moderate plant type, vigorous growth potential, dark green leaves, purple sheaths and apiculus, weak tillering capacity, high tiller-to-panicle formation rate, relatively large numbers of panicles and grains, high seed setting rate, partial panicle enclosure, low tolerance to lodging, and good maturing color. Tested in the late-season rice trials of Jiangxi in 2003 and 2004, it yielded 6.57 t/ha in 2003, higher than the yield of the control variety Shanyou 63 by 1.40%, and 6.82 t/ha in 2004, lower than the yield of the control variety Jinyou 207 by 1.80%, respectively. It is 109.7 d in whole growth duration, which is 2.0 d shorter than that of Jinyou 207 and 99.8 cm in plant height, has 2.61 million effective panicles per hectare and 129.4 spikelets per panicle with 104.0 filled spikelets per panicle. It is 80.4% in seed setting rate and 26.5 g in 1,000-grain weight, resistant to rice blast, Grade 0 resistant to seedling blast, and Grade 5 resistant to leaf blast and panicle blast.

Suitable areas－Late-season rice in Jiangxi where there is no severe occurrence of rice blast, early- or late-season rice in central Guangxi, and late-season rice in northern Guangxi. By 2015, the accumulated planting area exceeded 81,300 ha.

Ⅱ Youming 86

Institution (s)－Sanming Academy of Agricultural Sciences

Parental lines－Ⅱ－32A/Minghui 86

Approvals－State 2001 (GS2001012), Fujian in 2001 (MS2001009), Guizhou in 2000 (QPS228), super rice recognition by the MOA in 2005

Characters－It is a three-line late-maturing mid-season *indica* hybrid rice with a moderately compact plant type, thick and strong stems, tolerance to fertilizer, resistance to lodging, intermediate tillering capacity, thick and upright flag leaves, large panicles with large numbers of grains, high seed setting rate, good maturing color, high resistance to cold, high yield potential, wide adaptation, intermediate susceptibility to rice blast, and susceptibility to bacterial blight and rice planthopper. Tested in the late-maturing mid-season *indica* rice trials of southern China in 1999 and 2000, it yielded 9.48 t/ha and 8.48 t/ha, higher than the yields of the control variety Shanyou 63 by 8.19% and 3.15%, respectively. It is 150.8 d in whole growth duration, which is 3.7 d longer than that of Shanyou 63. It is 100 – 115 cm in plant height and 25.6 cm in panicle length, with 17 – 18 leaves on the main stem, 2.43 million effective panicles per hectare including 163.6 spikelets per panicle, and is 81.8% in seed setting rate and 28.2 g in 1,000-grain weight.

Suitable areas – Single-cropping mid-season rice in Guizhou, Yunnan, Sichuan, Chongqing, Hunan, Hubei, Zhejiang, and Shanghai, the Yangtze River basin in Anhui and Jiangsu, southern Henan and Hanzhong of Shaanxi. By 2015, the accumulated planting area exceeded 1,370,700 ha.

II Youhang 1

Institution (s) – Rice Research Institute of Fujian Academy of Agricultural Sciences

Parental lines – II – 32A/Hang 1

Approvals – State in 2005 (GS2005023), Fujian in 2004 (MS2004003), super rice recognition by the MOA in 2005

Characters – It is a three-line mid-season *indica* hybrid rice with a moderate plant type, thick and strong stems, intermediate tillering capacity, vigorous growth potential, long and wide flag leaves, moderate growth duration, good maturing color, high yield potential, intermediate susceptibility to rice blast, susceptibility to bacterial blight, high susceptibility to brown planthopper, and average grain quality. Tested in the late-maturing mid-season *indica* rice trials of the middle and lower reaches of the Yangtze River in 2003 and 2004, it yielded 8.33 t/ha, higher than the yield of the control variety Shanyou 63 by 5.13%. It is 135.8 d in whole growth duration, which is 2.7 d longer than that of Shanyou 63. It is 127.5 cm in plant height, has 2.49 million effective panicles per hectare, is 26.2 cm of panicle length, and has 165.4 spikelets per panicle. It is 77.9% in seed setting rate and 27.8 g in 1,000-grain weight.

Suitable areas – Single-cropping mid-season rice along the Yangtze River in Fujian, Jiangxi, Hunan, Hubei, Anhui, Zhejiang, and Jiangsu, excluding the Wuling Mountains; southern Henan where there is no severe occurrence of bacterial blight; and late-season rice in Fujian where there is no severe occurrence of rice blast. By 2015, the accumulated planting area exceeded 690,700 ha.

Teyouhang 1

Institution (s) – Rice Research Institute of Fujian Academy of Agricultural Sciences

Parental lines – LongtepuA/Hang 1

Approvals – Guangdong in 2008 (YS2008020), State in 2005 (GS2005007), Zhejiang in 2004 (ZS2004015), Fujian in 2003 (MS2003002), super rice recognition by the MOA in 2005

Characters – It is a three-line *indica* hybrid rice with a moderate plant type, relatively long flag leaves, thick and strong stems, high tolerance to lodging, intermediate tillering capacity, early maturity, good maturing color, high yield potential, intermediate tolerance to cold during booting and flowering, high or intermediate susceptibility to rice blast, normal to intermediate susceptibility to bacterial blight, high susceptibility to brown planthopper, and average grain quality. Tested in the late-maturing mid-season *indica* high-yielding rice trials of the upper reaches of the Yangtze River in 2002 and 2003, it yielded 8.88 t/ha, higher than the yield of the control variety Shanyou 63 by 5.50%. It is 150.5 d in whole growth duration, which is 2.6 d shorter than that of Shanyou 63. It is 112.7 cm in plant height and 24.4 cm in panicle length, with 2.355 million effective panicles and 166.1 spikelets per panicle, 83.9% in seed setting rate and 28.4 g in 1,000-grain weight.

Suitable areas – Single-cropping mid-season rice in low-to-medium altitude *indica* rice areas in Yunnan, Guizhou and Chongqing, excluding the Wuling Mountains; hilly rice areas in Pingba of Sichuan, southern Shaanxi where there is no severe occurrence of rice blast; late-season rice in Fujian where there is no severe occurrence of rice blast; early-season rice in Guangdong excluding the northern part; late-season rice in south-central and southwest Guangdong; and single-cropping rice in south-central Zhejiang. By 2015, the accumulated planting area exceeded 315,300 ha.

D You 527

Institution (s) – Rice Research Institute of Sichuan Agricultural University

Parental lines – D62A/Shuhui 527

Approvals – Honghe of Yunnan in 2005 (DTS [Honghe] 200503), State in 2003 (GS2003005), Shaanxi Introduction in 2003 (SY2003002), Fujian in 2002 (MS2002002), Sichuan in 2001 (CS135), Guizhou in 2000 (QPS242), super rice recognition by the MOA in 2005

Characters – It is a three-line late-maturing mid-season *indica* hybrid rice with a moderately compact plant type, thick and strong stems, dark green leaves, high tillering capacity, vigorous growth potential, intermediate panicle size, good maturing color, purple leaf sheaths and apiculus, high yield potential with high stability, wide adaptation, above-average grain quality, intermediate resistance to rice blast, susceptibility to bacterial blight and rice brown planthopper, and low susceptibility to rice false smut. Tested in the late-maturing mid-season *indica* rice trials of the Yangtze River basin in 2000, it yielded 8.57 t/ha, higher than the yield of the control variety Shanyou 63 by 4%. Tested in the late-maturing mid-season *indica* high-quality rice trials in 2001, it yielded 9.17 t/ha in the upper Yangtze areas, higher than the yield of the control variety Shanyou 63 by 4.48%, and 9.67t/ha in the middle and lower Yangtze areas, higher than the yield of the control variety Shanyou 63 by 6.31%, with a whole growth duration 4 d longer than that of Shanyou 63. It has 17 – 18 leaves on the main stems, is 114 – 120 cm in plant height and 25 cm in panicle length, and has 2.70 million effective panicles per hectare with 150 spikelets per panicle. It is 80% in seed setting rate and 30 g in 1,000-grain weight.

Suitable areas-Single-cropping mid-season rice of the Yangtze River Basin in Sichuan, Chongqing, Hubei, Hunan, Zhejiang, Jiangxi, Anhui, Shanghai, and Jiangsu, excluding the Wuling Mountains, and areas below 1,100 m in altitude in Yunnan and Guizhou, and Xinyang of Henan and Hanzhong of Shaanxi where there is no severe occurrence of bacterial blight; and mid-season or late-season rice in Fujian. By 2015, the accumulated planting area exceeded 1,640,000 ha.

Xieyou 527

Institution (s) – Rice Research Institute of Sichuan Agricultural University

Parental lines – XieqingzaoA/Shuhui 527

Approvals – State in 2004 (GS2004008), Hubei in 2004 (ES2004007), Shaoguan of Guangdong in 2004 (SS200402), Sichuan in 2003 (CS2003003), super rice recognition by the MOA in 2005

Characters – It is a three-line mid-season *indica* hybrid rice with a moderate plant type, long, wide and upright flag leaves, dark green leaves, high tillering capacity, vigorous growth potential, relatively high seed setting rate, scattered seeding on panicles, high grain weight, slender grains with awns, good maturing color, weak tolerance to cold, moderate growth duration, high yield potential, high to intermediate susceptibility to rice blast, high or normal susceptibility to bacterial blight, high susceptibility to brown planthopper, and average grain quality. Tested in the late-maturing mid-season *indica* rice trials of the upper reaches of the Yangtze River in 2002 and 2003, it yielded 8.93 t/ha, higher than the yield of the control variety Shanyou 63 by 6.12%. It is 153.2 d in whole growth duration, which is 0.1 d longer than that of Shanyou 63. It is 111.2 cm in plant height, has 2.55 million effective panicles per hectare, and is 24.6 cm in panicle length and has 139.2 spikelets per panicle. It is 82.7% in seed setting rate and 32.3 g in 1,000-grain weight.

Suitable areas – Single-cropping mid-season rice in low-to-medium altitude rice areas in Yunnan, Guizhou, Hubei, and Chongqing, excluding the Wuling Mountains; and rice areas in Pingba of Sichuan and southern Shaanxi where there is no severe occurrence of rice blast and bacterial blight. By 2015, the accumulated planting area exceeded 130,700 ha.

II You 162

Institution (s) – Rice Research Institute of Sichuan Agricultural University

Parental lines – II – 32A/Shuhui 162

Approvals – Ningde of Fujian in 2001 (MS2002J0K [Ningde]), Hubei in 2001 (ES008-2001), state in 2000 (GS2000003), Zhejiang in 1999 (ZPSZ195), Sichuan in 1997 (CS [97] 64), super rice recognition by the MOA in 2005

Characters – It is a three-line late-maturing mid-season *indica* hybrid rice with a compact plant type, good tillering capacity, vigorous growth potential, dark green leaves, high tiller-to-panicle formation rate, large panicles with large numbers of grains, good maturing color, high yield potential with high stability, intermediate susceptibility to rice blast, high susceptibility to bacterial blight, and vulnerability to sheath blight, rice false smut and rice sheath rot. Tested in the late-maturing mid-season *indica* rice trials of Sichuan in 1995 and 1996, it yielded 7.18 t/ha, higher than the yield of the control variety Shanyou 63 by 5.39% with a whole growth duration 3 – 4 d longer than that Shanyou 63. It is 120 cm in plant height with 150 – 180 spikelets per panicle, 80% in seed setting rate and 28 g in 1,000-grain weight, with Grade 4 – 5 resistance to leaf blast, Grade 0 – 3 resistance to panicle blast, and a better grain quality than Shanyou 63. Tested in mid-season rice trials of Hubei in 1997 and 1998, it yielded 9.11 t/ha, higher than the yield of the control variety Shanyou 63 by 7.50%.

Suitable areas – Single-cropping mid-season rice in southwest China and the Yangtze River Basin where there is no severe occurrence of bacterial blight. By 2015, the accumulated planting area exceeded 883,300 ha.

Ⅱ You 7

Institution (s) – Rice and Sorghum Research Institute of Sichuan Academy of Agricultural Sciences

Parental lines – Ⅱ – 32A/Luhui 17

Approvals – Sanming of Fujian in 2004 (MS2004G04 [Sanming]), Chongqing in 2001 (YNF [2001] 369), Sichuan in 1998 (CS82), super rice recognition by the MOA in 2005

Characters – It is a three-line late-maturing mid-season *indica* hybrid rice with high tolerance to cold in the seedling stage, thick and strong stems, above-average tillering capacity, high tolerance to lodging, uniform panicle layer, good maturing color, and better resistance than Shanyou 63. Tested in the late-maturing mid-season *indica* rice trials of Sichuan in 1996 and 1997, it yielded 8.71 t/ha, higher than the yield of the control variety Shanyou 63 by 3.85%, with a whole growth duration of 151 d. It is 115 cm in plant height, 25.7 cm in panicle length with 150 spikelets per panicle and 130 filled grains per panicle, and 27.5 g in 1,000-grain weight.

Suitable areas – Mid-season rice in areas below 800 m in altitude in Sichuan and similar ecological zones in Chongqing, and Sanming of Fujian where there is no severe occurrence of rice blast. By 2015, the accumulated planting area exceeded 990,000 ha.

Ⅱ You 602

Institution (s) – Rice and Sorghum Research Institute of Sichuan Academy of Agricultural Sciences

Parental lines – Ⅱ – 32A/Luhui 602

Approvals – State in 2004 (GS2004004), Sichuan in 2002 (CS2002030), super rice recognition by the MOA in 2005

Characters – It is a three-line mid-season *indica* hybrid rice with high tillering capacity, vigorous growth potential, high tolerance to cold, good maturing color, moderate growth duration, high yield potential, high susceptibility to rice blast, susceptibility to bacterial blight, intermediate susceptibility to brown planthopper, and average grain quality. Tested in the late-maturing *indica* mid-season rice trials of the upper reaches of the Yangtze River in 2001 and 2002, it yielded 8.86 t/ha, higher than the yield of the control variety Shanyou 63 by 4.74%, with a whole growth duration of 155.7 d, which is 2.4 d longer than that of Shanyou 63. It is 110.6 cm in plant height and 24.6 cm in panicle length, and has 2.445 million effective panicles per hectare and 150.5 spikelets per panicle. It is 82.4% in seed setting rate and 29.7 g in 1,000-grain weight.

Suitable areas – Single-cropping mid-season rice in areas of low-to-medium altitude in Yunnan, Guizhou, and Chongqing, excluding the Wuling Mountains; and Pingba of Sichuan and southern Shaanxi where there is no severe occurrence of rice blast and bacterial blight. By 2015, the accumulated planting area exceeded 410,700 ha.

Tianyou 998 (Tianfengyou 998)

Institution (s) – Rice Research Institute of Guangdong Academy of Agricultural Sciences

Parental lines – Tianfeng A/Guanghui 998

Approvals – State in 2006 (GS2006052), Jiangxi in 2005 (GS2005041), Guangdong in 2004 (YS2004008), super rice recognition by the MOA in 2005

Characters – It is a three-line late-season *indica* hybrid rice with a moderate plant type, vigorous growth potential, upright and dark green leaves, large numbers of panicles and grains, high seed setting rate, moderate growth duration, high yield potential, good grain quality, high susceptibility to rice blast, and susceptibility to bacterial blight. Tested in the mid- to late-maturing late-season *indica* rice trials of the middle and lower reaches of the Yangtze River in 2004 and 2005, it yielded 7.69 t/ha, higher than the yield of the control variety Shanyou 46 by 6.28%, with a whole growth duration of 117.7 d, which is 0.6 d shorter than that of Shanyou 46. It has 2.94 million effective panicles per hectare, is 98.0 cm in plant height and 21.1 cm in panicle length, and has 136.5 spikelets per panicle. It is 81.2% in seed setting rate and 25.2 g in 1,000-grain weight, with Grade 3 grain quality as per the national standard *High Quality Paddy*.

Suitable areas – Late-season rice in the double-cropping rice areas in north-central Guangxi, north-central Fujian, Jiangxi, south-central Hunan, and southern Zhejiang where there is no severe occurrence of rice blast and bacterial blight; early- or late-season rice in Guangdong. By 2015, the accumulated planting area exceeded 1,746,700 ha.

II You 084

Institution (s) – Jiangsu Hilly Areas Zhejiang Institute of Agricultural Sciences

Parental lines – II – 32A/Zhenhui 084

Approvals – State in 2003 (GS2003054), Jiangsu in 2001 (SS200103), super rice recognition in 2005

Characters – It is a three-line mid-season *indica* hybrid rice with a good plant type, thick and strong stems, high tillering capacity, high resistance to lodging, good maturing color, high susceptibility to rice blast and brown planthopper, susceptibility to sheath blight, intermediate susceptibility to bacterial blight, and good grain quality. Tested in the late-maturing mid-season *indica* rice trials of the southern China in 2000 and 2001, it yielded 8.41 t/ha and 9.73 t/ha, higher than the yields of the control variety Shanyou 63 by 1.9% and 6.89%, respectively, with a whole growth duration of 142.4 d, which is 3.1 d longer than that of Shanyou 63. It is 121.4 cm in plant height and 23.3 cm in panicle length, has 2.55 million effective panicles per hectare and 160.3 spikelets per panicle, and is 86% in seed setting rate and 27.8 g in 1,000-grain weight.

Suitable areas – Single-cropping mid-season rice in the Yangtze River basin of Jiangxi, Fujian, Anhui, Zhejiang, Jiangsu, Hubei, and Hunan, excluding the Wuling Mountains; and Xinyang of Henan where there is no severe occurrence of rice blast. By 2015, the accumulated planting area exceeded 1,448,000 ha.

II You 7954

Institution (s) – Institute of Crops and Nuclear Technology Utilization of Zhejiang Academy of Agricultural Sciences

Parental lines – II – 32A/Zhehui 7954

Approvals – State in 2004 (GS2004019), Zhejiang in 2002 (ZPSZ378), super rice recognition by the MOA in 2005

Characters – It is a three-line mid-season *indica* hybrid rice with a moderate plant type, dark green leaves, vigorous growth potential, intermediate tillering capacity, large panicles with large numbers of grains, high seed setting rate, intermediate maturing color, moderate growth duration, high yield potential, normal or intermediate susceptibility to rice blast, intermediate susceptibility to bacterial blight, high susceptibility to brown planthopper, and average grain quality. Tested in the late-maturing mid-season *indica* rice trials of the middle and lower reaches of the Yangtze River in 2002 and 2003, it yielded 8.52 t/ha, higher than the yield of the control variety Shanyou 63 by 9.01%, with a whole growth duration of 136.3 d, which is 3.0 d longer than that of Shanyou 63. It is 118.9 cm in plant height and 23.9 cm in panicle length, has 2.355 million effective panicles per hectare and 174.1 spikelets per panicle, and is 78.3% in seed setting rate and 27.3 g in 1,000-grain weight.

Suitable areas – Single-cropping mid-season rice of the Yangtze River Basin in Fujian, Jiangxi, Hunan, Hubei, Anhui, Zhejiang, and Jiangsu, excluding the Wuling Mountains; southern Henan where there is no severe occurrence of rice blast; and late-season rice in Wenzhou, Hangzhou, and Jinhua of Zhejiang. By 2015, the accumulated planting area exceeded 650,000 ha.

Section 2 Two-line *Indica* Hybrid Rice

Y Liangyou 900

Institution (s) – Biocentury Seeds Co., Ltd.

Parental lines – Y58/R900

Approvals – State in 2016 (GS2016044, south China riceareas), Guangdong in 2016 (YS2016021), State in 2015 (GS2015034, middle and lower Yangtze rice areas), super rice recognition in 2007

Characters – It is a two-line late-maturing mid-season *indica* hybrid rice with a moderate growth duration, large panicles with large numbers of grains, high yield potential, good grain quality, susceptibility to rice blast, bacterial blight and brown planthopper, high tolerance to lodging, weak or intermediate tolerance to cold, and intermediate tolerance to heat. Tested in the late-maturing mid-season *indica* rice trials of the middle and lower reaches of the Yangtze River in 2013 and 2014, it yielded 9.38 t/ha, higher than the yield of the control variety Fengliangyou 4 by 5.9%, with a whole growth duration of 140.7 d, which is 2.7 d longer than that of Fengliangyou 4. It is 119.7 cm in plant height and 27.2 cm in panicle length, has 238.2 spikelets per panicle and 2.235 million effective panicles per hectare, and is 78.3% in seed setting rate and 24.4 g in 1,000-grain weight. Tested in the photosensitive late-maturing *indica* rice trials of southern China in 2013 and 2014, it yielded 7.68 t/ha, higher than the yield of the control variety Boyou 998 by 6.0%, with a whole growth duration of 114.0 d, which is 2.1 d longer than that of Boyou 998. In 2014, it yielded an average of 15.09 t/ha in a 6.67-ha demo field in Longhui of Hunan, which is the first time that a super hybrid rice variety achieved the target yield of 15 t/ha.

Suitable areas – Single-cropping mid-season rice along the Yangtze River in Jiangxi, Hunan, Hubei, Anhui, Zhejiang, and Jiangsu, excluding the Wuling Mountains; northern Fujian and southern Henan; late-season rice in double-cropping rice areas in Hainan, south-central and southwest Guangdong plains, southern Guangxi, and southern Fujian where there is no severe occurrence of rice blast; and early-season rice in Guangdong excluding the northern part. By 2015, the accumulated planting area exceeded 12,000 ha.

Longliangyou-Huazhan

Institution (s) – Yuan Longping High-tech Agriculture Co., Ltd. and China National Rice Research Institute

Parental lines – Longke 638S/Huazhan

Approvals – State in 2017 (GS20170008, South China rice areas and the upper Yangtze rice areas), State in 2016 (GS2016045, the Wuling Mountains), Fujian in 2016 (MS2016028), state in 2015 (GS2015026, the middle and lower Yangtze rice areas), Hunan in 2015 (XS2015014), Jiangxi in 2015 (GS2015003), super rice recognition by the MOA in 2007

Characters – It is a two-line late-maturing mid-season *indica* hybrid rice with a moderate growth duration, good plant type, straight and stiff flag leaves, high tillering capacity, high yield potential with high stability, good grain quality, good resistance to diseases and adverse conditions, and wide adaptation. Tested in the late-maturing mid-season *indica* rice trials of the middle and lower reaches of the Yangtze River in 2013 and 2014, it yielded 9.70 t/ha, higher than the yield of the control variety Fengliangyou 4 by 8.4%, with a whole growth duration of 140.1 d, which is 2.0 d longer than that of Fengliangyou 4. It is 121.1 cm in plant height and 24.5 cm in panicle length, has 193.0 spikelets per panicle and 2.715 million effective panicles per hectare, and is 81.9% in seed setting rate and 23.8 g in 1,000-grain weight. Tested in the mid-season *indica* rice trials of the Wuling Mountains in 2013 and 2014, it yielded 9.20 t/ha, higher than the yield of the control variety Ⅱ You 264 by 6.59%, with a whole growth duration of 149.3 d, which is 1.5 d longer than that of Ⅱ You 264. It has intermediate resistance to rice blast and Grade 3 grain quality as per the national standard *High Quality Paddy*. Tested in the photosensitive late-maturing *indica* rice trials of southern China in 2014 and 2015, it yielded 7.66 t/ha, higher than the yield of the control variety Boyou 998 by 8.2%, with a whole growth duration of 115 d, which is 1.5 d longer than that of Boyou 998. Tested in the late-maturing mid-season *indica* rice trials of the upper reaches of the Yangtze River in 2014 and 2015, it yielded 9.39 t/ha, higher than the yield of the control variety F You 498 by 3.6%, with a whole growth duration of 157.9 d, which is 3.6 d longer than that of F You 498.

Suitable areas – Single-cropping mid-season rice in areas along the Yangtze River in Jiangxi, Hunan, Hubei, Anhui, Zhejiang and Jiangsu, northern Fujian and southern Henan, southern Shaanxi, hilly Pingba of Sichuan, Guizhou, low-to-medium altitude *indica* rice areas in Yunnan and areas in Chongqing below 800 m in altitude; late-season rice in double-cropping rice areas in Guangdong, excluding the northern part, southern Guangxi, Hainan, and southern Fujian. By 2015, the accumulated planting area exceeded 16,000 ha.

Shenliangyou 8386

Institution (s) – Guangxi Zhaohe Seeds Co., Ltd.

Parental lines – Shen 08S/R1386

Approvals – Guangxi in 2015 (GS2015007), super rice recognition by the MOA in 2017

Characters – It is a thermo-sensitive two-line *indica* rice hybrid. Tested in the early-season late-maturing rice trials of southern Guangxi in 2013 and 2014, it yielded 8.55 t/ha, higher than the yield of the control variety Teyou 63 by 7.71%, with a whole growth duration of 128.8 d, which is 3.0 d longer than that of Teyou 63. It has 15 – 16 leaves on the main stem, short, narrow and upright flag leaves and 2.40 million effective panicles per hectare. It is 112.4 cm in plant height, 25.1 cm in panicle length with 180 spikelets per panicle, 87.5% in seed setting rate and 25.4 g in 1,000-grain weight, with susceptibility to rice blast and intermediate susceptibility to bacterial blight.

Suitable areas – Early-season rice in southern Guangxi.

Y Liangyou 1173

Institution (s) – National Space Breeding Engineering Research Center (South China Agricultural University), Hunan Hybrid Rice Research Center

Parental lines – Y58S/Hanghui 1173

Approvals – Guangdong in 2015 (YSD 2015016), super rice recognition by the MOA in 2017

Characters – It is a thermo-sensitive two-line hybrid rice with a moderately compact plant type, intermediate to high tillering capacity, long panicles with a large number of grains, intermediate to high tolerance to lodging, intermediate tolerance to coldduring booting and flowering, resistance to rice blast, and susceptibility to bacterial blight. Tested in the early-season rice trials of Guangdong in 2013 and 2014, it yielded 7.33 t/ha and 7.15 t/ha, higher than the yields of the control variety Tianyou 122 by 15.30% and 12.86%, respectively, with a whole growth duration of 125 d, which is 3 d longer than that of Tianyou 122. It is 107.6 – 100.59 cm in plant height and 26.3 – 26.7 cm in panicle length, has 179 – 180 spikelets per panicle and 2.475 – 2.595 million effective panicles per hectare, and is 83.3% – 83.4% in seed setting rate and 20.4 – 20.7 g in 1,000-grain weight.

Suitable areas – Early-or late-season rice in Guangdong, excluding the northern part and single-cropping rice in the northern part of Guangdong.

Huiliangyou 996

Institution (s) – Hefei Keyuan Research Institute of Agricultural Sciences and Rice Research Institute of Anhui Academy of Agricultural Sciences

Parental lines – 1892S/R996

Approvals – State in 2012 (GS2012021), super rice recognition by the MOA in 2016

Characters – It is a two-line *indica* rice hybrid. Tested in the late-maturing mid-season *indica* rice trials of the middle and lower reaches of the Yangtze River in 2009 and 2010, it yielded 8.64 t/ha, higher than the yield of the control variety Ⅱ You 838 by 6.0%, with a whole growth duration of 132.4 d, which is 1.2 d shorter than that of Ⅱ You 838. It has 2.385 million effective panicles per hectare, is 113.6 cm in plant height and 24.0 cm of panicle length, has 180.7 spikelets per panicle, and is 80.1%

in seed setting rate and 26. 8 g in 1,000-grain weight, highly susceptible to rice blast and brown planthopper, and susceptible to bacterial blight.

Suitable areas – Single-cropping mid-season rice in areas along the Yangtze River in Jiangxi, Hunan, Hubei, Anhui, Zhejiang, and Jiangsu, excluding the Wuling Mountains, and northern Fujian and southern Henan where there is no severe occurrence of rice blast and bacterial blight. By 2015, the accumulated planting area exceeded 108,000 ha.

Shenliangyou 870

Institution (s) – Guangdong Zhaohua Seeds Co., Ltd. and Shenzhen Zhaonong Agricultural Sciences Co., Ltd.

Parental lines – Shen 08S/P5470

Approvals – Guangdong in 2014 (YS2014037), super rice recognition by the MOA in 2016

Characters – It is a thermo-sensitive two-line hybrid rice with a moderately compact plant type, weak to intermediate tillering capacity, high tolerance to lodging, intermediate tolerance to cold during booting and flowering, good maturing color, and high yield potential. Tested in the late-season rice trials of Guangdong in 2012 and 2013, it yielded 7. 45 t/ha and 6. 70 t/ha, higher than the yields of the control variety Yuejing Simiao 2 by 9. 9% and 8. 19%, respectively, with a whole growth duration of 117 d, which is the same as that of Yuejing Simiao 2. It is 96. 0 – 97 cm in plant height, 23. 5 – 24. 3 cm in panicle length, with 149 – 152 spikelets per panicle, 2. 25 – 2. 46 million effective panicles, resistant to rice blast, susceptible to bacteria blight, 83. 0% – 84. 2% in seed setting rate, and 26. 2 – 26. 7 g in 1,000-grain weight, with Grade 3 grain quality as per relevant national and provincial standards.

Suitable areas – Early-or late-season rice in Guangdong, excluding the northern part. By 2015, the accumulated planting area exceeded 27,300 ha.

H Liangyou 991

Institution (s) – Guangxi Zhaohe Seeds Co., Ltd.

Parental lines – HD9802S/R991

Approvals – Guangxi in 2011 (GS2011017), super rice recognition by the MOA in 2015

Characters – It is a thermo-sensitive two-line hybrid rice with a moderate plant type, upright flag leaves, thick and strong stems, high tillering capacity, and good maturing color. Tested in the mid-maturing late-season rice trials of central and northern Guangxi in 2009 and 2010, it yielded 7. 23 t/ha, higher than the yield of the control variety Zhongyou 838 by 6. 79%, with a whole growth duration of 108 d, similar to that of Zhongyou 838. It has 2. 55 million effective panicles per hectare, is 116. 9 cm in plant height and 22. 4 cm in panicle length, has 153. 8 spikelets per panicle, and is 77. 7% of seed setting rate, 24. 0 g in 1,000-grain weight, susceptible or intermediately susceptible to rice blast, and intermediate or highly susceptible to bacterial blight.

Suitable areas – Early- or late-season rice in central Guangxi; and late-season rice in northern Guangxi or early-season rice in southern Guangxi based on local growth conditions. By 2015, the accumulated planting area exceeded 79,300 ha.

N Liangyou 2

Institution (s) – Changsha Nianfeng Seeds Co., Ltd. and Hunan Hybrid Rice Research Center

Parental lines – N118S/R302

Approvals – Hunan in 2013 (XS2013010), super rice recognition by the MOA in 2015

Characters – It is a two-line late-maturing mid-season *indica* hybrid rice with a compact plant type, upright leaves, vigorous growth potential, and good maturing color. Tested in the late-maturing mid-season rice trials of Hunan in 2011 and 2012, it yielded 9.54 t/ha, higher than the yield of the control variety Y Liangyou 1 by 3.21%, with a whole growth duration of 141.8 d. It is 118.9 cm in plant height, has 2.3265 million effective panicles per hectare and 185.15 spikelets per panicle, is 84.97% in seed setting rate and 27.04 g in 1,000-grain weight, susceptible to rice blast, intermediately susceptible to bacterial blight and rice false smut, intermediately tolerant to heat or cold, with Grade 3 grain quality as per the national standard *High Quality Paddy*.

Suitable areas – Mid-season rice in the hilly regions of Hunan where there is no severe occurrence of rice blast.

Liangyou 616

Institution (s) – Fujian Nongjia Seeds Co., Ltd. of China National Seed Group, and Rice Research Institute of Fujian Academy of Agricultural Sciences

Parental lines – Guangzhan 63 – 4S/Fuhui 616

Approvals – Fujian in 2012 (MS2012003), super rice recognition in 2014

Characters – It is a two-line mid-season *indica* hybrid rice with a moderate plant type, large panicles with large number of grains, relatively high grain weight, and good maturing color. Tested in the mid-season rice trials of Fujian in 2009 and 2010, it yielded 9.46 t/ha and 9.07 t/ha, higher than the yields of the control variety Youming 86 by 6.31% and 13.88%, respectively, with a whole growth duration of 143.0 d, which is 1.3 d longer than that of Youming 86. It has 1.95 million effective panicles per hectare, is 127 cm in plant height and 26.5 cm in panicle length, has 182.9 spikelets per panicle, and is 86.61% in seed setting rate and 30.9 g in 1,000-grain weight, with intermediate susceptibility to rice blast and good grain quality.

Suitable areas – Mid-season rice in Fujian where there is no severe occurrence of rice blast. By 2015, the accumulated planting area exceeded 35,300 ha.

Liangyou 6

Institution (s) – Hubei Jingchu Seeds Co., Ltd.

Parental lines – HD9802S/Zaohui 6

Approvals – State in 2011 (GS2011003), super rice recognition by the MOA in 2014

Characters – It is a two-line *indica* rice hybrid with a compact plant type, vigorous growth potential and good maturing color. Tested in the late-maturing early-season *indica* rice trials of the middle and lower reaches of the Yangtze River in 2008 and 2009, it yielded 7.83 t/ha, higher than the yield of the control variety Jinyou 402 by 4.1%, with a whole growth duration of 112.7 d, which is 1.7 d shorter than that of Jinyou 402. It has 13 – 14 leaves on the main stem, is 94.6 cm in plant height and 19.9 cm in panicle length, has 127. 2 spikelets per panicle and 2. 955 million effective panicles per hectare, and is 88. 4% in seed setting rate and 25. 1 g in 1,000-grain weight, with high susceptibility to rice blast, brown planthopper, and whitebacked planthopper, susceptibility to bacterial blight, and Grade 3 grain quality as per the national standard *High Quality Paddy*.

Suitable areas – Early-season rice in the double-cropping rice areas in Jiangxi, Hunan, Hubei, northern Guangxi, northern Fujian, and south-central Zhejiang where there is no severe occurrence of rice blast and bacterial blight.

Guangliangyou 272

Institution (s) – Cereal Crops Research Institute of Hubei Academy of Agricultural Sciences

Parental lines – Guangzhan 63-4S/R7272

Approvals – Hubei in 2012 (ES2012003), super rice recognition by the MOA in 2014

Characters – It is a two-line late-maturing mid-season *indica* hybrid rice with a moderate plant type, high tillering capacity, thick and strong stems, partially exposed internodes, dark green leaves, long, wide and straight flag leaves, uniform panicle layer, intermediate-to-large panicle size with a large number of spikelets per panicle, good maturing color, green stems and yellow grains in maturity. Tested in the mid-season rice trials of Hubei in 2010 and 2011, it yielded 9. 07 t/ha, higher than the yield of the control variety Yangliangyou 6 by 1. 11%, with a whole growth duration of 139. 8 d, which is 2. 2 d shorter than that of Yangliangyou 6. It has 2. 415 million effective panicles per hectare, is 122. 9 cm in plant height and 25. 2 cm in panicle length, has 174. 5 spikelets per panicle and 144. 2 filled grains per panicle, is 82. 6% in seed setting rate and 28. 6 g in 1,000-grain weight, highly susceptible to rice blast and intermediately susceptible to bacterial blight, with Grade 2 grain quality as per the national standard *High Quality Paddy*.

Suitable areas – Mid-season rice in Hubei, excluding the southwestern part, where there is no severe occurrence of rice blast. By 2015, the accumulated planting area exceeded 32,000 ha.

C Liangyou-Huazhan

Institution (s) – Hunan Jinse Nonghua Seeds Co., Ltd.

Parental lines – C815S/Huazhan

Approvals – State in 2016 (GS2016002, south China rice areas), Hunan in 2016 (XS2016008), State in 2015 (GS2015022, middle and lower Yangtze River rice areas), Jiangxi in 2015 (GS2015008), State in 2013 (GS2013003, upper Yangtze River rice areas), Hubei in 2013 (ES2013008), super rice recognition by the MOA in 2014

Characters – It is a two-line mid-maturing mid-season *indica* hybrid rice with a moderate plant type, vigorous growth potential, high tillering capacity, upright leaves, large number of effective panicles, intermediate-to-large panicle size with a large number of grains per panicle, low grain weight, good maturing color, partially exposed internodes, and average tolerance to lodging. Trials in various regions show that it has intermediate tolerance to heat during heading, weak tolerance to cold, intermediate-to-high resistance to rice blast, intermediate to normal susceptibility to bacterial blight, intermediate susceptibility to rice false smut, and high susceptibility to brown planthopper and whitebacked planthopper. Tested in the mid-season *indica* rice trials of the upper reaches of the Yangtze River in 2010 and 2011, it yielded 9.04 t/ha, higher than the yield of the control variety Ⅱ You 838 by 4.8%, with a whole growth duration of 157.2 d, which is 0.7 d shorter than that of Ⅱ You 838. It is 101.8 cm in plant height and 23.0 cm in panicle length with 202.2 spikelets per panicle and 2.475 million effective panicles per hectare, 79.3% in seed setting rate and 23.7 g in 1,000-grain weight, with Grade 3 grain quality as per the national standard *High Quality Paddy*. Tested in the late-maturing mid-season *indica* rice trials of the middle and lower reaches of the Yangtze River in 2013 and 2014, it yielded 9.63 t/ha, higher than the yield of the control variety Fengliangyou 4 by 8.7%, with a whole growth duration of 136.1 d, which is 1.8 d shorter than that of Fengliangyou 4. Tested in the early-season *indica* rice trials of south China in 2013 and 2014, it yielded 7.63 t/ha, higher than the yield of the control variety Tianyou 998 by 6.7%, with a whole growth duration of 123.3 d, which is 0.8 d longer than that of Tianyou 998.

Suitable areas – Single-cropping mid-season rice in areas along the Yangtze River, excluding the Wuling Mountains, and early-season rice in south China where there is no severe occurrence of rice blast. By 2015, the accumulated planting area exceeded 147,300 ha.

Liangyou 038

Institution (s) – Jiangxi Tianya Seed Industry Co., Ltd.

Parental lines – 03S/R828

Approvals – Jiangxi 2010 (GS2010006), super rice recognition by the MOA in 2014

Characters – It is a two-line *indica* hybrid rice with a moderate plant type, short and wide flag leaves, vigorous growth potential, high tillering capacity, large numbers of panicles and grains per panicle, high seed setting rate, and good maturing color. Tested in the mid-season rice trials of Jiangxi in 2008 and 2009, it yielded 8.55 t/ha, higher than the yield of the control variety Ⅱ You 838 by 8.84%, with a whole growth duration of 122.6 d, which is 1.8 d shorter than that of Ⅱ You 838. It is 124.1 cm in plant height, has 2.325 million effective panicles per hectare and 163.5 spikelets per panicle with 138.4 filled grains per panicle, is 84.6% in seed setting rate and 28.0 g in 1,000-grain weight, with high susceptibility to rice blast.

Suitable areas – Mid-season rice in Jiangxi where there is no severe occurrence of rice blast. By 2015, the accumulated planting area exceeded 24,000 ha.

Y Liangyou 5867 (Shenliangyou 5867)

Institution (s) – Jiangxi Keyuan Seeds Co., Ltd. and Tsinghua Shenzhen Longgang Institute of China National Hybrid Rice R&D Center

Parental lines – Y58S/R674

Approvals – State in 2012 (GS2012027), Zhejiang in 2011 (ZS2011016), Jiangxi in 2010 (GS2010002), super rice recognition by the MOA in 2014

Characters – It is a two-line mid-season *indica* hybrid rice with a moderate plant height, relatively compact plant type, upright flag leaves, intermediate tillering capacity, large panicle size, high seed setting rate, high grain weight, high yield potential, intermediate tolerance to heat during heading, and good maturing color. Trials in various regions show that it ranges resistance to high susceptibility to rice blast and from intermediate resistance to intermediate susceptibility to bacterial blight. It is highly susceptible to brown planthopper. Tested in the late-maturing *indica* rice trials of the middle and lower reaches of the Yangtze River in 2009 and 2010, it yielded 8.67 t/ha, higher than the yield of the control variety Ⅱ You 838 by 5.0%; with a whole growth duration of 137.8 d, 3.9 d of longer than that of Ⅱ You 838. It has 2.565 million effective panicles per hectare, is 120.8 cm in plant height and 27.7 cm in panicle length, has 161.1 spikelets per panicle, and is 81.2% in seed setting rate and 27.7 g in 1,000-grain weight, with Grade 3 grain quality as per the national standard *High Quality Paddy*.

Suitable areas – Single-cropping mid-season rice in areas along the Yangtze River in Jiangxi, Hunan, Hubei, Anhui, Zhejiang, and Jiangsu, excluding the Wuling Mountains, and northern Fujian and southern Henan. By 2015, the accumulated planting area exceeded 290,000 ha.

Y Liangyou 2

Institution (s) – Hunan Hybrid Rice Research Center

Parental lines – Y58S/Yuanhui 2

Approvals – Anhui in 2014 (WS2014016), State in 2013 (GS2013027), Honghe of Yunnan in 2012 (DTS [Honghe] 2012017), Hunan in 2011 (XS2011020), super rice recognition by the MOA in 2014

Characters – It is a two-line mid-season *indica* hybrid rice and a representative variety of Phase Ⅲ of China's Super Rice Project. It has a moderate plant type, upright and slightly curved top-three leaves, high tillering capacity, good maturing color, and high tolerance to heat and cold. Tested in the mid-maturing mid-season *indica* rice trials of the middle and lower reaches of the Yangtze River in 2011 and 2012, it yielded 9.23 t/ha, higher than the yield of the control variety Fengliangyou 4 by 4.7%, with a whole growth duration of 139.1 d, which is 2.2 d longer than that of Fengliangyou 4. It is 122.6 cm in plant height and 28.3 cm in panicle length with 198.5 spikelets per panicle and 2.565 million effective panicles per hectare, is 78.9% in seed setting rate and 24.8 g in 1,000-grain weight. It is highly susceptible to rice blast and brown planthopper and susceptible to bacterial blight, with Grade 3 grain quality as per the national standard *High Quality Paddy*. Tested in a 6.67-ha demo field in Yanggu'ao township, Longhui, Hunan, in 2011, it had a yield of 13.90 t/ha, achieving the Phase Ⅲ yield target of 13.5 t/ha.

Suitable areas – Single-cropping mid-season rice in areas along the Yangtze River in Jiangxi, Hunan, Hubei, Anhui, Zhejiang, and Jiangsu, excluding the Wuling Mountains, and northern Fujian and southern Henan. By 2015, the accumulated planting area exceeded 145,300 ha.

Y Liangyou 087

Institution (s) – Nanning Wode Crops Research Institute, Hunan Hybrid Rice Research Center, and Guangxi Nanning Oumiyuan Agricultural Sciences Co., Ltd.

Parental lines – Y58S/R087

Approvals – Guangdong in 2015 (YS2015049), Guangxi in 2010 (GS2010014), super rice recognition by the MOA in 2013

Characters – It is a thermo-sensitive two-line hybrid rice with a moderately compact plant type, intermediate tillering capacity, intermediate-to-high tolerance to lodging, intermediate tolerance to cold, high yield potential, and susceptibility to rice blast and bacterial blight. Tested in the late-maturing early-season rice trials of southern Guangxi in 2008 and 2009, it yielded 8.08 t/ha, higher than the yield of the control variety Teyou 63 by 2.86%, with a whole growth duration of 128 d, which is 2 − 3 d longer than that of Teyou 63. It has 2.535 million effective panicles per hectare, is 117.2 cm in plant height and 24.0 cm in panicle length with 157.9 spikelets per panicle, 79.0% in seed setting rate and 26.0 g in 1,000-grain weight. Tested in the late-season rice trials of Guangdong in 2013 and 2014, it yielded 6.79 t/ha and 7.54 t/ha, higher than the yields of the control variety Yuejing Simiao 2 by 8.94% and 6.87%, respectively, with a whole growth duration of 115 − 119 d, which is 1 − 3 d longer than that of Yuejing Simiao 2. It has Grade 3 grain quality as per relevant national and provincial standards.

Suitable areas − Early-season rice in southern Guangxi and early-season or mid-season rice in other areas of Guangxi, depending on local conditions; and early- or late-season rice in Guangdong, excluding the northern part. By 2015, the accumulated planting area exceeded 57,300 ha.

Zhunliangyou 608

Institution (s) − Hunan Longping Seeds Industry Co., Ltd.

Parental lines − ZhunS/R608

Approvals − Hubei in 2015 (ES2015005), Hunan in 2010 (XS2010018 and XS2010027), state in 2009 (GS2009032), super rice recognition by the MOA in 2012

Characters − It is a two-line *indica* hybrid rice with a moderate plant height and plant type, intermediate tillering capacity, thick stems, slightly exposed and curved internodes, wide and thick leaves, curved and straight flag leaves, uniform panicle layer, medium panicle size with evenly distributed grains on panicles, good maturing color, and high yield potential. Tested in the mid- and late-maturing late-season *indica* rice trials of the middle and lower reaches of the Yangtze River in 2007 and 2008, it yielded 7.80 t/ha, higher than the yield of the control variety Shanyou 46 by 8.80%, with a whole growth duration of 119.0 d, which is 1.1 d longer than that of Shanyou 46. It is 108.9 cm in plant height and 24.1 cm in panicle length with 137.1 spikelets per panicle and 2.445 million effective panicles per hectare, 82% in seed setting rate and 31.0 g in 1,000-grain weight, highly susceptible to rice blast, bacterial blight, and brown planthopper, with high grain quality. Tested in the late-maturing mid-season rice trials of Hunan in 2008 and 2009, it yielded 8.03 t/ha, lower than the yield of the control variety Ⅱ You 58 by 1.67%, with a whole growth duration of 141 d. It is highly tolerant to cold and heat. Tested in the mid-season rice trials of Hubei in 2012 and 2013, it yielded 9.50 t/ha, higher than the yield of the control variety Fengliangyouxiang 1 by 5.76%, with a whole growth duration of 131.3 d, which is 2.2 d longer than that of Fengliangyouxiang 1.

Suitable areas − Late-season rice in double-cropping rice areas in north-central Guangxi, northern Guangdong, north-central Fujian, south-central Jiangxi, south-central Hunan, and southern Zhejiang where there is no severe occurrence of rice blast and bacterial blight; single-cropping late-season or mid-season rice in Hunan where there is no severe occurrence of rice blast; and mid-season rice in Hubei, excluding the southwestern part. By 2015, the accumulated planting area exceeded 192,700 ha.

Shenliangyou 5814

Institution (s) - Tsinghua Shenzhen Longgang Institute of China National Hybrid Rice R&D Center

Parental lines - Y58S/Bing 4114

Approvals - State in 2007 (GS20170013, upper Yangtze River rice areas), Hainan in 2013 (QS2013001), Chongqing in 2011 (YY2011007), State in 2009 (GS2009016, middle and lower Yangtze River rice areas), Guangdong in 2008 (YS2008023), super rice recognition by the MOA in 2012

Characters - It is a two-line late-maturing mid-season *indica* hybrid rice with a moderate plant type, straight and stiff leaves, intermediate tillering capacity, thick and strong stems, intermediate-to-high tolerance to lodging, moderate growth duration, good maturing color, high tolerance to cold, good grain quality, high and stable yield potential, and wide adaptation. Tested in the late-maturing mid-season *indica* rice trials of the middle and lower reaches of the Yangtze River in 2007 and 2008, it yielded 8.81 t/ha, higher than the yield of the control variety Ⅱ You 838 by 4.22%, with a whole growth duration of 136.8 d, which is 1.8 d longer than that of the control variety Ⅱ You 838. It is 124.3 cm in plant height, has 2.58 million effective panicles per hectare, is 26.5 cm in panicle length, has 171.4 spikelets per panicle, is 84.1% in seed setting rate and 25.7 g in 1,000-grain weight, with intermediate susceptibility to rice blast and bacterial blight, high susceptibility to brown planthopper and Grade 2 grain quality as per the national standard *High Quality Paddy*. Tested in the late-maturing mid-season *indica* rice trials of the upper reaches of the Yangtze River in 2014 and 2015, it yielded 9.36 t/ha, higher than the yield of the control variety F You 498 by 3.4%, with a whole growth duration of 158.7 d, which is 4.7 d longer than that of F You 498.

Suitable areas - Single-cropping mid-season rice in areas along the Yangtze River, excluding the Wuling Mountains, in Jiangxi, Hunan, Hubei, Anhui, Zhejiang and Jiangsu, northern Fujian and southern Henan; hilly areas in Pingba of Sichuan and Guizhou, excluding the Wuling Mountains, low-to-medium altitude *indica* rice areas in Yunnan, regions below 800 m in altitude in Chongqing, and southern Shaanxi; and late-maturing rice in Guangdong (excluding the northern part), and all areas of Hainan. By 2015, the accumulated planting area exceeded 1,152,000 ha.

Guangliangyouxiang 66

Institution (s) - Hubei Agro-Tech Extension and Service Center, Xiaonan Agricultural Bureau of Xiaogan city, and Hubei Zhongxiang Agricultural Technology Co. Ltd.

Parental lines - Guangzhan 63 - 4S/Xianghui 66

Approvals - State in 2012 (GS2012028), Henan in 2011 (YS2011004), Hubei in 2009 (ES2009005), super rice recognition by the MOA in 2012

Characters - It is a two-line mid-season *indica* hybrid rice with a relatively compact plant type, intermediate plant height, vigorous growth potential, high tillering capacity, thick stems, partially exposed internodes, dark green leaves, intermediately long, upright and stiff flag leaves, intermediate-to-large panicle size with large number of spikelets per panicle, and good maturing color. Tested in the mid-season rice trials of Hubei in 2007 and 2008, it yielded 9.03 t/ha, higher than the yield of the control variety Yangliangyou 6 by 2.64%, with a whole growth duration of 137.9 d, which is 0.6 d shorter than that of

Yangliangyou 6. The grain quality was rated as Grade 2 as per the national standard *High Quality Paddy*. Tested in the late-maturing mid-season *indica* rice trials of the middle and lower reaches of the Yangtze River in 2009 and 2010, it yielded 8. 33 t/ha, higher than the yield of the control variety Ⅱ You 838 by 2. 2% , with a whole growth duration of 138. 8 d, which is 5. 2 d longer than that of Ⅱ You 838. It has 2. 325 million effective panicles per hectare, is 128. 1 cm in plant height and 25. 3 cm in panicle length with 166. 1 spikelets per panicle, is 76. 1% in seed setting rate and 29. 8 g in 1,000-grain weight, is susceptible to rice blast and brown planthopper and intermediately susceptible to bacterial blight, with Grade 3 grain quality as per the national standard *High Quality Paddy*.

Suitable areas – Mid-season rice in areas along the Yangtze River in Jiangxi, Hunan, Hubei, south-central Anhui and Zhejiang, excluding the Wuling Mountains, northern Fujian and southern Henan where there is no severe occurrence of rice blast and bacterial blight. By 2015, the accumulated planting area exceeded 378,000 ha.

Lingliangyou 268

Institution (s) – Hunan Yahua Research Institute of Seeds Sciences

Parental lines – Xiangling 628S/Hua 268

Approvals – State in 2008 (GS2008008), super rice recognition by the MOA in 2011

Characters – It is a two-line early-season *indica* hybrid rice with a moderate plant type, thick and strong stems, short and upright leaves, moderate growth duration and high yield potential. Tested in the late-maturing early-season *indica* rice trails of the middle and lower reaches of the Yangtze River in 2006 and 2007, it yielded 7. 80 t/ha, higher than the yield of the control variety Jinyou 402 by 5. 63% , with a whole growth duration of 112. 2 d, which is 0. 3 d longer than that of Jinyou 402. It is 87. 7 cm in plant height and 19. 0 cm in panicle length, has 104. 7 spikelets per panicle and 3. 42 million effective panicles, is 87. 1% in seed setting rate and 26. 5 g in 1,000-grain weight, susceptible to rice blast and bacterial blight, intermediately resistant to brown planthopper and whitebacked planthopper, with average grain quality.

Suitable areas – Early-season rice in double-cropping rice areas in Jiangxi, Hunan, and northern Fujian, and south-central Zhejiang where there is no severe occurrence of rice blast and bacterial blight. By 2015, the accumulated planting area exceeded 173,300 ha.

Huiliangyou 6

Institution (s) – Rice Research Institute of Anhui Academy ofAgricultural Sciences

Parental lines – 1892S/Yangdao 6 selection

Approvals – State in 2012 (GS2012019), Anhui in 2008 (WS2008003), super rice recognition by the MOA in 2011

Characters – It is a two-line *indica* rice hybrid with intermediate-to-long, wide and upright flag leaves, large number of spikelets per panicle, and short awns. Tested in the late-maturing mid-season *indica* rice trials of the middle and lower reaches of the Yangtze River in 2009 and 2010, it yielded 8. 67 t/ha, higher than the yield of the control variety Ⅱ You 838 by 6. 4% , with a whole growth duration of 135. 1 d, which is 1. 5 d longer than that of Ⅱ You 838. It is 118. 5 cm in plant height, has 2. 415 million effective panicles per hectare, is 23.1 cm in panicle length, has 173.2 spikelets per panicle, is 80.8% in seed setting rate and 27.2 g in 1,000-grain weight, and is highly susceptible to rice blast and brown planthopper, susceptible to bacterial blight and averagely tolerant to heat during heading.

Suitable areas – Single-cropping mid-season rice in areas along the Yangtze River in Jiangxi, Hunan, Hubei, Anhui, Zhejiang and Jiangsu, excluding the Wuling Mountains, northern Fujian, southern Henan where there is no severe occurrence of rice blast and bacterial blight. By 2015 the accumulated planting area exceeded 130,000 ha.

Guiliangyou 2

Institution (s) – Rice Research Institute of Guangxi Academy of Agricultural Sciences

Parental lines – Guike-2S/Guihui 582

Approvals – Guangxi in 2008 (GS2008006), super rice recognition by the MOA in 2010

Characters – It is a thermo-sensitive two-line rice hybrid with a compact plant type, short and straight leaves, and good maturing color. Tested in the late-maturing early-season rice trials of southern Guangxi in 2006 and 2007, it yielded 7.67 t/ha, higher than the yield of the control variety Teyou 63 by 8.32%, with a whole growth duration of 124 d, which is 4 d shorter than that of Teyou 63. It is 112.2 cm in plant height, has 2.835 million effective panicles per hectare, is 23.2 cm in panicle length with 158 spikelets per panicle, 83% in seed setting rate and 21.6 g in 1,000-grain weight, with Grade 6 resistance to seedling blast, Grade 7 resistance to panicle blast, 46.2% in panicle blast induced loss index, 6.8 in the composite index of resistance to rice blast, Grade 7 resistance to bacterial blight Ⅳ and Grade 5 resistance to bacterial blight Ⅴ.

Suitable areas – Early-season rice in southern Guangxi. By 2015, the accumulated planting area exceeded 114,000 ha.

Fengliangyouxiang 1

Institution (s) – Hefei Fengle Seed Co., Ltd.

Parental lines – Guangzhan 63S/Fengxianghui 1

Approvals – State in 2007 (GS2007017), Anhui in 2007 (WPS07010622), Hunan in 2006 (XS2006037), Jiangxi in 2006 (GS2006022), super rice recognition by the MOA in 2009

Characters – It is a two-line *indica* hybrid rice with a loose plant type, straight and stiff flag leaves, average tillering capacity, good maturing color, early maturity, high yield potential, good grain quality, high susceptibility to rice blast, susceptibility to bacterial blight, and high tolerance to heat and cold. Tested in the late-maturing mid-season *indica* rice trials of the middle and lower reaches of the Yangtze River in 2005 and 2006, it yielded 8.53 t/ha, higher than the yield of the control variety Ⅱ You 838 by 6.17%, with a whole growth duration of 130.2 d, which is 3.5 d shorter than that of Ⅱ You 838. It is 116.9 cm in plant height, has 2.43 million effective panicles per hectare, is 23.8 cm in panicle length, has 168.6 spikelets per panicle, and is 82.0% in seed setting rate and 27.0 g in 1,000-grain weight.

Suitable areas – Mid-season rice in areas along the Yangtze River in Jiangxi, Hunan, Hubei, Anhui, Zhejiang and Jiangsu, excluding the Wuling Mountains, northern Fujian, southern Henan where there is no severe occurrence of rice blast and bacterial blight. By the 2015, the accumulated planting area exceeded 970,000 ha.

Yangliangyou 6

Institution (s) – Jiangsu Lixiahe Area Institute of Agricultural Sciences

Parental lines – Guangzhan 63 – 4S/93 – 11

Approvals – State in 2005 (GS2005024), Hubei in 2005 (ES2005005), Shaanxi in 2005 (SS2005003), Henan in 2004 (YS2004006), Jiangsu in 2003 (SS200302), Guizhou in 2003 (QS2003002), super rice recognition by the MOA in 2009

Characters – It is a two-line mid-season *indica* hybrid rice with a moderately compact plant type, thick and strong stems, straight and stiff flag leaves, high tolerance to lodging, vigorous growth potential, high tillering capacity, moderate growth duration, good maturing color, large panicle size with large number of grains per panicle, slender grains with short-to-medium awns, good grain quality, high yield potential with high stability, susceptibility to rice blast, intermediate susceptibility to bacterial blight and sheath blight, intermediate susceptibility to brown planthopper, and average tolerance to cold. Tested in the late-maturing mid-season *indica* rice trials of the middle and lower reaches of the Yangtze River in 2002 and 2003, it yielded 8.34 t/ha, higher than the yield of the control variety Shanyou 63 by 6.34%, with a whole growth duration of 134.1 d, which is 0.7 d longer than that of Shanyou 63. It is 120.6 cm in plant height, has 2.49 million effective panicles, is 24.6 in panicle length, has 167.5 spikelets per panicle, and is 78.3% in seed setting rate and 28.1 g in 1,000-grain weight. Tested in the rice trials of Jiangsu in 2001 and 2002, it yielded 9.51 t/ha, higher than the yield of the control variety Shanyou 63 by 5.69%, with a whole growth duration of 142 d, which is 1 – 2 d longer than that of Shanyou 63, and Grade 3 grain quality as per the national standard *High Quality Paddy*.

Suitable areas – Single-cropping mid-season rice in areas along the Yangtze River in Fujian, Jiangxi, Hunan, Hubei, Anhui, Zhejiang, and Jiangsu, excluding the Wuling Mountains, southern Henan where there is no severe occurrence of rice blast; and late-maturing *indica* rice areas in Guizhou. By 2015, the accumulated planting area exceeded 2,668,700 ha.

Luliangyou 819

Institution (s) – Hunan Yahua Research Institute of Seeds Sciences

Parental lines – Lu 18S/Hua 819

Approvals – State in 2008 (GS2008005), Hunan in 2008 (XS2008002), super rice recognition by the MOA in 2009

Characters – It is a two-line early-season *indica* hybrid rice with a moderate plant type, average tillering capacity, intermediate tolerance to fertilizer, moderate growth duration, high yield potential, susceptibility to rice blast, bacterial blight and whitebacked plant hopper, intermediate susceptibility to brown planthopper, low tolerance to lodging and average grain quality. Tested in the early-season early-to-medium maturing *indica* rice trials of the middle and lower reaches of the Yangtze River in 2006 and 2007, it yielded 7.62 t/ha, higher than the yield of the control variety Zhe 733 by 8.08%, with a whole growth duration of 107.2 d, which is 0.9 d shorter than that of Zhe 733. It is 87.2 cm in plant height, has 3.375 million effective panicles per hectare, is 19.6 cm in panicle length, has 109.5 spikelets per panicle, and is 83. 1% in seed setting rate and 26. 8 g in 1,000-grain weight.

Suitable areas - Early-season rice in double-cropping rice areas in Jiangxi, Hunan, Hubei, Anhui and Zhejiang where there is no severe occurrence of rice blast and bacterial blight. By 2015, the accumulated planting area exceeded 46,000 ha.

Xinliangyou 6380

Institution (s) - Rice Research Institute of Nanjing Agricultural University and Jiangsu Zhongjiang Seed Co., Ltd.

Parental lines - 03S/D208

Approvals - State in 2008 (GSD 2008012), Jiangsu in 2007 (SS200703), super rice recognition by the MOA in 2007

Characters - It is a two-line mid-season *indica* hybrid rice with a moderate plant type, tall plants, stiff and strong stems, high tolerance to lodging, straight and stiff leaves, intermediate tillering capacity, large panicle size and high yield potential. Tested in the late-maturing mid-season *indica* rice trials of the middle and lower reaches of the Yangtze River in 2006 and 2007, it yielded 8. 89 t/ha, higher than the yield of the control variety Ⅱ You 838 by 7. 56%, with a whole growth duration of 130. 4 d, which is 2. 8 d shorter than that of Ⅱ You 838. It has 2. 34 million effective panicles per hectare, is 124. 9 cm in plant height and 25. 4 cm in panicle length, has 168. 6 spikelets per panicle, and is 86. 2% in seed setting rate and 28. 2 g in 1,000-grain weight, with high susceptibility to rice blast, intermediate susceptibility to bacterial blight, susceptibility to brown planthopper, and average grain quality.

Suitable areas - Single-cropping mid-season rice in areas along the Yangtze River in Jiangxi, Hunan, Hubei, Anhui, Zhejiang and Jiangsu, excluding the Wuling Mountains, northern Fujian and southern Henan where there is no severe occurrence of rice blast. By 2015, the accumulated planting area exceeded 291,300 ha.

Wandao 186 (Fengliangyou 4)

Institution (s) - Hefei Fengle Seed Co., Ltd.

Parental lines - Feng 39S/Yandao 4 selection

Approvals - State in 2009 (GS2009012), Anhui in 2006 (WPS06010501), super rice recognition by the MOA in 2007

Characters - It is a two-line mid-season *indica* hybrid rice with a moderate plant type, straight and stiff leaves, high tillering capacity, vigorous growth potential, good maturing color, moderate growth duration, high yield potential and good grain quality. Tested in the late-maturing mid-season *indica* rice trials of the middle and lower reaches of the Yangtze River in 2007 and 2008, it yielded 9. 10 t/ha, higher than the yield of the control variety Ⅱ You 838 by 7. 04%, with a whole growth duration of 135. 3 d, which is 0. 1 d longer than that of Ⅱ You 838. It has 2. 415 million effective panicles, is 124. 8 cm in plant height ad 24. 2 cm in panicle length, has 180. 6 spikelets per panicle, and is 79. 7% in seed setting rate and 28. 2 g in 1,000-grain weight, highly susceptible to rice blast and brown planthopper and susceptible to bacterial blight, with Grade 2 grain quality as per the national standard *High Quality Paddy*.

Suitable areas - Single-cropping mid-season rice in areas along the Yangtze River in Jiangxi, Hunan,

Hubei, Anhui, Zhejiang and Jiangsu, excluding the Wuling Mountains, and northern Fujian and southern Henan where there is no severe occurrence of rice blast and bacterial blight. By 2015, the accumulated planting area exceeded 826,700 ha.

Y Liangyou 1

Institution (s) – Hunan Hybrid Rice Research Center

Parental lines – Y58S/93 – 11

Approvals – Guangdong in 2015 (YS2015047), State in 2013 (GS2013008, the upper Yangtze rice areas), State in2008 (GS2008001, the middle and lower Yangtze rice areas), Chongqing Introduction in 2008 (YY2008001), Hunan in 2006 (XS2006036), super rice recognition in 2006

Characters – It is a thermo-sensitive two-line late-maturing mid-season *indica* rice hybrid and a representative variety of Phase Ⅱ of China's Super Rice Project. It has a good plant type, straight and curve leaves, moderate growth duration, high yield potential with high stability, wide adaptation, good grain quality, high tolerance to heat, high susceptibility to rice blast, susceptibility to bacterial blight and brown planthopper, and intermediate susceptibility to whitebacked planthopper. Tested in the late-maturing mid-season *indica* rice trials of Hunan in 2004 and 2005, it had a yield of 9.52 t/ha, which was 8.8% higher than that of the control variety Liangyoupeijiu. Tested in the late-maturing mid-season *indica* rice trials of the middle and lower reaches of the Yangtze River in 2005 and 2006, it had an average yield of 8.44 t/ha, which was 3.95% higher than that of the control variety Ⅱ You 838 with a whole growth duration 0.3 d longer than that of the control variety. Tested in the late-maturing early-season *indica* rice trials of south China in 2006 and 2007, it yielded 7.56 t/ha, higher than the yield of the control variety Ⅱ You 128 by 3.32%, with a whole growth duration 0.1 d longer than that of the control variety. Tested in the mid-season *indica* rice trials of the upper reaches of the Yangtze River in 2010 and 2011, it yielded 8.74 t/ha, higher than the yield of the control variety Ⅱ You 838 by 2.6%, with a whole growth duration 2.6 d longer than that of Ⅱ You 838, with Grade 3 grain quality as per the national standard *High Quality Paddy*.

Suitable areas – Mid-season rice in single-cropping rice areas in low-to-medium altitude in Yunnan and Chongqing, hilly rice areas in Pingba of Sichuan and southern Shaanxi, along the Yangtze River in Jiangxi, Hunan, Hubei, Anhui, Zhejiang and Jiangsu, excluding the Wuling Mountains, and northern Fujian and southern Henan where there is no severe occurrence of rice blast and bacterial blight; early-season rice in the double-cropping rice areas in Hainan, southern Guangxi, and southern Fujian where there is no severe occurrence of rice blast; early- or late-season rice in Guangdong, excluding the northern part. By 2015, the accumulated planting area exceeded 2,153,300 ha.

Zhuliangyou 819

Institution (s) – Hunan Yahua Research Institute of Seeds Sciences

Parental lines – Zhu 1S/Hua 819

Approvals – Jiangxi in 2006 (GS2006004), Hunan in 2005 (XS2005010), super rice recognition in 2006

Characters – It is a two-line early-season mid-maturing *indica* hybrid rice with a moderate plant type, short growth period, stable high yield potential, average grain quality, and intermediate susceptibility to rice blast and bacterial blight. Tested in the early-season mid-maturing rice trials of Hunan in 2003 and 2004, it yielded 7.06 t/ha, higher than the yield of the control variety Xiangzaoxian 13 by 10.06%, with a whole growth duration of 106 d, which is 0.8 d shorter than that of Xiangzaoxian. It is 82 cm of plant height, has 3.54 million effective panicles per hectare and 109.6 spikelets per panicle, and is 79.8% in seed setting rate and 24.7 g in 1,000-grain weight.

Suitable areas – Early-season rice in double-cropping rice areas in Hunan and Jiangxi where there is severe occurrence of rice blast. By 2015, the accumulated planting area exceeded 478,000 ha.

Liangyou 287

Institution (s) – School of Life Sciences, Hubei University

Parental lines – HD9802S/R287

Approvals – Guangxi in 2006 (GS2006003), Hubei in 2005 (ES2005001), super rice recognition in 2006.

Characters – It is a two-line early-season medium-to-late maturing *indica* hybrid rice with high thermo-sensitivity, moderate plant type, fairly thick and strong stems, short, stiff and curve flag leaves, moderate growth duration, green leaves and grains yellow at maturity without premature senescence, and good grain quality. Tested in the early-season rice trials of Hubei in 2003 and 2004, it yielded 6.87 t/ha, lower than the yield of the control variety Jinyou 402 by 2.21%, with a whole growth duration of 113.0 d, which is 4.0 d shorter than that of Jinyou 402. It is 85.5 cm in plant height, has 3.18 million effective panicles per hectare, is 19.3 cm in panicle length with 110 – 138 spikelets per panicle and 84 – 113 filled grains per panicle, 79.3% in seed setting rate and 25.31 g in 1,000-grain weight, with high susceptibility to panicle neck blast, susceptibility to bacterial blight, and Grade 1 grain quality as per the national standard *High Quality Paddy*.

Suitable areas – Early-season rice in Hubei where there is no severe occurrence of rice blast and early- or late-season rice in central and northern Guangxi. By 2015, the accumulated planting area exceeded 627,300 ha.

Peizataifeng

Institution (s) – College of Agriculture, South China Agricultural University

Parental lines – Peiai 64S/Taifengzhan

Approvals – Jiangxi in 2006 (GS2006044), State in 2005 (GS2005002), Guangdong in 2004 (YS2004013), super rice recognition by the MOA in 2006

Characters – It is a thermo-sensitive two-line *indica* rice hybrid. Cultured as early-season rice in south China, it has moderate growth duration, high tillering capacity, good maturing color, relatively high yield potential, good grain quality, susceptibility to rice blast and high susceptibility to bacterial blight. Tested in the early-season rice trials of Guangdong in 2002 and 2003, it yielded 7.48 t/ha and 6.83 t/ha, higher than the yields of the control variety of Peizashuangqi by 7.36% and 8.63%, respectively. Tested in the early-season, high grain quality rice trials of south China in 2003 and 2004, it yielded 7.98 t/ha, higher than the yield of the control variety Yuexiangzhan by 3.29%, with a whole growth duration of

125. 8 d, which is 2. 5 d longer than that of Yuexiangzhan. It is 107. 7 cm in plant height, has 2. 76 million effective panicles per hectare, is 23. 3 cm in panicle length, has 176. 0 spikelets per panicle, and is 80. 1% in seed setting rate and 21. 2 g in 1,000-grain weight.

Suitable areas – Early rice in the double-cropping rice areas in Hainan, south-central Guangxi, and southern Fujian where there is no severe occurrence of rice blast and bacterial blight; late-season rice in Guangdong, early-season rice in Guangdong excluding the northern part; and late-season rice in Jiangxi where there is no severe occurrence of rice blast. By 2015, the accumulated planting area exceeded 330,700 ha.

Xinliangyou 6 (Wandao 147)

Institution (s) – Anhui Win-All Agricultural Hi-Tech Institute

Parental lines – Xin'an S/Anxuan 6

Approvals – State in 2007 (GS2007016), Jiangsu in 2006 (SS200602), Anhui in 2005 (WPS05010460), super rice recognition by the MOA in 2006

Characters – It is a two-line mid-season *indica* rice hybrid cultured as mid-season rice in the single-cropping rice areas of the middle and lower reaches of the Yangtze River. It has early maturity, moderate plant type, high yield potential, good grain quality, high susceptibility to rice blast, and intermediate susceptibility to bacterial blight. Tested in the mid-season *indica* rice trials of Anhui in 2003 and 2004, it yielded 8. 30 t/ha and 9. 49 t/ha, which is 10. 93% and 9. 3% higher than the yields of the control variety Shanyou 63, respectively. Tested in the late-maturing mid-season *indica* rice trials of the middle and lower reaches of the Yangtze River in 2005 and 2006, it yielded 8. 59 t/ha, higher than the yield of the control variety Ⅱ You 838 by 5. 71%, with a whole growth duration of 130 d, which is 3. 0 d shorter than that of Ⅱ You 838. It has 2. 425 million effective panicles, is 118. 7 cm in plant height and 23. 2 cm in panicle length with 169. 5 spikelets per panicle, 81. 2% in seed setting rate and 27. 7 g in 1,000-grain weight.

Suitable areas – Single-cropping mid-season rice in areas along the Yangtze River in Jiangxi, Hunan, Hubei, Anhui, Zhejiang and Jiangsu, excluding the Wuling Mountains, northern Fujian and southern Henan where there is no severe occurrence of rice blast. By 2015, the accumulated planting area exceeded 2,060,000 ha.

Liangyoupeijiu

Institution (s) – Cereal Crops Research Institute of Jiangsu Academy of Agricultural Sciences and Hunan Hybrid Rice Research Center

Parental lines – Peiai 64S/93 – 11

Approvals – State in 2001 (GS2001001), Hubei in 2001 (ES006—2001), Guangxi in 2001 (GS2001117), Fujian in 2001 (MS2001007), Shaanxi in 2001 (SS429), Hunan in 2001 (XPS300), Jiangsu in 1999 (SZSZ 313), super rice recognition by the MOA in 2005

Characters – It is a two-line late-maturing mid-season *indica* rice hybrid, a pioneer hybrid and the first hybrid to achieve the yield target of 10. 5 t/ha of Phase I of China's Super Rice Project in 2000. It has wide adaptation with the largest planting area among all two-line hybrid rice varieties in China so far. The yield is similar with that of Shanyou 63 in various rice yield trials and production trials in south China, but with a higher yield potential at a high fertilizer level. Its whole growth duration is 150 d, which is 3 -4 d longer than that of Shanyou 63. It has a good plant type, with high tillering capacity, good tolerance to lodging, good maturing color, average tolerance to cold in the middle and late growth stages, good grain quality, susceptibility to rice blast, intermediate susceptibility to bacterial blight, 16 - 17 leaves on the main stems, 110 -120 cm plants, 22. 8 cm panicles, 160 -220 spikelets per panicle, 76%-86% seed setting rate and 26. 2 g 1,000-grain weight.

Suitable areas - Single-cropping rice in Guizhou, Yunnan, Sichuan, Chongqing, Hunan, Hubei, Jiangxi, Anhui, Jiangsu, Zhejiang and Shanghai, and Xinyang of Henan and Hanzhong of Shaanxi. By 2015, the accumulated planting area exceeded 6,031,300 ha.

Zhunliangyou 527

Institution (s) - Hunan Hybrid Rice Research Center and Rice Research Institute of Sichuan Agricultural University

Parental lines - ZhunS/Shuhui 527

Approvals - State in 2006 (GS2006004, south China riceareas), Fujian in 2006 (MS2006024), State in 2005 (GS2005026, the middle and lower Yangtze and the Wuling Mountains), Guizhou Introduction in 2005 (QY2005001), Chongqing Introduction in 2005 (YY2005001), Hunan in 2003 (XS006—2003), super rice recognition in 2005

Characters - It is a two-line mid-season *indica* rice hybrid, iconic for Phase Ⅱ of China Super Rice Project and the first hybrid to achieve the yield target of 12 t/ha in 2004. It has moderate growth duration, high yield potential, good grain quality, average resistance to rice blast and bacteria blight, high susceptibility to brown planthopper, average tolerance to lodging, high tolerance to cold in the late growth stage, and wide adaptation. Tested in the late-maturing mid-season high-quality *indica* rice trials of the middle and lower reaches of the Yangtze River in 2003 and 2004, it yielded 8. 53 t/ha, higher than the yield of the control variety Shanyou 63 by 7. 09%, with a whole growth duration of 134. 3 d, which is 1. 1 d longer than that of the control variety. It is 123. 1 cm in plant height, has 2. 58 million effective panicles per hectare, is 26. 1 cm in panicle length, has 134. 1 spikelets per panicle, and is 84. 6% in seed setting rate and 31. 9 g in 1,000-grain weight, with Grade 3 grain quality as per the national standard *High Quality Paddy*. Tested in the mid-season *indica* rice trials of the Wuling Mountains in 2003 and 2004, it yielded 8. 87 t/ha, higher than the yield of the control variety Ⅱ You 58 by 7. 0% with a whole growth duration 2. 5 d shorter than that of Ⅱ You 58. Tested in the early-season *indica* rice trials of south China in 2004 and 2005, it yielded 9. 11 t/ha and 6. 94 t/ha, higher than the yields of the control varieties Yuexiangzhan and Ⅱ You 128 by 14. 51% and 3. 40%, respectively.

Suitable areas - Single-cropping mid-season rice in Fujian, Jiangxi, Hunan, Hubei, Anhui, Zhejiang, Jiangsu, Guizhou, Chongqing and southern Henan where there is no severe occurrence of rice blast and bacterial blight; early-season rice in the double-cropping rice areas in Hainan, southern Guangxi, south-central Guangdong where there is no severe occurrence of rice blast and bacterial blight. By 2015, the accumulated planting area exceeded 552,700 ha.

Section 3 *Indica-Japonica* Intersubspecific Three-line Hybrid Rice

Yongyou 2640

Institution (s) - Ningbo Seeds Co., Ltd.

Parental lines - Yongjing 26A/F7540

Approvals - Fujian in 2016 (MS2016022), Jiangsu in 2015 (SS201507), Zhejiang in 2013 (ZS2013024) super rice recognition by the MOA in 2017

Characters - It is a three-line *indica-japonica* hybrid rice, with a moderate plant type, high tolerance to lodging, weak photo sensitivity, average tillering capacity, large panicle size with large number of grains, good maturing color, high yield potential, good grain quality, and intermediate resistance to rice blast. Tested in the extremely early-maturing late-season *japonica* rice trials of Zhejiang in 2010 and 2011, it yielded 7.76 t/ha, higher than the yield of the control variety of Xiushui 417 by 10.9%, with a whole growth duration of 125.7 d, longer than that of Xiushui 417 by 2.7 d. It is 96.0 cm in plant height, has 2.865 million effective panicles per hectare, is 57.8% in tiller-to-panicle formation rate, 19.1 cm in panicle length with 189.4 spikelets per panicle and 143.5 filled grains per panicle, 75.9% in seed setting rate and 24.4 g in 1,000-grain weight. Tested in the rice trials of Jiangsu in 2011 and 2013, it yielded 9.54 t/ha, higher than the yield of the control variety Jiuyou 418 by 7.2%, with a whole growth duration of 149 d, shorter than that of Jiuyou 418 by 5.2 d. The loss caused by neck blast was rated as Grade 3 and the composite index of resistance was 3.25, indicating intermediate susceptibility to bacterial blight, and resistance to sheath blight and rice stripe virus. The grain quality was rated Grade 3 as per the national standard *High Quality Paddy*.

Suitable areas - Early-maturing late-season rice in areas to the south of the Qiantang River in Zhejiang; mid-season rice in Huaibei of Jiangsu and central Jiangsu, Putian of Fujian where there is no severe occurrence of rice blast.

Yongyou 538

Institution (s) - Ningbo Seeds Co., Ltd.

Parental lines - Yongjing 3A/F7538

Approvals - Zhejiang in 2013 (ZS2013022), super rice recognition in 2015

Characters - It is a three-line *indica-japonica japonica*-closed single-cropping hybrid rice with long growth duration, moderate plant height, thick and strong stems, high tolerance to lodging, long, upright and curve flag leaves, large panicle size with large number of grains, large number of spikelets per panicle, high yield potential, intermediate resistance to rice blast and bacterial blight and susceptibility to brown planthopper. Tested in the single-cropping late-season *japonica* rice trials of Zhejiang in 2011 and 2012, it yielded 10.78 t/ha, higher than the yield of the control variety of Jiayou 2 by 26.3%, with 153.5 d of whole growth duration, which is 7.3 d longer than that of the control variety. It has 2.10 million effective panicles per hectare, is 64.6% in tiller-to-panicle formation rate, 114.0 cm in plant height, 20.8 cm in panicle length with 289.2 spikelets per panicle and 239.2 filled grains per panicle, 84.9% in seed setting rate and 22.5 g in 1,000-grain weight.

Suitable areas – Single-cropping late-season rice in Zhejiang. By 2015, the accumulated planting area exceeded 54,700 ha.

Chunyou 84 (Chunyou 684)

Institution (s) – China National Rice Research Institute and Zhejiang Nongke Seed Co., Ltd.

Parental lines – Chunjiang 16A/C84

Approvals – Zhejiang in 2013 (ZSD 2013020), super rice recognition by the MOA in 2015

Characters – It is a three-line *indica-japonica japonica*-closed single-cropping rice hybrid. It has long growth duration, moderate plant height, compact plant type, vigorous growth potential, stick and strong stems, high tolerance to lodging, large panicles with large number of spikelets per panicle, high yield potential, intermediate resistance to rice blast, and susceptibility to bacterial blight and brown planthopper. Tested in the single-cropping late-season *japonica* rice trials of Zhejiang in 2010 and 2011, it yielded 10.29 t/ha, higher than the yield of the control variety Jiayou 2 by 22.9%, with a whole growth duration of 156.7 d, which is 9.2 d longer than that of the control variety. It has 2.10 million effective panicles per hectare, is 79.0% in tiller-to-panicle formation rate, 120.0 cm in plant height, 18.7 cm in panicle length with 244.9 spikelets and 200.1 filled grains per panicle, 83.6% in seed setting rate and 25.2 g in 1,000-grain weight.

Suitable areas – Single-cropping late-season rice in Zhejiang. By 2015, the accumulated planting area exceeded 26,700 ha.

Zheyou 18 (Zheyou 818)

Institution (s) – Institute of Crops and Nuclear Technology Utilization of Zhejiang Academy of Agricultural Sciences, Zhejiang Nongke Seed Co., Ltd., and Shanghai Institutes for Biological Sciences of Chinese Academy of Sciences

Parental lines – Zhe 04A/Zhehui 818

Approvals – Zhejiang in 2012 (ZS2012020), super rice recognition by the MOA in 2015

Characters – It is a three-line *indica-japonica japonica-closed* single-cropping hybrid rice with a compact plant type, stiff and straight flag leaves, intermediate plant height, thick and strong stems, high tolerance to lodging, weak-to-intermediate tillering capacity, large panicle size with large number of spikelets per panicle, high yield potential, intermediate susceptibility to rice blast and bacterial blight, and susceptibility to brown planthopper. Tested in the single-season *indica-japonica* hybrid rice trials of Zhejiang in 2010 and 2011, it yielded 9.93 t/ha, higher than the yield of the control variety Yongyou 9 by 7.8%, with a whole growth duration of 153.6 d, which is 1.0 d longer than that of the control variety. It is 122.0 cm in plant height, has 1.95 million effective panicles per hectare, is 64.0% in tiller-to-panicle formation rate, 20.5 cm in panicle length with 306.1 spikelets per panicle and 233.0 filled grains per panicle, 76.3% in seed setting rate and 23.2 g in 1,000-grain weight.

Suitable areas – Single-cropping rice in Zhejiang

Yongyou 15

Institution (s) – Crops Research Institute of Ningbo Academy of AgriculturalSciences and Ningbo Seeds Co., Ltd.

Parental lines – Yongjing 4A (formerly known as "Jingshuang A")/F8002 (formerly known as "F5032")

Approvals – Fujian in 2013 (MS2013006), Zhejiang in 2012 (ZS2012017), super rice recognition by the MOA in 2013

Characters – It is a three-line *indica-japonica indica*-closed hybrid rice with a moderate plant type, straight, stiff and slightly curled flag leaves, big plant height, strong and thick stems, high tolerance to lodging, low tillering capacity, large panicle size with large number of spikelets per panicle, large number of primary branches, green stems and yellow grains at maturity, high yield potential, good grain quality, resistance to rice blast, and susceptibility to bacterial blight and brown planthopper. Tested in the single-season *indica* hybrid rice trials of Zhejiang in 2008 and 2009, it had an average yield of 8.96 t/ha, higher than that of the control variety Liangyoupeijiu by 8.6%, with a whole growth duration of 138.7 d, which is 3.1 d longer than that of the control variety. It is 127.9 cm in plant height, has 1.785 million effective panicles per hectare, is 60.8% in tiller-to-panicle formation rate, 24.8 cm in panicle length with 235.1 spikelets per panicle and 184.4 filled grains per panicle, 78.5% in seed setting rate and 28.9 g in 1,000-grain weight.

Suitable areas – Single-cropping rice in Zhejiang, mid-season rice in Fujian where there is no severe occurrence of rice blast. By 2015, the accumulated planting area exceeded 168,000 ha.

Yongyou 12

Institution (s) – Crops Research Institute of Ningbo Academy of Agricultural Sciences, Ningbo Seeds Co., Ltd., and Shangyu Shunda Seeds Co., Ltd.

Parental lines – Yongjing 2A/F5032

Approvals – Zhejiang in 2010 (ZS2010015), super rice recognition by the MOA in 2011

Characters – It is a three-line late-maturing *indica-japonica* hybrid rice with high sensitivity to day-length, long growth duration, compact plant type, upright, stiff and curve flag leaves, tall plants, stiff and strong stems, high tolerance to lodging, intermediate tillering capacity, large panicle size with large number of spikelets per panicle, loose branches at panicle bases, high yield potential, average grain quality, intermediate resistance to rice blast and rice stripe virus, intermediate susceptibility to bacterial blight, and susceptibility to brown planthopper. Tested in the single-cropping late-season *japonica* rice trials of Zhejiang in 2007 and 2008, it yielded 8.48 t/ha, higher than the yield of the control variety Xiushui 09 by 16.2%, with a whole growth duration of 154.1 d, which is 7.3 d longer than that of the control variety. It is 120.9 cm in plant height, has 1.845 million effective panicles per hectare, is 57.1% in tiller-to-panicle formation rate, 20.7 cm in panicle length with 327 spikelets per panicle and 236.8 filled grains per panicle, 72.4% in seed setting rate and 22.5 g in 1,000-grain weight.

Suitable areas – Single-cropping rice in the south areas of Qiantang River of Zhejiang. By 2015, the accumulated planting area exceeded 203,300 ha.

Yongyou 6

Institution (s) – Crops Research Institute of Ningbo Academy of Agricultural Sciences and Ningbo Seeds Co., Ltd.

Parental lines – Yongjing 2A/K4806

Approvals – Fujian in 2007 (MS2007020), Zhejiang in 2005 (ZS2005020), super rice recognition by the MOA in 2013

Characters – It is a three-line *indica-japonica* hybrid rice with a tall and large plant type, stick and strong stems, stiff and straight leaves, large panicle size with large number of spikelets per panicle, intermediate resistance to rice blast and bacterial blight, susceptibility to rice planthopper, good grain quality and high seed production yield. Tested in the single-cropping *japonica* hybrid rice trials of Zhejiang in 2002 and 2003, it yielded 8.75 t/ha and 8.15 t/ha, higher than the yields of the control varieties Xiushui 63 and Yongyou 3 by 11.4% and 6.6%, respectively, with a whole growth duration of 156 d, which is 4.7 d longer than that of Xiushui 63 and 10.1 d longer than that of Yongyou 3, respectively. It has 2.01 million effective panicles per hectare, 210.1 filled grains per panicle, and is 72.9% in seed setting rate and 24.7 g in 1,000-grain weight.

Suitable areas – Single-cropping late-season rice in the south-central of Zhejiang, areas of Fujian where there is nosevere occurrence of rice blast. By 2015, the accumulated planting area exceeded 280,000 ha.

References

[1] CHENG SHIHUA, LIAO XIYUAN, MIN SHAOKAI. Studies on China's super hybrid rice: background, objectives, and thoughts on related issues[J]. China Rice, 1998 (1):3–5.

[2] YUAN LONGPING. Breeding seeds for super high yielding hybrid rice[J]. Hybrid Rice, 1997, 12 (6):1–6.

Chapter 20

Super Hybrid Rice-Awards and Achievements

Hu Zhongxiao

After more than 20 years of development, China has made world-renowned achievements in super hybrid rice research and application, with a large number of super hybrid rice varieties planted on a large scale and a series of high-yielding cultivation and related technologies applied to promote the development of super hybrid rice. The research and application of super hybrid rice has played an important role in China's rice production and made an important contribution to China's abundant grain production over the years, exerting significant influence at home and abroad. A series of achievements have been made in the breeding of super hybrid rice parents, and the selection and application of varieties. More than 10 achievements have won national science and technology awards, among which the two-line hybrid rice technology research and application project won the special prize of the State Scientific and Technological Progress Award in 2013, with its outcomes playing an important role in attaining the goals of all four phases of China's super rice project. This is the second time hybrid rice research has won a national-level science and technology award following the State Technological Invention Award granted to *indica* hybrid rice research in 1981. Then, in 2017, Yuan Longping's hybrid rice innovation team won the Innovation Team Award of the State Scientific and Technological Progress Award, the only rice research team to win such an honor so far. Based on public information from the Ministry of Science and Technology, this chapter presents national science and technology awards won by projects and teams of super hybrid rice research and application from 2000 to 2017, including three State Technological Invention Award winners and 12 Scientific and Technological Progress Award winners (including winner of the Innovation Team Award).

Section 1 State Technological Invention Award

Ⅰ. Technology System for the Breeding and Application of Two-line Super Hybrid Rice Liangyoupeijiu

Liangyoupeijiu is a two-line late-maturing mid-season *indica* rice hybrid bred jointly by the Food Crop Research Institute of Jiangsu Academy of Agricultural Sciences and Hunan Hybrid Rice Research Center (HHRRC) with Peiai 64S, which has a *japonica* kindship, and mid-season *indica* variety 9311 as the parents. It was approved at the national level and by the provincial authorities of Hubei, Guangxi, Fujian, Shaanxi, Hunan and Jiangsu successively, and recognized as a super hybrid rice variety by the Ministry of Agriculture (MOA) in 2005. It is the pioneer super hybrid rice variety in China. In 2000, it was the first to achieve the first-phase yield target of 10.5 t/ha of China's super rice project. It is also the two-line hybrid rice variety with the largest planting area so far. As of 2015, its total planting area had exceeded 6.0313 million hectares. It has an ideal plant type, and the plant type of a similar combination, Liangyou E32, was published in the journal *Science* as an ideal plant type model for super hybrid rice. Liangyoupeijiu has achieved an excellent combination of high quality, super-high yield and high resistance, which is a landmark achievement in the application of two-line hybrid rice in production. The breeding of Liangyoupeijiu represents innovations in the theory of hybrid rice breeding and seed production technology, and has secured two patents and one new plant variety right, and backed the publishing of three monographs and more than 50 research papers, greatly promoting progress in hybrid rice development in China.

"Technology System for the Breeding and Application of Two-line Super Hybrid Rice Liangyoupeijiu" won the second prize of the State Technological Invention Award in 2004. The awardees are Zou Jiangshi, Lv Chuangen, Lu Xinggui, Gu Fulin, Wang Cailin and Quan Yongming, and the awarded institutions were Jiangsu Academy of Agricultural Sciences, China National Hybrid Rice R&D Center and Hubei Academy of Agricultural Sciences.

Ⅱ. Late-stage Functional Super Hybrid Rice Breeding Technology and its Application

Through the introduction and utilization of excellent foreign germplasm resources and technologies, China National Rice Research Institute (CNRRI) established a technology system for late-stage functional super hybrid rice breeding, aiming at improving the photosynthetic function in the late growth stages of rice, and successively bred a number of new super hybrid rice varieties that have set world records for high rice yields, such as Guodao 1 and Guodao 6 (Nei 2 You 6). As of 2010, more than 10 rice research institutes and seed companies have used this patented technology to produce more than 20 new rice hybrids, with a cumulative planting area of 9,333,300 ha.

"Late-stage Functional Super Hybrid Rice Breeding Technology and its Application" won the second prize of the State Technological Invention Awards in 2011. The awardees are Cheng Shihua, Cao Liyong, Zhuang Jieyun, Zhan Xiaodeng, Ni Jianping and Wu Weiming; and the awarded institution is CNRRI.

Ⅲ. New Technology of Dual-purpose Rice PTGMS Line C815S Selection and Seed Production

C815S is a dual-purpose *indica* PTGMS line developed by Hunan Agricultural University (HAU) using the dual-purpose GMS material 5SH038 (F_6 of Anxiang S/Xiandang//02428) as the female parent and Peiai 64S as the male parent. It was selected under short daylength and low temperature conditions in Hainan, long daylength and low temperature conditions in summer of Changsha, and short daylength and lowtemperature conditions in autumn of Changsha plus water pool with artificially controlled temperature, through 10 generations in five years of directional breeding. It passed the technical evaluation of Hunan in 2004. This PTGMS line has a high combining ability. As of 2017, it had been used to develop 33 rice hybrids approved at the provincial level or above. Among them, the super hybrid rice C-Liangyouhuazhan had been planted with a cumulative area of more than 147,300 ha as of 2015.

"New Technology of Dual-purpose Rice GMS Line C815S Selection and Seed Production" won the second prize of the State Technological Invention Award in 2012. The awardees are Chen Liyun, Tang Wenbang, Xiao Yinghui, Liu Guohua, Deng Huabing and Lei Dongyang, and the awarded institution is HAU.

Section 2 State Scientific and Technological Progress Award

Ⅰ. Breeding and Application of the Dual-purpose PTGMS Line Peiai 64S

Peiai 64S is an *indica* PTGMS line developed through a series of crosses and backcrosses at the HHRRC by using Nongken 58S as the female parent and Peiai 64, a progeny of *indica-javanica* cross of Peidi/Aihuangmi//Ce64, as the male parent. It passed the technical evaluation of Hunan in 1991. Peiai 64S has broad and high compatibility. As of 2017, it had been used to develop more than 60 rice hybrids approved at the provincial level or above, including three super rice hybrids, namely Liangyoupeijiu, Peizataifeng and Peiliangyou 3076.

"Breeding and Application of the Dual-purpose PTGMS Line Peiai 64S won the first prize of the State Scientific and Technological Progress Award in 2001. The awardees are Luo Xiaohe, Li Renhua, Bai Delang, Zhou Chengshu, Chen Liyun, Qiu Zhizhong, Luo Zhibin, Liao Cuimeng, Liu Jianbin, Yi Junzhang, Wang Feng, He Jiang, Qin Xiyin, Xue Guangxing and Liu Jianfeng, and the awarded institutions are HHRRC, HAU, Yuan Longping Agricultural High-Tech Co. Ltd. (LPHT), Rice Research Institute of Guangdong Academy of Agricultural Sciences, Hybrid Rice Research Center of Guangxi Academy of Agricultural Sciences, and Crop Breeding and Cultivation Research Institute of Chinese Academy of Agricultural Sciences.

Ⅱ. Breeding and Application of the High-quality *Indica* CMS Line Jin 23A

Jin 23A is a mid-maturing early-season *indica* WA-type CMS line developed at Changde Agriculture Science and Research Academy. An F_5 progeny was selected from the cross of Feigai B/Ruanmi M (a Yunnan local soft rice variety) and crossed with the high quality rice variety Huangjin 3, and then a progeny plant from the three-way cross was backcrossed with V20A to develop the CMS line Jin 23A. The CMS line is thermo-sensitive with a compact plant type, high tillering capacity, high tiller-to-panicle formation rate, complete pollen abortion, early and concentrated flowering time, high stigma exsertion rate, high seed production yield and high combining ability. As of 2017, it had been used to breed more than 160 validated hybrid rice combinations, including four super rice hybrids, Jinyou 299, Jinyou 458, Jinyou 527 and Jinyou 785.

"Breeding and Application of the High-quality *Indica* CMS Line Jin 23A" won the second prize of the State Scientific and Technological Progress Award in 2002. The awardees are Li Yiliang, Xia Shengping, Jia Xianyong, Yang Nianchun, Zhang Deming, Wang Zebin, Zeng Geqi, Zhang Zhengguo, Pang Huaju and Xu Chunfang, and the awarded institutions are Changde Agricultural Science and Research Institute, Hunan Seed Group Co. Ltd., Changde Bureau of Agriculture and Rural Affairs, Hunan Rice Research Institute and Hunan Jinjian Cereals Industry Co., Ltd.

Ⅲ. Breeding and Application of Shuhui 162, a Hybrid Rice Restorer Line with High Combining Ability and Excellent Traits

Shuhui 162 is a restorer line with the traits of high combining ability, high resistance and wide adaptability, developed at Sichuan Agricultural University (SAU). It was developed by combining conventional breeding technology with biotechnology and integrating plant type breeding with hybrid heterosis. It involved introducing excellent foreign rice resources and creating a new breeding method that combines compound hybrid with anther culture. Specifically, Milyang 46, a rice variety introduced from South Korea, was used as the female parent to cross with an F_8 progeny from the cross of 707/Minghui 63, then the F_1 was anther-cultured and the progenies went through selection and testcrossing to develop the restorer line with good maturing color, no premature root senescence, strong stems with high tolerance to lodging, and kinship with Korean and African rice germplasm. The restorer line has high restorability, a broad restoring spectrum, high resistance to rice blast and high combining ability. It has been used to breed such hybrid rice combinations as D You 162, Ⅱ You 162 and Chiyou S162, which have been approved and applied in rice production.

"Breeding and Application of Shuhui 162, a Hybrid Rice Restorer Line with High Combining Ability and Excellent Traits" won the second prize of the State Scientific and Technological Progress Award in 2003. The awardees are Wang Xudong, Zhou Kaida, Wu Xianjun, Li Ping, Li Shigui, Gao Keming, Ma Yuqing, Ma Jun, Long Bin and Chen Yongchang, and the awarded institution is SAU.

Ⅳ. High-quality, Multi-resistance and High-yield Mid-season *Indica* Yangdao 6 (9311) and Its Application

The F_1 progeny from the cross of Yangdao 4 and 3021 (a mid-season *indica* rice variety) was treated with ^{60}Co-γ irradiation mutagenesis, and then the mutant went through directional breeding to develop the new mid-season conventional *indica* rice variety Yangdao 6 at the Institute of Agricultural Sciences of Lixiahe Region, Jiangsu. It passed variety validations of Jiangsu, Anhui and Hubei successively. This variety has good grain quality, high resistance to bacterial blight and rice blast, high heat tolerance, and high cold tolerance at the seedling stage. It is not only a mid-season *indica* rice variety, but also an important two-line restorer line. It has been used to breed three two-line mid-season *indica* super hybrid rice combinations, namely, Liangyoupeijiu, Y Liangyou 1 and Yangliangyou 6.

"High-quality, Multi-resistance and High-yield Mid-season Indica Yangdao 6 (9311) and Its Application" won the second prize of the State Scientific and Technological Progress Award in 2004. The awardees are Zhang Hongxi, Dai Zhengyuan, Xu Maolin, Li Aihong, Huang Niansheng, Liu Xiaobin, Lu Kaiyang, Wang Xinguo, Ji Jian'an and Hu Qingrong, and the awarded institution is the Institute of Agricultural Sciences in Lixiahe Region of Jiangsu province.

Ⅴ. Breeding of Super Rice Xieyou 9308, Basic Research on Super High Yield Physiology, and Demonstration and Promotion of Integrated Production Technology

Xieyou 9308 is a late-season *indica* rice hybrid developed at CNRRI by crossing Xieqingzao A and 9308. It can be planted as single-cropping rice in areas of low-to-medium altitudes in central and southern Zhejiang province, or as late-season rice in the double-cropping areas of Wenzhou city, Zhejiang province.

"Breeding of Super Rice Xieyou 9308, Basic Research on Super High Yield Physiology, and Demonstration and Promotion of Integrated Production" won the second prize of the State Scientific and Technological Progress Award in 2004. The awardees are Cheng Shihua, Chen Shenguang, Min Shaokai, Zhu Defeng, Wang Xi, Sun Yongfei, Ye Shuguang, Zhao Jianqun, Lv Hefa and Zheng Jiacheng, and the awarded institutions include CNRRI, Agricultural Bureau of Xinchang county of Zhejiang, Wenzhou Seed Company of Zhejiang, Agriculture Department of Zhejiang, Agriculture Bureau of Yueqing city of Zhejiang, and Agriculture Technology Extension Service Center of Zhuji city of Zhejiang.

Ⅵ. Discovery and Application of Indonesia Paddy-type Male Sterile Cytoplasm

HHRRC, CNRRI and other institutes used the WA-type CMS as the identification materials to search for new male sterile cytoplasm in cultivated rice varieties (restorer lines). Through hybridization and gene recombination between restorer lines and maintainer lines, 10 new sources of male sterile cytoplasm, including Indonesian Paddy No. 6, were discovered, which greatly enriched the types of sterile cytoplasm sources in hybrid rice. Indonesian Paddy No. 6 was used to breed three new CMS lines, i. e., Ⅱ-32A, You 1A and Zhong 9A, which have such good traits as high-yielding hybrids, good grain quality, high seed production yield and low seed production costs. Hybrids derived from Indonesia Paddy cytoplasm CMS set the world record of rice yield at that time (18.47 t/ha). The hybrid rice quality was improved significantly, with main indicators up to the Grade 2 or Grade 3 requirements as per relevant national standards. The yield of seed production was increased significantly, setting the high yield world record of seed production (6.6 t/ha). The cost of seed production is 4-6 yuan per kilogram with a more than 50% cost reduction. The breeding and application of Indonesia Paddy-type hybrid rice has raised the production of hybrid rice in China to a new level in terms of seed production yield, rice quality and hybrid yield, ushering in a new era of high yield in hybrid rice seed production, and playing a role in ensuring China's food security.

One of the Indonesia Paddy CMS lines, Ⅱ-32A, was developed at the HHRRC. Ⅱ-32B was a progeny from the cross of Zhenshan 97A/IR665, and then it was used as the recurrent parent to cross and backcross with Zhending 28A to develop Ⅱ-32A after multi-generations of backcrosses. It is one of the most important three-line CMS lines in China with vigorous growth potential, high combining ability, good flowering habits, high stigma exsertion rate and high outcrossing seed setting rate. Up to 2017, more than 200 hybrids derived from the CMS line have passed variety validations and put into production. There are eight super rice hybrids derived from Ⅱ-32A, namely Ⅱ Youming 86, Ⅱ Youhang 1, Ⅱ Youhang 2, Ⅱ You 162, Ⅱ You 7, Ⅱ You 602, Ⅱ You 084 and Ⅱ You 7954.

Zhong 9A is another early-season *indica* CMS line developed at the CNRRI with Indonesia Paddy cytoplasm, for which a progeny of You IB/L301//Feigai B as the male parent to backcross with You IA in multiple generations. It passed the technical validation in September 1997. The CMS line has high combining ability and 122 rice hybrids had been derived from the CMS line, passed variety validations and used in rice production by 2017. It was also used to breed two super hybrid rice varieties, i. e., Guodao 1 and Zhong 9 You 8012, with a total planting area of over 559,300 ha by 2015.

"Discovery and Application of Indonesia Paddy-type Male Sterile Cytoplasm" won the first prize of the State Scientific and Technological Progress Award in 2005. The awardees are Zhang Huilian, Deng Yingde, Peng Yingcai, Shen Xihong, Gan Mingfu, Fang Hongmin, Yi Junzhang, Shen Yuexin, Zhang Guoliang, Chen Jinjie, Xiong Wei and He Guowei, and the awarded institutions include CNRRI, HHRRC, Sichuan Seed Station, Jiangxi Seed Management Station, Hefei Fengle Seed Industry Co., Ltd., Anhui Quanyin Agricultural High-tech Research Institute, and Guangdong Crop Heterosis Development and Utilization Center.

Ⅶ. Creation and Application of the *Indica-Japonica* Hybrid Rice Restorer Line Luhui 17 with Heat Tolerance and High Combining Ability

Luhui 17 was developed from the cross of widely compatible *japonica* rice 02428 and Gui 630 at the Rice and Sorghum Research Institute of Sichuan Academy of Agricultural Sciences through testcrossing and selections. It has been used to breed five rice hybrids in production, i. e., Xieyou 117, Chuannong 2, B You 817, K You 17 and Ⅱ You 7.

"Creation and Application of the *Indica-Japonica* Hybrid Rice Restorer Line Luhui 17 with Heat Tolerance and High Combining Ability" won the second prize of the State Scientific and Technological Progress Award in 2005. The awardees are Kuang Haochi, Zheng Jiakui, Zuo Yongshu, Li Yun, Liu Guomin, Chen Guoliang, Xu Fuxian, Jiang Kaifeng, Xiong Hong and Liu Ming, and the awarded institutions include the Rice and Sorghum Research Institute of Sichuan Academy of Agricultural Sciences and Sichuan Academy of Agricultural Sciences.

Ⅷ. Breeding and Application of Backbone Parent Shuhui 527 and Heavy-panicle Hybrid Rice

Shuhui 527 was selected by pedigree breeding from the cross of Hui 1318 as the female parent, which was a progeny from the cross of Gui 630/Gu 154/IR1544-28-2-3, and 88-R3360 as the male parent, which was a restorer line with high grain quality derived from the cross of Fu 36-2/IR24, at the Rice Research Institute of SAU. Shuhui 527 has a large number of grains per panicle, high seed setting rate, high grain weight and high restorability. It has been used for more than 40 validated rice hybrids, including five super rice hybrids, i. e., D You 527, Xieyou 527, Zhunliangyou 527, Yifeng 8 and Jinyou 527.

"Breeding and Application of Backbone Parent Shuhui 527 and Heavy-panicle Hybrid Rice" won the second prize of the State Scientific and Technological Progress Award in 2009. The awardees are Li Shigui, Ma Jun, Li Ping, Li Hanyun, Zhou Kaida, Gao Keming, Wang Yuping, Tao Shishun, Wu Xianjun and Zhou Mingjing, and the awarded institutions are SAU and Southwest University of Science and Technology.

Ⅸ. Research and Application of Two-line Hybrid Rice Technology

The fertility of two-line hybrid rice is controlled by the nuclear genes, so the advantages are having no restoring and maintaining relationship, free mating, simplified breeding and seed production procedures, low cost, high utilization rate of rice germplasm and high probability of selection and breeding of superior combinations. After more than 20 years of research and development, the technology has established a new way to effectively utilize two-line hybrid heterosis based on PTGMS lines, solved the main limiting factors of three-line hybrid rice, and brought the utilization of rice heterozygous advantage to a new level, with innovations and breakthroughs in the following seven aspects.

(1) A complete hybrid rice breeding system was established; the hybrid rice breeding strategy of going from the three-line method to the two-line method and then to the one-line method and moving from intervarietal to intersubspecific and then to distant heterosis utilization was proposed; the relationship between fertility transition and light and temperature changes was clarified; and the mechanism of how day-length and temperature act on sterile lines in the temperature-sensitive period and sensitive parts of sterile lines was explored.

(2) The practical selection theory that PTGMS lines have a sterility threshold temperature below 23.5°C was proposed, and the practical PTGMS line selection and identification technology for a sterility threshold temperature below 23.5°C was proposed.

(3) A two-line super hybrid rice breeding technology that combines morphological improvement with intersubspecific heterosis utilization and the use of distant favorable genes was established. Using this breeding technology, China achieved the goals of the first three phases of its super rice project, yielding 10.5 t/ha, 12.0 t/ha and 13.5 t/ha respectively in 2000, 2004 and 2012. It also realized the sound combination between super-high yield, good rice quality, and high resistance in super hybrid rice. As of 2017, 36 two-line hybrids have been recognized as super rice varieties by the MOA.

(4) The systems of two-line hybrid rice seed production meteorological analysis, decision-making and high-yielding seed production technology have been established. The seed production specifications have been formulated. The average yield of seed production can reach 3.16 t/ha, which is 16.5% higher than the yield of the three-line method.

(5) Three sets of high-yielding and stable reproduction technology systems of two-line sterile lines were developed, namely winter reproduction in Hainan at low latitude, summer and autumn reproduc-

tion at regular environment temperature with cold water irrigation, and natural low-temperature summer reproduction at a high altitude. The average yield of reproduction can reach 5.80 t/ha, which is 153.4% higher than that of the three-line hybrid rice reproduction method, and the cost is reduced by 50%, solving the difficulties in the reproduction of sterile lines.

(6) The mechanism of the genetic drift in the fertility threshold temperature of PTGMS lines was clarified. The production procedure of core seeds and foundation seeds of PTGMS lines was created, thereby effectively preventing the shift of the fertility threshold temperature of PTGMS lines and ensuring the safety of seed production.

(7) The technical bottlenecks of two-line hybrid, i. e. *japonica* rice breeding and seed production, were broken through, promoting the development of *japonica* hybrid rice.

As of 2012, two-line hybrid rice had been planted in 16 provinces across the country. From 2005 to 2012, two-line hybrid rice ranked first in terms of annual planting area among all hybrid rice varieties for eight consecutive years. As of 2012, two-line hybrid rice cultivated by the project team had accumulated a total planting area of 33.2667 million hectares, the total rice yield was 235.82 billion kilograms, an increase of 11.099 billion kilograms, the total output value was 577.759 billion yuan, up by 27.193 billion yuan, helping ensure food security in China.

Two-line hybrid rice is China's first international scientific and technological achievement with independent intellectual property rights in the field of hybrid rice. It provided a new theory and technical method for the genetic improvement of crops and consolidated China's leading position in the world in hybrid rice research and application. Based on the theory and experience of two-line hybrid rice cultivation, research on two-line oilseed rape, sorghum and wheat has also succeeded, providing new methods for the utilization of heterosis that is difficult to utilize by the three-line method. Through technology transfer and cooperation, the two-line hybrid rice technology has been applied in the United States, and the yield has increased by more than 20%. Two-line hybrid rice provides the core technology support for China's seed industry to explore the international market and participate in the international competition in seed technology.

"Research and Application of Two-line Hybrid Rice Technology" won the special prize of the State Scientific and Technological Progress Award in 2013. The awardees are Yuan Longping, Shi Mingsong, Deng Huafeng, Lu Xinggui, Zou Jiangshi, Luo Xiaohe, Wang Shouhai, Yang Zhenyu, Mou Tongmin, Wang Feng, Chen Liangbi, He Haohua, Qin Xiyin, Liu Aimin, Yin Jianhua, Wan Banghui, Li Chengquan, Sun Zongxiu, Peng Huipu, Cheng Shihua, Pan Xigan, Yang Jubao, You Aiqing, Zeng Hanlai, Lv Chuangen, Wu Xiaojin, Deng Guofu, Zhou Guangqia, Huang Zonghong, Liu Yibai, Feng Yunqing, Yao Kemin, Wang Yanjun, Wang Dezheng, Zhu Yingguo, Liao Yilong, Liang Manzhong, Chen Dazhou, Su Xuejun, Xiao Cenglin, Yin Huaqi, Liao Fuming, Yuan Qianhua, Li Xinxin, Tong Zhe, Zhou Chengshu, Guo Mingqi, Yang Qinghua, Xu Xiaohong, and Zhu Renshan. The awarded institutions include HHRRC, Food Crops institute of Hubei Academy of Agricultural Sciences, Jiangsu Academy of Agricultural Sciences, Rice Research Institute of Anhui Academy of Agricultural Sciences, Huazhong Agricultural University, Wuhan University, Rice Research Institute of Guangdong Academy of Agricultural Sciences, Hunan Normal University, Jiangxi Agricultural University, Rice Research Institute of Guangxi Zhuang Autonomous Region Academy of Agricultural Sciences, CNRRI, LPHT, Rice Research Institute of Jiangxi Academy of Agricultural Sciences, South China Agricultural University, Rice Research Institute of Fujian Academy of Agricultural Sciences, Guizhou Rice Research Institute, Beijing Jinsenonghua Seed Industry Co. Ltd. and Hunan Institute of Meteorological Sciences.

X. Key Technology for Super-high Yielding Rice Cultivation and Regional Integrated Application

Super hybrid rice varieties generally have large biomass, large panicles and many grains, which are significantly different from the growth characteristics and yield formation patterns of ordinary rice varieties. However, due to a mismatch between cultivation techniques and the rice varieties, the techniques of ordinary rice production cannot give full play to the high yield and efficient production potential of super rice varieties. At the same time, rice production is in urgent need of mechanization and simplification through transformation and upgrading to further enhance the overall rice cultivation technological level and improve rice production and efficiency. To this end, CNRRI organized researchers from Yangzhou University, Jiangxi Agricultural University, HAU and Jilin Academy of Agricultural Sciences in typical ecological zones for super rice planting to carry out research on super rice cultivation technology, focusing on the common laws and key technologies for achieving high yield with super rice and carry out region-based integrated application according to specific local conditions, launching high yield demonstration in major rice areas in China, so as to provide technical support for large area production of super rice in China.

The researchers compared the growth characteristics and yield formation of different types of super rice and common rice varieties in different regions and seasons, revealed the high-yielding growth characteristics of super rice varieties, and clarified the common laws behind the formation of high yield. It was proposed that the biological basis of high yield of super rice is to stabilize matter production in the early stage and increase the matter production from jointing to heading and from heading to maturity. It was clarified that sufficient total number of spikelets in a population is the basis of a large sink for high yield formation of super rice. The number of grains per panicle is less related to the number of primary branches per panicle than to the number of secondary branches per panicle, so the formation of large panicles depends mainly on increasing the number of secondary branches per panicle. The above clarification provided a the oretical basis for promoting large panicle formation in high-yielding super rice cultivation. The nitrogen, phosphorus and potassium requirements of super rice varieties under high yield conditions were clarified, revealing the characteristics of high nitrogen production efficiency and high nitrogen uptake in the middle and late stages of super rice varieties, providing important theoretical guidance for quantitative fertilization of super rice varieties in high yield cultivation. Based on the growth and development rules of super rice, the factors affecting super rice yield increase in major rice regions were analyzed and it was clarified that to stabilize the number of panicles and increase the number of grains per panicle is the way to increase the yield of super rice. Comparing the growth characteristics and yield performance of different super rice planting methods (manual transplanting, machine transplanting, seedling scattering and direct seedling), the team clarified the high-yielding growth pattern of super rice varieties, and proposed practical indicators for the construction of high-yielding populations, such as basic seedling number, tiller-to-panicle formation rate, effective panicle number, leaf area index at heading, and population spikelets. The method of quantitative super rice fertilization, which is region-, variety-, and season-specific and with increased panicle fertilizer, was proposed; and the generic key high-yielding cultivation technology was created, emphasizing early development of sufficient seedlings in the early stage, strengthening stems to expand the "sink" in the middle stage, and ensuring abundant "source" to promote grain filling in the late stage. In addition, based on the transition of rice planting from the traditional manual transplanting to seedling scattering and machine transplanting, it was proposed that high-yielding planting methods of super rice varieties in different rice areas should be combined with the specific planting methods, and the layout of super rice varieties should be improved. This reduced the risk of improper planting methods for super rice varieties and provided sound basis for important decisions regarding the promotion of super rice in large areas.

This achievement provided the production technology for the large-scale promotion of super rice varieties in China, as well as the theory and practices regarding yield formation pattern, high-yielding plant types and yield increase pathways, contributing to the selection and breeding of super rice varieties. The application of super rice high-yield cultivation techniques and variety packages accelerated the formation of super rice varieties, and 17 sets of super rice high-yielding cultivation techniques were integrated and adapted to rice areas and cultivation methods. Cultivation technical protocols for super rice varieties were produced and approved by the MOA. The research team prepared more than 100 cultivation pattern maps, developed eight local production technology standards, published 10 monographs on this topic, including *Super Rice Variety Cultivation Technology* and *Super Rice Variety Cultivation Technology Pattern Maps*, and established a high-yielding cultivation technology system for super rice in China's major rice regions, creatively providing important technical support for the large-scale promotion of super rice. Super rice varieties were promoted and applied in the major rice production areas in South China, Southwest China, the middle and lower reaches of the Yangtze River and North China. The results of trials and demonstration around the country show that the large scale application of regionalized high-yielding cultivation techniques for super rice achieved yield increases of 756 to 1,098 kg/ha, with an average yield increase of 8.4%–13.1%, compared with traditional cultivation techniques. From 2011 to 2013, the research achievement was applied to 7.9273 million hectares of rice land with 895.5 kg/ha in yield increase and 6.4 million tons of paddy produced, delivering 11.65 billion yuan in yield and efficiency increase and 2.09 billion yuan in cost saving, totaling 13.74 billion yuan. It can be seen that the application of super rice high-yielding cultivation technology increased rice yields and made an important contribution to ensuring China's food security.

"Key Technology for Super-high Yielding Rice Cultivation and Regional Integrated Application" won the second prize of the State Scientific and Technological Progress Award in 2014. The awardees are Zhu Defeng, Zhang Hongcheng, Pan Xiaohua, Zou Yingbin, Hou Ligang, Huang Qing, Zheng Jiaguo, Wu Wenge, Chen Huizhe and Huo Zhongyang. The awarded institutions include CNRRI,

Yangzhou University, Jiangxi Agricultural University, HAU, Jilin Academy of Agricultural Sciences, Rice Research Institute of Guangdong Academy of Agricultural Sciences, and Crop Research Institute of Sichuan Academy of Agricultural Sciences.

Ⅺ. Breeding and Demonstration of New Double-cropping Super Rice Varieties in Jiangxi

The research addressed the technical difficulty in balancing between early maturity and high yield, high quality and high yield, high yield and stable yield in rice production in Jiangxi and other double-cropping rice areas, and made a major breakthrough in double-cropping super rice breeding theory, variety selection, technology integration and demonstration and promotion. The project team proposed the idea of ideal plant type, balance between panicle and grain, reasonable relationship between root and canopy, balance between "source" and "sink", well-matched heterosis, and comprehensive improvement, and used it as the guidance for the breeding of new double-cropping rice varieties. This became an important theory guiding the breeding of double-cropping rice varieties. Nine backbone double-cropping super rice parents were developed, each with distinctive features, the planting area of early-season super rice varieties derived from these parents accounted for 79.4% of the total planting area of early-season super rice and 65.4% of the total planting area of late-season super rice in Jiangxi. A total of 21 new double-cropping hybrid rice varieties were bred from these parents. Among them, the varieties Ganxin 688, Wufengyou T025 and Jinyou 458 were recognized as super rice varieties by the MOA; Ganxin 688 became the first super rice variety of Jiangxi province; Gangxin 688 and Jinyou 458 were listed as leading varieties of the MOA; Wufengyou T025 and Jinyou 458 passed the national validations; and Wufengyou T025 has been the hybrid rice variety with the largest planting area in Jiangxi since 2010. The project compiled four sets of high-yielding and high-efficiency seed production and cost-saving cultivation techniques for double-cropping super rice, established 215 6.67-ha demo plots, 108 66.7-ha demo plots, and 56 667-ha demo fields, achieving the goal of bringing and increase of 1,500 yuan/ha in benefits from double-cropping rice thorough cost saving and efficiency improvement in super rice production. Jiangxi thus reinforced its leading position in double-cropping rice development in China.

"Breeding and Demonstration of New Double-cropping Super Rice Varieties in Jiangxi province" won the second prize of the State Scientific and Technological Progress Award in 2016. The awardees are He Haohua, Cai Yaohui, Fu Junru, Yin Jianhua, He Xiaopeng, Xiao Yeqing, Cheng Feihu, Zhu Changlan, Hu Lanxiang and Chen Xiaorong; and the awarded institutions include Jiangxi Agricultural University, Rice Research Institute of Jiangxi Academy of Agricultural Sciences, Agriculture Technology Extension Service Center of Jiangxi, Jiangxi Modern Seed Industry Co., Ltd. and Jiangxi Dazhong Seed Industry Co., Ltd.

Ⅻ. Yuan Longping Hybrid Rice Innovation Team

Yuan Longping hybrid rice innovation team, aiming at meeting the needs of the national food security strategy, has been perseverant and innovative over the years. Twenty-one years after its inception, the team, with Yuan Longping, Deng Qiyun and Deng Huafeng as the leaders, has gathered many young and middle-aged experts as the mainstay, with a total of 85 members, 45 of whom have senior professional titles. With an average age of 42 years old the team has a complete discipline portfolio and a reasonable structure, and has remained in a world-leading position in its field.

Focuses on key scientific problems in hybrid rice research, the team has overcame a series of technical difficulties and ensured China's leading position in hybrid rice breeding and cultivation in the world. The team created two-line hybrid rice theory and technology, promoted the rapid development of two-line hybrid heterosis utilization in China, created the system of super hybrid rice breeding technology that combined morphological improvement with hybrid heterosis utilization, and was the first to achieve the yield goals of the first, second, third and fourth phases of China's super rice project, and set the world record of 15.40 t/ha in average yield in 6.67-ha demo plots. It also created breakthrough core parents such as Annong S-1, Peiai 64S and Y58S, which provided breeding resources for 80% of the two-line hybrid rice in China. The team developed a total of 93 hybrid rice varieties now in large-scale production in China, including Jinyou 207, Y Liangyou 1 and Y Liangyou 900, with a cumulative planting area of 53.3333 million hectares. The team created industrialized technology systems for safe seed production and nitrogen-saving and high-efficiency green cultivation of super hybrid rice, greatly promoting the development of China's seed industry. The team was awarded 11 national-level science and technology awards, including one State Supreme Science and Technology Award, one special prize of the State Scientific and Technological Progress Award, and two first prizes of the State Scientific and Technological Progress Award.

The team has five R&D platforms spearheaded by the State Key Laboratory of Hybrid Rice, sustainable talent echelon and financial investment, a nationwide outreach and the basic conditions for international development. In the future, the team will continue to pursue theoretical and technological innovations, strengthen cooperation between the industry, the academia and education institutions, breed more new hybrid rice varieties with high quality, multiple resistances, wide adaptation and high yield to meet market demand, promote industrialization, and strive to realize the dream of spreading hybrid rice to the world.

Yuan Longping hybrid rice innovation team won the Innovation Team Award of the State Scientific and Technological Progress Award in 2017. The key team members are Yuan Longping, Deng Qiyun, Deng Huafeng, Zhang Yuzhu, Ma Guohui, Xu Qiusheng, Yang Hehua, Qi Shaowu, Peng Jiming, Zhao Bingran, Yuan Dingyang, Li Xinqi, Wang Weiping, Wu Jun, Li Li, et al. The team's main support institutions are HHRRC and Hunan Academy of Agricultural Sciences.